NONLINEAR THEORY OF ELASTICITY

NORTH-HOLLAND SERIES IN

APPLIED MATHEMATICS AND MECHANICS

VOLUME 36

NORTH-HOLLAND-AMSTERDAM • NEW YORK • OXFORD • TOKYO

NONLINEAR THEORY OF ELASTICITY

by

A. I. LURIE

Translated from the Russian
by
K. A. Lurie

1990

NORTH-HOLLAND-AMSTERDAM • NEW YORK • OXFORD • TOKYO

ISBN: 0 444 87439 9

Publishers:

ELSEVIER SCIENCE PUBLISHERS B.V.
P.O. Box 1991
1000 BZ Amsterdam
The Netherlands

Sole distributors for the U.S.A. and Canada:

ELSEVIER SCIENCE PUBLISHING COMPANY. INC.
655 Avenue of the Americas
New York, N.Y. 10010
U.S.A.

Library of Congress Cataloging-in-Publication Data

Lur'e, A. I. (Anatoliĭ Isakovich), 1901-
[Nelineĭnaîa teoriîa uprugosti. English]
Nonlinear theory of elasticity / by A.I. Lurie ; translated from the Russian by K.A. Lurie.
p. cm. -- (North-Holland series in applied mathematics and mechanics ; v. 36)
Translation of: Nelineĭnaîa teoriîa uprugosti.
Includes bibliographical references and index.
ISBN 0-444-87439-9
1. Elasticity. 2. Nonlinear theories. I. Title. II. Series.
QA931.L95513 1990
531'.382--dc20 90-49215
CIP

PRINTED IN THE NETHERLANDS

PUBLISHERS' NOTE

This book had just been put into print when there came the grievous news of the death of its author, the distinguished scholar in the field of mechanics, Professor Anatolii Isakovich Lurie.

His last work was a tremendous feat in both human and scientific terms. Gravely ill, overcoming terrible sufferings, he hastened to complete the manuscript. It was his fate not to live to see the proofs: they have been read by his colleagues and pupils.

Everything created by Anatolii Isakovich - a scholar and a teacher - bears the imprint of his remarkable personality. Tireless and passionate in research, uncompromising in everything, he personified generosity and open-heartedness and deep human and professional decency.

This book is the last testimony of the unsparing talent of its author. The Publishers are proud of having introduced his works to readers for a half-century.

Anatolii Isakovich Lurie

1901–1980

FOREWORD

This book deals with the Theory of Elasticity which is a branch of the mechanics of a deformable solid. A revision of the fundamentals of continuum mechanics is not among the purposes of this book, the major emphasis being placed on the investigation of the process of deformation within the framework of the generally accepted model of a medium which, in this case, is an elastic body.

The first two chapters contain an account of the concepts of stress and strain, which are universal concepts throughout continuum mechanics. It would be pointless to reproduce this material here, the Contents and the introductory remarks accompanying each chapter furnish the reader with all the necessary information. The "direct" language of tensor calculus, which is the natural vehicle for continuum mechanics, is used without reference to component representations which hinder the perception of basic concepts and operation with them.

For readers familiar with the component representations (which cannot be avoided in special problems), their use will not cause difficulties; one needs only some skills in elementary algebraic transforms. For support, the author could draw on his many years of experience in teaching.

Chapter three attempts to determine the place of the Theory of Elasticity in the mechanics of materials in general. The fourth and the fifth chapters describe the behaviour of an elastic body. That is to say, an account is given of the constitutive equations arising from the specific strain-energy function. In these chapters the stationary value principle is also outlined and some space is devoted to a discussion of the inequalities arising from the requirements of monotonicity and strong ellipticity. It is quite probable that in the next few years great progress will be made in this "fundamental unsolved problem" in connection with problems of the existence of those solutions in nonlinear elasticity which are not treated here.

Chapter six deals with problems of the equilibrium of a compressible elastic body. It is assumed that the reader will disregard at a first reading purely mathematical questions and restrict himself to a detailed study of "the second order effects".

The most important achievements of the nonlinear theory of elasticity are those associated with investigations into incompressible (rubberlike) materials. These are treated in Chapter 7.

A convincing treatment of stability is not possible in linear theory. Progress here may come about by considering a small deformation superposed on an initially stressed body. This material is presented in Chapter 8.

Chapter 9, the final chapter, contains an account of fundamental relations of thermodynamics applied to an elastic medium. Whether the attention given to these fundamental physical premises is justified, is debatable. The alternative method for constructing the constitutive equations was based on the principles of thermodynamics but in the subsequent chapters there was insufficient space for the equations of thermoelasticity.

Careful study of Appendices I and III is recommended even for a competent reader in order to become accustomed to the notation used throughout this book. Appendix II contains the rules of differentiation with respect to a tensor argument used throughout the text.

There is as yet no standard notation in the nonlinear theory of elasticity. In particular, the notation used in the book has not been made to accord with that adopted in the foreign literature, where substantial discrepanceis are also observed. The reader will find apparently mutually contradictory forms of one and the same relationship both in this book and in foreign publications.

Intermediate calculations are carried through with necessary completeness. "One link in a sequence of transformations can be restored, two links cannot"*) (Littlewood).

I would like to address my warm thanks for their help and criticisms to my dear friends - E.L. Gurvich, V.V. Yeliseev, P.A. Zhilin, L.M. Zubov, V.A. Pal'mov, V.A. Pupirev, V.S. Chernina.

I clearly recognize the deficiencies of this work; no author is ever free of that feeling. Feci, quod potui.

The Author

*) Inverse translation from the Russian. - The translator.

CONTENTS

CHAPTER 1

DEFORMATION OF A CONTINUOUS MEDIUM

1. MATERIAL COORDINATES. SPATIAL COORDINATES

In the mechanics of a system containing a finite or countable set of material points, each point is designated by some number which is retained in the course of its motion. This possibility no longer exists when one considers a continuous medium, i.e. a continuous set of elements called particles, material points or "body-points". In this case, the elements are distinguished by the introduction of some continuously varying quantities. To this end, we fix some configuration of a three-dimensional continuous medium and ascribe to each particle M of it a set of three quantities q^1, q^2, q^3, i.e. its "number" which remains the same in the course of its motion. The particles in the immediate neighbourhood of M are described by the numbers $q^1 + \delta q^1$, $q^2 + \delta q^2$, $q^3 + \delta q^3$, where δq^1, δq^2, δq^3 are infinitely small. The position of a particle in this fixed configuration (which will be called below the reference v-configuration) is prescribed by the radius vector $\underline{r}$, the function

$$\underline{r} = \underline{r}(q^1,q^2,q^3) \tag{1}$$

is assumed to be continuous and differentiable as many times as necessary. The triplet (q^1,q^2,q^3) is called the material coordinates of the particle M. For example, q^s may be the Cartesian coordinates (designated as a^s) of the position occupied by a particle in the reference configuration of a medium in the frame $OXYZ$

$$\underline{r} = \underline{i}_1 a^1 + \underline{i}_2 a^2 + \underline{i}_3 a^3 \ . \tag{2}$$

This definition is of course not compulsory. Nothing prevents us from using curvilinear coordinates for q^s, e.g. cylindrical (r,φ,z) or spherical (R,ϑ,λ). It is essential only that there should exist a one-to-one correspondence between the set of triplets q^s and that of particles M.

The motion of the medium determines the position of each particle at any instant t of time. This motion is prescribed by the radius vector

$$\underline{R} = \underline{R}(q^1,q^2,q^3;\, t) \ . \tag{3}$$

This determines an actual V_t-configuration, i.e. a smooth mapping at the instant t onto the domain in Euclidean space E_3 is determined by the prescription of the region in which material coordinates may be ascribed admissible values.

It is not necessary to regard the reference v-configuration as being one of the actual configurations corresponding to some fixed instant t (e.g. $t = 0$) but this assumption is, however, often convenient in many respects. For this case,

$$\underline{R}(q^1,q^2,q^3\,;\,0) = \underline{r}(q^1,q^2,q^3)\ . \tag{4}$$

In the Cartesian coordinate system $OXYZ$ (the same for all configurations)

$$\underline{R} = \underline{i}_s x^s(q^1,q^2,q^3\,;\,t)\ , \qquad \underline{r} = \underline{i}_s a^s(q^1,q^2,q^3) \tag{5}$$

and, in particular, for $q^s = a^s$

$$x^s(a^1,a^2,a^3\,;\,0) = a^s \qquad (s = 1,2,3)\ . \tag{6}$$

In order that the mappings (1) and (3) be one-to-one, the equations

$$a^s = a^s(q^1,q^2,q^3)\ , \qquad x^s = x^s(q^1,q^2,q^3\,;\,t) \tag{7}$$

should be solvable for q^k. To this end, it is necessary and sufficient that the Jacobians

$$\sqrt{g} = \left|\frac{\partial a^s}{\partial q^k}\right| = \frac{D(a^1,a^2,a^3)}{D(q^1,q^2,q^3)}\ , \qquad \sqrt{G} = \left|\frac{\partial x^s}{\partial q^k}\right| = \frac{D(x^1,x^2,x^3)}{D(q^1,q^2,q^3)} \tag{8}$$

be non-zero within the admissible domain of q^s; the Jacobians will be positive for a suitable order of numeration of coordinates.

The well-known property of Jacobians implies that their ratio

$$\sqrt{\frac{G}{g}} = \left|\frac{\partial x^s}{\partial a^k}\right| = \frac{D(x^1,x^2,x^3)}{D(a^1,a^2,a^3)} = \frac{dV}{dv} \tag{9}$$

equals that of the elementary volume dV of particles in their actual configuration to the volume dv of the same particles in their reference configuration.

Any function $\Psi(q^1,q^2,q^3\,;\,t)$ of material coordinates and time, be it scalar, vector or tensor, may be considered as dependent upon (a^k,t) or (x^k,t) because of Equations (7). Instead of projections of $\underline{r}$, $\underline{R}$ onto the coordinate axes, one may consider components of these vectors along any vector basis and assume the notation

$$\Psi(q^s\,;\,t) = \Psi(\underline{r}\,;\,t)\ , \qquad \Psi(q^s,t) = \Psi(\underline{R}\,;\,t)\ . \tag{10}$$

These functions determine the fields of the quantity Ψ in the v- and V_t-configurations, respectively. The first representation is called referential, and the second spatial. One often comes across the term "Lagrangian" for $\Psi(\underline{r},t)$ as well as "Eulerian" for $\Psi(\underline{R},t)$; however, this terminology will not be used below. C. Truesdell points out that historically it is not justified because Euler used the "Lagrangian" and D'Alembert the "Eulerian" description.

The referential description implies following the motion of some fixed particle q^s, the spatial description the observation of a process developing with time at some given place. "This description seems perfectly suited to studies of fluids, where often a rapidly deforming mass comes from no-one knows where and goes no-one knows whither, so that we may prefer to consider what is going on here and now. However convenient kinematically, the spatial description is awkward in applications of the principles of continuum mechanics, since in fact it is what is happening to the body, not to the region of space occupied by the body, which enters the laws of dynamics." (C. Truesdell.)

For static problems, the time dependence of $\underline{R}$ vanishes. Instead of V_t-, one now consders the actual V-configuration. A clear distinction between configurations is a necessary prerequisite for a proper understanding of continuum mechanics.

2. VECTOR BASES

The original and dual vector bases (I, Par. 1) are represented for the reference configuration by triplets of vectors (v-bases)

$$\underline{r}_s = \frac{\partial \underline{r}}{\partial q^s}, \quad \underline{r}^s = g^{sk}\underline{r}_k = 1/2\,\varepsilon^{skt}\underline{r}_k \times \underline{r}_t, \tag{1}$$

the unit (metric) tensor and the symbols of Levi-Civita being expressed by the formulas

$$\underline{E} = g_{sk}\underline{r}^s\underline{r}^k = g^{sk}\underline{r}_s\underline{r}_k = \underline{r}_s\underline{r}^s = \underline{r}^s\underline{r}_s, \qquad \underline{\varepsilon} = \varepsilon^{skt}\underline{r}_s\underline{r}_k\underline{r}_t = \varepsilon_{skt}\underline{r}^s\underline{r}^k\underline{r}^t \tag{2}$$

in terms of their components

$$\begin{aligned} g_{sk} &= \underline{r}_s \cdot \underline{r}_k, \quad g^{sk} = \underline{r}^s \cdot \underline{r}^k, \quad g^s_k = \underline{r}^s \cdot \underline{r}_k = \delta^s_k; \\ \varepsilon^{skt} &= \underline{r}^s \cdot (\underline{r}^k \times \underline{r}^t), \quad \varepsilon_{skt} = \underline{r}_s \cdot (\underline{r}_k \times \underline{r}_t). \end{aligned} \tag{3}$$

For the actual configuration, the corresponding capital letters are used (the V_t-bases)

$$\underline{R}_s = \frac{\partial \underline{R}}{\partial q^s}, \quad \underline{R}^s = G^{sk}\underline{R}_k = 1/2\,\underline{\epsilon}^{skt}\underline{R}_k \times \underline{R}_t, \tag{4}$$

$$\underline{E} = G_{sk}\underline{R}^s\underline{R}^k = G^{sk}\underline{R}_s\underline{R}_k = \underline{R}_s\underline{R}^s = \underline{R}^s\underline{R}_s \; ;$$
$$\underline{\epsilon} = \epsilon^{skt}\underline{R}_s\underline{R}_k\underline{R}_t = \epsilon_{skt}\underline{R}^s\underline{R}^k\underline{R}^t \; . \tag{5}$$

The components of vectors and tensors in v-bases are designated by a zero superscript. For V_t-bases, one should use the superscript "t" but we will not do so since usually there is no need to fix the instant of time. For example,

$$\underline{a} = \overset{0}{a}_s\underline{r}^s = \overset{0}{a}{}^s\underline{r}_s = a_s\underline{R}^s = a^s\underline{R}_s \; ,$$
$$\underline{\underline{Q}} = \overset{0}{q}_{mn}\underline{r}^m\underline{r}^n = \overset{0}{q}{}^{mn}\underline{r}_m\underline{r}_n = q_{mn}\underline{R}^m\underline{R}^n = q^{mn}\underline{R}_m\underline{R}_n \tag{6}$$

and so on. The same system of notation is used to designate the Christoffel symbols, nabla operators and covariant differentiation:

$$\overset{0}{[rt,s]} = \frac{1}{2}\left(\frac{\partial g_{rs}}{\partial q^t} + \frac{\partial g_{ts}}{\partial q^r} - \frac{\partial g_{rt}}{\partial q^s}\right), \quad \overset{0}{\left\{ {s \atop rt} \right\}} = g^{sm}\overset{0}{[rt,m]} \; ,$$
$$[rt,s] = \frac{1}{2}\left(\frac{\partial G_{rs}}{\partial q^t} + \frac{\partial G_{ts}}{\partial q^r} - \frac{\partial G_{rt}}{\partial q^s}\right), \quad \left\{ {s \atop rt} \right\} = G^{sm}[rt,m] \; , \tag{7}$$

$$\overset{0}{\underline{\nabla}} = \underline{r}^s\frac{\partial}{\partial q^s}, \qquad \overset{0}{\underline{\underline{\nabla a}}} = \underline{r}^s\frac{\partial \underline{a}}{\partial q^s} = \underline{r}^s\underline{r}^k\overset{0}{\nabla}_s a_k \; ,$$
$$\underline{\nabla} = \underline{R}^s\frac{\partial}{\partial q^s}, \qquad \underline{\underline{\nabla a}} = \underline{R}^s\frac{\partial \underline{a}}{\partial q^s} = \underline{R}^s\underline{R}^k\nabla_s a_k \; , \tag{8}$$

so that

$$\frac{\partial \underline{a}}{\partial q^s} = \underline{r}_s \cdot \overset{0}{\underline{\underline{\nabla a}}} = \overset{0}{\underline{\underline{\nabla a}}}{}^T \cdot \underline{r}_s = \underline{R}_s \cdot \underline{\underline{\nabla a}} = \underline{\underline{\nabla a}}^T \cdot \underline{R}_s \; . \tag{9}$$

Simultaneous representations of tensors in two different configurations, i.e. the reference and actual configurations, are also admissible, e.g.

$$\underline{\underline{Q}} = q^{\alpha}_{m}\underline{r}_{\alpha}\underline{R}^m = q_{\alpha m}\underline{r}^{\alpha}\underline{R}^m = q_{m\alpha}\underline{R}^m\underline{r}^{\alpha} \; . \tag{10}$$

Here, for the reference configuration, we have used Greek indices.

Vector bases are represented in the Cartesian coordinate system by the formulas

$$\underline{r}_s = \underline{i}_m\frac{\partial a^m}{\partial q^s}, \quad \underline{r}^s = \underline{i}^m\frac{\partial q^s}{\partial a^m} \; ; \qquad \underline{R}_s = \underline{i}_m\frac{\partial x^m}{\partial q^s}, \quad \underline{R}^s = \underline{i}^m\frac{\partial q^s}{\partial x^m} \; . \tag{11}$$

The position vector $\underline{R}$ in the actual configuration and the vector $\underline{r}$ in the reference configuarion are sometimes represented in a reference and actual configuration bases by

$$\underline{R} = \underline{r}_s \chi^s = \underline{r}^s \chi_s , \qquad \underline{r} = \underline{R}_s \rho^s = \underline{R}^s \rho_s \tag{12}$$

with the corresponding expressions

$$\underline{R}_t = \underline{r}_s \overset{0}{\nabla}_t \chi^s , \qquad \underline{r}_t = \underline{R}^s \nabla_t \rho_s . \tag{13}$$

3. DEFORMATION GRADIENTS

In accordance with the definition of vector bases and nabla-operators, one can form tensors which represent gradients of vectors associated with a particle at a point in the actual and reference configurations

$$\overset{0}{\underline{\underline{\nabla R}}} = \underline{r}^s \frac{\partial \underline{R}}{\partial q^s} = \underline{r}^s \underline{R}_s , \qquad \underline{\underline{\nabla r}} = \underline{R}^s \frac{\partial \underline{r}}{\partial q^s} = \underline{R}^s \underline{r}_s \tag{1}$$

and their transposes

$$\overset{0}{\underline{\underline{\nabla R}}}{}^T = \underline{R}_s \underline{r}^s , \qquad \underline{\underline{\nabla r}}{}^T = \underline{r}_s \underline{R}^s . \tag{2}$$

These tensors allow the following representations in terms of their Cartesian components

$$\begin{aligned} \overset{0}{\underline{\underline{\nabla R}}} &= \underline{i}^s \underline{i}_k \frac{\partial x^k}{\partial a^s} , \qquad \underline{\underline{\nabla r}} = \underline{i}^s \underline{i}_k \frac{\partial a^k}{\partial x^s} ; \\ \overset{0}{\underline{\underline{\nabla R}}}{}^T &= \underline{i}_k \underline{i}^s \frac{\partial x^k}{\partial a^s} , \qquad \underline{\underline{\nabla r}}{}^T = \underline{i}_k \underline{i}^s \frac{\partial a^k}{\partial x^s} \end{aligned} \tag{3}$$

and in view of (1.9)

$$\det \overset{0}{\underline{\underline{\nabla R}}} = \left| \frac{\partial x^k}{\partial a^s} \right| = \sqrt{\frac{G}{g}} , \qquad \det \underline{\underline{\nabla r}} = \left| \frac{\partial a^k}{\partial x^s} \right| = \sqrt{\frac{g}{G}} . \tag{4}$$

The tensor $\underline{\underline{\nabla r}}$ is inverse of $\overset{0}{\underline{\underline{\nabla R}}}$, and $\overset{0}{\underline{\underline{\nabla R}}}$ is inverse to $\underline{\underline{\nabla r}}$

$$\underline{\underline{\nabla r}} = (\overset{0}{\underline{\underline{\nabla R}}})^{-1} , \qquad \overset{0}{\underline{\underline{\nabla R}}} = (\underline{\underline{\nabla r}})^{-1} , \tag{5}$$

which is a direct consequence of their definition (1)

$$\underline{\underline{\nabla r}} \cdot \underline{\underline{\overset{0}{\nabla} R}} = \underline{R}^s \underline{r}_s \cdot \underline{r}^k \underline{R}_k = \underline{R}^s \underline{R}_s = \underline{E} \ .$$

From here it follows that

$$d\underline{R} = d\underline{r} \cdot \underline{\underline{\overset{0}{\nabla} R}} = \underline{\underline{\overset{0}{\nabla} R}}^T \cdot d\underline{r} \ , \qquad d\underline{r} = d\underline{R} \cdot \underline{\underline{\nabla r}} = \underline{\underline{\nabla r}}^T \cdot d\underline{R} \ . \tag{6}$$

For example,

$$d\underline{r} \cdot \underline{\underline{\overset{0}{\nabla} R}} = dq^k \underline{r}_k \cdot \underline{r}^s \underline{R}_s = dq^k \underline{R}_k = d\underline{R} \ .$$

We may also represent the differentials dq^s as

$$dq^s = \underline{R}^s \cdot d\underline{R} = \underline{r}^s \cdot d\underline{r} \ . \tag{7}$$

The relationships

$$\frac{\partial}{\partial q^s} = \underline{r}_s \cdot \underline{\overset{0}{\nabla}} = \underline{R}_s \cdot \underline{\nabla} \tag{8}$$

give rise to the formulas connecting the nabla-operators $\underline{\overset{0}{\nabla}}$, $\underline{\nabla}$

$$\underline{\overset{0}{\nabla}}(\ ) = \underline{\underline{\overset{0}{\nabla} R}} \cdot \underline{\nabla}(\ ) \ , \qquad \underline{\nabla}(\ ) = \underline{\underline{\nabla r}} \cdot \underline{\overset{0}{\nabla}}(\ ) \tag{9}$$

applied to a vector $\underline{a}$ and to a tensor $\underline{\underline{Q}}$ of the second rank. For the vector $\underline{a}$ we also have

$$\underline{\underline{\overset{0}{\nabla} a}}^T = \underline{\underline{\nabla a}}^T \cdot \underline{\underline{\overset{0}{\nabla} R}}^T \ , \qquad \underline{\underline{\nabla a}}^T = \underline{\underline{\overset{0}{\nabla} a}}^T \cdot \underline{\underline{\nabla r}}^T \ . \tag{10}$$

Equations (2.12) imply that

$$\underline{\underline{\overset{0}{\nabla} R}} = \underline{r}^t \underline{r}_s \overset{0}{\nabla}_t \chi^s = \underline{r}^t \underline{r}^s \overset{0}{\nabla}_t \chi_s \ , \qquad \underline{\underline{\nabla r}} = \underline{R}^t \underline{R}_s \nabla_t \chi^s = \underline{R}^t \underline{R}^s \nabla_t \chi_s \ . \tag{11}$$

Note also the obvious relations

$$\underline{\underline{\nabla R}} = \underline{R}^s \underline{R}_s = \underline{E} \ , \qquad \underline{\underline{\overset{0}{\nabla} r}} = \underline{r}^s \underline{r}_s = \underline{E} \ . \tag{12}$$

The concept of a deformation gradient is used throughout the whole subsequent text.

4. CAUCHY-GREEN AND ALMANSI STRAIN MEASURES

Deformation of a continuum is characterized by a change of the magnitudes and directions of vectors

$$d\underline{r} = \overset{0}{\underline{e}}\,|d\underline{r}| = \overset{0}{\underline{e}}\,ds\,, \qquad d\underline{R} = \underline{e}\,|d\underline{R}| = \underline{e}\,dS \tag{1}$$

taken at the point $M(q^1,q^2,q^3)$ whose position in the v- and V-configurations is given by the position vectors $\underline{r}$ and $\underline{R}$. According to (3.6)

$$d\underline{R}\cdot d\underline{R} = dS^2 = d\underline{r}\cdot\overset{0}{\nabla}\underline{R}\cdot\overset{0}{\nabla}\underline{R}^T\cdot d\underline{r} = d\underline{r}\cdot\underline{G}\cdot d\underline{r} = ds^2\overset{0}{\underline{e}}\cdot\underline{G}\cdot\overset{0}{\underline{e}}\,, \tag{2}$$

$$d\underline{r}\cdot d\underline{r} = ds^2 = d\underline{R}\cdot\nabla\underline{r}\cdot\nabla\underline{r}^T\cdot d\underline{R} = d\underline{R}\cdot\underline{g}\cdot d\underline{R} = dS^2\underline{e}\cdot\underline{g}\cdot\underline{e}\,. \tag{3}$$

Here we have introduced two tensors: the Cauchy-Green strain measure $\underline{G}$

$$\overset{0}{\nabla}\underline{R}\cdot\overset{0}{\nabla}\underline{R}^T = \underline{G} = \underline{r}^s\underline{R}_s\cdot\underline{R}_k\underline{r}^k = G_{sk}\underline{r}^s\underline{r}^k \tag{4}$$

and the Almansi strain measure $\underline{g}$

$$\nabla\underline{r}\cdot\nabla\underline{r}^T = \underline{g} = \underline{R}^s\underline{r}_s\cdot\underline{r}_k\underline{R}^k = g_{sk}\underline{R}^s\underline{R}^k. \tag{5}$$

The Cauchy-Green measure refers to the v-basis, and the Almansi measure to the V_t-basis. The covariant components of these measures are equal, respectively, to those of the metric tensor $\underline{E}$ in the V_t- and v-bases. It would be a serious mistake to identify either $\underline{G}$ or $\underline{g}$ with $\underline{E}$. Transformation to the contravariant components of these tensors is performed according to the general rules applied to their convariant components in the v- and V_t-bases:

$$G^{sk} = g^{sm}g^{kn}G_{mn}\,, \qquad g^{sk} = G^{sm}G^{kn}g_{mn}\,, \tag{6}$$

this calculation does not yield contravariant components of $\underline{E}$.

The determinant of a product of tensors equals the product of their determinants /cf. (I.6.4)/; for this reason and in view of (3.4) and (4), (5) we have

$$\det\underline{G} = \frac{G}{g}\,, \qquad \det\underline{g} = \frac{g}{G}\,, \tag{7}$$

(while $\det\underline{E} = 1$). Note that the symbols G and g denote determinants of covariant components of $\underline{E}$ /cf. (I.1.9), (I.4.14), (2.5)/.

The tensors $\underline{G}$ and $\underline{g}$ are positive-definite since the quadratic forms

$(\frac{dS}{ds})^2$, $(\frac{ds}{dS})^2$ are positive-definite. The principal values G_k, g_k of these tensors are therefore also positive. Denoting by $\overset{0}{\underline{e}}_k$, $\underline{e}_k$ the principal directions of $\underline{G}$ and $\underline{g}$, we may write /cf. (I.9.14)/

$$\underline{G} = G_1 \overset{0}{\underline{e}}_1 \overset{0}{\underline{e}}{}^1 + G_2 \overset{0}{\underline{e}}_2 \overset{0}{\underline{e}}{}^2 + G_3 \overset{0}{\underline{e}}_3 \overset{0}{\underline{e}}{}^3 , \qquad \underline{g} = g_1 \underline{e}_1 \underline{e}^1 + g_2 \underline{e}_2 \underline{e}^2 + g_3 \underline{e}_3 \underline{e}^3 \tag{8}$$

(of course, $\overset{0}{\underline{e}}{}^s = \overset{0}{\underline{e}}_s$, $\underline{e}^s = \underline{e}_s$). Now, assuming in (2) and (3) that $\overset{0}{\underline{e}} = \overset{0}{\underline{e}}_k$ and $\underline{e} = \underline{e}_k$, we obtain the relationships

$$(\frac{dS}{ds})_k = \sqrt{G_k} = 1 + \delta_k , \qquad (\frac{ds}{dS})_k = \sqrt{g_k} = 1 + \Delta_k = (1 + \delta_k)^{-1} . \tag{9}$$

These formulas show that a unit segment oriented along the principal direction $\overset{0}{\underline{e}}_k$ of the Cauchy measure in the reference configuration acquires the length $1 + \delta_k$ in the V_t-configuration. At the same time, a unit segment oriented along the principal direction $\underline{e}_k$ of the Almansi measure in an actual configuration, possesses in the reference configuration the length $1 + \Delta_k$, so that we may call parameters δ_k and Δ_k the principal stretches

$$\delta_k = \frac{dS_k - ds_k}{ds_k} , \qquad \Delta_k = \frac{ds_k - dS_k}{dS_k} .$$

The following notation is used below

$$G_k = v_k^2 , \qquad v_k = 1 + \delta_k ; \qquad g_k = G_k^{-1} = v_k^{-2} , \tag{10}$$

where obviously $-1 < \delta_k < \infty$ and $0 < \Delta_k < \infty$ are the limiting values corresponding to the zero and infinite lengths of a rod.

Substituting into (2) and (3), we have

$$\overset{0}{\underline{e}} = \overset{0}{\underline{e}}_m = \frac{\underline{r}_m}{|\underline{r}_m|} = \frac{\underline{r}_m}{\sqrt{g_{mm}}} , \qquad \underline{e} = \underline{e}_m = \frac{\underline{R}_m}{|\underline{R}_m|} = \frac{\underline{R}_m}{\sqrt{G_{mm}}} , \tag{11}$$

that is considering unit segments along the unit vectors of the bases, we arrive at the formulas

$$(\frac{dS}{ds})_m = \left(\frac{\underline{r}_m \cdot \underline{G} \cdot \underline{r}_m}{g_{mm}}\right)^{1/2} = \sqrt{\frac{G_{mm}}{g_{mm}}} , \qquad (\frac{ds}{dS})_m = \left(\frac{\underline{R}_m \cdot \underline{g} \cdot \underline{R}_m}{G_{mm}}\right)^{1/2} = \sqrt{\frac{g_{mm}}{G_{mm}}} , \tag{12}$$

which provide a geometric interpretation of the diagonal components of the strain measures $\underline{G}$ and $\underline{g}$.

To find the geometric meaning of non-diagonal components, consider at the point M in the v- and V_t-configurations the vectors $d\mathbf{r}'$, $d\mathbf{r}''$ and $d\mathbf{R}'$, $d\mathbf{R}''$, respectively. According to (1), (2), (3), we obtain

$$d\mathbf{R}' = d\mathbf{r}' \cdot \overset{0}{\nabla\mathbf{R}} = \mathbf{e}'\,dS'\,, \qquad d\mathbf{R}'' = \overset{0}{\nabla\mathbf{R}}{}^{T} \cdot d\mathbf{r}'' = \mathbf{e}''\,dS''\,,$$

$$d\mathbf{R}' \cdot d\mathbf{R}'' = \mathbf{e}' \cdot \mathbf{e}''\,dS'\,dS'' = d\mathbf{r}' \cdot \overset{0}{\nabla\mathbf{R}} \cdot \overset{0}{\nabla\mathbf{R}}{}^{T} \cdot d\mathbf{r}'' = ds'\,ds''\,\overset{0}{\mathbf{e}}{}' \cdot \mathbf{G} \cdot \overset{0}{\mathbf{e}}{}''\,,$$

where, of course, $d\mathbf{r}' = \overset{0}{\mathbf{e}}{}'\,ds'$, $d\mathbf{r}'' = \overset{0}{\mathbf{e}}{}''\,ds''$. We have

$$\mathbf{e}' \cdot \mathbf{e}'' = \frac{ds'\,ds''}{dS'\,dS''}\,\overset{0}{\mathbf{e}}{}' \cdot \mathbf{G} \cdot \overset{0}{\mathbf{e}}{}'' = (\overset{0}{\mathbf{e}}{}' \cdot \mathbf{G} \cdot \overset{0}{\mathbf{e}}{}'\,\overset{0}{\mathbf{e}}{}'' \cdot \mathbf{G} \cdot \overset{0}{\mathbf{e}}{}'')^{-\frac{1}{2}}\,\overset{0}{\mathbf{e}}{}' \cdot \mathbf{G} \cdot \overset{0}{\mathbf{e}}{}'' \tag{13}$$

and similarly

$$\overset{0}{\mathbf{e}}{}' \cdot \overset{0}{\mathbf{e}}{}'' = \frac{dS'\,dS''}{ds'\,ds''}\,\mathbf{e}' \cdot \mathbf{g} \cdot \mathbf{e}'' = (\mathbf{e}' \cdot \mathbf{g} \cdot \mathbf{e}'\,\mathbf{e}'' \cdot \mathbf{g} \cdot \mathbf{e}'')^{-\frac{1}{2}}\,\mathbf{e}' \cdot \mathbf{g} \cdot \mathbf{e}''\,. \tag{14}$$

Directing $\overset{0}{\mathbf{e}}{}'$, $\overset{0}{\mathbf{e}}{}''$ along the unit vectors $\mathbf{r}_s$, $\mathbf{r}_t$ of the v-configuration and denoting by φ_{st} the angle formed by these vectors in the actual configuration, we may write /cf. (12)/

$$\cos\varphi_{st} = \sqrt{\frac{g_{ss}g_{tt}}{G_{ss}G_{tt}}}\,\frac{\mathbf{r}_s}{\sqrt{g_{ss}}} \cdot \mathbf{G} \cdot \frac{\mathbf{r}_t}{\sqrt{g_{tt}}} = \frac{G_{st}}{\sqrt{G_{ss}G_{tt}}}\,. \tag{15}$$

Similarly, from (14), we obtain

$$\cos\overset{0}{\varphi}_{st} = \frac{g_{st}}{\sqrt{g_{ss}g_{tt}}}\,, \tag{16}$$

where $\overset{0}{\varphi}_{st}$ denotes the angle formed in the v-configuration by the unit vectors $\mathbf{R}_s$, $\mathbf{R}_t$ of the V_t-configuration.

<u>The formulas relating the unit vectors $\overset{0}{\mathbf{e}}$, $\mathbf{e}$ at the point M in the v- and V_t-configurations</u>. We have

$$d\mathbf{R} = \mathbf{e}\,dS = d\mathbf{r} \cdot \overset{0}{\nabla\mathbf{R}} = \overset{0}{\mathbf{e}}\,ds \cdot \overset{0}{\nabla\mathbf{R}}\,, \qquad \mathbf{e} = \overset{0}{\mathbf{e}} \cdot \overset{0}{\nabla\mathbf{R}}\,\frac{ds}{dS} = \overset{0}{\nabla\mathbf{R}}{}^{T} \cdot \overset{0}{\mathbf{e}}\,\frac{ds}{dS}$$

and similarly

$$d\mathbf{r} = \overset{0}{\mathbf{e}}\,ds = d\mathbf{R} \cdot \nabla\mathbf{r} = \mathbf{e}\,dS \cdot \nabla\mathbf{r}\,, \qquad \overset{0}{\mathbf{e}} = \mathbf{e} \cdot \nabla\mathbf{r}\,\frac{dS}{ds} = \nabla\mathbf{r}^{T} \cdot \mathbf{e}\,\frac{dS}{ds}\,.$$

We arrive at the relationships

$$\mathbf{e} = \frac{\overset{0}{\mathbf{e}} \cdot \overset{0}{\nabla}\mathbf{R}}{(\overset{0}{\mathbf{e}} \cdot \mathbf{G} \cdot \overset{0}{\mathbf{e}})^{1/2}} = \frac{\overset{0}{\nabla}\mathbf{R}^{T} \cdot \overset{0}{\mathbf{e}}}{(\overset{0}{\mathbf{e}} \cdot \mathbf{G} \cdot \overset{0}{\mathbf{e}})^{1/2}}, \qquad \overset{0}{\mathbf{e}} = \frac{\mathbf{e} \cdot \nabla\mathbf{r}}{(\mathbf{e} \cdot \mathbf{g} \cdot \mathbf{e})^{1/2}} = \frac{\nabla\mathbf{r}^{T} \cdot \mathbf{e}}{(\mathbf{e} \cdot \mathbf{g} \cdot \mathbf{e})^{1/2}}. \tag{17}$$

In particular, if $\mathbf{e}$ is a unit vector directed along $\mathbf{r}_s$, then by (11), (12) and (3.1)

$$\frac{\overset{0}{\mathbf{e}} \cdot \overset{0}{\nabla}\mathbf{R}}{(\overset{0}{\mathbf{e}} \cdot \mathbf{G} \cdot \overset{0}{\mathbf{e}})^{1/2}} = \frac{1}{\sqrt{g_{ss}}}\sqrt{\frac{g_{ss}}{G_{ss}}}\, \mathbf{r}_s \cdot \overset{0}{\nabla}\mathbf{R} = \frac{\mathbf{R}_s}{\sqrt{G_{ss}}} = \mathbf{e}_s .$$

In other words, the vector at the left-hand side of $(17)_1$ is simply a unit vector directed along $\mathbf{R}_s$, while the vector $\overset{0}{\mathbf{e}}$ in $(17)_2$ is a unit vector along $\mathbf{r}_s$. In particular, for the principal directions $\mathbf{e}_k$, $\overset{0}{\mathbf{e}}_k$ of the measures $\mathbf{G}$ and $\mathbf{g}$, we obtain

$$\mathbf{e}_k = \frac{\overset{0}{\mathbf{e}}_k \cdot \overset{0}{\nabla}\mathbf{R}}{\sqrt{G_k}}, \qquad \overset{0}{\mathbf{e}}_k = \frac{\mathbf{e}_k \cdot \nabla\mathbf{r}}{\sqrt{g_k}} = \sqrt{G_k}\, \mathbf{e}_k \cdot \nabla\mathbf{r} . \tag{18}$$

By virtue of (3.11), (4) and (5) the strain measures $\mathbf{G}$ and $\mathbf{g}$ may also be represented in the form

$$\mathbf{G} = \overset{0}{\nabla}\mathbf{R} \cdot \overset{0}{\nabla}\mathbf{R}^{T} = \mathbf{r}^{s}\mathbf{r}^{t}\, \overset{0}{\nabla}_{s} \chi_{k}\, \overset{0}{\nabla}_{t} \chi^{k}, \qquad G_{st} = \overset{0}{\nabla}_{s} \chi_{k}\, \overset{0}{\nabla}_{t} \chi^{k}, \tag{19}$$

$$\mathbf{g} = \nabla\mathbf{r} \cdot \nabla\mathbf{r}^{T} = \mathbf{R}^{s}\mathbf{R}^{t}\, \nabla_{s} \rho_{k}\, \nabla_{t} \rho^{k}, \qquad g_{st} = \nabla_{s} \rho_{k}\, \nabla_{t} \rho^{k} . \tag{20}$$

In particular, for a Cartesian coordinate system

$$\mathbf{G} = G_{\langle st\rangle}\mathbf{i}^{s}\mathbf{i}^{t}, \qquad G_{\langle st\rangle} = \frac{\partial x^{k}}{\partial a^{s}} \frac{\partial x_{k}}{\partial a^{t}},$$
$$\mathbf{g} = g_{\langle st\rangle}\mathbf{i}^{s}\mathbf{i}^{t}, \qquad g_{\langle st\rangle} = \frac{\partial a^{k}}{\partial x^{s}} \frac{\partial a_{k}}{\partial x^{t}}, \tag{21}$$

unless otherwise specified. The Cartesian components of $\mathbf{Q}$ are given by $q_{\langle mn\rangle}$.

The displacement of a continuous medium is called rigid if its reference configuration is transformed into the actual configuration by the law of displacement of a rigid body

$$\mathbf{R} = \mathbf{R}_0 + (\mathbf{r} - \mathbf{r}_0) \cdot \mathbf{O} . \tag{22}$$

Here $\mathbf{O}$ denotes a proper orthogonal tensor which is the same for all particles of a medium, $\mathbf{R}_0$ and $\mathbf{r}_0$ denote the position vectors of the point O and $\mathbf{u}_0 = \mathbf{R}_0 - \mathbf{r}_0$ its displacement.

By (22) we have

$$\underset{-}{R}_s = \underset{-}{r}_s \cdot \underset{-}{O} , \qquad \overset{0}{\underset{--}{\nabla R}} = \underset{-}{r}^s \underset{-}{r}_s \cdot \underset{-}{O} = \underset{-}{O} , \qquad \overset{0}{\underset{--}{\nabla R}} \cdot \overset{0}{\underset{--}{\nabla R}}^T = \underset{-}{G} = \underset{-}{E} . \tag{23}$$

The Cauchy measure has thus turned out to be a unit tensor for a rigid displacement; this result could have been foreseen since for any direction $\underset{-}{e}$ we have (cf. (2))

$$\left(\frac{dS}{ds}\right)^2 = \overset{0}{\underset{-}{e}} \cdot \underset{-}{G} \cdot \overset{0}{\underset{-}{e}} = \overset{0}{\underset{-}{e}} \cdot \underset{-}{E} \cdot \overset{0}{\underset{-}{e}} = \overset{0}{\underset{-}{e}} \cdot \overset{0}{\underset{-}{e}} = 1 ,$$

i.e. the distance between any two particles of a medium remains the same. One should now prove the inverse, namely, that the condition $\underset{-}{G} = \underset{-}{E}$ holds only for a rigid displacement. By virtue of Ricci's theorem (III.2.17), (III.5.11), we have

$$\underset{--}{\nabla G} = \underset{--}{\nabla E} = 0 , \qquad \nabla_k G_{st} = (\nabla_k \underset{-}{R}_s) \cdot \underset{-}{R}_t + \underset{-}{R}_s \cdot \nabla_k \underset{-}{R}_t = 0 .$$

Adding two similar relations obtained by a cyclic permutation of the indices kst and subtracting the third from the sum of other two, we obtain

$$(\nabla_k \underset{-}{R}_s) \cdot \underset{-}{R}_t = 0 .$$

Now, taking Cartesian coordinates in the reference configuration as material coordinates, we arrive at the system of equations

$$\frac{\partial^2 x^m}{\partial a^k \partial a^s} \frac{\partial x_m}{\partial a^t} = 0 , \qquad (t = 1,2,3)$$

whose determinant (1.9) equals unity

$$\left| \frac{\partial x^m}{\partial a^t} \right| = \sqrt{G} = \sqrt{\det \underset{-}{E}} = 1 , \tag{24}$$

since we are considering a proper orthogonal transformation.

Thus the condition $\underset{-}{G} = \underset{-}{E}$ has lead to zero values for all the second derivatives of the Cartesian coordinates of the actual configuration with respect to the coordinates of the reference configuration

$$\frac{\partial^2 x^m}{\partial a^s \partial a^t} = 0 , \qquad \frac{\partial x^m}{\partial a^s} = \lambda_s^{\cdot m} , \qquad x^m = \lambda_s^{\cdot m} a^s + c^m , \qquad \underset{-}{R} = \underset{-}{r} \cdot \underset{-}{\Lambda} + \underset{-}{c} . \tag{25}$$

This type of linear transformation of a reference configuration into an actual configuration (Cartesian into Cartesian coordinates) is called an affine transformation. This is performed by a constant tensor $\underline{\Lambda}$. But

$$\overset{0}{\underline{\nabla R}} = \underline{\Lambda}\ , \qquad \underline{G} = \underline{E} = \underline{\Lambda}\cdot\underline{\Lambda}^{T}\ , \qquad \underline{\Lambda}^{T} = \underline{\Lambda}^{-1}\ , \qquad \underline{\Lambda} = \underline{O}\ ,$$

because the equality between transposed and inverse tensor defines an orthogonal tensor. By virtue of (24), this tensor is also properly orthogonal.

The formula (25) can now be reduced to the form (22) which is the desired result.

5. THE INVERSE TENSORS OF THE CAUCHY-GREEN AND ALMANSI STRAIN MEASURES

These tensors are introduced by the relationships /cf. (3.5)/

$$\underline{G}^{-1} = (\overset{0}{\underline{\nabla R}}\cdot\overset{0}{\underline{\nabla R}}^{T})^{-1} = \underline{\nabla r}^{T}\cdot\underline{\nabla r} = \underline{r}_{s}\underline{R}^{s}\cdot\underline{R}^{k}\underline{r}_{k} = G^{sk}\underline{r}_{s}\underline{r}_{k}\ , \tag{1}$$

$$\underline{g}^{-1} = (\underline{\nabla r}\cdot\underline{\nabla r}^{T})^{-1} = \overset{0}{\underline{\nabla R}}^{T}\cdot\overset{0}{\underline{\nabla R}} = \underline{R}_{s}\underline{r}^{s}\cdot\underline{r}^{k}\underline{R}_{k} = g^{sk}\underline{R}_{s}\underline{R}_{k}\ . \tag{2}$$

The first of these tensors is determined in the v-, the second in the V_t-basis, and their contravariant components are equal to the contravariant components of a unit tensor $\underline{E}$ in the V_t- and v-bases, respectively.

The principal values, and also the invariants, are the same for $\underline{G}$ and $\underline{g}^{-1}$, and for $\underline{g}$ and $\underline{G}^{-1}$. This is because the principal values of $\underline{A}\cdot\underline{A}^{T}$ and $\underline{A}^{T}\cdot\underline{A}$ are equal. At the same time, the principal values of $\underline{G}$ and $\underline{G}^{-1}$, and of $\underline{g}$ and $\underline{g}^{-1}$ are inverse to one another while the principal directions of these tensors coincide /cf. (I.9.14), (I.9.15)/. The tensor $\underline{g}^{-1}$ (used throughout the text) is called Finger's measure and is denoted by

$$\underline{F} = \underline{g}^{-1} = \overset{0}{\underline{\nabla R}}^{T}\cdot\overset{0}{\underline{\nabla R}}\ . \tag{3}$$

The four strain measures introduced so far are now given in the form

$$\underline{G} = G_1\overset{0}{\underline{e}}_1\overset{0}{\underline{e}}^1 + G_2\overset{0}{\underline{e}}_2\overset{0}{\underline{e}}^2 + G_3\overset{0}{\underline{e}}_3\overset{0}{\underline{e}}^3\ , \tag{4}$$

$$\underline{g} = \frac{1}{G_1}\underline{e}_1\underline{e}^1 + \frac{1}{G_2}\underline{e}_2\underline{e}^2 + \frac{1}{G_3}\underline{e}_3\underline{e}^3\ , \tag{5}$$

$$\underline{G}^{-1} = \frac{1}{G_1}\overset{0}{\underline{e}}_1\overset{0}{\underline{e}}^1 + \frac{1}{G_2}\overset{0}{\underline{e}}_2\overset{0}{\underline{e}}^2 + \frac{1}{G_3}\overset{0}{\underline{e}}_3\overset{0}{\underline{e}}^3 \tag{6}$$

$$\underline{F} = \underline{g}^{-1} = G_1\underline{e}_1\underline{e}^1 + G_2\underline{e}_2\underline{e}^2 + G_3\underline{e}_3\underline{e}^3\ . \tag{7}$$

The invariants of these tensors are determined by the formulas

$$
\begin{aligned}
&I_1(\underline{G}) = I_1(\underline{F}) = G_1 + G_2 + G_3 = g^{sk}G_{sk} \,, \\
&I_2(\underline{G}) = I_2(\underline{F}) = G_1G_2 + G_2G_3 + G_3G_1 = \frac{I_1(\underline{G}^{-1})}{I_3(\underline{G}^{-1})} = I_3(\underline{G})g_{sk}G^{sk} \,, \\
&I_3(\underline{G}) = I_3(\underline{F}) = G_1G_2G_3 = \frac{G}{g} \,, \\
&I_1(\underline{g}) = I_1(\underline{G}^{-1}) = \frac{1}{G_1} + \frac{1}{G_2} + \frac{1}{G_3} = \frac{I_2(\underline{G})}{I_3(\underline{G})} = g_{sk}G^{sk} \,, \\
&I_2(\underline{g}) = I_2(\underline{G}^{-1}) = \frac{I_1(\underline{G})}{I_3(\underline{G})} = \frac{1}{G_1G_2} + \frac{1}{G_2G_3} + \frac{1}{G_3G_1} = \frac{g}{G}\, g^{sk}G_{sk} \,, \\
&I_3(\underline{g}) = I_3(\underline{G}^{-1}) = \frac{1}{G_1G_2G_3} = \frac{g}{G} \,.
\end{aligned}
\tag{8}
$$

Here we used Equations (I.7.12), (I.9.8).

6. ORTHOGONAL TENSORS ACCOMPANYING DEFORMATION. THE LEFT AND RIGHT STRETCH TENSORS. THE HENCKY STRAIN MEASURE

The deformation gradient $\overset{0}{\nabla}\underline{\underline{R}}$ is a non-singular tensor, and therefore may be represented in polar form (I.Par.12.1)

$$
\overset{0}{\nabla}\underline{\underline{R}} = \underline{U}\cdot\underline{O} = \underline{O}\cdot\underline{V} \,, \tag{1}
$$

where $\underline{U}$, $\underline{V}$ denote some positive-definite symmetric tensors and $\underline{O}$ denotes an orthogonal tensor accompanying deformation.

In view of (1),

$$
\overset{0}{\nabla}\underline{\underline{R}}^{T} = \underline{O}^{T}\cdot\underline{U} = \underline{V}\cdot\underline{O}^{T} \tag{2}
$$

and the Cauchy-Green and Finger measures have the form

$$
\underline{G} = \overset{0}{\nabla}\underline{\underline{R}}\cdot\overset{0}{\nabla}\underline{\underline{R}}^{T} = \underline{U}\cdot\underline{O}\cdot\underline{O}^{T}\cdot\underline{U} = \underline{U}^{2} \,, \qquad \underline{F} = \overset{0}{\nabla}\underline{\underline{R}}^{T}\cdot\overset{0}{\nabla}\underline{\underline{R}} = \underline{V}\cdot\underline{O}^{T}\cdot\underline{O}\cdot\underline{V} = \underline{V}^{2} \,, \tag{3}
$$

so that

$$
\underline{U} = \underline{G}^{\frac{1}{2}} \,, \qquad \underline{V} = \underline{F}^{\frac{1}{2}} \,; \qquad \underline{O} = \underline{G}^{-\frac{1}{2}}\cdot\overset{0}{\nabla}\underline{\underline{R}} = \overset{0}{\nabla}\underline{\underline{R}}\cdot\underline{F}^{-\frac{1}{2}} \,. \tag{4}
$$

The tensors $\underline{U}$ and $\underline{V}$ are called the left and right stretch tensors. Their principal values in view of (4.10) are equal to $\sqrt{G_k} = v_k$, and their principal directions are prescribed by the normalized triplets $\overset{0}{\underline{e}}_s$ and $\underline{e}_s$ of unit vectors. The orthogonal tensor transforming one of these triplets into another is given by cf. (1.8.12),

$$\underline{O} = \overset{0}{\underline{e}}_s \underline{e}^s , \qquad \underline{O}^T = \underline{e}_s \overset{0}{\underline{e}}{}^s , \tag{5}$$

in view of (4.18), these expressions can be transformed to the following equivalent forms

$$\underline{O} = \left(\frac{\underline{e}_1 \overset{0}{\underline{e}}{}^1}{\sqrt{G_1}} + \frac{\underline{e}_2 \overset{0}{\underline{e}}{}^2}{\sqrt{G_2}} + \frac{\underline{e}_3 \overset{0}{\underline{e}}{}^3}{\sqrt{G_3}} \right) \cdot \overset{0}{\underline{\underline{\nabla R}}} = \underline{G}^{-\frac{1}{2}} \cdot \overset{0}{\underline{\underline{\nabla R}}} = \underline{U}^{-1} \cdot \overset{0}{\underline{\underline{\nabla R}}} , \tag{6}$$

$$\underline{O}^T = (\sqrt{G_1}\, \underline{e}_1 \underline{e}^1 + \sqrt{G_2}\, \underline{e}_2 \underline{e}^2 + \sqrt{G_3}\, \underline{e}_3 \underline{e}^3) \cdot \underline{\underline{\nabla r}} = \underline{F}^{\frac{1}{2}} \cdot \underline{\underline{\nabla r}} = \underline{V} \cdot \underline{\underline{\nabla r}} . \tag{7}$$

Equation (6) follows directly from (4). Equations (1) and (3.5) lead to the relationships

$$\underline{\underline{\nabla r}} = (\overset{0}{\underline{\underline{\nabla R}}})^{-1} = \underline{O}^T \cdot \underline{U}^{-1} = \underline{V}^{-1} \cdot \underline{O}^T , \qquad \underline{\underline{\nabla r}}^T = \underline{U}^{-1} \cdot \underline{O} = \underline{O} \cdot \underline{V}^{-1} , \tag{8}$$

which yield representation (7). Note also the formulas

$$\underline{V} = \underline{O}^T \cdot \underline{U} \cdot \underline{O} , \qquad \underline{U} = \underline{O} \cdot \underline{V} \cdot \underline{O}^T ; \qquad \underline{F} = \underline{O}^T \cdot \underline{G} \cdot \underline{O} , \qquad \underline{G} = \underline{O} \cdot \underline{F} \cdot \underline{O}^T . \tag{9}$$

The deformation process described by a Cauchy-Green measure can be characterized by a change of lengths of the unit segments oriented along the principal directions to the values of the components of the vector v_k and by a subsequent rotation of the triplet $\overset{0}{\underline{e}}_s$ into that of $\underline{e}_s$.

The orthogonal tensor accompanying deformation are denoted by $\underline{O}^{\times}$, $\underline{O}^{\times T}$ if it becomes necessary to distinguish them from other orthogonal tensors.

The tensors $\underline{G}$ and $\underline{g}$ (or $\underline{F}$) may be represented in terms of the derivatives of the position vectors $\underline{R}$ and $\underline{r}$. Evaluation of tensors $\underline{U}$, $\underline{V}$, that is taking the square root of a tensor, inevitably requires representation of the principal values and directions of $\underline{G}$ and $\underline{F}$ in terms of the same variables which is only possible when the components of G_{st} and g^{st} can be made explicit.

The right stretch tensor appears in the definition of the Hencky strain measure

$$\underline{H} = \ln \underline{V} , \qquad \underline{V} = \exp \underline{H} = \underline{e}_1 \underline{e}^1 \sqrt{G_1} + \underline{e}_2 \underline{e}^2 \sqrt{G_2} + \underline{e}_3 \underline{e}^3 \sqrt{G_3} , \tag{10}$$

which arises in a number of applications. The principal values and the first invariant of Hencky measure are equal to

$$H_k = \ln\sqrt{G_k} = \ln v_k \,, \qquad I_1(\underline{H}) = \ln\sqrt{G_1 G_2 G_3} = \ln\sqrt{\frac{G}{g}} \,. \tag{11}$$

Note. The eigenvectors $\overset{0}{\underline{e}}_s$, $\underline{e}_s$ are determined to within their sign provided that the eigenvalues U_k of the tensor $\underline{U}$ are simple. For this reason, Equation (5) determines eight different orthogonal tensors when all U_k are distinct, and an infinite number of such tensors when some U_k are multiple. In Equation (5) it is assumed that some three eigenvectors $\overset{0}{\underline{e}}_k$ are already chosen; the eigenvectors $\underline{e}_k$ of $\underline{V}$ are then determined by Equation (4.18).

7. STRAIN TENSORS

We introduce a vector

$$\underline{u} = \underline{R} - \underline{r} \tag{1}$$

for displacement from the reference into the actual configuration; the two coincide for $\underline{u} = \underline{0}$. Under this condition, the strain measures reduce to a unit tensor while the corresponding strain tensors turn out to be zero tensors. Linearization of these tensors is therefore simplified for suitably small displacement gradients.

We introduce the following representations of the deformation gradients /cf. (3.12)/

$$\overset{0}{\underline{\underline{\nabla R}}} = \overset{0}{\underline{\underline{\nabla r}}} + \overset{0}{\underline{\underline{\nabla u}}} = \underline{E} + \overset{0}{\underline{\underline{\nabla u}}} \,, \qquad \underline{\underline{\nabla r}} = \underline{\underline{\nabla R}} - \underline{\underline{\nabla u}} = \underline{E} - \underline{\underline{\nabla u}} \tag{2}$$

which enables us to transform the expressions for strain measures as follows

$$\underline{G} = (\underline{E} + \overset{0}{\underline{\underline{\nabla u}}}) \cdot (\underline{E} + \overset{0}{\underline{\underline{\nabla u}}}{}^T) = \underline{E} + 2[\overset{0}{\underline{\varepsilon}}(\underline{u}) + \tfrac{1}{2}\overset{0}{\underline{\underline{\nabla u}}} \cdot \overset{0}{\underline{\underline{\nabla u}}}{}^T] = \underline{E} + 2\underline{C} \,, \tag{3}$$

$$\underline{g} = (\underline{E} - \underline{\underline{\nabla u}}) \cdot (\underline{E} - \underline{\underline{\nabla u}}^T) = \underline{E} - 2[\underline{\varepsilon}(\underline{u}) - \tfrac{1}{2}\underline{\underline{\nabla u}} \cdot \underline{\underline{\nabla u}}^T] = \underline{E} - 2\underline{A} \,, \tag{4}$$

$$\underline{G}^{-1} = (\underline{E} - \underline{\underline{\nabla u}}^T) \cdot (\underline{E} - \underline{\underline{\nabla u}}) = \underline{E} - 2[\underline{\varepsilon}(\underline{u}) - \tfrac{1}{2}\underline{\underline{\nabla u}}^T \cdot \underline{\underline{\nabla u}}] = \underline{E} - 2\underline{C}' \,, \tag{5}$$

$$\underline{F} = (\underline{E} + \overset{0}{\underline{\underline{\nabla u}}}{}^T) \cdot (\underline{E} + \overset{0}{\underline{\underline{\nabla u}}}) = \underline{E} + 2[\overset{0}{\underline{\varepsilon}}(\underline{u}) + \tfrac{1}{2}\overset{0}{\underline{\underline{\nabla u}}}{}^T \cdot \overset{0}{\underline{\underline{\nabla u}}}] = \underline{E} + 2\underline{A}' \,. \tag{6}$$

Here $\overset{0}{\underline{\varepsilon}}(\underline{u})$ and $\underline{\varepsilon}(\underline{u})$ denote linear strain tensors (III.2.11) in the v- and V_t-bases, respectively

$$\overset{0}{\underline{\varepsilon}}(\underline{u}) = \tfrac{1}{2}(\overset{0}{\underline{\underline{\nabla u}}} + \overset{0}{\underline{\underline{\nabla u}}}{}^T) \,, \qquad \underline{\varepsilon}(\underline{u}) = \tfrac{1}{2}(\underline{\underline{\nabla u}} + \underline{\underline{\nabla u}}^T) \,. \tag{7}$$

The symmetric tensors $\underline{C}$, $\underline{A}$, $\underline{C}'$ and $\underline{A}'$ are strain tensors, $\underline{C}$ denotes the

Cauchy-Green tensor and $\mathbf{A}$ the Almansi strain tensor. Definitions (3)-(6) lead to the following relationships

$$\mathbf{C} = \tfrac{1}{2}(\mathbf{G} - \mathbf{E}) = \tfrac{1}{2}(G_{sk} - g_{sk})\mathbf{r}^s\mathbf{r}^k , \qquad \mathbf{A} = \tfrac{1}{2}(\mathbf{E} - \mathbf{g}) = \tfrac{1}{2}(G_{sk} - g_{sk})\mathbf{R}^s\mathbf{R}^k , \tag{8}$$

$$\mathbf{C}' = \tfrac{1}{2}(\mathbf{E} - \mathbf{G}^{-1}) = \tfrac{1}{2}(g^{sk} - G^{sk})\mathbf{r}_s\mathbf{r}_k , \qquad \mathbf{A}' = \tfrac{1}{2}(\mathbf{F} - \mathbf{E}) = \tfrac{1}{2}(g^{sk} - G^{sk})\mathbf{R}_s\mathbf{R}_k . \tag{9}$$

The covariant components of $\mathbf{C}$ and $\mathbf{A}$ are seen to be the same; a similar conclusion holds for the contravariant components of $\mathbf{C}'$ and $\mathbf{A}'$. Both pairs are of course components of different tensors since $\mathbf{C}$ and $\mathbf{C}'$ are given in the v-basis while $\mathbf{A}$ and $\mathbf{A}'$ are prescribed in the V_t-basis. For example, the contravariant components of $\mathbf{C}$ and $\mathbf{A}$ are equal to

$$C^{st} = g^{sm}g^{tn}C_{mn} , \qquad A^{st} = G^{sm}G^{tn}A_{mn} .$$

Making use of (III.2.4), (I.14.12), (I.11.4), we transform the tensor $\nabla\mathbf{u} \cdot \nabla\mathbf{u}^{\mathrm{T}}$ in the following way

$$\nabla\mathbf{u} \cdot \nabla\mathbf{u}^{\mathrm{T}} = (\boldsymbol{\varepsilon} - \boldsymbol{\Omega}) \cdot (\boldsymbol{\varepsilon} + \boldsymbol{\Omega}) = \boldsymbol{\varepsilon}^2 + \boldsymbol{\varepsilon} \cdot \boldsymbol{\Omega} - \boldsymbol{\Omega} \cdot \boldsymbol{\varepsilon} - \boldsymbol{\Omega}^2 =$$

$$= \boldsymbol{\varepsilon}^2 + \boldsymbol{\varepsilon} \cdot \boldsymbol{\Omega} + (\boldsymbol{\varepsilon} \cdot \boldsymbol{\Omega})^{\mathrm{T}} - \boldsymbol{\Omega}^2 = \boldsymbol{\varepsilon}^2 - \boldsymbol{\omega} \times \boldsymbol{\varepsilon} - (\boldsymbol{\omega} \times \boldsymbol{\varepsilon})^{\mathrm{T}} + \mathbf{E}\boldsymbol{\omega} \cdot \boldsymbol{\omega} - \boldsymbol{\omega}\boldsymbol{\omega} . \tag{10}$$

Here $\boldsymbol{\Omega}$ denotes the skew part of $\nabla\mathbf{u}^{\mathrm{T}}$ and $\boldsymbol{\omega}$ is the vector accompanying the tensor $\boldsymbol{\Omega}$. This transformation allows us to reduce the expressions for the Cauchy-Green and Almansi strain tensors to the form

$$\mathbf{C} = \overset{0}{\boldsymbol{\varepsilon}} + \tfrac{1}{2}[\overset{0}{\boldsymbol{\varepsilon}}{}^2 - \overset{0}{\boldsymbol{\omega}} \times \overset{0}{\boldsymbol{\varepsilon}} - (\overset{0}{\boldsymbol{\omega}} \times \overset{0}{\boldsymbol{\varepsilon}})^{\mathrm{T}} + \mathbf{E}\overset{0}{\boldsymbol{\omega}} \cdot \overset{0}{\boldsymbol{\omega}} - \overset{0}{\boldsymbol{\omega}}\overset{0}{\boldsymbol{\omega}}] , \tag{11}$$

$$\mathbf{A} = \boldsymbol{\varepsilon} - \tfrac{1}{2}[\boldsymbol{\varepsilon}^2 - \boldsymbol{\omega} \times \boldsymbol{\varepsilon} - (\boldsymbol{\omega} \times \boldsymbol{\varepsilon})^{\mathrm{T}} + \mathbf{E}\boldsymbol{\omega} \cdot \boldsymbol{\omega} - \boldsymbol{\omega}\boldsymbol{\omega}] . \tag{12}$$

Similarly, in view of

$$\nabla\mathbf{u}^{\mathrm{T}} \cdot \nabla\mathbf{u} = \boldsymbol{\varepsilon}^2 + \boldsymbol{\omega} \times \boldsymbol{\varepsilon} + (\boldsymbol{\omega} \times \boldsymbol{\varepsilon})^{\mathrm{T}} + \mathbf{E}\boldsymbol{\omega} \cdot \boldsymbol{\omega} - \boldsymbol{\omega}\boldsymbol{\omega} , \tag{13}$$

we may write down the expressions for $\mathbf{C}'$ and $\mathbf{A}'$. For example,

$$\mathbf{A}' = \overset{0}{\boldsymbol{\varepsilon}}(\mathbf{u}) + \tfrac{1}{2}[\overset{0}{\boldsymbol{\varepsilon}}{}^2 + \overset{0}{\boldsymbol{\omega}} \times \overset{0}{\boldsymbol{\varepsilon}} + (\overset{0}{\boldsymbol{\omega}} \times \overset{0}{\boldsymbol{\varepsilon}})^{\mathrm{T}} + \mathbf{E}\overset{0}{\boldsymbol{\omega}} \cdot \overset{0}{\boldsymbol{\omega}} - \overset{0}{\boldsymbol{\omega}}\overset{0}{\boldsymbol{\omega}}] . \tag{14}$$

In Cartesian coordinates, the components of $\mathbf{C}$ and $\mathbf{A}$ are represented by the formulas

$$C_{<st>} = \overset{0}{\varepsilon}_{<st>} + \frac{1}{2}\frac{\partial u_k}{\partial a^s}\frac{\partial u^k}{\partial a^t}\,, \qquad A_{st} = \varepsilon_{<st>} - \frac{1}{2}\frac{\partial u_k}{\partial x^s}\frac{\partial u^k}{\partial x^t}\,. \tag{15}$$

In curvilinear coordinates, we obtain corresponding representations with the aid of the rules of covariant differentiation (III.Par.5); for orthogonal curvilinear coordinates, see (III.Par.7).

Cartesian components of $\underset{-}{C}'$ and $\underset{-}{A}'$ are given by

$$C'_{<st>} = \varepsilon_{<st>} - \frac{1}{2}\frac{\partial u_s}{\partial x^k}\frac{\partial u_t}{\partial x_k}\,, \qquad A'_{<st>} = \overset{0}{\varepsilon}_{<st>} + \frac{1}{2}\frac{\partial u_s}{\partial a^k}\frac{\partial u_t}{\partial a_k} \tag{16}$$

(in Cartesian coordinates, there is no need to use the zero superscript to indicate components of the vector $\underset{-}{u}$).

<u>The invariants of strain tensors</u>. These invariants can easily be expressed in terms of those of particular strain measures $I_k(\underset{-}{G}) = I_k(\underset{-}{F})$ and $I_k(\underset{-}{g})$. To this end, one should represent powers of the tensors up to their cubes in terms of powers of the measures:

$$\underset{-}{C} = \frac{1}{2}(\underset{-}{G} - \underset{-}{E})\,, \qquad \underset{-}{C}^2 = \frac{1}{4}(\underset{-}{G} - 2\underset{-}{G} + \underset{-}{E})\,, \qquad \underset{-}{C}^3 = \frac{1}{8}(\underset{-}{G}^3 - 3\underset{-}{G}^2 + 3\underset{-}{G} - \underset{-}{E})\,. \tag{17}$$

It remains now to use the expressions (I.7.9) and (I.7.11) for the invariants of powers of a tensor in terms of its own invariants. We obtain then

$$I_1(\underset{-}{C}) = \frac{1}{2}[I_1(\underset{-}{G}) - 3]\,, \qquad I_2(\underset{-}{C}) = \frac{1}{4}[I_2(\underset{-}{G}) - 2I_1(\underset{-}{G}) + 3]\,,$$
$$I_3(\underset{-}{C}) = \frac{1}{8}[I_3(\underset{-}{G}) - I_2(\underset{-}{G}) + I_1(\underset{-}{G}) - 1] \tag{18}$$

and in precisely the same way

$$I_1(\underset{-}{A}) = \frac{1}{2}(3 - I_1(\underset{-}{g}))\,, \qquad I_2(\underset{-}{A}) = \frac{1}{4}[I_2(\underset{-}{g}) - 2I_1(\underset{-}{g}) + 3]\,,$$
$$I_3(\underset{-}{A}) = \frac{1}{8}[-I_3(\underset{-}{g}) + I_2(\underset{-}{g}) - I_1(\underset{-}{g}) + 1]\,. \tag{19}$$

The inverse relationships have the form

$$I_1(\underset{-}{G}) = 2I_1(\underset{-}{C}) + 3\,, \qquad I_2(\underset{-}{G}) = 4I_2(\underset{-}{C}) + 4I_1(\underset{-}{C}) + 3\,,$$
$$I_3(\underset{-}{G}) = 1 + 2I_1(\underset{-}{C}) + 4I_2(\underset{-}{C}) + 8I_3(\underset{-}{C})\,, \tag{20}$$

$$I_1(\underline{g}) = 3 - 2I_1(\underline{A}) , \qquad I_2(\underline{g}) = 3 - 4I_1(\underline{A}) + 4I_2(\underline{A}) ,$$
$$I_3(\underline{g}) = 1 - 2I_1(\underline{A}) + 4I_2(\underline{A}) - 8I_3(\underline{A}) . \tag{21}$$

The relationships between the invariants of tensors $\underline{A}$ and $\underline{C}$ are more complex. We may arrive at these formulas using the relationships (5.8) for the invariants of the measures. For example,

$$I_1(\underline{g}) = 3 - 2I_1(\underline{A}) = \frac{I_2(\underline{G})}{I_3(\underline{G})}$$

and in view of (20)

$$I_1(\underline{A}) = \frac{I_1(\underline{C}) + 4I_2(\underline{C}) + 12I_3(\underline{C})}{1 + 2I_1(\underline{C}) + 4I_2(\underline{C}) + 8I_3(\underline{C})} .$$

These expressions as well as the preceding formulas are rather unwieldy and for this reason strain measures should be preferred to tensors. "Nothing is gained by the introduction of displacements in place of coordinates*), but much is lost in the sense of brevity and elegance of results" (Kirchhoff).

For a rigid displacement, the strain tensors are zero tensors. This stands to reason and of course follows from (4.23), (8), (9).

As to the linear strain tensor, it is by no means zero for a rigid displacement. In fact, according to (7) and (4.22),

$$\overset{0}{\underline{\varepsilon}}(u) = \tfrac{1}{2}(\overset{0}{\underline{\nabla u}} + \overset{0}{\underline{\nabla u}}{}^{T}) = \tfrac{1}{2}(\overset{0}{\underline{\nabla R}} + \overset{0}{\underline{\nabla R}}{}^{T} - 2\underline{E}) = \tfrac{1}{2}(\underline{O} + \underline{O}^{T} - 2\underline{E}) . \tag{22}$$

At the same time,

$$\overset{0}{\underline{\nabla u}} \cdot \overset{0}{\underline{\nabla u}}{}^{T} = (\underline{O} - \underline{E}) \cdot (\underline{O}^{T} - \underline{E}) = 2\underline{E} - (\underline{O} + \underline{O}^{T})$$

and in view of (3)

$$2\underline{C} = \underline{O} + \underline{O}^{T} - 2\underline{E} + 2\underline{E} - (\underline{O} + \underline{O}^{T}) = \underline{0} .$$

8. DILATATION. THE ORIENTED ELEMENTARY AREA

1. The dilatation is defined as the ratio of the difference of elementary volumes in the actual and reference configurations to the elementary volume in the reference configuration. According to (1.9) and (7.20),

---

*) Here, "instead of a position vector".

$$\mathcal{D} = \frac{dV - dv}{dv} = \sqrt{\frac{G}{g}} - 1 = \sqrt{I_3(\underset{\sim}{G})} - 1 = [1 + 2I_1(\underset{\sim}{C}) + 4I_2(\underset{\sim}{C}) + 8I_3(\underset{\sim}{C})]^{\frac{1}{2}} - 1 \ . \qquad (1)$$

This expression may also be represented as

$$\mathcal{D} = \sqrt{G_1G_2G_3} - 1 = v_1v_2v_3 - 1 = (1+\delta_1)(1+\delta_2)(1+\delta_3) - 1 =$$

$$= [(1+2C_1)(1+2C_2)(1+2C_3)]^{\frac{1}{2}} - 1 \ , \qquad (2)$$

here C_k denote the principal values of the Cauchy-Green strain tensor.

2. Using the notation of (4.1), we may define the vectors of the oriented area in the reference and actual configurations by the following expressions

$$\underset{\sim}{n}\,do = d\underset{\sim}{r}' \times d\underset{\sim}{r}'' = \overset{0}{\underset{\sim}{e}}{}' \times \overset{0}{\underset{\sim}{e}}{}''\,ds'\,ds'' \ , \qquad \underset{\sim}{N}\,dO = d\underset{\sim}{R}' \times d\underset{\sim}{R}'' = \underset{\sim}{e}' \times \underset{\sim}{e}''\,dS'\,dS'' \ . \qquad (3)$$

Equations (4.17) and (3.1) enable us to represent the expression for $\underset{\sim}{N}\,dO$ in the form

$$\underset{\sim}{N}\,dO = \overset{0}{\underset{\sim}{e}}{}' \cdot \overset{0}{\nabla}\underset{\sim}{R} \times \overset{0}{\nabla}\underset{\sim}{R}^{T} \cdot \overset{0}{\underset{\sim}{e}}{}''\,ds'\,ds'' = \overset{0}{\underset{\sim}{e}}{}' \cdot \underset{\sim}{r}^{s}\underset{\sim}{R}_s \times \underset{\sim}{R}_k\underset{\sim}{r}^{k} \cdot \overset{0}{\underset{\sim}{e}}{}''\,ds'\,ds'' \ . \qquad (4)$$

In view of (I.2.4), (I.2.3) and (2.2), (2.3), we have

$$\underset{\sim}{R}_s \times \underset{\sim}{R}_k = \epsilon_{skt}\underset{\sim}{R}^{t} = \sqrt{\frac{G}{g}}\,\varepsilon_{skt}\underset{\sim}{R}^{t} = \sqrt{\frac{G}{g}}\,\varepsilon_{skt}\nabla\underset{\sim}{r} \cdot \underset{\sim}{r}^{t} = \sqrt{\frac{G}{g}}\,\nabla\underset{\sim}{r} \cdot (\underset{\sim}{r}_s \times \underset{\sim}{r}_k)$$

substituting this expression into (4), we obtain

$$\underset{\sim}{N}\,dO = \sqrt{\frac{G}{g}}\,\nabla\underset{\sim}{r} \cdot (\overset{0}{\underset{\sim}{e}}{}' \cdot \underset{\sim}{r}^{s}\underset{\sim}{r}_s \times \underset{\sim}{r}^{k}\underset{\sim}{r}_k \cdot \overset{0}{\underset{\sim}{e}}{}'')\,ds'\,ds'' = \sqrt{\frac{G}{g}}\,\nabla\underset{\sim}{r} \cdot (\overset{0}{\underset{\sim}{e}}{}' \times \overset{0}{\underset{\sim}{e}}{}'')\,ds'\,ds'' \ .$$

We have thus arrived at the following relationship between the parameters of the oriented elementary area in the actual configuration and in the reference configuration

$$\underset{\sim}{N}\,dO = \sqrt{\frac{G}{g}}\,\nabla\underset{\sim}{r} \cdot \underset{\sim}{n}\,do = \sqrt{\frac{G}{g}}\,\underset{\sim}{n} \cdot \nabla\underset{\sim}{r}^{T}\,do \ , \qquad (5)$$

this formula will often be used below.

The inverse relationship has the form (see (3.5))

$$\underset{\sim}{n}\,do = \sqrt{\frac{g}{G}}\,\overset{0}{\nabla}\underset{\sim}{R} \cdot \underset{\sim}{N}\,dO = \sqrt{\frac{g}{G}}\,\underset{\sim}{N} \cdot \overset{0}{\nabla}\underset{\sim}{R}^{T}\,dO \ . \qquad (6)$$

These formulas together with the definitions of strain measures (5.1) and (5.2) give rise to the relationships

$$\frac{dO}{do} = \sqrt{\frac{G}{g}}\,(\underline{n}\cdot\underline{\underline{\nabla r}}^{T}\cdot\underline{\underline{\nabla r}}\cdot\underline{n})^{\frac{1}{2}} = \sqrt{\frac{G}{g}}\,(\underline{n}\cdot\underline{G}^{-1}\cdot\underline{n})^{\frac{1}{2}}\,,\qquad \frac{do}{dO} = \sqrt{\frac{g}{G}}\,(\underline{N}\cdot\underline{F}\cdot\underline{N})^{\frac{1}{2}}\,. \tag{7}$$

We see that the measures $\underline{G}^{-1}$ and $\underline{F}$ determine the ratios of the elementary areas in the V_t- and v-configurations just as the measures $\underline{G}$ and $\underline{g}$ determine the ratios of lengths of the elementary line segments.

The unit vectors $\underline{N}$ and $\underline{n}$ are connected by the relationships

$$\underline{N} = (\underline{n}\cdot\underline{G}^{-1}\cdot\underline{n})^{-\frac{1}{2}}\,\underline{\underline{\nabla r}}\cdot\underline{n}\,,\qquad \underline{n} = (\underline{N}\cdot\underline{F}\cdot\underline{N})^{-\frac{1}{2}}\,\underline{\underline{\overset{0}{\nabla}R}}\cdot\underline{N} \tag{8}$$

or, in component representation,

$$N_s = (G^{kt}\overset{0}{n}_k\overset{0}{n}_t)^{-\frac{1}{2}}\,\overset{0}{n}_s\,,\qquad \overset{0}{n}_s = (g^{kt}N_kN_t)^{-\frac{1}{2}}\,N_s\,. \tag{9}$$

9. DIFFERENTIATION OF THE CAUCHY-GREEN AND FINGER MEASURES

The invariants $I_k(\underline{G})$ are differentiated with respect to the argument $\underline{G}$ according to the formulas (II.3.2), and with respect to the deformation gradient according to (II.3.5)

$$\begin{aligned} I_1(\underline{G})_{\underline{\underline{\overset{0}{\nabla}R}}} &= 2\underline{E}\cdot\underline{\underline{\overset{0}{\nabla}R}} = 2\underline{\underline{\overset{0}{\nabla}R}}\,,\\ I_2(\underline{G})_{\underline{\underline{\overset{0}{\nabla}R}}} &= 2(\underline{E}I_1(\underline{G})-\underline{G})\cdot\underline{\underline{\overset{0}{\nabla}R}} = 2\underline{\underline{\overset{0}{\nabla}R}}\cdot(\underline{E}I_1(\underline{G})-\underline{F})\,,\\ I_3(\underline{G})_{\underline{\underline{\overset{0}{\nabla}R}}} &= 2I_3(\underline{G})\,\underline{\underline{\nabla r}}^{T}\,. \end{aligned} \tag{1}$$

From the latter relationship, in view of

$$\sqrt{I_3(\underline{G})}_{G} = \tfrac{1}{2}\sqrt{I_3(\underline{G})}\,\underline{G}^{-1}\,, \tag{2}$$

we obtain

$$\sqrt{I_3(\underline{G})}_{\underline{\underline{\overset{0}{\nabla}R}}} = \sqrt{I_3(\underline{G})}\,\underline{\underline{\nabla r}}^{T}\,. \tag{3}$$

The divergence of this tensor is equal to zero. In fact, according to (III.6.4) and (III.4.8) we see that

$$\begin{aligned} \overset{0}{\underline{\nabla}}\cdot\sqrt{I_3(\underline{G})}\,\underline{\underline{\nabla r}}^{T} &= \frac{1}{\sqrt{g}}\frac{\partial}{\partial q^s}\sqrt{I_3(\underline{G})}\sqrt{g}\,\underline{r}^s\cdot\underline{\underline{\nabla r}}^{T} = \frac{1}{\sqrt{g}}\frac{\partial}{\partial q^s}\sqrt{G}\,\underline{R}^s = \\ &= \sqrt{\frac{G}{g}}\left(-\left\{{s \atop st}\right\}\underline{R}^t + \frac{\partial\ln\sqrt{G}}{\partial q^s}\,\underline{R}^s\right) = \underline{0}\,. \end{aligned}$$

We have thus arrived at the Piola identity

$$\overset{0}{\underline{\nabla}} \cdot \sqrt{I_3(\underline{G})}\, \overset{0}{\underline{\underline{\nabla r}}}{}^{T} = \overset{0}{\underline{\nabla}} \cdot \sqrt{I_3(\underline{G})}_{\overset{0}{\underline{\underline{\nabla R}}}} = \underline{0} \,. \tag{4}$$

By virtue of (II.3.4), we have

$$\left(\varphi_{\overset{0}{\underline{\underline{\nabla R}}}}\right)^{T} = \varphi_{\overset{0}{\underline{\underline{\nabla R}}}{}^{T}}$$

and, in view of (1) and because of the equality of the invariants $I_k(\underline{G})$ and $I_k(\underline{F})$,

$$I_1(\underline{F})_{\overset{0}{\underline{\underline{\nabla R}}}{}^{T}} = 2\overset{0}{\underline{\underline{\nabla R}}}{}^{T} \,, \qquad I_2(\underline{F})_{\overset{0}{\underline{\underline{\nabla R}}}{}^{T}} = 2\overset{0}{\underline{\underline{\nabla R}}}{}^{T} \cdot (\underline{E} I_1(\underline{F}) - \underline{G}) \,,$$

$$I_3(\underline{F})_{\overset{0}{\underline{\underline{\nabla R}}}{}^{T}} = 2 I_3(\underline{F})\, \underline{\underline{\nabla r}} \,. \tag{5}$$

The expression for the derivative of the tensor $\underline{G}$ with respect to $\overset{0}{\underline{\underline{\nabla R}}}$ is more complex. According to (I.15.9), we have

$$\begin{aligned}\underline{G}_{\overset{0}{\underline{\underline{\nabla R}}}} &= \overset{0}{\underline{\underline{\nabla R}}} \cdot \underline{C}_{III} + \underline{r}^{s}\underline{r}^{t} \cdot \overset{0}{\underline{\underline{\nabla R}}}{}^{T} \underline{r}_{s}\underline{r}_{t} = \overset{0}{\underline{\underline{\nabla R}}} \cdot \underline{C}_{III} + \underline{r}^{s}\underline{r}^{k}\underline{r}_{s}\underline{r}_{t}\underline{r}^{t} \cdot \underline{R}_{k} = \\ &= \overset{0}{\underline{\underline{\nabla R}}} \cdot \underline{C}_{III} + \underline{r}^{s}\underline{r}^{k}\underline{r}_{s}\underline{R}_{k} = \overset{0}{\underline{\underline{\nabla R}}} \cdot \underline{C}_{III} + \underline{C}_{II} \cdot \overset{0}{\underline{\underline{\nabla R}}} \,.\end{aligned}$$

Repeated application of (I.15.9) results in the relationship

$$\underline{G}_{\overset{0}{\underline{\underline{\nabla R}}}} = (\underline{C}_{II} + \underline{C}_{III}) \cdot \overset{0}{\underline{\underline{\nabla R}}} \,, \qquad \underline{g}_{\underline{\underline{\nabla r}}} = (\underline{C}_{II} + \underline{C}_{III}) \cdot \underline{\underline{\nabla r}} \tag{6}$$

where the latter formula is obtained in a similar manner to the former.

This formula is generalized to the differentiation of any tensor with respect to the deformation gradient. Following (II.4.11) and (I.15.3), we get

$$\underline{Q}_{\overset{0}{\underline{\underline{\nabla R}}}} = \underline{Q}_{\underline{G}} \cdot\cdot\, \underline{C}_{II} \cdot\cdot\, \underline{G}_{\overset{0}{\underline{\underline{\nabla R}}}} = \underline{Q}_{\underline{G}} \cdot\cdot\, (\underline{C}_{II} + \underline{C}_{III}) \cdot \overset{0}{\underline{\underline{\nabla R}}}$$

and making use of the definition of a derivative with respect to a symmetric tensor, we obtain

$$\underline{Q}_{\underline{G}} = \tfrac{1}{2}\underline{r}_{m}\underline{r}_{n} \frac{\partial \overset{0}{q}{}^{mn}}{\partial G_{st}} (\underline{r}_{s}\underline{r}_{t} + \underline{r}_{t}\underline{r}_{s}) \,,$$

$$(\underline{r}_{s}\underline{r}_{t} + \underline{r}_{t}\underline{r}_{s}) \cdot\cdot\, (\underline{C}_{II} + \underline{C}_{III}) = 2(\underline{r}_{s}\underline{r}_{t} + \underline{r}_{t}\underline{r}_{s}) \,,$$

$$\underline{\underline{Q}}_{\underline{G}} \cdot\cdot (\underline{C}_{II} + \underline{C}_{III}) = \underline{r}_m \underline{r}_n \frac{\partial \overset{0}{q}{}^{mn}}{\partial G_{st}} (\underline{r}_s \underline{r}_t + \underline{r}_t \underline{r}_s) = 2\underline{\underline{Q}}_{\underline{G}} ,$$

$$\underline{\underline{Q}}_{\overset{0}{\underline{\nabla}}\underline{R}} = 2\underline{\underline{Q}}_{\underline{G}} \cdot \overset{0}{\underline{\nabla}}\underline{R} , \qquad \underline{\underline{Q}}_{G} = \tfrac{1}{2}\underline{\underline{Q}}_{\overset{0}{\underline{\nabla}}\underline{R}} \cdot \underline{\nabla}\underline{r} . \tag{7}$$

In particular, for $\underline{\underline{Q}} = \underline{G}$. $\underline{G}_{\underline{G}} = \frac{1}{2}(\underline{C}_{II} + \underline{C}_{III})$, we arrive at $(6)_I$. In view of (II.4.16), we have also

$$\underline{F}_{\overset{0}{\underline{\nabla}}\underline{R}} = (\overset{0}{\underline{\nabla}}\underline{R}^T \cdot \overset{0}{\underline{\nabla}}\underline{R})_{\overset{0}{\underline{\nabla}}\underline{R}} = (\overset{0}{\underline{\nabla}}\underline{R}^T \cdot \underline{r}_t \underline{r}_s + \underline{r}_s \underline{r}_t \cdot \overset{0}{\underline{\nabla}}\underline{R})\, \underline{r}^t \underline{r}^s = (\underline{R}_t \underline{r}_s + \underline{r}_s \underline{R}_t)\, \underline{r}^t \underline{r}^s \tag{8}$$

and by (7), provided that the tensors $\underline{F}$ and $\underline{G}$ are coaxial,

$$\underline{F}_{\underline{G}} = \tfrac{1}{2}\underline{F}_{\overset{0}{\underline{\nabla}}\underline{R}} \cdot \underline{\nabla}\underline{r} = \tfrac{1}{2}(\underline{R}_t \underline{R}^s + \underline{R}^s \underline{R}_t)\, \underline{r}^t \underline{r}_s . \tag{9}$$

Equations (3) and (7) also show that

$$[\sqrt{I_3(\underline{G})}\, \underline{\underline{Q}}]_{\overset{0}{\underline{\nabla}}\underline{R}} = \sqrt{I_3(\underline{G})}\, (\underline{\underline{Q}}\, \underline{\nabla}\underline{r}^T + 2\underline{\underline{Q}}_{\underline{G}} \cdot \overset{0}{\underline{\nabla}}\underline{R}) . \tag{10}$$

In the component representation, recalling (3.11), and the definitions (II.4.7) and (6), we obtain

$$\underline{G}_{\overset{0}{\underline{\nabla}}\underline{R}} = \underline{r}^m \underline{r}^k \underline{r}_q \underline{r}^p \frac{\partial G_{mk}}{\partial \overset{0}{\nabla}_q X^p} = \underline{r}^s \underline{r}^t (\underline{r}_s \underline{r}_n \overset{0}{\nabla}_t X^n + \underline{r}_t \underline{r}_n \overset{0}{\nabla}_s X^n)$$

and therefore

$$\frac{\partial G_{mk}}{\partial \overset{0}{\nabla}{}_q X^p} = \underline{r}_p \underline{r}^q \underline{r}_k \underline{r}_m \cdots\cdot \underline{r}^s \underline{r}^t (\underline{r}_s \underline{r}_n \overset{0}{\nabla}_t X^n + \underline{r}_t \underline{r}_n \overset{0}{\nabla}_s X^n) .$$

The final relationships are given by

$$\frac{\partial G_{mk}}{\partial \overset{0}{\nabla}_q X^p} = \delta^q_m \overset{0}{\nabla}_k X_p + \delta^q_k \overset{0}{\nabla}_m X_p , \qquad \frac{\partial g_{mk}}{\partial \nabla_q \rho^p} = \delta^q_m \nabla_k \rho_p + \delta^q_k \nabla_m \rho_p . \tag{11}$$

We will also need an expression for the derivative $\underline{\nabla}\underline{r}^T_{\overset{0}{\underline{\nabla}}\underline{R}}$. We have

$$\underline{\nabla}\underline{r}^T_{\overset{0}{\underline{\nabla}}\underline{R}^T} = -\underline{\nabla}\underline{r}^T \cdot \underline{r}_s \underline{r}_t \cdot \underline{\nabla}\underline{r}^T \underline{r}^s \underline{r}^t = -\underline{r}_m \underline{R}^m \cdot \underline{r}_s g_{tn} \underline{R}^n \underline{r}^s \underline{r}^t = -\underline{r}_m \underline{R}^n \underline{R}^m \underline{r}_n$$

and further

$$\underline{\nabla}\underline{r}^T_{\overset{0}{\underline{\nabla}}\underline{R}} = \underline{\nabla}\underline{r}^T_{\overset{0}{\underline{\nabla}}\underline{R}^T} \cdot\cdot \underline{C}_{II} \cdot\cdot \overset{0}{\underline{\nabla}}\underline{R}^T_{\overset{0}{\underline{\nabla}}\underline{R}} = \underline{\nabla}\underline{r}^T_{\overset{0}{\underline{\nabla}}\underline{R}^T} \cdot\cdot \underline{C}_{II} \cdot\cdot \underline{C}_{III} = \underline{\nabla}\underline{r}^T_{\overset{0}{\underline{\nabla}}\underline{R}^T} \cdot\cdot \underline{C}_{II} ,$$

so that

$$\underline{\underline{\nabla r}}_{0}{}^{T}{}_{\underline{\underline{\nabla R}}} = -\underline{r}_m \underline{R}^t \underline{r}_t \underline{R}^m . \tag{12}$$

10. VARIATION OF THE STATE OF STRAIN

The material particles in their actual V_t-configuration are given, at some fixed instant t, a displacement $\eta\underline{w}(q^1,q^2,q^3)$ and to terms linear with respect to the small parameter η, variations (increments) are sought for the quantities characterizing the state of strain. In other words, we determine the configuration $\overset{\times}{V}_t$ for which the position of a particle is assigned by the radius-vector

$$\overset{\times}{\underline{R}} = \underline{R} + \eta\underline{w} . \tag{1}$$

As before, the radius vectors $\underline{r}$ and $\underline{R}$ prescribe the position of the particle in the reference v- and in the actual V_t-configurations. The variation $\delta\underline{R} = \underline{R}_s\delta q^s$ of the vector $\underline{R}$ describes the change produced by the displacement of the particle $M(q^k)$ to a new position $M(q^k + \delta q^k)$. The difference $\overset{\times}{\underline{R}} - \underline{R}$ determines the change in position of this particle caused by the imposed field $\eta\underline{w}$ of displacements. This difference will be denoted as

$$\overset{\times}{\delta}\underline{R} = \overset{\times}{\underline{R}} - \underline{R} = \eta\underline{w} \tag{2}$$

and we may write the formal derivative (which will be written as a dotted quantity) as

$$\dot{\underline{R}} = \lim_{\eta\to 0} \frac{\overset{\times}{\underline{R}} - \underline{R}}{\eta} = \underline{w} . \tag{3}$$

The vector basis in the $\overset{\times}{V}_t$-configuration is given by a triplet of vectors (see (1))

$$\begin{aligned} \overset{\times}{\underline{R}}_s &= \underline{R}_s + \eta\frac{\partial\underline{w}}{\partial q^s} = \underline{R}_s\cdot(\underline{E} + \eta\underline{\underline{\nabla w}}) = (\underline{E} + \eta\underline{\underline{\nabla w}}^T)\cdot\underline{R}_s , \\ \dot{\underline{R}}_s &= \underline{R}_s\cdot\underline{\underline{\nabla w}} = \underline{\underline{\nabla w}}^T\cdot\underline{R}_s . \end{aligned} \tag{4}$$

The dual basis can easily be determined from the relationship which expresses the invariance of a unit tensor. Discarding terms of the order η^2, we get

$$\underline{E} = \overset{\times}{\underline{R}}{}^s\overset{\times}{\underline{R}}_s = \overset{\times}{\underline{R}}{}^s\underline{R}_s + \overset{\times}{\underline{R}}{}^s\eta\underline{R}_s\cdot\underline{\underline{\nabla w}} = \overset{\times}{\underline{R}}{}^s\underline{R}_s + \eta\underline{R}^s\underline{R}_s\cdot\underline{\underline{\nabla w}} = \overset{\times}{\underline{R}}{}^s\underline{R}_s + \eta\underline{\underline{\nabla w}} ,$$

so that

$$\underline{R}^{\times s}\underline{R}_s = \underline{E} - \eta\underline{\underline{\nabla w}}\,, \qquad \underline{R}^{\times s} = \underline{R}^s - \eta\underline{\underline{\nabla w}}\cdot\underline{R}^s\,,$$
$$\underline{R}^{s\bullet} = -\underline{R}^s\cdot\underline{\underline{\nabla w}}^T = -\underline{\underline{\nabla w}}\cdot\underline{R}^s\,. \tag{5}$$

The nabla-operator in the $V_t^\times$-configuration can now be determined as

$$\underline{\nabla}^\times = \underline{R}^\times\frac{\partial}{\partial q^s} = (\underline{R}^s - \eta\underline{\underline{\nabla w}}\cdot\underline{R}^s)\frac{\partial}{\partial q^s}\,, \qquad \underline{\nabla}^\bullet = -\underline{\underline{\nabla w}}\cdot\underline{\nabla}\,. \tag{6}$$

<u>The deformation gradients in</u> $V_t^\times$. From (4) and (5) we get

$$(\overset{0}{\underline{\underline{\nabla R}}})^\times = \underline{r}^s\underline{R}_s^\times = \overset{0}{\underline{\underline{\nabla R}}} + \eta\overset{0}{\underline{\underline{\nabla R}}}\cdot\underline{\underline{\nabla w}}\,; \qquad (\overset{0}{\underline{\underline{\nabla R}}})^\bullet = \overset{0}{\underline{\underline{\nabla R}}}\cdot\underline{\underline{\nabla w}} = \overset{0}{\underline{\underline{\nabla w}}}\,,$$
$$(\overset{0}{\underline{\underline{\nabla R}}}{}^T)^\bullet = \underline{\underline{\nabla w}}^T\cdot\overset{0}{\underline{\underline{\nabla R}}}{}^T = \overset{0}{\underline{\underline{\nabla w}}}{}^T\,, \tag{7}$$

$$(\underline{\underline{\nabla r}})^\times = \underline{R}^{s\times}\underline{r}_s = \underline{\underline{\nabla r}} - \eta\underline{\underline{\nabla w}}\cdot\underline{\underline{\nabla r}}\,; \qquad (\underline{\underline{\nabla r}})^\bullet = -\underline{\underline{\nabla w}}\cdot\underline{\underline{\nabla r}}\,,$$
$$(\underline{\underline{\nabla r}}^T)^\bullet = -\underline{\underline{\nabla r}}^T\cdot\underline{\underline{\nabla w}}^T\,. \tag{8}$$

<u>Strain measures in</u> $V_t^\times$. In view of (7), (8), and following the easily checked rule

$$(\Phi\Psi)^\bullet = \Phi^\bullet\Psi + \Phi\Psi^\bullet\,, \tag{9}$$

we get

$$\underline{G}^\bullet = (\overset{0}{\underline{\underline{\nabla R}}}\cdot\overset{0}{\underline{\underline{\nabla R}}}{}^T)^\bullet = \overset{0}{\underline{\underline{\nabla R}}}\cdot(\underline{\underline{\nabla w}} + \underline{\underline{\nabla w}}^T)\cdot\overset{0}{\underline{\underline{\nabla R}}}{}^T = 2\overset{0}{\underline{\underline{\nabla R}}}\cdot\underline{\varepsilon}(\underline{w})\cdot\overset{0}{\underline{\underline{\nabla R}}}{}^T\,, \tag{10}$$

$$\underline{F}^\bullet = (\overset{0}{\underline{\underline{\nabla R}}}{}^T\cdot\overset{0}{\underline{\underline{\nabla R}}})^\bullet = \underline{F}\cdot\underline{\underline{\nabla w}} + \underline{\underline{\nabla w}}^T\cdot\underline{F}\,. \tag{11}$$

The position function of a particle $M(q^k)$ in the $V_t^\times$-configuration is represented by the expansion

$$\Phi(\underline{R}^\times) = \Phi(\underline{R}) + (\underline{R}^\times - \underline{R})\cdot\underline{\nabla}\Phi + \ldots = \Phi(\underline{R}) + \eta\underline{w}\cdot\underline{\nabla}\Phi + \ldots$$

and, in the usual notation,

$$\Phi^\bullet = \lim_{\eta\to 0}\frac{1}{\eta}[\Phi(\underline{R}^\times) - \Phi(\underline{R})] = \underline{w}\cdot\underline{\nabla}\Phi\,. \tag{12}$$

Applied to a vector $\underline{a}(\underline{R})$ and to a tensor $\underline{\underline{Q}}(\underline{R})$, these formulas will read as follows

$$\underline{a}(\underline{R})^{\bullet} = \underline{w} \cdot \underline{\underline{\nabla a}} = \underline{\underline{\nabla a}}^{T} \cdot \underline{w} , \qquad \underline{\underline{Q}}(\underline{R})^{\bullet} = \underline{w} \cdot \underline{\underline{\nabla Q}} . \tag{13}$$

The quantity $\Phi^{\bullet}$ should be called the convected derivative since it determines the change in the value of Φ associated with the particle $M(q^k)$, this change being induced by the displacement of this particle by the vector field $\underline{w}$ from the place $\underline{R}$ to the place $\overset{\times}{\underline{R}}$.

The convected derivative of a function of a tensor argument $\underline{\underline{Q}}$ is introduced in accordance with the definition of a derivative with respect to a tensor argument; we have

$$\delta^{\times}\Phi = \Phi_{\underline{\underline{Q}}} \cdot\cdot\, \delta^{\times}\underline{\underline{Q}}^{T} = \Phi_{\underline{\underline{Q}}} \cdot\cdot\, \eta\underline{\underline{Q}}^{T} ; \qquad \dot{\Phi}(\underline{\underline{Q}}) = \Phi_{\underline{\underline{Q}}} \cdot\cdot\, \underline{\underline{Q}}^{T} . \tag{14}$$

<u>The convected derivatives of the invariants</u> $(\underline{G})$. In view of (4) and (10), we have

$$I_1(\underline{G})^{\bullet} = \underline{E} \cdot\cdot\, \underline{G}^{\bullet} = 2\underline{E} \cdot\cdot\, \overset{0}{\underline{\underline{\nabla R}}} \cdot \underline{\varepsilon}(\underline{w}) \cdot \overset{0}{\underline{\underline{\nabla R}}}{}^{T} = 2\overset{0}{\underline{\underline{\nabla R}}}{}^{T} \cdot \overset{0}{\underline{\underline{\nabla R}}} \cdot\cdot\, \underline{\varepsilon}(\underline{w}) = 2I_1(\underline{F} \cdot \underline{\varepsilon}) , \tag{15}$$

$$I_2(\underline{G})^{\bullet} = [\underline{E}I_1(\underline{G}) - \underline{G}] \cdot\cdot\, \underline{G}^{\bullet} = 2I_1(\underline{G})\, I_1(\underline{F} \cdot \underline{\varepsilon}) - 2\underline{F}^{2} \cdot\cdot\, \underline{\varepsilon} =$$

$$= 2[I_1(\underline{F})\, I_1(\underline{F} \cdot \underline{\varepsilon}) - I_1(\underline{F}^{2} \cdot \underline{\varepsilon})] , \tag{16}$$

$$I_3(\underline{G})^{\bullet} = I_3(\underline{G})\underline{G}^{-1} \cdot\cdot\, \underline{G}^{\bullet} = 2I_3(\underline{G})\, \overset{0}{\underline{\underline{\nabla R}}}{}^{T} \cdot \underline{G}^{-1} \cdot \overset{0}{\underline{\underline{\nabla R}}} \cdot\cdot\, \underline{\varepsilon} =$$

$$= 2I_3(\underline{G})\underline{E} \cdot\cdot\, \underline{\varepsilon} = 2I_3(\underline{G})\, I_1(\underline{\varepsilon}) = 2I_3(\underline{G})\, \underline{\nabla} \cdot \underline{w} . \tag{17}$$

Here we used the rules (I.7.16) of contraction and formulas (II.3.1) and (II.3.3) for the derivatives of the invariants. Following (17), we also obtain

$$\sqrt{I_3(\underline{G})}^{\bullet} = \sqrt{\frac{G}{g}}^{\bullet} = \sqrt{\frac{G}{g}}\, \underline{\nabla} \cdot \underline{w} . \tag{18}$$

<u>The convected derivative of an elementary area</u>. In view of (8.5), (18), (8), we get

$$(\underline{N}\,dO)^{\bullet} = \left(\sqrt{\frac{G}{g}}\, \underline{n} \cdot \underline{\underline{\nabla r}}^{T} do\right)^{\bullet} = \sqrt{\frac{G}{g}}\, (\underline{n} \cdot \underline{\underline{\nabla r}}^{T}\underline{\nabla} \cdot \underline{w} - \underline{n} \cdot \underline{\underline{\nabla r}}^{T} \cdot \underline{\underline{\nabla w}}^{T})\, do =$$

$$= \underline{N}\,dO \cdot (\underline{\underline{E\nabla}} \cdot \underline{w} - \underline{\underline{\nabla w}}^{T}) = (\underline{\underline{E\nabla}} \cdot \underline{w} - \underline{\underline{\nabla w}}) \cdot \underline{N}\,dO . \tag{19}$$

We will also make use of the following relations

$$[\underline{T}(\underline{F})]^{\bullet} = \underline{T}_{\underline{F}} \cdot\cdot \underline{F}^{\bullet} = \underline{T}_{\underline{F}} \cdot\cdot (\underline{F} \cdot \underline{\nabla w} + \underline{\nabla w}^T \cdot \underline{F}) \ , \qquad (20)$$

$$[\underline{P}(\overset{0}{\underline{\nabla R}})]^{\bullet} = \underline{P}_{\overset{0}{\underline{\nabla R}}} \cdot\cdot \overset{0}{\underline{\nabla R}}{}^{T\bullet} = \underline{P}(\overset{0}{\underline{\nabla R}}) \cdot\cdot \overset{0}{\underline{\nabla w}}{}^T \ , \qquad (21)$$

$$(\underline{N} \cdot \underline{T}\, dO)^{\bullet} = (\underline{N}\, dO)^{\bullet} \cdot \underline{T} + \underline{N}\, dO \cdot \underline{T}^{\bullet} = \underline{N}\, dO \cdot (\underline{T}^{\bullet} + \underline{T\nabla} \cdot \underline{w} - \underline{\nabla w}^T \cdot \underline{T}) \ . \qquad (22)$$

<u>The convected derivative of a ratio of elementary areas</u>. According to (18), (8.7), (8) and (8.5), we have

$$(\frac{dO}{do})^{\bullet} = \left[\sqrt{\frac{G}{g}}\,(\underline{n} \cdot \underline{G}^{-1} \cdot \underline{n})^{\frac{1}{2}}\right]^{\bullet} = \frac{dO}{do}\,\underline{\nabla} \cdot \underline{w} +$$

$$+ \tfrac{1}{2}\sqrt{\frac{G}{g}}\,(\underline{n} \cdot \underline{G}^{-1} \cdot \underline{n})^{-\frac{1}{2}}\,(\underline{n} \cdot \underline{G}^{-1} \cdot \underline{n})^{\bullet} \ ;$$

$$(\underline{n} \cdot \underline{G}^{-1} \cdot \underline{n})^{\bullet} = -\underline{n} \cdot [\underline{\nabla r}^T \cdot \underline{\nabla w}^T \cdot \underline{\nabla r} + \underline{\nabla r}^T \cdot \underline{\nabla w} \cdot \underline{\nabla r}] \cdot \underline{n} =$$

$$= -2\underline{n} \cdot (\underline{\nabla r}^T \cdot \underline{\varepsilon} \cdot \underline{\nabla r}) \cdot \underline{n} = -2\,(\frac{dO}{do})^2\,\frac{g}{G}\,\underline{N} \cdot \underline{\varepsilon} \cdot \underline{N} \ ,$$

so that

$$\tfrac{1}{2}\sqrt{\frac{G}{g}}\,(\underline{n} \cdot \underline{G}^{-1} \cdot \underline{n})^{-\frac{1}{2}}\,(\underline{n} \cdot \underline{G}^{-1} \cdot \underline{n})^{\bullet} = -\frac{dO}{do}\,\underline{N} \cdot \underline{\varepsilon} \cdot \underline{N} \ .$$

We thus obtain

$$(\frac{dO}{do})^{\bullet} = \frac{dO}{do}\,(\underline{\nabla} \cdot \underline{w} - \underline{N} \cdot \underline{\varepsilon} \cdot \underline{N}) \ . \qquad (23)$$

11. THE VARIATION OF AN ORTHOGONAL TENSOR ACCOMPANYING DEFORMATION

Making use of (10.9), we perform the variation of the relationship determining the principal values q_s and directions $\underline{e}_s$ of a symmetric tensor $\underline{\underline{Q}}$;

$$\underline{\underline{Q}} \cdot \underline{e}_s = \sum_s q_s \underline{e}_s \ , \qquad \underline{\underline{Q}}^{\bullet} \cdot \underline{e}_s + \underline{\underline{Q}} \cdot \underline{e}_s^{\bullet} = \sum_s (q_s^{\bullet} \underline{e}_s + q_s \underline{e}_s^{\bullet}) \qquad (1)$$

or, after scalar multiplication by $\underline{e}_k$,

$$\underline{e}_k \cdot \underline{\underline{Q}}^{\bullet} \cdot \underline{e}_s + \sum_k q_k \underline{e}_k \cdot \underline{e}_s^{\bullet} = \sum_s (q_s^{\bullet} \underline{e}_k \cdot \underline{e}_s + q_s \underline{e}_k \cdot \underline{e}_s^{\bullet}) \ .$$

For $k = s$, $\underline{e}_s \cdot \underline{e}_s = 1$, $\underline{e}_s \cdot \underline{e}_s^{\bullet} = 0$; for $k \neq s$, $\underline{e}_k \cdot \underline{e}_s = 0$. We get

$$q_s^{\bullet} = \underline{e}_s \cdot \underline{\underline{Q}}^{\bullet} \cdot \underline{e}_s \ , \qquad \underline{e}_s^{\bullet} \cdot \underline{e}_k = \frac{\underline{e}_k \cdot \underline{\underline{Q}}^{\bullet} \cdot \underline{e}_s}{q_s - q_k} \ , \qquad \underline{e}_s^{\bullet} = \sum_k{}' \frac{\underline{e}_k \cdot \underline{\underline{Q}}^{\bullet} \cdot \underline{e}_s}{q_s - q_k}\,\underline{e}_k \qquad (2)$$

(a prime means omission of the terms $k = s$). We have thus obtained the formulas for convected derivatives of the principal values and principal directions of a tensor.

On applying this to the Cauchy-Green strain measure, we have by virtue of (10.10), (6.1) and (6.2)

$$\dot{G}_s = 2\overset{0}{\underline{e}}_s \cdot \overset{0}{\underline{\underline{\nabla R}}} \cdot \underline{\varepsilon} \cdot \overset{0}{\underline{\underline{\nabla R}}}{}^T \cdot \overset{0}{\underline{e}}_s = 2\overset{0}{\underline{e}}_s \cdot \underline{U} \cdot \underline{O} \cdot \underline{\varepsilon} \cdot \underline{O}^T \cdot \underline{U} \cdot \overset{0}{\underline{e}}_s =$$

$$= 2\sqrt{G_s}\,\overset{0}{\underline{e}}_s \cdot \underline{O} \cdot \underline{\varepsilon} \cdot \underline{O}^T \cdot \overset{0}{\underline{e}}_s \sqrt{G_s}\,,$$

here we have made use of the formula $\overset{0}{\underline{e}}_s \cdot \underline{U} = \overset{0}{\underline{e}}_s \sqrt{G_s}$. From (6.5) we have

$$\dot{G}_s = 2G_s \underline{e}_s \cdot \underline{\varepsilon} \cdot \underline{e}_s\,. \tag{3}$$

The vector $\overset{0\,\bullet}{\underline{e}}_s$ is represented in a similar way

$$\overset{0\,\bullet}{\underline{e}}_s = 2\sum_k{}' \frac{\overset{0}{\underline{e}}_k \cdot \overset{0}{\underline{\underline{\nabla R}}} \cdot \underline{\varepsilon} \cdot \overset{0}{\underline{\underline{\nabla R}}}{}^T \cdot \overset{0}{\underline{e}}_s}{G_s - G_k}\,\overset{0}{\underline{e}}_k = 2\sum_k{}' \frac{\sqrt{G_s G_k}}{G_s - G_k}\,\underline{e}_k \cdot \underline{\varepsilon} \cdot \underline{e}_s \overset{0}{\underline{e}}_k\,. \tag{4}$$

The expression (6.6) for an orthogonal tensor measure of deformation enables us to suggest the following representation for its convected derivative

$$\dot{\underline{O}} = \left[-\sum_s \frac{\dot{G}_s}{2G_s^{3/2}}\,\overset{0}{\underline{e}}_s\overset{0}{\underline{e}}_s + \sum_s \frac{1}{\sqrt{G_s}}\,(\overset{0\,\bullet}{\underline{e}}_s\overset{0}{\underline{e}}_s + \overset{0}{\underline{e}}_s\overset{0\,\bullet}{\underline{e}}_s)\right] \cdot \overset{0}{\underline{\underline{\nabla R}}} +$$

$$+ \sum_s \frac{\overset{0}{\underline{e}}_s\overset{0}{\underline{e}}_s}{\sqrt{G_s}} \cdot (\overset{0}{\underline{\underline{\nabla R}}})^\bullet = \left[-\sum_s \frac{1}{\sqrt{G_s}}\,\underline{e}_s \cdot \underline{\varepsilon} \cdot \underline{e}_s\overset{0}{\underline{e}}_s\overset{0}{\underline{e}}_s + \right.$$

$$\left. + 2\sum_s \sum_k{}' \frac{\sqrt{G_k}}{G_s - G_k}\,\underline{e}_k \cdot \underline{\varepsilon} \cdot \underline{e}_s(\overset{0}{\underline{e}}_k\overset{0}{\underline{e}}_s + \overset{0}{\underline{e}}_s\overset{0}{\underline{e}}_k)\right] \cdot \overset{0}{\underline{\underline{\nabla R}}} + \underline{O} \cdot \underline{\underline{\nabla w}}\,. \tag{5}$$

Here we have used (3), (4) and (10.7). The double sum is transformed to

$$\sum_s \sum_k{}' \frac{\underline{e}_k \cdot \underline{\varepsilon} \cdot \underline{e}_s}{G_s - G_k}\,(\sqrt{G_k} - \sqrt{G_s})\overset{0}{\underline{e}}_k\overset{0}{\underline{e}}_s = -\sum_s \sum_k{}' \frac{\underline{e}_k \cdot \underline{\varepsilon} \cdot \underline{e}_s}{\sqrt{G_s} + \sqrt{G_k}}\,\overset{0}{\underline{e}}_k\overset{0}{\underline{e}}_s$$

which allows the sums in the brackets in (5) to be combined. We thus arrive at the desired representation of the convected derivative

$$\dot{\underline{O}} = -2\sum_s \sum_k \frac{\underline{e}_k \cdot \underline{\varepsilon} \cdot \underline{e}_s}{\sqrt{G_s} + \sqrt{G_k}}\,\overset{0}{\underline{e}}_k\overset{0}{\underline{e}}_s \cdot \overset{0}{\underline{\underline{\nabla R}}} + \underline{O} \cdot \underline{\underline{\nabla w}} =$$

$$= -2\underline{\underline{\nabla r}}^T \cdot \sum_s \sum_k \frac{\underline{e}_k \cdot \underline{\varepsilon} \cdot \underline{e}_s}{\sqrt{G_s} + \sqrt{G_k}}\,\sqrt{G_s G_k}\,\underline{e}_k\underline{e}_s + \underline{O} \cdot \underline{\underline{\nabla w}}\,. \tag{6}$$

Here we have used Equations (4.18). We also have

$$\underline{O}^{T\bullet} = -2 \sum_s \sum_k \frac{\underline{e}_k \cdot \underline{\varepsilon} \cdot \underline{e}_s}{\sqrt{G_s} + \sqrt{G_k}} \sqrt{G_s G_k}\, \underline{e}_s \underline{e}_k \cdot \underline{\nabla r} + \underline{\nabla w}^T \cdot \underline{O}^T . \tag{7}$$

By virtue of the representation (6.5) for the tensor $\underline{O}$, it is easy to check the condition

$$\underline{O}^{\bullet} \cdot \underline{O}^T + \underline{O} \cdot \underline{O}^{T\bullet} = \underline{0} . \tag{8}$$

We also note the relationships

$$\underline{O}^{\bullet} \cdot\cdot \overset{0}{\underline{\nabla R}}{}^T = \overset{0}{\underline{\nabla R}} \cdot\cdot \underline{O}^{T\bullet} = \underline{0} . \tag{9}$$

In fact, we have

$$\underline{O}^{\bullet} \cdot\cdot \overset{0}{\underline{\nabla R}}{}^T = \underline{O}^{\bullet} \cdot\cdot \underline{O}^T \cdot \underline{U} = (\underline{O}^{\bullet} \cdot \underline{O}^T) \cdot\cdot \underline{U} = \underline{0} ,$$

this is because tensor $\underline{O}^{\bullet} \cdot \underline{O}^T$ is skew symmetric (see (8)) and the tensor $\underline{U}$ is symmetric.

The convected derivatives of the stretch tensors $\underline{U}$ and $\underline{V}$ can be calculated from (6) and (7), i.e.

$$\underline{\dot{U}} = (\overset{0}{\underline{\nabla R}} \cdot \underline{O}^T)^{\bullet} = \overset{0}{\underline{\nabla R}} \cdot (\underline{\nabla w} \cdot \underline{O}^T + \underline{O}^{T\bullet}) ,$$

$$\underline{\dot{V}} = (\overset{0}{\underline{\nabla R}}{}^T \cdot \underline{O})^{\bullet} = \tag{10}$$

$$= \underline{\nabla w}^T \cdot \underline{V} + \overset{0}{\underline{\nabla R}}{}^T \cdot \underline{O}^{\bullet} = \underline{\nabla w}^T \cdot \underline{V} + \underline{V} \cdot \underline{\nabla w} - 2 \sum_s \sum_k \frac{\underline{e}_k \cdot \underline{\varepsilon} \cdot \underline{e}_s}{V_s + V_k} V_s V_k \underline{e}_k \underline{e}_s .$$

From (9), we also have

$$I_1(\underline{U})^{\bullet} = I_1(\underline{U}^{\bullet}) = \underline{\nabla w} \cdot\cdot \underline{O}^T \cdot \overset{0}{\underline{\nabla R}} = \underline{\nabla w} \cdot\cdot \underline{V} . \tag{11}$$

12. THE SECOND DERIVATIVE OF A SCALAR FUNCTION OF A TENSOR ARGUMENT

The strain measure $\underline{G}^{\times}$ in the $V_t^{\times}$-configuration is given by

$$\underline{G}^{\times} = \overset{0}{\underline{\nabla R}}{}^{\times} \cdot \overset{0}{\underline{\nabla R}}{}^{T\times} = \overset{0}{\underline{\nabla R}} \cdot (\underline{E} + \eta \underline{\nabla w}) \cdot (\underline{E} + \eta \underline{\nabla w}^T) \cdot \overset{0}{\underline{\nabla R}}{}^T =$$

$$= \underline{G} + 2\eta \overset{0}{\underline{\nabla R}} \cdot \underline{\varepsilon} \cdot \overset{0}{\underline{\nabla R}}{}^T + \eta^2 \overset{0}{\underline{\nabla R}} \cdot \underline{\nabla w} \cdot \underline{\nabla w}^T \cdot \overset{0}{\underline{\nabla R}}{}^T . \tag{1}$$

We should therefore bear in mind the necessity of preserving the term $\eta^2\overset{0}{\underline{\nabla R}}\cdot\underline{\nabla w}\cdot\underline{\nabla w}^T\cdot\underline{\nabla R}^T$ in the expression for the factor $\delta\underline{G}^T$ in the first derivative when using formula (II.4.23) to calculate the second derivative of $\varphi(\underline{G}^\times)$ to terms of order η^2. Formula (II.4.23) is now reduced to

$$\varphi(\underline{G}^\times) - \varphi(\underline{G}) = \varphi_{\underline{G}}(\underline{G})\,..\,(2\eta\overset{0}{\underline{\nabla R}}\cdot\underline{\varepsilon}\cdot\overset{0}{\underline{\nabla R}}{}^T + \eta^2\overset{0}{\underline{\nabla R}}\cdot\underline{\nabla w}\cdot\underline{\nabla w}^T\cdot\overset{0}{\underline{\nabla R}}{}^T) +$$

$$+ \tfrac{1}{2}\,2\eta\overset{0}{\underline{\nabla R}}\cdot\underline{\varepsilon}\cdot\overset{0}{\underline{\nabla R}}{}^T\,..\,\varphi_{\underline{GG}}(\underline{G})\,..\,2\eta\overset{0}{\underline{\nabla R}}\cdot\underline{\varepsilon}\cdot\overset{0}{\underline{\nabla R}}{}^T + \ldots = \eta\varphi_{\underline{G}}(\underline{G})\,..\,\underline{G}^{\bullet} +$$

$$+ \frac{\eta^2}{2}\,[2\varphi_{\underline{G}}(\underline{G})\,..\,\overset{0}{\underline{\nabla R}}\cdot\underline{\nabla w}\cdot\underline{\nabla w}^T\cdot\overset{0}{\underline{\nabla R}}{}^T + \underline{G}^{\bullet}\,..\,\varphi_{\underline{GG}}(\underline{G})\,..\,\underline{G}^{\bullet}] + \ldots\,, \tag{2}$$

here $\underline{G}^{\bullet}$ is given by (10.10). This necessity vanishes when calculating the second derivative of a scalar function of the argument $\overset{0}{\underline{\nabla R}}$; for this case, according to (II.4.23), we have

$$\delta\overset{0}{\underline{\nabla R}}{}^T = \eta\underline{\nabla w}^T\cdot\overset{0}{\underline{\nabla R}}{}^T\,;$$

$$\varphi(\overset{0}{\underline{\nabla R}}{}^\times) - \varphi(\overset{0}{\underline{\nabla R}}) = \eta\varphi_{\overset{0}{\underline{\nabla R}}}\,..\,\underline{\nabla w}^T\cdot\overset{0}{\underline{\nabla R}}{}^T + \frac{\eta^2}{2}\,\underline{\nabla w}^T\cdot\overset{0}{\underline{\nabla R}}{}^T\,..\,\varphi_{\overset{0}{\nabla R}\overset{0}{\nabla R}}\,..\,\underline{\nabla w}^T\cdot\overset{0}{\underline{\nabla R}}{}^T + \ldots\,. \tag{3}$$

For example, for $\varphi(\underline{G}) = I_1(\underline{G})$ we have $\varphi_{\underline{G}} = \underline{E}$, $\varphi_{\underline{GG}} = \underline{0}$ and, in view of (2),

$$I_1(\underline{G}^\times) = I_1(\underline{G}) + 2\eta\underline{E}\,..\,\overset{0}{\underline{\nabla R}}\cdot\underline{\varepsilon}\cdot\overset{0}{\underline{\nabla R}}{}^T + \eta^2\underline{E}\,..\,\overset{0}{\underline{\nabla R}}\cdot\underline{\nabla w}\cdot\underline{\nabla w}^T\cdot\overset{0}{\underline{\nabla R}}{}^T =$$

$$= I_1(\underline{G}) + 2\eta\underline{F}\,..\,\underline{\varepsilon} + \eta^2\underline{F}\,..\,\underline{\nabla w}\cdot\underline{\nabla w}^T \tag{4}$$

of course, this expression follows immediately from (1), For

$$\varphi(\underline{G}) = \sqrt{I_3(\underline{G})}\,, \qquad \varphi_{\underline{G}}(\underline{G}) = \tfrac{1}{2}\sqrt{I_3(\underline{G})}\,\underline{G}^{-1}\,,$$

$$\varphi_{\underline{GG}}(\underline{G}) = \tfrac{1}{4}\sqrt{I_3(\underline{G})}\,\underline{G}^{-1}\underline{G}^{-1} + \tfrac{1}{2}\sqrt{I_3(\underline{G})}\,\underline{G}^{-1}_{\underline{G}}$$

and in view of (2) we have

$$\sqrt{I_3(\underline{G}^\times)} - \sqrt{I_3(\underline{G})} = \sqrt{I_3(\underline{G})}\,\{\tfrac{1}{2}\eta\underline{G}^{-1}\,..\,\underline{G}^{\bullet} +$$

$$+ \tfrac{1}{2}\eta^2\,[\tfrac{1}{4}\underline{G}^{\bullet}\,..\,\underline{G}^{-1}\cdot\underline{G}^{-1}\,..\,\underline{G}^{\bullet} + \underline{G}^{-1}\,..\,\overset{0}{\underline{\nabla R}}\cdot\underline{\nabla w}\cdot\underline{\nabla w}^T\cdot\overset{0}{\underline{\nabla R}}{}^T + \tfrac{1}{2}\underline{G}^{\bullet}\,..\,\underline{G}^{-1}_{\underline{G}}\,..\,\underline{G}^{\bullet}]\}\,.$$

Taking account of (II.4.20) and the rules of contraction (I.7.16), we arrive at the relationships

$$\tfrac{1}{2}\underline{G}^{-1} \cdot\cdot\, 2\overset{0}{\underline{\underline{\nabla R}}} \cdot \underline{\varepsilon} \cdot \overset{0}{\underline{\underline{\nabla R}}}{}^{T} = \overset{0}{\underline{\underline{\nabla R}}}{}^{T} \cdot \underline{G}^{-1} \cdot \overset{0}{\underline{\underline{\nabla R}}} \cdot\cdot\, \underline{\varepsilon} = \underline{E} \cdot\cdot\, \underline{\varepsilon} = \underline{\nabla} \cdot \underline{w} \ ,$$

$$\underline{G}^{-1} \cdot\cdot\, \overset{0}{\underline{\underline{\nabla R}}} \cdot \underline{\underline{\nabla w}} \cdot \underline{\underline{\nabla w}}{}^{T} \cdot \overset{0}{\underline{\underline{\nabla R}}}{}^{T} = \underline{E} \cdot\cdot\, \underline{\underline{\nabla w}} \cdot \underline{\underline{\nabla w}}{}^{T} = \underline{\underline{\nabla w}} \cdot\cdot\, \underline{\underline{\nabla w}}{}^{T} \ ,$$

$$\tfrac{1}{4}\underline{G}^{\bullet} \cdot\cdot\, \underline{G}^{-1}\underline{G}^{-1} \cdot\cdot\, \underline{G}^{\bullet} = \tfrac{1}{4}(\underline{G}^{\bullet} \cdot\cdot\, \underline{G}^{-1})^{2} = (\underline{\nabla} \cdot \underline{w})^{2} \ ,$$

$$\tfrac{1}{2}\underline{G}^{\bullet} \cdot\cdot\, \underline{G}^{-1}_{\underline{G}} \cdot\cdot\, \underline{G}^{\bullet} =$$

$$= -\overset{0}{\underline{\underline{\nabla R}}} \cdot \underline{\varepsilon} \cdot \overset{0}{\underline{\underline{\nabla R}}}{}^{T} \cdot\cdot\, \underline{G}^{-1} \cdot \underline{r}^{t}\underline{r}^{s} \cdot \underline{G}^{-1}(\underline{r}_{s}\underline{r}_{t} + \underline{r}_{t}\underline{r}_{s}) \cdot\cdot\, \overset{0}{\underline{\underline{\nabla R}}} \cdot \underline{\varepsilon} \cdot \overset{0}{\underline{\underline{\nabla R}}}{}^{T} =$$

$$= -\underline{\varepsilon} \cdot\cdot\, \overset{0}{\underline{\underline{\nabla R}}}{}^{T} \cdot \underline{G}^{-1} \cdot \underline{r}^{t}\underline{r}^{s} \cdot \underline{G}^{-1} \cdot \overset{0}{\underline{\underline{\nabla R}}}[\overset{0}{\underline{\underline{\nabla R}}}{}^{T} \cdot \underline{r}_{s}\underline{r}_{t} \cdot \overset{0}{\underline{\underline{\nabla R}}} + \overset{0}{\underline{\underline{\nabla R}}}{}^{T} \cdot \underline{r}_{t}\underline{r}_{s} \cdot \overset{0}{\underline{\underline{\nabla R}}}] \cdot\cdot\, \underline{\varepsilon} =$$

$$= -\underline{\varepsilon} \cdot\cdot\, \underline{R}^{t}\underline{R}^{s}(\underline{R}_{s}\underline{R}_{t} + \underline{R}_{t}\underline{R}_{s}) \cdot\cdot\, \underline{\varepsilon} = -\underline{\varepsilon} \cdot\cdot\, (\underline{C}_{III} + \underline{C}_{II}) \cdot\cdot\, \underline{\varepsilon} = -2\underline{\varepsilon} \cdot\cdot\, \underline{\varepsilon} = -2I_{1}(\underline{\varepsilon}^{2}) \ .$$

Here we have also used expressions (3.1) for the deformation gradients and those of (I.15.4) for contraction of isotropic tensors with a tensor of the second rank. We thus arrive at equation

$$\sqrt{I_3(\underline{G}^{\times})} - \sqrt{I_3(\underline{G})} = \sqrt{I_3(\underline{G})}\left\{\eta I_1(\underline{\varepsilon}) + \frac{\eta^2}{2}[I_1^2(\underline{\varepsilon}) + \underline{\underline{\nabla w}} \cdot\cdot\, \underline{\underline{\nabla w}}{}^{T} - 2I_1(\underline{\varepsilon}^2)]\right\} =$$

$$= \sqrt{I_3(\underline{G})}\,\{\eta\underline{\nabla} \cdot \underline{w} + \tfrac{1}{2}\eta^2[(\underline{\nabla} \cdot \underline{w})^2 - \underline{\underline{\nabla w}} \cdot\cdot\, \underline{\underline{\nabla w}}]\} \ . \tag{5}$$

From here we deduce the formula for the increment in the volume of the body produced by the field of displacements $\eta\underline{w}$, this increment is calculated to terms of order η^2:

$$V^{\times} - V = \eta \iiint\limits_{V} \underline{\nabla} \cdot \underline{w}\, dV + \tfrac{1}{2}\eta^2 \iiint\limits_{V} [(\underline{\nabla} \cdot \underline{w})^2 - \underline{\underline{\nabla w}} \cdot\cdot\, \underline{\underline{\nabla w}}]\, dV \ . \tag{6}$$

13. KINEMATIC RELATIONS

The equation of motion (1.3) of a particle $M(q^1, q^2, q^3)$ allows us to determine its velocity vector

$$\underline{v} = \frac{\partial}{\partial t}\underline{R}(q^1, q^2, q^3, t) \ . \tag{1}$$

Substituting material coordinates for coordinates in the actual V_t-configuration (see Par. 1), we arrive at the following representation

$$\underline{v} = \underline{v}(x^1, x^2, x^3 \,;\, t) = \underline{v}(\underline{R} \,;\, t) \ . \tag{2}$$

We have thus performed a transition from the material ("Lagrangian") description of motion to the spatial ("Eulerian") description, as it was put earlier in Par. 1. Formula (2) gives the field of velocities within the medium. In Equation (1), we follow the motion of a given particle, in Equation (2) we observe the motion "now, at this place".

Let us assume that within the field (2) of velocities, there is prescribed a scalar field

$$\varphi = \varphi(q^1, q^2, q^3, t) = \varphi(\underline{R}; t) \ . \tag{3}$$

Differentiation of this expression yields, in view of (3.7),

$$d\varphi = \frac{\partial \varphi}{\partial t} dt + \frac{\partial \varphi}{\partial q^s} dq^s = \frac{\partial \varphi}{\partial t} dt + d\underline{R} \cdot \underline{R}^s \frac{\partial \varphi}{\partial q^s} = \frac{\partial \varphi}{\partial t} dt + d\underline{R} \cdot \underline{\nabla}\varphi \ . \tag{4}$$

The first term determines the increment of φ at some fixed place for the time interval dt - the local increment, the second term has arisen because, for this time interval, the particle $M(q^k)$ carrying the value of the scalar φ, has moved from the position $\underline{R}$ to that of $\underline{R} + d\underline{R} = \underline{R} + \underline{v}dt$, at which the increment in the scalar field is $\underline{v} \cdot \underline{\nabla}\varphi dt$. In other words, we must calculate the convective transport produced by the field of velocities. Using the notation of Par. 10, we may write

$$\overset{\times}{\delta}\varphi = dt\, \underline{v} \cdot \underline{\nabla}\varphi \ , \qquad \varphi^{\bullet} = \underline{v} \cdot \underline{\nabla}\varphi \tag{5}$$

here the field $\underline{v}$ is used instead of that of $\underline{w}$, and η is replaced by dt.

Equation (4) now has the form

$$\dot{\varphi} = \frac{\partial \varphi}{\partial t} + \varphi^{\bullet} \ . \tag{6}$$

This expression determines the "material derivative" (the terms "individual" and "substantial" are equally applicable) of φ with respect to time, this derivative being equal to a sum of the local and convected derivatives. The dot above the symbol is a simple way of indicating differentiation with respect to time, but it is more usual to use the symbols d/dt, D/Dt.

The material derivative of a vector $\underline{a}$ and a tensor $\underline{\underline{Q}}$ may be introduced in a similar way, i.e.

$$\dot{\underline{a}}(\underline{R}; t) = \frac{\partial \underline{a}}{\partial t} + \underline{v} \cdot \underline{\nabla a} = \frac{\partial \underline{a}}{\partial t} + \underline{\nabla a}^T \cdot \underline{v} \ , \qquad \dot{\underline{\underline{Q}}}(\underline{R}; t) = \frac{\partial \underline{\underline{Q}}}{\partial t} + \underline{v} \cdot \underline{\nabla \underline{Q}} \ . \tag{7}$$

In particular, for $\underline{a} = \underline{v}$ the first relationship determines the field of accelerations $\underline{b}$

$$\underline{b} = \frac{\partial \underline{v}}{\partial t} + \underline{v}\cdot\underline{\underline{\nabla v}} = \frac{\partial \underline{v}}{\partial t} + \underline{\underline{\nabla v}}^T\cdot\underline{v} \ . \tag{8}$$

The notation (10.3) for the vector $\underline{R}^{\bullet}$ is the same as that for the definition (1) for the velocity vector; for this reason the remaining formulas of Par. 10 retain their meaning when $\underline{w}$ is substituted for $\underline{v}$ and a dot is placed above the symbol, this is, of course, provided that the differentiated quantities depend on $\underline{R}$ (but not on its derivatives with respect to q^s) and do not depend explicitly on $\underline{t}$.

For $\underline{a}$ and $\underset{\approx}{Q}$ not depending explicitly on time, Equations (7) reiterate the definition (10.13) of the convected derivative when we consider the convective transport produced by the field $\underline{w}$. The time derivatives of the quantities which depend only upon time (not upon position) can also be designated by a dot above the symbol. Due to lack of space non-stationary fields, i.e. fields which explicitly depend on time, are not dealt with here and so we do not differentiate between material and convected derivatives.

The time derivative of the deformation gradient prescribed by its polar representation (6.1) is given by (10.7)

$$(\overset{0}{\underline{\underline{\nabla R}}})^{\bullet} = \overset{0}{\underline{\underline{\nabla R}}}\cdot\underline{\underline{\nabla v}} = (\underline{U}\cdot\underline{O})^{\bullet} = \dot{\underline{U}}\cdot\underline{O} + \underline{U}\cdot\dot{\underline{O}} \ .$$

Therefore we have

$$\underline{\underline{\nabla v}} = \underline{\underline{\nabla r}}\cdot(\dot{\underline{U}}\cdot\underline{O} + \underline{U}\cdot\dot{\underline{O}}) = \underline{O}^T\cdot(\underline{U}^{-1}\cdot\dot{\underline{U}}\cdot\underline{O} + \dot{\underline{O}}) \ . \tag{9}$$

The linear strain tensor $\underline{\varepsilon}(\underline{v})$ for the velocity vector (the "deformation of velocity") is traditionally designated by $\underline{D}$, and a vorticity tensor (a spin of $\underline{v}$) is denoted by $\underline{W}$. In view of (9), we obtain

$$\underline{D} = \tfrac{1}{2}(\underline{\underline{\nabla v}} + \underline{\underline{\nabla v}}^T) = \tfrac{1}{2}\underline{O}^T\cdot(\underline{U}^{-1}\cdot\dot{\underline{U}} + \dot{\underline{U}}\cdot\underline{U}^{-1})\cdot\underline{O} \ , \tag{10}$$

$$\underline{W} = \tfrac{1}{2}(\underline{\underline{\nabla v}}^T - \underline{\underline{\nabla v}}) = \dot{\underline{O}}^T\cdot\underline{O} + \tfrac{1}{2}\underline{O}^T\cdot(\dot{\underline{U}}\cdot\underline{U}^{-1} - \underline{U}^{-1}\cdot\dot{\underline{U}})\cdot\underline{O} \ . \tag{11}$$

Here we have used Equation (11.8).

14. THE MATERIAL DERIVATIVE OF AN INTEGRAL AND THE LAW OF CONSERVATION OF MASS

We now consider the integral of the quantity prescribed in the V_t-configuration of a continuous medium occupying the volume V

$$\Psi = \iiint_V \varphi\, dV \ , \qquad \underline{c} = \iiint_V \underline{a}\, dV \ , \qquad \underline{P} = \iiint_V \underset{\approx}{Q}\, dV \ . \tag{1}$$

In the reference configuration the particles contained in V completely filled the volume v; we are thus able to eliminate dV using Equation (I.1.9) and so represent expressions (1) in terms of integrals evaluated over volume v

$$\Psi = \iiint\limits_{v} \sqrt{\frac{G}{g}}\, \varphi\, dv\,, \qquad \underline{c} = \iiint\limits_{v} \sqrt{\frac{G}{g}}\, \underline{a}\, dv\,, \qquad \underline{P} = \iiint\limits_{v} \sqrt{\frac{G}{g}}\, \underline{\underline{Q}}\, dv\,. \tag{2}$$

The volume V is variable because this is so for its boundary surface O; this point complicates calculation of the material derivatives of (1). This obstacle may, however, be overcome so long as we turn to the expressions (2). It now suffices only to evaluate the material derivatives only for the integrands themselves. In view of (10.18) and (13.6) we obtain

$$\dot{\Psi} = \iiint\limits_{v} \sqrt{\frac{G}{g}}\, (\varphi \underline{\nabla} \cdot \underline{v} + \dot{\varphi})\, dv = \iiint\limits_{v} \sqrt{\frac{G}{g}} \left(\varphi \underline{\nabla} \cdot \underline{v} + \underline{v} \cdot \underline{\nabla} \varphi + \frac{\partial \varphi}{\partial t} \right) dv =$$

$$= \iiint\limits_{V} \left(\frac{\partial \varphi}{\partial t} + \underline{\nabla} \cdot \varphi \underline{v} \right) dV \tag{3}$$

and similarly

$$\dot{\underline{c}} = \iiint\limits_{v} \sqrt{\frac{G}{g}} \left(\underline{a} \underline{\nabla} \cdot \underline{v} + \underline{v} \cdot \underline{\nabla} \underline{a} + \frac{\partial \underline{a}}{\partial t} \right) dv = \iiint\limits_{V} \left(\frac{\partial \underline{a}}{\partial t} + \underline{\nabla} \cdot \underline{v}\underline{a} \right) dV\,, \tag{4}$$

$$\dot{\underline{P}} = \iiint\limits_{v} \sqrt{\frac{G}{g}} \left(\underline{\underline{Q}} \underline{\nabla} \cdot \underline{v} + \underline{v} \cdot \underline{\nabla} \underline{\underline{Q}} + \frac{\partial \underline{\underline{Q}}}{\partial t} \right) dv = \iiint\limits_{V} \left(\frac{\partial \underline{\underline{Q}}}{\partial t} + \underline{\nabla} \cdot \underline{v}\underline{\underline{Q}} \right) dV\,. \tag{5}$$

In fact,

$$\underline{\nabla} \cdot \underline{v}\underline{\underline{Q}} = (\underline{\nabla} \cdot \underline{v})\underline{\underline{Q}} + \underline{R}^s \cdot \underline{v} \frac{\partial \underline{\underline{Q}}}{\partial q^s} = (\underline{\nabla} \cdot \underline{v})\underline{\underline{Q}} + \underline{v} \cdot \underline{\nabla} \underline{\underline{Q}}\,.$$

Provided that the Gauss-Ostrogradskii transform is applicable (III.Par.8), these formulas can be represented in the following way

$$\dot{\Psi} = \iiint\limits_{V} \frac{\partial \varphi}{\partial t}\, dV + \iint\limits_{O} \underline{N} \cdot \underline{v} \varphi\, dO\,, \qquad \dot{\underline{c}} = \iiint\limits_{V} \frac{\partial \underline{a}}{\partial t}\, dV + \iint\limits_{O} \underline{N} \cdot \underline{v}\underline{a}\, dO\,,$$

$$\dot{\underline{P}} = \iiint\limits_{V} \frac{\partial \underline{\underline{Q}}}{\partial t}\, dV + \iint\limits_{O} \underline{N} \cdot \underline{v}\underline{\underline{Q}}\, dO\,. \tag{6}$$

Equation (3) can be applied to the derivation of the equation of continuity which expresses the mass conservation law. The mass of a continuous medium occupying the volume V in its actual V_t-configuration, is determined by its integral

$$m = \iiint\limits_V \rho(q^1, q^2, q^3;\ t)\,dV = \iiint\limits_V \rho(\underline{R};\ t)\,dV\ . \tag{7}$$

Here, we have denoted the density, i.e. the mass of a unit volume of the medium which is a positive function of the material coordinates and time by ρ. The mass m is the sum of the masses of all the material particles filling the volume V. In the reference configuration the same particles occupy the volume v, their masses m_0 being distributed over this volume with density $\rho_0(q^1, q^2, q^3)$. From the law of conservation of mass,

$$m = m_0\ , \qquad \dot{m} = 0 \tag{8}$$

and by (7) and (3)

$$\left(\iiint\limits_V \rho(\underline{R};\ t)\,dV\right)^{\bullet} = \iiint\limits_V \left(\frac{\partial \varphi}{\partial t} + \underline{\nabla} \cdot \rho\underline{v}\right) dV = 0\ , \tag{9}$$

the volume V being arbitrarily prescribed. We thus arrive at the equation of continuity

$$\frac{\partial \rho}{\partial t} + \underline{\nabla} \cdot \rho\underline{v} = 0\ . \tag{10}$$

This relationship can also be obtained by expressing the conservation of the masses of the elementary volumes dv and dV

$$dm_0 = \rho_0 dv = dm = \rho\, dV\ , \qquad \frac{\rho_0}{\rho} = \frac{dV}{dv} = \sqrt{\frac{G}{g}}\ . \tag{11}$$

In fact differentiation with respect to time shows that

$$-\frac{\rho_0}{\rho^2}\dot{\rho} = \left(\sqrt{\frac{G}{g}}\right)^{\bullet} = \sqrt{\frac{G}{g}}\,\underline{\nabla} \cdot \underline{v}\ , \qquad \dot{\rho}(\underline{R};\ t) = \frac{\partial \rho}{\partial t} + \underline{v} \cdot \underline{\nabla}\rho\ , \tag{12}$$

and

$$-\frac{\rho_0}{\rho}\left(\frac{\partial \rho}{\partial t} + \underline{v} \cdot \underline{\nabla}\rho\right) = \rho\sqrt{\frac{G}{g}}\,\underline{\nabla} \cdot \underline{v}\ , \qquad \frac{\partial \rho}{\partial t} + \underline{v} \cdot \underline{\nabla}\rho + \rho\underline{\nabla} \cdot \underline{v} = 0\ ,$$

which is the desired result. In the time-independent case $\rho = \rho(\underline{R})$ and Equation (10) is reduced to the form

$$\underline{\nabla} \cdot \rho\underline{v} = 0\ ; \tag{13}$$

for $\rho = \rho_0 = \text{const.}$ the divergence of the velocity vanishes:

$$\underline{\nabla} \cdot \underline{v} = 0\ , \tag{14}$$

this condition expresses the incompressibility of an incompressible medium.

Given the condition of incompressibility for a compressible medium one may write

$$\frac{\partial\rho}{\partial t} + \underline{v}\cdot\underline{\nabla}\rho = 0 \;.$$

The material derivative of the circulation of a vector along a material contour Γ. By Equation (3.6) and by the definition (III.Par.8) of the circulation of a vector we have

$$C = \oint_\Gamma d\underline{R}\cdot\underline{a} = \oint_\gamma d\underline{r}\cdot\overset{0}{\underline{\underline{\nabla R}}}\cdot\underline{a} \;,$$

$$\dot{C} = \oint_\gamma d\underline{r}\cdot(\overset{0}{\underline{\underline{\nabla R}}}\cdot\underline{a})^{\bullet} = \oint_\gamma d\underline{r}\cdot\overset{0}{\underline{\underline{\nabla R}}}\cdot[\underline{\underline{\nabla v}}\cdot\underline{a}+\dot{\underline{a}}] = \oint_\Gamma d\underline{R}\cdot[\underline{\nabla}(\underline{v}\cdot\underline{a}) - (\underline{\underline{\nabla a}})\cdot\underline{v}+\dot{\underline{a}}] \;,$$

where γ denotes the contour in the reference configuration formed by the same particles as the contour Γ. At the same time, according to (III.2.3) we have

$$d\underline{R}\cdot\underline{\nabla}(\underline{v}\cdot\underline{a}) = d(\underline{v}\cdot\underline{a}) \;, \qquad \oint_\Gamma d\underline{R}\cdot\underline{\nabla}(\underline{v}\cdot\underline{a}) = 0$$

provided that the scalar $\underline{v}\cdot\underline{a}$ is single-valued. We thus have

$$\dot{C} = \oint d\underline{R}\cdot[\dot{\underline{a}} - (\underline{\underline{\nabla a}})\cdot\underline{v}] \;. \tag{15}$$

In particular, for $\underline{a} = \underline{v}$ (see III.3.2)

$$\underline{\nabla}\frac{v^2}{2} = \tfrac{1}{2}\underline{\nabla}(\underline{v}\cdot\underline{v}) = (\underline{\underline{\nabla v}})\cdot\underline{v} \;, \qquad \oint_\Gamma d\underline{R}\cdot(\underline{\underline{\nabla v}})\cdot\underline{v} = \tfrac{1}{2}\oint_\Gamma d\underline{R}\cdot\underline{\nabla}v^2 = \tfrac{1}{2}\oint_\Gamma dv^2 = 0 \;.$$

Hence we arrive at Kelvin's theorem:

$$C = \oint_\Gamma d\underline{R}\cdot\underline{v} \;, \qquad \dot{C} = \oint_\Gamma d\underline{R}\cdot\dot{\underline{v}} = \oint_\Gamma d\underline{R}\cdot\underline{b} \tag{16}$$

i.e. the time derivative of the circulation integral of the velocity along the material contour equals the circulation integral written in terms of the acceleration.

15. RIGID MOTIONS. FRAME-INDIFFERENT TENSORS

The stresses arising within solid bodies are produced as a result of deformations of these bodies. Therefore, if we superpose a rigid motion onto the actual motion of a body, that is if we add a motion which does not produce deformation, and if, moreover, all the remaining parameters (e.g. the temperature) retain their values, then we must also expect the stresses to remain unchanged. At the same time it is clear that, for a rigid rotation of the body, its stress tensor rotates together with the body which means that its principal axes exhibit rotation whereas its principal values remain the same. In other words, the stress tensor turns out to be "frozen" into the body. Quantities which manifest similar properties under rigid motion will from here on be called frame-indifferent.

Consider two motions $\underline{R}(q^1,q^2,q^3,t)$ and $\underline{R}'(q^1,q^2,q^3,t)$ of a deformable body. Neither of these motions will be assumed rigid in general. These motions will be regarded as differing from one another by a rigid motion if at any instant they are connected by the relationship

$$\underline{R}'(q^1,q^2,q^3;\ t) = \underline{c}(t) + [\underline{R}(q^1,q^2,q^3;\ t) - \underline{R}_0(t)]\cdot\underline{O}(t)\ . \tag{1}$$

Here $\underline{c}(t)$ denotes the position occupied by $\underline{R}_0$ in the motion $\underline{R}'$, and $\underline{O}(t)$ denotes a proper orthogonal tensor. If relationship (1) can be established between any two motions of a body, then this body is called absolutely rigid.

Definition: A tensor field of the k-th rank $\underline{A}(q^1,q^2,q^3,\underline{R})$ is called *frame-indifferent* if it is transformed according to the law

$$\underline{A}'(\underline{R}') = A^{i_1\ldots i_k}(\underline{R})\underline{e}_{i_1}\cdot\underline{O}(t)\ldots\underline{e}_{i_k}\cdot\underline{O}(t) \tag{2}$$

when a rigid motion is superposed. Here we have denoted by $\underline{O}(t)$ the same orthogonal tensor as in Equation (1), and an arbitrary basis by $\underline{e}_i$. In particular, a scalar is frame-indifferent provided that condition

$$\varphi'(\underline{R}') = \varphi(\underline{R})\ .$$

holds.

Density is a typical example of a frame-indifferent scalar. This function depends on the motion $\underline{R}$ since this motion results in a change of volume. But the volumes are the same for the two motions $\underline{R}$ and $\underline{R}'$, and so are the densities. Kinetic energy is an example of a scalar function which is not frame-indifferent, but this is not the case for the stored energy.

The vector $\underline{a}$ is frame-indifferent provided that

$$\underline{a}'(\underline{R}') = \underline{a}(\underline{R})\cdot\underline{O}(t) = \underline{O}^T(t)\cdot\underline{a}(\underline{R})\ . \tag{3}$$

From (1) it follows immediately that the basis vectors $\underline{R}_s$ are frame-indifferent. In fact, since $\underline{c}(t)$ and $\underline{O}(t)$ are the same for all particles of the body, we have

$$\underline{R}'_s = \underline{R}_s \cdot \underline{O} = \underline{O}^T \cdot \underline{R}_s \; , \tag{4}$$

which means that the basis vectors are "frozen" into the body.

We now determine the components of a frame-indifferent vector in the motions $\underline{R}$ and $\underline{R}'$. It is easy to see (cf. (3), (4)), that

$$\underline{a}'(\underline{R}') \cdot \underline{R}'_s = \underline{a}(\underline{R}) \cdot \underline{R}_s \; .$$

This equality means that the position of the vector $\underline{a}'$ referred to basis $\underline{R}'$ is exactly the same as that of the vector $\underline{a}$ referred to $\underline{R}_s$. In exactly the same way it follows from (2) and (4) that the components of a frame-indifferent tensor referred to basis $\underline{R}_s$ remain the same when a rigid motion is superposed.

According to (2), a frame-indifferent tensor $\underline{\underline{Q}}$ of the second rank is determined by the relationship

$$\underline{\underline{Q}}'(\underline{R}') = q^{st}\underline{R}_s \cdot \underline{O}\underline{R}_t \cdot \underline{O} = \underline{O}^T \cdot q^{st}\underline{R}_s\underline{R}_t \cdot \underline{O} \; ,$$

so that

$$\underline{\underline{Q}}'(\underline{R}') = \underline{O}^T \cdot \underline{\underline{Q}} \cdot \underline{O} \; .$$

The frame-indifferent quantities are "frozen" into the basis: the numerical values of their components are preserved together with their orientation with respect to the vector bases $\underline{R}_s$ and $\underline{R}'_s$. The fields of $\underline{a}$ and $\underline{\underline{Q}}$ move and rotate together with the basis, and the person observing the frame-indifferent fields does not know with which of the bases he is himself associated.

Differentiating with respect to t, we obtain

$$\underline{v}' = \dot{\underline{c}}(t) + \underline{O}^T \cdot \underline{v} + \dot{\underline{O}}^T \cdot (\underline{R} - \underline{R}_0) = \dot{\underline{c}} + \underline{O}^T \cdot \underline{v} + \dot{\underline{O}}^T \cdot \underline{O} \cdot (\underline{R}' - \underline{c}) \; ,$$

so that

$$\underline{v}' - \underline{O}^T \cdot \underline{v} = \dot{\underline{c}} + \dot{\underline{O}}^T \cdot \underline{O} \cdot (\underline{R}' - \underline{c}) \; . \tag{6}$$

The velocity vector is not frame-indifferent. This follows from (3) since the right-hand side of (6) is not zero. This result is obvious because the velocity vector of a moving particle is not "frozen" into the medium.

The skew symmetric tensor

$$\underline{\Omega} = \dot{\underline{O}}^T \cdot \underline{O} = -\underline{O}^T \cdot \dot{\underline{O}} = -\underline{\Omega}^T \tag{7}$$

represents the angular velocity tensor (the spin) of a primed system with regard to a non-primed system and the corresponding angular velocity vector is denoted by $\underline{\tilde{\omega}}$. The formula (6) now becomes

$$\underline{v}' = \dot{\underline{c}} + \underline{O}^T \cdot \underline{v} + \underline{\Omega} \cdot (\underline{R}' - \underline{c}) = \underline{O}^T \cdot \underline{v} + \underline{\tilde{\omega}} \times (\underline{R}' - \underline{c}) + \dot{\underline{c}} \; . \tag{8}$$

Differentiating Equations (1) and (8) with respect to q^s, we obtain

$$\underline{R}'_s = \underline{R}_s \cdot \underline{O} \, , \qquad \underline{O} = \underline{R}^s \underline{R}'_s \, , \qquad \underline{O} \cdot \underline{R}'^k = \underline{R}^s \delta_s^k = \underline{R}^k \, ,$$

$$\underline{R}'^s = \underline{O}^T \cdot \underline{R}^s \, , \qquad \underline{R}_s \underline{R}'^s = \underline{O}^T \, ,$$

$$\frac{\partial \underline{v}'}{\partial q^s} = \frac{\partial \underline{v}}{\partial q^s} \cdot \underline{O} + \underline{\Omega} \cdot \underline{R}'_s \, ,$$

$$\underline{\nabla}' \underline{v}' = \underline{R}'^s \frac{\partial \underline{v}}{\partial q^s} \cdot \underline{O} + \underline{R}'^s \underline{\Omega} \cdot \underline{R}'_s = \underline{O}^T \cdot \underline{R}^s \frac{\partial \underline{v}}{\partial q^s} \cdot \underline{O} - \underline{R}'^s \underline{R}'_s \cdot \underline{\Omega}$$

and, finally,

$$\underline{\nabla}' \underline{v}' = \underline{O}^T \cdot \underline{\nabla v} \cdot \underline{O} - \underline{\Omega} \, , \qquad (\underline{\nabla}' \underline{v}')^T = \underline{O}^T \cdot \underline{\nabla v}^T \cdot \underline{O} + \underline{\Omega} \; . \tag{9}$$

<u>The velocity gradient - a non frame-indifferent tensor</u>. By (13.10) and (13.11) we get

$$\underline{D}' = \underline{O}^T \cdot \underline{D} \cdot \underline{O} \, , \qquad \underline{W}' = \underline{O}^T \cdot \underline{W} \cdot \underline{O} + \underline{\Omega} \tag{10}$$

the rate of strain tensor is thus frame-indifferent which is not the case for the vorticity tensor. Equations (10) are a re-statement of Zorawski's theorem.

The transformation formulas for the deformation gradients are given by

$$\underline{\overset{0}{\nabla}R}' = \underline{r}^s \underline{R}'_s = \underline{r}^s \underline{R}_s \cdot \underline{O} = \underline{\overset{0}{\nabla}R} \cdot \underline{O} \, , \qquad \underline{\overset{0}{\nabla}R}'^T = \underline{O}^T \cdot \underline{\overset{0}{\nabla}R}^T \, , \tag{11}$$

$$\underline{\nabla}' \underline{r} = \underline{R}'^s \underline{r}_s = \underline{O}^T \cdot \underline{R}^s \underline{r}_s = \underline{O}^T \cdot \underline{\nabla r} \, , \qquad \underline{\nabla}' \underline{r}^T = \underline{\nabla r}^T \cdot \underline{O} \tag{12}$$

(here we have taken into account the fact that the reference configuration is not transformed). From these equations, we deduce the transformation formulas for the strain measures

$$\underline{G}' = \underline{\overset{0}{\nabla}R}' \cdot \underline{\overset{0}{\nabla}R}'^T = \underline{\overset{0}{\nabla}R} \cdot \underline{O} \cdot \underline{O}^T \cdot \underline{\overset{0}{\nabla}R}^T = \underline{\overset{0}{\nabla}R} \cdot \underline{\overset{0}{\nabla}R}^T = \underline{G} \, , \tag{13}$$

$$\underline{F}' = \underline{\overset{0}{\nabla}R}'^T \cdot \underline{\overset{0}{\nabla}R}' = \underline{O}^T \cdot \underline{\overset{0}{\nabla}R}^T \cdot \underline{\overset{0}{\nabla}R} \cdot \underline{O} = \underline{O}^T \cdot \underline{F} \cdot \underline{O} \; . \tag{14}$$

It can be seen that the Cauchy-Green measure is not frame-indifferent whereas the Finger measure is. From these formulas and from Equations (6.4) we deduce that the right stretch tensor is frame-indifferent which is not the case for the left stretch tensor. The same conclusion may be arrived at by a consideration of the polar representation of the deformation gradient

$$\overset{0}{\nabla}\underline{R} = \underline{U} \cdot \underline{O}^{\times} = \overset{0}{\nabla}\underline{R}' \cdot \underline{O}^{T} = \underline{U}' \cdot \underline{O}^{\times\prime} \cdot \underline{O}^{T} ,$$

here, in accordance with an earlier convention, a cross superscript designates an orthogonal transformation accompanying deformation (Par. 6). But the polar representation of $\overset{0}{\nabla}\underline{R}$ is unique, and therefore

$$\underline{U} = \underline{U}' , \qquad \underline{O}^{\times\prime} = \underline{O}^{\times} \cdot \underline{O} . \tag{15}$$

This equation together with Equation (11) yields

$$\overset{0}{\nabla}\underline{R} = \underline{O}^{\times} \cdot \underline{V} = \overset{0}{\nabla}\underline{R}' \cdot \underline{O}^{T} = \underline{O}^{\times\prime} \cdot \underline{V}' \cdot \underline{O}^{T} = \underline{O}^{\times} \cdot \underline{O} \cdot \underline{V}' \cdot \underline{O}^{T} , \qquad \underline{V} = \underline{O} \cdot \underline{V}' \cdot \underline{O}^{T}$$

which confirms the frame-indifference of $\underline{V}$

$$\underline{V}' = \underline{O}^{T} \cdot \underline{V} \cdot \underline{O} . \tag{16}$$

It follows from the definition (2) of a frame-indifferent tensor and from that of (II.7.1) for an isotropic tensor function, that the value of any isotropic function of any number of frame-indifferent tensors is itself a frame-indifferent tensor. Of course, the function of a non frame-indifferent tensor may also be isotropic.

16. THE OBJECTIVE DERIVATIVE OF A TENSOR

Let $\underline{a}$ and $\underset{\approx}{Q}$ be frame-indifferent. Their time derivatives are, however, non frame-indifferent, e.g.

$$\dot{\underline{a}}' = (\underline{a} \cdot \underline{O})^{\bullet} = \dot{\underline{a}} \cdot \underline{O} + \underline{a} \cdot \dot{\underline{O}} .$$

The relationships describing the intrinsic properties of continuous media must be equally valid in all bases, no one of which may be regarded as superior. This consideration requires that the definition of the time derivative of a frame-indifferent quantity presupposes the frame-indifference of the derivative itself. This requirement can actually be satisfied in a number of ways.

Making use of Equations (15.7), (15.10) and (15.3), we can transform (1) to obtain the relationship

$$\dot{\underline{a}}' = \dot{\underline{a}} \cdot \underline{O} - \underline{a} \cdot \underline{O} \cdot \underline{\Omega} = \dot{\underline{a}} \cdot \underline{O} - \underline{a} \cdot \underline{O} \cdot (\underline{W}' - \underline{O}^T \cdot \underline{W} \cdot \underline{O}) =$$

$$= (\dot{\underline{a}} + \underline{a} \cdot \underline{W}) \cdot \underline{O} - \underline{a}' \cdot \underline{W}'$$

which means that the vector is frame-indifferent (cf. (15.3))

$$\check{\underline{a}} = \dot{\underline{a}} + \underline{a} \cdot \underline{W} = (\dot{\underline{a}}' + \underline{a}' \cdot \underline{W}') \cdot \underline{O}^T = \dot{\underline{a}} - \underline{W} \cdot \underline{a} \ , \tag{2}$$

this vector is called the Jaumann-Noll objective derivative of a vector $\underline{a}$.

For a second rank tensor a similar calculation is more complex.

$$\dot{\underline{\underline{Q}}}' = (\underline{O}^T \cdot \underline{\underline{Q}} \cdot \underline{O})^{\bullet} = \underline{O}^T \cdot \dot{\underline{\underline{Q}}} \cdot \underline{O} + \dot{\underline{O}}^T \cdot \underline{\underline{Q}} \cdot \underline{O} + \underline{O}^T \cdot \underline{\underline{Q}} \cdot \dot{\underline{O}} =$$

$$= \underline{O}^T \cdot \dot{\underline{\underline{Q}}} \cdot \underline{O} + \underline{\Omega} \cdot \underline{O}^T \cdot \underline{\underline{Q}} \cdot \underline{O} - \underline{O}^T \cdot \underline{\underline{Q}} \cdot \underline{O} \cdot \underline{\Omega} = \underline{O}^T \cdot \dot{\underline{\underline{Q}}} \cdot \underline{O} + \underline{W}' \cdot \underline{\underline{Q}}' -$$

$$- \underline{\underline{Q}}' \cdot \underline{W}' - \underline{O}^T \cdot \underline{W} \cdot \underline{O} \cdot \underline{O}^T \cdot \underline{\underline{Q}} \cdot \underline{O} + \underline{O}^T \cdot \underline{\underline{Q}} \cdot \underline{O} \cdot \underline{O}^T \cdot \underline{W} \cdot \underline{O} =$$

$$= \underline{O}^T \cdot \dot{\underline{\underline{Q}}} \cdot \underline{O} + \underline{W}' \cdot \underline{\underline{Q}}' - \underline{\underline{Q}}' \cdot \underline{W}' - \underline{O}^T \cdot \underline{W} \cdot \underline{\underline{Q}} \cdot \underline{O} + \underline{O}^T \cdot \underline{\underline{Q}} \cdot \underline{W} \cdot \underline{O}$$

or

$$\dot{\underline{\underline{Q}}}' - \underline{W}' \cdot \underline{\underline{Q}}' + \underline{\underline{Q}}' \cdot \underline{W}' = \underline{O}^T \cdot (\dot{\underline{\underline{Q}}} - \underline{W} \cdot \underline{\underline{Q}} + \underline{\underline{Q}} \cdot \underline{W}) \cdot \underline{O} \ ,$$

from which the objective character of the Jaumann-Noll derivative of a tensor of the second rank follows

$$\check{\underline{\underline{Q}}} = \dot{\underline{\underline{Q}}} - \underline{W} \cdot \underline{\underline{Q}} + \underline{\underline{Q}} \cdot \underline{W} \ . \tag{3}$$

Alternative definitions of the objective derivative may be obtained if we introduce into the expression for the Jaumann-Noll derivative or eliminate from it the frame-different terms $\underline{D} \cdot \underline{a}$, $\underline{D} \cdot \underline{\underline{Q}}$ or $\underline{a} \cdot \underline{D}$, $\underline{\underline{Q}} \cdot \underline{D}$. For example, substituting $\underline{W}$ for $\underline{D} - \underline{\underline{\nabla v}}$, we get

$$\check{\underline{a}} = \dot{\underline{a}} + \underline{a} \cdot \underline{D} - \underline{a} \cdot \underline{\underline{\nabla v}} \ , \qquad \check{\underline{\underline{Q}}} = \dot{\underline{\underline{Q}}} - \underline{D} \cdot \underline{\underline{Q}} + \underline{\underline{\nabla v}} \cdot \underline{\underline{Q}} + \underline{\underline{Q}} \cdot \underline{D} - \underline{\underline{Q}} \cdot \underline{\underline{\nabla v}} \ .$$

Upon discarding the frame-invariant terms, we arrive at the Oldroyd derivatives

$$\check{\underline{a}} = \dot{\underline{a}} - \underline{a} \cdot \underline{\underline{\nabla v}} \ , \qquad \check{\underline{\underline{Q}}} = \dot{\underline{\underline{Q}}} + \underline{\underline{\nabla v}} \cdot \underline{\underline{Q}} - \underline{\underline{Q}} \cdot \underline{\underline{\nabla v}} \ , \tag{4}$$

or, alternatively, by addition of the terms $2\underline{a} \cdot \underline{D}$, $2\underline{\underline{Q}} \cdot \underline{D}$, we obtain

$$\overset{\Delta}{\underline{a}} = \dot{\underline{a}} + \underline{a} \cdot \underline{\underline{\nabla v}}^T \ , \qquad \overset{\Delta}{\underline{\underline{Q}}} = \dot{\underline{\underline{Q}}} + \underline{\underline{\nabla v}} \cdot \underline{\underline{Q}} + \underline{\underline{Q}} \cdot \underline{\underline{\nabla v}}^T \ . \tag{5}$$

The definition proposed by Truesdell is based upon a consideration of the vector $\underline{N}\cdot\underline{T}\,dO$ ($\underline{N}\,dO$ being an oriented elementary area, and $\underline{T}$ - a frame-indifferent tensor of second order. By (4) and (10.22) we obtain

$$(\underline{N}\cdot\underline{T}\,dO)^{\check{}} = (\underline{N}\cdot\underline{T}\,dO)^{\bullet} - \underline{N}\cdot\underline{T}\cdot\underline{\nabla v}\,dO =$$

$$= \underline{N}\,dO\cdot(\dot{\underline{T}} + \underline{T\nabla}\cdot\underline{v} - \underline{\nabla v}^{T}\cdot\underline{T} - \underline{T}\cdot\underline{\nabla v}) \qquad (6)$$

which enables us to determine the Truesdell objective derivative using the expression

$$\underline{T}^{\nabla} = \dot{\underline{T}} + \underline{T\nabla}\cdot\underline{v} - \underline{\nabla v}^{T}\cdot\underline{T} - \underline{T}\cdot\underline{\nabla v}\ . \qquad (7)$$

In these formulas, a dot above the symbol denotes the material derivatives

$$\dot{\underline{a}} = \frac{\partial\underline{a}}{\partial t} + \underline{v}\cdot\underline{\nabla a}\ , \qquad \dot{\underline{\underline{Q}}} = \frac{\partial\underline{\underline{Q}}}{\partial t} + \underline{v}\cdot\underline{\nabla\underline{Q}}\ . \qquad (8)$$

The component representation of $\underline{\underline{Q}}$ is given by any of the formulas

$$\dot{\underline{\underline{Q}}} = \underline{R}_s\underline{R}_t\left[\frac{\partial q^{st}}{\partial t} + v^r\frac{\partial q^{st}}{\partial q^r} + v^q\left(\begin{Bmatrix}s\\rq\end{Bmatrix}q^{rt} + \begin{Bmatrix}t\\qr\end{Bmatrix}q^{sr}\right)\right] =$$

$$= \underline{R}^s\underline{R}_t\left[\frac{\partial q_s^{\cdot t}}{\partial t} + v^r\frac{\partial q_s^{\cdot t}}{\partial q^r} + v^q\left(-\begin{Bmatrix}r\\sq\end{Bmatrix}q_r^{\cdot t} + \begin{Bmatrix}t\\qr\end{Bmatrix}q_s^{\cdot r}\right)\right]\ .$$

At the same time,

$$\underline{\nabla v}^T\cdot\underline{\underline{Q}} - \underline{\underline{Q}}\cdot\underline{\nabla v} = \underline{R}^s\underline{R}_t\left[q_r^{\cdot t}\frac{\partial v^r}{\partial q^s} - q_s^{\cdot r}\frac{\partial v^t}{\partial q^r} + v^q\left(\begin{Bmatrix}r\\sq\end{Bmatrix}q_r^{\cdot t} - \begin{Bmatrix}t\\qr\end{Bmatrix}q_s^{\cdot r}\right)\right]\ ,$$

$$\underline{\nabla v}^T\cdot\underline{\underline{Q}} + \underline{\underline{Q}}\cdot\underline{\nabla v} = \underline{R}_s\underline{R}_t\left[q^{rt}\frac{\partial v^s}{\partial q^r} + q^{sr}\frac{\partial v^t}{\partial q^r} + v^q\left(\begin{Bmatrix}s\\rq\end{Bmatrix}q^{rt} + \begin{Bmatrix}t\\qr\end{Bmatrix}q^{st}\right)\right]\ .$$

We thus arrive at the following expressions for the Oldroyd and Truesdell objective derivatives

$$\check{\underline{\underline{Q}}} = \underline{R}^s\underline{R}_t\left(\frac{\partial q_s^{\cdot t}}{\partial t} + v^r\frac{\partial q_s^{\cdot t}}{\partial q^r} + q_r^{\cdot t}\frac{\partial v^r}{\partial q^s} - q_s^{\cdot r}\frac{\partial v^t}{\partial q^r}\right)\ , \qquad (9)$$

$$\underline{\underline{Q}}^{\nabla} = \underline{R}_s\underline{R}_t\left(\frac{\partial q^{st}}{\partial t} + v^r\frac{\partial q^{st}}{\partial q^r} - q^{rt}\frac{\partial v^s}{\partial q^r} - q^{sr}\frac{\partial v^t}{\partial q^r}\right) + \underline{\underline{Q}}\,\frac{\partial(v^r\sqrt{G})}{\sqrt{G}\,\partial q^r}\ , \qquad (10)$$

these formulas do not include the Christoffel symbols.

17. REFERENCE WITH RESPECT TO THE PRESENT CONFIGURATION. THE RIVLIN-ERICKSEN TENSORS

In the foregoing, the reference configuration was associated with a time origin $t = 0$ and this was reflected in the notation for the deformation gradient (3.1). Nothing prevents us, however, from taking for the reference configuration the actual configuration V_τ at the instant τ, this assumption also to be indicated in the notation. Equations (3.1), (6.5) would then take the form

$$\overset{0}{\nabla \mathrm{R}} = \mathrm{R}^s(0)\, \mathrm{R}_s(t) \;, \qquad \underline{0}^\times = \mathrm{e}_s(0)\, \mathrm{e}^s(t) \tag{1}$$

and the corresponding notation for the Cauchy-Green and Finger strain measures would be $\overset{0}{\mathrm{G}}(t)$, $\overset{0}{\mathrm{F}}(t)$ instead of G, F. We need not reconstruct the notation for the quantities whose definition is not connected with the reference configuration, e.g. for the velocity vector v, its gradient $\nabla\mathrm{v}$, the deformation D and the vorticity W.

With the explicit indication of the reference configuration, the deformation gradients are represented by the formulas

$$\overset{\tau}{\nabla \mathrm{R}}(t) = \mathrm{R}^s(\tau)\, \mathrm{R}_s(t) \;, \qquad \overset{\tau}{\nabla \mathrm{R}}(t)^{\mathrm{T}} = \mathrm{R}_s(t)\, \mathrm{R}^s(\tau) \;, \tag{2}$$

and if we change places τ and t,

$$\overset{t}{\nabla \mathrm{R}}(\tau) = \mathrm{R}^s(t)\, \mathrm{R}_s(\tau) \;, \qquad \overset{t}{\nabla \mathrm{R}}(\tau)^{\mathrm{T}} = \mathrm{R}_s(\tau)\, \mathrm{R}^s(t) \;.$$

The formulas replacing Equations (3.5), (3.12) will be written as

$$\overset{\tau}{\nabla \mathrm{R}}(t) \cdot \overset{t}{\nabla \mathrm{R}}(\tau) = \mathrm{E} \;, \qquad \overset{t}{\nabla \mathrm{R}}(t) = \mathrm{E} \;. \tag{3}$$

The Cauchy-Green strain measure is now represented by the expression

$$\overset{\tau}{\mathrm{G}}(t) = \mathrm{R}^s(\tau)\, \mathrm{R}_s(t) \cdot \mathrm{R}_k(t)\, \mathrm{R}^k(\tau) \tag{4}$$

which, in view of the relationships

$$\mathrm{R}_s(t) = \mathrm{R}_s(0) \cdot \overset{0}{\nabla \mathrm{R}}(t) = [\overset{0}{\nabla \mathrm{R}}(t)]^{\mathrm{T}} \cdot \mathrm{R}_s(0) \;,$$

$$\mathrm{R}^s(\tau)\, \mathrm{R}_s(0) = [\overset{0}{\nabla \mathrm{R}}(\tau)]^{-1} \;, \qquad \mathrm{R}_s(0)\, \mathrm{R}^s(\tau) = [\overset{0}{\nabla \mathrm{R}}(\tau)^{\mathrm{T}}]^{-1}$$

can be transformed to

$$\overset{\tau}{\underline{G}}(t) = [\overset{0}{\underline{\underline{\nabla R}}}(\tau)]^{-1} \cdot \underline{G}(t) \cdot [\overset{0}{\underline{\underline{\nabla R}}}(\tau)^{T}]^{-1} , \qquad \underline{G}(t) = \overset{0}{\underline{\underline{\nabla R}}}(\tau) \cdot \overset{\tau}{\underline{G}}(t) \cdot \overset{0}{\underline{\underline{\nabla R}}}(\tau)^{T} . \tag{5}$$

Here we returned to the notation $\underline{G} = \overset{0}{\underline{G}}$ of the Cauchy-Green strain measure. A similar calculation for the Finger measure yields

$$\underline{F}(t) = \overset{\tau}{\underline{\underline{\nabla R}}}(t)^{T} \cdot \underline{F}(\tau) \cdot \overset{\tau}{\underline{\underline{\nabla R}}}(t) . \tag{6}$$

For these transformations, $\underline{R}_s(0) = \underline{r}_s$, $\underline{R}^s(0) = \underline{r}^s$.

Further, we consider the tensors

$$\left[\frac{d^n}{d\tau^n} \overset{t}{\underline{Q}}(\tau) \right]_{\tau=t} , \qquad n = 0,1,2,\ldots ,$$

which depend only on t, that is, are not associated with the reference configuration ("deprived of memory"); the definition of the quantities $\underline{G}(t)$, $\overset{0}{\underline{G}}(t)$ includes "memory" of the reference configuration.

Examples of quantities which are independent of deformation history are given by the tensors

$$\underline{L}_1 = \left[\frac{d}{d\tau} \overset{t}{\underline{\underline{\nabla R}}}(\tau) \right]_{\tau=t} = \underline{R}^s(t) \left[\frac{d}{d\tau} \underline{R}_s(\tau) \right]_{\tau=t} = \underline{R}^s \frac{\partial \underline{v}}{\partial q^s} = \underline{\underline{\nabla v}} , \tag{7}$$

$$\underline{L}_2 = \left[\frac{d^2}{d\tau^2} \overset{t}{\underline{\underline{\nabla R}}}(\tau) \right]_{\tau=t} = \underline{R}^s(t) \left[\frac{d^2}{d\tau^2} \underline{R}_s(\tau) \right]_{\tau=t} = \underline{R}^s \frac{\partial \underline{b}}{\partial q^s} = \underline{\underline{\nabla b}} , \tag{8}$$

etc. These quantities are also related to the Rivlin-Ericksen tensors

$$\underline{\ni}_n = \left[\frac{d^n}{d\tau^n} \overset{t}{\underline{G}}(\tau) \right]_{\tau=t} . \tag{9}$$

We get

$$\underline{\ni}_1 = \left[\frac{d}{d\tau} \overset{t}{\underline{\underline{\nabla R}}}(\tau) \cdot \overset{t}{\underline{\underline{\nabla R}}}(\tau)^{T} \right]_{\tau=t} = \left[\underline{R}^s(t) \frac{\partial \underline{v}(\tau)}{\partial q^s} \cdot \overset{t}{\underline{\underline{\nabla R}}}(\tau)^{T} + \right.$$

$$\left. + \overset{t}{\underline{\underline{\nabla R}}}(\tau) \cdot \frac{\partial \underline{v}(\tau)}{\partial q^s} \underline{R}^s(t) \right]_{\tau=t} = \underline{\underline{\nabla v}}(t) \cdot \overset{t}{\underline{\underline{\nabla R}}}(t) + \overset{t}{\underline{\underline{\nabla R}}}(t) \, \underline{\underline{\nabla v}}(t)^{T}$$

and by (4)

$$\underline{\ni}_1 = \underline{\underline{\nabla v}} + \underline{\underline{\nabla v}}^{T} = 2\underline{D} . \tag{10}$$

Similarly, we obtain

$$\underset{-2}{\ni} = \nabla b + \nabla b^T + 2\nabla v \cdot \nabla v^T \ . \tag{11}$$

Returning to (10), we also have

$$2D = \left[\frac{d}{d\tau}\overset{t}{G}(\tau)\right]_{\tau=t} = \left[\frac{d}{d\tau}[\overset{t}{U}(\tau)]^2\right]_{\tau=t} =$$

$$= \overset{t}{U}(t)\cdot\left[\frac{d}{d\tau}\overset{t}{U}(\tau)\right]_{\tau=t} + \left[\frac{d}{d\tau}\overset{t}{U}(\tau)\right]_{\tau=t}\cdot\overset{t}{U}(t) = 2\left[\frac{d}{d\tau}\overset{t}{U}(\tau)\right]_{\tau=t}$$

and the alternative representation of the rate of strain tensor is given by

$$D = \left[\frac{d}{d\tau}\overset{t}{U}(\tau)\right]_{\tau=t} \ . \tag{12}$$

By virtue of (7), we have

$$\nabla v = \left[\frac{d}{d\tau}\overset{t}{\nabla R}(\tau)\right]_{\tau=t} = \left[\frac{d}{d\tau}\overset{t}{U}(\tau)\cdot\overset{t\times}{O}(\tau)\right]_{\tau=t} = D + \left[\frac{d}{d\tau}\overset{t\times}{O}(\tau)\right]_{\tau=t} =$$

$$= D + e_s(t)\left[\frac{d}{d\tau}e^s(\tau)\right]_{\tau=t} = D + e_s(t)\,\dot{e}^s(t)$$

(we recall here that e_s denotes the same vector as e^s). At the same time

$$e_s(t)\,\dot{e}^s(t) + \dot{e}_s(t)\,e^s(t) = [e_s(t)\,e^s(t)]^{\cdot} = 0 \ .$$

We get

$$W = \tfrac{1}{2}(\nabla v^T - \nabla v) = -e_s(t)\,\dot{e}^s(t) = \dot{e}_s(t)\,e^s(t) = -\left[\frac{d}{d\tau}\overset{t\times}{O}(\tau)\right]_{\tau=t} ,$$

so that

$$\dot{e}_s(t) = W\cdot e_s(t) = -e_s(t)\cdot W \ .$$

These formulas determine the time derivatives of the vectors of an orthogonal basis $\overset{\times}{O}$; the time dependence of these vectors is stipulated by the vorticity field. The formulas (14) and (I.11.3) also allow the following representation

$$\dot{e}_s = \tfrac{1}{2}(\nabla\times v)\times e_s(t) \ ,$$

this because $\underline{\nabla} \times \underline{v}$ is the vector associated with the tensor $2\underline{W}$.

The component representation of a frame-indifferent tensor in the primed basis $\underline{Q}'^{\times}$ is given by the expression

$$\underline{Q}' = q^{st} \underline{e}'_s \underline{e}'_t$$

and for an observer associated with this basis the time derivative of $\underline{Q}'$ in the same basis denoted by $\check{\underline{Q}}'$ is equal to

$$\check{\underline{Q}}' = \dot{q}^{st} \underline{e}'_s \underline{e}'_t \ .$$

The "fixed" observer will observe rotation of the basis $\underline{e}'_s$ in the presence of the vorticity field. The time derivative determined by this observer will be equal to

$$\dot{\underline{Q}}' = \dot{q}^{st} \underline{e}'_s \underline{e}'_t + q^{st} (\underline{e}'_s \underline{e}'_t)^{\cdot} = \dot{q}^{st} \underline{e}'_s \underline{e}'_t + q^{st} (\underline{W}' \cdot \underline{e}'_s \underline{e}'_t - \underline{e}'_s \underline{e}'_t \cdot \underline{W}') =$$

$$= \check{\underline{Q}}' + \underline{W}' \cdot \underline{Q}' - \underline{Q}' \cdot \underline{W}' \ .$$

Discarding the primes, we get

$$\check{\underline{Q}} = \dot{\underline{Q}} - \underline{W} \cdot \underline{Q} + \underline{Q} \cdot \underline{W} \ .$$

We thus have shown that the Jaumann-Noll derivative (16.3) is simply the time derivative of a tensor with respect to the rotating system of axes. This derivative is determined by an observer following the behaviour of a tensor within a rotating field of velocities.

18. DETERMINATION OF AN ABSOLUTE POSITION BY PRESCRIBED STRAIN MEASURE

We suppose that the reference configuration is prescribed, so that we know the vector $\underline{r}(q^1, q^2, q^3)$. From this vector we construct the vector bases $\underline{r}_s$, $\underline{r}^s$, the components g_{sk}, g^{sk} of the metric tensor, and the Christoffel symbols $[st^0, r]$, $\left\{ {}^{\,0}_{\,r} \atop st \right\}$. A positive matrix

$$G_{st} = \underline{R}_s \cdot \underline{R}_t \ . \tag{1}$$

of covariant components of the Cauchy-Green strain measure is also prescribed. This matrix is used to determine the inverse matrix G^{st} (of contravariant components of the metric tensor in the basis of V_t-configuration) and the Christoffel symbols $[st, r]$, $\left\{ r \atop st \right\}$. We aim to determine the basis $\underline{R}_1, \underline{R}_2, \underline{R}_3$ of the actual configuration and the position vector $\underline{R}(q^1, q^2, q^3)$ in this configuration.

The problem is reduced to a system of differential equations (III.4.2)

$$\frac{\partial \underline{R}_s}{\partial q^t} = \left\{{q \atop st}\right\} \underline{R}_q \qquad \left(\left\{{q \atop st}\right\} = \left\{{q \atop ts}\right\}\right), \tag{2}$$

as has been explained in III.Par.10, this system is completely integrable, provided that the Riemann-Christoffel curvature tensor (or the Ricci tensor)

$$^4\underline{R} = R_{ktsr}\underline{R}^k\underline{R}^t\underline{R}^s\underline{R}^r , \tag{3}$$

vanishes identically. This tensor is evaluated from matrix (1) with the aid of Equations (III.10.15); in these formulas, we now assume that the symbols G_{sk}, G^{sk} are substituted for g_{sk}, g^{sk} and it is required that six components (III.10.21) of the Ricci tensor be equal to zero.

By Equation (2) we have

$$\frac{\partial \underline{R}_s}{\partial q^t} = \frac{\partial \underline{R}_t}{\partial q^s} , \tag{4}$$

and the relationship

$$d\underline{R} = \underline{R}_s dq^s$$

is integrable; the vector $\underline{R}(q^1,q^2,q^3)$ is determined by a quadrature as long as the basis vectors $\underline{R}_s$ are determined by Equations (2).

The statement of a problem will remain the same if we superpose (above the Cauchy measure (1)) the tensor $\underline{E}$ which does not produce deformation. We can thus assume that $\underline{G}(M_0) = \underline{E}$ at some fixed particle M_0.

Because the tensor (3) of curvature vanishes identically, there exists a unique Cartesian system of coordinates $OXYZ$ for which the position vectors are determined by the expressions

$$\underline{r} = \underline{i}_s a^s , \qquad \underline{R} = \underline{i}_s x^s(a^1,a^2,a^3) \tag{5}$$

and we may consider a_s as material coordinates q^s of a particle.

The system (2) can now be represented as follows

$$\frac{\partial x^m}{\partial a^s} = z^m_s , \qquad \frac{\partial z^m_s}{\partial a^t} = \left\{{q \atop st}\right\} z^m_q . \tag{6}$$

This system of eighteen partial differential equations is divided into three systems each containing six equations ($m = 1,2,3$). The system is "completely integrable" since its conditions of integrability are satisfied identically by virtue of (4).

It can be shown that there exists a solution to these systems which is three times continuously differentiable and which takes the initial values

$$z_s^m = \lambda_s^{\cdot m} \quad \text{and} \quad a^k = a_0^k \tag{7}$$

prescribed in advance.*) In the neighbourhood of these values, we have

$$x^m = \lambda_s^{\cdot m}(a^s - a_0^s) + x_0^m . \tag{8}$$

In other words, in the neighbourhood of $M_0(a_0^k)$ the position vector $\underline{R}$ can be represented in the form

$$\underline{R} = \underline{i}_m \lambda_s^{\cdot m}(a^s - a_0^s) + \underline{R}_0 = (\underline{r} - \underline{r}_0) \cdot \underline{\Lambda} + \underline{R}_0 , \qquad \underline{\Lambda} = \lambda_s^{\cdot m} \underline{i}^s \underline{i}_m , \tag{9}$$

here $\underline{\Lambda}$ denotes a constant tensor, and $\underline{R}_0$ the vector-radius of M_0. Therefore we have in the same neighbourhood

$$\underline{R}_s = \underline{r}_s \cdot \underline{\Lambda} , \qquad \overset{0}{\underline{\nabla}}\underline{R} = \underline{r}^s \underline{r}_s \cdot \underline{\Lambda} = \underline{\Lambda} , \qquad \overset{0}{\underline{\nabla}}\underline{R}^T = \underline{\Lambda}^T , \qquad \underline{G} = \underline{\Lambda} \cdot \underline{\Lambda}^T .$$

But the tensor $\underline{G}$ is prescribed, and for this reason the tensor $\underline{\Lambda}$ is not arbitrary, but its nine components are connected by six relationships. Assuming $\underline{G}(a_0^1, a_0^2, a_0^3) = \underline{E}$, we may satisfy these constraints if we set

$$\underline{E} = \underline{\Lambda} \cdot \underline{\Lambda}^T , \qquad \underline{\Lambda} = \underline{O} ; \qquad \underline{R} = (\underline{r} - \underline{r}_0) \cdot \underline{O} + \underline{R}_0 . \tag{10}$$

As could have been foreseen, the reference configuration is transformed into the actual configuration to within an arbitrary rigid displacement of the medium (Par.4). The solution includes six parameters: three components of $\underline{R}_0 = \underline{R}(M_0)$ and three independent parameters prescribing rotation of the medium at the point M_0.

---

*) E. Cartan. Leçons sur la Geomterie des Espaces. Paris, Gauthier-Villars et S^{ie}, Editeurs, 1928. The theorem is proved assuming that the functions $G_{st}(a^1, a^2, a^3)$ are twice continuously differentiable with respect to all variables a^k within some simply-connected domain. In "Cours d'Analyse Mathematique" by E. Goursat the existence of holomorphic solutions is proved for completely integrable systems whose right-hand sides are also holomorphic.

In a similar way the problem of determining the position vector $\underline{r}(q^1,q^2,q^3)$ in the reference configuration can be solved when the Almansi measure

$$g_{st} = \underline{r}_s \cdot \underline{r}_t \ . \tag{11}$$

is prescribed for the actual configuration.

The vector basis $\underline{r}_s$ is determined by the system of equations

$$\frac{\partial \underline{r}_s}{\partial q^t} = \overset{0}{\left\{ {q \atop st} \right\}} \underline{r}_q \ , \tag{12}$$

which is integrable provided that the six independent components (III.10.21) of the Ricci tensor vanish identically. Of course, in Equations (III.10.21) one should substitute $[ks,m]$ for $[\overset{0}{ks},m]$. The position vector is determined by a quadrature from the relationship

$$d\underline{r} = \underline{r}_s dq^s \ . \tag{13}$$

19. TENSORS OF AFFINE DEFORMATION

For a single-valued and twice differentiable transformation of material coordinates q^s to some new coordinates $\check{q}^r$

$$\check{q}^r = \check{q}^r(q^1,q^2,q^3) \tag{1}$$

the Christoffel symbols are transformed by Equations (III.4.9). Representing these formulas in the vector basis of the actual configuration and in that of the reference configuration and subtracting the two equations from one another, we arrive at the relationship

$$\left\{ {\check{q} \atop st} \right\} - \left\{ {\check{q} \atop st} \right\}^0 = c^m_t c^r_s \check{c}^q_k \left(\left\{ {k \atop rm} \right\} - \left\{ {k \atop rm} \right\}^0 \right) \ , \tag{2}$$

where the symbols c^m_t, c^q_k are determined only by transformation (1) and are in no way dependent upon the choice of the vector bases. The differences of the Christoffel symbols in the actual and reference configurations are now seen to be connected by the transformation (III.4.10) and therefore represent the components of a tensor of the third rank

$$A^q_{st} = \left\{ {q \atop st} \right\} - \overset{0}{\left\{ {q \atop st} \right\}} \ . \tag{3}$$

This tensor can be determined both in the actual and reference configurations with the aid of the representations

$$\underline{A} = \underline{R}^s \underline{R}^t \underline{R}_q A^q_{st} \, , \qquad \underline{A}^{\mathrm{T}} = \underline{R}_q \underline{R}^s \underline{R}^t A^q_{st} \, , \tag{4}$$

$$\overset{0}{\underline{A}} = \underline{r}^s \underline{r}^t \underline{r}_q A^q_{st} \, , \qquad \overset{0}{\underline{A}}{}^{\mathrm{T}} = \underline{r}_q \underline{r}^s \underline{r}^t A^q_{st} \, , \tag{5}$$

which are symmetric with respect to the lower indices s, t. The symbol "T" means substitution of the triad $\underline{R}^s \underline{R}^t \underline{R}_q$ (or $\underline{r}^s \underline{r}^t \underline{r}_q$) for that of $\underline{R}_q \underline{R}^s \underline{R}^t$ or $\underline{r}_q \underline{r}^s \underline{r}^t$). Equations (4), (5) determine different tensors possessing the same components (the tensors are "isomers").

We will call affine the transformation of the reference configuration into the actual configuration for which the Cartesian coordinates of the actual configuration are linear functions of the reference configuration

$$x^s = \lambda^s_{\cdot t} (a^t - \overset{0}{a}{}^t) + \overset{0}{x}{}^s \, .$$

Alternatively,

$$\underline{R} = (\underline{r} - \underline{r}_0) \cdot \underline{\Lambda} + \underline{R}_0 \, , \tag{6}$$

where $\underline{\Lambda}$ is a constant tensor, that is a tensor whose Cartesian components are constant. The covariant derivative of $\underline{\Lambda}$ is equal to zero (for arbitrary choice of q^s).

Equation (6) gives rise to the relationships

$$\underline{R}_s = \underline{r}_s \cdot \underline{\Lambda} \, , \qquad \frac{\partial \underline{R}_s}{\partial q^t} = \left\{ {q \atop st} \right\} \underline{R}_q = \frac{\partial \underline{r}_s}{\partial q^t} \cdot \underline{\Lambda} = \overset{0}{\left\{ {q \atop st} \right\}} \underline{r}_q \cdot \underline{\Lambda} = \overset{0}{\left\{ {q \atop st} \right\}} \underline{R}_q \, ,$$

so that

$$\left(\left\{ {q \atop st} \right\} - \overset{0}{\left\{ {q \atop st} \right\}} \right) \underline{R}_q = \underline{0} \, , \qquad A^q_{st} = 0 \, . \tag{7}$$

The tensors (4) and (5) are thus equal to zero if the transformation of the reference configuration into the actual configuration does not differ from an affine transformation. These tensors characterise the "non-affinity" of the transformation and are therefore called tensors of affine deformation (A.P. Norden).

Equations (3) may be reformulated with the aid of the covariant derivatives of the Cauchy-Green and Almansi strain measures. We have

$$\left\{ {q \atop st} \right\} = G^{qr}[st,r] = \tfrac{1}{2}G^{qr}\left(\frac{\partial G_{tr}}{\partial q^s} + \frac{\partial G_{rs}}{\partial q^t} - \frac{\partial G_{st}}{\partial q^r} \right). \tag{8}$$

But for the reference configuration's basis

$$\overset{0}{\nabla}_s G_{tr} = \frac{\partial G_{tr}}{\partial q^s} - \overset{0}{\left\{ {m \atop st} \right\}} G_{mr} - \overset{0}{\left\{ {m \atop sr} \right\}} G_{tm}.$$

From this and from two similar equations obtained by a cyclic permutation of indices str, we get

$$\overset{0}{\nabla}_s G_{tr} + \overset{0}{\nabla}_t G_{rs} - \overset{0}{\nabla}_r G_{st} = \frac{\partial G_{tr}}{\partial q^s} + \frac{\partial G_{rs}}{\partial q^t} - \frac{\partial G_{st}}{\partial q^r} - 2\overset{0}{\left\{ {m \atop st} \right\}} G_{mr}.$$

Taking now into account that

$$G^{qr}G_{mr}\overset{0}{\left\{ {m \atop st} \right\}} = \overset{0}{\left\{ {q \atop st} \right\}},$$

and substituting the resulting expressions into Equation (8), we arrive at the desired representation

$$A^q_{st} = \tfrac{1}{2}G^{qr}(\overset{0}{\nabla}_s G_{tr} + \overset{0}{\nabla}_t G_{sr} - \overset{0}{\nabla}_r G_{st}) \tag{9}$$

and, similarly,

$$A^q_{st} = -\tfrac{1}{2}g^{qr}(\nabla_s g_{tr} + \nabla_t g_{sr} - \nabla_r g_{st}). \tag{10}$$

Recalling that, by the Ricci theorem (III.5.11),

$$\nabla_k G_{st} = 0, \qquad \overset{0}{\nabla}_k g_{st} = 0;$$

we may also express (9) and (10) in terms of the covariant derivatives of the components of the strain tensors (7.8)

$$A^q_{st} = G^{qr}(\overset{0}{\nabla}_s c_{tr} + \overset{0}{\nabla}_t c_{sr} - \overset{0}{\nabla}_r c_{st}), \qquad A^q_{st} = g^{qr}(\nabla_s a_{tr} + \nabla_t a_{rs} - \nabla_r a_{st}). \tag{11}$$

The use of tensors of affine deformation allows us to avoid the introduction of Christoffel symbols into differential operations with tensors. We start from the formulas of differentiation of the deformation gradients

$$\frac{\partial}{\partial q^k}\,\underline{r}^s\underline{R}_s = \left\{ {q \atop sk} \right\} \underline{r}^s\underline{R}_q - \left\{ {\overset{0}{s} \atop kq} \right\} \underline{r}^q\underline{R}_s = A^q_{sk}\underline{r}^k\underline{R}_s = A^s_{qk}\underline{r}^q\underline{r}_s \cdot \overset{0}{\underline{\underline{\nabla R}}} = \underline{r}_k \cdot \overset{0}{\underline{A}} \cdot \overset{0}{\underline{\underline{\nabla R}}} \,.$$

We thus have

$$\frac{\partial}{\partial q^k}\,\overset{0}{\underline{\underline{\nabla R}}} = \underline{r}_k \cdot \overset{0}{\underline{A}} \cdot \overset{0}{\underline{\underline{\nabla R}}} \,, \qquad \overset{00}{\underline{\underline{\nabla\nabla R}}} = \overset{0}{\underline{A}} \cdot \overset{0}{\underline{\underline{\nabla R}}} \,. \tag{12}$$

Similarly, we get

$$\frac{\partial}{\partial q^k}\,\overset{0}{\underline{\underline{\nabla R}}}^{\mathrm{T}} = \overset{0}{\underline{\underline{\nabla R}}}^{\mathrm{T}} \cdot \overset{0}{\underline{A}}^{\mathrm{T}} \cdot \underline{r}_k \,, \qquad \overset{00}{\underline{\underline{\nabla\nabla R}}}^{\mathrm{T}} = \underline{r}^k \overset{0}{\underline{\underline{\nabla R}}}^{\mathrm{T}} \cdot \overset{0}{\underline{A}}^{\mathrm{T}} \cdot \underline{r}_k \,, \tag{13}$$

$$\frac{\partial}{\partial q^k}\,\underline{\underline{\nabla r}} = -\underline{R}_k \cdot \underline{A} \cdot \underline{\underline{\nabla r}} \,, \qquad \underline{\underline{\nabla\nabla r}} = -\underline{A} \cdot \underline{\underline{\nabla r}} \,, \tag{14}$$

$$\frac{\partial}{\partial q^k}\,\underline{\underline{\nabla r}}^{\mathrm{T}} = -\underline{\underline{\nabla r}}^{\mathrm{T}} \cdot \underline{A}^{\mathrm{T}} \cdot \underline{R}_k \,, \qquad \underline{\underline{\nabla\nabla r}}^{\mathrm{T}} = -\underline{R}^k \underline{\underline{\nabla r}}^{\mathrm{T}} \cdot \underline{A}^{\mathrm{T}} \cdot \underline{R}_k \,. \tag{15}$$

Making use of these formulas, we obtain representations for derivatives of the Cauchy-Green and Almansi strain measures

$$\frac{\partial \underline{G}}{\partial q^k} = \underline{r}_k \cdot \overset{0}{\underline{A}} \cdot \underline{G} + \underline{G} \cdot \overset{0}{\underline{A}}^{\mathrm{T}} \cdot \underline{r}_k \,, \qquad \frac{\partial \underline{g}}{\partial q^k} = -(\underline{R}_k \cdot \underline{A} \cdot \underline{g} + \underline{g} \cdot \underline{A}^{\mathrm{T}} \cdot \underline{R}_k) \,. \tag{16}$$

Taking contractions of the components of the tensors of affine deformation over the upper and lower indices, we obtain the relationships

$$A^k_{ks} = \frac{\partial \ln \sqrt{G}}{\partial q^s} - \frac{\partial \ln \sqrt{g}}{\partial q^s} = \frac{\partial}{\partial q^s} \ln \sqrt{\frac{G}{g}} \,. \tag{17}$$

Contraction over the lower indices of the tensors of affine deformation determines the vectors

$$G^{ts}\underline{R}_q A^q_{ts} = \underline{E} \cdot\cdot \underline{A} = \underline{A}^{\mathrm{T}} \cdot\cdot \underline{E} \,, \qquad g^{ts}\underline{r}_q A^q_{ts} = \underline{E} \cdot\cdot \overset{0}{\underline{A}} = \overset{0}{\underline{A}} \cdot\cdot \underline{E} \,. \tag{18}$$

<u>Divergence of the Finger measure</u>. From the above formulas, we get

$$\frac{\partial}{\partial q^k}\,\underline{F} = \frac{\partial}{\partial q^k}\,\overset{0}{\underline{\underline{\nabla R}}}^{\mathrm{T}} \cdot \overset{0}{\underline{\underline{\nabla R}}} = \overset{0}{\underline{\underline{\nabla R}}}^{\mathrm{T}} \cdot (\overset{0}{\underline{A}}^{\mathrm{T}} \cdot \underline{r}_k + \underline{r}_k \cdot \overset{0}{\underline{A}}) \cdot \overset{0}{\underline{\underline{\nabla R}}} =$$

$$= \overset{0}{\underline{\underline{\nabla R}}}^{\mathrm{T}} \cdot (\underline{r}_q\underline{r}^s + \underline{r}^s\underline{r}_q) \cdot \overset{0}{\underline{\underline{\nabla R}}} A^q_{sk} = g^{ms}(\underline{R}_q\underline{R}_m + \underline{R}_m\underline{R}_q) A^q_{sk} \,, \tag{19}$$

$$\underline{\nabla}\cdot\underline{F} = \delta^k_q \underline{R}^s\cdot\underline{F}A^q_{sk} + g^{ks}\underline{r}_q A^q_{sk}\cdot\overset{0}{\underline{\underline{\nabla R}}} = \underline{F}\cdot\underline{R}^s\frac{\partial}{\partial q^s}\ln\sqrt{\frac{G}{g}} + \overset{0}{\underline{\underline{\nabla R}}}{}^T\cdot(\overset{0}{\underline{A}}{}^T\cdot\cdot\,\underline{E})$$

or, equivalently,

$$\underline{\nabla}\cdot\underline{F} = \underline{F}\cdot\underline{\nabla}\ln\sqrt{\frac{G}{g}} + (\underline{E}\cdot\cdot\,\overset{0}{\underline{A}})\cdot\overset{0}{\underline{\underline{\nabla R}}} = \underline{F}\cdot\underline{\nabla}\ln\sqrt{\frac{G}{g}} + \overset{0}{\underline{\underline{\nabla R}}}{}^T\cdot(\overset{0}{\underline{A}}{}^T\cdot\cdot\,\underline{E})\ . \tag{20}$$

We also need the expression for $\underline{\nabla}\cdot\underline{F}^2$. We have

$$\underline{\nabla}\cdot\underline{F}^2 = \underline{F}\cdot\underline{\nabla}\cdot\underline{F} + \underline{F}\cdot\cdot\,\underline{\underline{\nabla F}}\ , \qquad \underline{F}\cdot\underline{\nabla}\cdot\underline{F} = \underline{F}^2\cdot\underline{\nabla}\ln\sqrt{\frac{G}{g}} + \underline{F}\cdot\overset{0}{\underline{\underline{\nabla R}}}{}^T\cdot(\overset{0}{\underline{A}}{}^T\cdot\cdot\,\underline{E})\ ,$$

$$\underline{F}\cdot\cdot\,\underline{\underline{\nabla F}} = (\underline{F}\cdot\cdot\,\underline{R}^k\underline{R}_q\underline{R}^s + \underline{F}\cdot\cdot\,\underline{R}^k\underline{F}\cdot\underline{R}^s\underline{R}_q)A^q_{sk} =$$

$$= \underline{F}\cdot\cdot\,\underline{R}_q\underline{R}^k\underline{R}^s A^q_{sk}\cdot\underline{F} + \underline{F}\cdot\cdot\,\underline{F}\cdot\underline{R}^k\underline{R}^s\underline{R}_q A^q_{sk} = \underline{F}\cdot\cdot\,(\underline{A}^T\cdot\underline{F} + \underline{F}\cdot\underline{A})$$

and thus

$$\underline{\nabla}\cdot\underline{F}^2 = \underline{F}^2\cdot\underline{\nabla}\ln\sqrt{\frac{G}{g}} + \underline{F}\cdot\overset{0}{\underline{\underline{\nabla R}}}{}^T\cdot(\underline{A}\cdot\cdot\,\underline{E}) + \underline{F}\cdot\cdot\,(\underline{A}^T\cdot\underline{F} + \underline{F}\cdot\underline{A}^T)\ . \tag{21}$$

Application of tensors of affine deformation enables us to establish the connection between the differential operations with functions of the deformation gradient (or of the strain measures) and derivatives with respect to these measures. For example, for a scalar function we have

$$\delta\psi(\overset{0}{\underline{\underline{\nabla R}}}) = \overset{0}{\underline{\nabla}}\psi\cdot\underline{r}_k\delta q^k = \psi_{\overset{0}{\underline{\underline{\nabla R}}}}\cdot\cdot\,\delta\overset{0}{\underline{\underline{\nabla R}}}{}^T = \psi_{\overset{0}{\underline{\underline{\nabla R}}}}\cdot\cdot\,\overset{0}{\underline{A}}{}^T\cdot\underline{r}_k\delta q^k\ ,$$

so that

$$\overset{0}{\underline{\nabla}}\psi = \psi_{\overset{0}{\underline{\underline{\nabla R}}}}\cdot\cdot\,\overset{0}{\underline{\underline{\nabla R}}}{}^T\cdot\overset{0}{\underline{A}}{}^T = \psi_{\overset{0}{\underline{\underline{\nabla R}}}}\cdot\overset{0}{\underline{\underline{\nabla R}}}{}^T\cdot\cdot\,\overset{0}{\underline{A}}{}^T = 2\psi_{\underline{G}}\cdot\underline{G}\cdot\cdot\,\overset{0}{\underline{A}}{}^T\ . \tag{22}$$

Analogously, for a tensor of second rank we have by virtue of (9.7)

$$\frac{\partial\underline{\underline{Q}}}{\partial q^k} = \underline{\underline{Q}}_{\overset{0}{\underline{\underline{\nabla R}}}}\cdot\cdot\,\overset{0}{\underline{\underline{\nabla R}}}{}^T\cdot\overset{0}{\underline{A}}{}^T\cdot\underline{r}_k\ ,$$

$$\overset{0}{\underline{\nabla}}\cdot\underline{\underline{Q}} = \underline{r}^k\cdot(\underline{\underline{Q}}_{\overset{0}{\underline{\underline{\nabla R}}}}\cdot\overset{0}{\underline{\underline{\nabla R}}}{}^T\cdot\cdot\,\overset{0}{\underline{A}}{}^T)\cdot\underline{r}_k = 2\underline{r}^k\cdot(\underline{\underline{Q}}_{\underline{G}}\cdot\underline{G}\cdot\cdot\,\overset{0}{\underline{A}}{}^T)\cdot\underline{r}_k\ . \tag{23}$$

The differential equation (12) for the deformation gradient reproduced in the form of a system of the first order

$$\overset{0}{\underline{\nabla}}\underline{R} = \underline{Z}\,, \qquad \overset{0}{\underline{\nabla}}\underline{Z} = \overset{0}{\underline{A}} \cdot \underline{Z}\,, \tag{24}$$

is immediately transformed to the form (18.6) - it suffices to accept the Cartesian coordinates a^s of the reference configuration for material coordinates, to represent $\underline{Z}$ in the form of (3) and to take into account that the Christoffel symbols in the reference configuration vanish for $q^s = a^s$. The problem of Par. 18 is thus reduced to determination of the tensor $\underline{Z}$ by (24) and to the subsequent evaluation of the position vector $\underline{R}$ from its total differential

$$d\underline{R} = d\underline{r} \cdot \underline{Z}\,. \tag{25}$$

CHAPTER 2

STRESS IN A CONTINUOUS MEDIUM

1. BODY AND SURFACE FORCES

In this chapter we will consider a body (a continuous medium) mostly in the actual configuration. As before, the volume of the body and its boundary surface will be designated by V, O in this configuration, and by v, o in the reference configuration. The notation ρ, ρ_0 for densities in the actual and reference configurations will also be preserved. By the law of conservation mass we have

$$dm = \rho\, dV = \rho_0\, dv \,, \tag{1}$$

so that

$$\frac{\rho_0}{\rho} = \frac{dV}{dv} = \sqrt{\frac{G}{g}} \,. \tag{2}$$

The external forces represent the effect produced upon the particles of the body B by the surrounding bodies. These forces are subdivided into the body and surface forces.

The body forces act upon each particle of a medium. The resultant vector of the body forces per unit mass will be denoted by $\underline{k}$. The body force applied to the elementary mass occupying the volume V is equal to

$$\underline{k}\, dm = \rho \underline{k}\, dV = \rho_0 \underline{k}\, dv \tag{3}$$

here, $\rho\underline{k}$ represents the force acting upon a unit volume of the medium. The resultant vector and the resulting moment of the body forces referred to the origin of the vector $\underline{R}$ are

$$\iiint_V \rho\underline{k}\, dV = \iiint_v \rho_0\underline{k}\, dv \,, \qquad \iiint_V \underline{R} \times \rho\underline{k}\, dV = \iiint_v \underline{R} \times \rho_0\underline{k}\, dv \,. \tag{4}$$

An example of a body force is the gravitational force

$$\underline{k} = -\underline{i}_3\, g \tag{5}$$

where $\underline{i}_3$ is a unit vector in an upward direction and g denotes the acceleration due to gravity. If the particles are fixed in a system of axes which rotate uniformly about some fixed axis (in a Galilean system), then the "centrifugal" force

$$\underline{k} = -\underline{\omega} \times (\underline{\omega} \times \underline{R}) = \omega^2 h \underline{e} \tag{6}$$

must be regarded as a body force. Here, $\underline{\omega}$ denotes a constant angular velocity vector, $\underline{R}$ is the position vector of a particle in a system of axes rotating with this angular velocity (the origin of $\underline{R}$ is placed on the axis of rotation), h is the radius of the circular path of a rotating particle and $\underline{e}$ is the unit vector directed from the centre of this circle along its radius. The Coriolis force does not appear in (6) because the particles of the medium are fixed within the system of rotating axes. The "tangential" force of inertia is also absent since $\underline{\omega}$ is constant.

For potential forces,

$$\underline{k} = -\underline{\nabla}\pi \ . \tag{7}$$

Here π denotes the potential energy of the field of body forces. For the gravitational field and for the field of centrifugal forces, we have, respectively,

$$\pi = g\underline{i}_3 \cdot \underline{R} \ , \qquad \pi = -\tfrac{1}{2}\,|\underline{\omega} \times \underline{R}|^2 = \tfrac{1}{2}\underline{\omega} \cdot (\underline{\underline{RR}} - \underline{\underline{ER}} \cdot \underline{R}) \cdot \underline{\omega} \ . \tag{8}$$

The forces of interaction between the particles contained in the volume considered, e.g. the Newtonian forces of attraction, are not regarded as body forces. According to the accepted definition, the fields of the body forces are produced by the bodies surrounding the volume considered. This volume is assumed to be sufficiently small so that the mutual attraction of its particles is negligible compared with the gravitational field of the Earth.

The forces distributed throughout the surface O of the volume V are the external surface forces. The surface force related to a unit area of O is denoted by $\underline{t}$, the resultant vector and the resultant moment of these forces are given by (cf. (1.8.9))

$$\begin{aligned} \iint_O \underline{f}\, dO &= \iint_o \sqrt{\frac{G}{g}}\,(\underline{n} \cdot \underline{G}^{-1} \cdot \underline{n})^{\frac{1}{2}}\, \underline{f}\, do \ , \\ \iint_O \underline{R} \times \underline{f}\, dO &= \iint_o \sqrt{\frac{G}{g}}\,(\underline{n} \cdot \underline{G}^{-1} \cdot \underline{n})^{\frac{1}{2}}\, \underline{R} \times \underline{f}\, do \ . \end{aligned} \tag{9}$$

The formula

$$\underline{f} = \underline{f} \cdot \underline{\underline{NN}} + (\underline{N} \times \underline{f}) \times \underline{N} \ , \tag{10}$$

where $\underline{N}$ is the unit vector of the outward normal to O, yields the decomposition of the surface force into its normal and tangential components.

The pressure exerted by a fluid upon a submerged body is another example of a surface force. This is given by

$$\underline{f} = -\rho \underline{N} \ , \tag{11}$$

Another example is that of the reaction force over the surface of contact between a body and other bodies supporting it.

For "dead loading" the magnitude and direction of the contact force $\underline{f}\, dO$ remain constant when applied at the same point M of the body*)

$$\underline{f}\, dO = \underline{f}^0\, do \ , \qquad \underline{f} = \underline{f}^0 \sqrt{\frac{g}{G}}\, (\underline{n} \cdot \underline{G}^{-1} \cdot \underline{n})^{-\frac{1}{2}} \ . \tag{12}$$

Here, $\underline{f}^0$ denotes the contact force in the reference configuration. The force associated with pressure (11) provides an example of "follow-up" loading.

The incremental work done when a pressure uniformly distributed over the surface of a body (p = const) produces a virtual displacement $\delta\underline{R}$ is given by

$$- \iint_O p\underline{N} \cdot \delta\underline{R}\, dO = -p \iiint_V \underline{\nabla} \cdot \delta\underline{R}\, dV = -p \iiint_v \delta \sqrt{\frac{G}{g}}\, dv =$$

$$= -p\delta \iiint_v \sqrt{\frac{G}{g}}\, dv = -\delta p\, V$$

here, we have used the Gauss-Ostrogradskii transformation and Eqs. (1.10.18) and (1.10.2). We get

$$- \iint_O p\underline{N} \cdot \delta\underline{R}\, dO = -\delta\Pi \ , \qquad \Pi = pV \ , \tag{13}$$

in which Π is the potential energy due to the pressure uniformly distributed over the surface of the body.

The pressure exerted by a fluid on a submerged body varies according to the law $p = \gamma z$ where γ is a specific weight (the weight per unit volume) of the fluid, and z is the vertical coordinate in the negative sense of a point of O measured from the horizontal surface of the fluid.

The incremental work done by the virtual displacement $\delta\underline{R}$ is now given by

---

*) For example, if initially, the force vector makes an acute angle with the normal $\underline{N}$ to the surface, then, after the body has been rotated through 180°, this angle becomes obtuse (the vector $\underline{N}$ being converted to $-\underline{N}$).

$$-\gamma \iint\limits_{O} z\underline{N} \cdot \delta\underline{R}\, dO = -\gamma \iiint\limits_{V} \underline{\nabla} \cdot z\delta\underline{R}\, dV = -\gamma \iiint\limits_{v} \sqrt{\frac{G}{g}}\, \underline{\nabla} \cdot z\delta\underline{R}\, dv =$$

$$= -\gamma \iiint\limits_{v} \sqrt{\frac{G}{g}} \left(\underline{R}^{s} \frac{\partial z}{\partial q^{s}} \cdot \delta\underline{R} + z\underline{\nabla} \cdot \delta\underline{R} \right) dv =$$

$$= -\gamma \iiint\limits_{v} \left(\sqrt{\frac{G}{g}}\, \delta z + z\delta \sqrt{\frac{G}{g}} \right) dv =$$

$$= -\gamma\delta \iiint\limits_{v} \sqrt{\frac{G}{g}}\, z\, dv = -\gamma\delta \iiint\limits_{V} z\, dV = -\gamma\delta\, V z_{C} ,$$

where z_C is the coordinate of the centre of mass of the submerged portion of the volume V. We have thus arrived at the formulas

$$\delta\Pi = \delta\gamma z_{C} V , \qquad \Pi = \gamma V z_{C} . \tag{14}$$

For "dead loading" of the entire surface,

$$\iint\limits_{O} \underline{f} \cdot \delta\underline{R}\, dO = \iint\limits_{o} \underline{f}^{0} \cdot \delta\underline{R}\, do = \delta \iint\limits_{o} \underline{f}^{0} \cdot \underline{R}\, do = \delta \iint\limits_{O} \underline{f} \cdot \underline{R}\, dO ,$$

$$\Pi = - \iint\limits_{O} \underline{f} \cdot \underline{R}\, dO . \tag{15}$$

The notion of a "force tensor" is introduced by the expression

$$\underline{B} = \iiint\limits_{V} \rho \underline{k}\,\underline{R}\, dV + \iint\limits_{O} \underline{f}\underline{R}\, dO . \tag{16}$$

Twice the skew-symmetric part of this tensor is given by the resultant moment $\underline{m}^{0}$ of the external forces which are the body and contact forces. In fact, from (I.14.19) we have

$$\underline{B} - \underline{B}^{T} = \iiint\limits_{V} \rho (\underline{k}\underline{R} - \underline{R}\underline{k})\, dV + \iint\limits_{O} (\underline{f}\underline{R} - \underline{R}\underline{f})\, dO =$$

$$= \underline{E} \times \left(\iiint\limits_{V} \rho \underline{R} \times \underline{k}\, dV + \iint\limits_{O} \underline{R} \times \underline{f}\, dO \right) = \underline{E} \times \underline{m}^{0} . \tag{17}$$

If the body and contact loadings are "dead", then their potential energy is given by the first invariant of the force tensor

$$I_{1}(\underline{B}) = -\Pi = \iiint\limits_{V} \rho \underline{k} \cdot \underline{R}\, dV + \iint\limits_{O} \underline{f} \cdot \underline{R}\, dO . \tag{18}$$

The calculation of the incremental work produced by a uniform pressure on some part $O_\times$ of the surface O is more complex. The closed contour on O which defined $O_\times$ is denoted by $\Gamma_\times$. By Stokes' theorem (III.9.11) and in view of (III.3.4) and (I.10.19) we may write

$$\oint_{\Gamma_\times} d\underline{R} \cdot (\underline{R} \times \delta\underline{R}) = \iint_{O_\times} \underline{N} \cdot \underline{\nabla} \times (\underline{R} \times \delta\underline{R})\, dO =$$

$$= \iint_{O_\times} \underline{N} \cdot [\delta\underline{R} \cdot \underline{\nabla R} - \underline{\delta R \nabla} \cdot \underline{R} - \underline{R} \cdot \underline{\nabla}\delta\underline{R} + \underline{R\nabla} \cdot \delta\underline{R}]\, dO =$$

$$= -3 \iint_{O_\times} \underline{N} \cdot \delta\underline{R}\, dO + \iint_{O_\times} \underline{N} \cdot \delta\underline{R}\, dO + \iint_{O_\times} \underline{N}\, dO \cdot (\underline{E\nabla} \cdot \delta\underline{R} - \underline{\nabla}\delta\underline{R}^{T}) =$$

$$= -3 \iint_{O_\times} \underline{N} \cdot \delta\underline{R}\, dO + \delta \iint_{O_\times} \underline{N} \cdot \underline{R}\, dO \,.$$

The expression for the incremental work is thus reduced to

$$-p \iint_{O_\times} \underline{N} \cdot \delta\underline{R}\, dO = -\frac{1}{3}\, p\delta \iint_{O_\times} \underline{N} \cdot \underline{R}\, dO + \frac{1}{3}\, p \int_{\Gamma_\times} d\underline{R} \cdot (\underline{R} \times \delta\underline{R}) \,. \tag{19}$$

The loading of part of the surface by a uniform pressure cannot be represented by a potential field. An exception arises in the case of a clamped edge: for this case $\delta\underline{R} = \underline{0}$, and by (19)

$$-p \iint_{O_\times} \underline{N} \cdot \delta\underline{R}\, dO = \frac{1}{3}\, p\delta \iint_{O_\times} \underline{N} \cdot \underline{R}\, dO = -\delta\Pi \,, \qquad \Pi = \frac{1}{3} \iint_{O_\times} p\underline{N} \cdot \underline{R}\, dO \,. \tag{20}$$

This result can also be obtained for the case of a rigid contour constrained to move in its plane.

2. THE CAUCHY STRESS TENSOR

The stress tensor is one of the principal and fundamental concepts of continuum mechanics. There are many ways of introducing the stress tensor but these all include, in essence, the following assumptions.

1. The equations of momentum balance and moment of momentum for a system of a finite number of material points and for a solid body are generalized in a natural way applicable to deformable bodeis. These are Euler's laws of dynamics.

2. If we prescribe within the body an arbitrary volume V defined by some closed surface O, then the influence of the remaining part of the body upon V reduces to a prescription on O, some vector field called the stress vector $\underline{t}_{\underline{N}}$ and which represents the surface density of the force vector. The force vector

$d\underline{P}$ acting upon an infinitesimal surface element dO is then calculated from the formula $d\underline{P} = \underline{t}_{\underline{N}} dO$. It is assumed that $\underline{t}_{\underline{N}}$ may vary continuously with respect to the spatial coordinates.

The second assumption excludes the so-called polar media for which, beside the stress vector on the surface O there also acts the vector of the moment stresses $\underline{u}_{\underline{N}}$ such that the moment $d\underline{M}$ applied to dO is determined by the formula $d\underline{M} = \underline{u}_{\underline{N}} dO$. We do not consider polar media in the present context.

The following analysis is based upon the assumptions given above.

For the volume V we choose a cylinder of radius a and length l. A unit vector $\underline{N}$ will be assumed to be oriented along the axis of the cylinder. For this body we may write the momentum balance equation

$$\frac{d}{dt}\int_0^l\int_0^a\int_0^{2\pi} \rho\underline{v}r\,d\varphi\,dr\,dz = \int_0^l\int_0^a\int_0^{2\pi} \rho\underline{k}\,dV + \int_0^l\int_0^{2\pi} \underline{t}_{\underline{r}}r\,d\varphi\,dz + \int_0^a\int_0^{2\pi} \underline{t}_{-\underline{N}}r\,d\varphi\,dr + \int_0^a\int_0^{2\pi} \underline{t}_{\underline{N}}r\,d\varphi\,dr\ . \qquad (1)$$

The volume integral on the left-hand side of this equality represents the resultant momentum vector of the cylinder. The right-hand side, which is the vector of the resultant of the forces acting, includes the volume integral of the body forces, the integral over the curved surface ($r = a$) of the cylinder and two integrals evaluated over its ends $z = 0$ and $z = 1$. The vector $\underline{N}$ is the outward normal to the face $z = 1$, and the vector $(-\underline{N})$ is that for the face $z = 0$. The stress vector acting on the infinitely small area $ds\underline{N}$ is denoted by $\underline{t}_{\underline{N}}$. As the length l of the cylinder tends to zero, the left-hand side and the first two integrals on the right-hand side of (1) vanish and the last two integrals are evaluated over the opposite sides of the same area $z = 0$. Equation (1) now takes the form

$$\int_0^a\int_0^{2\pi} (\underline{t}_{-\underline{N}} + \underline{t}_{\underline{N}})\,r\,d\varphi\,dr = 0\ .$$

Applying the first mean value theorem to this equation and dividing through by πa we obtain

$$\underline{t}_{-\underline{N}}\Big|_{\substack{r=r_1\\ \varphi=\varphi_1}} + \underline{t}_{\underline{N}}\Big|_{\substack{r=r_2\\ \varphi=\varphi_2}} = \underline{0}\ , \qquad 0 \le r_1, r_2 \le a\ , \qquad 0 \le \varphi_1, \varphi_2 \le 2\pi\ ,$$

As the radius a of the cylinder tends to zero, we obtain the first property of the stress vector $\underline{t}_{\underline{N}}$

$$\underline{t}_{\underline{N}} = -\underline{t}_{-\underline{N}} \ , \tag{2}$$

here, both sides are evaluated at the same point. Since the orientation of the cylinder's axis as well as the point $r = 0$, $z = 0$ are completely arbitrary, we conclude that Equation (2) is satisfied at all points and for all elementary areas of the body considered.

We now apply the momentum balance equation to the elementary tetrahedron formed by the vectors $\underline{R}_1 dq^1$, $\underline{R}_2 dq^2$, $\underline{R}_3 dq^3$. The sides of this tetrahedron are represented by the triangles $\underline{R}_1 \times \underline{R}_2 dq^1 dq^2$, $\underline{R}_2 \times \underline{R}_3 dq^2 dq^3$ and $\underline{R}_3 \times \underline{R}_1 dq^3 dq^1$ and by a plane passing through the points $\underline{R}_1 dq^1$, $\underline{R}_2 dq^2$, $\underline{R}_3 dq^3$. The integral of a unit outward vector evaluated over the closed (piecewise smooth) surface is evidently equal to zero. In fact, by virtue of the Gauss-Ostrogradskii theorem, we have

$$\iint_O \underline{N} dO = \iint_O \underline{N} \cdot \underline{E} dO = \iiint_V \underline{\nabla} \cdot \underline{E} dV = \underline{0} \ .$$

Taking the surface of the tetrahedron as O, we obtain

$$\underline{N} dO = \underline{R}^1 \frac{\underline{R}_2 dq^2 \times \underline{R}_3 dq^3}{|\underline{R}^1|} + \underline{R}^2 \frac{\underline{R}_3 dq^3 \times \underline{R}_1 dq^1}{|\underline{R}^2|} + \underline{R}^3 \frac{\underline{R}_1 dq^1 \times \underline{R}_2 dq^2}{|\underline{R}^3|} \ . \tag{3}$$

Here the outward normal to the surface of the tetrahedron corresponding to the vectors $\underline{R}_2 dq^2$, $\underline{R}_3 dq^3$ is given by $(-\underline{R}^1 |\underline{R}^1|^{-1})$. From Equation (3) we obtain the following three relationships

$$|\underline{R}_m dq^m \times \underline{R}_n dq^n| = \underline{N} \cdot \underline{R}_p |\underline{R}^p| dS \ , \qquad m \neq n \ , \ n \neq p \ , \ m \neq p \ . \tag{4}$$

There is of course no summation over the repeated indices in these formulas.

The stress vector acting on the infinitely small area with the normal $\underline{R}^m |\underline{R}^m|^{-1}$ is denoted by $\underline{t}_{(m)}$. This vector acts within the areas $q^m = \text{const}$; it characterizes the influence of that part of the medium which occupies the half-space corresponding to the increasing values of q^m. The momentum balance equation for the elementary tetrahedron can be written as

$$\underline{t}_{\underline{N}} dS + \underline{t}_{-1} |\underline{R}_2 \times \underline{R}_3| \, |dq^2 dq^3| + \underline{t}_{-2} |\underline{R}_3 \times \underline{R}_1| \, |dq^3 dq^1| +$$

$$+ \underline{t}_{-3} |\underline{R}_1 \times \underline{R}_2| \, |dq^1 dq^2| = \left(\frac{d}{dt} \rho \underline{v} - \rho \underline{k}\right) (\underline{R}_1 dq^1 \times \underline{R}_2 dq^2) \cdot \underline{R}_3 dq^3 \ . \tag{5}$$

The left-hand side of this equation is of second order, while the right-hand side is of third order; therefore, it can be considered to be zero to within the second order terms. By virtye of (2) and (4), we reduce (5) to the form

$$\underline{t}_{\underline{N}} = \underline{N} \cdot \sum_{m=1}^{3} \underline{R}_m \underline{t}_{(m)} \sqrt{G^{mm}} , \qquad |\underline{R}^m| \equiv \sqrt{G^{mm}} . \tag{6}$$

The vectors $\underline{t}_{\underline{N}}$ and $\underline{N}$ do not depend upon the choice of the coordinate system. Consequently, it is also true for the vector sum of the dyads on the right-hand side, though each of these dyads is of course dependent upon that choice. This is possible only provided that the right-hand side of (6) is expressed as the product of the vector $\underline{N}$ and a tensor of the second rank, so that the quantity $\underline{R}_m \underline{t}_m \sqrt{G^{mm}}$ represents a contravariant tensor whose "vector components" are determined by

$$\underline{t}^m = \underline{t}_{(m)} \sqrt{G^{mm}} .$$

Note that the vectors $\underline{t}_{(m)}$ represent the physical components of stress, these vectors act, however, on areas whose value and orientation depend upon the choice of coordinates.

By Equation (6) we have thus introduced a tensor of the second rank

$$\underline{T} = \underline{R}_m \underline{t}^m , \qquad \underline{t}^m = \underline{t}_{(m)} \sqrt{G^{mm}} , \tag{7}$$

which is called the Cauchy stress tensor. This tensor is a function of points of the body and is independent of the area chosen within the body. The meaning of this tensor is that the stress vector on the area with the normal $\underline{N}$ is determined by the Cauchy formula

$$\underline{t}_{\underline{N}} = \underline{N} \cdot \underline{T} . \tag{8}$$

The tensor $\underline{T}$ which is a function of position in the actual configuration, describes the state of stress at this point. Relationship (8) is of fundamental importance in continuum mechanics in general, it was formulated by Cauchy in the period 1823-1828. The tensor $\underline{T}$ is the Cauchy stress tensor.

The stress tensor is traditionally prescribed either by its contravariant or by mixed components

$$\underline{T} = t^{sk} \underline{R}_s \underline{R}_k = t^{s}_{\cdot k} \underline{R}_s \underline{R}^k = t_s^{\cdot k} \underline{R}^s \underline{R}_k . \tag{9}$$

In Paragraph 3 we shall show that this tensor is symmetric

$$\underline{T} = \underline{T}^{\mathrm{T}} , \qquad t^{sk} = t^{ks} , \qquad t^{s}_{\cdot k} = t_s^{\cdot k} = t^s_k , \qquad \underline{N} \cdot \underline{T} = \underline{T} \cdot \underline{N} . \tag{10}$$

A normal stress acting on the area $\underline{N}dO$ is denoted by $\sigma_{\underline{N}}$ and is given by

$$\sigma_{\underline{N}} = t_{\underline{N}}\,\underline{N} = \underline{N}\cdot\underline{T}\cdot\underline{N} = t^{sk}N_sN_k \tag{11}$$

this expression being a quadratic form in the components of $\underline{N}$. The component of $\underline{t}_{\underline{N}}$ lying in the plane tangential to the elementary area is denoted by $\underline{\tau}_{\underline{N}}$ and is given by

$$\underline{\tau}_{\underline{N}} = \underline{t}_{\underline{N}} - \underline{N}\sigma_{\underline{N}} = \underline{N}\times(\underline{t}_{\underline{N}}\times\underline{N}) = \underline{N}\cdot\underline{T}\cdot(\underline{E}-\underline{NN}) \ . \tag{12}$$

The square of the absolute value of $\underline{t}_{\underline{N}}$ is given by the quadratic form

$$\underline{t}_{\underline{N}}\cdot\underline{t}_{\underline{N}} = \underline{N}\cdot\underline{T}\cdot\underline{T}\cdot\underline{N} = \underline{N}\cdot\underline{T}^2\cdot\underline{N} \ , \tag{13}$$

and the square of the total tangential stress is

$$\underline{\tau}_{\underline{N}}\cdot\underline{\tau}_{\underline{N}} = \underline{t}_{\underline{N}}\cdot\underline{t}_{\underline{N}} - \sigma_{\underline{N}}^2 = \underline{N}\cdot\underline{T}^2\cdot\underline{N} - \underline{N}\cdot\underline{T}\cdot\underline{NN}\cdot\underline{T}\cdot\underline{N} =$$

$$= \underline{N}\cdot\underline{T}\cdot(\underline{E}-\underline{NN})\cdot\underline{T}\cdot\underline{N} \ . \tag{14}$$

The stress tensor can be expressed in terms of its principal values ("principal stresses") σ_s and principal directions $\underline{e}_s = \underline{e}^s$ (the principal axes of stress)

$$\underline{T} = \sigma_1\underline{e}_1\underline{e}^1 + \sigma_2\underline{e}_2\underline{e}^2 + \sigma_3\underline{e}_3\underline{e}^3 \ . \tag{15}$$

These formulas can now be rewritten as

$$\underline{t}_{\underline{N}} = \sigma_1N_1\underline{e}^1 + \sigma_2N_2\underline{e}^2 + \sigma_3N_3\underline{e}^3 \ , \tag{16}$$

$$\sigma_{\underline{N}} = \sigma_1N_1^2 + \sigma_2N_2^2 + \sigma_3N_3^2 \ , \tag{17}$$

$$\underline{t}_{\underline{N}}^2 = \sigma_1^2N_1^2 + \sigma_2^2N_2^2 + \sigma_3^2N_3^2 \ , \tag{18}$$

$$\underline{\tau}_{\underline{N}}^2 = 4(N_2^2N_3^2\tau_1^2 + N_3^2N_1^2\tau_2^2 + N_1^2N_2^2\tau_3^2) \ . \tag{19}$$

Here, τ_1, τ_2, τ_3 denote the absolute values of half-differences of the principal stresses

$$\tau_1 = \tfrac{1}{2}|\sigma_2-\sigma_3| \ , \qquad \tau_2 = \tfrac{1}{2}|\sigma_3-\sigma_1| \ , \qquad \tau_3 = \tfrac{1}{2}|\sigma_1-\sigma_2| \tag{20}$$

- the "principal tangential stresses". According to (19), these stresses are produced on the areas bisecting the angle between the principal planes. The value of $\tau^2_{\underline{N}}$ is stationary on these areas.

On the area whose unit normal $\underline{N}_x$ forms equal angles with the principal axes of stress (i.e. on the so-called "octahedral area" where $N^2_{xk} = \frac{1}{3}$) we have

$$\sigma_{\underline{N}_x} = \frac{1}{3}(\sigma_1 + \sigma_2 + \sigma_3) = \frac{1}{3}I_1(\underline{T}) \ , \tag{21}$$

$$t^2_{\underline{N}_x} = \frac{1}{3}(\sigma_1^2 + \sigma_2^2 + \sigma_3^2) = \frac{1}{3}I_1(\underline{T}^2) \ , \tag{22}$$

$$\tau^2_{\underline{N}_x} = \frac{4}{9}(\tau_1^2 + \tau_2^2 + \tau_3^2) = \frac{1}{3}[I_1(\underline{T}^2) - \frac{1}{3}I_1^2(\underline{T})] =$$

$$= \frac{2}{3}[\frac{1}{3}I_1^2(\underline{T}) - I_2(\underline{T})] = -\frac{2}{3}I_2(\mathrm{dev}\,\underline{T}) \ ,$$

so that

$$-I_2(\mathrm{dev}\,\underline{T}) = \frac{2}{3}(\tau_1^2 + \tau_2^2 + \tau_3^2) = \frac{1}{6}[(\sigma_1 - \sigma_2)^2 + (\sigma_2 - \sigma_3)^2 + (\sigma_3 - \sigma_1)^2] \ . \tag{23}$$

These formulas provide a mechanical interpretation of invariants of the stress tensor. Its first invariant is proportional to the normal stress on the octahedral area, and the second invariant of the deviatoric stress tensor is proportional to the square of the total tangential stress on the same area.

We will see below that for an isotropic elastic material, the principal axes of stress are parallel to those of the Almansi strain measure (or the Finger measure $\underline{F} = \underline{g}^{-1}$). The area $\overset{1}{do}$ of a face of a unit cube in the reference configuration characterized by the normal $\overset{0}{\underline{e}}_1$ coinciding with the principal direction of the Cauchy-Green measure $\underline{G}$ (and of that of $\underline{G}^{-1}$) acquires the value $\overset{1}{dO}$ in the actual configuration. By virtue of (1.8.7), (1.5.1), (1.8.2) we have

$$\overset{1}{dO} = \overset{1}{do}\sqrt{\frac{G}{g}}\,(\overset{0}{\underline{e}}_1 \cdot \underline{G}^{-1} \cdot \overset{0}{\underline{e}}_1)^{1/2} = \sqrt{\frac{G}{g}}\,\frac{\overset{1}{do}}{\sqrt{G_1}} = \frac{\sqrt{I^3(G)}}{v_1}\,\overset{1}{do} = \frac{\sqrt{I_3(\underline{G})}}{1 + \delta_1}\,\overset{1}{do} \ ,$$

where δ_k denote the principal stretches. The force acting on a unit area in the reference configuration is equal to

$$t_1 = \sigma_1 \frac{\overset{1}{dO}}{\overset{1}{do}} = \frac{\sqrt{I_3(G)}}{v_1}\,\sigma_1 = v_2 v_3 \sigma_1 = (1 + \delta_2)(1 + \delta_3)\sigma_1 \ . \tag{24}$$

The quantities

$$t_k = \sigma_k \frac{\sqrt{I_3(G)}}{v_k} = \sigma_k \frac{\sqrt{I_3(G)}}{1+\delta_k} \tag{25}$$

are distinguished from the principal stresses. They are called the "principal forces".

The "stresses" in a traditional sense are determined as the "physical" components of the tensor $\underline{T}$.

The unit vectors normal to the surfaces q^s = const as well as the vectors of force related to a unit area on these surfaces, are designated by $\overset{s}{\underline{N}}$, $\overset{s}{\underline{t}}$ (below q^s will be thought of as having Cartesian coordinates)

$$\overset{s}{\underline{N}} = \frac{\underline{R}^s}{|\underline{R}^s|}, \qquad \overset{s}{\underline{t}} = \overset{s}{\underline{N}} \cdot \underline{T}$$

by (9), we have

$$\overset{s}{\underline{t}} = \frac{\underline{R}^s}{|\underline{R}^s|} \cdot t^{kt}\underline{R}_k\underline{R}_t = \frac{t^{st}}{|\underline{R}^s|}\underline{R}_t = \sum_{t=1}^{3} t^{st}\frac{|\underline{R}_t|}{|\underline{R}^s|}\underline{e}_t ,$$

where $\underline{e}_t$ denote the unit vectors along the coordinate axes of q^t.

The physical components of $\underline{T}$ designated by σ_{st} are equal to projections of $\overset{s}{\underline{t}}$ in these directions

$$\sigma_{st} = \overset{s}{\underline{t}} \cdot \underline{e}_t = t^{st}\frac{|\underline{R}_t|}{|\underline{R}^s|} = t^{st}\sqrt{\frac{G_{tt}}{G^{ss}}}, \qquad t^{st} = \sigma_{st}\sqrt{\frac{G^{ss}}{G_{tt}}} . \tag{26}$$

Here G_{tt}, G^{ss} denote the diagonal components of the strain measures $\underline{G}$ and $\underline{G}^{-1}$. Of course the quantities σ_{st} are not the components of a tensor and obviously $\sigma_{st} = \sigma_{ts}$ since $G^{-1}_{ss} = G^{ss}$.

Cartesian coordinates are an exception. For these $\underline{e}_t = \underline{i}_t$ and the components of the stress tensor do not differ from its physical components. The notation σ_x, σ_y, σ_z for normal stresses, as well as that of τ_{xy}, τ_{yz}, τ_{zx} for tangential stresses is universally adopted, so that

$$t^{11} = \sigma_x , \quad t^{22} = \sigma_y , \quad t^{33} = \sigma_z , \quad t^{12} = \tau_{xy} , \quad t^{23} = \tau_{yz} , \quad t^{31} = \tau_{zx} . \tag{27}$$

In a cylindrical coordinate system R, Φ, Z

$$G_{11} = G^{11} = 1 , \quad G_{33} = G^{33} = 1 , \quad G_{22} = R^2 , \quad G^{22} = R^{-2}$$

and by (26)

$$t^{11} = \sigma_R\,,\quad t^{33} = \sigma_Z\,,\quad t^{31} = \tau_{ZR}\,,\quad t^{22} = \frac{\sigma_\Phi}{R^2}\,,\quad t^{12} = \frac{\tau_{R\Phi}}{R}\,,\quad t^{23} = \frac{\tau_{\Phi Z}}{R}\,. \tag{28}$$

3. THE EQUATIONS OF MOTION OF A CONTINUOUS MEDIUM

We consider within a continuous medium a volume $V^\times$ bounded by an arbitrary surface $O^\times$. Whatever this portion of the volume V is (the surface $O^\times$ may in part belong to the boundary O of V), the contact force on $O^\times$ is determined by the Cauchy equation (2.8)

$$\underline{f} = \underline{N}\cdot\underline{T} = \underline{t}_{\underline{N}}\,. \tag{1}$$

The Eulerian laws of dynamics formulated for the volume V have the form of balance equations for momentum and moment of momentum

$$\begin{aligned} \frac{d}{dt}\iiint_{V^\times} \rho\underline{v}\,dV &= \iiint_{V^\times} \rho\underline{k}\,dV + \iint_{V^\times} \underline{t}_{\underline{N}}\,dO\,,\\ \frac{d}{dt}\iiint_{V^\times} \rho\underline{R}\times\underline{v}\,dV &= \iiint_{V^\times} \underline{R}\times\rho\underline{k}\,dV + \iint_{V^\times} \underline{R}\times\underline{t}_{\underline{N}}\,dO\,. \end{aligned} \tag{2}$$

Substituting for $\underline{t}_{\underline{N}}$ in the surface integrals from (1) and applying the Gauss-Ostrogradskii transform, we obtain by virtue of (III.8.4) and (III.8.5) that

$$\iint_{O^\times} \underline{N}\cdot\underline{T}\,dO = \iiint_{V^\times} \underline{\nabla}\cdot\underline{T}\,dV\,,\qquad \iint_{O^\times} \underline{R}\times\underline{N}\cdot\underline{T}\,dO = \iiint_{V^\times} (\underline{R}\times\underline{\nabla}\cdot\underline{T} - 2\underline{\omega})\,dV\,, \tag{3}$$

where $\underline{\omega}$ denotes the vector corresponding to the skew-symmetric part of $\underline{T}$, this vector being equal to zero if the tensor $\underline{T}$ is symmetric. Substituting these results into (2), we arrive at the relationships

$$\begin{aligned} &\iiint_{V^\times} \left(\rho\frac{d\underline{v}}{dt} - \rho\underline{k} - \underline{\nabla}\cdot\underline{T}\right)dV = \underline{0}\,,\\ &\iiint_{V^\times} \left[\underline{R}\times\left(\rho\frac{d\underline{v}}{dt} - \rho\underline{k} - \underline{\nabla}\cdot\underline{T}\right) + 2\underline{\omega}\right]dV = \underline{0}\,. \end{aligned} \tag{4}$$

Because the volume $V^\times$ is arbitrary the integrands should be zero. The first equation leads to the Cauchy equation of motion

$$\underline{\nabla}\cdot\underline{T} + \rho\underline{k} = \rho\frac{d\underline{v}}{dt}\,; \tag{5}$$

and the second expresses the symmetry of the stress tensor

$$\underline{\omega} = 0 \Rightarrow \underline{T} = \underline{T}^T . \tag{6}$$

The relationship (1) related to the surface O of the volume V yields the boundary condition at this surface. For the equilibrium configurations of the volume V, the right-hand side of (5) is equal to zero. For this special case, the dynamic equations are called static equations.

Returning now to the expression (1.16) for the force tensor, we obtain, after the elimination of $\underline{f}$ by (1), that

$$\underline{B} = \iiint\limits_V \rho\underline{k}\,\underline{R}\,dV + \iint\limits_O \underline{N}\cdot\underline{\underline{TR}}\,dO = \iiint\limits_V (\rho\underline{k} + \underline{\nabla}\cdot\underline{T})\underline{R}\,dV + \iiint\limits_V \underline{T}\,dV ,$$

since

$$\underline{\nabla}\cdot\underline{\underline{TR}} = (\underline{\nabla}\cdot\underline{T})\underline{R} + \underline{R}^s\cdot\underline{\underline{TR}}_s = (\underline{\nabla}\cdot\underline{T})\underline{R} + \underline{T} .$$

We thus have

$$\underline{B} = \iiint\limits_V \rho\underline{k}\,\underline{R}\,dV + \iint\limits_O \underline{\underline{fR}}\,dO = \iiint\limits_V \underline{T}\,dV = V\underline{T}_{(m)} , \qquad \underline{T}_{(m)} = \frac{1}{V}\iiint\limits_V \underline{T}\,dV . \tag{7}$$

Here, $\underline{T}_{(m)}$ denotes the mean value of the stress tensor evaluated for the volume V. The symmetry of the stress tensor is also a direct consequence of (1.17) and (7)

$$\underline{B} - \underline{B}^T = \underline{E}\times\underline{m}^O = \iiint\limits_V (\underline{T} - \underline{T}^T)\,dV = \underline{0} , \qquad \underline{T} = \underline{T}^T \tag{8}$$

because this relationship is applicable to any volume $V^{\times}$ contained in the volume V.

These formulas also show that for the resultant vectors of force $\rho\underline{k}$ and $\underline{f}$ equal to zero and for non-zero resultant moments of these vectors, we are able to satisfy the necessary conditions of equilibrium if we impart a finite rotation to the volume and keep the directions of the forces acting constant (or rotate the force vectors maintaining the orientation of the body). In fact, rotation, characterized by an orthogonal tensor, induces the following transformation of the force tensor

$$\underline{B}' = \iiint\limits_V \rho\underline{k}\,\underline{R}'\,dV + \iint\limits_O \underline{\underline{fR}}'\,dO = \underline{B}\cdot\underline{O} , \tag{9}$$

here we have used the relationship $\underline{R}' = \underline{R}\cdot\underline{O}$. By (8) tensor $\underline{O}$ satisfies the condition

$$\underline{B}' - \underline{B}'^{T} = \underline{B} \cdot \underline{O} - \underline{O}^{T} \cdot \underline{B}^{T} = \underline{0} \ . \tag{10}$$

At the same time, assuming that the tensor $\underline{B}$ is non-singular ($\det \underline{B} \neq 0$) and making use of its polar representation

$$\underline{B} = (\underline{B} \cdot \underline{B}^{T})^{\frac{1}{2}} \cdot \underline{\tilde{O}} \ , \qquad \underline{B}^{T} = \underline{\tilde{O}}^{T} \cdot (\underline{B} \cdot \underline{B}^{T})^{\frac{1}{2}} \ , \tag{11}$$

Equation (10) may be satisfied if we assume that

$$\underline{O} = \underline{\tilde{O}}^{T} \ .$$

This rotation satisfies all the static equations because the resultant vector of all forces retains its zero value.

Note. The relationships of the type (7) are generalized by substitution of dyads $\underline{\underline{kR}}$, $\underline{\underline{fR}}$ for tensors having the form of $\underline{\underline{k\Psi}}(\underline{R})$, $\underline{\underline{f\Psi}}(\underline{R})$ where $\underline{\Psi}$ denotes a tensor of arbitrary rank. We have

$$\iint\limits_{O} \underline{\underline{f\Psi}}(\underline{R})\, dO = \iint\limits_{O} \underline{N} \cdot \underline{\underline{T\Psi}}(\underline{R})\, dO = \iiint\limits_{V} \underline{\nabla} \cdot \underline{\underline{T\Psi}}\, dV =$$

$$= \iiint\limits_{V} [(\underline{\nabla} \cdot \underline{T})\, \underline{\Psi}(\underline{R}) + \underline{T} \cdot \underline{\underline{\nabla\Psi}}]\, dV \tag{12}$$

and by (5)

$$\iiint\limits_{V} \rho \underline{k}\, \underline{\Psi}\, dV + \iint\limits_{O} \underline{\underline{f\Psi}}\, dO = \iiint\limits_{V} \underline{T} \cdot \underline{\underline{\nabla\Psi}}\, dV \ . \tag{13}$$

Assuming for example that $\underline{\Psi} = \underline{\underline{RR}}$, we arrive at the relationship

$$\iiint\limits_{V} \rho \underline{k}\, \underline{\underline{RR}}\, dV + \iint\limits_{O} \underline{\underline{fRR}}\, dO = \iiint\limits_{V} \underline{T} \cdot (\underline{\underline{ER}} + \underline{R}^{s} \underline{\underline{RR}}_{s})\, dV \ , \tag{14}$$

which reduces to a system of eighteen equations determining all the moments of Cartesian components of the stress tensor to the first order

$$\iiint\limits_{V} (t_{<sq>} x^{r} + t_{<sr>} x^{q})\, dV = \iiint\limits_{V} \rho k_{<s>} x^{q} x^{r}\, dV + \iint\limits_{O} f_{<s>} x^{q} x^{r}\, dO \ . \tag{15}$$

This system is divided into three easily solved systems. If we assume that $\underline{\Psi} = \underline{\underline{RRR}}$, we would obtain as many as 30 equations for the 36 moments of the second order.

4. THE TENSOR OF STRESS FUNCTIONS

It is known that any tensor with zero divergence may be thought of as the curl of some different tensor, this is because the divergence of a curl of any tensor vanishes identically:

$$\underline{\nabla} \cdot \underline{L} = \underline{0} , \qquad \underline{L} = \underline{\nabla} \times \underline{M} , \qquad \underline{\nabla} \cdot \underline{L} = \underline{\nabla} \cdot (\underline{\nabla} \times \underline{M}) = \underline{0} . \tag{1}$$

If $\underline{L}$ is a symmetric tensor ($\underline{L} = \underline{L}^T$), then the tensor $\underline{M}$ should be subjected to the condition

$$(\underline{\nabla} \times \underline{M})^T = \underline{\nabla} \times \underline{M} . \tag{2}$$

This condition may be satisfied by (III.6.23), (III.6.27) if we assume that

$$\underline{M} = (\underline{\nabla} \times \underline{\underline{Q}})^T , \qquad \underline{\nabla} \times \underline{M} = \underline{\nabla} \times (\underline{\nabla} \times \underline{\underline{Q}})^T , \qquad \underline{L} = \mathrm{Ink}\,\underline{\underline{Q}} , \quad \underline{\underline{Q}} = \underline{\underline{Q}}^T . \tag{3}$$

From here it follows that the static equations of a continuous medium may be satisfied in the absence of a body force by the "incompatibility" tensor $\mathrm{Ink}\,\underline{\underline{Q}}$ for any twice differentiable symmetric tensor

$$\underline{\nabla} \cdot \underline{T} = \underline{0} , \quad \underline{T} = \underline{T}^T , \qquad \underline{T} = \mathrm{Ink}\,\underline{\underline{Q}} = \underline{\nabla} \times (\underline{\nabla} \times \underline{\underline{Q}})^T , \quad \underline{\underline{Q}} = \underline{\underline{Q}}^T . \tag{4}$$

In a Cartesian basis, assuming that

$$\underline{\underline{Q}} = Q^{<11>} \underline{i}_1 \underline{i}_1 + Q^{<22>} \underline{i}_2 \underline{i}_2 + Q^{<33>} \underline{i}_3 \underline{i}_3 ,$$

$$\underline{\underline{Q}} = Q^{<12>} (\underline{i}_1 \underline{i}_2 + \underline{i}_2 \underline{i}_1) + Q^{<23>} (\underline{i}_2 \underline{i}_3 + \underline{i}_3 \underline{i}_2) + Q^{<31>} (\underline{i}_3 \underline{i}_1 + \underline{i}_1 \underline{i}_3) , \tag{5}$$

we arrive, respectively, at the well-known representations of Maxwell and Morera for the stress tensor in terms of the stress functions.

For a plane tensor $\underline{T}$ with zero divergence whose components do not depend on the coordinate x^3 we obtain the following expression

$$\underline{T} = t^{\alpha\beta} \underline{R}_\alpha \underline{R}_\beta = \mathrm{Ink}\, \underline{i}_3 \underline{i}_3 U(q^1, q^3) \qquad (\alpha, \beta = 1,2) . \tag{6}$$

In Equation (III.6.26) we should now set

$$\underline{\nabla}^2 \underline{\underline{Q}} = \underline{i}_3 \underline{i}_3 \underline{\nabla}^2 U , \qquad \underline{\nabla} \cdot \underline{\underline{Q}} = \underline{R}^\alpha \cdot \frac{\partial U}{\partial q^\alpha} \underline{i}_3 = \underline{0} , \qquad I_1(\underline{\underline{Q}}) = U$$

and therefore

$$\underline{T} = \mathrm{Ink}\,\underline{\underline{Q}} = \mathrm{Ink}\,\underline{i}_3\underline{i}_3 U = -\underline{i}_3\underline{i}_3\nabla^2 U + \underline{\underline{E}}\nabla^2 U - \underline{\nabla\nabla} U \ .$$

We have thus obtained a well-known representation of the stress tensor with the aid of the Airy function

$$\underline{T} = \underline{E}_2\nabla^2 U - \underline{\nabla\nabla} U = \underline{R}^\alpha \underline{R}_\alpha \nabla^2 U - \underline{R}^\alpha \underline{R}^\beta \nabla_\alpha \nabla_\beta U \ . \tag{7}$$

In particular, for the Cartesian coordinates we have that

$$t^{11} = \sigma_x = \frac{\partial^2 U}{\partial x^{2\,2}} \ , \qquad t^{22} = \sigma_y = \frac{\partial^2 U}{\partial x^{1\,2}} \ , \qquad t^{12} = \tau_{12} = -\frac{\partial^2 U}{\partial x^1 \partial x^2} \ . \tag{8}$$

The expressions for the components of the stress tensor in terms of the remaining components Q^{11}, Q^{22}, Q^{12}, Q^{23}, Q^{31} of the tensor of the stress functions are now reduced to the form

$$t^{11} = \frac{\partial^2 Q^{22}}{\partial x^{3\,2}} \ , \qquad t^{22} = \frac{\partial^2 Q^{11}}{\partial x^{3\,2}} \ , \qquad t^{12} = \frac{\partial}{\partial x^3}\left(\frac{\partial Q^{23}}{\partial x^1} + \frac{\partial Q^{31}}{\partial x^2} - \frac{\partial Q^{12}}{\partial x^3}\right) \ , \tag{9}$$

$$t^{23} = -\frac{\partial^2 Q^{11}}{\partial x^2 \partial x^3} + \frac{\partial^2 Q^{12}}{\partial x^1 \partial x^3} + \frac{\partial}{\partial x^1}\left(\frac{\partial Q^{31}}{\partial x^2} - \frac{\partial Q^{23}}{\partial x^1}\right) \ , \tag{10}$$

$$t^{31} = -\frac{\partial^2 Q^{22}}{\partial x^3 \partial x^1} + \frac{\partial^2 Q^{12}}{\partial x^2 \partial x^3} + \frac{\partial}{\partial x^2}\left(\frac{\partial Q^{23}}{\partial x^1} - \frac{\partial Q^{31}}{\partial x^2}\right) \ ,$$

$$t^{33} = \frac{\partial^2 Q^{11}}{\partial x^{2\,2}} + \frac{\partial^2 Q^{22}}{\partial x^{1\,2}} - 2\,\frac{\partial^2 Q^{12}}{\partial x^1 \partial x^2} \ . \tag{11}$$

The components t^{31}, t^{23} of the stress tensor not depending on x^3 are represented in terms of the Prandtl stress function $\Phi(x^1, x^2)$ which arises in the problem of the torsion of a rod

$$t^{31} = \frac{\partial}{\partial x^2}\left(\frac{\partial Q^{23}}{\partial x^1} - \frac{\partial Q^{31}}{\partial x^2}\right) = \frac{\partial \Phi}{\partial x^2} \ , \qquad t^{23} = \frac{\partial}{\partial x^1}\left(\frac{\partial Q^{31}}{\partial x^2} - \frac{\partial Q^{23}}{\partial x^1}\right) = -\frac{\partial \Phi}{\partial x^1} \ . \tag{12}$$

For the Saint-Venant problem it is essential to consider the case of a linear dependence of the components Q^{11}, Q^{12}, Q^{22} upon x^3. We have then

$$t^{11} = 0 \ , \qquad t^{12} = 0 \ , \qquad t^{22} = 0$$

and the expression for the stress tensor takes the form

$$\underline{T} = \underline{i}_3\underline{\tau} + \underline{\tau}\underline{i}_3 + \underline{i}_3\underline{i}_3(\underline{\nabla}_0^2 I_1(\underline{\underline{Q}}) - \underline{\nabla}_0 \cdot \underline{\nabla}_0 \cdot \underline{\underline{Q}}) \ .$$

Here the vector $\underline{\tau}$ is determined by the formula

$$\underline{\tau} = \frac{\partial}{\partial x^3}(\underline{\nabla}_0 I_1(\underline{\underline{Q}}) + \underline{\nabla}_0 \cdot \underline{\underline{Q}}) - \underline{i}_3 \times \underline{\nabla}_0 \Phi \ ,$$

where $\underline{\nabla}_0$ denotes the nabla operator in two variables x^1, x^2

$$\underline{\underline{Q}} = \underline{i}_1\underline{i}_1 Q^{11} + \underline{i}_2\underline{i}_2 Q^{22} + (\underline{i}_1\underline{i}_2 + \underline{i}_2\underline{i}_1)Q^{12} \ .$$

5. ON POLAR MEDIA

So far it has been assumed that the influence of the bodies surrounding the volume V is expressed by a prescription of the external body and contact forces $\underline{k}$ and $\underline{f}$. For more general constructions, distributions of the external body and contact moments (torques) $\underline{\mu}$ and $\underline{\nu}$ are also permissible. In order to describe interactions among the particles of the medium which we have envisaged, it must be assumed that the distribution of influences of the particles filling the volume V_2 on the adjacent particles of the volume V_1 is statically equivalent not only to the resultant vector $\underline{t}_{\underline{N}} dO'$ of contact force acting on the elementary area $\underline{N}dO'$ but also to the resultant moment $\underline{m}_{\underline{N}} dO'$ evaluated with respect to some pole. The assumption that the moment may be considered to be of the same order of magnitude (dO') as the force, stems from the very notion of smallness within the model of a continuous medium.

The "infinitely small" volume itself represents a complicated congomeration of an enormous number of elementary particles, and the influences communicated across the area should be treated as the integrated actions of these particles.

We will confine our consideration to analysis of the equilibrium state (the static equations).

In order to describe the state of stress we introduce not only the stress tensor $\underline{T}$, but also the tensor $\underline{M}$ of couple-stress moments of stresses. In a similar fashion to (2.8), this tensor is determined by the relationship

$$\underline{N} \cdot \underline{M} = \underline{m}_{\underline{N}} \ , \qquad (1)$$

which is confirmed by a consideration of the equilibrium of moments (torques) distributed over the sides of the elementary tetrahedron.

The first relation of (3.2), which now expresses the vanishing of the resultant force acting on an arbitrarily defined volume in the medium is preserved as is also the static equation within the volume which follows from (3.2). However, the second relation in (3.2) should be expanded and to this end we represent it in the form

$$\iiint\limits_{V^\times} (\underline{R} \times \rho\underline{k} + \rho\underline{\mu})\, dV + \iint\limits_{O^\times} (\underline{R} \times \underline{t}_N + \underline{m}_N)\, dO = \underline{0} \,. \tag{2}$$

Eliminating $\underline{t}_N$ and $\underline{m}_N$ from the surface integral with the aid of (3.1) and (1) and transforming this integral into one evaluated over the volume $V^\times$, we obtain

$$\iiint\limits_{V^\times} [\underline{R} \times (\rho\underline{k} + \underline{\nabla} \cdot \underline{T}) - 2\underline{\omega} + \underline{\nabla} \cdot \underline{M} + \rho\underline{\mu}]\, dV = \underline{0} \,. \tag{3}$$

The static equations in the volume are now reduced to the form

$$\underline{\nabla} \cdot \underline{T} + \rho\underline{k} = \underline{0} \,, \qquad \underline{\nabla} \cdot \underline{M} + \rho\underline{\mu} = 2\underline{\omega} \,, \tag{4}$$

where by (I.14.17)

$$\underline{\omega} = \tfrac{1}{2}\underline{\epsilon} \cdot\cdot\, \underline{T} = \tfrac{1}{2} t^{sk} \underline{R}_s \times \underline{R}_k \,.$$

Materials whose behaviour displays the action of moments of stress are called polar materials. The theory of such materials was worked out by E. and F. Cosserat in 1909 and has been developed further over the last few decades in a considerable number of publications on the "moment theories of continuous medium". Due to lack of space these theories are not included here. In this section the Cauchy stress tensor will always be assumed symmetric.

Note 1. Example (4) may be obtained by the variational procedure if we include the second deformation gradients in the expression for the stored energy of an elastic medium (Toupin, 1964). This, however, is incompatible with the assumption (adopted below) that the elastic medium is of the class of "simple materials" (Chapter 3, Par. 1). It has also been shown (Gurtin, 1965) that the theory of "non-simple materials of the second order" requires the introduction of moments of stresses in order to avoid conflict with the principles of thermodynamics.

Note 2. In the multipolar theory of continuous media (materials of higher order than the second) moments of higher order are introduced in order to describe the interaction of particles across the oriented area.

6. ALTERNATIVE DEFINITIONS OF THE STRESS TENSOR

The description of the state of stress using the Cauchy tensor $\underline{T}$ determined in the actual configuration is natural and physically obvious. The main difficulty is that this configuration itself is usually not known in advance, while the reference configuration forms part of the initial data.

General formulations as well as approaches to some specific problems can be simplified by introduction of the Piola tensor $\underline{P}$ determined in the reference configuration and simply related to the Cauchy tensor $\underline{T}$.

We substitute the oriented area $\underline{N}dO$ in the initial relationship (2.3) for its representation (1.8.4) in terms of $\underline{n}do$ in the reference configuration. We then obtain

$$\underline{t}_N dO = \underline{N}\cdot\underline{T}dO = \sqrt{\frac{G}{g}}\,\underline{n}\cdot\underline{\underline{\nabla r}}^{T}\cdot\underline{T}do = \underline{n}\cdot\underline{P}do \; . \tag{1}$$

Here we have introduced the Piola tensor

$$\underline{P} = \sqrt{\frac{G}{g}}\,\underline{\underline{\nabla r}}^{T}\cdot\underline{T}\,, \qquad \underline{P}^{T} = \sqrt{\frac{G}{g}}\,\underline{T}\cdot\underline{\underline{\nabla r}} \; . \tag{2}$$

The inverse relationships have the form

$$\underline{T} = \sqrt{\frac{g}{G}}\,\underline{\underline{\overset{0}{\nabla}R}}^{T}\cdot\underline{P} = \sqrt{\frac{g}{G}}\,\underline{P}^{T}\cdot\underline{\underline{\overset{0}{\nabla}R}} \; . \tag{3}$$

The resultant external force acting on the arbitrary volume $\overset{\times}{V}$ can be expressed in terms of the tensor $\underline{P}$ by the relationship

$$\iiint_{\overset{\times}{V}} \rho\underline{k}\,dV + \iint_{\overset{\times}{O}} \underline{t}_N\,dO = \iiint_{\overset{\times}{v}} \rho_0\underline{k}\,dv + \iint_{\overset{\times}{o}} \underline{n}\cdot\underline{P}\,do = \iiint_{\overset{\times}{v}} (\rho_0\underline{k} + \underline{\overset{0}{\nabla}}\cdot\underline{P})\,dv \; .$$

The requirement that this force be zero for any $\overset{\times}{V}$ yields the static equation within the volume, this equation containing the Piola tensor

$$\underline{\overset{0}{\nabla}}\cdot\underline{P} + \rho_0\underline{k} = \underline{0} \; . \tag{4}$$

The nabla operator is defined here in the reference configuration. Equation (1) now has the form

$$\underline{n}\cdot\underline{P} = \underline{f}\,\frac{dO}{do} \; . \tag{5}$$

This is the static equation on a surface. It is considerably simplified for "dead" contact forces (cf. (1.12))

$$\underline{n} \cdot \underline{P} = \underline{f}^{0} \; . \tag{6}$$

The Piola tensor is non-symmetric; its left product with $\overset{0}{\underline{\underline{\nabla R}}}{}^{T}$ is symmetric

$$\overset{0}{\underline{\underline{\nabla R}}}{}^{T} \cdot \underline{P} = \sqrt{\frac{G}{g}}\, \underline{T} = \underline{T}_{(O)} \; . \tag{7}$$

The tensor $\underline{T}_{(O)}$ has been used by several authors (Hamel, Kappus, Trefftz).

The static equations in the volume for tensors $\underline{T}$ and $\underline{P}$ can be represented as

$$\frac{\partial}{\partial q^{s}} \sqrt{G}\, t^{st} \underline{R}_{t} + \rho_0 \sqrt{g}\, \underline{k} = \underline{0} \; , \qquad \frac{\partial}{\partial q^{s}} \sqrt{g}\, p^{st} \underline{r}_{t} + \rho_0 \sqrt{g}\, \underline{k} = \underline{0} \; , \tag{8}$$

(cf. (III.6.4)), here p^{st} denotes the contravariant components of $\underline{P}$ in the reference basis. For the tensor $\underline{T}_{(O)}$ the static equations take the form

$$\frac{\partial}{\partial q^{s}} \sqrt{g}\, t^{st}_{(O)} \underline{R}_{t} + \rho_0 \sqrt{g}\, \underline{k} = \underline{0} \; , \tag{9}$$

in the volume, and

$$\underline{f}\, \frac{dO}{do} = \sqrt{\frac{g}{G}}\, \underline{N} \cdot \underline{T}_{(O)} \frac{dO}{do} = \underline{t}_{(O)\underline{N}} \; , \qquad \underline{t}_{(O)\underline{N}} = \underline{n} \cdot \underline{\underline{\nabla r}}^{T} \cdot \underline{T}_{(O)} \; . \tag{10}$$

on the surface. The force vector on the oriented area $\underline{N}dO$ was denoted by $\underline{t}_{(O)\underline{N}}$, this vector referred, however, to a unit area in the reference configuration.

We now consider the tensor $\underline{T}^{\vee}$ which is defined by the formulas

$$\underline{T}^{\vee} = \underline{\underline{\nabla r}}^{T} \cdot \underline{T} \cdot \underline{\underline{\nabla r}} \; , \qquad \underline{T} = \overset{0}{\underline{\underline{\nabla R}}}{}^{T} \cdot \underline{T}^{\vee} \cdot \overset{0}{\underline{\underline{\nabla R}}} \; , \qquad \underline{T}^{\vee} = \sqrt{\frac{g}{G}}\, \underline{P} \cdot \underline{\underline{\nabla r}} \; ,$$

$$\underline{P} = \sqrt{\frac{G}{g}}\, \underline{T}^{\vee} \cdot \overset{0}{\underline{\underline{\nabla R}}} \; . \tag{11}$$

This tensor is represented in terms of its contravariant components in the following way

$$\underline{T}^{\vee} = \underline{\underline{\nabla r}}^{T} \cdot t^{sk} \underline{R}_{s} \underline{R}_{k} \cdot \underline{\underline{\nabla r}} = t^{sk} \underline{r}_{s} \underline{r}_{k} \tag{12}$$

in the vector basis of the reference configuration, the contravariant components of $\underline{T}^{\vee}$ coincide with the same components of $\underline{T}$ in the actual configuration.

Substituting the deformation gradients in (12) for their polar representations, we get

$$\underline{T} = \underline{O}^{\times T} \cdot \underline{U} \cdot \underline{T}^{\vee} \cdot \underline{U} \cdot \underline{O}^{\times} ,$$

$$\underline{T}^{\vee} = \underline{U}^{-1} \cdot \underline{O}^{\times} \cdot \underline{T} \cdot \underline{O}^{\times T} \cdot \underline{U}^{-1} = \underline{O}^{\times} \cdot \underline{V}^{-1} \cdot \underline{T} \cdot \underline{V}^{-1} \cdot \underline{O}^{\times T} . \tag{13}$$

In this section the tensor $\underline{T}^{\vee}$ will be called "energetic" (Chapter 2, Par. 8); the term "reduced" is equally applicable.

Representing the tensors $\underline{P}$ and $\underline{T}$ in terms of their mixed components in the bases of both configurations (cf. (1.2.10))

$$\underline{P} = p^{\alpha}_{\cdot t} \underline{r}_{\alpha} \underline{R}^{t} , \qquad \underline{T} = t^{\alpha}_{\cdot t} \underline{R}_{\alpha} \underline{R}^{t} , \tag{14}$$

we get

$$p^{\alpha}_{\cdot t} \underline{r}_{\alpha} \underline{R}^{t} = \sqrt{\frac{G}{g}}\, \underline{\nabla r}^{T} \cdot \underline{T} = \sqrt{\frac{G}{g}}\, \underline{r}_{\beta} \underline{R}^{\beta} \cdot t^{\alpha}_{\cdot t} \underline{R}_{\alpha} \underline{R}^{t} = \sqrt{\frac{G}{g}}\, r_{\alpha} t^{\alpha}_{\cdot t} \underline{R}^{t} ,$$

$$p^{\alpha}_{\cdot t} = \sqrt{\frac{G}{g}}\, t^{\alpha}_{\cdot t} . \tag{15}$$

The tensors $\underline{P}$ and $\underline{T}^{\vee}$ are themselves very convenient auxiliary quantities which, however, do not directly determine the state of stress. To this end it is necessary to return to the Cauchy tensor $\underline{T}$ which is sometimes called the true stress tensor.

The static equations expressed in terms of the energetic stress tensor

$$\overset{0}{\underline{\nabla}} \cdot \underline{P} + \rho_0 \underline{k} = \overset{0}{\underline{\nabla}} \cdot \left(\sqrt{\frac{G}{g}}\, \underline{T}^{\vee} \cdot \overset{0}{\underline{\nabla R}} \right) + \rho_0 \underline{k} = \underline{0} , \tag{16}$$

can be transformed to (cf. (1.19.12))

$$\overset{0}{\underline{\nabla}} \cdot \left(\sqrt{\frac{G}{g}}\, \underline{T}^{\vee} \right) + \underline{r}_s \sqrt{\frac{G}{g}}\, \underline{T}^{\vee} \cdot\cdot (\overset{0}{\underline{A}} \cdot \underline{r}^{s}) + \rho_0 \underline{k} \cdot \underline{\nabla r} = \underline{0} , \tag{17}$$

and in component representation become

$$\overset{0}{\nabla}_s \sqrt{\frac{G}{g}}\, t^{sq} + \sqrt{\frac{G}{g}}\, t^{st} A^{q}_{st} + \rho_0 k^{q} = \underline{0} . \tag{18}$$

The tensor

$$\underline{T}^{\times} = \sqrt{\frac{G}{g}}\,\underline{T}^{\vee} = \underline{P}\cdot\underline{\nabla r} = t^{st}\underline{r}_s\underline{r}_t \tag{19}$$

is called the second Piola-Kirchhoff tensor.

7. THE INCREMENTAL WORK

The incremental work done by the external body and contact forces for a virtual displacement $\delta\underline{R}$ of the particles of a medium from their state of equilibrium in the actual configuration within the volume V is equal to

$$\delta' A_{(e)} = \iiint\limits_V \rho\underline{k}\cdot\delta\underline{R}\,dV + \iint\limits_O \underline{f}\cdot\delta\underline{R}\,dO\ . \tag{1}$$

The notation $\delta' A_{(e)}$ (and, later on, $\delta' a_{(e)}$) serves to remind one that a small quantity of order $\delta\underline{R}$ is implied but by no means implies the differentiation of a function of state for the medium. Also, the symbol $\delta' A_{(e)}$ denotes an incremental amount of work but not a variational quantity, the latter quantity does not exist in general.

The transformation of (1) can easily be performed by integration over the volume v in the reference configuration together with the introduction of the Piola tensor

$$\delta' A_{(e)} = \iiint\limits_v \rho_0\underline{k}\cdot\delta\underline{R}\,dv + \iint\limits_o \underline{n}\cdot\underline{P}\cdot\delta\underline{R}\,do = \iiint\limits_v (\rho_0\underline{k} + \overset{0}{\underline{\nabla}}\cdot\underline{P})\cdot\delta\underline{R}\,dv +$$
$$+ \iiint\limits_v \underline{P}\cdot\cdot(\overset{0}{\underline{\nabla}}\delta\underline{R})^{\mathrm{T}}dv = \iiint\limits_v \underline{P}\cdot\cdot\,\delta\overset{0}{\underline{\nabla R}}{}^{\mathrm{T}}dv\ . \tag{2}$$

Here we have used the transformation (III.3.10) for the product of a tensor and vector, the formulas (7.4), (7.5) and the commutative property of operations $\overset{0}{\underline{\nabla}}$ and δ

$$\overset{0}{\underline{\nabla}}\delta\underline{R} = \underline{r}^s\frac{\partial}{\partial q^s}\delta\underline{R} = \underline{r}^s\delta\underline{R}_s = \delta\underline{r}^s\underline{R}_s = \delta\overset{0}{\underline{\nabla R}}\ ,\qquad \overset{0}{\underline{\nabla}}\delta\underline{R} = \delta\overset{0}{\underline{\nabla R}} \tag{3}$$

(this is so because the reference configuration is not subjected to variation, we have $\delta\underline{r}^s = \underline{0}$). The operations $\underline{\nabla}$ and δ do not commute. By (1.3.9) we have

$$\underline{\nabla}\delta\underline{R} = \underline{\nabla r}\cdot\overset{0}{\underline{\nabla}}\delta\underline{R} = \underline{\nabla r}\cdot\delta\overset{0}{\underline{\nabla R}}\ ,\qquad (\underline{\nabla}\delta\underline{R})^{\mathrm{T}} = (\delta\overset{0}{\underline{\nabla R}}{}^{\mathrm{T}})\cdot\underline{\nabla r}^{\mathrm{T}}\ . \tag{4}$$

Returning to (2), we obtain

$$\delta' A_{(e)} = \iiint\limits_{v} \delta' a_{(e)} dv = \iiint\limits_{v} \underline{P} \cdot\cdot\, \delta\underline{\overset{0}{\nabla}\underline{R}}^{T} dv \ , \qquad \delta' a_{(e)} = \underline{P} \cdot\cdot\, \delta\underline{\overset{0}{\nabla}\underline{R}}^{T} \ . \tag{5}$$

Here $\delta' a_{(e)}$ denotes the incremental work done by the external forces related to a unit volume of the reference configuration (the incremental work per unit volume). It can also be represented in terms of the tensor $\underline{T}$, thus from (7.11) we have

$$\delta' a_{(e)} = \sqrt{\frac{G}{g}}\, \underline{T}^{\vee} \cdot \underline{\overset{0}{\nabla}\underline{R}} \cdot\cdot\, \delta\underline{\overset{0}{\nabla}\underline{R}}^{T} = \sqrt{\frac{G}{g}}\, \underline{T}^{\vee} \cdot\cdot\, \underline{\overset{0}{\nabla}\underline{R}} \cdot \delta\underline{\overset{0}{\nabla}\underline{R}}^{T} =$$

$$= \tfrac{1}{2} \sqrt{\frac{G}{g}}\, \underline{T}^{\vee} \cdot\cdot\, \left(\underline{\overset{0}{\nabla}\underline{R}} \cdot \delta\underline{\overset{0}{\nabla}\underline{R}}^{T} + (\delta\underline{\overset{0}{\nabla}\underline{R}}) \cdot \underline{\overset{0}{\nabla}\underline{R}}^{T}\right) \ .$$

We have used here rules (1.7.16) and (1.7.17) for contraction. Turning now to the definitions (1.4.4) and (1.7.8) of the Cauchy-Green strain measure and strain tensor we obtain

$$\delta' a_{(e)} = \tfrac{1}{2} \sqrt{\frac{G}{g}}\, \underline{T}^{\vee} \cdot\cdot\, \delta\underline{G} = \sqrt{\frac{G}{g}}\, \underline{T}^{\vee} \cdot\cdot\, \delta\underline{C} \ . \tag{6}$$

This formula provides justification for the name "energetic" being applied to the stress tensor $\underline{T}$. In the linear theory, the incremental work is defined as a contraction of the stress tensor and the linear strain tensor.

The incremental work can be expressed in terms of the Cauchy stress tensor. By virtue of (6.11) and (4) we have

$$\delta' a_{(e)} = \sqrt{\frac{G}{g}}\, \underline{\nabla \underline{r}}^{T} \cdot \underline{T} \cdot \underline{\nabla \underline{r}} \cdot \underline{\overset{0}{\nabla}\underline{R}} \cdot\cdot\, \delta\underline{\overset{0}{\nabla}\underline{R}}^{T} = \sqrt{\frac{G}{g}}\, \underline{T} \cdot\cdot\, (\delta\underline{\overset{0}{\nabla}\underline{R}}^{T}) \cdot \underline{\nabla \underline{r}}^{T} =$$

$$= \sqrt{\frac{G}{g}}\, \underline{T} \cdot\cdot\, (\underline{\nabla}\delta\underline{R})^{T} \ . \tag{7}$$

By the definition (1.5.2), (1.5.3) of the Finger deformation measure and in view of (4) we obtain

$$\delta\underline{F} = (\delta\underline{\overset{0}{\nabla}\underline{R}}^{T}) \cdot \underline{\overset{0}{\nabla}\underline{R}} + \underline{\overset{0}{\nabla}\underline{R}}^{T} \cdot \delta\underline{\overset{0}{\nabla}\underline{R}} = (\underline{\nabla}\delta\underline{R})^{T} \cdot \underline{F} + \underline{F} \cdot \underline{\nabla}\delta\underline{R} \ ,$$

$$\underline{\nabla}\delta\underline{R}^{T} = (\delta\underline{F} - \underline{F} \cdot \underline{\nabla}\delta\underline{R}) \cdot \underline{F}^{-1}$$

and substitution into (7) yields the relationship

$$\delta' a_{(e)} = \sqrt{\frac{G}{g}}\, (\underline{F}^{-1} \cdot \underline{T} \cdot\cdot\, \delta\underline{F} - \underline{F}^{-1} \cdot \underline{T} \cdot \underline{F} \cdot\cdot\, \underline{\nabla}\delta\underline{R}) \ ,$$

This can be simplified considerably when the tensors $\underline{T}$ and $\underline{F}$ commute (which is the case if the body is isotropic). We then have by (7)

$$\delta' a_{(e)} = \tfrac{1}{2}\sqrt{\frac{G}{g}}\,\underline{T}\cdot\cdot\,\underline{F}^{-1}\cdot\delta\underline{F}\ . \tag{8}$$

Recalling the definition (1.6.10) of the Hencky measure, we have from (8)

$$\delta' a_{e} = \sqrt{\frac{G}{g}}\,\underline{T}\cdot\cdot\,\delta\underline{H}\ . \tag{9}$$

This formula will be derived towards the end of this section. It provides the simplest definition of the incremental work per unit volume in terms of the Cauchy strain tensor, this definition being valid, however, only for an isotropic body.

Proceeding now to the analysis of motion, we substitute a virtual displacement $\delta\underline{R}$ for a true displacement $d\underline{R} = \underline{v}dt$ and following (1) we obtain

$$\frac{dA_{(e)}}{dt} = \iiint\limits_V \rho\underline{k}\cdot\underline{v}\,dV + \iint\limits_O \underline{t}_{\underline{N}}\cdot\underline{v}\,dO = \iiint\limits_V (\rho\underline{k} + \underline{\nabla}\cdot\underline{T})\cdot\underline{v}\,dV +$$

$$+ \iiint\limits_V \underline{T}\cdot\cdot\,\underline{\underline{\nabla v}}^{T}\,dV = \iiint\limits_V \rho\dot{\underline{v}}\cdot\underline{v}\,dV + \frac{1}{2}\iiint\limits_V \underline{T}\cdot\cdot\,(\underline{\underline{\nabla v}}^{T} + \underline{\underline{\nabla v}})\,dV \tag{10}$$

where we have used the equation of motion (3.5) and the symmetry of the tensor $\underline{T}$. The first term is transformed to

$$\iiint\limits_V \rho\dot{\underline{v}}\cdot\underline{v}\,dV = \iiint\limits_v \rho_0\dot{\underline{v}}\cdot\underline{v}\,dv = \frac{1}{2}\iiint\limits_v \rho_0\frac{d}{dt}\,\underline{v}\cdot\underline{v}\,dv = \frac{1}{2}\frac{d}{dt}\iiint\limits_v \rho_0\underline{v}\cdot\underline{v}\,dv = \dot{K}\ .$$

Here K denotes the kinetic energy of the particles filling the volume considered

$$K = \frac{1}{2}\iiint\limits_v \rho_0\underline{v}\cdot\underline{v}\,dv = \frac{1}{2}\iiint\limits_V \rho\underline{v}\cdot\underline{v}\,dV = \frac{1}{2}\iiint\limits_V \underline{v}\cdot\underline{v}\,dm\ . \tag{11}$$

The second term in (10) can be rewritten with the notation of (1.13.10)

$$N = \frac{1}{2}\iiint\limits_V \underline{T}\cdot\cdot\,(\underline{\underline{\nabla v}}^{T} + \underline{\underline{\nabla v}})\,dV = \iiint\limits_V \underline{T}\cdot\cdot\,\underline{D}\,dV\ , \tag{12}$$

and determines a quantity called the power. We have now arrived at the equation of work

$$\frac{dA_{(e)}}{dt} = \dot{K} + N \tag{13}$$

which means that the sum of the power and the material time derivative for the kinetic energy is equal to the work produced per unit time by the external body and contact forces.

The expression for the power may be given in terms of the Piola stress tensor as follows

$$N = \iiint_V \underline{T} \cdot\cdot \underline{\nabla v}^T dV = \iiint_V \sqrt{\frac{g}{G}}\, \underline{\nabla}^0 \underline{R}^T \cdot \underline{P} \cdot\cdot \underline{\nabla v}^T dV = \iiint_v \underline{P} \cdot\cdot \underline{\nabla v}^T \cdot \underline{\overset{0}{\nabla} R}^T dv$$

and by (1.3.9) we have

$$N = \iiint_V \underline{T} \cdot\cdot \underline{D}\, dV = \iiint_v \underline{P} \cdot\cdot \underline{\overset{0}{\nabla} v}^T dv \,. \tag{14}$$

<u>Note</u>. *Derivation of formula (9)*. By definition, the tensor $\exp \underline{M}$ is equal to

$$\exp \underline{M} = \sum_{k=0}^{\infty} \frac{\underline{M}^k}{k!} = \underline{E} + \underline{M} + \frac{1}{2!}\underline{M}^2 + \frac{1}{3!}\underline{M}^3 + \ldots$$

and the variation of this tensor allows the following representation

$$\delta \exp \underline{M} = \delta\underline{M} + \frac{1}{2!}(\underline{M} \cdot \delta\underline{M} + \delta\underline{M} \cdot \underline{M}) + \frac{1}{3!}(\underline{M}^2 \cdot \delta\underline{M} + \underline{M} \cdot \delta\underline{M} \cdot \underline{M} + \delta\underline{M} \cdot \underline{M}^2) + \ldots =$$

$$= \sum_{k=1}^{\infty} \frac{1}{k!} \sum_{s=0}^{k-1} \underline{M}^s \cdot \delta\underline{M} \cdot \underline{M}^{k-1-s} \,.$$

The tensors $\underline{M}$ and $\underline{Q}$ are assumed symmetric and coaxial. By the rule of contraction, we obtain

$$\underline{Q} \cdot\cdot \delta \exp \underline{M} = \underline{Q} \cdot\cdot (\delta\underline{M} + \underline{M} \cdot \delta\underline{M} + \frac{1}{2!}\underline{M}^2 \cdot \delta\underline{M} + \ldots) =$$

$$= \underline{Q} \cdot\cdot \left[\delta\underline{M} + \sum_{k=1}^{\infty} \frac{1}{k!} \underline{M}^k \cdot \delta\underline{M}\right] = \underline{Q} \cdot\cdot (\exp \underline{M}) \cdot \delta\underline{M} = \underline{Q} \cdot \exp \underline{M} \cdot\cdot \delta\underline{M} \,.$$

Returning now to Equation (8) and to the representation (1.6.10) of the Hencky measure, we have

$$\underline{F}^{-1} = \exp(-2\underline{H}) \,, \qquad \delta\underline{F} = \delta \exp 2\underline{H} \,,$$

$$\underline{T} \cdot\cdot \underline{F}^{-1} \cdot \delta\underline{F} = \underline{T} \cdot \exp(-2\underline{H}) \cdot \exp 2\underline{H} \cdot\cdot \delta 2\underline{H} = \underline{T} \cdot\cdot 2\delta\underline{H} \,,$$

which is the desired result.

CHAPTER 3

THE STATE EQUATIONS

1. THE SIMPLE BODY

The equations of motion of a continuous medium determine the divergence of the stress tensor in a given field of body forces and velocities, but they do not prescribe the state of stress of the medium itself. All processes (of motion and equilibrium) obey these equations which are the necessary conditions for these processes to take place. They are, however, not sufficient to provide a complete description because different materials (e.g. a piece of clay or a steel rod) exhibit different responses to the action of the same system of forces. The universal theorems of mechanics (which are valid for all materials) - the principles of conservation of momentum and moment of momentum - provide a basis for the derivation of the equations of motion but they must be supplemented by the physical laws which are adopted to express the diversity of behaviour of various materials. These laws are represented by the equations of state (also called the constitutive equations), which establish the connection between the stress tensor and the quantities describing the motion of particles, provided that we confine ourselves to purely a mechanical statement of a problem (thermal effects will be considered in Chapter 9). Experiments enable these laws to be established conclusively, but only at the final stage. A speculative consideration is inevitable if we aim to formulate the general principles of the construction of the state equations and of the classification of materials. Only if we start from a mathematical model for a sufficiently narrow class of materials may be obtain information about it experimentally, the latter step to be made "the later the better" (Signorini).

For the state equations, the notions "now" and "before" are essential, because elimination of time would constrict their content. For this reason, static considerations will be postponed until the later stages of analysis.

A very general description of the state of stress "now, at this place" or, in other words, at some given instant t of time for a particle denoted in the reference configuration by $\underline{r}(q^1,q^2,q^3)$ is given by prescription of the stress tensor $\underline{T}$ as a function of $\underline{r}(q^1,q^2,q^3)$, t and the variable

$$\tau = t - s \leq t , \qquad 0 \leq s < \infty , \tag{1}$$

this variable participates in description of the history of each particle of the body $\mathcal{B}$

$$\overset{\mathcal{B}}{\overline{\underline{R}(\underline{r}(q^1,q^2,q^3);\tau)}} \ . \tag{2}$$

The stress tensor $\underline{T}(\underline{r}(q^1,q^2,q^3);t)$ is regarded as a functional over the history (2), and a function of the material coordinates q^1, q^2, q^3 of a given particle M, and probably of the time t, which may enter an explicit representation

$$\underline{T}(\underline{R}(\underline{r};t)) = \mathcal{T}\overset{\mathcal{B}}{\overline{(\underline{R}(\underline{r}(q^1,q^2,q^3);\tau);q^1,q^2,q^3;t)}} \ . \tag{3}$$

The explicit dependence on t reflects an attempt to take into account the variability of the material's properties stipulated by thermal, chemical and other processes which occur in the material (e.g. aging of concrete). A more detailed description would be obtained by the introduction, instead of t, of the quantities describing these processes together with their own histories. Confining ourselves to purely mechanical formulations we will omit t from a number of the arguments considered.

It may be assumed that temperature and, possibly, other parameters of chemical or some other origin participate in the representation (3) of the state equation. However, in this chapter temperature does not appear as a parameter of state. This is because there is a wide range of problems which do not require a thermodynamic treatment. Questions of symmetry, the concept of a solid body, types of anisotropy, and the concept of an elastic fluid etc. constitute the main body of this chapter. The introduction of additional parameters, which would introduce additional material into the discussion, is considered inappropriate here since there exists a vast class of phenomena which can be adequately described without considering temperature. In particular, in the absence of chemical reactions, such a description is equally valid both for isothermal and adiabatic deformation processes. The more general problems which require thermodynamic considerations for their analysis, will be considered in Chapter 9.

Simplifying the problem still more we will assume that the medium is homogeneous or "materially uniform".*) This means that its state equation is formulated in exactly the same way for all particles of the body $\mathcal{B}$. Also, under this condition, Equation (3) will not contain the arguments q^1, q^2, q^3 of the position occupied by a particle in $\mathcal{B}$.

---

*) We will not discuss the difference between the two notions here.

Under these simplifying assumptions we have

$$\underline{T}(\underline{R}(\underline{r};t)) = \overline{\tau(\underline{R}(\underline{r}(q^1,q^2,q^3);\tau))}^{B} \ . \tag{4}$$

This representation of a constitutive equation is based on the principle of causality i.e. "the material does not know future but retains a memory of the past".

Of course, Equation (3) and its simplified form (4) represent only the starting points for further detailed discussion. The next step is provided by the principle of local action which states that the state of stress at the point $\underline{r}(q^1,q^2,q^3)$ is determined by the action produced upon M by the particles in its immediate neighbourhood only

$$|\underline{r}(\bar{q}^1,\bar{q}^2,\bar{q}^3) - \underline{r}(q^1,q^2,q^3)| < \varepsilon \ , \tag{5}$$

here, ε is assumed to be small. Instead of Equation (4) we may now write

$$\underline{T}(\underline{R}(\underline{r};t)) = T\ \overline{(\underline{R}(\underline{r}(q^1,q^2,q^3);\tau))}^{N} \ , \tag{6}$$

in which N indicates that the influence is produced on M only by its immediate neighbourhood. The argument of the functional in (6) is now represented as

$$\overline{\underline{R}(\underline{r}(q^1;q^2;q^3);\tau)}^{N} = \underline{R}(\underline{r}(q^1,q^2,q^3);\tau) + \Delta\underline{R} = d\underline{r}\cdot\overset{0}{\underline{\nabla}}\underline{R}(\underline{r}(q^1,q^2,q^3);\tau) +$$

$$+ \tfrac{1}{2}d\underline{r}\cdot\overset{00}{\underline{\nabla\nabla}}\underline{R}(\underline{r}(q^1,q^2,q^3);\tau)\cdot d\underline{r} + \dots \ . \tag{7}$$

The stress tensor is represented "now, at this place" by a functional over the history of the motion as well as over the first, second etc. deformation gradients

$$\underline{T}(\underline{R}(\underline{r};t)) = T(\underline{R}(\underline{r};\tau)\ ,\ \overset{0}{\underline{\nabla}}\underline{R}(\underline{r};\tau)\ ,\ \overset{00}{\underline{\nabla\nabla}}\underline{R}(\underline{r};\tau),\dots) \ . \tag{8}$$

It would, however, not be sensible to preserve the argument $\underline{R}(\underline{r},\tau)$ since the reaction of a particle to the action of its neighbours does not depend on the position occupied by this particle within a medium. Equation (8) is now replaced by

$$\underline{T}(\underline{R}(\underline{r};t)) = T(\overset{0}{\underline{\nabla}}\underline{R}(\underline{r};\tau)\ ,\ \overset{00}{\underline{\nabla\nabla}}\underline{R}(\underline{r};\tau),\dots) \ . \tag{9}$$

The highest order of the gradient entering the functional determines the "order" of the material. We will confine ourselves here to materials of the first order, i.e. to "simple bodies".

$$\underline{T}(\underline{R}(\underline{r};t)) = \mathcal{T}(\overset{0}{\underline{\underline{\nabla R}}}(\underline{r};\tau)) \; . \tag{10}$$

This class includes not only classical materials, i.e. elastic bodies and viscous fluids, but also some broader classes of materials. The "non-simple" materials are outside the scope of this book. Our earlier exclusion of polar media is consistent with the exclusion of second order materials.

2. THE PRINCIPLE OF MATERIAL FRAME-INDIFFERENCE

Following the definition of a frame-indifferent quantity (Chapter 1, Par. 15) we assume the invariance of the force vector $\underline{t}_{\underline{N}}$ acting on an area $\underline{N}dO$. The invariance of $\underline{N}$ is easily checked with the aid of Equations (1.8.8), (1.15.13) and (1.15.12)

$$\underline{N}' = (\underline{n}\cdot\underline{G}^{-1'}\cdot\underline{n})^{-\frac{1}{2}}\cdot\underline{\underline{\nabla r}}'\cdot\underline{n} = (\underline{n}\cdot\underline{G}^{-1}\cdot\underline{n})^{-\frac{1}{2}}\underline{O}^{T}\cdot\underline{\underline{\nabla r}}\cdot\underline{n} = \underline{O}^{T}\cdot\underline{N} = \underline{N}\cdot\underline{O} \; . \tag{1}$$

Therefore we have

$$\underline{t}'_{\underline{N}'} = \underline{N}'\cdot\underline{T}' = \underline{N}\cdot\underline{O}\cdot\underline{T}' = \underline{t}_{\underline{N}}\cdot\underline{O} \; , \qquad \underline{t}_{\underline{N}} = \underline{N}\cdot\underline{O}\cdot\underline{T}'\cdot\underline{O}^{T} = \underline{N}\cdot\underline{T}$$

which proves the invariance of the Cauchy stress tensor $\underline{T}$ since the vector $\underline{N}$ can be chosen arbitrarily

$$\underline{T}'(\underline{r};t) = \underline{O}^{T}\cdot\underline{T}(\underline{r};t)\cdot\underline{O} \tag{2}$$

a parallel statement for the Piola tensor is expressed by the equality

$$\underline{P}'(\underline{r};t) = \sqrt{\frac{G}{g}}\,\underline{\underline{\nabla r}}^{T'}\cdot\underline{T}' = \sqrt{\frac{G}{g}}\,\underline{\underline{\nabla r}}^{T}\cdot\underline{O}\cdot\underline{O}^{T}\cdot\underline{T}\cdot\underline{O} = \underline{P}(\underline{r};t)\cdot\underline{O} \; . \tag{3}$$

The material property expressed by the functional (1.10) cannot be dependent on the vector basis, i.e. the functional $\mathcal{T}$ over $\overset{0}{\underline{\underline{\nabla R}}}'$ in a primed basis has the same form as the functional over $\overset{0}{\underline{\underline{\nabla R}}}$ in a non-primed basis

$$T = \mathcal{T}(\overset{0}{\underline{\underline{\nabla R}}}'(\underline{r};\tau)) \; , \qquad \underline{T} = \mathcal{T}(\overset{0}{\underline{\underline{\nabla R}}}(\underline{r};\tau)) \; . \tag{4}$$

We have thus formulated the principle of material frame-indifference.

Let us give a simple illustration of this principle. Suppose that in the initial basis (in a fixed room) we measure the stretching of a string loaded by a weight at its free end. The "primed" observer moving vertically in a lift will observe that the same weight will produce a different elongation but will conclude that the material properties of the string have remained the same while the change in its length has arisen due to the acceleration of the lift which quantity can itself be determined by this change. Similarly, if we know the elongation of a vertically suspended loaded string, then, if one end of the same string, loaded with the same mass, is fixed to the axis of rotation of a rotating horizontal table, the change in this elongation enables the angular velocity of the table to be calculated. In the differential equation of motion the physical properties of the string, i.e. its Young modulus and density, remain unchanged.

From (1.15.11), (2) and (4) we have the relationship

$$\mathcal{T}(\overset{0}{\underline{\underline{\nabla R}}}'(\underline{r};\tau)) = \mathcal{T}(\overset{0}{\underline{\underline{\nabla R}}}(\underline{r};\tau)\cdot\underline{O}(\tau)) = \underline{O}^{T}(t)\,\mathcal{T}(\overset{0}{\underline{\underline{\nabla R}}}(\underline{r};\tau))\cdot\underline{O}(t) \ . \tag{5}$$

Making use of the polar representation (1.6.1) of the deformation gradient

$$\overset{0}{\underline{\underline{\nabla R}}}(\underline{r},\tau) = \underline{U}(\underline{r};\tau)\cdot\underline{O}^{\times}(\tau) \ , \tag{6}$$

where $\underline{O}^{\times}$ denotes a tensor representing an orthogonal deformation, we arrive at the equality

$$\mathcal{T}(\underline{U}(\underline{r};\tau)\cdot\underline{O}^{\times}(\tau)\cdot\underline{O}(\tau)) = \underline{O}^{T}(t)\cdot\mathcal{T}(\overset{0}{\underline{\underline{\nabla R}}}(\underline{r},\tau))\cdot\underline{O}(t) \ , \tag{7}$$

which holds for all orthogonal tensors $\underline{O}$. This allows us to conclude that for all values of t

$$\underline{O}(t) = \underline{O}^{\times T}(t) \ , \qquad \underline{O}^{T}(t) = \underline{O}^{\times}(t) \tag{8}$$

and Equation (7) is now transformed to

$$\underline{T}(\underline{r};t) = \mathcal{T}(\overset{0}{\underline{\underline{\nabla R}}}(\underline{r},\tau)) = \underline{O}^{\times T}(t)\cdot\mathcal{T}(\underline{U}(\underline{r};\tau))\cdot\underline{O}^{\times}(t) \ . \tag{9}$$

We have thus proved the fundamental property of simple materials which is that their behaviour is determined only by the strain history of a stretch tensor $\underline{U}$ while the history of rotations exerts no influence upon tensor $\underline{T}(\underline{r},t)$ "at this instant".

The principle of material frame-indifference provides us with a tool for determining the suitability of an *a priori* prescription of the stress tensor using quantities characterizing the motion of a medium. To illustrate this point, we will consider two examples.

If we were to incorporate the position vector $\underline{R}(\underline{r},t)$ into the functional

$$T(\underline{r};t) = T(\underline{R}(\underline{r};\tau), \overset{0}{\underline{\nabla}}\underline{R}(\underline{r};\tau)),$$

we would obtain by (1.15.1) and (5) that

$$T(\underline{R}'(\underline{r};\tau), \overset{0}{\underline{\nabla}}\underline{R}'(\underline{r};\tau)) = \underline{O}^{T}(t) \cdot T(\underline{R}(\underline{r};\tau), \overset{0}{\underline{\nabla}}\underline{R}(\underline{r};\tau)) \cdot \underline{O}(t) .$$

In particular, for the continuous motion of the primed frame

$$\underline{O} = \underline{O}^{T} = \underline{E}, \qquad \overset{0}{\underline{\nabla}}\underline{R}' = \overset{0}{\underline{\nabla}}\underline{R}, \qquad \underline{R}' = \underline{c} + \underline{R}$$

and the preceding equality would lead to the relationship

$$T(\underline{R}(\underline{r};\tau) + \underline{c}(\tau), \overset{0}{\underline{\nabla}}\underline{R}(\underline{r};\tau)) = T(\underline{R}(\underline{r};\tau), \overset{0}{\underline{\nabla}}\underline{R}(\underline{r};\tau)),$$

which cannot be satisfied for arbitrary $\underline{c}(\tau)$. The equation of state of a material, as has already been pointed out earlier, cannot depend on the position $\underline{R}$ of a particle. This obvious fact is given here as a consequence of the principle of material frame-indifference.

The second example is more interesting. The state equation

$$\underline{T} = \underline{H}(\underline{v},\underline{D},\underline{W},\rho) \tag{10}$$

is put forward here as a description of the motion of the fluid, i.e. the stress tensor is represented as a function of velocity $\underline{v}$ and its strain tensor $\underline{D}$ as well as of vorticity $\underline{W}$ and density ρ. The principle of material frame-indifference is now represented by the relationship

$$\underline{T}' = \underline{H}(\underline{v}',\underline{D}',\underline{W}',\rho') = \underline{O}^{T} \cdot \underline{H}(\underline{v},\underline{D},\underline{W},\rho) \cdot \underline{O}$$

which, together with (1.15.6), (1.15.10) and (1.15.7), implies

$$\underline{H}(\dot{\underline{c}} + \underline{O}^{T} \cdot \underline{v} + \dot{\underline{O}}^{T} \cdot \underline{R}, \underline{O}^{T} \cdot \underline{D} \cdot \underline{O}, \underline{O}^{T} \cdot \underline{W} \cdot \underline{O} + \dot{\underline{O}}^{T} \cdot \underline{O}, \rho) = \underline{O}^{T} \cdot \underline{H}(\underline{v},\underline{D},\underline{W},\rho) \cdot \underline{O} . \tag{11}$$

For a progressive motion of the primed frame we get

$$\underline{O} = \underline{E}, \qquad \dot{\underline{O}} = \underline{0}, \qquad \underline{H}(\dot{\underline{c}} + \underline{v},\underline{D},\underline{W},\rho) = \underline{H}(\underline{v},\underline{D},\underline{W},\rho),$$

which means (as could have been foreseen) that the velocity vector cannot enter representation (10). Assuming now that

$$\underline{T} = \underline{H}(\underline{D},\underline{W},\rho) \;, \qquad \underline{T}' = \underline{H}(\underline{O}^T\cdot\underline{D}\cdot\underline{O}\,,\,\underline{O}^T\cdot\underline{W}\cdot\underline{O}+\dot{\underline{O}}^T\cdot\underline{O}\,,\,\rho)$$

and considering the instanteneous rotational motion, we get

$$\underline{O} = \underline{E} \;, \qquad \dot{\underline{O}} \neq \underline{0} \;, \qquad \underline{H}(\underline{D},\underline{W}+\dot{\underline{O}}^T,\rho) = \underline{H}(\underline{D},\underline{W},\rho) \;,$$

thus excluding the argument $\underline{W}$ from the independent variables of $\underline{H}$.

The principle of material frame-indifference can now be represented as

$$\underline{T} = \underline{H}(\underline{D},\rho) \;, \qquad \underline{T}' = \underline{H}(\underline{O}^T\cdot\underline{D}\cdot\underline{O},\rho) = \underline{O}^T\cdot\underline{H}(\underline{D},\rho)\cdot\underline{O} \;. \tag{12}$$

According to (II.7.1), the symmetric tensor function $\underline{T}$ of the symmetric tensor $\underline{D}$ is isotropic, and by (II.7.7) it can be represented as a polynomial of the second degree

$$\underline{T} = \varphi_0\underline{E} + \varphi_1\underline{D} + \varphi_2\underline{D}^2 \tag{13}$$

with the coefficients φ depending both on ρ and the invariants of $\underline{D}$. In particular, if we permit only a linear dependence upon $\underline{D}$, we should assume that

$$\varphi_0 = -p(\rho) + \lambda(\rho)I_1(\underline{D}) = -p(\rho) + \lambda(\rho)\underline{\nabla}\cdot\underline{v} \;, \qquad \varphi_1 = 2\mu(\rho) \;, \qquad \varphi_2 = 0 \;,$$

which yields the classical Navier-Stokes equations of a fluid

$$\underline{T} = [-p(\rho) + \lambda(\rho)\underline{\nabla}\cdot\underline{v}]\underline{E} + 2\mu(\rho)\underline{D} \;. \tag{14}$$

For an incompressible fluid

$$\rho = \rho_0 = \text{const} \;, \qquad \underline{\nabla}\cdot\underline{v} = 0 \;, \qquad \underline{T} = -p\underline{E} + 2\mu\underline{D} \;, \tag{15}$$

but the pressure p can by no means be assumed constant. It is determined from the equation of motion combined with the equation of continuity

$$\underline{\nabla}\cdot\underline{T} + \rho_0\underline{k} = \rho_0\underline{b} \;, \qquad \underline{\nabla}\cdot\underline{v} = 0 \;,$$

or

$$-\underline{\nabla} p + \mu \nabla^2 \underline{v} + \rho_0 \underline{k} = \rho_0 \underline{b} \ , \qquad \underline{\nabla} \cdot \underline{v} = 0 \qquad (\underline{b} = \dot{\underline{v}}) \ , \tag{16}$$

since

$$\underline{\nabla} \cdot \underline{\underline{\nabla \dot{v}}}^T = \underline{\underline{\nabla\nabla}} \cdot \underline{v} = \underline{0} \ , \qquad \underline{\nabla} \cdot \underline{D} = \tfrac{1}{2} \nabla^2 \underline{v} \ .$$

3. ELASTIC MATERIALS

Following the principles of causality, local action and material frame-indifference, we arrived in Par. 2 at the state equation of a simple material in the form (2.9) which represents the stress tensor as a functional over the history of the stretch tensor. To simplify this relationship, we will omit below the tacitly implied dependence on the material coordinates of particles

$$\underline{T}(t) = \underline{O}^{\times T}(t) \mathcal{T}(\underline{U}(\tau)) \cdot \underline{O}^{\times}(t) \ . \tag{1}$$

Actually, we are only able to put forward more or less acceptable assumptions concerning strain history. The problem which now arises is that of the elimination of strain history, i.e. a transition from a functional to a function depending on the quantities determined "now", at the instant τ, not at that of $\tau = t - s$. The first step must be the representation of $\underline{U}(\tau)$ in the form

$$\underline{U}(\tau) = \underline{U}(t) + \underline{Q}(t,\tau) \ , \tag{2}$$

where the term $\underline{Q}(t,\tau)$ is outlined explicitly. Making use of the notation of Chapter 1, Par. 17 we substitute in the identity

$$\overset{0}{\underline{\underline{\nabla R}}}(\tau) = \underline{r}^s \underline{R}_s(\tau) = \underline{r}^s \underline{R}_s(t) \cdot \overset{t}{\underline{\underline{\nabla R}}}(\tau) = \overset{0}{\underline{\underline{\nabla R}}}(t) \cdot \overset{t}{\underline{\underline{\nabla R}}}(\tau)$$

the deformation gradients for their polar representations

$$\underline{U}(\tau) \cdot \underline{O}^{\times}(\tau) = \underline{U}(t) \cdot \underline{O}^{\times}(t) \cdot \overset{t}{\underline{U}}(\tau) \cdot \overset{t\times}{\underline{O}}(\tau) \ ,$$

$$\underline{U}(\tau) = \underline{U}(t) \cdot \underline{O}^{\times}(t) \cdot \overset{t}{\underline{U}}(\tau) \cdot \overset{t\times}{\underline{O}}(\tau) \cdot \underline{O}^{\times T}(\tau) \ .$$

By virtue of (1.12.12) we have

$$\overset{t\times}{\underline{O}}(\tau) = \underline{R}^s(t)\underline{R}_s(\tau) \ , \qquad \underline{O}^{\times T}(\tau) = \underline{R}^k(\tau)\underline{r}_k \ ,$$

$$\overset{t\times}{\underline{O}}(\tau) \cdot \underline{O}^{\times T}(\tau) = \underline{R}^s(t)\underline{r}_s = \underline{O}^{\times T}(t) \ ,$$

and

$$\underline{U}(\tau) = \underline{U}(t) \cdot \underline{O}^{\times}(t) \cdot \overset{t}{\underline{U}}(\tau) \cdot \underline{O}^{\times T}(t) \ . \tag{3}$$

The representation (2) now acquires the form

$$\underline{U}(\tau) = \underline{U}(t) + \underline{U}(t) \cdot \underline{O}^{\times}(t) \cdot (\overset{t}{\underline{U}}(\tau) - \underline{E}) \cdot \underline{O}^{\times T}(t) \ ,$$

$$\underline{Q}(t,\tau) = \underline{U}(t) \cdot \underline{O}^{\times}(t) \cdot (\overset{t}{\underline{U}}(\tau) - \underline{E}) \cdot \underline{O}^{\times T}(t) \ , \tag{4}$$

where $\underline{Q}(t,t) = \underline{0}$ because $\overset{t}{\underline{U}}(t) = \underline{E}$. The functional (1) can now be represented as

$$T(\underline{U}(\tau)) = T(\underline{U}(t) + \underline{Q}(t,\tau)) = T(\underline{U}(t)) + E(t,\tau) \ , \tag{5}$$

where

$$E(t,\tau) = T(\underline{U}(t) + \underline{Q}(t,\tau)) - T(\underline{U}(t)) \ . \tag{6}$$

For the state equation, we obtain the relationship

$$\underline{T}(t) = \underline{O}^{\times T} \cdot T(\underline{U}(t)) \cdot \underline{O}(t) + \underline{O}^{\times T}(t) \cdot E(t,\tau) \cdot \underline{O}^{\times}(t) \ , \tag{7}$$

where the first term represents a function (not a functional) determined for the actual configuration, and the second is a functional over the history of the motion (the functional of memory). The structure of the latter functional is determined by Equations (6) and (4).

Throughout this book we will assume the existence of materials satisfying the condition

$$\forall t - \tau \geq 0 \qquad E(t,\tau) = 0 \ , \tag{8}$$

which agrees with the relationship $E(t,t) = 0$ which follows from Equations (4) and (6).

This assumption excludes viscous and visco-elastic materials since the behaviour of such materials cannot be adequately described without taking into account the continuing effects of previous deformations. Neither do we need to know the sequence of the deformation processes since we do not consider plasticity either. We consider only materials totally deprived of "memory", and thus no question arises about their history. These are properties which characterize an elastic material.

We will now consider two situations. Firstly, we will be dealing with purely time independent processes for which the loaded body remains ar rest throughout so that the quantities determining its deformation are totally time-independent depending only on the material coordinates

$$\underline{U} = \underline{U}(q^1, q^2, q^3) \ .$$

The constitutive equation of a simple material now takes the form

$$\underline{T} = \underline{O}^{\times T} \cdot T(\underline{U}) \cdot \underline{O}^{\times} \ . \tag{9}$$

Secondly, we may confine ourselves to consideration of materials possessing an "instantly vanishing memory". The tensor $\underline{U}(t)$ preserves then its time dependence but the functional $E(t,\tau)$ of memory (which equals zero for $t = \tau$) is negligible as compared to the first term in (5).

For this case we have by (7)

$$\underline{T}(t) = \underline{O}^{\times T}(t) \cdot T(\underline{U}(t)) \cdot \underline{O}^{\times}(t) \ . \tag{10}$$

The materials obeying the state equations (9), (10) are called elastic materials. The theory of elasticity is that part of continuum mechanics concerned with the behaviour of these materials. It must be noted that the term elastic materials covers a range of materials which differ substantially.

The results based on the state equation (9) remain valid for all simple materials. Under static conditions, the term "theory of elasticity" is generally assumed to be much more widely applicable than is in fact the case.

The treatment afforded by state equation (10) is completely different since this implies an extremely idealized model of a material which is able to respond immediately to a deformation produced at the instant t irrespectively of its prior state. A study of the functional of memory shows that his idealization is acceptable as a "zero approximation" valid for sufficiently slow processes of deformation. The next approximation must be a function of the velocities of these processes.

Thus a simple material is "elastic" provided that its state of stress at any instant t does not depend on the strain history but is determined by the magnitude of the deformation at that instant. An "elastic" material immediately "forgets" the preceding process of deformation.

The strain measures are determined by a comparison of the actual and reference configurations. This gives rise to the statement that an elastic material possesses a perfect memory of a single configuration - the reference configuration. It is hardly possible to agree with this view since this configuration can be assigned arbitrarily and need not be associated with any specific physical properties. The arbitrariness of choice of this configuration often enables the state equation to be simplified and thus provides grounds for preferring some specifically chosen actual configuration.

In forms (9) and (10) of the state equation for an elastic material we substitute the orthogonal tensor $\overset{\times}{\underline{O}}$ by the expressions

$$\underline{O}^{\times} = \underline{U}^{-1} \cdot \overset{0}{\underline{\nabla R}} , \qquad \underline{O}^{\times T} = \overset{0}{\underline{\nabla R}}{}^{T} \cdot \underline{U}^{-1} \tag{11}$$

and we obtain the relationship

$$\underline{T} = \overset{0}{\underline{\nabla R}}{}^{T} \cdot \underline{U}^{-1} \cdot \mathcal{T}(\underline{U}) \cdot \underline{U}^{-1} \cdot \overset{0}{\underline{\nabla R}} = \overset{0}{\underline{\nabla R}}{}^{T} \cdot \underline{\Phi}(\underline{U}) \cdot \overset{0}{\underline{\nabla R}} . \tag{12}$$

Here we have introduced the notation

$$\underline{\Phi}(\underline{U}) = \underline{U}^{-1} \cdot \mathcal{T}(\underline{U}) \cdot \underline{U}^{-1} = \underline{G}^{-\frac{1}{2}} \cdot \mathcal{T}(\underline{G}^{\frac{1}{2}}) \cdot \underline{G}^{-\frac{1}{2}} = \underline{\Phi}(\underline{G}^{\frac{1}{2}}) . \tag{13}$$

Making use of the symbol

$$\underline{\Phi}(\underline{U}) = \underline{\Phi}(\underline{G}^{\frac{1}{2}}) = \underline{\Psi}(\underline{G}) \tag{14}$$

we may represent the state equation in the form

$$\underline{T} = \overset{0}{\underline{\nabla R}}{}^{T} \cdot \underline{\Psi}(\underline{G}) \cdot \overset{0}{\underline{\nabla R}} , \tag{15}$$

obtaining for the Piola tensor (2.6.2) and energetic tensor (2.6.11)

$$\underline{P} = \sqrt{\frac{G}{g}}\, \Phi(\underline{U}) \cdot \overset{0}{\underline{\nabla R}} = \sqrt{\frac{G}{g}}\, \Psi(\underline{G}) \cdot \overset{0}{\underline{\nabla R}} , \tag{16}$$

$$\underline{T}^{\vee} = \underline{\Phi}(\underline{U}) = \underline{\Psi}(\underline{G}) . \tag{17}$$

It should also be recalled that by (2.5) the principle of material frame-indifference for an elastic material can be expressed as

$$\mathcal{T}(\overset{0}{\underline{\nabla R}} \cdot \underline{O}) = \underline{O}^{T} \cdot \mathcal{T}(\overset{0}{\underline{\nabla R}}) \cdot \underline{O} , \qquad \mathcal{T}(\overset{0}{\underline{\nabla R}}) = \underline{O} \cdot \mathcal{T}(\overset{0}{\underline{\nabla R}} \cdot \underline{O}) \cdot \underline{O}^{T} . \tag{18}$$

4. THE SYMMETRY GROUP OF A MATERIAL

The restrictions which the functional equation (3.18) imposes on the dependence of the state equation on the deformation gradient follow from considerations of the invariance of the motions which can be compared in the primed and nonprimed basis. The reference configuration is thereby preserved. Analysis of the problems associated with the choice of this configuration enables a classification of simple elastic materials (a solid body, a fluid etc.) to be given and also provides a strict definition of the notion of isotropy. In these considerations, the actual configuration is assumed fixed.

We consider the reference configurations $\overset{1}{v}$ and $\overset{2}{v}$; in these, the position of a particle $M(q^1,q^2,q^3)$ is specified by the vectors $\overset{1}{\underline{r}}$, $\overset{2}{\underline{r}}$. The one-to-one dependence between the two configurations is represented by the relationships

$$\overset{2}{\underline{r}} = \overset{2}{\underline{r}}(\overset{1}{\underline{r}}) \,, \qquad \overset{1}{\underline{r}} = \overset{1}{\underline{r}}(\overset{2}{\underline{r}}) \tag{1}$$

of which the first determines the transformation $\overset{1}{v} \to \overset{2}{v}$, and the second the transformation $\overset{2}{v} \to \overset{1}{v}$. The nabla operators defined for these configurations are denoted by

$$\overset{1}{\underline{\nabla}} = \overset{1}{\underline{r}}{}^{s} \frac{\partial}{\partial q^{s}} \,, \qquad \overset{2}{\underline{\nabla}} = \overset{2}{\underline{r}}{}^{s} \frac{\partial}{\partial q^{s}} \tag{2}$$

and the gradients of the transformations $\overset{1}{v} \to \overset{2}{v}$, $\overset{2}{v} \to \overset{1}{v}$ are

$$\overset{1}{v} \to \overset{2}{v} : \overset{12}{\underline{\underline{\nabla r}}} = \overset{1}{\underline{r}}{}^{s} \overset{2}{\underline{r}}_{s} = \underline{S} \,; \qquad \overset{2}{v} \to \overset{1}{v} : \overset{21}{\underline{\underline{\nabla r}}} = \overset{2}{\underline{r}}{}^{s} \overset{1}{\underline{r}}_{s} = \underline{S}^{-1} \,. \tag{3}$$

The deformation gradients $\overset{k}{\underline{\underline{\nabla R}}}$ in the actual configuration are now represented by the formulas

$$\overset{1}{v} \to V_t : \overset{1}{\underline{\underline{\nabla R}}} = \overset{1}{\underline{r}}{}^{s} \underline{R}_{s} = \underline{S} \cdot \overset{2}{\underline{\underline{\nabla R}}} \,; \qquad \overset{2}{v} \to V_t : \overset{2}{\underline{\underline{\nabla R}}} = \overset{2}{\underline{r}}{}^{s} \underline{R}_{s} = \underline{S}^{-1} \cdot \overset{1}{\underline{\underline{\nabla R}}} \,. \tag{4}$$

The equation of state may be written in one of the following forms

$$\underline{T} = \overset{1}{T}(\overset{1}{\underline{\underline{\nabla R}}}) = \overset{1}{T}(\underline{S} \cdot \overset{2}{\underline{\underline{\nabla R}}}) = \overset{2}{T}(\overset{2}{\underline{\underline{\nabla R}}}) = \overset{2}{T}(\underline{S}^{-1} \cdot \overset{1}{\underline{\underline{\nabla R}}}) \,. \tag{5}$$

Here $\overset{1}{T}$, $\overset{2}{T}$ denote two different functions related to the same state of stress. In exactly the same way we may distinguish between two equations of one and the same curve which are formulated in different coordinate systems. We cannot, however, exclude the possibility that the two equations may preserve their form when the coordinate axes (which are at the same time the axes of symmetry of the curve) are rotated by 180°.

The question now arises of how to find the transformations of the reference configuration which preserve the functional dependence between the stress tensor and the deformation gradient, or in other words, which preserve the form of the state equation. The gradient of such a selected transformation $\overset{1}{v} \to \overset{2}{v}$ will be denoted by $\underline{H}$ (instead of $\underline{S}$), and the indices over the symbol T discarded so that $\overset{1}{T} = \overset{2}{T} = T$. Equation (5) must now be written in the form*)

$$\forall \overset{1}{\underline{\underline{\nabla R}}}, \overset{2}{\underline{\underline{\nabla R}}} : T(\overset{1}{\underline{\underline{\nabla R}}}) = T(\overset{2}{\underline{\underline{\nabla R}}}) = T(\underline{H} \cdot \overset{2}{\underline{\underline{\nabla R}}}) = T(\underline{H}^{-1} \cdot \overset{1}{\underline{\underline{\nabla R}}}) \ . \tag{6}$$

These formulas show that there exists no experiment on the state of stress which would enable us to discover whether or not the material had been subjected in its reference $\overset{1}{v}$- (or $\overset{2}{v}$-) configuration to the $\underline{H}$-transformation $\overset{1}{v} \to \overset{2}{v}$ (or the $\underline{H}^{-1}$-transformation $\overset{2}{v} \to \overset{1}{v}$).

The change of density can be found and the $\underline{H}$-transformation should leave the density unchanged

$$\det \underline{H} = \pm 1 \tag{7}$$

(we do not exclude transformation accompanied by inversion).

The simplest example is provided by a thought experiment in which a cube is subjected to forces which are constant in magnitude and direction. The faces of the cube normal to the axes XYZ will be denoted by 1, 2, 3; the loading is produced by a force parallel to the Y-axis. In the $\overset{1}{v}$-configuration, face 2 is loaded, and the $\overset{1}{v} \to \overset{2}{v}$ transformation is performed by rotating the cube by 90° about the Z-axis. As a result, face 1 of the cube will find itself under load and the state of deformation of the cube and, consequently, its state of stress may either be changed, or remain the same. In the latter case, the rotation does represent the $\underline{H}$-transformation whereas in the first case it does not. Rotation about the X-axis, when the face 3 becomes subjected to the action of load, leads to the same conclusion.

The $\underline{H}$-transformations constitute a group. This means that if the $\underline{H}_1$- and $\underline{H}_2$-transformations satisfy Equations (6) then the transformations $\underline{H}_1 \cdot \underline{H}_2$, $\underline{H}_2 \cdot \underline{H}_1$ are $\underline{H}$-transformations too. The $\underline{H}^{-1}$-transformation also belongs to the class of $\underline{H}$-transformations, which is certainly the case for a unit tensor $\underline{E}$.

The proof is obvious since it is sufficient to substitute $\overset{2}{\underline{\underline{\nabla R}}}$ for $\underline{H}_2 \cdot \overset{2}{\underline{\underline{\nabla R}}}$ ($\underline{H}_1 \cdot \overset{2}{\underline{\underline{\nabla R}}}$) in the first (second) relationship of the pair

$$\forall \overset{2}{\underline{\underline{\nabla R}}} : T(\overset{2}{\underline{\underline{\nabla R}}}) = T(\underline{H}_1 \cdot \overset{2}{\underline{\underline{\nabla R}}}) \ , \qquad T(\overset{2}{\underline{\underline{\nabla R}}}) = T(\underline{H}_2 \cdot \overset{2}{\underline{\underline{\nabla R}}}) \ . \tag{8}$$

---

*) The symbol $\forall x$ means "for all x".

This operation is admissible since Equations (8) are valid for all $\overset{2}{\underline{\nabla R}}$. We thus arrive at the equations

$$T(\underline{H}_2 \cdot \overset{2}{\underline{\nabla R}}) = T(\underline{H}_1 \cdot \underline{H}_2 \cdot \overset{2}{\underline{\nabla R}}) , \qquad T(\underline{H}_1 \cdot \overset{2}{\underline{\nabla R}}) = T(\underline{H}_2 \cdot \underline{H}_1 \cdot \overset{2}{\underline{\nabla R}})$$

and, by (8),

$$T(\overset{2}{\underline{\nabla R}}) = T(\underline{H}_1 \cdot \overset{2}{\underline{\nabla R}}) = T(\underline{H}_2 \cdot \overset{2}{\underline{\nabla R}}) = T(\underline{H}_1 \cdot \underline{H}_2 \cdot \overset{2}{\underline{\nabla R}}) = T(\underline{H}_2 \cdot \underline{H}_1 \cdot \overset{2}{\underline{\nabla R}})$$

which is the desired result.

The group of $\underline{H}$-transformations (denoted by g) is called a symmetry group of material; the elements of this group are the tensors $\underline{H}$:

$$\underline{H} \subset g \ .$$

The symbol u will be used to denote the group of all unimodular transformations, i.e. those preserving the volume. By (7), the symmetry group is a subgroup of the unimodular group

$$g \subset u \ . \tag{10}$$

Recalling that the $\underline{H}$-transformation is related to a particular reference υ-configuration we may use this to determine, by an $\underline{H}$-transformation, reference configurations which cannot be distinguished experimentally from the initial configuration.

We now consider a material possessing in its $\overset{1}{\upsilon}$-configuration the symmetry group $\overset{1}{g}$ and in the $\overset{2}{\upsilon}$-configuration the symmetry group $\overset{2}{g}$. By (5) and (4), we have

$$\forall \overset{1}{\underline{\nabla R}} : \underline{T} = \overset{1}{T}(\overset{1}{\underline{\nabla R}}) = \overset{2}{T}(\overset{2}{\underline{\nabla R}}) = \overset{2}{T}(\underline{S}^{-1} \cdot \overset{1}{\underline{\nabla R}}) \ , \tag{11}$$

and by (8),

$$\forall \overset{1}{\underline{\nabla R}}, \underline{H}_1 \subset \overset{1}{g} : \overset{1}{T}(\overset{1}{\underline{\nabla R}}) = \overset{1}{T}(\underline{H}_1 \cdot \overset{1}{\underline{\nabla R}}) \ . \tag{12}$$

The condition $\forall \overset{1}{\underline{\nabla R}}$ in (11) enables us to make the substitution $\overset{1}{\underline{\nabla R}} \to \underline{H}_1 \cdot \overset{1}{\underline{\nabla R}}$, so that

$$\overset{1}{T}(\underline{H}_1 \cdot \overset{1}{\underline{\nabla R}}) = \overset{2}{T}(\underline{S}^{-1} \cdot \underline{H}_1 \cdot \overset{1}{\underline{\nabla R}})$$

and this relationship can, by (11) and (12), be rewritten as

$$\overset{1}{T}(\overset{1}{\underline{\nabla R}}) = \overset{1}{T}(\underline{H}_1 \cdot \overset{1}{\underline{\nabla R}}) = \overset{2}{T}(\underline{S}^{-1} \cdot \overset{1}{\underline{\nabla R}}) = \overset{2}{T}(\underline{S}^{-1} \cdot \overset{1}{\underline{H}} \cdot \overset{1}{\underline{\nabla R}}) \ . \tag{13}$$

Substituting, we have, from (4),

$$\underline{S}^{-1}\cdot\overset{1}{\underline{\nabla R}} = \overset{2}{\underline{\nabla R}}\ ,\qquad \overset{1}{\underline{\nabla R}} = \underline{S}\cdot\overset{2}{\underline{\nabla R}}\ ,$$

and we obtain, from (13),

$$\forall\overset{2}{\underline{\nabla R}},\underline{H}_2\subset g\ :\ \overset{2}{T}(\overset{2}{\underline{\nabla R}}) = \overset{2}{T}(\underline{H}_2\cdot\overset{2}{\underline{\nabla R}}) = \overset{2}{T}(\underline{S}^{-1}\cdot\underline{H}_1\cdot\underline{S}\cdot\overset{2}{\underline{\nabla R}})\ . \tag{14}$$

We have now arrived at the relationship known as Noll's rule

$$\underline{H}_2 = \underline{S}^{-1}\cdot\underline{H}_1\cdot\underline{S}\ ,\qquad \underline{H}_1 = \underline{S}\cdot\underline{H}_2\cdot\underline{S}^{-1}\ . \tag{15}$$

In the notation of group theory this rule is expressed by the formulas

$$\overset{2}{g} = \underline{S}^{-1}\overset{1}{g}\underline{S}\ ,\qquad \overset{1}{g} = \underline{S}\overset{2}{g}\underline{S}^{-1}\ . \tag{16}$$

We have thus established a connection between the symmetry groups referred to the reference $\overset{1}{v}$- and $\overset{2}{v}$-configurations connected by the $\underline{S}$-transformation (3). Relationship (15) satisfies condition (10) and if $\overset{1}{g}\subset u$, then $\overset{2}{g}\subset u$. We find, in fact, that by (15)

$$\det\underline{H}_2 = \det\underline{S}^{-1}\cdot\det\underline{H}_1\cdot\det\underline{S} = \det\underline{H}_1 = \pm 1\ ,\qquad u = \underline{S}^{-1}u\underline{S}\ . \tag{17}$$

For the similarity transformation (with the coefficient α)

$$\overset{2}{\underline{r}} = \alpha\overset{1}{\underline{r}}\ ,\qquad \underline{S} = \overset{12}{\underline{\nabla r}} = \alpha\overset{1}{\underline{r}}^{s}\overset{1}{\underline{r}}_{s} = \alpha\underline{E}\ ,\qquad \underline{S}^{-1} = \alpha^{-1}\underline{E}$$

and by (16) we have

$$\overset{2}{g} = \alpha^{-1}\underline{E}\overset{1}{g}\underline{E}\alpha = \overset{1}{g} \tag{18}$$

the symmetry group thus remains the same. We will see below that this enables us to conclude that the similarity transformation preserves the intrinsic symmetry properties of a material. For the identity transformation ($\underline{S} = \underline{E}$) and for that of inversion ($\underline{S} = -\underline{E}$) we also have $\overset{2}{g} = \overset{1}{g}$ (cf. (16)).

The material is called triclinic if its symmetry group consists of two elements: $\overset{1}{g} = \{\underline{E},-\underline{E}\}$. This material of minimal symmetry preserves this property for all reference configurations

$$\overset{2}{g} = \underline{S}^{-1}\{\underline{E},-\underline{E}\}\underline{S} = \{\underline{E},-\underline{E}\} = \overset{1}{g}\ . \tag{19}$$

For materials of maximum symmetry (such as simple fluids) the symmetry group $\overset{1}{g} = u$ is a complete unimodular group not only for the $\overset{1}{v}$-configuration but for all configurations, which follows from (17)

$$\overset{2}{g} = \underline{S}^{-1} u \underline{S} = u = \overset{1}{g} \, . \tag{20}$$

5. ORTHOGONAL TRANSFORMATION. ISOTROPIC MATERIAL

An orthogonal tensor $\underline{O} \subset o$ is unimodular*) ($\det \underline{O} = \pm 1$), $o \subset g$. In other words, the set of $\underline{H}$-transformations of a reference v-configurations which cannot be discovered experimentally, may also include orthogonal transformations. For these transformations (which are of primary interest), Equation (4.6) can be rewritten in the form

$$\forall \underline{\underline{\overset{0}{\nabla R}}} : \underline{T} = T(\underline{\underline{\overset{0}{\nabla R}}}) = T(\underline{O}^T \cdot \underline{\underline{\overset{0}{\nabla R}}}) \tag{1}$$

and, by substituting $\underline{\underline{\overset{0}{\nabla R}}}$ for $\underline{\underline{\overset{0}{\nabla R}}} \cdot \underline{O}$, we have

$$\forall \underline{\underline{\overset{0}{\nabla R}}} : \underline{T} = T(\underline{\underline{\overset{0}{\nabla R}}} \cdot \underline{O}) = T(\underline{O}^T \cdot \underline{\underline{\overset{0}{\nabla R}}} \cdot \underline{O}) \, . \tag{2}$$

But in view of the principle of material frame-indifference written in the form (3.18), we have

$$T(\underline{\underline{\overset{0}{\nabla R}}} \cdot \underline{O}) = \underline{O}^T \cdot T(\underline{\underline{\overset{0}{\nabla R}}}) \cdot \underline{O} \, . \tag{3}$$

We have thus obtained the relationship

$$\forall \underline{\underline{\overset{0}{\nabla R}}}, \underline{O} \subset g : T(\underline{O}^T \cdot \underline{\underline{\overset{0}{\nabla R}}} \cdot \underline{O}) = \underline{O}^T \cdot T(\underline{\underline{\overset{0}{\nabla R}}}) \cdot \underline{O} \, , \tag{4}$$

which represents an orthogonal transformation belonging to the symmetry group in a certain reference v-configuration.

We have arrived at the definition of an isotropic material - "one of Cauchy's greatest discoveries" (Truesdell): a material is isotropic it there exists the reference v-configuration for which its symmetry group contains a complete orthogonal group o, i.e. Equation (4) is satisfied for all transformations belonging to this group

$$\forall \underline{\underline{\overset{0}{\nabla R}}}, \forall \underline{O} \subset o : T(\underline{O}^T \cdot \underline{\underline{\overset{0}{\nabla R}}} \cdot \underline{O}) = \underline{O}^T \cdot T(\underline{\underline{\overset{0}{\nabla R}}}) \cdot \underline{O} \, . \tag{5}$$

---

*) The symbol o is used to denote a complete group of orthogonal transformations.

This $\mathcal{V}$-configuration is called an undeformed state of the material. Under any orthogonal transformation the undistorted state remains unaltered. For an isotropic material, Equation (5) is satisfied identically for all $\underline{O} \subset \mathcal{O}$.

The $\underline{H}$-tensor in (4.6) may be any orthogonal tensor. Setting $\underline{H} = \underline{O}^{\times}$ where $\underline{O}^{\times}$ is the orthogonal tensor accompanying deformation we may rewrite (4.6) in the form

$$\underline{T}(\underline{\overset{0}{\nabla}R}) = \mathcal{T}(\underline{\overset{0}{\nabla}R}) = \mathcal{T}(\underline{O}^{\times T} \cdot \underline{\overset{0}{\nabla}R}) = \mathcal{T}(\underline{O}^{\times T} \cdot \underline{O}^{\times} \cdot \underline{V}) = \mathcal{T}(\underline{V}) \tag{6}$$

here we have used the definition (1.6.1) of the right stretch tensor.

Equation (6) demonstrates the admissibility of the substitution of $\underline{\overset{0}{\nabla}R}$ for $\underline{V}$ in Equation (5).

We obtain

$$\forall \underline{O} \subset g \ : \ \mathcal{T}(\underline{O}^{T} \cdot \underline{V} \cdot \underline{O}) = \underline{O}^{T} \cdot \mathcal{T}(\underline{V}) \cdot \underline{O} \tag{7}$$

this result means that the symmetric tensor function

$$\underline{T} = \underline{T}^{T} = \mathcal{T}(\underline{V}) \tag{8}$$

of a symmetric tensor argument $\underline{V}$ satisfies the definition (II.7.1) of an isotropic tensor function. For this reason, it can be represented by (II.7.7) as a quadratic polynomial over $\underline{V}$

$$\underline{T} = \chi_0\underline{E} + \chi_1\underline{V} + \chi_2\underline{V}^2 \tag{9}$$

its scalar coefficients being functions of the invariants $I_k(\underline{V}) = I_k(\underline{U})$

$$\chi_\Gamma = \chi_\Gamma(I_1(\underline{V}), I_2(\underline{V}), I_3(\underline{V})) \qquad (\Gamma = 0,1,2) \ . \tag{10}$$

For an undistorted reference configuration we have

$$\underline{V} = \underline{E} \ , \qquad I_1(\underline{V}) = I_2(\underline{V}) = 3 \ , \qquad I_3(\underline{V}) = 1$$

and by (9),

$$\underline{T} = -p\underline{E} \ , \qquad -p = \chi_0(3,3,1) + \chi_1(3,3,1) + \chi_2(3,3,1) \tag{11}$$

that is the state of stress in the undistorted configuration may only be either a field of uniform compression ($p > 0$) or of uniform tension ($p < 0$). For $p = 0$ the undistorted state is called the natural state (stress absent).

It should be specially emphasized that representation (9) is valid only in the case when the reference configuration is that of the undistorted state of the material whereas the state equations (3.12) and (3.15) preserve their form irrespective of the choice of reference configuration.

We have excluded from Equation (8) all rotations associated with deformation, which stands in full harmony with the idea of isotropy. It is also clear that the stress tensor $\underline{T}$ is expressed directly through the frame-indifferent tensor $\underline{V}$ (this may, of course, be substituted for the Finger tensor $\underline{F} = \underline{V}^2$), this tensor is quite naturally prescribed together with $\underline{T}$ in the vector basis of the actual configuration.

By (3.12) and (9) we have

$$\Phi(\underline{U}) = \underline{\underline{\nabla r}}^T \cdot (\chi_0\underline{E} + \chi_1\underline{V} + \chi_2\underline{V}^2) \cdot \underline{\underline{\nabla r}}$$

and according to (3.16), (3.17), the representations for the Piola tensor and energetic tensor take the form

$$\underline{P} = \sqrt{\frac{G}{g}}\,(\chi_0\underline{U}^{-1} + \chi_1\underline{E} + \chi_2\underline{U}) \cdot \underline{O}^{\times T} ,$$

$$\underline{T}^{\vee} = \chi_0\underline{U}^{-2} + \chi_1\underline{U}^{-1} + \chi_2\underline{E} .$$

6. THE SOLID BODY

For a solid body, there exist some "preferred" states and any change of shape (deformation) from those states influences the behaviour of a material under load. This behaviour may (or may not) be influenced by the rotation of a sample of the material under test. This qualitative description gives rise to the following definition: a simple material is solid if there exists a reference υ-configuration such that its symmetry group g is a subgroup of the orthogonal group. Contrary to (5.4), for this configuration (called the undistorted configuration), we have

$$g \subset o \ . \tag{1}$$

The term "undistorted configuration" possesses different meaning here and in the definition of isotropy (Par. 5) We will see in Par. 7 that for an isotropic solid body the two definitions are equivalent.

A simple solid body is anisotropic if its symmetry group g in an undistorted state represents a subgroup of the orthogonal group. The particular type of anisotropy is determined by the prescription of g. The elements of this group satisfy (5.4) for all $\overset{0}{\underline{\underline{\nabla R}}}$. This relationship is evidently satisfied by the group

$\{\underline{E},-\underline{E}\}$ which consists of two elements. This allows us to confine ourselves to the consideration of proper orthogonal transformations.

Many arbitrary rotation groups g may be proposed but only twelve groups provide a complete description of the symmetries inherent in real solid bodies.

Let υ, υ' be two undistorted configurations, and by the notation of Par. 4 we have

$$\underline{r}' = \underline{r}\cdot\underline{O}\ , \qquad \underline{r}'_s = \underline{r}_s\cdot\underline{O}\ , \qquad \underline{S} = \underline{r}^s\underline{r}'_s = \underline{r}^s\underline{r}_s\cdot\underline{O} = \underline{O}$$

and from Noll's rule (see (4.16)), we have

$$g' = \underline{O}^{\mathrm{T}} g \underline{O}\ . \tag{2}$$

Here g denotes the symmetry group of an undistorted state (the υ-configuration) and g' denotes the same group for the υ'-configuration, this is also undistorted since an orthogonal transformation is not accompanied by a change of shape. The groups g and g' are conjugate within the orthogonal group. The term "type of anisotropy" should be interpreted in a broad sense. This is determined by a symmetry group g' and not by a group g of special rotations directly associated with material symmetries.

For example, the symmetries of an orthotropic material are determined by the transformations (II.5.3)

$$g = \{\underline{O}^{\pi}_{\underline{c}_1}, \underline{O}^{\pi}_{\underline{c}_2}, \underline{O}^{\pi}_{\underline{c}_3}\}\ , \qquad \underline{O}^{\pi}_{\underline{c}_k} = 2\underline{c}_k\underline{c}_k - \underline{E}\ , \tag{3}$$

while its symmetry group would preferably be prescribed by the relationship (cf. (2))

$$g' = \underline{O}^{\mathrm{T}}\cdot\{\underline{O}^{\pi}_{\underline{c}_1}, \underline{O}^{\pi}_{\underline{c}_2}, \underline{O}^{\pi}_{\underline{c}_3}\}\cdot\underline{O} = \{2\underline{c}'_1\underline{c}'_1 - \underline{E}\, ,\ 2\underline{c}'_2\underline{c}'_2 - \underline{E}\, ,\ 2\underline{c}'_3\underline{c}'_3 - \underline{E}\} \tag{4}$$

which reflects the possibility of the introduction of any orthonormal basis $\underline{c}'_k = \underline{c}_k\cdot\underline{O}$ instead of $\underline{c}_1$, $\underline{c}_2$, $\underline{c}_3$. Equation (4) expresses the symmetry property of a material in its invariant form which is not associated with some particular choice of basis $\underline{c}_1$, $\underline{c}_2$, $\underline{c}_3$.

Proceeding to a more general consideration, we denote by g and $\hat{g}$ the symmetry groups of a solid material corresponding to two undistorted configurations υ and $\hat{\upsilon}$. According to Noll's rule (4.15), we have

$$\hat{\underline{O}} = \underline{S}^{-1}\cdot\underline{O}\cdot\underline{S}\ , \qquad \underline{S}\cdot\hat{\underline{O}} = \underline{O}\cdot\underline{S}\ , \tag{5}$$

here $\underline{S} = \underline{r}^s\hat{\underline{r}}_s$ is the deformation gradient generated by the transformation $\upsilon \to \hat{\upsilon}$.

As does any non-singular tensor, the tensor $\underline{S}$ allows a polar representation

$$\underline{S} = \underline{U} \cdot \underline{O}^{\times} = \underline{O}^{\times} \cdot \underline{V} , \qquad \underline{V} = \underline{O}^{\times T} \cdot \underline{U} \cdot \underline{O}^{\times} \tag{6}$$

we have hereby preserved the notation for the stretch tensors as well as for the orthogonal tensor accomapnying the deformation $\upsilon \to \hat{\upsilon}$. Of course, the values of $\underline{U}$, $\underline{V}$, $\underline{O}^{\times}$ differ from those associated with the deformation $\upsilon \to V_t$.

Now, in view of (5) and (6), we have

$$\underline{U} \cdot \underline{O}^{\times} \cdot \hat{\underline{O}} = \underline{O} \cdot \underline{U} \cdot \underline{O}^{\times} = \underline{O} \cdot \underline{O}^{\times} \cdot (\underline{O}^{\times T} \cdot \underline{U} \cdot \underline{O}^{\times}) = \underline{O} \cdot \underline{O}^{\times} \cdot \underline{V} \tag{7}$$

and by the uniqueness theorem for the polar representation

$$\tilde{\underline{O}} = \underline{O}^{\times} \cdot \hat{\underline{O}} = \underline{O} \cdot \underline{O}^{\times} , \qquad \underline{U} \cdot \tilde{\underline{O}} = \tilde{\underline{O}} \cdot \underline{V} , \qquad \hat{\underline{O}} = \underline{O}^{\times T} \cdot \underline{O} \cdot \underline{O}^{\times} . \tag{8}$$

The latter equation can be rewritten in group terms

$$\hat{g} = \underline{O}^{\times T} g \underline{O}^{\times} , \tag{9}$$

but now with a more general assumption about the transformation $\upsilon \to \hat{\upsilon}$, the groups $\hat{g}$ and g are thought of as symmetry groups within the orthogonal group of transformations.

Now, by (8) and (6)

$$\underline{U} = \tilde{\underline{O}} \cdot \underline{V} \cdot \tilde{\underline{O}}^{T} = \tilde{\underline{O}} \cdot \underline{O}^{\times T} \cdot \underline{U} \cdot \underline{O}^{\times} \cdot \tilde{\underline{O}}^{T} = \underline{O} \cdot \underline{O}^{\times} \cdot \underline{O}^{\times T} \cdot \underline{U} \cdot \underline{O}^{\times} \cdot \underline{O}^{\times T} \cdot \underline{O}^{T}$$

and we have thus arrived at the desired relationship

$$\underline{U} = \underline{O} \cdot \underline{U} \cdot \underline{O}^{T} , \qquad \underline{O}^{T} \cdot \underline{U} = \underline{U} \cdot \underline{O}^{T} , \qquad \underline{U} \cdot \underline{O} = \underline{O} \cdot \underline{U} . \tag{10}$$

This means that the tensor $\underline{O} \subset g$ yielding the transformation of the undistorted reference υ-configuration into the (also undistorted) reference $\hat{\upsilon}$-configuration commutes with the stretch tensor $\underline{U}$ associated with the deformation $\upsilon \to \hat{\upsilon}$. We can also show that the converse is true. Under condition (10), the tensor $\hat{\underline{O}}$ in Equation (5) is orthogonal ($\hat{\underline{O}} \cdot \hat{\underline{O}}^{T} = \underline{E}$). In other words, the symmetry group $\hat{g} \subset o$ provided that $g \subset o$. Thus, eliminating the tensor $\underline{S}$ from Equation (5) with the aid of expression (6), we obtain

$$\hat{\underline{O}} = \underline{O}^{\times T} \cdot \underline{U}^{-1} \cdot \underline{O} \cdot \underline{U} \cdot \underline{O}^{\times} = \underline{O}^{\times T} \cdot \underline{O} \cdot \underline{O}^{\times} , \qquad \hat{\underline{O}}^{T} = \underline{O}^{\times T} \cdot \underline{O}^{T} \cdot \underline{O}^{\times} ,$$

$$\hat{\underline{O}} \cdot \hat{\underline{O}}^{T} = \underline{O}^{\times T} \cdot \underline{O} \cdot \underline{O}^{\times} \cdot \underline{O}^{\times T} \cdot \underline{O}^{T} \cdot \underline{O}^{\times} = \underline{E} ,$$

which is the desired result.

Substituting now in the commutation relationship (10) the orthogonal tensor for its general representation (I.11.10), we obtain

$$\underline{U} \cdot \underline{O} = \underline{U} \cos\omega + (1 - \cos\omega) \underline{U} \cdot \underline{cc} - \underline{U} \times \underline{c} \sin\omega ,$$

$$\underline{O} \cdot \underline{U} = \underline{U} \cos\omega + (1 - \cos\omega) \underline{cc} \cdot \underline{U} - \underline{c} \times \underline{U} \sin\omega ,$$

so that

$$\underline{U} \cdot \underline{O} - \underline{O} \cdot \underline{U} = (\underline{U} \cdot \underline{cc} - \underline{cc} \cdot \underline{U})(1 - \cos\omega) + (\underline{c} \times \underline{U} - \underline{U} \times \underline{c}) \sin\omega = \underline{0}$$

or

$$\underline{U} \cdot \underline{cc} - \underline{cc} \cdot \underline{U} + (\underline{c} \times \underline{U} - \underline{U} \times \underline{c}) \operatorname{ctg} \frac{\omega}{2} = \underline{0} . \tag{11}$$

As in Chapter II, Par. 5 we will consider below the monoclinic, orthotropic, cubic and transversally isotropic symmetry groups.

The tensor $\underline{U}$ is represented in terms of its principal values and directions

$$\underline{U} = u_1 \underline{e}_1 \underline{e}_1 + u_2 \underline{e}_2 \underline{e}_2 + u_3 \underline{e}_3 \underline{e}_3 \tag{12}$$

and Equation (11) is transformed to

$$\sum_{k=1}^{3} \left[u_k \underline{e}_k \cdot \underline{c} (\underline{e}_k \underline{c} - \underline{c} \underline{e}_k) + u_k (\underline{c} \times \underline{e}_k \underline{e}_k + \underline{e}_k \underline{c} \times \underline{e}_k) \operatorname{ctg} \frac{\omega}{2} \right] = \underline{0} . \tag{13}$$

Suppose now that the vector $\underline{e}_k$ is parallel to one of the vectors $\underline{c}_m$ of the orthonormal basis $\underline{c}_1$, $\underline{c}_2$, $\underline{c}_3$. Substituting in Equation (13) $\underline{c}$ for $\underline{c}_m$ (where m is assumed fixed), we get

$$\underline{e}_k \cdot \underline{c}_m = \delta_{km} = \begin{cases} 0, & k \neq m, \\ 1, & k = m, \quad \underline{c}_m \times \underline{e}_k = e_{mkt} \underline{e}_t , \quad \underline{e}_k \times \underline{c}_m = -e_{mkt} \underline{e}_t . \end{cases}$$

The first group of terms in (13) is discarded and we arrive at the relationship

$$\operatorname{ctg}\frac{\omega}{2}\sum_{k=1}^{3} e_{mkt}(u_k - u_t)\underline{e}_k\underline{e}_t = \underline{0} \ . \tag{14}$$

In particular, for $m = 1$, $\underline{c}_1 = \underline{e}_1$ we have

$$\operatorname{ctg}\frac{\omega}{2}(u_2 - u_3)(\underline{e}_2\underline{e}_3 + \underline{e}_3\underline{e}_2) = \underline{0} \tag{15}$$

and for $\omega \neq \pi$

$$u_2 = u_3 \ , \qquad \underline{U} = (u_1 - u_2)\underline{e}_1\underline{e}_1 + u_2\underline{E} \tag{16}$$

which means that a transversally isotropic material subjected to stretching along one of the axes of transversal isotropy and to a subsequent similarity transformation with coefficient u_2 preserves its transversal isotropy in the new reference configuration. The material remains undistorted and its symmetries unchanged.

For

$$\omega = \pi \ , \qquad \operatorname{ctg}\frac{\omega}{2} = 0 \ , \qquad \underline{O} = \underline{O}^{\pi}_{\underline{e}_1}$$

Equation (15) is satisfied for a monoclinic material regardless of the principal values of the stretch tensor provided that one of its axes is directed along an axis of monoclinic symmetry. A monoclinic material remains the same under any stretching both along this axis and along the perpendicular directions. This material does not lose its inherent symmetries, neither does it acquire new ones.

The preservation of the symmetry group of an orthotropic material requires the coincidence of all three exes $\underline{c}_1$, $\underline{c}_2$, $\underline{c}_3$ of orthotropy with the principal directions of the stretch tensor ($\underline{c}_m = \underline{e}_m$, $m = 1,2,3$). Stretched along these directions, the material remains orthotropic.

For $\omega \neq \pi$ Equation (14) obtains the form

$$(u_2 - u_3)(\underline{e}_2\underline{e}_3 + \underline{e}_3\underline{e}_2) + (u_3 - u_1)(\underline{e}_3\underline{e}_1 + \underline{e}_1\underline{e}_3) + (u_1 - u_2)(\underline{e}_1\underline{e}_2 + \underline{e}_2\underline{e}_1) = \underline{0}$$

this condition is fulfilled only for a similarity transformation

$$u_1 = u_2 = u_3 \ , \qquad \underline{U} = \alpha\underline{E} \ .$$

In particular, only this transformation preserves the group of cubic symmetry.

The commutation theorem (10) certainly holds under the trivial condition

$\underline{U} = \alpha\underline{E}$ regardless of the type of anisotropy, and in particular for a triclinic material. Such a material can neither gain, nor lose symmetry.

The conditions determining a monoclinic material are $\omega = \pi$ and $u_2 \neq u_3$. We may, however, treat this differently. A transversally isotropic material becomes monoclinic (i.e. its group of symmetry is changed) when it is subjected to the stretchings u_2, u_3 along two perpendicular directions normal to the axis $\underline{c}_1 = \underline{e}_1$ of transverse isotropy. Quite similarly, a material with cubic symmetry becomes orthotropic when it is strained along three axes of cubic symmetry.

7. AN ISOTROPIC SOLID MATERIAL

By the definition of an isotropic material, its symmetry group g_v in any undistorted v-configuration is a subgroup of the group u of all unimodular transformations. At the same time, it contains within itself the group of all orthogonal transformations

$$o \subset g_v \subset u \ . \tag{1}$$

For a solid material there are undistorted configurations $\tilde{v}$ whose symmetry group $g_{\tilde{v}}$ is a subgroup of the orthogonal group

$$g_{\tilde{v}} \subset o \ . \tag{2}$$

In other words the undistorted state does not retain this property under any non-orthogonal transformation.

From (1) and (2) it follows that for a solid isotropic body there exist configurations v (undistorted according to the definition of "isotropy") and $\tilde{v}$ (undistorted according to the definition of "solidity") such that

$$g_v \supset o \ , \qquad g_{\tilde{v}} \subset o \ . \tag{3}$$

It is shown in group theory that (1) implies

$$\text{either } g_v = o \ , \qquad \text{or } g_v = u \ . \tag{4}$$

Assuming the latter ($g_v = u$) and following (4.20), we may determine the transformation $v \to \tilde{v}$ generated by the tensor $\underline{S}$ such that

$$g_{\tilde{v}} = \underline{S}^{-1} g_v \underline{S} = \underline{S}^{-1} u \underline{S} = u \tag{5}$$

which is in contradiction with the second condition in (3) since $o \subset u$. From

(6.9) it also follows that the groups g_v and $g_{\tilde{v}}$ are conjugate within the group of orthogonal transformations. But $g_v = o$ by the first relationship in (4) and therefore $g_{\tilde{v}} = o$.

We have just proved that the group of symmetry of an isotropic solid material in its undistorted v-configuration is a complete orthogonal group

$$g_v = o \ . \tag{6}$$

Any orthogonal transformation transfers an undistorted state into a state which is also undistorted. There exists no experiment on a prismatic sample of a solid isotropic material in its undistorted state which might enable us to say how the axis of this sample had been oriented within the body before the sample was prepared.

We recall here that the transformation of one undistorted v-configuration into another (also undistorted) $\hat{v}$-configuration generated by a tensor $\underline{S} = \underline{U} \cdot \underline{O}^{\times}$ can be realised under the condition (6.10) which is satisfied only for the similarity transformation $\underline{S} = \alpha\underline{E}$. An isotropic solid body remains the same only under this transformation. For example, the transformation (6.16) converts a solid isotropic material into a transverse isotropic material.

8. AN ELASTIC FLUID

A simple material is a simple fluid if, for a certain reference v-configuration, its symmetry group g_v is the unimodular group of transformations

$$g_v = u \ . \tag{1}$$

This definition already contains the assertion that a fluid is isotropic. Referring to (4.20), we may say further that the symmetry group of a fluid continues to be unimodular for any configuration. A fluid is deprived of "preferred" configurations. All its configurations are undistorted. We have already observed that an isotropic material is either a solid body or a fluid (cf. (7.4)).

The state equation of an elastic fluid as an isotropic material is given by Equation (3.5.9) but this is substantially simplified in this case. Firstly, any configuration may be thought of as a reference configuration so that $\underline{V} = \underline{E}$; secondly, only the variations in density are detectable. In other words, the expression for the stress tensor may include only the third invariant of $\underline{V}$

$$I_3(\underline{V}) = \sqrt{\frac{G}{g}} = \frac{\rho_0}{\rho} \ , \tag{2}$$

not the first and second invariants, otherwise we would find properties which a fluid does not possess. We thus find that the equation of state of a fluid must have the form

$$\underline{T} = -p(\rho)\underline{E} \tag{3}$$

which means that within an elastic fluid no tangential stresses can arise.

In Par. 3 we have excluded from consideration materials whose behaviour depends on the history of their motion. But a substantial part of the content of Pars. 4-7 can be extended to such (non-elastic) materials, including symmetry group, the definition of isotropy and solid body. Restriction of the analysis to only elastic materials has constricted the concept of a fluid: even the classical Navier-Stokes fluid has proved to be outside our consideration.

The concept of a material fully deprived of memory is no longer tenable here, and in Equation (3.3.7) we must retain the functional $E(t,\tau)$ of memory up to a certain approximation. For example, a Stokes' fluid "possesses memory" about the state immediately preceding its configuration at the instant t. On the other hand, in hydrostatics any simple fluid behaves in accordance with the state equation of an elastic fluid (3.3). This fully conforms to the definition of an elastic material under time-independent conditions.

In hydrodynamics, the model of an elastic fluid repeats an "ideal" fluid deprived of friction. This model is applicable to the same processes as classical "Eulerian" hydrodynamics, particularly to the motion of an incompressible fluid ρ = const. It would, however, be erroneous to assume that (3) now implies p = const. The point is that for ρ = const. the motion of a fluid is subjected to the equation of continuity (1.14.13)

$$\underline{\nabla} \cdot \underline{v} = 0$$

and this equation makes an incompressible fluid a "material with superposed constraints". The theory of such materials will be dealt with in Chapter 7.

Note. In order to complete the classification of simple materials, "liquid crystals" are included in this class. Contrary to the concept of a solid body, assertion (7.6) is not valid in this case since the symmetry group of a liquid crystal may contain non-orthogonal transformations. In this respect a liquid crystal is similar to a fluid within which no unimodular transformation can be found. A liquid crystal is distinguished from a fluid by the existence of orthogonal transformations which can be detected just as they are in a solid body. A liquid crystal is a fluid when and only when it is isotropic and when no orthogonal transformation can be detected within it. A liquid crystal is a material which is not deprived of memory.

Chapter 4

THE EQUATIONS OF NONLINEAR THEORY OF ELASTICITY AND THE STATEMENT OF PROBLEMS

1. THE SPECIFIC STORED ENERGY OF DEFORMATION

The theory of elasticity is discussed in this book as a purely mechanical discipline dealing with such notions as force, the state of stress, strain and the state equation. The fundamental concepts of thermoelasticity are discussed in Chapter 9 fully recognizing the extent to which continuum mechanics is enriched by the introduction of thermodynamical principles. This is because the author does not wish to further complicate a problem which is already fairly complex. Moreover, for many applications, thermal effects are of minor importance.

The concept of an elastic material defined in Chapter 3, Par. 3 as a simple material whose behaviour does not depend on the strain history (i.e. which is "deprived of memory"), is now augmented by the requirement of the existence of the stress potential, i.e. the Э-function of either the deformation gradient $\overset{0}{\nabla}\underline{\underline{R}}$ or other possible strain measures. The variation of this function is equal to the incremental work per unit volume of the external forces in the reference v-configuration

$$\delta Э = \delta' a_{(e)} \ . \tag{1}$$

In accordance with this definition, the Э-function represents a "measure of the stored energy". It will be referred to below as the specific stored energy of deformation (abbreviation s.e. whenever there is no possible ambiguity).

The stored energy of deformation of a body occupying the volume V is determined by the expression

$$\mathfrak{Z} = \iiint\limits_{v} Э \, dv \ . \tag{2}$$

The existence of the Э-function is associated with the property of the elastic medium to store the work done by the external forces on loading and to return the "stored energy" on unloading. The concept of s.e. may be associated with the thermodynamic quantities - the free energy (for isothermal processes) and the internal energy (for adiabatic processes). Essentially, relationship

(1) expresses the first law of thermodynamics as applied to these processes. A more general case will be considered in Chapter 9.

Two definitions of elasticity may be distinguished. One due to Cauchy who did not postulate the existence of a potential energy, and the other due to Green who did.*) An elastic material in the Green sense is (as suggested by Truesdell) also called hyperelastic. Many authors do not, however, see any reason for such a distinction. Following the latter, we will assume that an elastic material is at the same time hyperelastic.

Returning to (1) and recalling the definition of the incremental work (2.7.5), we obtain

$$\delta\text{Э} = \underline{P} \cdot\cdot\, \delta\overset{0}{\nabla}\underline{\underline{R}}^{T} \,. \tag{3}$$

For this reason, assuming that Э is a scalar function of the tensor argument $\overset{0}{\nabla}\underline{\underline{R}}$, we arrive, using the definition (II.2.7) of a derivative, at the fundamental relationship for the Piola tensor

$$\underline{P} = \frac{\partial \text{Э}}{\partial \overset{0}{\nabla}\underline{\underline{R}}} = \text{Э}_{\overset{0}{\nabla}\underline{\underline{R}}} \,. \tag{4}$$

Appealing once again to one of the derivations of Chapter 3, Par. 2, we may represent the frame-indifferent scalar $\text{Э}(\overset{0}{\nabla}\underline{\underline{R}})$ in the primed basis by the following expression

$$\text{Э}(\overset{0}{\nabla}\underline{\underline{R}}) = \text{Э}(\overset{0}{\nabla}\underline{\underline{R}}') = \text{Э}(\overset{0}{\nabla}\underline{\underline{R}} \cdot \underline{O}) = \text{Э}(\underline{U} \cdot \underline{O}^{\times} \cdot \underline{O}) = \text{Э}(\underline{U}) = \text{Э}(\underline{G}^{\frac{1}{2}}) \tag{5}$$

here we have made use of relationship (3.2.6). This now allows us below to treat the stored energy as a function of $\overset{0}{\nabla}\underline{\underline{R}}$, $\underline{U}$ and $\underline{V}$, this assumption being valid only for an isotropic material. In all cases, we retain the notation Э:

$$\text{Э} = \text{Э}(\overset{0}{\nabla}\underline{\underline{R}}) = \text{Э}(\underline{U}) = \text{Э}(\underline{G}) \,. \tag{6}$$

By (II.3.5) we also obtain

$$\underline{P} = \text{Э}_{\overset{0}{\nabla}\underline{\underline{R}}} = 2\text{Э}_{\underline{G}} \cdot \overset{0}{\nabla}\underline{\underline{R}} \tag{7}$$

furthermore, in view of (2.6.3), (2.6.11)

---

*) The notion of potential was introduced to mathematical physics by G. Green in his publications of 1839-1841.

$$\underline{T} = 2\sqrt{\frac{g}{G}}\, \overset{0}{\underline{\underline{\nabla R}}}{}^{T} \cdot \underline{\ni}_{G} \cdot \overset{0}{\underline{\underline{\nabla R}}} \,, \qquad \underline{T}^{\vee} = 2\sqrt{\frac{g}{G}}\, \underline{\ni}_{G} \,. \tag{8}$$

For an isotropic material in its undistorted reference configuration, the tensors $\underline{P}$ and $\underline{T}$ allow the following expressions in terms of $\ni$

$$\underline{P} = \ni_{\overset{0}{\underline{\underline{\nabla R}}}} = 2\overset{0}{\underline{\underline{\nabla R}}} \cdot \underline{\ni}_{F} \,, \qquad \underline{T} = \sqrt{\frac{g}{G}}\, \overset{0}{\underline{\underline{\nabla R}}} \cdot \underline{P} = 2\sqrt{\frac{g}{G}}\, \underline{F} \cdot \underline{\ni}_{F} \tag{9}$$

(here, we have used the rule for differentiation given by (II.3.6)).

These formulas, except for (9), can be applied to all elastic materials (in statics, to all simple materials). The specification of the functional dependence of the s.e. on its arguments generates a certain group of materials. The validity of the generally accepted law of dependence may be verified by a comparison of the results obtained by solving the simplest problems (extension, simple shear, etc.) with the experimental data. Criteria based upon some a priori notions on the behaviour of an elastic body under a load are also put forward (see Chapter 5, Pars. 9-13).

Note too that the transition from the basic relationship (3) to the representation (4) of the Piola tensor and the subsequent formulas (6)-(9) are valid only if the variation $\delta\overset{0}{\underline{\underline{\nabla R}}}{}^{T}$ of the deformation gradient is an independent quantity. On this point, see Chapter 7.

The vector bases of the two undistorted reference configurations v and $\hat{v}$ are given by the triplets of vectors $\underline{r}_s$ and $\hat{\underline{r}}_s$ which can be made coincident by an orthogonal transformation $\overset{\times}{\underline{O}} \subset \overset{\times}{o}$ prescribing the symmetry group g of material (Chapter 3, Pars. 4, 6). The gradients $\overset{0}{\underline{\underline{\nabla R}}}$ and $\hat{\underline{\underline{\nabla R}}}$ in the two bases are connected by the relationships

$$\hat{\underline{\underline{\nabla R}}} = \hat{\underline{r}}^{s}\underline{R}_{s} = \underline{O}^{\times T} \cdot \underline{r}^{s}\underline{R}_{s} = \underline{O}^{\times T} \cdot \overset{0}{\underline{\underline{\nabla R}}} \,, \qquad \hat{\underline{\underline{\nabla R}}}{}^{T} = \overset{0}{\underline{\underline{\nabla R}}}{}^{T} \cdot \underline{O}^{\times} \tag{10}$$

one should not confuse these formulas with those of (1.15.11) which prescribe the transformation of the actual configuration associated with a transition to the primed basis.

The Cauchy-Green and Finger strain measures are transformed, in contrast with Equations (1.15.13) and (1.15.14), by the formulas

$$\hat{\underline{G}} = \hat{\underline{\underline{\nabla R}}} \cdot \hat{\underline{\underline{\nabla R}}}{}^{T} = \underline{O}^{\times T} \cdot \underline{G} \cdot \underline{O}^{\times} \,, \qquad \hat{\underline{F}} = \hat{\underline{\underline{\nabla R}}}{}^{T} \cdot \hat{\underline{\underline{\nabla R}}} = \underline{F} \,. \tag{11}$$

The tensor $\hat{\underline{F}}$ is prescribed in the basis of the actual configuration, and it is therefore indifferent to alteration of the reference basis. The components $\hat{G}_{sk}$ of the rotated tensor $\hat{\underline{G}}$ in the $\hat{\underline{r}}_s$-basis are equal to the components G_{sk} of the tensor $\underline{G}$ in the $\underline{r}_s$-basis (cf. (I.8.11)). By (II.5.2) the function $\ni(\underline{G})$ of these components is isotropic in the subgroup $\underline{O}^{\times} \subset O^{\times}$ of orthogonal transformations

$$\ni(\underline{G}) = \ni(G_{11},\dots,G_{31}) = \ni(\hat{G}_{11},\dots,\hat{G}_{31}) = \ni(\hat{\underline{G}}) \ . \tag{12}$$

This confirms the conservation of the symmetry group of the material. One cannot discover in which of the reference configurations (v or $\hat{v}$) the measure $\underline{G}$ is prescribed.

In this section, we will consider a solid material to be as described in Chapter 3, Pars. 6, 7.

2. THE STATE EQUATIONS OF ORTHOTROPIC AND TRANSVERSALLY ISOTROPIC MATERIALS

Possible forms of the dependence of a scalar function φ on its arguments which are invariant with respect to certain subgroups of orthogonal transformations are considered in II, Par. 5. Following these ideas and identifying the φ-function with the specific stored energy of deformation $\ni$, we arrive, using (1.8), at the equations of state. We will confine ourselves to the cases of orthotropic and transversally isotropic materials.

For an orthotropic material, we have by (II.5.10)

$$\ni = \ni(G_{11},G_{22},G_{33},G_{12}^2,G_{23}^2,G_{31}^2,I_3(\underline{G})) \ , \tag{1}$$

where G_{sk} denote the components of $\underline{G}$ in the orthonormal basis $\underline{c}_1$, $\underline{c}_2$, $\underline{c}_3$ of the material's axes of symmetry.

By definition (II.2.8) of the derivative of a scalar function by a symmetric tensor, we have

$$\ni_{\underline{G}} = \frac{1}{2}\sum_{s=1}^{3}\sum_{k=1}^{3}\frac{\partial\ni}{\partial G_{sk}}(\underline{c}_s\underline{c}_k + \underline{c}_k\underline{c}_s) =$$

$$= \sum_{s=1}^{3}\frac{\partial\ni}{\partial G_{ss}}\underline{c}_s\underline{c}_s + \frac{\partial\ni}{\partial G_{12}^2}G_{12}(\underline{c}_1\underline{c}_2 + \underline{c}_2\underline{c}_1) + \frac{\partial\ni}{\partial G_{23}^2}G_{23}(\underline{c}_2\underline{c}_3 + \underline{c}_3\underline{c}_2) +$$

$$+ \frac{\partial\ni}{\partial G_{31}^2}G_{31}(\underline{c}_3\underline{c}_1 + \underline{c}_1\underline{c}_3) = \frac{\partial\ni}{\partial I_3}I_3\underline{G}^{-1} \ . \tag{2}$$

Now, bearing in mind that

$$\overset{0}{\underline{\underline{\nabla R}}}{}^{T}\cdot\underline{c}_s\underline{c}_k\cdot\overset{0}{\underline{\underline{\nabla R}}} = \underline{R}_s\underline{R}_k \ , \qquad \overset{0}{\underline{\underline{\nabla R}}}{}^{T}\cdot\underline{G}^{-1}\cdot\overset{0}{\underline{\underline{\nabla R}}} = \underline{E} \ ,$$

and that $\underline{c}_s = \underline{c}^s$ for a Cartesian basis,

$$\overset{0}{\underline{\nabla R}}^{T} \cdot \underline{c}_s = \underline{R}_q \underline{c}^q \cdot \underline{c}_s = \underline{R}_s \,, \qquad \underline{c}_k \cdot \overset{0}{\underline{\nabla R}} = \underline{R}_k \,,$$

we arrive via (1.8) at the state equation

$$\underline{T} = 2\sqrt{\frac{g}{G}}\left[I_3 \frac{\partial \ni}{\partial I_3}\underline{E} + \sum_{s=1}^{3} \frac{\partial \ni}{\partial G_{ss}} \underline{R}_s \underline{R}_s + \frac{\partial \ni}{\partial G_{12}^2} G_{12}(\underline{R}_1\underline{R}_2 + \underline{R}_2\underline{R}_1) + \right.$$

$$\left. + \frac{\partial \ni}{\partial G_{23}^2} G_{23}(\underline{R}_2\underline{R}_3 + \underline{R}_3\underline{R}_1) + \frac{\partial \ni}{\partial G_{31}^2} G_{31}(\underline{R}_3\underline{R}_1 + \underline{R}_1\underline{R}_3)\right] . \tag{3}$$

The position vector $\underline{r}$ in the reference configuration is determined here by its Cartesian coordinates a^s in the basis $\underline{c}_1$, $\underline{c}_2$, $\underline{c}_3$, these coordinates now represent the material coordinates of a particle

$$\underline{r} = \underline{c}_s a^s \,, \qquad \underline{r}_s = \underline{c}_s \,, \qquad \underline{R}_s = \frac{\partial \underline{R}}{\partial a^s} \,. \tag{4}$$

For a transverse isotropic material, we have by (II.5.28)

$$\ni = \ni(I_1(\underline{G}), I_2(\underline{G}), I_3(\underline{G}), L, M) \,, \qquad L = \underline{c} \cdot \underline{G} \cdot \underline{c} \,, \qquad M = \underline{c} \cdot \underline{G}^2 \cdot \underline{c} \,, \tag{5}$$

here $\underline{c}$ denotes the axis of transverse isotropy. By (II.3.3) and (II.3.8), (II.3.9) we get

$$\ni_{\underline{G}} = \left(\frac{\partial \ni}{\partial I_1} + I_1 \frac{\partial \ni}{\partial I_2}\right)\underline{E} - \frac{\partial \ni}{\partial I_2}\underline{G} + I_3 \frac{\partial \ni}{\partial I_3}\underline{G}^{-1} + \frac{\partial \ni}{\partial L}\underline{cc} + \frac{\partial \ni}{\partial M}(\underline{cc} \cdot \underline{G} + \underline{G} \cdot \underline{cc}) \tag{6}$$

and the state equation (1.8) is transformed to

$$\underline{T} = 2\sqrt{\frac{g}{G}}\left[I_3 \frac{\partial \ni}{\partial I_3}\underline{E} + \left(\frac{\partial \ni}{\partial I_1} + I_1 \frac{\partial \ni}{\partial I_2}\right)\underline{F} - \frac{\partial \ni}{\partial I_2}\underline{F}^2 + \right.$$

$$\left. + \frac{\partial \ni}{\partial L} \overset{0}{c}{}^{s} \overset{0}{c}{}^{k} \underline{R}_s \underline{R}_k + \frac{\partial \ni}{\partial M} \overset{0}{c}{}^{s} \overset{0}{c}{}^{k} (\underline{R}_s \underline{R}_k \cdot \underline{F} + \underline{F} \cdot \underline{R}_k \underline{R}_s)\right] . \tag{7}$$

Here $\overset{0}{c}_s$ denote the contravariant components of $\underline{c}$ in the vector basis $\underline{r}_s$ of the reference configuration. For material coordinates q^s, we have chosen Cartesian coordinates with the basis $\underline{c}_1$, $\underline{c}_2$, $\underline{c}_3$.

3. THE STATE EQUATION OF AN ELASTIC ISOTROPIC MATERIAL

The variety of form of the state equations in the literature is due to the fact that it is possible to choose various strain measures and to use definitions of a stress tensor which differ from one another.

The strain energy is represented as a function of the invariants (with respect to the full orthogonal group of transformation) of the chosen strain measure. The undistorted state serves as a reference configuration; by (3.5.11) the state of stress here is represented by an isotropic tensor describing either uniform compression or tension. In particular, the stress may be absent if the reference configuration is the natural one. The similarity transformation of the natural configuration leads to a new undistorted reference configuration, this, however, is no longer the natural configuration.

The $\ni$-function is now prescribed as a function of the invariants of the Cauchy-Green measure (or, equally, of the Finger measure) by the relationship (cf. (1.5.8))

$$\ni = \ni(I_1(\underline{G}), I_2(\underline{G}), I_3(\underline{G})) \, , \qquad I_k(\underline{G}) = I_k(\underline{F}) \, . \tag{1}$$

A different representation is associated with the invariants of the Almansi measure $\underline{g}$ (or $\underline{G}^{-1}$)

$$\ni = \ni(I_1(\underline{g}), I_2(\underline{g}), I_3(\underline{g})) \, , \qquad I_k(\underline{g}) = I_k(\underline{G}^{-1}) \, . \tag{2}$$

These invariants are connected with $I_k(\underline{G})$ by Equations (1.5.8).

The strain energy is determined up to some additive constant and may be set equal to zero for the reference configuration

$$\ni = \ni(I_1(\underline{E}), I_2(\underline{E}), I_3(\underline{E})) = \ni(3,3,1) = 0 \, . \tag{3}$$

The state equations of an isotropic elastic body in Finger's form follow from Equations (1), (1.8) combined with the formulas (II.3.3) for the differentiation of the invariants

$$\underline{T} = 2\sqrt{\frac{g}{G}}\,(\psi_0\underline{E} + \psi_1\underline{F} + \psi_2\underline{F}^2) = 2\sqrt{\frac{g}{G}}\,\underline{F} \cdot \frac{\partial \ni}{\partial \underline{F}} \, . \tag{4}$$

Here, we have introduced the notation $I_k(\underline{F})$ for the functions of the invariants and we have

$$\psi_0 = I_3 \frac{\partial \ni}{\partial I_3} \, , \qquad \psi_1 = I_1 \frac{\partial \ni}{\partial I_2} + \frac{\partial \ni}{\partial I_1} \, , \qquad \psi_2 = -\frac{\partial \ni}{\partial I_2} \, . \tag{5}$$

This representation should be preferred to that of (3.5.9) because the connection between tensor $\underline{V}$ and the position vector $\underline{R}$ is very complex. It must be recalled that Equation (3.5.9) was obtained without considering any connection with the strain energy of an "elastic", not necessarily a "hyperelastic" material. We may of course also arrive at the Equation (4) for an "elastic" material by repeating the derivation of Equation (3.5.9). However, for a "hyperelastic" material, the functions ψ_Γ are connected by the differential relationships

$$\frac{\partial}{\partial I_2}(\psi_1 + I_1\psi_2) = -\frac{\partial\psi_2}{\partial I_1}, \qquad I_3\frac{\partial}{\partial I_3}(\psi_1 + I_1\psi_2) = \frac{\partial\psi_0}{\partial I_1}, \qquad -I_3\frac{\partial\psi_2}{\partial I_3} = \frac{\partial\psi_0}{\partial I_2}, \tag{6}$$

which are obtained by elimination of $\ni$ from Equations (5) in the same way as for (II.7.18).

The representation of the state equation in terms of the Almansi strain measure can certainly be obtained by (II.3.7), (1.9) and (2)

$$\underline{T} = 2\sqrt{\frac{g}{G}}\,\underline{F}\cdot\ni_{\underline{F}} = -2\sqrt{\frac{g}{G}}\,\ni_{\underline{g}}\cdot\underline{g}\ .$$

We get

$$\underline{T} = 2\sqrt{\frac{g}{G}}\,(\psi_0'\underline{E} + \psi_1'\underline{g} + \psi_2'\underline{g}^2) \tag{7}$$

and from (5) we have

$$\psi_0' = -I_3'\frac{\partial\ni}{\partial I_3'}, \qquad \psi_1' = -\left(\frac{\partial\ni}{\partial I_1'} + I_1'\frac{\partial\ni}{\partial I_2'}\right), \qquad \psi_2' = \frac{\partial\ni}{\partial I_2'}\ ; \quad I_k' = I_k(\underline{g})\ . \tag{8}$$

The functions ψ_Γ' are of course connected by relationships of the type (6).

As can be seen from the state equation (4), the principal directions of the tensors $\underline{T}$ and $\underline{F}$ are the same. Their principal values are connected by the relationships

$$\sigma_k = 2\sqrt{\frac{g}{G}}\,(\psi_0 + \psi_1 v_k^2 + \psi_2 v_k^4) = \frac{2}{v_1 v_2 v_3}\sum_{\Gamma=0}^{2}\psi_\Gamma v_k^{2\Gamma}, \qquad (k = 1,2,3)\ . \tag{9}$$

Here, σ_k denote the principal stresses, and by (1.4.10) and (1.5.7)

$$v_k^2 = G_k = F_k = (1 + \delta_k)^2 \tag{10}$$

this Equation expresses the principal values of the Cauchy-Green and Finger strain measures. The coefficients ψ_Γ are also expressed in terms v_1^2, v_2^2, v_3^2

with the aid of the invariants

$$I_1(\underline{F}) = v_1^2 + v_2^2 + v_3^2, \quad I_2(\underline{F}) = v_1^2v_2^2 + v_2^2v_3^2 + v_3^2v_1^2, \quad I_3(\underline{F}) = v_1^2v_2^2v_3^2 . \tag{11}$$

The principal forces defined by (2.2.25) are given by the formulas

$$t_k = \frac{\sqrt{I_3}}{v_k}\sigma_k = \frac{v_1v_2v_3}{v_k}\sigma_k ;$$

$$t_1 = v_2v_3\sigma_1, \quad t_2 = v_3v_1\sigma_2, \quad t_3 = v_1v_2\sigma_3 \tag{12}$$

and by (9)

$$t_1 = 2v_1\left[\frac{\partial \ni}{\partial I_1} + (v_2^2+v_3^2)\frac{\partial \ni}{\partial I_2} + v_2^2v_3^2\frac{\partial \ni}{\partial I_3}\right] = t(v_1,v_2,v_3),$$

$$t_2 = 2v_2\left[\frac{\partial \ni}{\partial I_1} + (v_3^2+v_1^2)\frac{\partial \ni}{\partial I_2} + v_3^2v_1^2\frac{\partial \ni}{\partial I_3}\right] = t(v_2,v_3,v_1), \tag{13}$$

$$t_3 = 2v_3\left[\frac{\partial \ni}{\partial I_1} + (v_1^2+v_2^2)\frac{\partial \ni}{\partial I_2} + v_1^2v_2^2\frac{\partial \ni}{\partial I_3}\right] = t(v_3,v_1,v_2),$$

or, equivalently,

$$t_k = 2v_k\left[\frac{\partial \ni}{\partial I_1} + (I_1 - v^2)\frac{\partial \ni}{\partial I_2} + I_3v_k^{-2}\frac{\partial \ni}{\partial I_3}\right] . \tag{14}$$

Regarding the $\ni$-function as depending on the v_k-variables, we have

$$\frac{\partial I_1}{\partial v_k} = 2v_k, \quad \frac{\partial I_2}{\partial v_k} = 2v_k(I_1 - v_k^2), \quad \frac{\partial I_3}{\partial v_k} = 2v_k\frac{I_3}{v_k^2}, \tag{15}$$

so that

$$\frac{\partial \ni}{\partial v_k} = \sum_{s=1}^{3}\frac{\partial \ni}{\partial I_s}\frac{\partial I_s}{\partial v_k} = 2v_k\left[\frac{\partial \ni}{\partial I_1} + (I_k - v_k^2)\frac{\partial \ni}{\partial I_2} + \frac{I_3}{v_k^2}\frac{\partial \ni}{\partial I_3}\right] .$$

Comparing this with Equation (14), we arrive at the formulas

$$t_k = \frac{\partial \ni}{\partial v_k} = \frac{\partial \ni}{\partial \delta_k} . \tag{16}$$

This simple result could have been foreseen. In fact, the incremental work $\delta' a_e = \delta \ni$ of the tractions t_k distributed over the faces of a cube whose edges

$v_k = 1 + \delta_k$ are parallel to the principal directions of the Finger measure, on the virtual displacement $\delta v_k = \delta(1 + \delta_k)$ of the faces from their actual configuration, is equal to

$$\delta \ni = \sum_{k=1}^{3} \frac{\partial \ni}{\partial v_k} \delta v_k = \sum_{k=1}^{3} t_k \delta v_k ,$$

from which relationship Equation (16) follows immediately.

The representation of the Piola tensor for an isotropic elastic body can be reduced to the form

$$\underline{P} = 2\underline{\underline{\nabla r}}^T \cdot (\psi_0 \underline{E} + \psi_1 \underline{F} + \psi_2 \underline{F}^2) = 2(\psi_0 \underline{\underline{\nabla r}}^T + \psi_1 \underline{\underline{\nabla r}}^T \cdot \overset{0}{\underline{\underline{\nabla R}}}^T \cdot \overset{0}{\underline{\underline{\nabla R}}} +$$

$$+ \psi_2 \underline{\underline{\nabla r}}^T \cdot \overset{0}{\underline{\underline{\nabla R}}}^T \cdot \overset{0}{\underline{\underline{\nabla R}}} \cdot \overset{0}{\underline{\underline{\nabla R}}}^T \cdot \overset{0}{\underline{\underline{\nabla R}}}) = 2(\psi_0 \underline{\underline{\nabla r}}^T + \psi_1 \overset{0}{\underline{\underline{\nabla R}}} + \psi_2 \underline{G} \cdot \overset{0}{\underline{\underline{\nabla R}}}) ,$$

or

$$\underline{P} = 2(\psi_0 \underline{G}^{-1} + \psi_1 \underline{E} + \psi_2 \underline{G}) \cdot \overset{0}{\underline{\underline{\nabla R}}} . \tag{17}$$

We have used here the definition (2.6.2) of the Piola tensor together with those of (1.4.4) and (1.5.3) for the strain measures.

The elimination of $\underline{G}^{-1}$ using the Hamilton-Cayley theorem (I.9.22) leads to a different representation

$$\underline{P} = 2(\varphi_0 \underline{E} + \varphi_1 \underline{G} + \varphi_2 \underline{G}^2) \cdot \overset{0}{\underline{\underline{\nabla R}}} \tag{18}$$

with the coefficients

$$\varphi_0 = \frac{\partial \ni}{\partial I_1} + I_1 \frac{\partial \ni}{\partial I_2} + I_2 \frac{\partial \ni}{\partial I_3} , \qquad \varphi_1 = -\left(\frac{\partial \ni}{\partial I_2} + I_1 \frac{\partial \ni}{\partial I_3} \right) , \qquad \varphi_2 = \frac{\partial \ni}{\partial I_3} . \tag{19}$$

For the stress tensor derived from the energy equation we obtain by (2.6.11) the following representation

$$\underline{T}^{\vee} = 2\sqrt{\frac{g}{G}} [\psi_0 \underline{G}^{-1} + \psi_1 \underline{E} + \psi_2 \underline{G}] = 2\sqrt{\frac{g}{G}} [\varphi_0 \underline{E} + \varphi_1 \underline{G} + \varphi_2 \underline{G}^2] . \tag{20}$$

In the reference configuration, the stress tensor is isotropic

$$\underline{T} = 2\underline{E}(\psi_0 + \psi_1 + \psi_2)^0 = 2\underline{E}\left(\frac{\partial \ni}{\partial I_1} + 2 \frac{\partial \ni}{\partial I_2} + \frac{\partial \ni}{\partial I_3} \right)^0 = -p\underline{E} , \tag{21}$$

$$p = -2\left(\frac{\partial\ni}{\partial I_1} + 2\frac{\partial\ni}{\partial I_2} + \frac{\partial\ni}{\partial I_3}\right)^0 \tag{22}$$

a zero superscript means that, after differentiation, the invariants take the values $I_1 = I_2 = 3$, $I_3 = 1$.

Note also that by (11) and (16),

$$\sigma_k = \sqrt{\frac{g}{G}}\, v_k \frac{\partial\ni}{\partial v_k} = \sqrt{\frac{g}{G}}\, \frac{\partial\ni}{\partial \ln v_k} = e^{-I_1(\underline{H})} \frac{\partial\ni}{\partial \ln v_k}\,, \tag{23}$$

so that

$$\underline{T} = \sqrt{\frac{g}{G}}\, \frac{\partial\ni}{\partial \ln V} = e^{-I_1(\underline{H})} \frac{\partial\ni}{\partial \underline{H}}\,, \tag{24}$$

where $\underline{H}$ denotes the logarithmic strain measure (1.6.10) (the Hencky measure).

In view of (1.8) and of the definition (2.6.19) of the second Piola-Kirchhoff tensor we also have

$$\underline{T}^{\times} = 2\frac{\partial\ni}{\partial \underline{G}} = \frac{\partial\ni}{\partial \underline{C}}\,, \qquad t^{st} = \frac{\partial\ni}{\partial C_{st}}\,. \tag{25}$$

4. THE SIMILARITY TRANSFORMATION OF A REFERENCE CONFIGURATION

We consider two undistorted reference configurations connected by the similarity transformation

$$\hat{\underline{r}}(q^1,q^2,q^3) = \alpha\underline{r}(q^1,q^2,q^3)\,, \qquad \hat{\underline{r}}_s = \alpha\underline{r}_s\,. \tag{1}$$

The elementary volumes in the v- and $\hat{v}$-configurations are equal to

$$d\hat{v} = \hat{\underline{r}}_1 \cdot (\hat{\underline{r}}_2 \times \hat{\underline{r}}_3)\,dq^1 dq^2 dq^3\,, \qquad d\hat{v} = \alpha^3 dv\,. \tag{2}$$

For this reason,

$$\hat{\underline{r}}^1 = \frac{\hat{\underline{r}}_2 \times \hat{\underline{r}}_3}{\hat{\underline{r}}_1 \cdot (\hat{\underline{r}}_2 \times \hat{\underline{r}}_2)} = \frac{1}{\alpha}\,\frac{\underline{r}_2 \times \underline{r}_3}{\underline{r}_1 \cdot (\underline{r}_2 \times \underline{r}_3)} = \alpha^{-1}\underline{r}^1\,, \qquad \hat{\underline{r}}_2 = \alpha^{-1}\underline{r}^2\,, \qquad \hat{\underline{r}}^3 = \alpha^{-1}\underline{r}^3. \tag{3}$$

The position vector $\underline{R}$ in the actual configuration is of course the same, and therefore

$$\hat{\nabla}\underline{R} = \hat{\underline{r}}^s\underline{R}_s = \alpha^{-1}\overset{0}{\nabla}\underline{R}\,, \qquad \hat{\underline{G}} = \alpha^{-2}\underline{G}\,, \qquad \hat{\underline{F}} = \alpha^{-2}\underline{F} \tag{4}$$

and the relationships between the invariants are determined by the formulas

$$\hat{I}_1(\hat{\underline{F}}) = \alpha^{-2} I_1(\underline{F}) , \quad \hat{I}_2(\hat{\underline{F}}) = \alpha^{-4} I_2(\underline{F}) , \quad \hat{I}_3(\hat{\underline{F}}) = \alpha^{-6} I_3(\underline{F}) . \tag{5}$$

The strain energy is the same in the transformations $v \to V$, $\hat{v} \to V$ into the actual configuration

$$= \iiint_v \ni(I_1, I_2, I_3)\, dv = \iiint_{\hat{v}} \hat{\ni}(\hat{I}_1, \hat{I}_2, \hat{I}_3)\, d\hat{v} = \iiint_v \hat{\ni}(\alpha^{-2} I_1, \alpha^{-4} I_2, \alpha^{-6} I_3)\, \alpha^3\, dv$$

and, since the volume v is arbitrary,

$$\ni(I_1, I_2, I_3) = \alpha^3 \hat{\ni}(\alpha^{-2} I_1, \alpha^{-4} I_2, \alpha^{-6} I_3) . \tag{6}$$

For this reason

$$\frac{\partial \ni}{\partial I_1} = \alpha^3 \frac{\partial \hat{\ni}}{\partial I_1} = \alpha \frac{\partial \hat{\ni}}{\partial \hat{I}_1} , \quad \frac{\partial \ni}{\partial I_2} = \alpha^{-1} \frac{\partial \hat{\ni}}{\partial \hat{I}_2} , \quad \frac{\partial \ni}{\partial I_3} = \alpha^{-3} \frac{\partial \ni}{\partial \hat{I}_3}$$

and by (3.5)

$$\psi_0 = \alpha^3 \hat{\psi}_0 , \quad \psi_1 = \alpha \hat{\psi}_1 , \quad \psi_2 = \alpha^{-1} \hat{\psi}_2 . \tag{7}$$

The state equation (3.4) in the reference $\hat{v}$-configuration

$$\underline{T} = 2\hat{I}_3^{-\frac{1}{2}} (\hat{\psi}_0 \underline{E} + \hat{\psi}_1 \hat{F} + \hat{\psi}_2 \hat{\underline{F}}^2) ,$$

is now transformed to

$$\hat{\underline{T}} = 2\alpha^3 I_3^{-\frac{1}{2}} (\alpha^{-3} \psi_0 \underline{E} + \alpha^{-1} \psi_1 \alpha^{-2} \underline{F} + \alpha \psi_2 \alpha^{-4} \underline{F}^2)$$

and, as might be expected,

$$\hat{\underline{T}} = 2 I_3^{-\frac{1}{2}} (\psi_0 \underline{E} + \psi_1 \underline{F} + \psi_2 \underline{F}^2) = \underline{T} . \tag{8}$$

The stress tensor in the actual configuration is independent of the choice of the undistorted reference configuration.

5. VARIATION OF THE STRESS STATE

It is assumed that the field of virtual displacements $\eta \underline{w}(q^1, q^2, q^3)$, where η is a small parameter, is superposed on the actual configuration of the medium.

Up to terms linear in η the Cauchy stress tensor becomes

$$\underline{T}^{\times} = \underline{T} + \eta\dot{\underline{T}} , \qquad \dot{\underline{T}} = \lim_{\eta\to 0} \frac{1}{\eta} (\underline{T}^{\times} - \underline{T}) . \tag{1}$$

By the definition of Chapter 1, Par. 10, the tensor $\dot{\underline{T}}$ represents the convected derivative of $\underline{T}$ which does not differ from its material derivative since $\underline{T}$ does not depend explicitly on time (η is identified with δt).

Following (3.4) and bearing in mind the formulas of Chapter 1, Par. 10, we obtain

$$\dot{\sqrt{\frac{g}{G}}} = -\sqrt{\frac{g}{G}}\, \underline{\nabla} \cdot \underline{w} ,$$

$$\dot{\underline{T}} = -\underline{T}\underline{\nabla} \cdot \underline{w} + 2\sqrt{\frac{g}{G}}\, [\psi_1\dot{\underline{F}} + \psi_2(\underline{F} \cdot \dot{\underline{F}} + \dot{\underline{F}} \cdot \underline{F}) + (\underline{E}\dot{\psi}_0 + \underline{F}\dot{\psi}_1 + \underline{F}^2\dot{\psi}_2)] \tag{2}$$

and by (1.10.11)

$$2\sqrt{\frac{g}{G}}\, [\psi_1\dot{\underline{F}} + \psi_2(\underline{F} \cdot \dot{\underline{F}} + \dot{\underline{F}} \cdot \underline{F})] = 2\sqrt{\frac{g}{G}}\, [\psi_1(\underline{F} \cdot \underline{\nabla w} + \underline{\nabla w}^T \cdot \underline{F}) +$$

$$+ \psi_2\underline{F} \cdot (\underline{F} \cdot \underline{\nabla w} + \underline{\nabla w}^T \cdot \underline{F}) + \psi_2(\underline{F} \cdot \underline{\nabla w} + \underline{\nabla w}^T \cdot \underline{F}) \cdot \underline{F}] =$$

$$= 2\sqrt{\frac{g}{G}}\, [(\psi_1\underline{F} + \psi_2\underline{F}^2) \cdot \underline{\nabla w} + \underline{\nabla w}^T \cdot (\psi_1\underline{F} + \psi_2\underline{F}^2) + \psi_2\underline{F} \cdot (\underline{\nabla w} + \underline{\nabla w}^T) \cdot \underline{F}] .$$

We now introduce the linear strain tensor $\underline{\varepsilon}(\underline{w})$ and use (3.4) to transform the latter expression thus obtaining

$$\underline{T} \cdot \underline{\nabla w} + \underline{\nabla w}^T \cdot \underline{T} - 4\sqrt{\frac{g}{G}}\, [\psi_0\underline{\varepsilon}(\underline{w}) - \psi_2\underline{F} \cdot \underline{\varepsilon}(\underline{w}) \cdot \underline{F}] . \tag{3}$$

Proceeding now to the calculation of ψ_Γ $(\Gamma = 0,1,2)$, we obtain by virtue of (II.3.3) and (1.10.15)-(1.10.17)

$$\dot{\psi}_\Gamma = \frac{\partial\psi_\Gamma}{\partial I_1}\dot{I}_1 + \frac{\partial\psi_\Gamma}{\partial I_2}\dot{I}_2 + \frac{\partial\psi_\Gamma}{\partial I_3}\dot{I}_3 =$$

$$= 2\left[\frac{\partial\psi_\Gamma}{\partial I_1}\underline{F} + \frac{\partial\psi_\Gamma}{\partial I_2}(I_1\underline{F} - \underline{F}^2) + \frac{\partial\psi_\Gamma}{\partial I_3} I_3\underline{E}\right] \cdot\cdot\, \underline{\varepsilon}(\underline{w}) ,$$

or

$$\dot{\psi}_\Gamma = 2(\vartheta_{0\Gamma}\underline{E} + \vartheta_{1\Gamma}\underline{F} + \vartheta_{2\Gamma}\underline{F}^2) \cdot\cdot\, \underline{\varepsilon}(\underline{w}) = 2\sum_{N=0}^{2} \vartheta_{N\Gamma}\underline{F} \cdot\cdot\, \underline{\varepsilon}(\underline{w}) \tag{4}$$

(here we assume summation over $N = 0,1,2;\ \underline{F}^0 = \underline{E}$). The coefficients $\vartheta_{N\Gamma}$ are defined by the formulas

$$\vartheta_{0\Gamma} = I_3 \frac{\partial\psi_\Gamma}{\partial I_3}, \qquad \vartheta_{1\Gamma} = \frac{\partial\psi_\Gamma}{\partial I_1} + I_1 \frac{\partial\psi_\Gamma}{\partial I_2}, \qquad \vartheta_{2\Gamma} = -\frac{\partial\psi_\Gamma}{\partial I_2} \tag{5}$$

and the nature of the dependence of $\vartheta_{N\Gamma}$ on ψ_Γ is the same as that of ψ_Γ on $\ni$ and formulas (3.5) express $\vartheta_{N\Gamma}$ through the second derivatives of $\ni$ with respect to the invariants, This calculation shows that these derivatives are symmetric with respect to the indices N, Γ

$$\vartheta_{N\Gamma} = \vartheta_{\Gamma N}, \qquad N,\Gamma = 0,1,2\ . \tag{6}$$

Substitution into (2) results in the following representation of $\dot{\underline{T}}$

$$\dot{\underline{T}} = -\underline{T}\underline{\nabla}\cdot\underline{w} + \underline{\nabla}\underline{w}^T\cdot\underline{T} + \underline{T}\cdot\underline{\nabla}\underline{w} + 4\sqrt{\frac{g}{G}}\,[-\psi_0\underline{\varepsilon}(\underline{w}) + \psi_2\underline{F}\cdot\underline{\varepsilon}(\underline{w})\cdot\underline{F}] + \sum_\Gamma\sum_N \vartheta_{N\Gamma}\underline{F}^\Gamma\underline{F}^N \cdot\cdot\, \underline{\varepsilon}(\underline{w})\ . \tag{7}$$

Evidently, the tensor $\dot{\underline{T}}$ is symmetric. It is useful at this point to introduce the tensor

$$\underline{\theta} = \underline{T}\cdot\underline{\nabla}\underline{w} + 4\sqrt{\frac{G}{g}}\,[-\psi_0\underline{\varepsilon}(\underline{w}) + \psi_2\underline{F}\cdot\underline{\varepsilon}(\underline{w})\cdot\underline{F}] + \sum_\Gamma\sum_N \vartheta_{N\Gamma}\underline{F}^\Gamma\underline{F}^N \cdot\cdot\, \underline{\varepsilon}(\underline{w})\ , \tag{8}$$

so that

$$\dot{\underline{T}} = -\underline{T}\underline{\nabla}\cdot\underline{w} + \underline{\nabla}\underline{w}^T\cdot\underline{T} + \underline{\theta}\ . \tag{9}$$

The convected derivative of the Piola tensor is now calculated by (2.6.2) and (1.10.8)

$$\dot{\underline{P}} = \left(\sqrt{\frac{G}{g}}\,\underline{\nabla}\underline{r}^T\cdot\underline{T}\right)^{\cdot} = \left(\sqrt{\frac{G}{g}}\right)^{\cdot}\underline{\nabla}\underline{r}^T\cdot\underline{T} + \sqrt{\frac{G}{g}}\,(\underline{\nabla}\underline{r}^T)^{\cdot}\cdot\underline{T} + \sqrt{\frac{G}{g}}\,\underline{\nabla}\underline{r}^T\cdot\dot{\underline{T}} = \sqrt{\frac{G}{g}}\,\underline{\nabla}\underline{r}^T\cdot(\underline{T}\underline{\nabla}\cdot\underline{w} - \underline{\nabla}\underline{w}^T\cdot\dot{\underline{T}} + \dot{\underline{T}})$$

and by (9) we obtain

$$\underline{\dot{P}} = \sqrt{\frac{G}{g}}\, \underline{\underline{\nabla r}}^T \cdot \underline{\theta} \,. \tag{10}$$

The tensor $\underline{\dot{P}}$ is thus shown to be connected with $\underline{\theta}$ by precisely the same relationship as that which connects $\underline{P}$ and $\underline{T}$.

6. THE EQUILIBRIUM EQUATIONS FOR A VARIED STRESS STATE

The static equilibrium equations (2.6.4) for the Piola tensor enable us to avoid the difficulties associated with varying the basis of the actual configuration. We have

$$\overset{0}{\underline{\nabla}} \cdot (\underline{P} + \eta\underline{\dot{P}}) + \rho_0\underline{k} + \eta\rho_0\underline{k}_x = \underline{0}\,, \qquad \overset{0}{\underline{\nabla}} \cdot \underline{P} + \rho_0\underline{k} + \eta(\overset{0}{\underline{\nabla}} \cdot \underline{\dot{P}} + \rho_0\underline{k}_x) = \underline{0}\,.$$

We denote the variation of the body force by $\underline{k}_x$. Assuming that the actual configuration is that of equilibrium we arrive at the equation of equilibrium in the new configuration

$$\overset{0}{\underline{\nabla}} \cdot \underline{\dot{P}} + \rho_0\underline{k}_x = \underline{0} \tag{1}$$

this equation is formulated for the vector basis of the reference configuration. It is transformed to the basis of the actual configuration by (5.10), (III.3.11) and the Piola identity

$$\overset{0}{\underline{\nabla}} \cdot \sqrt{\frac{G}{g}}\, \underline{\underline{\nabla r}}^T \cdot \underline{\theta} = \underline{\theta}^T \cdot \overset{0}{\underline{\nabla}} \cdot \sqrt{\frac{G}{g}}\, \underline{\underline{\nabla r}}^T + \sqrt{\frac{G}{g}}\, \underline{\underline{\nabla r}} \cdot\cdot \overset{0}{\underline{\underline{\nabla \theta}}} =$$

$$= \sqrt{\frac{G}{g}}\, \underline{R}^s \underline{r}_s \cdot\cdot \underline{r}^k \frac{\partial \underline{\theta}}{\partial q^k} = \sqrt{\frac{G}{g}}\, \underline{\nabla} \cdot \underline{\theta} \,.$$

In view of (2.1.2), the equation of equilibrium in the varied state can also be reduced to the form

$$\underline{\nabla} \cdot \underline{\theta} + \rho\underline{k}_x = \underline{0} \,. \tag{2}$$

On the surface of the new volume, we have by (2.6.5)

$$\underline{n} \cdot (\underline{P} + \eta\underline{\dot{P}}) = \underline{f}\left[\frac{dO}{do} + \eta\left(\frac{dO}{do}\right)^{\cdot}\right] + \eta\underline{f}_x \frac{dO}{do} \tag{3}$$

where $\underline{f}_x$ denotes the variation of the body force. By (1.10.23) we obtain

$$\underline{n} \cdot \dot{\underline{P}} = \frac{dO}{do} [\underline{f}(\underline{\nabla} \cdot \underline{w} - \underline{N} \cdot \underline{\varepsilon}(\underline{w}) \cdot \underline{N}) + \underline{f}_x]$$

and in view of (5.10)

$$\underline{N} \cdot \underline{\theta} = \underline{f}(\underline{\nabla} \cdot \underline{w} - \underline{N} \cdot \underline{\varepsilon}(\underline{w}) \cdot \underline{N}) + \underline{f}_x \; . \tag{4}$$

In a varied state, the tensor $\underline{\theta}$ plays the role of the stress tensor $\underline{T}$. The equations of static equilibrium involve this tensor both within the volume and on the surface. It is assumed of course that the state of stress in the actual configuration is known in advance. The boundary value problem given by (2) and (4) is linear with respect to the vector $\underline{w}$. Its coefficients are constant provided that the transformation of the reference configuration into the actual is affine; for this case the tensor $\underline{T}$ is constant.

For loading by a constant surface pressure which is always directed along the normal to the surface, the constant force in the varied state will, in view of Equations (2.1.11) and (1.10.19) be determined by the relationship

$$-p\underline{N}\,dO - \eta p(\underline{N}\,dO)^{\cdot} = -p\underline{N}\,dO - \eta p\underline{N}\,dO \cdot (\underline{E\nabla} \cdot \underline{w} - \underline{\nabla w}^T) \tag{5}$$

and the equations of equilibrium (4) and (3) on the surface will take the form

$$\begin{aligned} \underline{N} \cdot \underline{\theta} &= -p\underline{N} \cdot (\underline{E\nabla} \cdot \underline{w} - \underline{\nabla w}^T) \; , \\ \underline{n} \cdot \dot{\underline{P}} &= -\sqrt{\frac{G}{g}}\, \underline{n} \cdot \underline{\nabla r}^T \cdot (\underline{E\nabla} \cdot \underline{w} - \underline{\nabla w}^T) \; . \end{aligned} \tag{6}$$

For the "dead" loading of a surface, the contact force acting on the area dO preserves in the varied state the same value and direction as in the actual configuration and so by (1.10.22) and (5.9) we have

$$(\underline{f}\,dO)^{\cdot} = (\underline{N} \cdot \underline{T}\,dO)^{\cdot} = \underline{N}\,dO \cdot (\dot{\underline{T}} + \underline{T\nabla} \cdot \underline{w} - \underline{\nabla w}^T) = \underline{N} \cdot \underline{\theta}\,dO = \underline{0} \; . \tag{7}$$

As it might be expected, the equations of equilibrium on a surface prove to be homogeneous with respect to the tensors $\underline{\theta}$ and $\dot{\underline{P}}$

$$\underline{N} \cdot \underline{\theta} = \underline{0} \; , \qquad \underline{n} \cdot \dot{\underline{P}} = \underline{0} \; . \tag{8}$$

7. THE RELAXATION TENSOR OF AN ISOTROPIC MEDIUM

The relaxation tensor is a tensor of the fourth rank defined as the derivative of the stress tensor with respect to the measure of deformation through which it is expressed. For example, for a linearly elastic isotropic body the stress tensor is a linear function of the linear strain tensor

$$\underline{T} = \lambda I_1(\underline{\varepsilon})\underline{E} + 2\mu\underline{\varepsilon}(\underline{u}) \tag{1}$$

and by (II.3.1), (II.4.13)

$$\underline{T} = \lambda \underline{\underline{EE}} + \mu(\underline{C}_{II} + \underline{C}_{III}) = \underline{\tau}^{pqrt}\underline{r}_p\underline{r}_q\underline{r}_r\underline{r}_t \,, \tag{2}$$

and the components of the relaxation tensor are represented by the expressions

$$\underline{\tau}^{pqrt} = \underline{r}^p\underline{r}^q\underline{r}^r\underline{r}^t \cdots\cdot [\underline{\underline{EE}}\lambda + (\underline{C}_{II} + \underline{C}_{III})\mu] =$$

$$= g^{pq}g^{rt}\lambda + (g^{rp}g^{tq} + g^{rq}g^{tp})\mu \,. \tag{3}$$

The tensor $\underline{T}_{\varepsilon}$ preserves its value under the permutation of indices within each pair (pq), (rt).

We will now proceed to the determination of the relaxation tensor by the Finger representation (3.4) of the Cauchy stress tensor. By (1.10.2) the convected derivative $\dot{\underline{T}}$ is defined as the contraction

$$\dot{\underline{T}} = \underline{\underline{T}}_F \cdot\cdot \dot{\underline{F}} \tag{4}$$

and it is required that the expression (5.2) for $\dot{\underline{T}}$ be reduced to this form.

The divergence of the vector $\underline{w}$ can be represented in the following form

$$\tfrac{1}{2}\underline{F}^{-1}\cdot\cdot\dot{\underline{F}} = \tfrac{1}{2}\underline{F}^{-1}\cdot\cdot(\underline{F}\cdot\underline{\underline{\nabla w}} + \underline{\underline{\nabla w}}^T\cdot\underline{F}) =$$

$$= \tfrac{1}{2}\underline{E}\cdot\cdot\underline{\underline{\nabla w}} + \tfrac{1}{2}\underline{F}\cdot\underline{F}^{-1}\cdot\cdot\underline{\underline{\nabla w}} = \underline{E}\cdot\cdot\underline{\varepsilon}(\underline{w}) = \underline{\nabla}\cdot\underline{w} \tag{5}$$

and the first term in (5.2) is substituted for the expression

$$-\underline{\underline{T\nabla}}\cdot\underline{w} = -\tfrac{1}{2}\underline{\underline{TF}}^{-1}\cdot\cdot\dot{\underline{F}} \,.$$

Now, by virtue of (I.15.4) and (I.15.7) we have the relationships

$$(\psi_1\underline{E} + \psi_2\underline{F})\cdot\dot{\underline{F}} = \tfrac{1}{2}(\psi_1\underline{E} + \psi_2\underline{F})\cdot(\underline{C}_{II} + \underline{C}_{III})\cdot\cdot\dot{\underline{F}} \,,$$

$$\dot{\underline{F}}\cdot\underline{F} = \underline{C}_{II}\cdot\cdot\underline{F}\cdot\dot{\underline{F}} = \tfrac{1}{2}\underline{C}_{II}\cdot\cdot[\underline{F}\cdot(\underline{C}_{II} + \underline{C}_{III})]\cdot\cdot\dot{\underline{F}} \,. \tag{6}$$

It remains only to refer to (II.3.3) and (5.5) and we may write the formulas

$$\dot{\psi}_\Gamma = (\psi_\Gamma)_{\underline{F}} \cdot\cdot\, \dot{\underline{F}} = [\vartheta_{0\Gamma}\underline{F}^{-1} + \vartheta_{1\Gamma}\underline{E} + \vartheta_{2\Gamma}\underline{F}] \cdot\cdot\, \dot{\underline{F}} = \sum_{N=0}^{2} \vartheta_{N\Gamma}\underline{F}^{N-1} \cdot\cdot\, \dot{\underline{F}} \,. \qquad (7)$$

We obtain

$$\dot{\underline{T}} = \Big\{ -\tfrac{1}{2}\,\underline{\underline{TF}}^{-1} + \sqrt{\frac{g}{G}}\,[(\psi_1\underline{E} + \psi_2\underline{F}) \cdot (\underline{C}_{II} + \underline{C}_{III}) +$$

$$+ \psi_2\underline{C}_{II} \cdot\cdot \{\underline{F} \cdot (\underline{C}_{II} + \underline{C}_{III})\}] + 2\sqrt{\frac{g}{G}} \sum_{\Gamma=0}^{2} \sum_{N=0}^{2} \vartheta_{N\Gamma}\underline{F}^{\Gamma}\underline{F}^{N-1}\Big\} \cdot\cdot\, \dot{\underline{F}} \,. \qquad (8)$$

The tensor of the fourth order in the braces is, by (4), the relaxation tensor. The expanded expression for this tensor is obtained with the aid of (5.6) and we have

$$\underline{T}_{\underline{F}} = -\tfrac{1}{2}\,\underline{\underline{TF}}^{-1} + \sqrt{\frac{g}{G}}\,[(\psi_1\underline{E} + \psi_2\underline{F}) \cdot (\underline{C}_{II} + \underline{C}_{III}) +$$

$$+ \psi_2\underline{C}_{II} \cdot\cdot \{\underline{F} \cdot (\underline{C}_{II} + \underline{C}_{III})\}] + 2\sqrt{\frac{g}{G}}\,[\vartheta_{00}\underline{\underline{EF}}^{-1} + \vartheta_{11}\underline{\underline{FE}} + \vartheta_{22}\underline{F}^2\underline{F} +$$

$$+ \vartheta_{01}(\underline{\underline{EE}} + \underline{\underline{FF}}^{-1}) + \vartheta_{02}(\underline{\underline{EF}} + \underline{F}^2\underline{F}^{-1}) + \vartheta_{12}(\underline{\underline{FF}} + \underline{F}^2\underline{E})] \,. \qquad (9)$$

If the reference configuration is the natural one ($\underline{T} = \underline{0}$), we have

$$(\underline{T}_{\underline{F}})_0 = (\psi_1 + 2\psi_2)^0(\underline{C}_{II} + \underline{C}_{III}) + 2\underline{\underline{EE}}(\vartheta_{00} + \vartheta_{11} + \vartheta_{22} + 2\vartheta_{01} + 2\vartheta_{02} + 2\vartheta_{12})^0 \,. \qquad (10)$$

Substituting from (5.5) for $\vartheta^0_{N\Gamma}$, we have

$$(\vartheta_{00} + \vartheta_{11} + \vartheta_{22} + 2\vartheta_{01} + 2\vartheta_{02} + 2\vartheta_{12})^0 =$$

$$= \left(\frac{\partial\psi_0}{\partial I_3} + \frac{\partial\psi_1}{\partial I_1} + \frac{\partial\psi_1}{\partial I_2} - \frac{\partial\psi_2}{\partial I_2} + 2\,\frac{\partial\psi_1}{\partial I_3} - 2\,\frac{\partial\psi_0}{\partial I_2} \right)^0 .$$

Making use of representations (3.5) of the coefficients ψ_Γ in terms of the strain energy $\ni$, we may rewrite this expression as follows

$$\left[\left(\frac{\partial}{\partial I_2} + \frac{\partial}{\partial I_3} \right) \ni + \left(\frac{\partial}{\partial I_1} + 2\,\frac{\partial}{\partial I_2} + \frac{\partial}{\partial I_3} \right)^2 \ni \right]^0 =$$

$$= \left[\left(\frac{\partial}{\partial I_1} + 2\,\frac{\partial}{\partial I_2} + \frac{\partial}{\partial I_3} \right) (\psi_0 + \psi_1 + \psi_2) \right]^0 ,$$

because

$$\left[\left(\frac{\partial}{\partial I_1} + 2\frac{\partial}{\partial I_2} + \frac{\partial}{\partial I_3}\right)(\psi_0 + \psi_1 + \psi_2)\right]^0 =$$

$$= \left[\left(\frac{\partial}{\partial I_1} + 2\frac{\partial}{\partial I_2} + \frac{\partial}{\partial I_3}\right)\left(I_3\frac{\partial\ni}{\partial I_3} + \frac{\partial\ni}{\partial I_1} + I_1\frac{\partial\ni}{\partial I_2} - \frac{\partial\ni}{\partial I_2}\right)\right]^0 =$$

$$= \left(\frac{\partial\ni}{\partial I_2} + \frac{\partial\ni}{\partial I_3}\right)^0 + \left[\left(\frac{\partial}{\partial I_1} + 2\frac{\partial}{\partial I_2} + \frac{\partial}{\partial I_3}\right)^2\ni\right]^0 .$$

We thus have

$$(\underline{T}_{\underline{F}})^0 = (\psi_1 + 2\psi_2)^0(\underline{C}_{II} + \underline{C}_{III}) + 2\underline{EE}\left[\left(\frac{\partial}{\partial I_1} + 2\frac{\partial}{\partial I_2} + \frac{\partial}{\partial I_3}\right)(\psi_0 + \psi_1 + \psi_2)\right]^0 . \quad (11)$$

Comparing this with (2) and taking into account that in the linear approximation

$$\underline{F} = \underline{E} + 2\underline{\varepsilon}(\underline{u}) \; ,$$

we may express the Lamé coefficients of the state equation of a linearly elastic body in terms of the coefficients ψ_Γ and the values of their derivatives in the natural reference configuration. Bearing in mind also that by (3.22)

$$(\psi_0 + \psi_1 + \psi_2)^0 = \left[\left(\frac{\partial}{\partial I_1} + 2\frac{\partial}{\partial I_2} + \frac{\partial}{\partial I_3}\right)\ni\right]^0 = \underline{0} \; ,$$

we arrive at the formulas

$$\mu = 2(\psi_1 + 2\psi_2)^0 = 2(\psi_2 - \psi_0)^0 =$$

$$= 2\left[\left(\frac{\partial}{\partial I_1} + \frac{\partial}{\partial I_2}\right)\ni\right]^0 = -2\left[\left(\frac{\partial}{\partial I_2} + \frac{\partial}{\partial I_3}\right)\ni\right]^0 , \quad (12)$$

$$\lambda = 4\left[\left(\frac{\partial}{\partial I_1} + 2\frac{\partial}{\partial I_2} + \frac{\partial}{\partial I_3}\right)(\psi_0 + \psi_1 + \psi_2)\right]^0 =$$

$$= 4\left[\left(\frac{\partial}{\partial I_2} + \frac{\partial}{\partial I_3}\right)\ni\right]^0 + 4\left[\left(\frac{\partial}{\partial I_1} + 2\frac{\partial}{\partial I_2} + \frac{\partial}{\partial I_3}\right)^2\ni\right]^0 . \quad (13)$$

From Equations (12), (13) it also follows that

$$\lambda + 2\mu = 4\left[\left(\frac{\partial}{\partial I_1} + 2\,\frac{\partial}{\partial I_2} + \frac{\partial}{\partial I_3}\right)^2 \ni\right]^0 . \tag{14}$$

8. THE REPRESENTATION OF THE RELAXATION TENSOR WITH REFERENCE TO THE PROPER BASIS OF A STRESS TENSOR

In Equation (7.9), the stress tensor, as well as the Finger and Almansi strain measures are substituted for their component representations

$$\underline{T} = t^{sk}\underline{R}_s\underline{R}_k , \qquad \underline{F} = g^{sk}\underline{R}_s\underline{R}_k , \qquad \underline{F}^{-1} = \underline{g} = g_{sk}\underline{R}^s\underline{R}^k . \tag{1}$$

The component representation of $\underline{T}_{\underline{F}}$ takes the form

$$\underline{T}_{\underline{F}} = \underline{\tau}^{qpst}\underline{R}_q\underline{R}_p\underline{R}_s\underline{R}_t , \tag{2}$$

so that

$$\tau^{qpst} = \underline{R}^p\underline{R}^q \cdot\cdot\, \underline{T}_{\underline{F}} \cdot\cdot\, \underline{R}^t\underline{R}^s . \tag{3}$$

By (7.9) we get

$$\tau^{qpst} = -\tfrac{1}{2}\, t^{qp} g_{mn} G^{ms} G^{nt} + 2\sqrt{\frac{g}{G}}\,[\vartheta_{00} G^{pq} G^{nt} G^{ms} g_{mn} +$$

$$+ \vartheta_{11} g^{qp} G^{ts} + \vartheta_{22} g^{qm} g^{np} G_{mn} g^{st} + \vartheta_{01}(G^{pq} G^{st} + g^{pq} g_{mn} G^{tm} G^{sn}) +$$

$$+ \vartheta_{02}(G^{pq} g^{st} + g^{qn} g^{mp} G_{mn} G^{kt} G^{ts} g_{kt}) + \vartheta_{12}(g^{pq} g^{st} + g^{qm} g^{pn} G_{mn} G^{ts})] +$$

$$+ \sqrt{\frac{g}{G}}\,[\psi_1(G^{qs}G^{pt} + G^{qt}G^{ps}) + \psi_2(g^{qs}G^{pt} + g^{qt}G^{ps} + g^{pt}G^{qs} + g^{ps}G^{qt})] . \tag{4}$$

The relaxation tensor preserves its value under the permutation of indices within any pair (qp), (st). By (II.1.6), the number of its components is 21. This follows from (7.9), (4) as well as from the definition (II.4.5) of the derivative of a tensor by a tensor argument

$$\underline{T}_{\underline{F}} = \underline{R}_p\underline{R}_q\underline{R}^s\underline{R}^t\,\frac{\partial t^{pq}}{\partial g^{st}} = \tau^{pq\cdot\cdot}_{\cdot\cdot st}\,\underline{R}_p\underline{R}_q\underline{R}^s\underline{R}^t .$$

The above expression (4) can be represented in a comparatively simple way by the orthonormal eigenvector basis $\underline{e}_1$, $\underline{e}_2$, $\underline{e}_3$ of tensors (1). Recall that their principal values in this basis are, respectively, σ_q, v_q^2, v_q^{-2}.

We have

$$G^{mn} = G_{mn} = \begin{cases} 0\,, & m \neq n\,, \\ 1\,, & m = n\,, \end{cases} \qquad g^{mn} = \begin{cases} 0\,, & m \neq n\,, \\ v_m^2\,, & m = n\,, \end{cases} \qquad g_{mn} = \begin{cases} 0\,, & m \neq n\,, \\ v_m^{-2}\,, & m = n \end{cases} \tag{5}$$

and the following components are different from zero: a) the three components

$$\tau^{ssss} = -\frac{1}{2}\,\frac{\sigma_s}{v_s^2} + \frac{2}{v_1 v_2 v_3}\left[\frac{\vartheta_{00}}{v_s^2} + \vartheta_{11} v_s^2 + \vartheta_{22} v_s^6 + 2\vartheta_{01} + \right.$$
$$\left. + 2\vartheta_{02} v_s^2 + 2\vartheta_{12} v_s^4 + \psi_1 + 2\psi_2 v_s^2\right], \tag{6}$$

b) the six components

$$\tau^{sskk} = -\frac{1}{2}\,\frac{\sigma_s}{v_k^2} + \frac{2}{v_1 v_2 v_3}\left[\frac{\vartheta_{00}}{v_k^2} + \vartheta_{11} v_s^2 + \vartheta_{22} v_s^4 v_k^2 + \vartheta_{01}\left(1 + \frac{v_s^2}{v_k^2}\right) + \right.$$
$$\left. + \vartheta_{02}\left(v_k^2 + \frac{v_s^4}{v_k^2}\right) + \vartheta_{12}(v_s^2 v_k^2 + v_s^4)\right] \qquad (\tau^{sskk} \neq \tau^{kkss})\,, \tag{7}$$

c) the three components

$$\tau^{sksk} = \frac{1}{v_1 v_2 v_3}\,[\psi_1 + \psi_2(v_s^2 + v_k^2)]\,, \qquad \tau^{sksk} = \tau^{ksks} = \tau^{kssk}\,. \tag{8}$$

Making, in all, 12 non-zero components in this basis.

These relationships may be simplified considerably and expressed more explicitly by making use of the representation (3.9) of the principal stresses σ_s through the principal values v_s^2 of the Finger (or Cauchy-Green) strain measure.

Referring to Equations (3.15), (5.5) we get

$$\frac{\partial \sigma_s}{\partial v_k^2} = -\frac{1}{2}\,\frac{\sigma_s}{v_k^2} + \frac{2}{v_1 v_2 v_3}\left[\sum_{\Gamma=0}^{2} v_s^{2\Gamma}\left(\frac{\partial \psi_\Gamma}{\partial I_1} + I_1\,\frac{\partial \psi_\Gamma}{\partial I_2} - v_k^2\,\frac{\partial \psi_\Gamma}{\partial I_2} + \frac{I_3}{v_k^2}\,\frac{\partial \psi_\Gamma}{\partial I_3}\right) + \right.$$
$$\left. + \begin{cases} \psi_1 + 2\psi_2\,, & k = s \\ 0\,, & k \neq s \end{cases}\right] =$$
$$= -\frac{1}{2}\,\frac{\sigma_s}{v_k^2} + \frac{2}{v_1 v_2 v_3}\left[\left(\frac{\vartheta_{00}}{v_k^2} + \vartheta_{01} + \vartheta_{02} v_k^2\right) + \left(\vartheta_{10}\,\frac{v_s^2}{v_k^2} + \vartheta_{11} v_s^2 + \vartheta_{12} v_s^2 v_k^2\right) + \right.$$

$$+ \left(\vartheta_{20} \frac{v_s^4}{v_k^2} + \vartheta_{21} v_s^4 + \vartheta_{22} v_s^4 v_k^2 \right) + \begin{cases} \psi_1 + 2\psi_2 , & k = s \\ 0 , & k \neq s \end{cases}$$

these expressions reproducting Equations (6) and (7). We thus have

$$\tau^{ssss} = \frac{\partial \sigma_s}{\partial v_s^2} , \qquad \tau^{sskk} = \frac{\partial \sigma_s}{\partial v_k^2} . \tag{9}$$

These components of the relaxation tensor may be called the tangential moduli of longitudinal and transverse extension because $\partial\sigma_s/\partial v_k^2$ is the slope of a tangent to the curve $\sigma_s(v_1^2, v_2^2, v_3^2)$ on the plane (v_k^2, σ_s).

In view of (3.9), formula (8) may be given in the form

$$\tau^{sksk} = \frac{\sigma_s - \sigma_k}{2(v_s^2 - v_k^2)} \tag{10}$$

this value must be interpreted in terms of the "principal shear stresses" and "principal shear strains".

9. THE RELAXATION TENSOR

By (1.4), the relaxation tensor $\underline{P}_{\underline{\underline{\overset{0}{\nabla}R}}}$ is represented as the second derivative of the strain energy with respect to the deformation gradient

$$\underline{P}_{\underline{\underline{\overset{0}{\nabla}R}}} = \frac{\partial^2 \ni}{\partial \underline{\underline{\overset{0}{\nabla}R}} \partial \underline{\underline{\overset{0}{\nabla}R}}} = \ni_{\underline{\underline{\overset{0}{\nabla}R\overset{0}{\nabla}R}}} . \tag{1}$$

Making use of the representation (1.3.11) of the deformation gradient and bearing in mind the definition (II.4.4) of a derivative of a tensor with respect to a tensor argument, we arrive at the component representation

$$\underline{P} = \ni_{\underline{\underline{\overset{0}{\nabla}R}}} = \underline{r}_m \underline{r}^n \frac{\partial \ni}{\partial \overset{0}{\nabla}_m x^n} , \qquad \underline{P}_{\underline{\underline{\overset{0}{\nabla}R}}} = \underline{r}_m \underline{r}^n \underline{r}_p \underline{r}^q \frac{\partial^2 \ni}{\partial \overset{0}{\nabla}_p x^q \partial \overset{0}{\nabla}_m x^n} , \tag{2}$$

and for the Cartesian basis

$$\underline{P}_{\underline{\underline{\overset{0}{\nabla}R}}} = \underline{i}_m \underline{i}^n \underline{i}_p \underline{i}^q \frac{\partial^2 \ni}{\partial \dfrac{\partial x^q}{\partial a^p} \partial \dfrac{\partial x^n}{\partial a^m}} . \tag{3}$$

Alternative representations follow from the rules for the calculation of derivative of a tensor product and for the substitution of the argument under differentiation with respect to a tensor argument (see (II.4.10), (II.4.12))

$$\underline{P} = 2\ni_{\underline{G}} \cdot \overset{0}{\underline{\underline{\nabla R}}}\,, \qquad \underline{P}_{0\atop \underline{\underline{\nabla R}}} = 2\ni_{\underline{G}} \cdot \underline{C}_{II} + 2\left(\frac{\partial}{\partial \overset{0}{\underline{\underline{\nabla R}}}}\ni_{\underline{G}} \cdot\cdot\, \underline{r}_s \underline{r}_t\right) \cdot \overset{0}{\underline{\underline{\underline{\nabla R r}}}}{}^t \underline{r}^s =$$

$$= \underline{P} \cdot \underline{\underline{\nabla r}} \cdot \underline{C}_{II} + 2\{[\ni_{\underline{\underline{GG}}} \cdot\cdot\, (\underline{C}_{II} + \underline{C}_{III}) \cdot \overset{0}{\underline{\underline{\nabla R}}}] \cdot\cdot\, \underline{r}_s \underline{r}_t\} \cdot \overset{0}{\underline{\underline{\underline{\nabla R r}}}}{}^t \underline{r}^s\,.$$

We have also used here rules (II.4.13), Table (I.15.3) and Equation (1.9.6). We thus arrive at a relationship which does not assume the medium to be isotropic

$$\underline{P}_{0\atop \underline{\underline{\nabla R}}} = \underline{P} \cdot \underline{\underline{\nabla r}} \cdot \underline{C}_{II} + 2[\ni_{\underline{\underline{GG}}} \cdot\cdot\, (\underline{r}^q \underline{r}_t + \underline{r}_t \underline{r}^q)] \cdot \overset{0}{\underline{\underline{\underline{\nabla R r}}}}{}^t \underline{R}_q\,. \tag{4}$$

For an isotropic elastic body

$$\ni_{\underline{G}} = \varphi_0 \underline{E} + \varphi_1 \underline{G} + \varphi_2 \underline{G}^2\,, \tag{5}$$

where φ_Γ are defined by Equations (3.19). Reproducing the calculation of Par. 7, we obtain

$$\ni_{\underline{\underline{GG}}} = \sum_{N=0}^{2} \sum_{\Gamma=0}^{2} \zeta_{N\Gamma} \underline{G}^N \underline{G}^\Gamma + \tfrac{1}{2}(\varphi_1 \underline{E} + \varphi_2 \underline{G}) \cdot (\underline{C}_{II} + \underline{C}_{III}) +$$

$$+ \tfrac{1}{2}\varphi_2 \underline{C}_{II} \cdot\cdot\, \{\underline{G} \cdot (\underline{C}_{II} + \underline{C}_{III})\}\,, \tag{6}$$

where the coefficients $\zeta_{N\Gamma}$ are expressed in terms of φ_Γ by the formulas (3.19) defining φ_Γ in terms of $\ni$

$$\zeta_{0\Gamma} = \left(\frac{\partial}{\partial I_1} + I_1 \frac{\partial}{\partial I_2} + I_2 \frac{\partial}{\partial I_3}\right)\varphi_\Gamma\,,$$

$$\zeta_{1\Gamma} = -\left(\frac{\partial}{\partial I_2} + I_1 \frac{\partial}{\partial I_3}\right)\varphi_\Gamma\,, \qquad \zeta_{2\Gamma} = \frac{\partial \varphi_\Gamma}{\partial I_3}\,. \tag{7}$$

Substituting (6) into (4), we get

$$\underline{P}_{0\atop \underline{\underline{\nabla R}}} = \underline{P} \cdot \underline{\underline{\nabla r}} \cdot \underline{C}_{II} + 4\overset{0}{\underline{\underline{\nabla R}}} \cdot \sum_{N=0}^{2} \sum_{\Gamma=0}^{2} \zeta_{N\Gamma} \underline{F}^N \overset{0}{\underline{\underline{\nabla R}}} \cdot \underline{F}^\Gamma +$$

$$+ 2(\varphi_1 \underline{E} + \varphi_2 \underline{G}) \cdot [\underline{C}_{II} \cdot\cdot\, \{\overset{0}{\underline{\underline{\nabla R}}}{}^T \cdot (\underline{C}_{II} + \underline{C}_{III})\}] \cdot \overset{0}{\underline{\underline{\nabla R}}} +$$

$$+ 2\varphi_2 \underset{\sim}{C}_{II} \cdot\cdot [\overset{0}{\nabla R}{}^T \cdot \underset{\sim}{G} \cdot (\underset{\sim}{C}_{II} + \underset{\sim}{C}_{III})] \cdot \overset{0}{\nabla R} \tag{8}$$

it is now possible (just as in (7.9)) to represent $\underset{\sim}{P}_{\underset{\nabla R}{0}}$ in an invariant form which does not include the basis vectors.

For a natural reference configuration, repeating the calculation from Par. 7 yields

$$\underset{\sim}{P}_{\underset{\nabla R}{0}} = 4\underset{\sim}{EE}\left[\left(\frac{\partial}{\partial I_1} + 2\frac{\partial}{\partial I_2} + \frac{\partial}{\partial I_3}\right)(\varphi_0 + \varphi_1 + \varphi_2)\right]^0 -$$

$$- 2(\underset{\sim}{C}_{II} + \underset{\sim}{C}_{III})\left[\left(\frac{\partial}{\partial I_2} + \frac{\partial}{\partial I_3}\right)\ni\right]^0 \tag{9}$$

and comparison with (4.7.2) leads to the relationships

$$\lambda = 4\left[\left(\frac{\partial}{\partial I_1} + 2\frac{\partial}{\partial I_2} + \frac{\partial}{\partial I_3}\right)(\varphi_0 + \varphi_1 + \varphi_2)\right]^0 ,$$

$$\mu = -2\left[\left(\frac{\partial}{\partial I_2} + \frac{\partial}{\partial I_3}\right)\ni\right]^0 , \tag{10}$$

which again yield (4.7.12) and (4.7.13).

The formula expressing $\underset{\sim}{P}_{\underset{\nabla R}{0}}$ in terms of $\underset{\sim}{T}_F$ may be obtained by referring once again to rules (II.4.10) and (II.4.12)

$$\underset{\sim}{P}_{\underset{\nabla R}{0}} = \left(\sqrt{\frac{G}{g}}\,\nabla r^T \cdot \underset{\sim}{T}\right)_{\underset{\nabla R}{0}} = \sqrt{\frac{G}{g}}\,\nabla r^T \cdot \underset{\sim}{T}_{\underset{\nabla R}{0}} + \left[\left(\sqrt{\frac{G}{g}}\,\nabla r^T\right)_{\underset{\nabla R}{0}} \cdot\cdot\, \underset{\sim}{r}_s \underset{\sim}{r}_t\right] \cdot \underset{\sim}{T}\underset{\sim}{r}^t \underset{\sim}{r}^s .$$

Referring to the formulas of Chapter 1, Par. 9, we arrive at the relationship

$$\underset{\sim}{P}_{\underset{\nabla R}{0}} = \underset{\sim}{P}\nabla r^T - \underset{\sim}{r}_s \underset{\sim}{r}^t \cdot \underset{\sim}{P} \underset{\sim}{r}_t \underset{\sim}{r}^s \cdot \nabla r^T + \sqrt{\frac{G}{g}}\,\nabla r^T \cdot \underset{\sim}{T}_F \cdot\cdot (\underset{\sim}{R}_t \underset{\sim}{r}_s + \underset{\sim}{r}_s \underset{\sim}{R}_t) \underset{\sim}{r}^t \underset{\sim}{r}^s , \tag{11}$$

which can also be reduced to the form

$$\underset{\sim}{P}_{\underset{\nabla R}{0}} = \underset{\sim}{P}\nabla r^T - \underset{\sim}{r}_s \underset{\sim}{P}^T \underset{\sim}{R}^s + \sqrt{\frac{G}{g}}\,\nabla r^T \cdot \underset{\sim}{T}_F \cdot\cdot (\underset{\sim}{C}_{II} + \underset{\sim}{C}_{III}) \cdot\cdot \overset{0}{\nabla R}{}^T \cdot \underset{\sim}{C}_{II} . \tag{12}$$

10. EQUATIONS OF MOTION AND EQUILIBRIUM FOR AN ISOTROPIC ELASTIC BODY

Here we shall consider analogs of the equations of the linear theory of elasticity in terms of displacements which are obtained by substitution of the stress tensor for its representation in terms of the linear strain tensor, the latter tensor being itself expressed in terms of the displacement vector. In non-linear theory the problem becomes more complicated because it is possible to determine the state of stress with the aid of various strain tensors (Cauchy, Piola), and these strain tensors allow various representations either in terms of the strain measures (Cauchy-Green, Finger, Almansi) or the deformation gradient. Instead of the displacement vector, it is preferable to make use of the position vector in the actual configuration.

The divergence of the Piola tensor is given by the expression (1.19.23)

$$\overset{0}{\underline{\nabla}} \cdot \underline{P} = \underline{r}^k \cdot \underline{P}_{\overset{0}{\underline{\nabla}}\underline{R}} \cdot\cdot \overset{0}{\underline{\nabla}}\underline{R}^{\mathrm{T}} \cdot \overset{0}{\underline{A}}{}^{\mathrm{T}} \cdot \underline{r}_k \,, \tag{1}$$

in view of (1.19.23) this can be transformed to

$$\overset{0}{\underline{\nabla}} \cdot \underline{P} = \underline{r}^k \cdot \underline{P}_{\overset{0}{\underline{\nabla}}\underline{R}} \cdot\cdot \underline{r}_k \cdot \overset{00}{\underline{\nabla\nabla}}\underline{R}^{\mathrm{T}} = \underline{r}^k \cdot \underline{P}_{\overset{0}{\underline{\nabla}}\underline{R}} \cdot\cdot \frac{\partial}{\partial q^k} \overset{0}{\underline{\nabla}}\underline{R}^{\mathrm{T}} \,. \tag{2}$$

Substituting the deformation gradient by its coordinate representation, we arrive at three equations of motion

$$\underline{r}_q \underline{r}^k \cdot\cdot \underline{P}_{\overset{0}{\underline{\nabla}}\underline{R}} \cdot\cdot \underline{r}_s \underline{r}^t \overset{0}{\nabla}_k \overset{0}{\nabla}_t \chi^s + \rho_0 k_q = \rho_0 \overset{0}{b}_q \,, \tag{3}$$

which are linear with respect to the second derivatives of the components in the reference basis of the position vector in the actual configuration. This system clearly possesses a very complex structure. The first covariant derivatives of χ^s enter nonlinearly in it, both through the relaxation tensor $\underline{P}_{\overset{0}{\underline{\nabla}}\underline{R}}$ and through the extended representations of the second covariant derivatives.

Substituting for the relaxation tensor in (3) the expressions of (9.2), we arrive at the equations of motion of an anisotropic elastic medium

$$\frac{\partial^2 \ni}{\partial \overset{0}{\nabla}_k \chi^s \, \partial \overset{0}{\nabla}_t \chi^q} \overset{0}{\nabla}_k \overset{0}{\nabla}_t \chi^s + \rho_0 k_q = \rho_0 \overset{0}{b}_q \qquad (q = 1,2,3) \,. \tag{4}$$

In (1) we now substitute the relaxation tensor $\underline{P}_{\overset{0}{\underline{\nabla}}\underline{R}}$ by its representation in terms of $\underline{T}_{\underline{F}}$ causing the terms containing $\underline{P}$ to disappear. In fact, referring to (1.19.17) we obtain

$$\underline{r}^k \cdot \underline{P\nabla r}^{\mathrm{T}} \cdot\cdot \overset{0}{\underline{\nabla}}\underline{R}^{\mathrm{T}} \cdot \overset{0}{\underline{A}}{}^{\mathrm{T}} \cdot \underline{r}_k = \underline{P}^{\mathrm{T}} \cdot \underline{r}^k \underline{\nabla r}^{\mathrm{T}} \cdot\cdot \overset{0}{\underline{\nabla}}\underline{R}^{\mathrm{T}} \cdot A^q_{mk} \underline{r}_q \underline{r}^m =$$

$$= \underline{P}^T \cdot \underline{r}^k A^q_{qk} = \underline{P}^T \cdot \overset{0}{\underline{\nabla}} \ln \sqrt{\frac{G}{g}} \ ,$$

$$\underline{r}^k \cdot \underline{r}_s \underline{r}^t \cdot \underline{P}\underline{r}_t \underline{r}^s \cdot \underline{\nabla r}^T \cdot\cdot \overset{0}{\underline{\nabla R}}{}^T \cdot \overset{0}{\underline{A}}{}^T \cdot \underline{r}_k = \underline{P}^T \cdot \underline{r}^t \underline{r}_t \underline{r}^k \cdot\cdot \underline{r}_q \underline{r}^s A^q_{st} =$$

$$= \underline{P}^T \cdot \underline{r}^s A^q_{qs} = \underline{P}^T \cdot \overset{0}{\underline{\nabla}} \ln \sqrt{\frac{G}{g}} \ ,$$

which is the desired result. Also, making use of (1.19.12), (1.19.13) and (9.11), we have

$$\underline{r}^k \cdot \underline{P}_{\overset{0}{\underline{\nabla R}}} \cdot\cdot \overset{0}{\underline{\nabla R}}{}^T \cdot \overset{0}{\underline{A}}{}^T \cdot \underline{r}_k = \sqrt{\frac{G}{g}}\, \underline{R}^k \cdot \underline{T}_{\underline{F}} \cdot\cdot (\underline{R}_t \underline{r}_s + \underline{r}_s \underline{R}_t) \underline{r}^t \underline{r}^s \cdot\cdot \underline{R}_m \underline{r}^q A^m_{qk} =$$

$$= \sqrt{\frac{G}{g}}\, \underline{R}^k \cdot \underline{T}_{\underline{F}} \cdot\cdot (\overset{0}{\underline{\nabla R}}{}^T \cdot \underline{r}^q \underline{R}_m + \underline{R}_m \underline{r}^q \cdot \overset{0}{\underline{\nabla R}}) A^m_{qk} =$$

$$= \sqrt{\frac{G}{g}}\, \underline{R}^k \cdot \underline{T}_{\underline{F}} \cdot\cdot \left(\overset{0}{\underline{\nabla R}}{}^T \frac{\partial \overset{0}{\underline{\nabla R}}}{\partial q^k} + \frac{\partial \overset{0}{\underline{\nabla R}}{}^T}{\partial q^k} \cdot \overset{0}{\underline{\nabla R}} \right) .$$

The equation of motion of an isotropic elastic body now takes the form

$$\underline{R}^k \cdot \underline{T}_{\underline{F}} \cdot\cdot \frac{\partial \underline{F}}{\partial q^k} + \rho \underline{k} = \rho \underline{b} \ . \tag{5}$$

The same equation could have been obtained if we had started from an equation of motion formulated with the aid of the Cauchy stress tensor $\underline{T}$. By (1.19.19), (1.19.12) and (1.9.8) we have

$$\underline{\nabla} \cdot \underline{T} = \underline{R}^k \cdot \frac{\partial \underline{T}}{\partial q^k} = \underline{R}^k \cdot \left(\underline{T}_{\overset{0}{\underline{\nabla R}}} \cdot\cdot \overset{0}{\underline{\nabla R}}{}^T \cdot \overset{0}{\underline{A}}{}^T \right) \cdot \underline{r}_k = \underline{R}^k \cdot \underline{T}_{\underline{F}} \cdot\cdot \underline{C}_{II} \cdot\cdot \underline{F}_{\overset{0}{\underline{\nabla R}}} \cdot\cdot \underline{R}_m \underline{r}^q A^m_{qk} =$$

$$= \underline{R}^k \cdot \underline{T}_{\underline{F}} \cdot\cdot (\underline{R}_m \underline{r}^q \cdot \overset{0}{\underline{\nabla R}} + \overset{0}{\underline{\nabla R}}{}^T \cdot \underline{r}^q \underline{R}_m) A^m_{qk} = \underline{R}^k \cdot \underline{T}_{\underline{F}} \cdot\cdot \frac{\partial \underline{F}}{\partial q^k} \ ,$$

as required.

Substituting the relaxation tensor by its component representation

$$\underline{T}_{\underline{F}} = \tau^{kn\cdot\cdot}_{st} \underline{R}_k \underline{R}_n \underline{R}^s \underline{R}^t \tag{6}$$

and using the above representation for $\partial \underline{F}/\partial q^k$, we obtain the equation of motion in component form

$$(\tau^{kn\cdot\cdot}_{st} + \tau^{kn\cdot\cdot}_{ts}) g^{qt} A^s_{qk} + \rho k^n = \rho b^n \qquad (n = 1,2,3) \ . \tag{7}$$

This equation can be expressed in terms of the displacement vector. To this end, we perform the transformations

$$\underline{R}_k = \underline{r}_k + \underline{R}_t \nabla_k u^t \,, \qquad \frac{\partial \underline{R}_k}{\partial q^q} = \left\{ \begin{matrix} 0 \\ t \\ kq \end{matrix} \right\} \underline{r}_t + \underline{R}_t \nabla_q \nabla_k u^t \,,$$

which allow us to represent A^s_{qk} in the form

$$A^s_{qk} = \left\{ \begin{matrix} s \\ qk \end{matrix} \right\} - \left\{ \begin{matrix} 0 \\ s \\ qk \end{matrix} \right\} = \underline{R}^s \frac{\partial \underline{R}_k}{\partial q^q} - \left\{ \begin{matrix} 0 \\ s \\ qk \end{matrix} \right\} = \nabla_q \nabla_k u^s - \left\{ \begin{matrix} 0 \\ t \\ kq \end{matrix} \right\} \nabla_t u^s \,.$$

Substitution into (7) leads to the equations

$$(\tau^{kn\cdot\cdot}_{st} + \tau^{kn\cdot\cdot}_{ts}) g^{qt} \left[\nabla_q \nabla_k u^s - \left\{ \begin{matrix} 0 \\ r \\ kq \end{matrix} \right\} \nabla_r u^s \right] + \rho k^n = \rho b^n \,. \tag{8}$$

An alternative form of the equations of motion of an isotropic elastic medium is obtained by substituting the Cauchy stress tensor for its representation (3.4) in terms of the Finger strain measure

$$\underline{\nabla} \cdot \underline{T} + \rho \underline{k} = \underline{\nabla} \cdot 2 \sqrt{\frac{g}{G}} \, (\psi_0 \underline{E} + \psi_1 \underline{F} + \psi_2 \underline{F}^2) + \rho \underline{k} = \rho \underline{b}$$

or, in extended form,

$$\underline{T} \cdot \underline{\nabla} \ln \sqrt{\frac{g}{G}} + 2 \sqrt{\frac{g}{G}} \left[\sum_{\Gamma=0}^{2} \underline{F}^{\Gamma} \cdot \underline{\nabla} \psi_{\Gamma} + \psi_1 \underline{\nabla} \cdot \underline{F} + \psi_2 \underline{\nabla} \cdot \underline{F}^2 \right] + \rho \underline{k} = \rho \underline{b} \,.$$

By (1.19.22) and (1.3.10) we have

$$\underline{\nabla} \psi_{\Gamma} = \sum_{s=1}^{3} \frac{\partial \psi_{\Gamma}}{\partial I_s} \underline{\nabla} I_s = 2 \sum_{s=1}^{3} \frac{\partial \psi_{\Gamma}}{\partial I_s} \underline{\underline{\nabla r}} \cdot (I_s)_{\underline{G}} \cdot \underline{G} \cdot\cdot \overset{0}{\underline{A}}{}^{T} =$$

$$= 2 \underline{\underline{\nabla r}} \cdot (\vartheta_{1\Gamma} \underline{G} + \vartheta_{2\Gamma} \underline{G}^2 + \vartheta_{0\Gamma} \underline{E}) \cdot\cdot \overset{0}{\underline{A}}{}^{T} = 2 \sum_{N=0}^{2} \vartheta_{N\Gamma} \underline{F}^{N-1} \cdot \overset{0}{\underline{\underline{\nabla R}}}{}^{T} \cdot\cdot \overset{0}{\underline{A}}{}^{T} \,.$$

Making use of the formulas (1.19.20) and (1.19.21) and eliminating $\underline{T}$ with the aid of (3.4), we arrive at the equation

$$2 \sum_{\Gamma=0}^{2} \sum_{N=0}^{2} \vartheta_{N\Gamma} \underline{F}^{\Gamma} \underline{F}^{N-1} \cdot \overset{0}{\underline{\underline{\nabla R}}}{}^{T} \cdot\cdot \overset{0}{\underline{A}}{}^{T} - \psi_0 \underline{\nabla} \ln \sqrt{\frac{G}{g}} +$$

$$+ (\psi_1 \underline{E} + \psi_2 \underline{F}) \cdot \overset{0}{\underline{\underline{\nabla R}}}{}^{T} \cdot (\overset{0}{\underline{A}}{}^{T} \cdot\cdot \underline{E}) + \psi_2 \underline{F} \cdot\cdot (\underline{A}^{T} \cdot \underline{F} + \underline{F} \cdot \underline{A}^{T}) + \tfrac{1}{2} \rho_0 \underline{k} = \tfrac{1}{2} \rho_0 \underline{b} \,.$$

11. ELLIPTICITY OF THE EQUATIONS OF EQUILIBRIUM

We will now consider the equations of equilibrium - the equations of Par. 10 do not include the acceleration vector $\underline{b}$.

As in III, Par. 11, we consider in the undistorted reference v-configuration of an elastic medium some arbitrarily chosen smooth surface S. The Gaussian coordinates on this surface will be denoted by u^{α} ($\alpha = 1,2$), and the coordinate measured along the normal $\underline{n}$ to the surface will be designated as ζ. The material coordinates of a particle within the volume v will be denoted by q^{s}.

It is assumed that in this volume there is prescribed the function $\varphi(q^1,q^2,q^3)$ which is twice continuously differentiable with respect to the coordinates u^{α} on the surface S and that its normal derivative $\partial\varphi/\partial\zeta$ is also continuously differentiable with respect to u^{α}. Using the notation (III.11.21) we have

$$\left[\frac{\partial^2\varphi}{\partial u^{\alpha}\partial\zeta}\right]_{\zeta=0} = 0 .$$

From (III.11.26) the discontinuity of the second derivative with respect to ζ on S is defined under these conditions by the expression

$$\left[\frac{\partial^2\varphi}{\partial\zeta^2}\right]_{\zeta=0} = \underline{n}\cdot[\overset{00}{\underline{\nabla\nabla}}\varphi]_{\zeta=0}\cdot\underline{n} .$$

Alternatively this may be written in the form

$$[\overset{00}{\underline{\nabla\nabla}}\varphi]_{\zeta=0} = \underline{r}^k\underline{r}^t[\overset{0}{\nabla}_k\overset{0}{\nabla}_t\varphi]_{\zeta=0} = \underline{nn}\left[\frac{\partial^2\varphi}{\partial\zeta^2}\right]_{\zeta=0} \tag{1}$$

and in component form as

$$[\overset{0}{\nabla}_k\overset{0}{\nabla}_t\varphi]_{\zeta=0} = n_k n_t\left[\frac{\partial^2\varphi}{\partial\zeta^2}\right]_{\zeta=0} , \tag{2}$$

where n_k denotes the covariant components of $\underline{n}$ in the basis $\underline{r}_k$. Equation (2) determines the "weak" discontinuities of φ across S, that is the discontinuities of its second derivatives.

Applied to the functions χ^s in the equations (10.3) of equilibrium, Equation (2) enables three linear homogeneous equations to be written

$$\underline{r}_q\underline{r}^k\cdot\cdot\,\underline{P}_{0\,\underline{\nabla R}}\cdot\cdot\,\underline{r}_s\underline{r}^t n_k n_t\left[\frac{\partial^2\chi^s}{\partial\zeta^2}\right]_{\zeta=0} = 0 \qquad (q = 1,2,3) \tag{3}$$

for the unknown discontinuities of the second derivatives of χ^s by ζ.

Following Knowles and Sternberg*) (J.K. Knowles and E. Sternberg, 1975), we shall call the system of equations (10.3) elliptic if, in the reference v-configuration, there does not exist any weak discontinuity surface of the function χ^s, that is provided that the Equations (3) allow only trivial (zero) solutions, which means that the determinant of the matrix of their coefficients is non-zero

$$\det \| \underline{r}_q \underline{n} \cdot\cdot \underset{\underline{\nabla R}}{\underline{P}_0} \cdot\cdot \underline{r}_s \underline{n} \| \neq 0 \ . \tag{4}$$

With this matrix (as with any other) we can associate the quadratic form of three variables m^s - the components of a unit vector $\underline{m}$ in the $\underline{r}_s$-basis

$$m^q \underline{r}_q \underline{n} \cdot\cdot \underset{\underline{\nabla R}}{\underline{P}_0} \cdot\cdot m^s \underline{r}_s \underline{n} = \underline{mn} \cdot\cdot \underset{\underline{\nabla R}}{\underline{P}_0} \cdot\cdot \underline{mn} \ . \tag{5}$$

The system of equations of equilibrium is called strongly elliptic if this form is positive definite

$$\underline{mn} \cdot\cdot \underset{\underline{\nabla R}}{\underline{P}_0} \cdot\cdot \underline{mn} > 0 \tag{6}$$

for any dyad $\underline{mn}$ of unit vectors. For a strongly elliptic system not only the determinant (4) of the matrix of coefficients should be non-zero, but this matrix should also satisfy all the conditions of Sylvester's theorem. Of course, a strongly elliptic system is also elliptic.

Applied to the equations of equilibrium in the form (10.4), the requirement of strong ellipticity is equivalent to the condition that for any n_k the matrix

$$\left\| \frac{\partial^2 \ni}{\partial \overset{0}{\nabla}_t \chi^s \, \partial \overset{0}{\nabla}_k \chi^q} n_k n_t \right\| \ , \tag{7}$$

satisfies Sylvester's conditions, or, equally, the inequality

$$\frac{\partial^2 \ni}{\partial \overset{0}{\nabla}_t \chi^s \, \partial \overset{0}{\nabla} \chi^q} n_k n_t m^s m^q > 0 \ . \tag{8}$$

Lack of ellipticity presupposes the possibility of discontinuities in the solutions to the equations of equilibrium of an elastic body arising on certain

---

*) See also V.I. Smirnov, A Course of Higher Mathematics (in Russian), Vol. IV, Par. 161; Nauka, Moscow, 1951.

surfaces. It would be difficult to reconcile this with the idea of the physical properties ascribed to an elastic material. At the same time, there are no serious reasons for excluding such a possibility, for sufficiently large deformations, for example. Strong ellipticity represents an additional, more restrictive requirement. We will see below that it corresponds to some properties of an elastic medium which are assumed *a priori* and which do not follow immediately from its definition as a simple material which has no memory and which has the property of being able to store the work done on it by external forces. Strong ellipticity is a property of a material determined by its specific strain energy. For this energy being specified, strong ellipticity may be present for certain values of $\overset{0}{\nabla}\mathbf{R}$ and absent for other values.

The matrix corresponding to the determinant (4) is transformed in the case of an isotropic material by the substitution of $\mathbf{P}_{\overset{0}{\nabla}\mathbf{R}}$ for the relaxation tensor $\mathbf{T}_F$ using (9.12). This substitution results in the disappearance of the first two terms of (9.12)

$$\mathbf{r}_q\mathbf{n} \cdot\cdot\, \mathbf{P} \cdot \nabla\mathbf{r}^T \cdot\cdot\, \mathbf{r}_s\mathbf{n} = \mathbf{r}_q\mathbf{n} \cdot\cdot\, \mathbf{P}\mathbf{R}^m \cdot \mathbf{r}_s n_m \,,$$

$$\mathbf{r}_q\mathbf{n} \cdot\cdot\, \mathbf{r}_m\mathbf{P}^T\mathbf{R}^m \cdot\cdot\, \mathbf{r}_s\mathbf{n} = n_m\mathbf{r}_q \cdot \mathbf{P}^T \cdot \mathbf{n}\mathbf{R}^m \cdot \mathbf{r}_s = \mathbf{r}_q\mathbf{n} \cdot\cdot\, \mathbf{P}\mathbf{R}^m \cdot \mathbf{r}_s n_m \,,$$

which is the desired result. We thus arrive at the matrix

$$\sqrt{I_3}\, \mathbf{r}_q\mathbf{n} \cdot\cdot\, \nabla\mathbf{r}^T \cdot \mathbf{T}_F \cdot\cdot\, (\mathbf{R}_t\mathbf{r}_p + \mathbf{r}_p\mathbf{R}_t)\mathbf{r}^t\mathbf{r}^p \cdot\cdot\, \mathbf{r}_s\mathbf{n} =$$

$$= \sqrt{I_3}\, \mathbf{r}_q\mathbf{n} \cdot \nabla\mathbf{r}^T \cdot\cdot\, \mathbf{T}_F \cdot\cdot\, (\overset{0}{\nabla}\mathbf{R}^T \cdot \mathbf{n}\mathbf{r}_s + \mathbf{r}_s\overset{0}{\nabla}\mathbf{R}^T \cdot \mathbf{n}) \,,$$

which becomes transformed to

$$\mathbf{r}_q \cdot [\sqrt{I_3}\, \mathbf{n} \cdot \nabla\mathbf{r}^T \cdot (\mathbf{T}_F + \mathbf{T}_F \cdot\cdot\, \mathbf{C}_{II}) \cdot \overset{0}{\nabla}\mathbf{R}^T \cdot \mathbf{n}] \cdot \mathbf{r}_s = \mathbf{r}_q \cdot \mathbf{Q} \cdot \mathbf{r}_s \,, \tag{9}$$

where $\mathbf{T}_F \cdot\cdot\cdot\, \mathbf{C}_{II}$ is the tensor $\mathbf{T}_F$ with the third and fourth indices transposed

$$\mathbf{T}_F \cdot\cdot\, \mathbf{C}_{II} = \tau^{pqst}\mathbf{R}_p\mathbf{R}_q\mathbf{R}_s\mathbf{R}_t \cdot\cdot\, \mathbf{C}_{II} = \tau^{pqts}\mathbf{R}_p\mathbf{R}_q\mathbf{R}_s\mathbf{R}_t \,. \tag{10}$$

The system of equilibrium equations is thus seen to be elliptic provided that

$$\det\mathbf{Q} \neq 0 \,, \qquad \mathbf{Q} = \sqrt{I_3}\, \mathbf{n} \cdot \nabla\mathbf{r}^T \cdot (\mathbf{T}_F + \mathbf{T}_F \cdot\cdot\, \mathbf{C}_{II}) \cdot \overset{0}{\nabla}\mathbf{R}^T \cdot \mathbf{n} \tag{11}$$

in other words, the determinant of the covariant components of $\mathbf{Q}$ must be nonzero (by (I.5.1) its sign does not differ from that of $\det\mathbf{Q}$).

It is preferable to substitute the normal vector $\underline{n}$ to the surface in the reference configuration for the vector $\underline{N}$ in the actual configuration. By (1.8.8)

$$\underline{n}\cdot\underline{\nabla r}^{T} = (\underline{n}\cdot\underline{G}^{-1}\cdot\underline{n})^{\frac{1}{2}}\underline{N}\ , \qquad \overset{0}{\underline{\nabla R}}{}^{T}\cdot\underline{n} = (\underline{n}\cdot\underline{G}^{-1}\cdot\underline{n})^{\frac{1}{2}}\underline{F}\cdot\underline{N}\ . \tag{12}$$

Discarding the positive quadratic form $\sqrt{I_3}\,(\underline{n}\cdot\underline{G}^{-1}\cdot\underline{n})$ which does not influence the sign of $\underline{\underline{Q}}$ we arrive at the rèpresentation in the actual configuration

$$\underline{\underline{Q}} = \underline{N}\cdot(\underline{\underline{T}}_F + \underline{\underline{T}}_F\cdot\cdot\,\underline{C}_{II})\cdot\underline{F}\cdot\underline{N}\ . \tag{13}$$

Substituting in (13) for the tensor $\underline{\underline{T}}_F$ its expression (7.9), we arrive at the expanded expression for the tensor $\underline{\underline{Q}}$

$$\begin{aligned}\underline{\underline{Q}} = {} & (\vartheta_{00} - \tfrac{1}{2}\psi_0)\,\underline{NN} + \vartheta_{11}\underline{N}\cdot\underline{FF}\cdot\underline{N} + \vartheta_{22}\underline{N}\cdot\underline{F}^2\underline{F}^2\cdot\underline{N} + \\ & + \vartheta_{01}(\underline{NF}\cdot\underline{N} + \underline{N}\cdot\underline{FN}) + \vartheta_{02}(\underline{N}\cdot\underline{F}^2\underline{N} + \underline{NF}^2\cdot\underline{N}) + \\ & + \vartheta_{12}(\underline{N}\cdot\underline{FF}^2\cdot\underline{N} + \underline{N}\cdot\underline{F}^2\underline{F}\cdot\underline{N}) + \tfrac{1}{2}(\psi_1\underline{E} + \psi_2\underline{F})\,\underline{N}\cdot\underline{F}\cdot\underline{N} + \\ & + \tfrac{1}{2}\psi_2(\underline{N}\cdot\underline{FF}\cdot\underline{N} + \underline{EN}\cdot\underline{F}^2\cdot\underline{N})\ .\end{aligned}$$

This representation shows that the tensor $\underline{\underline{Q}}$ is symmetric. Its eigenvalues are proportional to the squares of the velocities of propagation of plane waves in the direction of $\underline{N}$ through a pre-stressed elastic medium (for the case when the transformation of the reference configuration to the actual configuration is affine). This makes it reasonable to call $\underline{\underline{Q}}$ an acoustical tensor (see Chapter 8, Par. 7). The velocities are real whenever the system is strongly elliptic.

12. THE ACOUSTICAL TENSOR OF AN ELASTIC MEDIUM

For a linearly elastic medium, we have by (7.2) and (7.11) - (7.13)

$$(\underline{\underline{T}}_F)^0 = \tfrac{1}{2}[\mu(\underline{C}_{II} + \underline{C}_{III}) + \lambda\underline{EE}] = \tfrac{1}{2}\underline{T}_\varepsilon\ , \qquad (\underline{\underline{T}}_F + \underline{\underline{T}}_F\ \ \underline{C}_{II})^0 = \underline{T}_\varepsilon$$

and $\underline{F} = \underline{E}$ in (11.13). The representation of the acoustic tensor is reduced to the form

$$\underline{\underline{Q}} = \underline{N}\cdot\underline{T}_\varepsilon\cdot\underline{N} = (\lambda+\mu)\underline{NN} + \mu\underline{EN}\cdot\underline{N} = (\lambda+2\mu)\underline{NN} + \mu(\underline{t}_1\underline{t}_1 + \underline{t}_2\underline{t}_2)\ , \tag{1}$$

because

$$\underline{N}\cdot\underline{C}_{II}\cdot\underline{N} = N_s \underline{R}_t \underline{R}^s N^t = \underline{NN} \,, \qquad \underline{N}\cdot\underline{C}_{III}\cdot\underline{N} = N_s \underline{E} N^s = \underline{EN}\cdot\underline{N} \,,$$

and the unit tensor $\underline{E}$ is represented in the orthonormal base $\underline{N}$, $\underline{t}_1$, $\underline{t}_2$ by the expression

$$\underline{E} = \underline{NN} + \underline{t}_1\underline{t}_1 + \underline{t}_2\underline{t}_2 \,.$$

The matrix of the tensor $\underline{Q}$ is diagonal, and its determinant is

$$\det \underline{Q} = \mu^2(\lambda + 2\mu) \,. \tag{2}$$

The equilibrium equations of the linear theory are elliptic provided that

$$\mu \neq 0 \,, \qquad \lambda + 2\mu \neq 0 \,. \tag{3}$$

They are strongly elliptic if the Sylvester criteria are fulfilled which for this case reduce to the inequalities

$$\mu > 0 \,, \qquad \lambda + 2\mu = 2\mu \frac{1-\nu}{1-2\nu} > 0 \rightarrow -\infty < \nu < \frac{1}{2} \,. \tag{4}$$

The eigenvalues of the tensor $\underline{Q}$ are equal to $k_1 = \lambda + 2\mu$, $k_2 = k_3 = \mu$, and the conditions (4) of strong ellipticity guarantee the existence of strain extension and shear waves (propagating with the velocities $\sqrt{(\lambda + 2\mu)/\rho}$ and $\sqrt{\mu/\rho}$) respectively in a linearly elastic medium.

It is well known that for the specific strain energy to be positive it is necessary and sufficient that the following inequalities hold

$$\mu > 0 \,, \qquad 3\lambda + 2\mu = 2\mu \frac{1+\nu}{1-2\nu} > 0 \rightarrow -1 < \nu < \frac{1}{2} \,. \tag{5}$$

These conditions guarantee strong ellipticity. However, the inverse statement is not true: e.g. for $\lambda < -2\mu/3$, $3\lambda + 2\mu < 0$ but $\lambda + 2\mu > 0$.

The analysis for a non-linear elastic medium is of course more complicated. Using the component representation (8.2) of the relaxation tensor, we have by (11.13)

$$\underline{Q} = N_s N_k g^{qk}(\tau^{stpq} + \tau^{stqp})\underline{R}_t \underline{R}_p \tag{6}$$

and in the basis of the principal directions of the Finger measure

$$\underline{Q} = N_s N_q v_q^2(\tau^{stpq} + \tau^{stqp})\underline{e}_t \underline{e}_p \,, \qquad Q^{tp} = N_s N_q v_q^2(\tau^{stpq} + \tau^{stqp}) \,. \tag{7}$$

Of the 12 components of the relaxation tensor, only 4 are substantially different

$$\tau^{1111}, \tau^{1122}, \tau^{1212}, \tau^{1313} . \tag{8}$$

We obtain

$$Q^{11} = 2(N_1^2\tau^{1111}v_1^2 + N_2^2\tau^{2121}v_2^2 + N_3^2\tau^{3131}v_3^2) ,$$

$$Q^{12} = 2N_1N_2(v_2^2\tau^{1122} + v_1^2\tau^{1212}) , \qquad Q^{21} = 2N_2N_1(v_1^2\tau^{2211} + v_2^2\tau^{2121}) .$$

Here we have used rule (8.8) for the transposition of indices. Referring to (8.9) and (8.10), we arrive at the expressions

$$Q^{11} = v_1 \frac{\partial\sigma_1}{\partial v_1} N_1^2 + v_2^2 \frac{\sigma_1 - \sigma_2}{v_1^2 - v_2^2} N_2^2 + v_3^2 \frac{\sigma_1 - \sigma_3}{v_1^2 - v_3^2} N_3^2 , \tag{9}$$

$$Q^{12} = N_1N_2 \left(v_2 \frac{\partial\sigma_1}{\partial v_2} + v_1^2 \frac{\sigma_1 - \sigma_2}{v_1^2 - v_2^2}\right) , \qquad Q^{21} = N_2N_1 \left(v_1 \frac{\partial\sigma_2}{\partial v_1} + v_2^2 \frac{\sigma_1 - \sigma_2}{v_1^2 - v_2^2}\right) \tag{10}$$

and it is not difficult to verify that $Q^{12} = Q^{21}$ - the acoustical tensor is symmetric. In fact, by (3.12) and (3.16) we have

$$v_2 \frac{\partial\sigma_1}{\partial v_2} = \frac{1}{v_3} \frac{\partial^2 \ni}{\partial v_2 \partial v_1} - \sigma_1 , \qquad v_1 \frac{\partial\sigma_2}{\partial v_1} = \frac{1}{v_3} \frac{\partial^2 \ni}{\partial v_1 \partial v_2} - \sigma_2$$

and, as expected,

$$Q^{12} - Q^{21} = N_1N_2 \left[-\sigma_1 + \sigma_2 + (v_1^2 - v_2^2) \frac{\sigma_1 - \sigma_2}{v_1^2 - v_2^2} \right] = 0 ,$$

The other components of $\underset{\sim}{Q}$ are obtained by cyclic permutation of indices in Equations (9), (10).

From the expressions obtained for the components Q^{mn} it is evident that the structure of $\det \underset{\sim}{Q}$ is fairly complicated and it is hardly possible to obtain from it any precise statements about strongly elliptic materials.

We will confine our discussion to the case in which the normal vector $\underline{N}$ is directed along one of the principal axes of the Finger measure. Assume, for example, that $\underline{N} = \underline{e}_1$, $N_1 = 1$, $N_2 = N_3 = 0$. The tensor $\underset{\sim}{Q}$ then becomes diagonal

$$Q^{11} = v_1 \frac{\partial \sigma_1}{\partial v_1}\,, \qquad Q^{22} = v_1^2 \frac{\sigma_2 - \sigma_1}{v_2^2 - v_1^2}\,, \qquad Q^{33} = v_1^2 \frac{\sigma_3 - \sigma_1}{v_3^2 - v_1^2}\,. \tag{11}$$

The Sylvester conditions for strong ellipticity reduce to

$$\frac{\partial \sigma_1}{\partial v_1} > 0\,, \qquad \frac{\sigma_1 - \sigma_2}{v_1^2 - v_2^2} > 0\,, \qquad \frac{\sigma_1 - \sigma_3}{v_1^2 - v_3^2} > 0\,. \tag{12}$$

Repeating this argument, we arrive, in the first place, at the requirement that the principal stresses are monotonic increasing with respect to the corresponding principal strains

$$\frac{\partial \sigma_s}{\partial v_s} > 0\,, \qquad s = 1,2,3\,. \tag{13}$$

In the second place, we see that the signs of the principal stress differences and of the corresponding principal strains should be the same

$$\operatorname{sgn}(\sigma_s - \sigma_k) = \operatorname{sgn}(v_s - v_k)\;; \qquad s,k = 1,2,3\,, \quad s \neq k\,, \tag{14}$$

these are the "$\mathcal{BE}$-criteria" (Baker-Ericksen).

As stated in Par. 11, the notion of strong ellipticity accords well with the idea of "how the material should behave". More detailed information may be elicited from a consideration of specific materials by an explicit prescription of their functions $\ni(I_1,I_2,I_3)$.

13. ON THE POSING OF EQUILIBRIUM BOUNDARY-VALUE PROBLEMS

It is assumed that the undistorted reference v-configuration is known. As in the linear theory, the first boundary-value problem consists in the determination in the actual V-configuration - of the position vector $\underline{R}(q^1,q^2,q^3)$ of any particle $M(q^1,q^2,q^3)$ of the body - by either this vector or the displacement $\underline{u} = \underline{R} - \underline{r}$ prescribed on the boundary surface O in the same configuration

$$\text{on } O\text{:}\;\; \underline{R} = \underline{R}(q^1,q^2,q^3) = \underline{r}(q^1,q^2,q^3) + \underline{u}(q^1,q^2,q^3)\,. \tag{1}$$

The desired solution should satisfy the equilibrium equations in one of the forms given in Par. 10 ($\underline{b} = \underline{0}$).

The formulation of the second boundary-value problem is more complicated. It would be unrealistic to require that the static equilibrium equations be fulfilled on the surface O without any resctrictions as to the class of surface tractions. We do not know in advance either the surface or the directions of the forces prescribed upon it. For this reason consideration is usually confined to "dead" and "progressive" loadings.

For a dead loading, we have by (2.1.12) and (1.8.7)

$$\text{on } O:\ \underline{N}\cdot\underline{T} = \underline{f} = \underline{f}^0 \frac{do}{dO} = \sqrt{\frac{g}{G}}\,\underline{f}^0(\underline{n}\cdot\underline{G}^{-1}\cdot\underline{n})^{-\frac{1}{2}}\,; \qquad \text{on } o:\ \underline{n}\cdot\underline{P} = \underline{f}^0\,, \tag{2}$$

where the contact force $\underline{f}^0$ is prescribed on the surface o of a body in its reference configuration. The solution must satisfy the equations of equilibrium in the volume V of the actual configuration which means that the resultant of the external forces and the couple produced must both be equal to zero

$$\iiint\limits_V \rho\underline{k}\,dV + \iint\limits_O \underline{f}\,dO = \underline{0}\,, \qquad \iiint\limits_V \underline{R}\times\rho\underline{k}\,dV + \iint\limits_O \underline{R}\times\underline{f}\,dO = \underline{0}\,. \tag{3}$$

The reference configuration must also be in equilibrium,

$$\iiint\limits_v \rho_0\underline{k}\,dv + \iint\limits_o \underline{f}^0\,do = \underline{0}\,, \qquad \iiint\limits_v \underline{r}\times\rho_0\underline{k}\,dv + \iint\limits_o \underline{r}\times\underline{f}^0\,do = \underline{0}\,. \tag{4}$$

The first condition (3) is immediately transformed to the first condition (4) - this follows from the conservation law of mass $\rho\,dV = \rho_0 dv$ and from the definition (2) of "dead" loading. In other words, vanishing of the resultant vector of external forces in the reference configuration guarantees its vanishing in the actual configuration. This statement cannot be made with regard to the resultant moment of external forces. In the actual configuration, this moment will be zero provided that the compatibility condition is fulfilled (Signorini)

$$\iiint\limits_v (\underline{R}-\underline{r})\times\rho_0\underline{k}\,dv + \iint\limits_o (\underline{R}-\underline{r})\times\underline{f}^0 do = \underline{0}$$

or

$$\iiint\limits_v \underline{R}\times\rho_0\underline{k}\,dv + \iint\limits_o \underline{R}\times\underline{f}^0 do = \underline{0}\,, \tag{5}$$

which follows immediately from (3) and (4).

As in the linear theory, prescription of the external forces determines the position vector $\underline{R}$ up to the rigid displacement [see (1.4.22)]. This arbitrariness may be used to satisfy condition (5) assuming that the force tensor (2.1.16) is non-singular; this has been proved in Chapter 3, Par. 3. Complications may arise if the force tensor $\underline{B}$ is singular ($\det\underline{B} = 0$). On this point see Chapter

6, Par. 11.

A real example of a "follow-up" loading is given by the hydrostatic pressure which remains directed along the normal to the deformed surface of a body. The equation of equilibrium on a surface is then presented in the form

$$\text{on } O: \; -p\underline{N} = \underline{N} \cdot \underline{T} \; ; \qquad \text{on } o: \; -\sqrt{\frac{G}{g}} \; \underline{n} \cdot \underline{\nabla r}^{T} = \underline{n} \cdot \underline{P} \; . \tag{6}$$

It would be erroneous to expect the solution of the boundary-value problems of the non-linear theory of elasticity to be unique. This statement is easily demonstrated by some simple examples.

Let the inner surface of a hollow cylinder be fixed, and the outer surface subjected to rotation through angle φ_0 in the first experiment, and through $\varphi_0 + 2\pi$ in the second (rotation is carried out using a friction band acting upon the outer surface of the cylinder). The states of stress in the two experiments differ markedly while the boundary conditions (1) are the same. All the points on the outer surface have the same position vector $\underline{R}$.

The non-uniqueness of the solution to the second boundary-value problem is illustrated by the example of an everted semi-spherical cupola when its outer and inner surfaces in the reference configuration become the inner and outer surfaces in the actual configuration. External forces are absent in both configurations although the reference configuration may well be the natural one, a state of stress exists in the actual configuration. A similar example is furnished by the problem of the eversion of a hollow cylinder (see Chapter 7, Par. 12).

"Dead" forces are applied to the upper and lower faces O_1 and O_2 of a prismatic rod parallel to its generators. These forces are directed upward on O_1 and downward on O_2, so that the rod is stretched. For the second thought experiment, the rod is rotated through 180°: the face O_1 becomes the lower, and the face O_2 the upper but the direction of the forces remains constant since they are "dead". The rod is now compressed. This is an example of the non-uniqueness of the solution to the second boundary value problem under "dead" surface loading. We do not mean of course, that the rod is fixed in some loading device because the loading produced by such a device is not "dead"; it preserves its direction regardless of the orientation of the rod which is permanently either compressed or stretched.

14. METHODS OF ANALYSIS OF EQUILIBRIUM BOUNDARY-VALUE PROBLEMS FOR A NONLINEARLY ELASTIC BODY

A "frontal assault" upon the boundary-value problems of the nonlinear theory of elasticity described in Par. 13 is, as a rule, unrewarding. For small but finite values of the gradient $\overset{0}{\nabla}u = \overset{0}{\nabla}R - E$ of the displacement vector u from the natural reference configuration, it is possible to construct solutions to the second boundary-value problem in the form of a power series in $|\overset{0}{\nabla}u|$ in which the linear theory solution plays the role of an initial approximation. The procedure for constructing such series has been worked out by Signorini (1940) and their convergence and uniqueness (under certain restrictions) investigated by Stoppelli (Stoppelli, 1954) and Van Buren (W. van Buren, 1968). At each stage of this process one has to determine solutions to the linear equations under the "body and contact" forces determined by the preceding approximations which are assumed known. At all stages of this approximation process the resultant vector of these forces turns out to be zero, and difficulties arise in connection with the "compatibility conditions" (13.5) because it cannot be excluded in advance that the force tensor B may prove to be singular.

The effective determination of "second order effects", that is terms of second degree with respect to $|\overset{0}{\nabla}u|$, is itself fairly complicated and attempts to proceed further along these lines lead to expressions for "forces" in the right-hand sides of the linear equations which are difficult to envisage.

It must be recalled that for three-dimensional problems in the linear theory of elasticity (except for the cases of a half-space and sphere) there is no way of representing effectively a solution to the second boundary value problem for an arbitrary law of distribution of the body and contact forces. This excludes the possibility of finding the stress state for second order effects, only certain integral characteristics of this state are determinable. Planar problems are more accessible because the methods of the theory of functions applied to problems in linear elasticity theory are effective for a vast class of domains regardless of the way in which body and contact forces are prescribed. This has not only enabled us to obtain solutions to nonlinear problems for second order effects, but also to proceed, in a number of cases, to fourth order terms (see numerous papers by Yu.I. Koifman and others). We should also mention here the investigation of the behviour of a material in a vicinity of the end of a straight crack carried out in the framework of a nonlinear plane problem (J.K. Knowles, E. Sternberg, 1975).

Some experience in solving nonlinear problems for large deformations has been gained by the application of a "semi-inverse" method which has also provided the first outstanding achievements of linear theory. At the first stage of the procedure some hypothetical form $R(r(q^1,q^2,q^3))$ of the transformation of the reference configuration to the actual configuration is prescribed which contains the functions of the material coordinates sought. In the second stage this form is used to construct an expression for the strain measure and for this expression (by virtue of the state equation of a material) there is determined the stress tensor (either the Cauchy tensor $\underline{T}$ or the Piola tensor $\underline{P}$). In the third stage, from the equations of equilibrium within the volume and on its boundary surface, distributions of the body and contact forces are found which should be compatible with the assumed prescription of the position vector $\underline{R}$. It is necessary that the body and contact forces so determined agree with their prescribed values, i.e. either be constant (weight) or proportional to the distance from a certain axis (contrifugal force). It is usually taken that $\underline{k} = \underline{0}$ when it is assumed in advance that the stress state is produced by contact forces and that the influence of body forces is negligible. For this case, the choice of $\underline{R}$ is subject to the condition $\underline{\nabla} \cdot \underline{T} = \underline{0}$ (or $\overset{0}{\underline{\nabla}} \cdot \underline{P} = \underline{0}$). All that remains now is the requirement that the arbitrary (as yet undetermined) quantities entering the proposed representation of $\underline{R}$ be so chosen that the contact forces calculated following this procedure correspond to the problem as formulated. The success of this latter step is of course connected with the particular way in which the position vector $\underline{R}$, which should be consistent with the intuitively envisaged properties of the stress state, its symmetries etc., is prescribed. One may often expect a large measure of success in determining the contact forces on a substantial part of the surface O in the actual configuration but not over the entire surface. Over the remaining part one must remain content with the coincidence of the resultant and the couple of the distributions of the contact forces so obtained with their known values. An example of this is given by a sufficiently long prismatic body on whose lateral and end faces these conditions are usually satisfied. The theory of torsion provides a classic example of this.

It must be recalled that in the linear theory of elasticity, a version of the semi-inverse method is used which presupposes at its initial stage the prescription of a set of possible static stress states, that is the state satisfying the static equilibrium equations within the volume and on its surface. It is later verified that this state satisfies the Beltrami-Michell equations which guarantee that the linear strain tensor calculated from the usual stress tensor allows us to determine the displacement vector $\underline{u}$. In nonlinear theory this approach is hampered by the fact that inversion of the equation of state, that is the determination of the strain measure in terms of the stress tensor, is practically non-realizable (II, Par. 8) and not unique. In the nonlinear theory, an analog of the Beltrami-Mitchell equations can only be used for some exceptional cases (Par. 17). For this reason, the second version of the semi-inverse method may consist of the prescription of the strain measure satisfying the requirement that the corresponding Ricci tensor is zero (III.10.21). The stress tensor is then constructed using this measure and the equation of state. This tensor should describe one of a set of possible static stress states and its divergence should be zero if no account is taken of body forces and the surface forces are determined by its derivative along the normal. There is of course no reason to expect such a procedure not to require special prescription of the coefficients of the constitutive equation as functions of the invariants of the strain measure (e.g. the coefficients $\psi_\Gamma(I_1,I_2,I_3)$ in (4.3.4)). This also means that the forms of representation of the contact forces depend on the expressions for those coefficients, in other words, they cannot be represented in a universal form valid for any form of strain energy function $\ni(I_1,I_2,I_3)$ on its arguments.

The formulas for the contact forces are different for different materials. For some of these, the static equilibrium equations hold within the volume for the initially prescribed strain measure. For other materials, some different measure might be required.

One cannot exclude, however, the fact that these difficulties may not arise at all. Those deformations which guarantee that in this method static equilibrium holds for any material, i.e. regardless of the specific description of $\ni(I_1,I_2,I_3)$ or equivalently of $\psi_\Gamma(I_1,I_2,I_3)$, we will call "general". For this case the formulas which determine the contact forces giving rise to such a "general" deformation would be expressed only in terms of $\ni$. These formulas are general because they enable the forces for any material to be calculated. The numerical values of these forces would certainly be different for different materials. We will also call general the solution $\underline{R}(q^1,q^2,q^3)$ to the equations of the theory of elasticity which is determined by a general deformation. It must be recalled that these definitions assume the absence of body forces.

15. GENERAL SOLUTIONS OF EQUATIONS OF THE NONLINEAR THEORY OF ELASTICITY. THE THEOREM OF ERICKSEN

It is quite plain that the solution which determines the affine transformation of the reference configuration to the actual configuration is a general one

$$\underline{R} = \underline{r} \cdot \underline{\Lambda}\,, \qquad x^s = a^k \Lambda_k^{\cdot s}\,, \tag{1}$$

here, $\underline{\Lambda}$ denotes a constant tensor, $\det \underline{\Lambda} \neq 0$. The strain measures are also constant

$$\overset{0}{\underline{\nabla}}\underline{R} = \underline{\Lambda}\,, \quad \overset{0}{\underline{\nabla}}\underline{R}^{\mathrm{T}} = \underline{\Lambda}^{\mathrm{T}}\,, \quad \underline{G} = \underline{\Lambda} \cdot \underline{\Lambda}^{\mathrm{T}}\,, \quad \underline{F} = \underline{\Lambda}^{\mathrm{T}} \cdot \underline{\Lambda}\,, \tag{2}$$

the same is true for their invariants, including $I_3 = (\det \underline{\Lambda})^2$. By (4.3.4), the stress tensor $\underline{T}$ is represented as a quadratic polynomial with constant coefficients

$$\underline{T} = 2(\det \underline{\Lambda})^{-1}\left[I_3 \frac{\partial \ni}{\partial I_3} \underline{E} + \left(\frac{\partial \ni}{\partial I_1} + I_1 \frac{\partial \ni}{\partial I_2}\right)\underline{F} - \frac{\partial \ni}{\partial I_2} \underline{F}^2\right] . \tag{3}$$

In the absence of body forces, the static equilibrium equation $\underline{\nabla} \cdot \underline{T} = \underline{0}$ is satisfied within the volume, the same equation on the surface determines the distribution of the contact forces

$$\underline{f} = \underline{N} \cdot \underline{T} = 2(\det \underline{\Lambda})^{-1}\left[I_3 \frac{\partial \ni}{\partial I_3} \underline{N} + \left(\frac{\partial \ni}{\partial I_1} + I_1 \frac{\partial \ni}{\partial I_2}\right)\underline{N} \cdot \underline{F} - \frac{\partial \ni}{\partial I_2} \underline{N} \cdot \underline{F}^2\right] , \tag{4}$$

which produce the required transformation.

The theorem of Ericksen (J.L. Ericksen, 1955) asserts that there exist no general solutions to the problem of the equilibrium of an isotropic elastic body which would differ from those produced by the affine transformation (1). The following proof is due to Shield (R.T. Shield, 1971).

First, the following two auxiliary propositions will be established:

a) for $\ni(I_1, I_2, I_3) = I_1$, the equations of static equilibrium are satisfied if $\underline{R}$ is a harmonic vector. In fact, by (II.3.5) we have

$$\underline{P} = \ni_{\overset{0}{\nabla}\underline{R}} = (I_1)_{\overset{0}{\nabla}\underline{R}} = 2\overset{0}{\nabla}\underline{R} , \qquad \overset{0}{\underline{\nabla}} \cdot \underline{P} = 2\overset{0}{\underline{\nabla}} \cdot \overset{0}{\underline{\nabla}}R = \overset{0}{\nabla}{}^2\underline{R} = \underline{0} . \tag{5}$$

b) If we now assume that $\ni(I_1, I_2, I_3) = I_3$, the equations of static equilibrium are satisfied if I_3 = const. Referring to Equations (1.9.1), (1.9.4), we have

$$(\sqrt{I_3})_{\overset{0}{\nabla}\underline{R}} = \sqrt{I_3}\,\underline{\nabla r}^T , \qquad \overset{0}{\underline{\nabla}} \cdot \sqrt{I_3}\,\underline{\nabla r}^T = \underline{0} , \qquad (I_3)_{\overset{0}{\nabla}\underline{R}} = 2\sqrt{I_3}\,(\sqrt{I_3})_{\overset{0}{\nabla}\underline{R}} \tag{6}$$

and therefore (making use of the rule $\underline{\nabla} \cdot \varphi\underline{Q} = \varphi\underline{\nabla} \cdot \underline{Q} + \underline{Q}^T \cdot \underline{\nabla}\varphi$)

$$\underline{P} = \ni_{\overset{0}{\nabla}\underline{R}} = (I_3)_{\overset{0}{\nabla}\underline{R}} ,$$

$$\overset{0}{\underline{\nabla}} \cdot \underline{P} = 2\overset{0}{\underline{\nabla}} \cdot \sqrt{I_3}\,(\sqrt{I_3})_{\overset{0}{\nabla}\underline{R}} =$$

$$= 2\sqrt{I_3}\,\overset{0}{\underline{\nabla}} \cdot \sqrt{I_3}\,\underline{\nabla r}^T + 2\sqrt{I_3}\,\underline{\nabla r} \cdot \overset{0}{\underline{\nabla}}\sqrt{I_3} = 2\sqrt{I_3}\,\underline{\nabla}\sqrt{I_3} = \underline{\nabla} I_3 = \underline{0} , \tag{7}$$

which means that I_3 = const (since the gradient of this scalar is zero).

If we now let $\text{Э} = \sqrt{I_3}\,\tilde{\text{Э}}(I_1, I_2, I_3)$ and the stress tensors calculated by Э and $\tilde{\text{Э}}$ satisfy the equations of static equilibrium, we have

$$\underline{P} = \text{Э}_{\underline{\underline{\overset{0}{\nabla}R}}} = \sqrt{I_3}\,\tilde{\text{Э}}_{\underline{\underline{\overset{0}{\nabla}R}}} + (\sqrt{I_3})_{\underline{\underline{\overset{0}{\nabla}R}}}\,\tilde{\text{Э}} \,,$$

$$\underline{\underline{\overset{0}{\nabla}P}} = \sqrt{I_3}\,\underline{\overset{0}{\nabla}}\cdot\tilde{\text{Э}}_{\underline{\underline{\overset{0}{\nabla}R}}} + \tilde{\text{Э}}^{T}_{\underline{\underline{\overset{0}{\nabla}R}}}\cdot\underline{\overset{0}{\nabla}}\sqrt{I_3} + \tilde{\text{Э}}\underline{\overset{0}{\nabla}}\cdot(\sqrt{I_3})_{\underline{\underline{\overset{0}{\nabla}R}}} + (\sqrt{I_3})^{T}_{\underline{\underline{\overset{0}{\nabla}R}}}\cdot\underline{\overset{0}{\nabla}}\tilde{\text{Э}} \,.$$

The first term is by assumption equal to zero, the second vanishes due to (7), and the third due to (6). We thus arrive at the equation

$$\underline{\overset{0}{\nabla}}\cdot\underline{P} = \sqrt{I_3}\,\underline{\underline{\nabla r}}\cdot\underline{\overset{0}{\nabla}}\tilde{\text{Э}} = \sqrt{I_3}\,\underline{\nabla}\tilde{\text{Э}} = \underline{0}\,, \qquad \tilde{\text{Э}} = \text{const}\,, \tag{8}$$

which means that all the invariants are constant. Therefore

$$\underline{\overset{0}{\nabla}}{}^2 I_1 = 0\,. \tag{9}$$

At the same time,

$$\underline{G} = \underline{\underline{\overset{0}{\nabla}R}}\cdot\underline{\underline{\overset{0}{\nabla}R}}^{T}\,, \qquad I_1(\underline{G}) = \underline{\underline{\overset{0}{\nabla}R}}\cdot\cdot\,\underline{\underline{\overset{0}{\nabla}R}}^{T} = \sum_{s=1}^{3}\sum_{k=1}^{3}\left(\frac{\partial x^k}{\partial a^s}\right)^2,$$

$$\underline{\overset{0}{\nabla}}{}^2 I_1(\underline{G}) = \sum_{r=1}^{3}\sum_{k=1}^{3}\sum_{s=1}^{3}\frac{\partial^2}{\partial a^{r2}}\left(\frac{\partial x^k}{\partial a^s}\right)^2 =$$

$$= 2\sum_{s=1}^{3}\sum_{k=1}^{3}\frac{\partial x^k}{\partial a^s}\,\frac{\partial}{\partial a^s}\,\underline{\overset{0}{\nabla}}{}^2 x^k + 2\sum_{r=1}^{3}\sum_{s=1}^{3}\sum_{k=1}^{3}\left(\frac{\partial^2 x^k}{\partial a^r\partial a^s}\right)^2 = 0\,.$$

But the first term is zero by (5) and so

$$\frac{\partial^2 x^k}{\partial a^r\partial a^s} = 0\,, \qquad k = 1,2,3\,. \tag{10}$$

The functions x^k are thus linear functions of a^s, and the transformation of the reference configuration to the actual configuration is affine which is the desired result.

A linear transformation is of course also a general solution in the case of an incompressible medium ($I_3 = 1$), but in this case Ericksen's theorem is no longer valid in its original form. A linear transformation proves to be not the only general solution (see Chapter 7, Pars. 8 and 9).

16. THE STATIONARY VALUE PRINCIPLE FOR THE STORED ENERGY OF A SYSTEM

Here, as in Pars. 17, 18, we establish the variational principles of the nonlinear theory of elasticity. We show that the equations of static equilibrium as well as the boundary conditions on the surface O of a body in its actual configuration follow from these statements. At the same time the possibility arises of applying direct methods of the Ritz and Bubnov-Galerkin type, as well as of their modification by L.V. Kantorovich, to the formulation of approximate solutions to the problems of the nonlinear theory of elasticity.

The derivation of the stationary value principle of the stored energy of deformation of a system starts from the relationship

$$\delta\ni = \delta' a_{(e)} , \tag{1}$$

which expresses the equality of the strain energy function to the incremental work of the external body and contact forces produced by a virtual displacement $\delta\underline{R}$ of the particles of a body from their state of equilibrium in the actual configuration.

The integral form of the principle is as follows

$$\iiint_v \delta\ni \, dv - \iiint_V \rho\underline{k} \cdot \delta\underline{R} \, dV - \iint_O \underline{f} \cdot \delta\underline{R} \, dO = 0 . \tag{2}$$

The body forces are constant, they remain the same in both the considered equilibrium state and in the states to be compared with this. Therefore

$$\iiint_V \rho\underline{k} \cdot \delta\underline{R} \, dV = \iiint_v \rho_0\underline{k} \cdot \delta\underline{R} \, dv = \delta \iiint_v \rho_0\underline{k} \cdot \underline{R} \, dv = \delta \iiint_V \rho\underline{k} \cdot \underline{R} \, dV . \tag{3}$$

The sign of the variation has been interchanged with the sign of the integral because the reference configuration is, of course, also preserved under variation.

It is assumed that a "dead" surface loading $\underline{f}^0$ is prescribed on the region o_1 of the surface of a body in its reference configuration, and on the region $o_2 = o - o_1$ a displacement vector is prescribed. Then, as has already been stated,

$$\text{on } o_2 : \ \delta\underline{R} = \underline{0} ; \qquad \text{on } o_1 : \ \delta\underline{f}^0 = \underline{0} . \tag{4}$$

We thus have

$$\iint_O \underline{f} \cdot \delta\underline{R} \, dO = \iint_{O_1} \underline{f} \cdot \delta\underline{R} \, dO + \iint_{O_2} \underline{f} \cdot \delta\underline{R} \, dO = \iint_{o_1} \underline{f}^0 \cdot \delta\underline{R} \, do = \delta \iint_{o_1} \underline{f}^0 \cdot \underline{R} \, do \tag{5}$$

and relationship (2) is reduced to the form

$$\delta\left[\iiint\limits_{v} (\ni - \rho_0 \underline{k} \cdot \underline{R})\, dv - \iint\limits_{o_1} \underline{f}^0 \cdot \underline{R}\, do\right] = 0 \tag{6}$$

or, after transition to integration in the actual configuration,

$$\delta\left[\iiint\limits_{V} \left(\sqrt{\frac{g}{G}}\, \ni - \rho \underline{k} \cdot \underline{R}\right) dv - \iint\limits_{O_1} \underline{f}^0 \frac{do}{dO} \cdot \underline{R}\, dO\right] = 0 , \tag{7}$$

where do/dO should be substituted by its value in (1.8.7).

The quantity in the brackets

$$W_1 = \iiint\limits_{v} (\ni - \rho_0 \underline{k} \cdot \underline{R})\, dv - \iint\limits_{o_1} \underline{f}^0 \cdot \underline{R}\, do = \iiint\limits_{V} \left(\sqrt{\frac{g}{G}}\, \ni - \rho \underline{k} \cdot \underline{R}\right) dV - \iint\limits_{O_1} \underline{f}^0 \frac{do}{dO} \cdot \underline{R}\, dO , \tag{8}$$

is called the potential energy of a system ("the body forces plus the external forces"). It represents a functional of the position vector $\underline{R}$ of a particle in the actual configuration. Equation (6) (or (7)) expresses the stationary value principle (theorem) of this functional. In the equilibrium state, of all the admissible fields of the position vectors $\underline{R}$ (i.e., of vectors which have the prescribed value on O_2), the field adopted is that for which the functional W_1 assumes its stationary value ($\delta W_1 = 0$).

It is to be recalled that the stationary value is defined as that value of the functional which acquires an increment ΔW_1 of order higher than $\delta\underline{R}$ whenever the position vector $\underline{R}$ is subjected to the variation $\delta\underline{R}$. It is shown in linear theory that $\Delta W_1 > 0$ (for $\ni > 0$), i.e. that the stationary value of the functional is a minimum. There is no basis for such an assertion in nonlinear theory.

The differential equations and the natural boundary conditions of the variational stationary value problem of the functional W_1 represent the equations of equilibrium in the volume and on the surface in one of the forms given in Par. 10.

In fact, recalling (4.1.3), (2.6.2), (II.2.7), (1.3.10) and (III.3.10), we have

$$\iiint\limits_{v} \delta\ni\, dv = \iiint\limits_{v} \ni_{\overset{0}{\underline{\underline{\nabla R}}}} \cdot\cdot\, \delta\overset{0}{\underline{\underline{\nabla R}}}{}^{T} dv = \iiint\limits_{v} \underline{P} \cdot\cdot\, \overset{0}{\underline{\nabla}}\delta\underline{R}^{T} dv =$$

$$= \iiint\limits_{v} \sqrt{\frac{G}{g}}\, \underline{\underline{\nabla r}}^{T} \cdot \underline{T} \cdot\cdot\, \overset{0}{\underline{\nabla}}\delta\underline{R}^{T} dv = \iiint\limits_{V} \underline{T} \cdot\cdot\, \overset{0}{\underline{\nabla}}\delta\underline{R}^{T} \cdot \underline{\underline{\nabla r}}^{T} dV =$$

$$= \iiint_V \underline{T} \cdot\cdot \underline{\nabla}\delta\underline{R}^T \, dV = \iiint_V [\underline{\nabla} \cdot (\underline{T} \cdot \delta\underline{R}) - (\underline{\nabla} \cdot \underline{T}) \cdot \delta\underline{R}] \, dV =$$

$$= \iint_O \underline{N} \cdot \underline{T} \cdot \delta\underline{R} \, dO - \iiint_V \delta\underline{R} \cdot (\underline{\nabla} \cdot \underline{T}) dV \tag{9}$$

this allows us to write (2) in the form

$$- \iiint_V (\underline{\nabla} \cdot \underline{T} + \rho\underline{k}) \cdot \delta\underline{R} \, dV + \iint_{O_1} (\underline{N} \cdot \underline{T} - \underline{f}) \cdot \delta\underline{R} \, dO + \iint_{O_2} (\underline{N} \cdot \underline{T} - \underline{f}) \cdot \delta\underline{R} \, dO = 0 \, . \tag{10}$$

According to a fundamental lemma of the calculus of variations and because $\delta\underline{R}$ in the volume V is arbitrary, we have

$$\underline{\nabla} \cdot \underline{T} + \rho\underline{k} = \underline{0} \, . \tag{11}$$

The surface integral over O_2 vanishes since on this part of the surface the vector $\underline{R}$ is prescribed, $\delta\underline{R} = \underline{0}$. On O_1, vector $\underline{R}$ is arbitrary, and therefore

$$\underline{N} \cdot \underline{T} = \underline{f}^0 \, . \tag{12}$$

Note that in these transformations, we have used the notation $\underline{P}$ or $\underline{T}$ purely in order to simplify the representation and to avoid repetition of the calculation carried out in Par. 10. The specific strain energy is expressed in terms of the strain measure and so long as we have introduced the principle of stationary values of the potential energy of a system as a primary assumption of the mechanics of an elastic body, we are free to make no mention of the state of stress at all, i.e. "to be unaware of its existence".

If the body forces may be represented in terms of a potential and the contact forces are "dead", then the functional W_1 takes the form

$$W_1 = \iiint_v (\ni + \rho_0\pi) dv - \iint_{o_1} \underline{f}^0 \cdot \underline{R} \, do \, , \tag{13}$$

since $\underline{k} \cdot \delta\underline{R} = -\underline{\nabla}\pi \cdot \delta\underline{R} = -\delta\pi$ from (1.1.7) and from the definition of a gradient of a scalar (III.2.1).

If, furthermore, a uniform pressure p is included in the external forces distributed over the whole surface of the body, then by (2.1.13)

$$W_1 = \iiint_v \left(\ni + \rho_0\pi + p\sqrt{\frac{G}{g}}\right) dv - \iint_{o_1} \underline{f}^0 \cdot \underline{R} \, do \, . \tag{14}$$

17. THE PRINCIPLE OF STATIONARY COMPLEMENTARY ENERGY

This principle stands in contrast to the principle of stationary potential energy of a system. The problem here is that we have to choose from a set of possible static stress states, the state which is actually realized within an elastic body, whereas the principle of stationary potential energy makes no mention of stresses but rather expresses a property of the actual field of displacements.

The functional the stationary values of which are under consideration must be expressed in terms of the stress tensor or its invariants if the medium is isotropic. Geometric quantities should not enter into it. This does not cause difficulties in linear theory because the expression for a linear strain tensor in terms of the stress tensor $\underline{T}$ is known and this enables us to express the specific strain energy directly in terms of the stress tensor. In nonlinear theory, this procedure requires the inversion of the equation of state of the material and, as we have already mentioned in Par. 14 and in II, Par. 8, this procedure is practically unrealizable in the general case (for any material). However, the procedure for deriving the stationary value principle of additional work requires the assumption that inversion is carried out. It is assumed that equation

$$\underline{P} = \exists_{\underline{\underline{\overset{0}{\nabla}R}}} = \underline{P}(\underline{\underline{\overset{0}{\nabla}R}}) \tag{1}$$

has been solved with respect to $\underline{\underline{\overset{0}{\nabla}R}}$

$$\underline{\underline{\overset{0}{\nabla}R}} = \underline{\underline{\overset{0}{\nabla}R}}(\underline{P}) , \qquad \exists(\underline{\underline{\overset{0}{\nabla}R}}) = \exists(\underline{P}) . \tag{2}$$

The generating function of the Legendre transformation inverse to (1) is defined as

$$\exists_{\times}(\underline{P}) = \underline{P} \cdot\cdot \underline{\underline{\overset{0}{\nabla}R}}^{T}(\underline{P}) - \exists(\underline{\underline{\overset{0}{\nabla}R}}(\underline{P})) . \tag{3}$$

In fact, by (1) we have

$$\delta\exists_{\times}(\underline{P}) = \delta\underline{P} \cdot\cdot \underline{\underline{\overset{0}{\nabla}R}}^{T}(\underline{P}) + \underline{P} \cdot\cdot \delta\underline{\underline{\overset{0}{\nabla}R}}(\underline{P}) - \frac{\partial\exists}{\partial\underline{\underline{\overset{0}{\nabla}R}}} \cdot\cdot \delta\underline{\underline{\overset{0}{\nabla}R}}^{T} =$$

$$= \delta\underline{P} \cdot\cdot \underline{\underline{\overset{0}{\nabla}R}}^{T}(\underline{P}) = \underline{\underline{\overset{0}{\nabla}R}}(\underline{P}) \cdot\cdot \delta\underline{P}^{T} , \tag{4}$$

so that by the definition of the derivative of a scalar by a tensor argument we have

$$\overset{0}{\underline{\nabla}}\underline{R}(\underline{P}) = (\ni_{\times})_{\underline{P}} , \tag{5}$$

which is the desired result; the function $\ni_{\times}(\underline{P})$ is called the specific additional work. This relationship is an analog of Castigliano's theorem in the linear theory of elasticity.

We will now consider a possible static stress state. For this state the Piola tensor $\underline{P}$ satisfies the equations of static equilibrium in the volume v and on that part of the surface o_1 where a "dead" surface loading is prescribed

$$\text{in } v\text{: } \overset{0}{\underline{\nabla}} \cdot \underline{P} + \rho_0 \underline{k} = \underline{0} \text{ ;} \qquad \text{on } o_1\text{: } \underline{n} \cdot \underline{P} = \underline{f}^0 . \tag{6}$$

For the possible static states which are compared with the realizable equilibrium state the following conditions hold

$$\text{in } v\text{: } \overset{0}{\underline{\nabla}} \cdot (\underline{P} + \delta\underline{P}) + \rho_0 \underline{k} = \underline{0} \text{ ;} \qquad \text{on } o_1\text{: } \underline{n} \cdot (\underline{P} + \delta\underline{P}) = \underline{f}^0 ,$$

since the body and contact forces on o_1 remain unchanged. We thus have

$$\text{in } v\text{: } \overset{0}{\underline{\nabla}} \cdot \delta\underline{P} = \underline{0} \text{ ;} \qquad \text{on } o_1\text{: } \underline{n} \cdot \delta\underline{P} = \underline{0} . \tag{7}$$

We now consider the functional W_2, called the additional work, over the possible static strain tensors $\underline{P}$

$$W_2 = \iiint\limits_{v} \ni_{\times}(\underline{P})\, dv - \iint\limits_{o_2} \underline{R} \cdot \underline{f}^0 do . \tag{8}$$

Here $\underline{f}^0$ should be regarded as the reactions of the constraints which ensure that the position vector $\underline{R}$ is fixed on that part o_2 of the surface.

Referring now to the definition (4) and bearing in mind that on o_2 the vector $\underline{R}$ is not varied, we have by (1)

$$\delta W_2 = \iiint\limits_{v} \overset{0}{\underline{\nabla}}\underline{R}^T \cdot\cdot\, \delta\underline{P}\, dv - \iint\limits_{o_2} \underline{R} \cdot \delta\underline{f}^0 do =$$

$$= \iiint\limits_{v} [\overset{0}{\underline{\nabla}} \cdot (\delta\underline{P} \cdot \underline{R}) - (\overset{0}{\underline{\nabla}} \cdot \delta\underline{P}) \cdot \underline{R}]\, dv - \iint\limits_{o_2} \underline{R} \cdot \delta\underline{f}^0 do =$$

$$= - \iiint\limits_{v} (\overset{0}{\underline{\nabla}} \cdot \delta\underline{P}) \cdot \underline{R}\, dv + \iint\limits_{o_1} \underline{n} \cdot \delta\underline{P} \cdot \underline{R}\, do + \iint\limits_{o_2} (\underline{n} \cdot \delta\underline{P} - \delta\underline{f}^0) \cdot \underline{R}\, do . \tag{9}$$

But on o_2

$$\underline{n} \cdot \delta \underline{P} = \delta \underline{f}^{0} \tag{10}$$

and, from (7), we arrive at the principle of stationary complementary energy

$$\delta W_2 = 0\ , \qquad W_2 = \iiint\limits_{v} [\underline{P} \cdot\cdot \overset{0}{\nabla}\underline{R}^{T} - \ni(\overset{0}{\nabla}\underline{R}(\underline{P}))]\, dv - \iint\limits_{o_2} \underline{n} \cdot \underline{P} \cdot \underline{R}\, do \tag{11}$$

- functional W_2 preserves its stationary value for all possible Piola tensors (i.e. those satisfying (6)).

Bearing in mind that the curl of the gradient of any vector is a zero tensor, we deduce from (5) that

$$\overset{0}{\underline{\nabla}} \times \overset{0}{\nabla}\underline{R} = \overset{0}{\underline{\nabla}} \times (\ni_{\times})_{\underline{P}} = \underline{0}\ . \tag{12}$$

This relationship is an analog of the compatibility conditions of linear theory - the Beltrami-Michell equations. In this theory the principle of minimum complementary energy distinguishes from a number of statically admissible stress states that state which is realized, i.e. enables the displacement vector to be determined. It would be natural to expect that in nonlinear theory, the principle of stationary complementary energy performs the same role.*)

To verify this statement we shall construct the Euler equations for a variational calculus problem of the stationary values of the functional (11) subject to constraints (7).

Introducing the Lagrange multiplier $\underline{\lambda}$ (a vector), we have

$$\delta W_2 = \iiint\limits_{v} \overset{0}{\nabla}\underline{R}^{T} \cdot\cdot\, \delta\underline{P}\, dv - \iint\limits_{o_2} \underline{R} \cdot \delta\underline{f}^{0} do + \iiint\limits_{v} \underline{\lambda} \cdot \overset{0}{\underline{\nabla}} \cdot \delta\underline{P}\, dv = 0\ .$$

Repeating transformation (III.3.10), we get

$$\iiint\limits_{v} \underline{\lambda} \cdot \overset{0}{\underline{\nabla}} \cdot \delta\underline{P}\, dv = \iiint\limits_{v} [\overset{0}{\underline{\nabla}} \cdot (\delta\underline{P} \cdot \underline{\lambda}) - (\overset{0}{\nabla}\lambda)^{T} \cdot\cdot\, \delta\underline{P}]\, dv =$$

$$= \iint\limits_{o_1} \underline{\lambda} \cdot \underline{n} \cdot \delta\underline{P}\, do + \iint\limits_{o_2} \underline{\lambda} \cdot \underline{n} \cdot \delta\underline{P}\, do - \iiint\limits_{v} \overset{0}{\nabla}\lambda^{T} \cdot\cdot\, \delta\underline{P}\, dv =$$

---

*) There is no reason to assert in advance that the function $\overset{0}{\nabla}\underline{R}(\underline{P})$ of the argument $\underline{P}$ entering Equations (2)-(5) is itself a gradient, i.e. that it satisfies condition (12). The symbol $\overset{0}{\nabla}\underline{R}(\underline{P})$ should be regarded in these formulas simply as a notation for a tensor function of the argument $\underline{P}$. If, however, we ascribe the sense of a gradient to this symbol as has been done in (9), then we arrive at the principle of additional work in the form (11). The proof that the above symbol may actually be treated as a gradient is given below.

$$= \iint_{o_2} \underline{\lambda} \cdot \underline{n} \cdot \delta\underline{P}\, do - \iiint_{v} \overset{0}{\underline{\underline{\nabla\lambda}}}{}^{T} \cdot\cdot\, \delta\underline{P}\, dv \ ,$$

since the integral over o_1 vanishes in view of (7). Referring to (10), we obtain

$$\delta W_2 = \iiint_{v} (\overset{0}{\underline{\underline{\nabla R}}}{}^{T} - \overset{0}{\underline{\underline{\nabla\lambda}}}{}^{T}) \cdot\cdot\, \delta\underline{P}\, dv + \iint_{o_2} (\underline{\lambda} - \underline{R}) \cdot \underline{n} \cdot \delta\underline{P}\, do = 0 \ . \tag{13}$$

By a suitable choice of the three components of $\underline{\lambda}$ we may treat all nine components of $\delta\underline{P}$ as independent functions connected by the three equations (7) in the volume v. We thus arrive at the desired results

$$\text{in } v\text{:}\ \overset{0}{\underline{\underline{\nabla R}}} = \overset{0}{\underline{\underline{\nabla\lambda}}} \ ; \qquad \text{on } o_2\text{:}\ \underline{\lambda} = \underline{R} \ . \tag{14}$$

It is thus proved that the tensor $\overset{0}{\underline{\underline{\nabla R}}}$ determined by (5) actually represents the gradient of the vector, i.e. that Equation (5) is integrable

$$\overset{0}{\underline{\underline{\nabla R}}} = \underline{r}^{s}\underline{R}_{s} = \frac{\partial \text{Э}_{\times}}{\partial \underline{P}} \ , \qquad \underline{R}_{s} = \underline{r}_{s} \cdot \frac{\partial \text{Э}_{\times}}{\partial \underline{P}} \ , \qquad \underline{R} = \int_{M_0}^{M} d\underline{r} \cdot \frac{\partial \text{Э}_{\times}}{\partial \underline{P}} + \underline{R}(M_0) \tag{15}$$

here, the integration is performed along any curve connecting the two points $M(q^1,q^2,q^3)$ and $M_0(q_0^1,q_0^2,q_0^3)$.

In some papers which followed the publication of that of L.M. Zubov doubts were raised regarding the assumption of the inversion of (1) which formed the basis for the derivation of the principle of stationary complementary energy. For instance, in two papers on the Zubov principle, W.T. Koiter (1973, 1975), pointed out that application of this principle gives rise to ambiguities. L.M. Zubov has proved using the polar form of the Piola tensor that representation (2) of the strain measure in terms of the Piola tensor possesses, in general, as many as four branches, but that as long as the deformations are not too great (i.e. that the angle of rotation of the basis of the principal axes of the Cauchy strain measure does not exceed 70°), only one of these branches is realized (1976).

18. THE COMBINED STATIONARY VALUE PRINCIPLES OF REISSNER AND OF KHU-WASHIZU

In the stationary value principle proposed by Reissner (E. Reissner, 1950) the functional W_3 over the position vector $\underline{R}$ and the Piola tensor $\underline{P}$ is considered

$$W_3 = \iiint_{v} [\underline{P} \cdot\cdot\, \overset{0}{\underline{\underline{\nabla R}}}{}^{T} - \text{Э}_{\times}(\underline{P}) - \rho_0\underline{k} \cdot \underline{R}]\, dv - \iint_{o_1} \underline{f}^{0} \cdot \underline{R}\, do - \iint_{o_2} \underline{n} \cdot \underline{P} \cdot (\underline{R} - \underline{R}_{\times})\, do \ . \tag{1}$$

Here, $\ni_\times(\underline{P})$ denotes the specific complementary energy and $\underline{R}_\times$ is the value of the position vector prescribed on o_2. Using (17.3), an alternative form of the Reissner functional has the form

$$W_3 = \iiint_v [\ni(\underline{P}) - \rho_0\underline{k}\cdot\underline{R}]\,dv - \int_{o_1} \underline{f}^0\cdot\underline{R}\,do - \iint_{o_2} \underline{n}\cdot\underline{P}\cdot(\underline{R}-\underline{R}_\times)\,do \ . \tag{2}$$

Varying W_3, we get

$$\delta W_3 = \iiint_v [\delta\underline{P}\cdot\cdot\overset{0}{\underline{\nabla}}\underline{R}^T + \underline{P}\cdot\cdot\delta\overset{0}{\underline{\nabla}}\underline{R}^T - (\ni_\times)_{\underline{P}}\cdot\cdot\delta\underline{P}^T - \rho_0\underline{k}\cdot\delta\underline{R}]\,dv -$$

$$- \iint_{o_1} \underline{f}^0\cdot\delta\underline{R}\,do - \iint_{o_2} \underline{n}\cdot\delta\underline{P}\cdot(\underline{R}-\underline{R}_\times)\,do - \iint_{o_2} \underline{n}\cdot\underline{P}\cdot\delta\underline{R}\,do \ .$$

Substituting the surface integral for the volume integral, we obtain

$$\iint_{o_2} \underline{n}\cdot\underline{P}\cdot\delta\underline{R}\,do = \iint_{o} \underline{n}\cdot\underline{P}\cdot\delta\underline{R}\,do - \iint_{o_1} \underline{n}\cdot\underline{P}\cdot\delta\underline{R}\,do =$$

$$= \iiint_v [(\overset{0}{\underline{\nabla}}\cdot\underline{P})\cdot\delta\underline{R} + \underline{P}\cdot\cdot\delta\overset{0}{\underline{\nabla}}\underline{R}^T]\,dv - \iint_{o_1} \underline{n}\cdot\underline{P}\cdot\delta\underline{R}\,do$$

which allows us to represent δW_3 in the form

$$\delta W_3 = -\iiint_v \left[\left(\frac{\partial\ni_\times}{\partial\underline{P}} - \overset{0}{\underline{\nabla}}\underline{R}\right)\cdot\cdot\delta\underline{P}^T + (\overset{0}{\underline{\nabla}}\cdot\underline{P}+\rho_0\underline{k})\cdot\delta\underline{R}\right]dv +$$

$$+ \iint_{o_1} (\underline{n}\cdot\underline{P}-\underline{f}^0)\cdot\delta\underline{R}\,do - \iint_{o_2} \underline{n}\cdot\delta\underline{P}\cdot(\underline{R}-\underline{R}_\times)\,do = 0 \ . \tag{3}$$

Since $\delta\underline{R}$ and $\delta\underline{P}$ are arbitrary within the volume, we obtain the relationships

$$\overset{0}{\underline{\nabla}}\cdot\underline{P} + \rho_0\underline{k} = \underline{0} \ , \qquad \frac{\partial\ni_\times}{\partial\underline{P}} = \overset{0}{\underline{\nabla}}\underline{R} \ . \tag{4}$$

On the part o_1 of the surface the variation $\delta\underline{R}$ is arbitrary, on the part o_2 the same is true for $\delta\underline{P}$. We thus have

$$\text{on } o_1\text{: } \underline{n}\cdot\underline{P} = \underline{f}^0 \ ; \qquad \text{on } o_2\text{: } \underline{R} = \underline{R}_\times \ . \tag{5}$$

When using Reissner's functional, we need not concern ourselves over the choice of the tensor $\underline{P}$ within the set of the statically admissible tensors. This fact represents an advantage which Reissner's principle enjoys over the principle of complementary energy. As for the latter principle, Equation (17.5) still holds. The Piola tensor determined from Reissner's principle satisfies the state equation of the material.

As long as we know the expression for the position vector $\underline{R}$ in the actual configuration as well as that for the tensor $\underline{P}$, which expressions contain a number of functions of the material coordinates and a number of constant parameters, we may construct an expression for the functional W_3. These functions and parameters are then determined from the Euler equation and from the natural boundary conditions of the corresponding variational problem. This "double approximation" ($\underline{R}$ and $\underline{P}$) is sucessfully applied to the linear problems. In the nonlinear theory, the Euler equations for the approximating functions are of course nonlinear. The difficulties associated with the representation of the specific additional work $\ni_x(\underline{P})$ still of course remain when we attempt to construct the functional W_3.

This problem has been overcome in the stationary value principle put forward by Khu and Washizu (K. Washizu, 1955).

Here, the functional is given by the expression

$$W_4 = \iiint_v [\ni(\underline{C}) - \underline{P}^T \cdot\cdot (\underline{C} - \overset{0}{\underline{\nabla}}\underline{R}^T) - \rho_0 \underline{k} \cdot \underline{R}]\, dv - \\ - \iint_{o_1} \underline{f}^0 \cdot \delta\underline{R}\, do - \iint_{o_2} \underline{n} \cdot \delta\underline{P} \cdot (\underline{R} - \underline{R}_x)\, do \,. \tag{6}$$

where $\underline{C}$ denotes a tensor of the second rank, $\ni(\underline{C})$ represents the expression for the specific strain energy where $\overset{0}{\underline{\nabla}}\underline{R}$ is substituted for $\underline{C}$. The functional W_4 contains three independent quantities $\underline{P}$, $\underline{C}$ and $\underline{R}$. From $\underline{R}$ we construct the expression for $\overset{0}{\underline{\nabla}}\underline{R}^T$.

The calculation of δW_4 substantially repeats that of (3) and we get

$$\delta W_4 = \iiint_v \left[\left(\frac{\partial \ni}{\partial \underline{C}} - \underline{P}\right) \cdot\cdot\, \delta\underline{C}^T - (\overset{0}{\underline{\nabla}} \cdot \underline{P} + \rho_0 \underline{k}) \cdot \delta\underline{R} - \\ - (\underline{C} - \overset{0}{\underline{\nabla}}\underline{R}) \cdot\cdot\, \delta\underline{P}\right] dv + \iint_{o_1} (\underline{n} \cdot \underline{P} - \underline{f}^0) \cdot \delta\underline{R}\, do - \iint_{o_2} \underline{n} \cdot \delta\underline{P} \cdot (\underline{R} - \underline{R}_x)\, do = 0 \,. \tag{7}$$

Because the variations $\delta\underline{P}$, $\delta\underline{C}$, $\delta\underline{R}$ are independent, the Euler equations and the natural boundary conditions are reduced to the form

$$\underline{P} = \frac{\partial \ni}{\partial \underline{C}}\,, \quad \underline{C} = \overset{0}{\underline{\nabla}}\underline{R}\,, \quad \overset{0}{\underline{\nabla}} \cdot \underline{P} + \rho_0 \underline{k} = \underline{0}\,, \tag{8}$$

$$\text{on } O_1 \colon \underline{n} \cdot \underline{P} = \underline{f}^0 \ ; \qquad \text{on } O_2 \colon \underline{R} = \underline{R}_x \ . \tag{9}$$

The expressions (8) contain both the equilibrium equations and the constitutive equation.

19. THE HAMILTON-OSTROGRADSKII PRINCIPLE

We call the Hamilton action the functional over the position vector $\underline{R}(q^1,q^2,q^3;t)$ which is evaluated for a system in the course of its motion:

$$S = \int_{t_0}^{t_1} (K - W)dt \ . \tag{1}$$

Here, K denotes the kinetic energy

$$K = \frac{1}{2}\iiint_V \rho \underline{v} \cdot \underline{v}\, dV = \frac{1}{2}\iiint_v \rho_0 \underline{v} \cdot \underline{v}\, dv \ , \tag{2}$$

and W the potential energy of a system given by (16.8)

$$W = \iiint_v (\ni - \rho_0 \underline{k} \cdot \underline{R})\, dv - \iint_{O_1} \underline{f}^0 \cdot \underline{R}\, do = \iiint_V \left(\sqrt{\frac{g}{G}} \ni - \rho \underline{k} \cdot \underline{R} \right) dV - \iint_{O_1} \underline{f}^0 \frac{do}{dO} \cdot \underline{R}\, dO \ . \tag{3}$$

The actual motion $\underline{R}(q^1,q^2,q^3;t)$ is to be compared with that of $\underline{R} + \delta\underline{R}$ where $\delta\underline{R}$ denotes a virtual displacement mentally imparted to a particle to remove it from its instanteneous position to some infinitely close point (a synchronous variation); $\delta\underline{R}$ is a continuous differentiable function of the material coordinates and the time. The operations of differentiation and variation are assumed interchangeable

$$(\delta\underline{R})^{\cdot} = \delta\dot{\underline{R}} \ . \tag{4}$$

At the limits of the interval $[t_0,t_1]$, the variations $\delta\underline{R}$ are assumed to vanish; the motions being compared coincide at these instants for all particles of the medium:

$$\delta\underline{R}(q^1,q^2,q^3;t_0) = \underline{0} \ , \qquad \delta\underline{R}(q^1,q^2,q^3;t_1) = \underline{0} \ . \tag{5}$$

The property of motion contained in the Hamilton-Ostrogradskii principle, is formulated as follows: from all possible motions (i.e. those permitted by the constraints) within the time interval $[t_0,t]$ and which coincide at t_0 and t, the actual motion is distinguished by the Hamilton action attaining a stationary value, the variation thus being zero

$$\delta S = 0 \ . \tag{6}$$

We have to proof that this requirement yields the equations of motion given in Par. 10. These include not only those of Chapter 2, Par. 6, but also the state equations of the medium.

Making use of the properties (4) and (5) of variation, we get

$$\delta K = \frac{1}{2}\iiint_v \rho_0\, dv \int_{t_0}^{t_1} \delta\underline{v}\cdot\underline{v}\, dt \ ; \qquad \frac{1}{2}\int_{t_0}^{t_1} \delta\underline{v}\cdot\underline{v}\, dt = \int_{t_0}^{t_1} (\delta\dot{\underline{R}})\cdot\underline{v}\, dt =$$

$$= \int_{t_0}^{t_1} (\delta\underline{R})^{\bullet}\cdot\underline{v}\, dt = \delta\underline{R}\cdot\underline{v}\Big|_{t_0}^{t_1} - \int_{t_0}^{t_1} \dot{\underline{v}}\cdot\delta\underline{R}\, dt = -\int_{t_0}^{t_1} \underline{b}\cdot\delta\underline{R}\, dt \ ,$$

so that

$$\delta\int_{t_0}^{t_1} K\, dt = -\int_{t_0}^{t_1} dt \iiint_v \rho_0\underline{b}\cdot\delta\underline{R}\, dv \ . \tag{7}$$

The calculation of the variation of the functional W is reduced to the transformations already carried out in Par. 16.

Because the force $\underline{f}^0$ is prescribed on o_1 and the position vector $\underline{R}$ on o_2 is known, we get

$$-\delta W = \iiint_v (\overset{0}{\underline{\nabla}}\cdot\underline{P} + \rho_0\underline{k})\cdot\delta\underline{R}\, dv + \iint_{o_1} (\underline{f}^0 - \underline{n}\cdot\underline{P})\cdot\delta\underline{R}\, do \ , \tag{8}$$

where we have used the equation of state

$$\underline{P} = (\ni)_{\overset{0}{\underline{\nabla}}\underline{R}} \tag{9}$$

throughout the calculation.

The condition (6) now takes the form

$$\delta S_\times = \int_{t_0}^{t_1} dt \left[\iiint_v (-\rho_0\underline{b} + \overset{0}{\underline{\nabla}}\cdot\underline{P} + \rho_0\underline{k})\cdot\delta\underline{R}\, dv + \iint_{o_1} (\underline{f}^0 - \underline{n}\cdot\underline{P})\cdot\delta\underline{R}\, do\right] = 0 \tag{10}$$

and by the usual argument we arrive at the anticipated equations of motion and the boundary conditions

$$\text{in } v\colon \overset{0}{\underline{\nabla}} \cdot \underline{P} + \rho_0 \underline{k} = \rho_0 \underline{b} \; ; \qquad \text{on } o_1\colon \underline{n} \cdot \underline{P} = \underline{f}^0 \; . \tag{11}$$

It must be stressed that the Piola tensor $\underline{P}$ is determined here by (9).

The kinematic boundary condition on o_2 is assumed to hold because of the way in which the functional S is prescribed since the vector $\underline{R}$ contained in S must satisfy this principle. This is reflected in the formulation of the principle (where only those motions permitted by the constraints are regarded as admissible).

Reversing the calculation, we may easily arrive from Equations (11) at the stationary value principle of the Hamilton action. Formal transformation enables Equations (11) to be written as follows

$$\text{in } V\colon \underline{\nabla} \cdot \underline{T} + \rho \underline{k} = \rho \underline{b} \; ; \qquad \text{on } O_1\colon \underline{N} \cdot \underline{T} = \underline{f}^0 \frac{do}{dO} = \underline{f} \; . \tag{12}$$

The difference $K - W$ is known as the Lagrangian L of a system and we have

$$L = K - W \, , \qquad S = \int_{t_0}^{t_1} L \, dt \tag{13}$$

and Equation (10) can be represented as

$$\delta S = \int_{t_0}^{t_1} \delta L \, dt = \int_{t_0}^{t_1} dt \left[\iiint_V (-\rho \underline{b} + \underline{\nabla} \cdot \underline{T} + \rho \underline{k}) \cdot \delta \underline{R} \, dV + \right.$$

$$\left. + \iint_{O_1} (\underline{f} - \underline{N} \cdot \underline{T}) \cdot \delta \underline{R} \, dO \right] = 0 \; . \tag{14}$$

Note. In our representation of the potential energy W the contact forces were assumed "dead", that is possessing a potential. This assumption may be sacrificed if we write the expression for the incremental work done by those forces in the form

$$\delta' A_{(e)} = \iint_{o_1} \underline{f} \cdot \delta \underline{R} \, do \tag{15}$$

and eliminate the corresponding terms from (3). Determining $\delta S_\times$ in the same way as for δS using (10), but with the modified expression for W, we obtain

$$\delta S_\times = \int_{t_0}^{t_1} dt \left[\iiint_v (-\rho_0 \underline{b} + \overset{0}{\underline{\nabla}} \cdot \underline{P} + \rho_0 \underline{k}) \cdot \delta \underline{R} \, dv - \iint_{o_1} \underline{n} \cdot \underline{P} \cdot \delta \underline{R} \, do \right] ,$$

so that

$$\delta S_{\times} + \int_{t_0}^{t_1} \delta' A_{(e)} dt = 0 .$$

This relation may be called the Hamilton-Ostrogradskii principle but without the adjective "variational" since now there is no functional whose stationary values are to be considered.

CHAPTER 5

THE STATE EQUATIONS FOR A NONLINEARLY ELASTIC MATERIAL

1. ON THE CHOICE OF A STATE EQUATION FOR AN ISOTROPIC ELASTIC BODY

The quadratic equation (4.3.4) of state of an elastic isotropic homogeneous material is specified either by *a priori* prescription of an explicit expression for the specific strain energy $\ni$ as a function of the invariants of the strain measure or by the representation of the coefficients $\psi_\Gamma(I_1,I_2,I_3)$ of this law compatible with the existence of $\ni$ (for a hyperelastic body). Analysis of the simplest types of deformation (e.g. uniform compression, stretching and torsion) which permits comparison with experiment provides a basis for a conclusion as to whether or not the suggested representations for $\ni$ (or ψ_Γ) are valid for the material considered.

A particular version of $\ni$ (or ψ_Γ) is chosen on the basis of how closely it approximates to the equation of state of a linearly elastic body. For example, the prescription which leads to the appearance in the equation of state of terms which are quadratic in $\underline{\underline{\nabla u}}$ is compared with the representation of $\ni$ in linear theory using a quadratic form of the deformation gradient $\underline{\underline{\nabla u}}$ which has constant coefficients. An alternative approach is based upon the retention of the terms of the same order in those state equations which still retain their invariant form. An additional criterion is associated with the comparative tractability of the subsequent mathematical analysis. Finally, in contrast to the traditional approaches of continuum mechanics, the constitutive equations are constructed using statistical considerations; the relationships obtained are further corrected and complemented by experiment.

There is no lack of critical appraisal of the above approaches. For example, "nature does not give preference to representations of functional relationships based upon power series" (Truesdell); "wherever ideas do not suffice, they are replaced by words such as simplicity or accesability" (Rivlin). *)

One may hope to solve "main unsolved problem of continuum mechanics" (Truesdell) successfully if sufficiently general requirements regarding the mathematical structure of the constitutive equation can be postulated, requirements which allow us to either decline or accept the proposed version. On this point see Pars. 9-13.

---

*) Inverse translation from the Russian. (Note by the translator.)

2. SETH'S BODY. SIGNORINI'S BODY

It seems evident that the simplest attempt to describe the behaviour of a material subjected to large deformations would be associated with the substitution in the linear Hooke's law of the linear strain tensor $\underline{\varepsilon}$ by some finite strain tensor, e.g. the Almansi strain tensor $\underline{A}$ (1.7.8) or the corresponding strain measure $\underline{g}$ (Seth, 1935)

$$\underline{T} = \lambda \underline{E} I_1(\underline{A}) + 2\mu \underline{A} = \tfrac{1}{2}(3\lambda + 2\mu)\underline{E} - \tfrac{1}{2}(\lambda I_1(\underline{g})\underline{E} + 2\mu \underline{g}) \ . \tag{1}$$

Comparison with (4.3.7) yields

$$\psi_0' = \frac{1}{4}\sqrt{\frac{G}{g}}\,[3\lambda + 2\mu - \lambda I_1(\underline{g})] \ , \qquad \psi_1' = -\tfrac{1}{4}\sqrt{\frac{G}{g}}\,\mu \ , \qquad \psi_2' = 0 \ . \tag{2}$$

The integrability conditions (4.3.6) do not hold and the "quasilinear" law (1) is no good for describing the behaviour of a hyperelastic body. However, as it has been shown by Seth, this law makes it possible to take into account certain peculiarities of a non-linear medium, e.g. the finite value of the force causing the fracture of a sample (that is, the infinite increase of one of the principla stretches), and also the necessity to apply normal forces to produce simple shear deformation. For small values of the gradient of the displacement vector, the quantitative results cannot be very different from those prescribed by the linear theory but the quasilinear law imposes no restrcitions on displacements and rotations and for this reason it allows phenomena inaccesible to the linear theory to be examined.

Signorini (Signorini, 1943-48) has proposed a quadratic law for the dependence of the stress tensor $\underline{T}$ on the Almansi strain tensor, such a law being well-suited to the structure of the constitutive equation (4.3.7)

$$\underline{T} = (m_1 I_2' + m_2 I_1'^2 + m_3 I_1' + m_4)\underline{E} - (m_5 I_1' + m_6)\underline{g} + m_7 \underline{g}^2 \ . \tag{3}$$

Here, $I_k' = I_k(\underline{g})$. The constants m_k are subject to the requirement that $\ni$ exists, in other words they must be chosen subject to (4.3.8). These conditions lead to

$$\frac{\partial^2 \ni}{\partial I_2' \partial I_3'} = -\tfrac{1}{4}(I_3')^{-3/2} m_7 = -\tfrac{1}{2}(I_3')^{-3/2} m_1 \ ,$$

$$\frac{\partial^2 \ni}{\partial I_3' \partial I_1'} = -\tfrac{1}{4}(I_3')^{-3/2}[(m_5 - m_7)I_1' + m_6] = -\tfrac{1}{2}(I_3')^{-3/2}(2m_2 I_1' + m_3)$$

the third equation is satisfied identically. We get

$$m_7 = 2m_1 , \quad m_5 = 2m_1 + 4m_2 , \quad m_6 = 2m_3 .$$

The number of coefficients has now been reduced to four

$$\underline{T} = (m_1 I_2' + m_2 I_1'^2 + m_3 I_1' + m_4)\underline{E} - [(2m_1 + 4m_2)I_1' + 2m_3]\underline{g} + 2m_1 \underline{g}^2 . \tag{4}$$

The expression for the potential energy takes the form

$$\ni = (I_3')^{-1/2}(m_1 I_2' + m_2 I_1'^2 + m_3 I_1' + m_4) + \text{const.} .$$

In the reference configuration for a hydrostatic compression ($\underline{g} = \underline{E}$, $I_1' = I_2' = 3$, $I_3' = 1$)

$$\underline{T} = -\underline{E} , \quad -p = -m_1 - 3m_2 + m_3 + m_4 \tag{5}$$

and Equations (4) and (5) yield the representations

$$\underline{T} = [-p + m_1(I_2' + 1) + m_2(I_1'^2 + 3) + m_3(I_1' - 1)]\underline{E} -$$

$$- [(2m_1 + 4m_2)I_1' + 2m_3]\underline{g} + 2m_1\underline{g}^2 , \tag{6}$$

$$\ni = (I_3')^{-1/2}[m_1(I_2' + 1) + m_2(I_1'^2 + 3) + m_3(I_1' - 1) - p] + \text{const.} . \tag{7}$$

The number of constants is reduced to three if the undistorted reference state is the natural one ($p = 0$).

Going from the strain measure to $\underline{A}$, the Almansi tensor whose invariants are given by $j_k' = I_k(\underline{A})$ and introducing the new constants

$$4m_1 = c , \quad 4m_2 = \frac{1}{2}(\lambda + \mu - \frac{c}{2}) , \quad 4m_1 + 12m_2 + 2m_3 = \mu + \frac{c}{2} \tag{8}$$

in order to retain as closely as possible the notation of the linear theory, we have from (1.7.21)

$$\underline{T} = [\lambda j_1' + c j_2' + \frac{1}{2}(\lambda + \mu - \frac{c}{2}) j_1'^2]\underline{E} + 2[\mu - (\lambda + \mu + \frac{c}{2}) j_1']\underline{A} + 2c\underline{A}^2 , \tag{9}$$

$$\ni = \sqrt{\frac{G}{g}}\,[c j_2' + \frac{1}{2}(\lambda + \mu - \frac{c}{2}) j_1'^2 + (\mu + \frac{c}{2})(1 - j_1')] - (\mu + \frac{c}{2}) . \tag{10}$$

The reference configuration (for it, $\underline{A} = \underline{0}$) is natural, and the additional constant in (10) is so chosen that for this configuration, $\ni = 0$.

Signorini's law is a general and uniquely correct representation of the relationship determining the quadratic dependence of the components of the stress tensor on those of strain for a hyperelastic isotropic material. For a "simplified" quasilinear law $c = 0$ and we have

$$\underline{T} = [\lambda j_1' + \underline{\tfrac{1}{2}(\lambda+\mu)j_1'^2}]\underline{E} + 2[\mu - \underline{(\lambda+\mu)j_1'}]\underline{A} \,. \tag{11}$$

This law is distinguished from Seth's law, (1), by the underlined terms which ensure the existence of the potential $\ni$. The expression for $\ni$ can be written for the simplified law if we return to the invariants $I_k'(\underline{g})$ of the Almansi measure and we now have

$$8\ni\sqrt{\frac{g}{G}} = 9\lambda + 5\mu - 2(3\lambda+\mu)I_1' + (\lambda+\mu)I_1'^2 - 8\mu\sqrt{I_3'} =$$

$$= \frac{4}{9}\mu[(I_1'+3)I_1' - 18\sqrt{I_3'}] + (9\lambda+5\mu)(\frac{1}{3}I_1' - 1)^2 \,. \tag{12}$$

Moreover, this yields the sufficient conditions for $\ni$ to be positive. The quantity in the square brackets is positive. In fact, the inequality

$$I_3' = g_1g_2g_3 \le \frac{1}{3^3}(g_1+g_2+g_3)^3 = \frac{I_1'^3}{27} \,,$$

may be strengthened by omitting the negative term in the right-hand side

$$\frac{I_1'^3}{27} = \frac{1}{27\cdot 12}[I_1'^2(I_1'+3)^2 - I_1'^2(I_1'-3)^2] \,.$$

Hence

$$I_3' \le \frac{1}{27\cdot 12}I_1'^2(I_1'+3)^2 \,, \qquad (I_1'+3)I_1' - 18\sqrt{I_3'} > 0 \tag{13}$$

and by (12) the required conditions are reduced to

$$\mu > 0 \,, \qquad 9\lambda + 5\mu > 0 \qquad (\lambda > -\frac{5}{9}\mu) \,. \tag{14}$$

That the first condition is necessary follows from (12) for which $I_1' = 3$.

For a linearly elastic body, the range of admissible values of parameters λ and μ is given by

$$\mu > 0 \,, \qquad 3\lambda + 2\mu > 0 \tag{15}$$

this range is wider than that for a Signorini body. This may be explained by the fact that Inequalities (15) must ensure that $\ni$ is positive only for small strains. Introducing the quantity

$$\nu = \frac{1}{2}\,\frac{\lambda}{\lambda + \mu},$$

the Poisson ratio of linear theory, we may represent Inequalities (14) and (15) for a Signorini body and a linearly elastic medium respectively as follows

$$\mu > 0\,, \quad -\frac{5}{8} < \nu < \frac{1}{2}\,; \qquad \mu > 0\,, \quad -1 < \nu < \frac{1}{2}\,. \tag{16}$$

Note. In their 1939 paper N.V. Zvolinsky and P.M. Riz proposed an equation of state which included five constants

$$\underline{T} = [\lambda' j_1' + (B + \tfrac{1}{2}\lambda') j_1'^2 - (C + 3\lambda') j_2'] \underline{E} +$$

$$+ [2\mu' + (C + \lambda' - 2\mu') j_1'] \underline{A} + (D + 5\mu')\, \underline{A}^2\,.$$

This equation may be made to agree with that of Signorini, (9), if the parameters B, C, D are expressed in terms λ, μ and c by the equalities

$$B = \frac{1}{2}\left(\mu - \frac{c}{2}\right), \qquad C = -c - 3\lambda\,, \qquad D = 2c - 5\mu\,;$$

$$\lambda' = \lambda\,, \qquad \mu' = \mu\,.$$

3. MURNAGHAN'S MATERIAL

A representation of the specific strain energy in the form of a polynomial of the Cauchy-Green strain tensor whose constant coefficients were determined experimentally was used in numerous papers by Murnaghan (F.D. Murnaghan), the contents of which were reproduced in his monograph of 1951.

We start with the prescription of the specific stored energy Э which guarantees the appearance in the state equation of all terms quadratic with respect to the displacement gradient $\underline{\nabla u}$. Restricting the analysis to isotropic materials, the Э-function must be specified by (I.15.16), (I.15.14) which are scalar structures of the second and third order respectively. Addition of the linear terms yields

$$Э = -pI_1(\underline{C}) + \frac{1}{2}[\lambda I_1^2(\underline{C}) + 2\mu I_1(\underline{C}^2)] + \frac{1}{6}[\nu_1 I_1^3(\underline{C}) +$$

$$+ 6\nu_2 I_1(\underline{C}) I_1(\underline{C}^2) + 8\nu_3 I_1(\underline{C}^3)] = -pI_1(\underline{C}) + \frac{1}{2}(\lambda + 2\mu) I_1^2(\underline{C}) -$$

$$- 2\mu I_2(\underline{C}) + \frac{1}{3}(l + 2m) I_1^3(\underline{C}) - 2m I_1(\underline{C}) I_2(\underline{C}) + n I_3(\underline{C}) \tag{1}$$

which contains six constants. Here λ, μ and ν_1, ν_2, ν_3 are Lamé parameters of the second and the third order respectively and l, m and n are the Murnaghan constants connected with ν_1, ν_2 and ν_3 by Equations (I.15.15).

Experiments associated with the problems of nonlinear acoustics have provided a considerable amount of data on the third order Lamé parameters of a number of materials. Some data on these parameters have been obtained on the basis of static tests carried out by Bridgman at very high pressures (P.W. Bridgman, 1948).

On substituting $I_k(\underline{C})$ by the invariants $I_k(\underline{G}) = I_k$ of the Cauchy-Green strain measure formula (1) is transformed to

$$\begin{aligned}\ni = \frac{1}{4}\Big[& (-3\lambda - 2\mu + \frac{9}{2}l + \frac{n}{2})I_1(\underline{G}) + \frac{1}{2}(\lambda + 2\mu - 3l - 2m)I_1^2(\underline{G}) + \\ & + (-2\mu + 3m - \frac{n}{2})I_2(\underline{G}) - mI_1(\underline{G})I_2(\underline{G}) + \frac{1}{6}(l + 2m)I_1^3(\underline{G}) + \\ & + \frac{n}{2}(I_3(\underline{G}) - 1) + \frac{9}{2}\lambda + 3\mu - \frac{9}{2}l\Big] \end{aligned} \tag{2}$$

(note that $p = 0$ if the initial state is the natural one).

We have by (4.3.5)

$$\begin{aligned}\psi_0 &= \frac{1}{8}nI_3 , \\ \psi_1 &= \frac{1}{4}\Big[(I_1 - 3)\lambda - 2\mu + \frac{1}{2}(I_1 - 3)^2 + m(I_1 - I_2) - \frac{n}{2}(I_1 - 1)\Big] , \\ \psi_2 &= \frac{1}{4}\Big[2\mu + m(I_1 - 3) + \frac{n}{2}\Big] . \end{aligned} \tag{3}$$

It is easy to verify that Equations (4.7.12), (4.7.13) hold.

In subsequent formulas, we will confine ourselves to terms which are quadratic with respect to the derivatives of the displacement vector. We have

$$\begin{aligned}\underline{F} &= \underline{E} + 2\overset{0}{\underline{\varepsilon}}(\underline{u}) + \overset{0}{\underline{\nabla u}}^T \cdot \overset{0}{\underline{\nabla u}} , \\ \underline{F}^2 &= \underline{E} + 4\overset{0}{\underline{\varepsilon}}(\underline{u}) + 4\overset{0}{\underline{\varepsilon}}{}^2(\underline{u}) + 2\overset{0}{\underline{\nabla u}}^T \cdot \overset{0}{\underline{\nabla u}} , \end{aligned} \tag{4}$$

$$\begin{aligned} I_1(\underline{F}) &= 3 + 2\vartheta + \chi , \\ I_2(\underline{F}) &= 3 + 4\vartheta + 2\chi + 2\kappa , \\ I_3(\underline{F}) &= 1 + 2\vartheta + \chi + 2\kappa \end{aligned} \tag{5}$$

here we used the notation

$$\vartheta = I_1(\overset{0}{\underline{\varepsilon}}(\underline{u})) = \overset{0}{\underline{\nabla}} \cdot \underline{u} \,, \qquad \chi = I_1(\overset{0}{\underline{\underline{\nabla u}}}^T \cdot \overset{0}{\underline{\underline{\nabla u}}}) = 2\overset{0}{\underline{\omega}} \cdot \overset{0}{\underline{\omega}} + I_1(\overset{0}{\underline{\varepsilon}}^2) \,,$$

$$\kappa = 2I_2(\overset{0}{\underline{\varepsilon}}) = \vartheta^2 - I_1(\overset{0}{\underline{\varepsilon}}(\underline{u})^2) \,,$$

$$I_3^{-1/2}(\underline{F}) = 1 - \vartheta - \frac{1}{2}\chi - \kappa + \frac{3}{2}\vartheta^2 \,, \tag{6}$$

$$\overset{0}{\underline{\varepsilon}}(\underline{u}) = \frac{1}{2}(\overset{0}{\underline{\underline{\nabla u}}} + \overset{0}{\underline{\underline{\nabla u}}}^T) \,,$$

by $\overset{0}{\underline{\omega}}(\underline{u})$ we denoted the vector associated with the skew part of $\overset{0}{\underline{\underline{\nabla u}}}$. To this approximation,

$$4\psi_0 = \frac{1}{2}n + \frac{1}{2}n(2\vartheta + \chi + 2\kappa) \,,$$

$$4\psi_1 = -2\mu - n + 2\vartheta(\lambda - m - \frac{n}{2}) + \chi(\lambda - m - \frac{n}{2}) - 2m\kappa + 2l\vartheta^2 \,, \tag{7}$$

$$4\psi_2 = 2\mu + 2m\vartheta + m\chi + \frac{1}{2}n$$

and the expression for the stress tensor may be written as

$$\underline{T} = (1 - \vartheta)\underline{T}^0 + 2\overset{0}{\underline{\varepsilon}} \cdot \underline{T}^0 + \underline{T}' \,. \tag{8}$$

Here, $\underline{T}^0$ denotes the stress tensor of linear theory

$$\underline{T}^0 = \lambda\vartheta\underline{E} + 2\mu\overset{0}{\underline{\varepsilon}}(\underline{u}) \,, \tag{9}$$

and the symbol $\underline{T}'$ is related to the group of terms

$$\underline{T}' = \frac{1}{2}\left[\lambda\chi + l\vartheta^2 - (m - \frac{n}{2})\kappa\right]\underline{E} + 2(m - \frac{n}{2})\overset{0}{\underline{\varepsilon}}\vartheta + n\overset{0}{\underline{\varepsilon}}^2 + \mu\overset{0}{\underline{\underline{\nabla u}}}^T \cdot \overset{0}{\underline{\underline{\nabla u}}} \,, \tag{10}$$

where $\overset{0}{\underline{\underline{\nabla u}}}^T \cdot \overset{0}{\underline{\underline{\nabla u}}}$ is given by (1.7.13). It is natural that the Murnaghan coefficients enter only the representation for $\underline{T}'$. At the same time, if we confine ourselves to terms quadratic with respect to the components of $\underline{C}$ in the expression (1) for the specific stored energy, in other words, if we assume that l, m, n equal zero, we would arrive, just as for a "Seth body", at a representation of $\underline{T}$ which differs from the expression of linear theory in that tensor $\underline{C}$ is used instead of $\overset{0}{\underline{\varepsilon}}$.

The values of the Murnaghan coefficients (λ, μ) and (ν_1, ν_2, ν_3) of the second and third order

$$l = \tfrac{1}{2}\nu_1 + \nu_2 , \qquad m = \nu_2 + 2\nu_3 , \qquad n = 4\nu_3$$

are given in the Tables 1-3 for various materials. These data have been communicated to the author by S.S. Sekoyan.

In the remainder of this section we shall examine the connection between the Murnaghan coefficients and the increments in the moduli λ, μ of the linear law which results from the superposition of a small perturbation on the natural reference configuration.

We therefore assume that the particles of the medium in the equilibrium state in their actual configuration are given a set of virtual displacements $\eta\underline{w}(q^1,q^2,q^3)$. Then by (8) we get

$$\dot{\underline{T}} = (1-\vartheta)\dot{\underline{T}}^0 - \dot{\vartheta}\underline{T}^0 + 2\overset{0}{\underline{\varepsilon}}{}^{\cdot} \cdot \underline{T}^0 + \dot{\underline{T}}' + 2\overset{0}{\underline{\varepsilon}} \cdot \dot{\underline{T}}^0 \tag{11}$$

and a calculation carried out according to the formulas of Chapter 1, Par. 10 shows that

$$\dot{\underline{T}}^0 = \lambda\overset{0}{\nabla} \cdot \underline{w} + 2\overset{0}{\underline{\varepsilon}}(\underline{w}) , \qquad \dot{\vartheta} = \overset{0}{\nabla} \cdot \underline{w} , \qquad \overset{0}{\underline{\varepsilon}}{}^{\cdot} = \overset{0}{\underline{\varepsilon}}(\underline{w}) ,$$

$$\dot{\chi} = \overset{0}{\nabla}\underline{u}^{T} \cdot\cdot \overset{0}{\nabla}\underline{w} + \overset{0}{\nabla}\underline{w}^{T} \cdot\cdot \overset{0}{\nabla}\underline{u} = 2I_1(\overset{0}{\nabla}\underline{u} \cdot \overset{0}{\nabla}\underline{w}^{T}) ,$$

$$(\vartheta^2)^{\cdot} = 2\vartheta\overset{0}{\nabla} \cdot \underline{w} , \qquad [I_1(\overset{0}{\underline{\varepsilon}}{}^2)]^{\cdot} = 2I_1(\overset{0}{\underline{\varepsilon}} \cdot \overset{0}{\underline{\varepsilon}}(w))$$

etc. Here we have put $\overset{0}{\underline{\varepsilon}} = \overset{0}{\underline{\varepsilon}}(\underline{u})$ and $\vartheta = \overset{0}{\nabla} \cdot \underline{u}$ instead of $\overset{0}{\underline{\varepsilon}}(\underline{w})$ and $\overset{0}{\nabla} \cdot \underline{w} = I_1(\overset{0}{\nabla}\underline{w})$.

Substitution into (8) and (10) yields

$$\begin{aligned}
\dot{\underline{T}} = {} & (1-\vartheta)\dot{\underline{T}}^0 - \underline{T}^0\overset{0}{\nabla} \cdot \underline{w} + 2\overset{0}{\underline{\varepsilon}}(\underline{w}) \cdot \underline{T}^0 + 2\overset{0}{\underline{\varepsilon}} \cdot \dot{\underline{T}}^0 + \\
& + \underline{E}\left\{ \lambda I_1(\overset{0}{\nabla}\underline{u} \cdot \overset{0}{\nabla}\underline{w}) + 2l\vartheta\overset{0}{\nabla} \cdot \underline{w} - (2m-n)\left[\vartheta\overset{0}{\nabla} \cdot \underline{w} - I_1(\overset{0}{\underline{\varepsilon}} \cdot \overset{0}{\underline{\varepsilon}}(\underline{w}) \right]\right\} + \\
& + (2m-n)\,[\overset{0}{\underline{\varepsilon}}\overset{0}{\nabla} \cdot \underline{w} + \vartheta\overset{0}{\underline{\varepsilon}}(\underline{w})] + n(\overset{0}{\underline{\varepsilon}} \cdot \overset{0}{\underline{\varepsilon}}(\underline{w}) + \overset{0}{\underline{\varepsilon}}(\underline{w}) \cdot \overset{0}{\underline{\varepsilon}}) + \\
& + \mu(\overset{0}{\nabla}\underline{u}^{T} \cdot \overset{0}{\nabla}\underline{w} + \overset{0}{\nabla}\underline{w}^{T} \cdot \overset{0}{\nabla}\underline{u}) \; .
\end{aligned} \tag{12}$$

We shall confine ourselves to the simplest case in which the actual configuration is a similarity transformation of the natural or refernce configuration. Then by (6) and (9) we have

$$\underline{R} = \underline{r}(1+\delta) , \qquad \underline{u} = \underline{r}\delta , \qquad \overset{0}{\nabla}\underline{u} = \underline{E}\delta , \qquad \vartheta = 3\delta ,$$

Table 1. Steels

Material	Reference	Density g/cm^3	In units of 10^{12} dyne/cm^2					Remarks
			λ	μ	ν_1	ν_2	ν_3	
Steel Rex 535	1	---	0,909	0,780	3,58	-2,65	-1,82	Nickel-bearing
Steel Rex 535	2	7,065	1,09± ±0,01	0,818± ±0,005	-1,75± ±0,50	-2,40± ±0,50	-1,69± ±0,15	---
Steel Hecla 37	2	7,823	1,11± ±0,01	0,821± ±0,005	-3,58± ±0,70	-2,82± ±0,30	-1,77± ±0,08	0.4% C; 0.3% Si; 0.8% Mn
Steel Hecla 17	2	7,825	1,10± ±0,01	0,820± ±0,005	-1,34± ±0,20	-2,61± ±0,20	-1,67± ±0,06	0.6% C; 0.2% Si; 0.8% Mn
Steel Hecla 138A	2	7,843	1,09± ±0,01	0,819± ±0,005	-3,23± ±0,50	-2,65± ±0,30	-1,77± ±0,10	0.4% C; 0.4% Cr; 0.5% Mo; 2.5% Ni
Steel Hecla ATV	2	8,065	0,87± ±0,02	0,716± ±0,03	0,34± ±0,20	-5,52± ±0,80	-1,00	3.6% Ni; 10% Cr; 1% Mn
Steel 3	3	---	---	---	-1,6	-2,8	-2,05	Annealed specimen
Steel 35XΓCA	4	7,748	1,10± ±0,002	0,804± ±0,002	-1,9± ±0,15	-2,3± ±0,1	-2,01± ±0,04	Hardening 880°C; tempering 350°C. HRC = 46
Steel 50XΓCM2Φ	4	7,785	1,129± ±0,002	0,803± ±0,002	-2,8± ±0,2	-2,7± ±0,1	-1,87± ±0,04	Hardening 1020°C; tempering 350-400°C for 1 hour; HRC = 50

Notation: 50-0.5% of carbon, X - 1% of chromium, -1% of manganese, C - 1% of silicon, M - 1% of molgbdenium, 20 - 2% of vanadium; A - improvable steel (subjected to any processing), HRC - Rockwell hardness.

Table 2. Aluminium and Copper Alloys

Material	Reference	Density g/cm^3	In units of 10^{12} $dyne/cm^2$					Remarks
			λ	μ	ν_1	ν_2	ν_3	
B 53 S	2	2,677	0,580± ±0,002	0,260± ±0,001	-2,49± ±0,60	-0,99± ±0,10	-0,60± ±0,05	2.8% Mg; 0.8% Mn; 0.1% Cr
D 54 S	2	2,719	0,491± ±0,002	0,260± ±0,001	-3,79± ±0,025	-1,98± ±0,09	-0,80± ±0,03	4.5% Mg; 0.8% Mn; 1% Cr
D 16	3	---	---	---	-1,3	-2,1	-0,92	
Brass Л-62	3	---	---	---	-3,0	-2,1	-1,25	Annealed specimen
Brass ЛС-59-1	3	---	---	---	4,8	1,7	-1,22	Annealed specimen
Phosphor bronze БрОФ	3	---	---	---	12,6	-1,5	-0,60	
Phosphor bronze БрОФ	3	---	---	---	12,1	-2,4	-0,42	Annealed specimen
БрОФ 65-0,15	4	8,921	1,133± ±0,002	0,417± ±0,002	-4,9± ±0,4	-1,67± ±0,08	-0,500± ±0,010	Annealing at 350°C, 1 hour
Beryllium bronze Бр 52	4	8,317	1,042± ±0,002	0,490± ±0,002	-4,0± ±0,3	-1,7± ±0,08	-0,600± ±0,010	Hardening 770-780°C, tempering 350°C, 1 hour. HRC = 36

Table 3. Miscellaneous Materials

Material	Reference	Density g/cm^3	In units of 10^{12} dyne/cm^2					Remarks
			λ	μ	ν_1	ν_2	ν_3	
Copper	5	---	1,07	0,477	-5,6± ±1,4	1,72± ±0,14	-3,98± ±0,05	
Magnesium	2	1,716	0,259± ±0,002	0,166± ±0,03	-0,654± ±0,01	-0,574± ±0,01	-0,421± ±0,04	96% Mg, 3%Al, 1% Zn
ARMCO iron	8	---	1,100± ±0,004	0,820± ±0,010	2,6± ±13,7	-4,8± ±6,2	-2,8± ±2,8	
Molybdenium	2	9,71	1,57± ±0,02	1,10± ±0,01	-0,51± ±0,50	-2,83± ±0,30	-0,93± ±0,04	Sintered from powder of 99,99% purity
		10,2	1,78	1,24	1,94	-3,98	-2,27	Recrystallized
Tungsten	2	14,16	0,750± 0,005	0,730± ±0,01	-2,15± ±0,30	-1,43± ±0,15	-1,24± ±0,02	Sintered from powder of 99,99% purity
		17,8	1,63	1,37	-4,29	-2,58	-2,67	Recrystallized
Acrylic plastic	6	1,160	0,390	0,186	-0,078	-0,070	0,047	
Fused quartz	7	2,203	0,15872	0,31261	0,54± ±0,13	0,93± ±0,08	-0,11± ±0,03	
Glass (pyrex)	8	---	0,1353± ±0,0003	0,275± ±0,003	2,6± ±3,7	-1,68± ±1,38	1,05± ±0,88	

$$\underline{T}^0 = (3\lambda + 2\mu)\underline{E}\delta\ , \qquad \chi = 3\delta^2\ , \qquad \kappa = 6\delta^2$$

and by (8) and (10)

$$\underline{T} = (3\lambda + 2\mu)\delta\underline{E} - \delta^2[\tfrac{1}{2}(3\lambda + 2\mu) - (9l + n)]\underline{E}\ . \tag{13}$$

When the set of displacements $\eta\underline{w}$ is superposed on this equilibrium state the convected derivative $\dot{\underline{T}}$ will, by (12), be represented as

$$\dot{\underline{T}} = \dot{\underline{T}}^0 - \underline{E}\delta\overset{0}{\underline{\nabla}}\cdot\underline{w}(3\lambda + 2\mu - 6l + 2m - n) + 2\overset{0}{\underline{\varepsilon}}(\underline{w})\delta(3\lambda + 2\mu + 3m - \frac{n}{2})\ . \tag{14}$$

Two special cases will now be considered.

Case (a) in which it is assumed that

$$\begin{aligned} &\underline{w} = \frac{1}{3}\underline{r}(1+\delta)\ , \qquad \overset{0}{\underline{\varepsilon}}(\underline{w}) = \frac{1}{3}(1+\delta)\underline{E}\ , \\ &\overset{0}{\underline{\nabla}}\cdot\underline{w} = 1 + \delta\ , \qquad \dot{\underline{T}}^0 = k(1+\delta)\underline{E}\ , \end{aligned} \tag{15}$$

where $k = \lambda + \frac{2}{3}\mu$ denotes the bulk modulus of linear theory. Substitution of (15) into (14) results in the relationship

$$\dot{\underline{T}} = -\dot{p}\underline{E}\ , \qquad \dot{p} = -\left[(1-\delta^2)k + 6\delta(1+\delta)(l + \frac{n}{9})\right] = -k_x(\delta)\ . \tag{16}$$

Here $\dot{p}$ denotes the pressure which is applied to produce a displacement $\underline{w}$ in a rod which is initially either compressed ($\delta < 0$) or stretched ($\delta > 0$). This yields

$$k_x(0) = k\ , \qquad k_x'(0) = 6(l + \frac{n}{9})\ . \tag{17}$$

This relationship shows that it is possible, in principle, to determine the value of $l + n/9$ from the slope of the tangent to the curve $k_x(\delta)$ in the reference configuration at the point $\delta = 0$. Most of the experimental data show that $k_x'(0) < 0$, i.e. that resistance of the material to change in density increases with increase in density.

For example, experiments with sodium carried out by Bridgman at pressures up to 100,000 atm led (Murnaghan) to the numerical values

$$k = 0.628\cdot 10^4\ \text{atm}\ , \qquad n + 9l = -406.2\cdot 10^4\ \text{atm}\ .$$

It can be seen that $|n + 9l| \gg k$, and under such loads non-linear terms must be taken into account.

Case (b) in which it is assumed that the initially compressed medium ($\delta < 0$) is subjected to simple shear in the plane OXY parallel to the axis OX. Letting s be the shear parameter we have

$$\underline{w} = \underline{i}_1(1+\delta)sa^2 , \quad \overset{0}{\underline{\nabla}}\underline{w} = \underline{i}_2\underline{i}_1(1+\delta)s , \quad \overset{0}{\underline{\nabla}}\cdot\underline{w} = 0 ,$$

$$\overset{0}{\underline{\varepsilon}}(\underline{w}) = \tfrac{1}{2}(\underline{i}_1\underline{i}_2+\underline{i}_2\underline{i}_1)(1+\delta)s , \qquad \dot{\underline{T}}^0 = \mu(\underline{i}_1\underline{i}_2+\underline{i}_2\underline{i}_1)(1+\delta)s . \tag{18}$$

By (14) we get

$$\dot{\underline{T}} = (\underline{i}_1\underline{i}_2+\underline{i}_2\underline{i}_1)\dot{\tau}_{12} ,$$

$$\dot{\tau}_{12} = \left[(1+\delta)\mu + (1+\delta)\delta(3k+3m-\frac{n}{2})\right]s = \mu_\times(\delta) , \tag{19}$$

so that

$$\mu_\times(0) = \mu , \qquad \mu'_\times(0) = 3(\lambda+\mu) + 3m - \frac{n}{2} . \tag{20}$$

Here $\dot{\tau}_{12}$ denotes the tangential stress producing the shear in the initially compressed material.

The values $k'_\times(0)$ and $\mu'_\times(0)$ which determine the changes in the bulk and shear moduli from those of the natural reference configuration are related in Equations (17) and (20) to the values, δ, of the relative elongations. Relating these values to the increase in volume

$$k_\times(\delta) = k_{\times\times}(3\delta) , \qquad \mu_\times(\delta) = \mu_{\times\times}(3\delta) ,$$

we obtain

$$k'_{\times\times}(0) = 2(l+\frac{n}{9}) , \qquad \mu'_{\times\times}(0) = \lambda + \mu + m - \frac{n}{6} . \tag{21}$$

The relationships (17), (20) and (21) are due to Wang (Wang, 1965).

4. THE TRANSFORMATION OF THE COEFFICIENTS OF THE STATE EQUATIONS OF SIGNORINI AND MURNAGHAN

When the reference configurations v and $\hat{v}$ are connected by the similarity transformation (4.4.3), the representations (2.7) may be written

$$\ni = (I'_3)^{-1/2}[m_1(I'_2+1) + m_2(I'^2_1+3) + m_3(I'_1-1) - p] , \tag{1}$$

$$\hat{\ni} = (\hat{I}'_3)^{-1/2}[\hat{m}_1(\hat{I}'_2 + 1) + \hat{m}_2(\hat{I}'^2_1 + 3) + \hat{m}_3(\hat{I}'_1 - 1) - \hat{p}] \ . \tag{2}$$

Here, $I'_k = I_k(\underline{g})$, and the invariants I'_k, $\hat{I}'_k$ are connected by the relationships

$$\hat{I}'_1 = K^2 I'_1 \ , \qquad \hat{I}'_2 = K^4 I'_2 \ , \qquad \hat{I}'_3 = K^6 I'_3 \tag{3}$$

different from Equations (4.4.5) related to the invariants $I_k(\underline{F})$ (here, the similarity coefficient α of (4.4) has been replaced by K).

From (4.4.6), (1) and (2) we obtain

$$\begin{aligned} \ni &= K^3\hat{\ni}, m_1(I'_2 + 1) + m_2(I'^2_1 + 3) + m_3(I'_1 - 1) - p = \\ &= \hat{m}_1(K^4 I'_2 + 1) + \hat{m}_2(K^4 I'^2_1 + 3) + \hat{m}_3(K^2 I'_1 - 1) - \hat{p} \end{aligned} \tag{4}$$

and this relationship is satisfied by the values $\hat{m}_k$ and $\hat{p}$ given by the formulas

$$\hat{m}_1 = K^{-4} m_1 \ , \qquad \hat{m}_2 = K^{-4} m_2 \ , \qquad \hat{m}_3 = K^{-2} m_3 \ ,$$

$$\hat{p} = p + \frac{1 - K^2}{K^4}[(m_1 + 3m_2)(1 + K^2) - m_3 K^2] \ .$$

Making use of the notation of (2.8) these equations reduce to the form

$$\left.\begin{aligned} \hat{\lambda} &= K^{-4}\lambda - \tfrac{1}{2}K^{-4}(3\lambda + \mu - \tfrac{c}{2})(1 - K^2) \ , \\ \hat{\mu} &= K^{-4}\mu + \tfrac{1}{2}K^{-4}(3\lambda + \mu - \tfrac{c}{2})(1 - K^2) \ , \qquad \hat{c} = K^{-4}c \ , \end{aligned}\right\} \tag{5}$$

$$\hat{p} = \frac{1}{8}\,\frac{1 - K^2}{K^4}\left[(1 + K^2)(\frac{c}{2} + 3\lambda + 3\mu) + K^2(6\lambda + 2\mu - c)\right] \ . \tag{6}$$

We have assumed that $p = 0$, the v-configuration is the natural one and that $\hat{p}$ is a uniform tension (compression for $\hat{p} < 0$) which transforms the v-configuration into that of $\hat{v}$.

Similarly for a Murnaghan material the application of Equations (3.2) and (4.4.5) leads to the Brillouin formulas (L. Brillouin, 1938)

$$\hat{n} = K^3 n \ , \qquad \hat{m} = K^3 m \ , \qquad \hat{l} = K^3 l \ ,$$

$$2\hat{\mu} = K[2\mu + (1 - K^2)(\frac{n}{2} - 3m)] \ ,$$

$$\hat{\lambda} + 2\hat{\mu} = K[\lambda + 2\mu + (1 - K^2)(2m + 3l)] \ , \tag{7}$$

$$\hat{p} = \frac{1-K^2}{K}\left[-\frac{3}{2}k + \frac{1}{4}(1-K^2)(n+9l)\right] \qquad (k = \lambda + \frac{2}{3}\mu) \ . \tag{8}$$

Equation (8) determines the state of uniform tension (compression for $\hat{p} < 0$) in the $\hat{v}$-configuration. If powers of $\delta = K - 1$ higher than the second are neglected, this formula agrees well with (3.13). Denoting the relative volume compression by θ

$$K^3 = \frac{\hat{v}}{v} = 1 - \theta \ , \qquad \theta = -\frac{\hat{v}-v}{v} \ , \tag{9}$$

and we have by (8)

$$-\hat{p} = (1-\theta)^{-1/3}\left\{[1 - (1-\theta)^{2/3}]\,\frac{3}{2}k - \frac{1}{4}[1 - (1-\theta)^{2/3}]\,(n+9l)\right\} =$$

$$= k\theta + \frac{1}{2}(k - l - \frac{n}{9})\theta^2 \ , \tag{10}$$

where the powers higher than the second are neglected. For metallic materials under uniform pressures not exceeding 10^4 atmospheres, the linear dependence

$$-\hat{p} = k\theta \tag{11}$$

still holds.

The temperature coefficient of linear expansion $\alpha(\vartheta)$ is given by the equation

$$d\hat{l} = \hat{l}\alpha(\vartheta)d\vartheta$$

and under the similarity transformation from isothermal state v to isothermal state $\hat{v}$ we have

$$\hat{l} = Kl \ , \qquad d\hat{l} = l dK = \hat{l}\,\frac{dK}{K} \ , \qquad \frac{d\hat{l}}{\hat{l}} = \frac{dK}{K} = \alpha(\vartheta)d\vartheta \ ,$$

so that

$$K = \exp\int_0^{\hat{\vartheta}} \alpha(\vartheta)d\vartheta \ , \tag{12}$$

where $\hat{\vartheta}$ is the temperature difference between the isothermal states v and $\hat{v}$. This formula together with (5) and (7) determines the temperature dependence of the coefficients to Signorini and Murnaghan materials.

For a weak temperature dependence and within a narrow range of temperatures we have

$$K = 1 + \alpha\hat{\vartheta} \ . \tag{13}$$

5. A QUASI-LINEAR JOHN MATERIAL

In linear elasticity theory the specific strain energy, $\ni$, is known to be expressed as a quadratic form of the principal values of a linear strain tensor

$$\ni = \tfrac{1}{2}[\lambda I_1^2(\underline{\varepsilon}) + 2\mu I_1(\underline{\varepsilon}^2)] = \tfrac{1}{2}[\lambda(\varepsilon_1+\varepsilon_2+\varepsilon_3)^2 + 2\mu(\varepsilon_1^2+\varepsilon_2^2+\varepsilon_3^2)] \ . \tag{1}$$

Following a similar representation of $\ni$ through the principal stretches δ_k determined by the formulas (1.4.10) of the theory of finite deformations

$$\ni = \tfrac{1}{2}[\lambda(\delta_1+\delta_2+\delta_3)^2 + 2\mu(\delta_1^2+\delta_2^2+\delta_3^2)] \ , \tag{2}$$

John has considered several problems concerning the propagation of waves in a non-linear elastic medium. From (1.6.4) and (1.4.10) the principal values, δ_k, of the tensor $\underline{U} - \underline{E}$ are

$$\delta_1 + \delta_2 + \delta_3 = s_1 = I_1(\underline{U}-\underline{E}) \ , \qquad \delta_1^2 + \delta_2^2 + \delta_3^2 = s_2 = I_1((\underline{U}-\underline{E})^2) \tag{3}$$

and (2) reduces to

$$\ni = \tfrac{1}{2}\lambda s_1^2 + \mu s_2 = \tfrac{1}{2}\lambda I_1^2(\underline{U}-\underline{E}) + \mu I_1((\underline{U}-\underline{E})^2) \ . \tag{4}$$

John called this material "harmonic" because with this definition of $\ni$ the planar problem is reduced to a non-linear boundary-value problem in the theory of harmonic functions (or functions of the complex variable). Such a material may be appropriately termed "quasi-linear material".

Referring to Equations (II.3.10), (II.4.11), (1.9.6) and (1.6.1) we get

$$I_1(\underline{U}-\underline{E})_{\overset{0}{\underline{\underline{\nabla R}}}} = I_1(\underline{U})_{\underline{G}} \cdot\cdot \underline{G}_{\overset{0}{\underline{\underline{\nabla R}}}} = \tfrac{1}{2}\underline{U}^{-1} \cdot\cdot (\underline{C}_{II} + \underline{C}_{III}) \cdot \overset{0}{\underline{\underline{\nabla R}}} = \underline{U}^{-1} \cdot \overset{0}{\underline{\underline{\nabla R}}} = \underline{O}^{\times} \ ,$$

$$I_1((\underline{U}-\underline{E})^2)_{\overset{0}{\underline{\underline{\nabla R}}}} = I_1(\underline{G})_{\overset{0}{\underline{\underline{\nabla R}}}} - 2I_1(\underline{U})_{\overset{0}{\underline{\underline{\nabla R}}}} = 2\overset{0}{\underline{\underline{\nabla R}}} - 2\underline{O}^{\times}$$

and the state equations for the Piola and Cauchy tensors take the form

$$\underline{P} = \ni_{\overset{0}{\underline{\underline{\nabla R}}}} = (\lambda s_1 - 2\mu)\underline{O}^{\times} + 2\mu\overset{0}{\underline{\underline{\nabla R}}} = [(\lambda s_1 - 2\mu)\underline{U}^{-1} + 2\mu\underline{E}] \cdot \overset{0}{\underline{\underline{\nabla R}}} \ ,$$

$$\underline{T} = \sqrt{\frac{g}{G}}\, \underline{\underline{\nabla R}}^{T} \cdot \underline{P} = \sqrt{\frac{g}{G}}\, [(\lambda s_1 - 2\mu)\underline{V} + 2\mu\underline{F}] =$$

$$= \sqrt{\frac{g}{G}}\, [\lambda(I_1(\underline{V}) - 3)\underline{E} + 2\mu(\underline{V}-\underline{E})] \cdot \underline{V} \ . \tag{5}$$

The specific strain energy $\ni$ of a quasi-linear material can be given in terms of the invariants of the symmetric tensor $(\underline{P}\cdot\underline{P}^{T})^{1/2}$. In fact, we have

$$\underline{P}\cdot\underline{P}^{T} = [(\lambda s_1 - 2\mu)\underline{O}^{\times} + 2\mu\overset{0}{\underline{\underline{\nabla R}}}]\cdot[(\lambda s_1 - 2\mu)\underline{O}^{\times T} + 2\mu\overset{0}{\underline{\underline{\nabla R}}}{}^{T}] =$$

$$= (\lambda s_1 - 2\mu)^2\underline{E} + 2\mu(\lambda s_1 - 2\mu)(\underline{O}^{\times}\cdot\overset{0}{\underline{\underline{\nabla R}}}{}^{T} + \overset{0}{\underline{\underline{\nabla R}}}\cdot\underline{O}^{\times T}) + 4\mu^2\underline{U}^2 =$$

$$= (\lambda s_1 - 2\mu)^2\underline{E} + 4\mu(\lambda s_1 - 2\mu)\underline{U} + 4\mu^2\underline{U}^2$$

or

$$\underline{P}\cdot\underline{P}^{T} = [(\lambda s_1 - 2\mu)\underline{E} + 2\mu\underline{U}]^2 , \qquad (\underline{P}\cdot\underline{P}^{T})^{1/2} = (\lambda s_1 - 2\mu)\underline{E} + 2\mu\underline{U} . \tag{6}$$

The first invariants of these tensors are equal to

$$I_1((\underline{P}\cdot\underline{P}^{T})^{1/2}) = I_1(\lambda s_1\underline{E} + 2\mu(\underline{U} - \underline{E})) = (3\lambda + 2\mu)s_1 ,$$

$$I_1(\underline{P}\cdot\underline{P}^{T}) = 3\lambda^2 s_1^2 + 4\mu\lambda s_1 I_1(\underline{U} - \underline{E}) + 4\mu^2 I_1((\underline{U} - \underline{E})^2) = \lambda(3\lambda + 4\mu)s_1^2 + 4\mu^2 s_2 . \tag{7}$$

From these we have

$$s_1 = \frac{I_1((\underline{P}\cdot\underline{P}^{T})^{1/2})}{3\lambda + 2\mu} , \qquad s_2 = \frac{1}{4\mu^2}\left[I_1(\underline{P}\cdot\underline{P}^{T}) - \frac{\lambda(3\lambda + 4\mu)}{(3\lambda + 2\mu)^2} I_1^2((\underline{P}\cdot\underline{P}^{T})^{1/2})\right] \tag{8}$$

and further by (4)

$$\ni = \frac{1}{4\mu}\left[I_1(\underline{P}\cdot\underline{P}^{T}) - \frac{\nu}{1 + \nu} I_1^2((\underline{P}\cdot\underline{P}^{T})^{1/2})\right] , \qquad \nu = \frac{\lambda}{2(\lambda + \mu)} . \tag{9}$$

We also have from (5)

$$\underline{P}\cdot\cdot\overset{0}{\underline{\underline{\nabla R}}}{}^{T} = [(\lambda s_1 - 2\mu)\underline{U}^{-1} + 2\mu\underline{E}]\cdot\overset{0}{\underline{\underline{\nabla R}}}\cdot\cdot\overset{0}{\underline{\underline{\nabla R}}}{}^{T} = I_1((\lambda s_1 - 2\mu)\underline{U} + 2\mu\underline{U}^2) =$$

$$= I_1((\lambda s_1 + 2\mu)(\underline{U} - \underline{E}) + 2\mu(\underline{U} - \underline{E})^2 + \lambda s_1\underline{E}) ,$$

so that by (4) and (8)

$$\underline{P}\cdot\cdot\overset{0}{\underline{\underline{\nabla R}}}{}^{T} = (\lambda s_1 + 2\mu)s_1 + 2\mu s_2 + 3\lambda s_1 = \lambda s_1^2 + 2\mu s_2 + (3\lambda + 2\mu)s_1 =$$

$$= 2\ni + I_1((\underline{P}\cdot\underline{P}^{T})^{1/2}) . \tag{10}$$

Now by virtue of (4.17.3) the specific complementary energy $\ni_\times$ is also seen to be expressed in terms of the invariants of the tensor $(\underline{P}\cdot\underline{P}^T)^{1/2}$

$$\ni_\times = \ni + I_1((\underline{P}\cdot\underline{P}^T)^{1/2}) = \frac{1}{4\mu}\left[I_1(\underline{P}\cdot\underline{P}^T) - \frac{\nu}{1+\nu} I_1^2((\underline{P}\cdot\underline{P}^T)^{1/2})\right] + $$

$$+ I_1((\underline{P}\cdot\underline{P}^T)^{1/2}) \ . \tag{11}$$

The relationship (4.17.5) enables us to determine the deformation gradient $\overset{0}{\underline{\underline{\nabla R}}}$. We have

$$\delta I_1(\underline{P}\cdot\underline{P}^T) = \delta(\underline{P}\cdot\cdot\underline{P}^T) = \underline{P}\cdot\cdot\delta\underline{P}^T + \delta\underline{P}\cdot\cdot\underline{P}^T = 2\underline{P}\cdot\cdot\delta\underline{P}^T \ ,$$

$$\delta I_1((\underline{P}\cdot\underline{P}^T)^{1/2}) = \tfrac{1}{2}(\underline{P}\cdot\underline{P}^T)^{-1/2}\cdot\cdot\,\delta\underline{P}\cdot\underline{P}^T = \tfrac{1}{2}(\underline{P}\cdot\underline{P}^T)^{-1/2}\cdot\cdot\,(\underline{P}\cdot\delta\underline{P}^T + (\delta\underline{P})\cdot\underline{P}^T) =$$

$$= \tfrac{1}{2}\{(\underline{P}\cdot\underline{P}^T)^{-1/2}\cdot\underline{P}\cdot\cdot\delta\underline{P}^T + [(\underline{P}\cdot\underline{P}^T)^{-1/2}]^T\cdot\underline{P}\cdot\cdot\delta\underline{P}^T\} = (\underline{P}\cdot\underline{P}^T)^{-1/2}\cdot\underline{P}\cdot\cdot\delta\underline{P}^T$$

where we have used rule (II.3.10) for differentiation and the symmetry of $(\underline{P}\cdot\underline{P}^T)^{-1/2}$.

Referring now to the definition (II.2.7) of the derivative of a tensor argument, we obtain

$$I_1(\underline{P}\cdot\underline{P}^T)_{\underline{P}} = 2\underline{P} \ , \qquad I_1((\underline{P}\cdot\underline{P}^T)^{1/2})_{\underline{P}} = (\underline{P}\cdot\underline{P}^T)^{-1/2}\cdot\underline{P} \tag{12}$$

and by (4.17.5)

$$\overset{0}{\underline{\underline{\nabla R}}} = (\ni_\times)_{\underline{P}} = \frac{1}{2\mu}\left\{\underline{E} + \left[2\mu - \frac{\nu}{1+\nu} I_1((\underline{P}\cdot\underline{P}^T)^{1/2})\right](\underline{P}\cdot\underline{P}^T)^{-1/2}\right\}\cdot\underline{P} \ . \tag{13}$$

It appears that for a quasi-linear material, the equation of state, (5), can be inverted.

For that deformation which preserves the directions of the principal axes of the strain measure

$$\underline{O}^\times = \underline{E} \ , \qquad \underline{U} = \overset{0}{\underline{\underline{\nabla R}}} = \overset{0}{\underline{\underline{\nabla u}}} + \underline{E} \ , \qquad s_1 = \overset{0}{\underline{\nabla}}\cdot\underline{u} \tag{14}$$

and for a quasilinear material, we have from (5),

$$\underline{P} = \lambda\overset{0}{\underline{\nabla}}\cdot\underline{u}\underline{\underline{E}} + 2\mu\overset{0}{\underline{\underline{\nabla u}}} \ . \tag{15}$$

The equilibrium equations (2.7.4) and (2.7.5) in the volume and on its surface take the form

$$(\lambda + 2\mu)\overset{0}{\underline{\nabla}}^2\underline{u} + \rho_0\underline{k} = 0 \,, \qquad \lambda\underline{n}\overset{0}{\underline{\nabla}}\cdot\underline{u} + 2\mu\underline{n}\cdot\overset{0}{\underline{\nabla}}\underline{u} = \underline{f}\,\frac{dO}{do}\,. \tag{16}$$

This is a special case of the equations of equilibrium of a linearly elastic body for $\overset{0}{\underline{\nabla}}\underline{u} = \overset{0}{\underline{\nabla}}\underline{u}^T$ (or $\overset{0}{\underline{\nabla}} \times \underline{u} = \underline{0}$). The non-linearity of the problem is stipulated by the right-hand side of the boundary condition for the case when the contact forces are not "dead".

The problems of the stretching of a rod, the axisymmetric deformation of a cylinder, the symmetric deformation of a sphere may all be reduced to Equations (16).

The acoustic tensor of a quasi-linear material. The principal stresses determined by Equation (5) are

$$\sigma_1 = \frac{1}{v_2 v_3}\,[(\lambda s_1 - 2\mu) + 2\mu v_1] \qquad \text{etc.}$$

From these we get

$$v_1\,\frac{\partial\sigma_1}{\partial v_1} = \frac{1}{v_2 v_3}\,(\lambda + 2\mu)v_1\,, \qquad v_2\,\frac{\partial\sigma_1}{\partial v_2} = -\,\frac{1}{v_2 v_3}\,[(\lambda s_1 - 2\mu) + 2\mu v_1 - \lambda v_2]\,,$$

$$\frac{\sigma_1 - \sigma_2}{v_1^2 - v_2^2} = \frac{1}{v_1 v_2 v_3}\left(\frac{\lambda s_1 - 2\mu}{v_1 + v_2} + 2\mu\right) = \frac{1}{v_1 v_2 v_3}\left[\lambda + 2\mu + \frac{\lambda v_3 - (3\lambda + 2\mu)}{v_1 + v_2}\right]$$

and substitution into Equations (4.12.9) and (4.12.10) determines the components of the acoustical tensor in the orthonormal trihedron $\underline{e}_1$, $\underline{e}_2$, $\underline{e}_3$ of the principal directions of the stress tensor $\underline{T}$ (or Finger's*) strain measure $\underline{F}$)

$$Q^{11} = \left[(\lambda + 2\mu)(v_1^2N_1^2 + v_2^2N_2^2 + v_3^2N_3^2) + \frac{\lambda v_3 - (3\lambda + 2\mu)}{v_1 + v_2}\,v_2^2N_2^2 + \right.$$

$$\left. + \frac{\lambda v_2 - (3\lambda + 2\mu)}{v_1 + v_3}\,v_3^2N_3^2\right],$$

$$Q^{12} = -v_1 v_2\,\frac{\lambda v_3 - (3\lambda + 2\mu)}{v_1 + v_2}\,N_1 N_2 = Q^{21} \qquad \text{etc.}$$

---

*) As in (4.12.9) and (4.12.10), we neglect the factor $\sqrt{I_3}$. This does not affect the sign of $\underline{Q}$.

Using the notation

$$\frac{\lambda v_1 - (3\lambda + 2\mu)}{v_2 + v_3} = A \ , \qquad \frac{\lambda v_2 - (3\lambda + 2\mu)}{v_3 + v_1} = B \ , \qquad \frac{\lambda v_3 - (3\lambda + 2\mu)}{v_1 + v_2} = C \ ,$$

$$v_1 N_1 = x \ , \qquad v_2 N_2 = y \ , \qquad v_3 N_3 = z$$

the acoustical tensor $\underline{\underline{Q}}$ may be given in the following form

$$\underline{\underline{Q}} = (\lambda + 2\mu)(x^2 + y^2 + z^2)\underline{\underline{E}} + \underline{e}_1\underline{e}_1(Cy^2 + Bz^2) + \underline{e}_2\underline{e}_2(Az^2 + Cx^2) +$$

$$+ \underline{e}_3\underline{e}_3(Bx^2 + Ay^2) - (\underline{e}_1\underline{e}_2 + \underline{e}_2\underline{e}_1)Cxy - (\underline{e}_2\underline{e}_3 + \underline{e}_3\underline{e}_2)Ayz -$$

$$- (\underline{e}_3\underline{e}_1 + \underline{e}_1\underline{e}_3)Bzx \ . \tag{17}$$

For vector $\underline{N}$ which is parallel to $\underline{e}_1$ we have $N_1 = 1$ and $N_2 = N_3 = 0$ and tensor $\underline{\underline{Q}}$ takes a diagonal form

$$\underline{\underline{Q}} = (\lambda + 2\mu)\underline{e}_1\underline{e}_1 + \frac{1}{v_1 + v_2}[(\lambda + 2\mu)(v_1 + v_2) + \lambda v_3 - (3\lambda + 2\mu)]\,\underline{e}_2\underline{e}_2 +$$

$$+ \frac{1}{v_1 + v_3}[(\lambda + 2\mu)(v_1 + v_3) + \lambda v_2 - (3\lambda + 2\mu)]\,\underline{e}_3\underline{e}_3 \ .$$

The inequalities

$$\lambda + 2\mu > 0 \ , \qquad (\lambda + 2\mu)(v_1 + v_2) + \lambda v_3 > 3\lambda + 2\mu \ ,$$

$$(\lambda + 2\mu)(v_1 + v_3) + \lambda v_2 > 3\lambda + 2\mu \tag{18}$$

are the conditions under which this tensor is positive. If we put $N_2 = 1$ and $N_3 = N_1 = 0$ in (17), we would of course obtain the additional inequaliy

$$(\lambda + 2\mu)(v_2 + v_3) + \lambda v_1 > 3\lambda + 2\mu \ . \tag{19}$$

Inequalities (18) and (19) are transformed to

$$(1 - 2q)(v_1 + v_2) + qv_3 > 1 \ , \qquad (1 - 2q)(v_2 + v_3) + qv_1 > 1 \ ,$$

$$(1 - 2q)(v_3 + v_1) + qv_2 > 1 \ , \tag{20}$$

where $q = \nu/(1+\nu)$ and $-\infty < q < 1/3$.

When only one component of $\underline{N}$ becomes zero, e.g. $N_3 = 0$, the tensor $\underset{\approx}{Q}$ takes the value

$$\underset{\approx}{Q} = [(\lambda+2\mu)(x^2+y^2) + Cy^2]\, \underline{e}_1\underline{e}_1 + [(\lambda+2\mu)(x^2+y^2) + Cx^2]\, \underline{e}_2\underline{e}_2 +$$

$$+ [(\lambda+2\mu)(x^2+y^2) + (Ay^2+Bx^2)]\, \underline{e}_3\underline{e}_3 - (\underline{e}_1\underline{e}_2 + \underline{e}_2\underline{e}_1)Cxy \ .$$

The conditions under which the component Q^{33} is positive are

$$\lambda + 2\mu + B > 0 \ , \qquad \lambda + 2\mu + A > 0$$

and these are already contained in (18) and (19). These two conditions equally imply positiveness of the minor

$$[(\lambda+2\mu)(x^2+y^2) + Cy^2][(\lambda+2\mu)(x^2+y^2) + Cx^2] - C^2x^2y^2 =$$

$$= (\lambda+2\mu)(x^2+y^2)^2(\lambda+2\mu+C)$$

as well as of the two minors associated with the diagonal elements.

It is easily verified that Equations (20) ensure that the diagonal terms and the diagonal minors of the matrix of the acoustic tensor (17) for any $\underline{N}$ are positive (x, y, z different from zero). This also ensures that $\det \underset{\approx}{Q}$ is positive

$$\det \underset{\approx}{Q} > 0 \ .$$

We have thus proved that Inequalities (20) are not only the necessary but also the sufficient conditions for the acoustic tensor to be positive, that is for the strong ellipticity of a quasilinear material

6. A BLATZ AND KO MATERIAL

The data obtained by Blatz and Ko in the course of their experiments on the deformation of a special rubber over a large range of loads show that the specific strain energy may be approximated by the following

$$\ni = \frac{1}{2}\mu\beta\left[I_1 + \frac{1}{\alpha}(I_3^{-\alpha} - 1) - 3\right] + \frac{1}{2}\mu(1-\beta)\left[I_2 I_3^{-1} + \frac{1}{\alpha}(I_3^{\alpha} - 1) - 3\right] , \tag{1}$$

$$\alpha = \frac{\lambda}{2\mu} = \frac{\nu}{1-2\nu} \ , \qquad I_k = I_k(\underset{\approx}{G}) \ . \tag{2}$$

Knowles and Sternberg (J.K. Knowles, E. Sternberg, 1975) used a simplified form of this relation in which they assumed the following values of the constants α, β and ν to be valid

$$\beta = 0\ , \qquad \nu = \tfrac{1}{4}\ , \qquad \alpha = \tfrac{1}{2}\ .$$

Using these, the expression for $\ni$ takes the form

$$\ni = \frac{1}{2}\,\mu\left(\frac{I_2}{I_3} + 2\sqrt{I_3} - 5\right) \tag{3}$$

and the state equations of the Piola and Cauchy tensors are given by

$$\begin{aligned} \underline{P} &= \ni_{\underset{\nabla R}{0}} = \frac{\mu}{I_3}\,[\underline{E}I_1 - \underline{G} + (I_3^{3/2} - I_2)\underline{G}^{-1}]\cdot \overset{0}{\underline{\nabla R}}\ , \\ \underline{T} &= \mu I_3^{-3/2}[\underline{E}(I_3^{3/2} - I_2) + I_1\underline{F} - \underline{F}^2]\ . \end{aligned} \tag{4}$$

In the linear approximation, in view of (3.4) and (3.5) we get

$$\underline{T} = \mu(\underline{E}\vartheta + 2\underline{\varepsilon})$$

this is a linearly elastic material for $\lambda = \mu$ ($\nu = 0.25$).

The principal stress σ_1 turns out to be

$$\frac{1}{\mu}\,\sigma_1 = 1 + I_3^{-3/2}[-(v_1^2v_2^2 + v_2^2v_3^2 + v_3^2v_1^2) + v_1^2(v_1^2 + v_2^2 + v_3^2) - v_1^4] = 1 - \frac{1}{\sqrt{I_3}v_1^2}\ ,$$

so that

$$1 - \frac{\sigma_1}{\mu} = \frac{1}{\sqrt{I_3}v_1^2}\ , \qquad 1 - \frac{\sigma_2}{\mu} = \frac{1}{\sqrt{I_3}v_2^2}\ , \qquad 1 - \frac{\sigma_3}{\mu} = \frac{1}{\sqrt{I_3}v_3^2}\ . \tag{5}$$

From this we have

$$\begin{aligned} \left(1 - \frac{\sigma_1}{\mu}\right)\left(1 - \frac{\sigma_2}{\mu}\right)\left(1 - \frac{\sigma_3}{\mu}\right) &= I_3^{-5/2}\ , \\ \sqrt{I_3} &= \left[\left(1 - \frac{\sigma_1}{\mu}\right)\left(1 - \frac{\sigma_2}{\mu}\right)\left(1 - \frac{\sigma_3}{\mu}\right)\right]^{-1/5}\ , \end{aligned} \tag{6}$$

so that, in view of (5),

$$v_s^2 = \left[\left(1-\frac{\sigma_1}{\mu}\right)\left(1-\frac{\sigma_2}{\mu}\right)\left(1-\frac{\sigma_3}{\mu}\right)\right]^{1/5}\left(1-\frac{\sigma_s}{\mu}\right)^{-1} \qquad (s = 1,2,3)\ . \tag{7}$$

The equation (4) of state thus proves to be invertible. Making use of the notation

$$D = \left(1-\frac{\sigma_1}{\mu}\right)\left(1-\frac{\sigma_2}{\mu}\right)\left(1-\frac{\sigma_3}{\mu}\right) = 1 - \frac{I_1(\underline{T})}{\mu} + \frac{I_2(\underline{T})}{\mu^2} - \frac{I_3(\underline{T})}{\mu^3}\,,$$

$$v_s^2 = D^{1/5}\left(1-\frac{\sigma_3}{\mu}\right)^{-1} \tag{8}$$

we obtain

$$I_1(\underline{G}) = D^{-4/5}\left(3 - \frac{2I_1(\underline{T})}{\mu} + \frac{I_2(\underline{T})}{\mu^2}\right),$$

$$I_2(\underline{G}) = D^{-3/5}\left(3 - \frac{I_1(\underline{T})}{\mu}\right),$$

$$I_3(\underline{G}) = D^{-2/5}\ .$$

By virtue of (3), the expression for $\ni$ may now be written in terms of the invariants of the Cauchy stress tensor

$$\ni = \frac{1}{2}\mu\left[5(D^{-1/5}-1) - D^{-1/5}\frac{I_1(\underline{T})}{\mu}\right]\ . \tag{9}$$

Since

$$\underline{P}\cdot\cdot\overset{0}{\underline{\nabla}}\underline{R}^T = \sqrt{I_3}\,I_1(\underline{T}) = D^{-1/5}I_1(\underline{T})$$

the expression (4.7.3) for the specific complementary energy reduces to the form

$$\ni_\times = \frac{1}{2}\mu\,[5(1-D^{-1/5}) + 3D^{-1/5}I_1(\underline{T})]\ . \tag{10}$$

The equation of state, (4), can be inverted to express $\underline{F}$ in terms of $\underline{T}$ as follows

$$\underline{F} = \sum_{s=1}^{3} v_3^2\underline{e}_s\underline{e}_s = D^{1/5}\sum_{s=1}^{3}\frac{\underline{e}_s\underline{e}_s}{1-\sigma_s/\mu} =$$

$$= D^{-4/5}\left[\underline{e}_1\underline{e}_1\left(1-\frac{\sigma_2}{\mu}\right)\left(1-\frac{\sigma_3}{\mu}\right) + \underline{e}_2\underline{e}_2\left(1-\frac{\sigma_3}{\mu}\right)\left(1-\frac{\sigma_1}{\mu}\right) + \right.$$

$$+ \underline{e}_3\underline{e}_3 \left(1 - \frac{\sigma_1}{\mu}\right)\left(1 - \frac{\sigma_2}{\mu}\right)\bigg] = D^{-4/5} \left[\underline{E}\left(1 - \frac{I_1(\underline{T})}{\mu}\right) + \frac{1}{\mu}\underline{T} + \frac{1}{\mu^2} I_3(\underline{T})\underline{T}^{-1}\right]$$

or

$$\underline{F} = D^{-4/5} \left[\underline{E}\left(1 - \frac{I_1(\underline{T})}{\mu} + \frac{I_2(\underline{T})}{\mu^2}\right) + \frac{1}{\mu}\underline{T}\left(1 - \frac{I_1(\underline{T})}{\mu}\right) + \frac{1}{\mu^2}\underline{T}^2\right] =$$

$$= -5\mu (D^{1/5})_{\underline{T}} \, . \tag{11}$$

The latter equation follows from (8) and from the rules for differentiation given in Chapter II, Par. 3

$$-\mu D_{\underline{T}} = \underline{E}\left(1 - \frac{I_1(\underline{T})}{\mu} + \frac{I_2(\underline{T})}{\mu^2}\right) + \frac{1}{\mu}\underline{T}\left(1 - \frac{I_1(\underline{T})}{\mu}\right) + \frac{\underline{T}^2}{\mu^2} = \underline{F}D^{4/5} \, . \tag{12}$$

The inverse formula proves to be fairly complicated and D should be eliminated using (8).

We now turn to the formulation of the acoustic tensor $\underline{\underline{Q}}$ for a Blatz and Ko material. From (4.11.14), (4.3.5) and (4.5.5) on discarding the positive factor μ, we have

$$\psi_0 = \frac{1}{2I_3}(-I_2 + I_3^{3/2}) \, , \qquad \psi_1 = \frac{1}{2}\frac{I_1}{I_3} \, , \qquad \psi_2 = -\frac{1}{2}\frac{1}{I_3} \, ,$$

$$\vartheta_{00} = \frac{1}{2I_3}(I_2 + I_3^{3/2}) \, , \qquad \vartheta_{10} = -\frac{1}{2}\frac{I_1}{I_3} \, , \qquad \vartheta_{02} = \frac{1}{2I_3} \, ,$$

$$\vartheta_{11} = \frac{1}{2I_3} \, , \qquad \vartheta_{22} = \vartheta_{12} = 0$$

and substitution into (4.11.14) yields

$$4I_3\underline{\underline{Q}} = 3I_2\underline{NN} - 2I_1(\underline{NF}\cdot\underline{N} + \underline{N}\cdot\underline{FN}) + 2(\underline{NF}^2\cdot\underline{N} + \underline{N}\cdot\underline{F}^2\underline{N}) +$$

$$+ \underline{N}\cdot\underline{FF}\cdot\underline{N} + (I_1\underline{E} - \underline{F})\underline{N}\cdot\underline{F}\cdot\underline{N} - \underline{EN}\cdot\underline{F}^2\cdot\underline{N} \, . \tag{13}$$

Tensor $\underline{\underline{Q}}$ is now expressed in terms of the principal values and directions of the measure $\underline{F}$

$$\underline{\underline{Q}} = \sum_{s=1}^{3} (2N_s^2 + N^2)\frac{\underline{e}_s\underline{e}_s}{v_s^2} + (\underline{e}_1\underline{e}_2 + \underline{e}_2\underline{e}_1)\left(\frac{1}{v_1^2} + \frac{1}{v_2^2}\right)N_1N_2 +$$

$$+ (\underline{e}_2\underline{e}_3 + \underline{e}_3\underline{e}_2)\left(\frac{1}{v_2^2} + \frac{1}{v_3^2}\right)N_2N_3 + (\underline{e}_3\underline{e}_1 + \underline{e}_1\underline{e}_3)\left(\frac{1}{v_3^2} + \frac{1}{v_1^2}\right)N_3N_1 \ . \qquad (14)$$

The introduction of the notation $\underline{e}_s = v_s\underline{e}'_s$ leads to the form obtained by Knowles and Sternberg

$$\underline{\underline{Q}} = \sum_{s=1}^{3} (2N_s^2 + N^2)\underline{e}'_s\underline{e}'_s + (\underline{e}'_1\underline{e}'_2 + \underline{e}'_2\underline{e}'_1)\left(\frac{v_1}{v_2} + \frac{v_2}{v_1}\right) +$$

$$+ (\underline{e}'_2\underline{e}'_3 + \underline{e}'_3\underline{e}'_2)\left(\frac{v_2}{v_3} + \frac{v_3}{v_2}\right) + (\underline{e}'_3\underline{e}'_1 + \underline{e}'_1\underline{e}'_3)\left(\frac{v_3}{v_1} + \frac{v_1}{v_3}\right) \ . \qquad (15)$$

It has been shown by Knowles and Sternberg that the matrix of the tensor $\underline{\underline{Q}}$ satisfies all conditions of Sylvester's theorem for all $\underline{N}$ and for those states of deformation which are restricted by

$$2 - \sqrt{3} < \frac{v_i}{v_j} < 2 + \sqrt{3} \ . \qquad (16)$$

For this range of deformations, the material is strongly elliptic.

It is also of interest to consider another form of a Blatz and Ko material for which $\beta = 1$ in Equation (1):

$$\ni = \frac{1}{2}\mu\left[I_1 + \frac{1}{\alpha}(I_3^{-\alpha} - 1) - 3\right] \ . \qquad (17)$$

For this "hypothetical" material, the Piola and Cauchy tensors are given by the expressions

$$\underline{P} = \ni_{\overset{0}{\underline{\underline{\nabla R}}}} = \mu(\overset{0}{\underline{\underline{\nabla R}}} - I_3^{-\alpha}\underline{\underline{\nabla r}}^T) \ , \qquad \underline{T} = \mu I_3^{-1/2}(\underline{F} - \underline{E}I_3^{-\alpha}) \qquad (18)$$

and to a linear approximation, they lead to the equation of state of the linear theory

$$\sigma_s = \lambda\vartheta + 2\mu\varepsilon_s \ , \qquad s = 1,2,3 \ . \qquad (19)$$

It can be seen from (4.3.5) and (4.5.5) that the only non-zero functions are

$$\phi_0 = -\frac{1}{2}\mu I_3^{-\alpha} \ , \qquad \phi_1 = \frac{1}{2}\mu \ , \qquad \vartheta_{00} = \frac{1}{2}\mu\alpha I_3^{-\alpha}$$

and in view of (4.11.14) the acoustical tensor may be written

$$\underset{\approx}{Q} = \frac{1}{4}\mu\,[\underline{\underline{EN}}\cdot\underline{F}\cdot\underline{N} + (2\alpha+1)I_3^{-\alpha}\underline{\underline{NN}}] \ . \tag{20}$$

For $-1/2 < \alpha < \infty$, that is $-\infty < \nu < 1/2$, the determinant $\det\underset{\approx}{Q}$ satisfies all the conditions of Sylvester's theorem for all vectors $\underline{N}$ and all positive tensors $\underline{F}$ which means that for all deformations the system of equations for this "hypothetical" material is strongly elliptic.

7. THE ENERGY OF DILATATION AND DISTORTION

V.A. Pal'mov (1976) proposed a generalization of the well-known expression for the specific potential energy as given by linear theory in the form of the sum of two terms which determine the energy of dilatation and the energy respectively.

The basis of the success of Pal'mov's proposal was the substitution of the Cauchy-Green and Finger strain measures by the measures

$$\underline{G}^{+} = I_3^{-1/3}\underline{G} \ , \qquad \underline{F}^{+} = I_3^{-1/3}\underline{F} \qquad (I_3 = I_3(\underline{G}) = I_3(\underline{F})) \ . \tag{1}$$

According to this definition, the principal values of these tensors are

$$G_k^{+} = F_k^{+} = I_3^{-1/3}G_k = I_3^{-1/3}F_k \qquad (G_k = F_k = v_k^2,\ k = 1,2,3) \tag{2}$$

and their third invariant is therefore equal to 1 - these tensors are insensitive to change of volume

$$I_3(\underline{G}^{+}) = I_3(\underline{F}^{+}) = G_1^{+}G_2^{+}G_3^{+} = (I_3^{-1/3})^3 G_1G_2G_3 = 1 \ . \tag{3}$$

Similar relationships for Almansi's measure $\underline{g}$ and the measure $\underline{G}^{-1}$ are given by

$$\underline{g}^{+} = I_3^{1/3}\underline{g} \ , \qquad (\underline{G}^{-1})^{+} = I_3^{1/3}\underline{G}^{-1} = (I_3')^{-1/3}\underline{G}^{-1} \ , \qquad I_3' = I_3(\underline{g}) = I_3^{-1} \ . \tag{4}$$

The specific strain energy, $\ni$, is now represented as the sum of the energy of dilatation, $\ni_{I}$, and the energy of distortion, $\ni_{II}$.

$$\ni = \ni_{I} + \ni_{II} \ , \qquad \ni_{I} = \ni_{I}(\sqrt{I_3}) \ , \qquad \ni_{II} = \ni_{II}(I_1(\underline{F}^{+}), I_2(\underline{F}^{+})) \ . \tag{5}$$

The stress tensor is given by (4.3.2)

$$\underline{T} = 2I_3^{-1/2}\ni_{\underline{F}}\cdot\underline{F} = 2I_3^{-1/2}(\ni_{I})_{\underline{F}}\cdot\underline{F} + 2I_3^{-1/2}(\ni_{II})_{\underline{F}}\cdot\underline{F} = \underline{T}_{I} + \underline{T}_{II} \ . \tag{6}$$

By the rules for differentiation given in II, Pars. 2, 3 we have

$$\ni_I(\sqrt{I_3})_{\underline{F}} = \frac{\partial \ni_I}{\partial \sqrt{I_3}} (\sqrt{I_3})_{\underline{F}} = \frac{1}{2} \ni_I'(\sqrt{I_3}) \sqrt{I_3}\, \underline{F}^{-1}$$

where the first term in (6) is

$$\underline{T}_I = \ni_I'(\sqrt{I_3}) \underline{E} \ . \tag{7}$$

The calculation of $\underline{T}_{II}$ is more complex. We have

$$(\ni_{II})_{\underline{F}} = \left\{ \left[\frac{\partial \ni_{II}}{\partial I_1(\underline{F}^+)} + I_1(\underline{F}^+) \frac{\partial \ni_{II}}{\partial I_2(\underline{F}^+)} \right] \underline{E} - \frac{\partial \ni_{II}}{\partial I_2(\underline{F}^+)} \underline{F}^+ \right\} \cdot\cdot \underline{F}^+_{\underline{F}}$$

where we have made use of rule (II.4.12). We also have from (II.4.9) and (II.4.18)

$$(\underline{F}^+)_{\underline{F}} = (\underline{F} I_3^{-1/3})_{\underline{F}} = -\frac{1}{3} I_3^{-1/3} \underline{F}\underline{F}^{-1} + I_3^{-1/3} \frac{1}{2} (\underline{C}_{II} + \underline{C}_{III})$$

whence

$$\underline{T}_{II} = 2 I_3^{-1/2} I_3^{-1/3} \left\{ \left[\frac{\partial \ni_{II}}{\partial I_1(\underline{F}^+)} + I_1(\underline{F}^+) \frac{\partial \ni_{II}}{\partial I_2(\underline{F}^+)} \right] \underline{F} - \frac{\partial \ni_{II}}{\partial I_2(\underline{F}^+)} \underline{F}^+ \cdot \underline{F} - \right.$$

$$\left. - \frac{1}{3} \left[\frac{\partial \ni_{II}}{\partial I_1(\underline{F}^+)} + I_1(\underline{F}^+) \frac{\partial \ni_{II}}{\partial I_2(\underline{F}^+)} \right] I_1(\underline{F}) \underline{E} + \frac{1}{3} \frac{\partial \ni_{II}}{\partial I_2(\underline{F}^+)} I_1(\underline{F}^+ \cdot \underline{F}) \underline{E} \right\}$$

or, from (1),

$$\underline{T}_{II} = 2 I_3^{-1/2} \left\{ \left[\frac{\partial \ni_{II}}{\partial I_1(\underline{F}^+)} + I_1(\underline{F}^+) \frac{\partial \ni}{\partial I_2(\underline{F}^+)} \right] \left(\underline{F}^+ - \frac{1}{3} \underline{E} I_1(\underline{F}^+) \right) - \right.$$

$$\left. - \frac{\partial \ni_{II}}{\partial I_2(\underline{F}^+)} \left[\underline{F}^{+2} - \frac{1}{3} \underline{E} I_1(\underline{F}^{+2}) \right] \right\} .$$

With a notation similar to that of (4.3.5)

$$\psi_1^+ = \frac{\partial \ni_{II}}{\partial I_1(\underline{F}^+)} + I_1(\underline{F}^+) \frac{\partial \ni}{\partial I_2(\underline{F}^+)} , \qquad \psi_2^+ = - \frac{\partial \ni_{II}}{\partial I_2(\underline{F}^+)} \tag{8}$$

and recalling the definition in I, Par. 13 of a deviator, we have

$$\underset{-}{T}_{II} = 2I_3^{-1/2}\,\mathrm{dev}\,(\overset{+}{\psi}_1\overset{+}{\underset{-}{F}} + \overset{+}{\psi}_2\underset{-}{F}^{+2}) \ . \tag{9}$$

Thus the Cauchy stress tensor may be represented as the sum of spherical and a deviatoric term

$$T = \mathfrak{I}_I'(\sqrt{I_3})\underset{=}{E} + 2\sqrt{\frac{g}{G}}\,\mathrm{dev}(\overset{+}{\psi}_1\overset{+}{\underset{-}{F}} + \overset{+}{\psi}_2\underset{-}{F}^{+2}) \ . \tag{10}$$

Substitution of a modified form of the Almansi measure yields the similar form

$$\underset{=}{T} = \mathfrak{I}_I'(\sqrt{I_3'})\underset{=}{E} + 2\sqrt{\frac{g}{G}}\,\mathrm{dev}(\psi_1'^{+}\overset{+}{\underset{=}{g}} + \psi_2'^{+}\underset{=}{g}^{+2}) \ , \tag{11}$$

where in the same way as for (4.3.8),

$$\psi_1'^{+} = -\left[\frac{\partial \mathfrak{I}_{II}}{\partial I_1(\overset{+}{\underset{=}{g}})} + I_1(\overset{+}{\underset{=}{g}})\frac{\partial \mathfrak{I}_{II}}{\partial I_2(\overset{+}{\underset{=}{g}})}\right] , \qquad \psi_2'^{+} = \frac{\partial \mathfrak{I}_{II}}{\partial I_2(\overset{+}{\underset{=}{g}})} \ . \tag{12}$$

To specify $\mathfrak{I}_{II}$, the following expression which is linear with respect to $I_1(\overset{+}{\underset{=}{F}})$ and $I_2(\overset{+}{\underset{=}{F}})$ may be used

$$\mathfrak{I}_{II} = C_1I_1(\overset{+}{\underset{=}{F}}) + C_2I_2(\overset{+}{\underset{=}{F}}) \qquad (C_1 > 0,\ C_2 > 0) \ . \tag{13}$$

This is similar to that suggested by Mooney for an incompressible medium (see Chapter 7, Par. 2). We have then

$$\underset{-}{T}_{II} = 2\sqrt{\frac{1}{I_3}}\,\mathrm{dev}\{[C_1 + C_2I_1(\overset{+}{\underset{=}{F}})]\,\overset{+}{\underset{=}{F}} - C_2\underset{-}{F}^{+2}\} \ . \tag{14}$$

The expression

$$\mathfrak{I}_I(\sqrt{I_3}) = -k(\ln\sqrt{I_3} - \sqrt{I_3} + 1) \ , \tag{15}$$

may serve as an acceptable approximation for $\mathfrak{I}_I$, so that

$$\underset{-}{T}_I = -k(I_3^{-1/2} - 1)\underset{=}{E} \ , \qquad I_1(\underset{=}{T}) = -3k(I_3^{-1/2} - 1) \ ,$$

$$q = \frac{1}{3}I_1(\underset{=}{T}) = -k\,\frac{\rho - \rho_0}{\rho_0} \ ,$$

where q denotes the mean normal stress.

8. THE TRIGONOMETRIC REPRESENTATION OF THE EQUATION OF STATE

Following I, Par. 13 the Cauchy stress tensor $\underline{T}$ may be written as the sum of a spherical and a deviatoric term

$$\underline{T} = \frac{1}{3}I_1(\underline{T})\underline{E} + \operatorname{dev}\underline{T} = \frac{1}{3}I_1(\underline{T})\underline{E} + \sum_{s=1}^{3} \sigma'_s \underline{e}_s \underline{e}_s \tag{1}$$

in which the principal values of $\operatorname{dev}\underline{T}$ are denoted by σ'_s. The deformed state is given by the logarithmic strain tensor $\underline{H}$, this tensor being coaxial with $\underline{T}$. In a similar way to (1) we have

$$\underline{H} = \frac{1}{3}I_1(\underline{H})\underline{E} + \operatorname{dev}\underline{H} = \frac{1}{3}I_1(\underline{H}) + \sum_{s=1}^{3} h'_s \underline{e}_s \underline{e}_s \; . \tag{2}$$

Here, h'_s denote the principal values of $\operatorname{dev}\underline{H}$. From (I.13.10) these values can be represented as

$$h'_1 = \sqrt{\frac{g_2}{3}}\sin\psi \; , \qquad h'_2 = \sqrt{\frac{g_2}{3}}\sin\left(\psi + \frac{2\pi}{3}\right) \; ,$$

$$h'_3 = \sqrt{\frac{g_2}{3}}\sin\left(\psi + \frac{4\pi}{3}\right) \; ; \qquad |\psi| < \frac{\pi}{6} \; , \tag{3}$$

where, from (I.13.4) and (I.13.5),

$$g_2 = -4I_2(\operatorname{dev}\underline{H}) = \frac{2}{3}\,[(h_1 - h_2)^2 + (h_2 - h_3)^2 + (h_3 - h_1)^2] =$$

$$= \frac{2}{3}\left[\left(\ln\frac{v_1}{v_2}\right)^2 + \left(\ln\frac{v_2}{v_3}\right)^2 + \left(\ln\frac{v_3}{v_1}\right)^2 \tag{4}$$

we have made use of (1.6.11). From the definition of a deviator we have

$$h'_s = h_s - \frac{1}{3}I_1(\underline{H}) = \ln v_s - \frac{1}{3}\ln v_1 v_2 v_3 \; ;$$

$$h'_1 = \frac{1}{3}\ln\frac{v_1^2}{v_2 v_3} \; , \qquad h'_2 = \frac{1}{3}\ln\frac{v_2^2}{v_3 v_1} \; , \qquad h'_3 = \frac{1}{3}\ln\frac{v_3^2}{v_1 v_2} \; . \tag{5}$$

From (3) and (5) we deduce the relationships

$$\sin\psi = \frac{1}{\sqrt{3g_2}}\ln\frac{v_1^2}{v_2 v_3} \; , \qquad \cos\psi = \frac{1}{\sqrt{g_2}}\ln\frac{v_2}{v_3} \tag{6}$$

the second of which has been obtained from the expression for h_2' following the substitution

$$\sin\left(\psi + \frac{2\pi}{3}\right) = -\frac{1}{2}\sin\psi + \frac{\sqrt{3}}{2}\cos\psi \ .$$

The value of ψ is found from (6). This parameter may also be found using (I.13.5) and (I.13.9), i.e.

$$g_3 = -\left(\frac{g_2}{3}\right)^{3/2} \sin 3\psi = 4I_3(\operatorname{dev} \underline{H}) \ . \tag{7}$$

Applying this to $\operatorname{dev}\underline{T}$, the formulas of I, Par. 13 lead to the form

$$\sigma_1' = \sqrt{\frac{G_2}{3}}\sin\chi \ , \qquad \sigma_2' = \sqrt{\frac{G_2}{3}}\sin\left(\chi + \frac{2\pi}{3}\right) \ ,$$

$$\sigma_3' = \sqrt{\frac{G_2}{3}}\sin\left(\chi + \frac{4\pi}{3}\right) \ , \qquad |\chi| < \frac{\pi}{6} \ , \tag{8}$$

$$G_2 = -4I_2(\operatorname{dev}\underline{T}) = \frac{2}{3}\left[(\sigma_1 - \sigma_2)^2 + (\sigma_2 - \sigma_3)^2 + (\sigma_3 - \sigma_1)^2\right] \ , \tag{9}$$

$$\sigma_1' = \frac{1}{3}(2\sigma_1 - \sigma_2 - \sigma_3) \ , \qquad \sigma_2' = \frac{1}{3}(2\sigma_2 - \sigma_3 - \sigma_1) \ ,$$

$$\sigma_3' = \frac{1}{2}(2\sigma_3 - \sigma_1 - \sigma_2) \ , \tag{10}$$

and analogs of Equations (6) and (7) are given by the relationships

$$\sin\chi = \frac{1}{\sqrt{3G_2}}(2\sigma_1 - \sigma_2 - \sigma_3) \ , \qquad \cos\chi = \frac{1}{\sqrt{G_2}}(\sigma_2 - \sigma_3) \ , \tag{11}$$

$$G_3 = -\left(\frac{G_2}{3}\right)^{3/2} \sin 3\chi = 4I_3(\operatorname{dev}\underline{T}) \ . \tag{12}$$

By (6) and (11) we now have

$$\sqrt{g_2 G_2}\cos\omega = (\sigma_2 - \sigma_3)\ln\frac{v_2}{v_3} + \frac{1}{3}(2\sigma_1 - \sigma_2 - \sigma_3)\ln\frac{v_1^2}{v_2 v_3} \ , \tag{13}$$

$$\sqrt{3 g_2 G_3}\sin\omega = (2\sigma_1 - \sigma_2 - \sigma_3)\ln\frac{v_2}{v_3} - (\sigma_2 - \sigma_3)\ln\frac{v_1^2}{v_2 v_3} \ . \tag{14}$$

Here

$$\omega = \chi - \psi$$

denotes the phase difference of the deviators of the Cauchy stress tensor and the Hencky strain measure introduced by (II.9.6).

Returning to Equations (4.3.16), (4.3.12), we have now

$$\delta \ni = \sum_{k=1}^{3} t_k \delta v_k = e^{I_1(\underline{H})} \sum_{k=1}^{3} \sigma_k \frac{\delta v_k}{v_k} =$$

$$= e^{I_1(\underline{H})} \sum_{k=1}^{3} \sigma_k \delta \ln v_k = e^{I_1(\underline{H})} \sum_{k=1}^{3} \sigma_k \delta h_k \, ,$$

and after identifying the spherical and deviatoric terms of tensors $\underline{T}$ and $\underline{H}$

$$\delta \ni = e^{I_1(\underline{H})} \sum_{k=1}^{3} \left[\sigma_k' + \frac{1}{3} I_1(\underline{T})\right]\left[\delta h_k' + \frac{1}{3} \delta I_1(\underline{H})\right] =$$

$$= e^{I_1(\underline{H})} \sum_{k=1}^{3} \sigma_k' \delta h_k' + \frac{1}{3} e^{I_1(\underline{H})} I_1(\underline{T}) \delta I_1(\underline{H}) \, ,$$

the latter being so because

$$\sum_{k=1}^{3} \sigma_k' = 0 \, , \qquad \sum_{k=1}^{3} \delta h_k' = 0 \, .$$

From (3) and an expression similar to that for σ_k', we have

$$\sum_{k=1}^{3} \sigma_k' \delta h_k' = \frac{\sqrt{G_2}}{3} \left\{ \sqrt{g_2} \left[\sin\chi \cos\psi + \sin\left(\chi + \frac{2\pi}{3}\right) \cos\left(\psi + \frac{2\pi}{3}\right) + \right.\right.$$

$$\left. + \sin\left(\chi + \frac{4\pi}{3}\right) \cos\left(\psi + \frac{4\pi}{3}\right)\right] \delta\psi + \left[\sin\chi \sin\psi + \right.$$

$$\left.\left. + \sin\left(\chi + \frac{2\pi}{3}\right) \sin\left(\psi + \frac{2\pi}{3}\right) + \sin\left(\chi + \frac{4\pi}{3}\right) \sin\left(\psi + \frac{4\pi}{3}\right)\right] \delta\sqrt{g_2} \right\}$$

and after the obvious simplifications this becomes

$$\sum_{k=1}^{3} \sigma_k' \delta h_k' = \frac{1}{2} \left(\sqrt{g_2} \sin\omega\delta\psi + \cos\omega\delta\sqrt{g_2}\right) \, . \tag{15}$$

The variation of the specific strain energy, treated here as a function of the quantities determining the Hencky measure is now reduced to the form

$$\delta\ni = \partial\tilde{\ni}(I_1(\underline{H}),\psi,\sqrt{g_2}) = e^{I_1(\underline{H})}\left[\frac{1}{3} I_1(\underline{T})\delta I_1(\underline{H}) + \right.$$

$$\left. + \frac{1}{2}\sqrt{G_2}(\sqrt{g_2}\sin\omega\delta\psi + \cos\omega\delta\sqrt{g_2})\right] . \tag{16}$$

We have thus arrived at the relationship between the invariant quantities which determine the tensors $\underline{T}$ and $\underline{H}$

$$\frac{\partial\tilde{\ni}}{\partial I_1(\underline{H})} = \frac{1}{3} I_1(\underline{T})e^{I_1(\underline{H})} , \qquad \frac{\partial\tilde{\ni}}{\partial\psi} = \frac{1}{2}\sqrt{G_2 g_2}\, e^{I_1(\underline{H})}\sin\omega ,$$

$$\frac{\partial\tilde{\ni}}{\partial\sqrt{g_2}} = \frac{1}{2}\sqrt{G_2}\, e^{I_1(\underline{H})}\cos\omega . \tag{17}$$

By (II.9.4) and (II.9.12) the constitutive equation for $\operatorname{dev}\underline{T}$ and its inverse take the form

$$\operatorname{dev}\underline{T} = \frac{1}{\cos 3\psi}\sqrt{\frac{G_2}{g_2}}\left\{\cos(\omega+3\psi)\operatorname{dev}\underline{H} - 2\sqrt{\frac{3}{g_2}}\left[(\operatorname{dev}\underline{H})^2 - \frac{1}{6}\underline{E}g_2\right]\sin\omega\right\} , \tag{18}$$

$$\operatorname{dev}\underline{H} = \frac{1}{\cos 3\chi}\sqrt{\frac{g_2}{G_2}}\left\{\cos(-\omega+3\chi)\operatorname{dev}\underline{T} + 2\sqrt{\frac{3}{G_2}}\left[(\operatorname{dev}\underline{T})^2 - \frac{1}{6}\underline{E}G_2\right]\sin\omega\right\} . \tag{19}$$

These equations are satisfied identically by the relationships given above.

9. THE MONOTONICITY CRITERION FOR THE COLEMAN-NOLL STRESS EQUATION OF STATE

A continuous function $f(x)$ is called monotonically increasing if, for any two values x, x_0 of the variable x within its range of definition, the following relationships hold

$$[f(x) - f(x_0)](x - x_0) > 0 , \qquad x \neq x_0 . \tag{1}$$

This definition can be generalized to tensor functions of a tensor argument as follows. $\underline{Q}(\underline{A})$ is called a monotonically increasing function of $\underline{A}$ if the condition

$$[\underline{Q}(\underline{A}^\times) - \underline{Q}(\underline{A})]\cdot\cdot(\underline{A}^{\times T} - \underline{A}^T) > 0 \tag{2}$$

is satisfied for all tensors $\underline{A}^\times$ having the form

$$\underline{A}^{\times} = \underline{A} \cdot \underline{S} , \qquad \underline{S} \neq \underline{E} . \tag{3}$$

It has been suggested that the monotonicity of the Piola tensor $\underline{P}(\overset{0}{\underline{\nabla R}})$ should be regarded as a criterion of the suitability of the following form of the constitutive equation

$$[\underline{P}(\overset{0}{\underline{\nabla R}}{}^{\times}) - \underline{P}(\overset{0}{\underline{\nabla R}})] \cdot\cdot (\overset{0}{\underline{\nabla R}}{}^{\times T} - \overset{0}{\underline{\nabla R}}{}^{T}) > 0 \qquad (\overset{0}{\underline{\nabla R}}{}^{\times} = \overset{0}{\underline{\nabla R}} \cdot \underline{S},\ \underline{S} \neq \underline{E}) . \tag{4}$$

1. We will show, however, that for $\underline{S} = \underline{O}$, where $\underline{O}$ is an orthogonal tensor, the principle of material frame indifference (Chapter 3, Par. 2) restricts the suitability of the criterion (4) to a narrow class of states of stress. In fact, according to this principle, we have in view of (3.2.3) and (2.6.2)

$$\underline{P}^{\times} = \underline{P} \cdot \underline{O} , \qquad \underline{P}^{\times} - \underline{P} = \sqrt{\frac{G}{g}}\, \underline{\nabla r}^{T} \cdot \underline{T} \cdot (\underline{O} - \underline{E}) .$$

Criterion (4) is now given by the inequality

$$\underline{\nabla r}^{T} \cdot \underline{T} \cdot (\underline{O} - \underline{E}) \cdot\cdot (\underline{O}^{T} - \underline{E}) \cdot \overset{0}{\underline{\nabla R}}{}^{T} = \underline{T} \cdot\cdot (2\underline{E} - \underline{O} - \underline{O}^{T}) =$$

$$= 2 [I_1(\underline{T}) - \frac{1}{2} \underline{T} \cdot\cdot (\underline{O} + \underline{O}^{T})] > 0$$

and by (I.11.10) we have

$$\frac{1}{2} (\underline{O} + \underline{O}^{T}) = \underline{E} \cos\omega + (1 - \cos\omega) \underline{ee} ,$$

$$I_1(\underline{T}) - \frac{1}{2} \underline{T} \cdot\cdot (\underline{O} + \underline{O}^{T}) = (1 - \cos\omega)(I_1(\underline{T}) - \underline{T} \cdot\cdot \underline{ee}) .$$

Replacing $\underline{T}$ by its representation in terms of its principal stresses and denoting the angles between the axis $\underline{e}$ of rotation and the principal directions of $\underline{T}$ by γ_s, we arrive at the inequality

$$\sigma_1\alpha_1 + \sigma_2\alpha_2 + \sigma_3\alpha_3 > 0 , \qquad \alpha_s = \frac{1}{2} \sin^2\gamma_s \tag{5}$$

for all

$$0 \le \alpha_s \le \frac{1}{2} , \qquad \alpha_1 + \alpha_2 + \alpha_3 = 1 .$$

This inequality is of course valid provided that

$$\min_s \sigma_s > 0 \qquad \sigma_1 > 0\,, \qquad \sigma_2 > 0\,, \qquad \sigma_3 > 0\,.$$

We will further assume that

$$\min_s \sigma_s = \sigma_3 < 0\,, \qquad \sigma_1 - \sigma_3 > 0\,, \qquad \sigma_2 - \sigma_3 > 0\,.$$

On eliminating α_3 Inequality (5) reduces to

$$(\sigma_1 - \sigma_3)\alpha_1 + (\sigma_2 - \sigma_3)\alpha_2 - |\sigma_3| > 0\,,$$

$$f(\alpha_1,\alpha_2) = \frac{\sigma_1 - \sigma_3}{|\sigma_3|}\alpha_1 + \frac{\sigma_2 - \sigma_3}{|\sigma_3|}\alpha_2 - 1 > 0\,.$$

In the α_1,α_2 plane the straight line $f(\alpha_1,\alpha_2) = 0$ cuts the axes at

$$\alpha_1^0 = \frac{|\sigma_3|}{\sigma_1 - \sigma_3}\,, \qquad \alpha_2^0 = \frac{|\sigma_3|}{\sigma_2 - \sigma_3}$$

and the region $f(\alpha_1,\alpha_2) > 0$ lies above this line but not necessarily within the triangle

$$0 \le \alpha_1 \le 0.5\,, \qquad 0 \le \alpha_2 \le 0.5\,,$$

bounded by the axes and the straight line $\alpha_1 + \alpha_2 = 0.5$. Therefore

$$\frac{|\sigma_3|}{\sigma_k - \sigma_3} < 0.5\,, \qquad |\sigma_3| < 0.5\sigma_k + 0.5|\sigma_3|\,, \qquad \sigma_k + \sigma_3 > 0 \qquad (k = 1,2)\,.$$

We have thus shown that the criterion given by Inequality (4) is applicable only for those materials whose state of stress is characterized by the fact that the sum of any two principal stresses is positive

$$\sigma_1 + \sigma_2 > 0\,, \qquad \sigma_2 + \sigma_3 > 0\,, \qquad \sigma_3 + \sigma_1 > 0$$

regardless of the character of the deformation $\overset{0}{\nabla}\underline{\underline{R}}$.

2. This forces us to restrict the class of tensors which are admissible in the sense of Inequality (4). The criterion suggested by Coleman and Noll proposes that only pure strain be chosen for the tensor $\underline{S}$, on other words, it is required that this tensor be symmetric and positive, that is

$$\underline{S} = \underline{S}^T\,, \qquad \underline{a}\cdot\underline{S}\cdot\underline{a} > 0 \tag{6}$$

for any vector $\underline{a} \neq \underline{0}$.

This condition should be supplemented by the requirement that the orthogonal tensor associated with the deformation be the same for the actual state and for the state with which it is being compared. We have then

$$\overset{0}{\underline{\underline{\nabla R}}}{}^{\times} = \underline{U}^{\times} \cdot \underline{O} = \underline{O} \cdot \underline{V}^{\times} , \qquad \underline{\underline{\nabla r}}^{\times} = \underline{V}^{\times -1} \cdot \underline{O}^{T} , \qquad \underline{\underline{\nabla r}}^{\times T} = \underline{O} \cdot \underline{V}^{\times -1} ,$$

$$\underline{P}^{\times} = \sqrt{\frac{G}{g}}^{\times} \underline{O} \cdot \underline{V}^{\times -1} \cdot \underline{T}^{\times}$$

and criterion (4) is as follows

$$\underline{O} \cdot \left(\sqrt{\frac{G}{g}}^{\times} \underline{V}^{\times -1} \cdot \underline{T}^{\times} - \sqrt{\frac{G}{g}} \underline{V}^{-1} \cdot \underline{T} \right) \cdot\cdot (\underline{V}^{\times} - \underline{V}) \cdot \underline{O}^{T} =$$

$$= \left(\sqrt{\frac{G}{g}}^{\times} \underline{V}^{\times -1} \cdot \underline{T}^{\times} - \sqrt{\frac{G}{g}} \underline{V}^{-1} \cdot \underline{T} \right) \cdot\cdot (\underline{V}^{\times} - \underline{V}) > 0 .$$

The tensors $\underline{V}^{\times}$ and $\underline{V}$ in this inequality have the same axes by definition, and this is also true for $\underline{T}$ and $\underline{V}$ because the medium is isotropic. Recalling the definition (4.3.12) of the principal forces, we have

$$\sqrt{\frac{G}{g}} \underline{V}^{-1} \cdot \underline{T} = \sqrt{I_3} \sum_{k=1}^{3} \underline{e}_k \underline{e}_k \frac{\sigma_k}{v_k} = \sum_{k=1}^{3} \underline{e}_k \underline{e}_k t_k$$

and the preceding inequality reduces to the form

$$\sum_{k=1}^{3} (t_k^{\times} - t_k)(v_k^{\times} - v_k) = (t_1^{\times} - t_1)(v_1^{\times} - v_1) + (t_2^{\times} - t_2)(v_2^{\times} - v_2) +$$

$$+ (t_3^{\times} - t_3)(v_3^{\times} - v_3) > 0 , \tag{7}$$

which includes the criteria discussed below (Par. 13) and which confirms the general concepts regarding isotropic elastic materials. The description of the behaviour of a material is by no means exhausted by Inequality (7) since the derivation of this inequality presupposes that the axes of the actual deformation and the deformation being compared with it coincide.

Inequality (7) shows that if $v_k^{\times} = v_k$ for some value of k, then $t_n^{\times} = t_n$ for at least one value of n. In other words, that to different v_k there correspond different values of t_k. The functions $v_k(t_1, t_2, t_3)$ are single-valued functions, and the Equations (5.3.13) which determine $t_k(v_1, v_2, v_3)$ are uniquely reversible if the Coleman-Noll criterion is satisfied.

In the differential form of Inequality (7), only the first powers of differences $\overset{\times}{v}_k - v_s = \Delta v_s$ are retained. We then have

$$\overset{\times}{t}_s - t_s = \sum_{k=1}^{3} \frac{\partial t_s}{\partial v_k} \Delta v_k , \qquad \sum_{s=1}^{3} \sum_{k=1}^{3} \frac{\partial t_s}{\partial v_k} \Delta v_k \Delta v_s \geq 0 . \tag{8}$$

It must be kept in mind that under the transition from a strict inequality (>) to its differential form the increment $\overset{\times}{t}_s - t_s$ evaluated to linear terms in Δv_k may turn out to be zero. For this reason the sign ">" is substituted for "≥".

We now examine the strenghtened criterion of monotonicity which eliminates the equality sign in (8). For this case, the matrix

$$\left\| \frac{\partial t_s}{\partial v_k} \right\|$$

is positive and its determinant satisfies the conditions of the Sylvester's theorem

$$\sum_{s=1}^{3} \sum_{k=1}^{3} \frac{\partial t_s}{\partial v_k} \Delta v_k \Delta v_s > 0 \tag{9}$$

or, referring to (4.3.16),

$$\sum_{s=1}^{3} \sum_{k=1}^{3} \frac{\partial^2 \ni}{\partial v_s \partial v_k} \Delta v_s \Delta v_k > 0 \tag{10}$$

the symmetric tensor $\nabla\nabla\ni$ is positive (its eigenvalues are positive).

3. For a linearly elastic body, the state of stress is compared with the natural state (stress absent)

$$\overset{0}{\underline{\underline{\nabla R}}} = \underline{E} , \qquad \underline{P}(\overset{0}{\underline{\underline{\nabla R}}}) = \underline{P}(\underline{E}) = \underline{0} , \qquad \overset{0}{\underline{\underline{\nabla R}}}{}^{\times} = \underline{E} + \underline{\varepsilon} = \underline{\underline{ES}} , \qquad \underline{S} = \underline{E} + \underline{\varepsilon} ,$$

here, $\underline{\varepsilon}$ denotes a linear strain tensor, $\underline{S} - \underline{S}^T$ is a positive tensor for small $\underline{\varepsilon}$. In linear theory, the Piola tensor is indistinguishable from the stress tensor $\underline{T}$. Using (4.7.2) Inequality (4) is reduced to the form

$$\underline{P}(\underline{E} + \underline{\varepsilon}) \cdot\cdot\, \underline{\varepsilon} = \underline{T}(\underline{E} + \underline{\varepsilon}) \cdot\cdot\, \underline{\varepsilon} = \underline{\varepsilon} \cdot\cdot\, \underline{T}_{\varepsilon} \cdot\cdot\, \underline{\varepsilon} =$$

$$= \underline{\varepsilon} \cdot\cdot\, [\lambda \underline{\underline{EE}} + \mu(\underline{C}_{II} + \underline{C}_{III})] \cdot\cdot\, \underline{\varepsilon} = \lambda I_1^2(\underline{\varepsilon}) + 2\mu I_1(\underline{\varepsilon}^2) =$$

$$= (\lambda + 2\mu)(\varepsilon_1^2 + \varepsilon_2^2 + \varepsilon_3^2) + 2\lambda(\varepsilon_1\varepsilon_2 + \varepsilon_2\varepsilon_3 + \varepsilon_3\varepsilon_1) > 0 \tag{11}$$

and demonstrates the positiveness of the (doubled) specific strain energy, Э, of a linearly elastic body. The determinant of the matrix of this quadratic form and its diagonal minors are equal to

$$4\mu^2(3\lambda + 2\mu) , \qquad \lambda + 2\mu , \qquad 4\mu(\lambda + \mu)$$

and the conditions of the Sylvester's theorem are satisfied if

$$\mu > 0 , \qquad \lambda + \mu > 0 , \qquad \lambda + 2\mu > 0 , \qquad 3\lambda + 2\mu > 0 .$$

The second and the third inequalities follow from the first and the fourth since $3\lambda + 2\mu + \mu = 3(\lambda + \mu)$, $3\lambda + 2\mu + 4\mu = 3(\lambda + 2\mu)$. The two inequalities

$$\mu > 0 , \qquad 3\lambda + 2\mu > 0 \tag{12}$$

are not only sufficient but also the necessary conditions of validity of the criterion (4) for a linearly elastic body. Naturally criterion (10) leads to the same conditions.

10. THE CRITERIA FOR INCREASE IN MAGNITUDE OF STRESS

When the configuration to be compared with the actual one is prescribed in a more general way than that given in Par. 9, one should expect the appearance of material properties which are not contained in criterion (9.7).

Let us assume

$$\underline{S} = \underline{E} + \tau\underline{D} , \qquad \underline{D} = \underline{D}^T , \qquad \overset{0}{\underline{\underline{\nabla R}}}{}^{\times} = \overset{0}{\underline{\underline{\nabla R}}} + \tau\overset{0}{\underline{\underline{\nabla R}}} \cdot \underline{D} ,$$

$$\overset{0}{\underline{\underline{\nabla R}}}{}^{\times T} - \overset{0}{\underline{\underline{\nabla R}}}{}^{T} = \tau\underline{D} \cdot \overset{0}{\underline{\underline{\nabla R}}}{}^{T} . \tag{1}$$

Here, the parameter τ is assumed to be so small that tensor $\underline{S}$ remains positive. Criterion (9.4) now yields

$$\tau[f(\tau) - f(0)] = \tau\underline{D} \cdot \overset{0}{\underline{\underline{\nabla R}}}{}^{T} \cdot\cdot [\underline{P}(\overset{0}{\underline{\underline{\nabla R}}} + \tau\overset{0}{\underline{\underline{\nabla R}}} \cdot \underline{D}) - \underline{P}(\overset{0}{\underline{\underline{\nabla R}}})] > 0 . \tag{2}$$

The quantity

$$f(0) = \underline{D} \cdot \overset{0}{\underline{\underline{\nabla R}}}{}^{T} \cdot\cdot \underline{P}(\overset{0}{\underline{\underline{\nabla R}}}) = \sqrt{\frac{G}{g}}\, \underline{D} \cdot \overset{0}{\underline{\underline{\nabla R}}}{}^{T} \cdot\cdot \underline{\underline{\nabla r}}^{T} \cdot \underline{T} = \sqrt{\frac{G}{g}}\, \underline{D} \cdot\cdot \underline{T} \tag{3}$$

is proportional to the magnitudes of the stresses in the actual configuration, and $f(\tau)$ is proportional to the magnitudes of the stresses in the configuration being compared. In the following approximation, which is linear with respect to τ, we have

$$\mathbf{P}(\overset{0}{\nabla}\mathbf{R} + \tau\overset{0}{\nabla}\mathbf{R}\cdot\mathbf{D}) = \mathbf{P}(\overset{0}{\nabla}\mathbf{R}) + \tau\,\mathbf{P}(\overset{0}{\nabla}\mathbf{R})_{\overset{0}{\nabla}\mathbf{R}} \cdot\cdot\, \mathbf{D}\cdot\overset{0}{\nabla}\mathbf{R}^{T} . \tag{4}$$

On dividing by τ^2 and taking it to the limit $\tau \to 0$, Inequality (2) reduces to a quadratic form in six variables $D_{mn} = D_{nm}$

$$f'(0) = \mathbf{D}\cdot\overset{0}{\nabla}\mathbf{R}^{T} \cdot\cdot\, \mathbf{P}_{\overset{0}{\nabla}\mathbf{R}} \cdot\cdot\, \mathbf{D}\cdot\overset{0}{\nabla}\mathbf{R}^{T} \geq 0 . \tag{5}$$

Substituting for $\mathbf{P}_{\overset{0}{\nabla}\mathbf{R}}$ from (4.9.11) and expressing $\mathbf{P}$ in terms of $\mathbf{T}$, we obtain

$$\sqrt{\frac{g}{G}}\, f'(0) = \mathbf{D}\cdot\cdot\,(\mathbf{TE} + \mathbf{R}_s\mathbf{TR}^s)\cdot\cdot\,\mathbf{D} + \mathbf{D}\cdot\cdot\,[\mathbf{T}_{\mathbf{F}}\cdot\cdot\,(\mathbf{R}_t\mathbf{r}_s + \mathbf{r}_s\mathbf{R}_t)(\mathbf{r}^t\mathbf{r}^s\cdot\cdot\,\mathbf{D}\cdot\overset{0}{\nabla}\mathbf{R}^{T})] .$$

After the transformation

$$(\mathbf{R}_t\mathbf{r}_s + \mathbf{r}_s\mathbf{R}_t)\mathbf{r}^t\mathbf{r}^s\cdot\cdot\,\mathbf{D}\cdot\overset{0}{\nabla}\mathbf{R}^{T} =$$

$$= (\mathbf{R}_t\mathbf{r}_s + \mathbf{r}_s\mathbf{R}_t)g^{tm}\mathbf{R}_m\mathbf{r}^s\cdot\cdot\,\mathbf{D} = (\mathbf{F}\cdot\mathbf{R}^m\mathbf{E}\mathbf{R}_m + \mathbf{r}_s\mathbf{F}\mathbf{r}^s)\cdot\cdot\,\mathbf{D}$$

Inequality (5) can be expressed in the following form

$$\sqrt{\frac{g}{G}}\, f'(0) = \mathbf{D}\cdot\cdot\,\mathbf{B}\cdot\cdot\,\mathbf{D} \geq 0 , \tag{6}$$

where the fourth rank tensor $\mathbf{B}$ turns out to be

$$\mathbf{B} = \mathbf{TE} + \mathbf{R}_s\mathbf{TR}^s + \mathbf{T}_{\mathbf{F}}\cdot\cdot\,(\mathbf{F}\cdot\mathbf{R}^m\mathbf{E}\mathbf{R}_m + \mathbf{r}_s\mathbf{F}\mathbf{r}^s) . \tag{7}$$

We also have

$$\mathbf{D}\cdot\cdot\,\mathbf{TE}\cdot\cdot\,\mathbf{D} = I_1(\mathbf{D}\cdot\mathbf{T})I_1(\mathbf{D}) ,$$

$$\mathbf{D}\cdot\cdot\,\mathbf{R}_s\mathbf{TR}^s\cdot\cdot\,\mathbf{D} = D_{ms}D^{sq}\mathbf{R}^m\mathbf{R}_q\cdot\cdot\,\mathbf{T} = I_1(\mathbf{D}^2\cdot\mathbf{T}) .$$

We now introduce the component representation for $\mathbf{T}_{\mathbf{F}}$, so that

$$\mathbf{D}\cdot\cdot\,\tau^{qpst}\mathbf{R}_q\mathbf{R}_p\mathbf{R}_s\mathbf{R}_t\cdot\cdot\,(\mathbf{F}\cdot\mathbf{R}^m\mathbf{E}\mathbf{R}_m + \mathbf{r}_s\mathbf{F}\mathbf{r}^s)\cdot\cdot\,\mathbf{D} =$$

$$= D_{pq}\tau^{qpst}(D_{sm}\underline{F}\cdot\cdot\underline{R}_t\underline{R}^m + D_{tm}\underline{F}\cdot\cdot\underline{R}_s\underline{R}^m) = 2D_{pq}D_{sm}\tau^{pqst}\underline{F}\cdot\cdot\underline{R}_t\underline{R}^m$$

here we have used the permutation rule $\tau^{qpst} = \tau^{pqts}$ and the symmetry of $\underline{D}$. Inequality (6) is thus reduced to the form

$$\frac{1}{2}\sqrt{\frac{g}{G}}\,f'(0) = \frac{1}{2}[I_1(\underline{D})I_1(\underline{D}\cdot\underline{T}) - I_1(\underline{D}^2\cdot\underline{T})] + \tau^{qpst}D_{pq}D_{sm}g^{mn}G_{nt} \geq 0$$

where ≥ should be substituted for > if we wish to obtain the strengthened criterion.

In the orthonormal basis of the principal directions of $\underline{T}$ and $\underline{F}$,

$$\underline{T} = \sum_{s=1}^{3}\sigma_s\underline{e}_s\underline{e}_s\,, \qquad \underline{F} = \sum_{s=1}^{3}v_s^2\underline{e}_s\underline{e}_s\,; \qquad g^{mn} = \begin{cases} v_m^2\,, & m = n\,,\\ 0\,, & m \neq n\,,\end{cases}$$

$$G_{nt} = \begin{cases} 1\,, & n = t\,,\\ 0\,, & n \neq t\,,\end{cases} \qquad \underline{D} = \sum_{s=1}^{3}\sum_{m=1}^{3}D_{sm}\underline{e}_s\underline{e}_m \tag{8}$$

and therefore

$$I_1(\underline{D})I_1(\underline{D}\cdot\underline{T}) - I_1(\underline{D}^2\cdot\underline{T}) = \sum_{k=1}^{3}D_{kk}\sum_{s=1}^{3}D_{ss}\sigma_s - \sum_{s=1}^{3}\sigma_s\sum_{k=1}^{3}D_{sk}^2 =$$

$$= (\sigma_1+\sigma_2)(D_{11}D_{22}-D_{12}^2) + (\sigma_2+\sigma_3)(D_{22}D_{33}-D_{23}^2) + (\sigma_3+\sigma_1)(D_{33}D_{11}-D_{31}^2)\,.$$

The second group in (7) contains terms which are proportional to the 12 non-zero components (4.8.9) and (4.8.10) of the elasticity tensor $\underline{T}_{\underline{F}}$

$$\sum_{q=1}^{3}D_{qq}^2v_q^2\frac{\partial\sigma_q}{\partial v_q^2} + D_{11}D_{22}\left(v_1^2\frac{\partial\sigma_2}{\partial v_1^2} + v_2^2\frac{\partial\sigma_1}{\partial v_2^2}\right) +$$

$$+ D_{22}D_{33}\left(v_2^2\frac{\partial\sigma_3}{\partial v_2^2} + v_3^2\frac{\partial\sigma_2}{\partial v_3^2}\right) + D_{33}D_{11}\left(v_3^2\frac{\partial\sigma_1}{\partial v_3^2} + v_1^2\frac{\partial\sigma_3}{\partial v_1^2}\right) +$$

$$+ D_{12}^2\frac{\sigma_1-\sigma_2}{v_1^2-v_2^2}(v_1^2+v_2^2) + D_{23}^2\frac{\sigma_2-\sigma_3}{v_2^2-v_3^2}(v_2^2+v_3^2) + D_{31}^2\frac{\sigma_3-\sigma_1}{v_3^2-v_1^2}(v_3^2+v_1^2)\,.$$

The quadratic form (6) of the six variables

$$D_{ss} = x_s\,, \qquad s = 1,2,3\,; \qquad\qquad x_4 = D_{12}\,, \quad x_5 = D_{23}\,, \quad x_6 = D_{31}$$

may now be written in the form

$$\frac{1}{2}\,\underline{D}\cdot\cdot\,\underline{B}\cdot\cdot\,\underline{D} = \sum_{s=1}^{3} x_s^2 v_s^2 \frac{\partial\sigma}{\partial v_s^2} + x_1x_2\left[v_1^2\frac{\partial\sigma_2}{\partial v_1^2} + v_2^2\frac{\partial\sigma_1}{\partial v_2^2} + \frac{1}{2}(\sigma_1+\sigma_2)\right] +$$

$$+ x_2x_3\left[v_2^2\frac{\partial\sigma_3}{\partial v_2^2} + v_3^2\frac{\partial\sigma_2}{\partial v_3^2} + \frac{1}{2}(\sigma_2+\sigma_3)\right] +$$

$$+ x_3x_1\left[v_3^2\frac{\partial\sigma_1}{\partial v_3^2} + v_1^2\frac{\partial\sigma_3}{\partial v_1^2} + \frac{1}{2}(\sigma_3+\sigma_1)\right] +$$

$$+ x_4^2\left[\frac{\sigma_1-\sigma_2}{v_1^2-v_2^2}(v_1^2+v_2^2) - \frac{1}{2}(\sigma_1+\sigma_2)\right] +$$

$$+ x_5^2\left[\frac{\sigma_2-\sigma_3}{v_2^2-v_3^2}(v_2^2+v_3^2) - \frac{1}{2}(\sigma_2+\sigma_3)\right] +$$

$$+ x_6^2\left[\frac{\sigma_3-\sigma_1}{v_3^2-v_1^2}(v_3^2+v_1^2) - \frac{1}{2}(\sigma_3+\sigma_1)\right]. \tag{9}$$

Here the principal stresses are replaced by the principal forces

$$v_1^2\frac{\partial\sigma_2}{\partial v_1^2} = \frac{1}{2}v_1\frac{\partial}{\partial v_1}\frac{t_2}{v_3v_1} = \frac{1}{2}\left(\frac{1}{v_3}\frac{\partial t_2}{\partial v_1} - \sigma_2\right);$$

$$v_1^2\frac{\partial\sigma_2}{\partial v_1^2} + v_2^2\frac{\partial\sigma_1}{\partial v_2^2} + \frac{1}{2}(\sigma_1+\sigma_2) = \frac{1}{2v_3}\left(\frac{\partial t_2}{\partial v_1} + \frac{\partial t_1}{\partial v_2}\right) = \frac{1}{v_3}\frac{\partial^2\ni}{\partial v_1\partial v_2} = \sqrt{\frac{g}{G}}\,v_1v_2\frac{\partial^2\ni}{\partial v_1\partial v_2}\,;$$

$$v_1^2\frac{\partial\sigma_1}{\partial v_1^2} = \frac{1}{2}v_1\frac{\partial}{\partial v_1}\frac{t_1}{v_2v_3} = \frac{1}{2}\sqrt{\frac{g}{G}}\,v_1^2\frac{\partial^2\ni}{\partial v_1^2}$$

etc. Using the notation

$$x_s' = \sqrt{\frac{g}{G}}\,v_sx_s \qquad (s = 1,2,3)\,,$$

$$A_1 = \frac{\sigma_1-\sigma_2}{v_1^2-v_2^2}(v_1^2+v_2^2) - \frac{1}{2}(\sigma_1+\sigma_2) \quad \text{etc.} \tag{10}$$

we arrive at the quadratic form

$$\Phi = \frac{1}{2} \sum_{s=1}^{3} \sum_{k=1}^{3} \frac{\partial^2 \backepsilon}{\partial v_s \partial v_k} x'_s x'_k + A_1 x_4^2 + A_5 x_5^2 + A_6 x_6^2 \ . \tag{11}$$

The matrix generating this form possesses the structure

$$\left\| \begin{matrix} \frac{\partial^2 \backepsilon}{\partial v_1^2} & \frac{\partial^2 \backepsilon}{\partial v_1 \partial v_2} & \frac{\partial^2 \backepsilon}{\partial v_1 \partial v_3} & & & \\ \frac{\partial^2 \backepsilon}{\partial v_2 \partial v_1} & \frac{\partial^2 \backepsilon}{\partial v_2^2} & \frac{\partial^2 \backepsilon}{\partial v_2 \partial v_3} & & 0 & \\ \frac{\partial^2 \backepsilon}{\partial v_3 \partial v_1} & \frac{\partial^2 \backepsilon}{\partial v_3 \partial v_2} & \frac{\partial^2 \backepsilon}{\partial v_3^2} & & & \\ & & & A_1 & & \\ & 0 & & & A_2 & \\ & & & & & A_3 \end{matrix} \right\| \tag{12}$$

and the condition that (11) should be positive-definite (i.e. the strengthened monotonicity criterion) is satisfied provided that the following inequalities hold

$$\frac{\partial^2 \backepsilon}{\partial v_1^2} > 0 \ , \qquad \frac{\partial^2 \backepsilon}{\partial v_1^2} \frac{\partial^2 \backepsilon}{\partial v_2^2} - \left(\frac{\partial^2 \backepsilon}{\partial v_1 \partial v_2} \right)^2 > 0 \ \text{etc.} \qquad \det \left\| \frac{\partial^2 \backepsilon}{\partial v_s \partial v_k} \right\| > 0 \ , \tag{13}$$

which again yield (9.10), and

$$A_\Gamma > 0 \ , \qquad \Gamma = 1,2,3 \ . \tag{14}$$

An alternative representation of Inequalities (14) is given by

$$A_{ab} = A_{ba} = \frac{\sigma_a - \sigma_b}{v_a^2 - v_b^2} (v_a^2 + v_b^2) - \frac{1}{2} (\sigma_a + \sigma_b) > 0 \ , \tag{15}$$

where $a,b = 1,2,3$ and $a \neq b$.

After an identical set of transformations involving the discarding of several positive factors, Inequality (15) is substituted by the equivalent condition

$$(v_a - v_b)(\sigma_a - \sigma_b) > 2\sigma_b \frac{(v_a - v_b)^2}{v_a^2 + 3v_b^2}(v_a + v_b) \qquad (a \neq b) , \tag{16}$$

whose left-hand side is positive when the *BE*-criterion (4.12.12) holds and in the right-hand side of which σ_b can be substituted for σ_a from (15). For this inequality to hold, it is sufficient that in the pair (σ_a, σ_b) there be at least one non-positive number.

All three Inequalities (14) hold therefore provided that within any pair (σ_1, σ_2), (σ_2, σ_3) or (σ_3, σ_1) there is at least one non-positive number, in other words, if the *BE*-criteria are satisfied two of the three principal stresses should be non-positive.

A similar consideration for the principal forces leads to the inequalities

$$(v_a - v_b)(t_a - t_b) > -t_b \frac{(v_a - v_b)^4}{v_a(v_a^2 + 3v_b^2)} \qquad (a \neq b) . \tag{17}$$

The *OF*-criterion ("ordered forces") will be derived in Chapter 5, Par. 13

$$(v_a + v_b)(t_a - t_b) > 0 \qquad (a \neq b) . \tag{18}$$

When this criterion is satisfied, for Inequality (17) to hold it is sufficient that in the pair (t_a, t_b) there should be at least one negative number and of the three principal forces, ar least two should be negative.

The principal stresses and forces have the same sign, and therefore two of the three principal stresses are always either non-positive or nonnegative.

We thus see that whenever the *BE*- and *OF*-criteria hold simultaneously, then all Inequalities (14) are satisfied. The inverse of course does not hold.

11. EMPIRICAL CRITERION

The equation of state

$$\underline{T} = 2\sqrt{\frac{g}{G}}\left[\underline{E}\left(I_2 \frac{\partial \ni}{\partial I_2} + I_3 \frac{\partial \ni}{\partial I_3}\right) + \underline{F}\frac{\partial \ni}{\partial I_1} - I_3 \frac{\partial \ni}{\partial I_2}\underline{F}^{-1}\right] , \tag{1}$$

which follows immediately on eliminating $\underline{F}^2$ from (4.3.4) using the Hamilton-Cayley theorem (I.9.22) enables Inequalities (10.14) to be written in the following form

$$\frac{1}{2}(v_1^2 + v_2^2)\left(\frac{\partial \ni}{\partial I_1} + 3v_3^2 \frac{\partial \ni}{\partial I_2}\right) - \left(I_2 \frac{\partial \ni}{\partial I_2} + I_3 \frac{\partial \ni}{\partial I_3}\right) > 0 \qquad \text{etc.} \tag{2}$$

These inequalities are satisfied whenever the following conditions hold

$$\frac{\partial \ni}{\partial I_1} \geq 0 \; , \qquad \frac{\partial \ni}{\partial I_2} \geq 0 \; , \qquad I_2 \frac{\partial \ni}{\partial I_2} + I_3 \frac{\partial \ni}{\partial I_3} \leq 0 \; . \tag{3}$$

These are called the "empirical" criteria. These conditions are of course sufficient but not necessary for the validity of Inequalities (10.14).

(a) For a simplified (two constants) Signorini material, we have from (2.12) that

$$\ni = \frac{1}{8} \sqrt{I_3} \left[9\lambda + 5\mu - 2(3\lambda + \mu) \frac{I_2}{I_3} + (\lambda + \mu) \left(\frac{I_2}{I_3} \right)^2 \right] ; \tag{4}$$

and the quantity in the rectangular brackets, which is equal to μ in the natural state, remains positive if $\mu > 0$ for any value of I_2/I_3. This follows from Inequalities (2.14) together with the condition

$$(9\lambda + 5\mu)(\lambda + \mu) - (3\lambda + \mu)^2 = 4\mu(2\lambda + \mu) > 0 \; , \qquad 2\lambda + \mu > 0 \; ,$$

since otherwise for $2\lambda + \mu < 0$ we would have simultaneously that $3\lambda + \mu < 0$ and $\lambda + \mu > 0$ for the values of ν from the interval given by (2.16) and the roots I_2/I_3 of the expression in brackets would then be negative.

The first of the conditions given by (3) is satisfied for equality. At the same time we have

$$I_2 \frac{\partial \ni}{\partial I_2} + I_3 \frac{\partial \ni}{\partial I_3} = \frac{1}{16} \sqrt{I_3} \left[(\lambda + \mu) \frac{I_2^2}{I_3^2} - 2(3\lambda + \mu) \frac{I_2}{I_3} + 9\lambda + 2\mu \right] =$$

$$= \frac{1}{2} (\ni + \mu) > 0 \tag{5}$$

which contradicts the third inequality. This casts doubt upon the appropriateness of the empirical criterion (3) but not that of a Signorini material per se.

(b) For a simplified version (6.3) of a Blatz and Ko material

$$\frac{\partial \ni}{\partial I_1} > 0 \; , \qquad I_2 \frac{\partial \ni}{\partial I_2} = \frac{1}{2} \mu \frac{I_2}{I_3} > 0 \; , \qquad I_2 \frac{\partial \ni}{\partial I_2} + I_3 \frac{\partial \ni}{\partial I_3} = \frac{1}{2} \mu \sqrt{I_3} > 0$$

the third condition of (3) is also violated. At the same time, from (10.10) and (6.5), Inequalities (10.14) are reduced to

$$\frac{1}{v_1^2} + \frac{1}{v_2^2} > \frac{2}{3} v_1 v_2 v_3 \; , \qquad \frac{1}{v_2^2} + \frac{1}{v_3^2} > \frac{2}{3} v_1 v_2 v_3 \; , \qquad \frac{1}{v_3^2} + \frac{1}{v_1^2} > \frac{2}{3} v_1 v_2 v_3 \; , \tag{6}$$

these evidently being satisfied within the cube bounded by $v_s \leq 3^{1/5} \approx 1.25$.

(c) The conclusions regarding the appropriateness or otherwise of the empirical criteria for a Murnaghan material are unreliable and contradictory. From (3.2) we have

$$4\frac{\partial \ni}{\partial I_1} = (I_1 - 3)\lambda + 2\mu(I_1 - 1) + \frac{1}{2}l(I_1 - 3)^2 + m(I_1^2 - 2I_1 - I_2) + \frac{n}{2},$$

$$4\frac{\partial \ni}{\partial I_2} = -2\mu - \frac{n}{2} - m(I_1 - 3), \qquad 4\frac{\partial \ni}{\partial I_3} = \frac{n}{2}$$

and the criteria of (3) in the natural configuration lead to the inequalities

$$4\mu + \frac{n}{2} > 0, \qquad 4\mu + n < 0, \qquad 6\mu + n > 0,$$

or (for $\mu > 0$)

$$1 < -\frac{n}{4\mu} < 1,5.$$

For two varieties of Hecla steel, acoustical measurements (R.T. Smith, R. Stern, R.W. Stephens, 1966) have shown that for Hecla 37 (in dyne/cm^2)

$$\lambda = 1,11 \pm 0,01, \qquad \mu = 0,821 \pm 0,005, \qquad l = -4,61 \pm 0,65,$$

$$m = -6,36 \pm 0,46, \qquad n = -7,08 \pm 0,32,$$

and for Hecla 17 (in 10^{12} dyne/cm^2)

$$\lambda = 1,10 \pm 0,01, \qquad \mu = 0,820 \pm 0,005, \qquad l = -3,28 \pm 0,30,$$

$$m = -5,95 \pm 0,32, \qquad n = -6,68 \pm 0,24.$$

It is difficult to draw from these data any conclusion regarding the appropriateness or otherwise of the criteria of (3).

12. CONVEXITY OF THE SPECIFIC STRAIN ENERGY

The Coleman-Noll monotonicity criterion may be associated with the property of convexity of the specific strain energy $\ni$ regarded as a function of the invariants $I_k(\underset{\sim}{G})$ of either the Cauchy-Green or Finger strain measure.

To elucidate the notion of convexity, it is sufficient to confine our consideration to a twice continuously differentiable function $F(x)$. This function is convex (from below) provided that its first derivative $f(x)$ increases monotonically. Introducing the parameter $\tau \subset [0,1]$ and assuming that

$$x(\tau) = x_0 + \tau(x_1 - x_0) \; , \tag{1}$$

we have by (9.1)

$$[f(x(\tau)) - f(x_0)][x(\tau) - x_0] = \tau(x_1 - x_0)\,[f(x(\tau)) - f(x_0)] > 0 \; ,$$

$$\frac{dF(x,\tau)}{d\tau} = \frac{dF(x(\tau))}{dx(\tau)}\,(x_1 - x_0) = f(x(\tau))\,(x_1 - x_0) \; . \tag{2}$$

These inequalities imply

$$\frac{dF(x,\tau)}{d\tau} - f(x_0)\,(x_1 - x_0) = \frac{1}{\tau}\,[f(x(\tau)) - f(x_0)]\,(x(\tau) - x_0) \tag{3}$$

and integration with respect to τ leads to the representation of the convex function in terms of its monotonically increasing derivative

$$F(x_1) - F(x_0) - f(x_0)\,(x_1 - x_0) = \int_0^1 \frac{d\tau}{\tau}\,[f(x(\tau)) - f(x_0)]\,(x(\tau) - x_0) \; . \tag{4}$$

For a function of a tensor argument, this definition corresponds to the relationship

$$\ni(\underline{\underline{\overset{0}{\nabla}R}}^{\times}) - \ni(\underline{\underline{\overset{0}{\nabla}R}}) - \underline{P}(\underline{\underline{\overset{0}{\nabla}R}}) \cdot\cdot (\underline{\underline{\overset{0}{\nabla}R}}^{T\times} - \underline{\underline{\overset{0}{\nabla}R}}^{T}) =$$

$$= \int_0^1 \frac{d\tau}{\tau}\,[\underline{P}(\underline{\underline{\overset{0}{\nabla}R}}(\tau)) - \underline{P}(\underline{\underline{\overset{0}{\nabla}R}})] \cdot\cdot [\underline{\underline{\overset{0}{\nabla}R}}^{T}(\tau) - \underline{\underline{\overset{0}{\nabla}R}}^{T}] > 0 \; , \tag{5}$$

where we have by (1)

$$\underline{\underline{\overset{0}{\nabla}R}}(\tau) = \underline{\underline{\overset{0}{\nabla}R}} + \tau(\underline{\underline{\overset{0}{\nabla}R}}^{\times} - \underline{\underline{\overset{0}{\nabla}R}}) \; , \tag{6}$$

where it is required that $\underline{\underline{\overset{0}{\nabla}R}}^{\times} = \underline{\underline{\overset{0}{\nabla}R}} \cdot \underline{S}$ and $\underline{S} = \underline{S}^{T}$ denotes a positive tensor. This condition is analogous to the requirement $x_1 - x_2 > 0$ in the definition of a monotonically increasing function.

Interchanging $\underline{\underline{\overset{0}{\nabla}R}}$ and $\underline{\underline{\overset{0}{\nabla}R}}^{\times}$ in (5), we have from (5)

$$\ni(\underline{\underline{\overset{0}{\nabla}R}}) - \ni(\underline{\underline{\overset{0}{\nabla}R}}^{\times}) > \underline{P}(\underline{\underline{\overset{0}{\nabla}R}}^{\times}) \cdot\cdot (\underline{\underline{\overset{0}{\nabla}R}}^{T} - \underline{\underline{\overset{0}{\nabla}R}}^{T\times}) \tag{7}$$

this, together with Inequalities (5) and (7), leads to the criterion of monotonicity as initially given in (9.4).

If the reference configuration is natural ($\overset{0}{\underline{\nabla R}} = \underline{E}$, $\overset{0}{\underline{\nabla R}}{}^{\times} = \underline{S}$) then, from (5),

$$\ni(\underline{S}) > \ni(\underline{E}) \tag{8}$$

which means that the specific strain energy is a minimum for the natural state. This has already been expressed by Inequality (10.13) which determines the convexity of a twice differentiable function.

An alternative representation of the criterion of monotonicity is obtained if we substitute the Piola tensor in (9.4) expressed in terms of the specific strain energy $\ni(\overset{0}{\underline{\nabla R}})$.

As in (10.1), we assume that

$$\overset{0}{\underline{\nabla R}}{}^{\times} = \overset{0}{\underline{\nabla R}} \cdot \underline{S} = \overset{0}{\underline{\nabla R}} \cdot (\underline{E} + \eta \underline{D})$$

where η denotes a small parameter. Referring now to the expansion

$$\ni(\overset{0}{\underline{\nabla R}} + \eta \overset{0}{\underline{\nabla R}} \cdot \underline{D}) = \ni(\overset{0}{\underline{\nabla R}}) + \ni_{\overset{0}{\underline{\nabla R}}} \cdot\cdot\, \eta \underline{D} \cdot \overset{0}{\underline{\nabla R}}{}^{T} + \frac{1}{2}\eta^2 \underline{D} \cdot \overset{0}{\underline{\nabla R}}{}^{T} \cdot\cdot\, \ni_{\overset{0}{\underline{\nabla R}}\overset{0}{\underline{\nabla R}}} \cdot\cdot\, \underline{D} \cdot \overset{0}{\underline{\nabla R}}{}^{T} + \ldots =$$

$$= \ni(\overset{0}{\underline{\nabla R}}) + \eta \overset{0}{\underline{\nabla R}}{}^{T} \cdot \underline{P} \cdot\cdot\, \underline{D} + \frac{1}{2}\eta^2 \underline{D} \cdot \overset{0}{\underline{\nabla R}}{}^{T} \cdot\cdot\, \underline{P}_{\overset{0}{\underline{\nabla R}}} \cdot\cdot\, \underline{D} \cdot \overset{0}{\underline{\nabla R}}{}^{T} + \ldots , \tag{9}$$

we once again obtain the relationship (10.3)

$$\left[\frac{d}{d\eta} \ni(\overset{0}{\underline{\nabla R}} + \eta \overset{0}{\underline{\nabla R}} \cdot \underline{D})\right]_{\eta=0} = \overset{0}{\underline{\nabla R}}{}^{T} \cdot \sqrt{\frac{G}{g}}\, \underline{\nabla r}^{T} \cdot \underline{T} \cdot\cdot\, \underline{D} = \sqrt{\frac{G}{g}}\, I_1(\underline{T} \cdot \underline{D}) . \tag{10}$$

At the same time

$$\underline{P}(\overset{0}{\underline{\nabla R}}{}^{\times}) = \underline{P}(\overset{0}{\underline{\nabla R}} + \eta \overset{0}{\underline{\nabla R}} \cdot \underline{D}) = \underline{P}(\overset{0}{\underline{\nabla R}}) + \eta \underline{D} \cdot \overset{0}{\underline{\nabla R}}{}^{T} \cdot\cdot\, \underline{P}_{\overset{0}{\underline{\nabla R}}} + \ldots ,$$

so that by (9)

$$\left\{\frac{1}{\eta^2}[\underline{P}(\overset{0}{\underline{\nabla R}}{}^{\times}) - \underline{P}(\overset{0}{\underline{\nabla R}})] \cdot (\overset{0}{\underline{\nabla R}}{}^{\times T} - \overset{0}{\underline{\nabla R}}{}^{T})\right\}_{\eta=0} = \left[\frac{d^2}{d\eta^2} \ni(\overset{0}{\underline{\nabla R}} + \eta \overset{0}{\underline{\nabla R}} \cdot \underline{D})\right]_{\eta=0} .$$

The monotonicity criterion (9.4) is now given by the inequality

$$\left[\frac{d^2}{d\eta^2} \ni(\overset{0}{\underline{\nabla R}} + \eta \overset{0}{\underline{\nabla R}} \cdot \underline{D})\right]_{\eta=0} > 0 . \tag{11}$$

13. THE ADDITIONAL INEQUALITIES OF NONLINEAR THEORY

The conclusions which follow from a consideration of the conditions of the ellipticity of the equilibrium equations enable us to formulate certain specific criteria ("additional inequalities") which are consistent with the expected properties of the state of stress in an elastic isotropic medium.

1. *TE*-, *ET*-inequalities (tension-extension, extension-tension). It is assumed here that two of the three principal elongations have the same values in the state under consideration as in the actual configuration. By (9.7) we arrive at the *TE*-inequalities for the principal forces

$$(\overset{\times}{t}_i - t_i)(\overset{\times}{v}_i - v_i) > 0 \qquad (i = 1,2,3\,, \quad v_j = \overset{\times}{v}_j\,, \quad j \neq i)\,. \tag{1}$$

The same inequalities hold for the principal stresses

$$(\overset{\times}{\sigma}_i - \sigma_i)(\overset{\times}{v}_i - v_i) > 0 \qquad (i = 1,2,3\,, \quad v_j = \overset{\times}{v}_j\,, \quad j \neq i)\,. \tag{2}$$

In fact, we have

$$t_1 = v_2 v_3 \sigma_1\,, \qquad \overset{\times}{t}_1 = \overset{\times}{v}_2 \overset{\times}{v}_3 \overset{\times}{\sigma}_1 = v_2 v_3 \overset{\times}{\sigma}_1\,,$$

$$(\overset{\times}{\sigma}_1 - \sigma_1)(\overset{\times}{v}_1 - v_1) = \frac{1}{v_2 v_3}(\overset{\times}{t}_1 - t_1)(\overset{\times}{v}_1 - v_1)\,,$$

because for the state being compared, $\overset{\times}{v}_2 = v_2$, $\overset{\times}{v}_3 = v_3$.

In the Inequalities (1) and (2)

$$t_s = t_s(v_1, v_2, v_3)\,, \qquad \sigma_s = \sigma_s(v_1, v_2, v_3) \qquad (s = 1,2,3)\,. \tag{3}$$

The first system of equations is uniquely solvable for

$$v_k = v_k(t_1, t_2, t_3)\,, \tag{4}$$

because, by (5.9.10), the Jacobian

$$\det \left\| \frac{\partial t_s}{\partial v_k} \right\| = \det \left\| \frac{\partial^2 \ni}{\partial v_k \partial v_s} \right\| > 0$$

is non-zero.

This allows us to substitute Inequalities (1) by the *ET*-inequalities

$$(\overset{\times}{t}_i - t_i)(\overset{\times}{v}_i - v_i) > 0\,, \qquad i = 1,2,3\,; \qquad \overset{\times}{t}_j = t_j\,, \qquad j \neq i\,. \tag{5}$$

However, there are no grounds for the assertion that the second system of equations (3) is also solvable with respect to v_k. For example, the state equation (3.8.3) for an elastic fluid

$$\sigma_1 = \sigma_2 = \sigma_3 = -p\left(\rho_0 \sqrt{\frac{g}{G}}\right) = -p\left(\frac{\rho_0}{v_1 v_2 v_3}\right) = f(\sqrt{I_3})$$

may be solved with respect to the product $v_1 v_2 v_3$ but not to the values of v_k taken separately. For the principal forces, the situation is different

$$t_i = \sigma_i \frac{\sqrt{I_3}}{v_i} = \frac{1}{v_i}\sqrt{I_3} f(\sqrt{I_3}) , \qquad (t_1 t_2 t_3)^{1/3} = I_3 f(\sqrt{I_3}) .$$

The latter equation enables us to determine $\sqrt{I_3}$ in terms of the product $t_1 t_2 t_3$, which makes it possible to determine v_i with the aid of the formulas

$$v_i = \frac{\varphi(t_1 t_2 t_3)}{t_i} , \qquad i = 1,2,3 . \tag{6}$$

2. The OF-inequality of "ordered forces"

$$(t_i - t_k)(v_i - v_k) > 0 , \qquad v_i \neq v_k ; \qquad i = 1,2,3 , \tag{7}$$

given above, expresses the commonly held idea that a larger deformation is associated with a greater force. Its derivation is based on a consideration of the state being compared for which

$$\overset{\times}{v}_1 = v_2 , \qquad \overset{\times}{v}_2 = v_3 , \qquad \overset{\times}{v}_3 = v_1 , \tag{8}$$

and on Equations (4.3.13) from which we have

$$t_1 = t(v_1, v_2, v_3) , \qquad \overset{\times}{t}_1 = t(v_2, v_3, v_1) = t_2 .$$

Inequalities (5) can now be represented in the form

$$(t_2 - t_1)(v_2 - v_1) > 0 , \qquad v_1 \neq v_2$$

the remaining two OF-inequalities are obtained in a similar way. It remains only to ensure that permutation of indices (8) does not violate conditions (9.6) of the derivation of Inequality (9.7). We have in fact

$$\overset{0}{\underline{\underline{\nabla R}}}{}^{\times} = \underline{O} \cdot \underline{V}^{\times} = \overset{0}{\underline{\underline{\nabla R}}} \cdot \underline{S} = \underline{O} \cdot \underline{V} \cdot \underline{S} , \qquad \underline{V}^{-1} \cdot \underline{V}^{\times} = \underline{S} ,$$

but now

$$\underline{V}^{\times} = \underline{e}_1\underline{e}_1 v_2 + \underline{e}_2\underline{e}_2 v_3 + \underline{e}_3\underline{e}_3 v_1 \ ,$$

$$\underline{S} = \underline{e}_1\underline{e}_1 \frac{v_2}{v_1} + \underline{e}_2\underline{e}_2 \frac{v_3}{v_2} + \underline{e}_3\underline{e}_3 \frac{v_1}{v_3} = \underline{V}^{\times} \cdot \underline{V}^{-1} = \underline{S}^{T}$$

and $\underline{S} = \underline{S}^T$ is a positive tensor, which is the desired result.

3. Together with the $\mathcal{OF}$-inequalities the $\mathcal{BE}$-inequalities (4.12.12) are also cited

$$(\sigma_i - \sigma_k)(v_i - v_k) > 0 \ , \qquad i \neq k \ ; \qquad i,k = 1,2,3 \ , \tag{9}$$

which follows from the conditions of strong ellipticity for the equations of the theory of elasticity. Alternatively, relations

$$\frac{\sigma_1 - \sigma_2}{v_1^2 - v_2^2} > 0 \ , \qquad \frac{\sigma_2 - \sigma_3}{v_2^2 - v_3^2} > 0 \ , \qquad \frac{\sigma_3 - \sigma_1}{v_3^2 + v_1^2} > 0 \tag{10}$$

may also be expressed, on the basis of the state equations (4.3.9), in the form

$$\frac{\partial \ni}{\partial I_1} + v_s^2 \frac{\partial \ni}{\partial I_2} > 0 \quad \text{or} \quad \frac{\partial \ni}{\partial I_1} + \frac{I_3}{v_i^2 v_k^2} \frac{\partial \ni}{\partial I_2} > 0 \ , \qquad v_i \neq v_k \ ; \qquad i,k = 1,2,3 \ .$$

A similar calculation for Inequalities (7) using (4.3.13) enables the former to be expressed in the form

$$(t_1 - t_2)(v_1 - v_2) = 2(v_1 - v_2)^2 \left[\frac{\partial \ni}{\partial I_1} + (v_3^2 - v_1 v_2) \frac{\partial \ni}{\partial I_2} - \frac{1}{v_1 v_2} I_3 \frac{\partial \ni}{\partial I_3} \right] =$$

$$= 2(v_1 - v_2)^2 \left[\frac{\partial \ni}{\partial I_1} + \frac{v_3^2}{v_1 v_2} (v_1^2 + v_2^2 + v_1 v_2) \frac{\partial \ni}{\partial I_2} - \right.$$

$$\left. - \frac{1}{v_1 v_2} \left(I_2 \frac{\partial}{\partial I_2} + I_3 \frac{\partial}{\partial I_3} \right) \right] > 0 \ . \tag{12}$$

Inequalities (11) and (12) are obviously confirmed by the empirical criteria (11.3) but the inverse certainly does not hold and criteria (11.3) by no means follow from the $\mathcal{BE}$- and $\mathcal{OF}$-inequalities. It must be recalled here (Chapter 5, Par. 10) that the simultaneous fulfillment of these inequalities guarantees the validity of Inequalities (10.14).

4. As an example of the TE-inequalities consider the case of uniform stretching or compression: $\overset{\times}{v}_1 = \overset{\times}{v}_2 = \overset{\times}{v}_3 = \overset{\times}{v}$. If the undistorted configuration is natural ($p = 0$, $t = 0$, $v = 1$), then by (1) and (2)

$$\overset{\times}{p}\,\overset{\times}{\delta} > 0\,, \qquad \overset{\times}{t}\,\overset{\times}{\delta} > 0 \qquad (\overset{\times}{\delta} = \overset{\times}{v} - 1)$$

and the tensile stresses (forces) increase the linear dimensions while the compressing stresses decrease them.

If, however, the undistorted state is not natural, then the inequalities

$$(\overset{\times}{p} - p)(\overset{\times}{v} - v) > 0\,, \qquad (\overset{\times}{t} - t)(\overset{\times}{v} - v) > 0$$

are not equivalent to one another. In fact,

$$\begin{aligned} (\overset{\times}{t} - t)(\overset{\times}{v} - v) &= (\overset{\times}{v} - v)(\overset{\times}{v}{}^2\,\overset{\times}{p} - v^2 p) = \\ &= (\overset{\times}{v} - v)\,[\overset{\times}{v}{}^2(\overset{\times}{p} - p) + p(\overset{\times}{v}{}^2 - v^2)] = \\ &= (\overset{\times}{v} - v)(\overset{\times}{p} - p)\left[\overset{\times}{v}{}^2 + \frac{p}{(\overset{\times}{p} - p)^2}(\overset{\times}{p} - p)(\overset{\times}{v} - v)(\overset{\times}{v} + v)\right], \\ (\overset{\times}{p} - p)(\overset{\times}{v} - v) &= \frac{\overset{\times}{v} - v}{v^2\overset{\times}{v}{}^2}\,(\overset{\times}{t}v^2 - t\overset{\times}{v}{}^2) = \\ &= \frac{1}{v^2\overset{\times}{v}{}^2}\,(\overset{\times}{v} - v)\,[(\overset{\times}{t} - t)v^2 - t(\overset{\times}{v}{}^2 - v^2)] = \\ &= \frac{1}{v^2\overset{\times}{v}{}^2}\,(\overset{\times}{v} - v)(\overset{\times}{t} - t)\left[v^2 - \frac{t}{(\overset{\times}{t} - t)^2}(\overset{\times}{t} - t)(\overset{\times}{v} - v)(\overset{\times}{v} + v)\right]. \end{aligned} \tag{13}$$

For the state of tension ($p > 0$), the second Equation (19) follows from the first, for the state of compression, the first Equation (13) is implied by the second.

CHAPTER 6

PROBLEMS OF THE NONLINEAR THEORY OF A COMPRESSIBLE ELASTIC MEDIUM

1. THE AFFINE TRANSFORMATION OF A REFERENCE CONFIGURATION

In Chapter 4, Par. 15 a proof of the Ericksen's theorem was given which asserts the non-existence of general solutions (that is, solutions preserving their form regardless of the prescribed form of the specific stored energy $\ni(I_1,I_2,I_3)$ of deformation) to the problems of the nonlinear theory of elasticity under a transformation of the reference configuration to the actual configuration which is different from the affine transformation.

In this chapter, unless stated to the contrary, we will assume the absence of the body forces and that the state of stress is due to the surface tractions. Under these conditions the constant stress tensor $\underline{T}$ is determined by the state equation (4.3.4), the principal forces and stresses being equal to

$$t_s = \frac{\partial\ni}{\partial v_s} = \frac{\sqrt{I_3}}{v_s}\sigma_s \qquad (s = 1,2,3) , \tag{1}$$

$$\sigma_s = \frac{v_s}{\sqrt{I_3}} t_s = 2\left\{\sqrt{I_3}\frac{\partial\ni}{\partial I_3} + \frac{v_s^2}{\sqrt{I_3}}\left[\frac{\partial\ni}{\partial I_1} + (I_1 - v_s^2)\frac{\partial\ni}{\partial I_2}\right]\right\} \qquad (s = 1,2,3) . \tag{2}$$

The simplest affine transformation is that of similarity. For this transformation,

$$\underline{R} = K\underline{r} , \qquad \overset{0}{\underline{\nabla}}\underline{R} = K\underline{r}^s\underline{r}_s = K\underline{E} , \qquad \underline{F} = K^2\underline{E} ,$$

$$v_s = v = K \qquad (s = 1,2,3)$$

and, assuming

$$\ni(v_1,v_2,v_3) = f(v) , \tag{4}$$

we have

$$\sum_{s=1}^{3}\frac{\partial\ni}{\partial v_s} = f'(v) = t_1 + t_2 + t_3 = 3t , \qquad t = \frac{1}{3}f'(v) . \tag{5}$$

According to (2), in the state of uniform compression the dependence of pressure ($p = -\sigma_s$, $s = 1,2,3$) on the density is given by the relationship

$$-p = \frac{1}{3v^2} f'(v) = -\frac{\rho^2}{\rho_0} \frac{d}{d\rho} f\left(\left(\frac{\rho_0}{\rho}\right)^{1/3}\right) . \tag{6}$$

The problems of uniaxial stretching (Chapter 6, Pars. 2-3), of a simple shear (Par. 4) and of a pure shear (Par. 5) provide examples of affine transformations.

2. THE UNIAXIAL STRETCHING OF A ROD

The axis of a prismatic rod is assumed coincident with OX, in which case $\sigma_2 = \sigma_3 = 0$. The principal values of the Finger measure are denoted by v^2, $\alpha^2 v^2$ and $\alpha^2 v^2$ and the invariants of this measure are equal to

$$I_1 = (1+2\alpha^2)v^2 , \qquad I_2 = (2+\alpha^2)\alpha^2 v^4 , \qquad I_3 = \alpha^4 v^6 . \tag{1}$$

By (1.2) we have

$$\sigma_1 = 2v\left[\frac{1}{v^2\alpha^2}\frac{\partial \ni}{\partial I_1} + 2\frac{\partial \ni}{\partial I_2} + v^2\alpha^2\frac{\partial \ni}{\partial I_3}\right] , \tag{2}$$

$$\sigma_2 = 2v\left[\frac{1}{v^2}\frac{\partial \ni}{\partial I_1} + (1+\alpha^2)\frac{\partial \ni}{\partial I_2} + v^2\alpha^2\frac{\partial \ni}{\partial I_3}\right] = 2v\psi(\alpha^2, v^2) = 0 . \tag{3}$$

In the natural reference configuration, $\underline{F} = \underline{E}$, $v = 1$ and $\alpha^2 = 1$ and in its neighbourhood

$$\psi(\alpha^2, v^2) = \left[(\alpha^2-1)\frac{\partial\psi}{\partial\alpha^2} + (v^2-1)\frac{\partial\psi}{\partial v^2}\right]_{\underline{F}=\underline{E}} + \ldots . \tag{4}$$

Referring to (4.7.12) and (4.7.13) we have by (3)

$$\left.\frac{\partial\psi}{\partial\alpha^2}\right|_{\underline{F}=\underline{E}} = \left\{\frac{\partial \ni}{\partial I_2} + v^2\frac{\partial \ni}{\partial I_3} + 2\left[\frac{\partial^2 \ni}{\partial I_1^2} + v^4(1+\alpha^2)^4\frac{\partial^2 \ni}{\partial I_2^2} + v^8\alpha^2\frac{\partial^2 \ni}{\partial I_3^2} + \right.\right.$$

$$\left.\left. + 2v^2(1+\alpha^2)\frac{\partial^2 \ni}{\partial I_1 \partial I_2} + 2\alpha^2(1+\alpha^2)v^6\frac{\partial^2 \ni}{\partial I_2 \partial I_3} + 2\frac{\partial^2 \ni}{\partial I_3 \partial I_1}v^4\alpha^2\right]\right\}_{\substack{v=1\\ \alpha=1}} =$$

$$= \left(\frac{\partial \ni}{\partial I_2} + \frac{\partial \ni}{\partial I_3}\right)^0 + 2\left[\left(\frac{\partial}{\partial I_1} + 2\frac{\partial}{\partial I_2} + \frac{\partial}{\partial I_3}\right)^2 \ni\right]^0 = \frac{1}{2}(\mu+\lambda) \tag{5}$$

and a similar calculation yields

$$\left(\frac{\partial\psi}{\partial v^2}\right)_{\underline{F}=\underline{E}} = \frac{1}{4}(3\lambda + 2\mu) \ . \tag{6}$$

By (4) we now get

$$\psi(\alpha^2, v^2) = \frac{1}{4}[(3\lambda + 2\mu)(v^2 - 1) + 2(\lambda + \mu)(\alpha^2 - 1)] + \dots \ . \tag{7}$$

In the linear approximation, denoting the longitudinal and transverse elongations by δ_1 and δ_2, we have

$$v^2 = 1 + 2\delta_1 \ , \qquad \alpha^2 v^2 = 1 + 2\delta_2$$

and by (7)

$$(3\lambda + 2\mu)\delta_1 + 2(\lambda + \mu)(\delta_2 - \delta_1) + \dots = 0 \ , \qquad \frac{\delta_2}{\delta_1} = -\frac{\lambda}{2(\lambda + \mu)} = -\nu \tag{8}$$

the latter relationship determines the Poisson ratio in the linear theory.

Since $\left(\frac{\partial\psi}{\partial\alpha^2}\right)_{\underline{F}=\underline{E}} \neq 0$, we may assert, referring to the implicit functions' theorem, that the Equation (3) is uniquely solvable with regard to α^2 at any rate within some vicinity of the initial state. Substitution of this value $\alpha^2(v^2)$ into (2) leads to the expression for $\sigma_1(v)$ - the diagram of stretching of a sample.

Application of the criterion (5.10.13) of monotonicity leads to a more general conclusion. Setting $v_2 = v_3 = v$, we get

$$t_2(v_1, v_2, v_3) = \left(\frac{\partial\ni(v_1, v_2, v_3)}{\partial v_2}\right)_{v_2=v_3=v} = \left(\frac{\partial\ni(v_1, v_2, v_3)}{\partial v_3}\right)_{v_2=v_3=v} = 0 \ . \tag{9}$$

This equation is solvable with respect to v provided that

$$\frac{\partial t_2}{\partial v} = \left(\frac{\partial^2\ni(v_1, v_2, v_3)}{\partial v_2^2} + \frac{\partial^2\ni(v_1, v_2, v_3)}{\partial v_3 \partial v_2}\right)_{v_2=v_3=v} \neq 0 \ . \tag{10}$$

Applying Sylvester's theorem to the determinant of the matrix (5.10.13), we have

$$\left(\frac{\partial^2 \ni}{\partial v_2^2}\right)_{v_2=v_3=v} = \left(\frac{\partial^2 \ni}{\partial v_3^2}\right)_{v_2=v_3=v} > 0 ,$$

$$\left[\frac{\partial^2 \ni}{\partial v_2^2}\frac{\partial^2 \ni}{\partial v_3^2} - \left(\frac{\partial^2 \ni}{\partial v_2 \partial v_3}\right)^2\right]_{v_2=v_3=v} = \left[\left(\frac{\partial^2 \ni}{\partial v_2^2}\right)^2 - \left(\frac{\partial^2 \ni}{\partial v_2 \partial v_3}\right)^2\right]_{v_2=v_3=v} > 0 .$$

These inequalities imply that

$$\left(\frac{\partial^2 \ni}{\partial v_2^2} > \left|\frac{\partial^2 \ni}{\partial v_2 \partial v_3}\right|\right)_{v_2=v_3=v} > 0 ,$$

which means that

$$\left(\frac{\partial^2 \ni}{\partial v_2^2} + \frac{\partial^2 \ni}{\partial v_2 \partial v_3}\right)_{v_2=v_3=v} > 0$$

irrespective of the sign of the mixed derivative, and, in view of (10), Equation (9) yields the solution $v = v_1$ which is the desired result.

3. THE UNIAXIAL STRETCHING IN SIGNORINI AND BLATZ AND KO MATERIALS AND IN A QUASI-LINEAR MATERIAL

1. Substituting for λ in terms of μ and ν from (2.8) for $c = 0$ (a simplified material of Signorini) allows us to reduce the expression (5.2.9) for the tensor of stress to the following form

$$\frac{1-2\nu}{2\mu}\underline{T} = (\nu j_1' + \frac{1}{4} j_1'^2)\underline{E} + (1 - 2\nu - j_1')\underline{A} . \tag{1}$$

Here $\underline{A}$ denotes the Almansi strain tensor $j_1' = I_1(\underline{A})$. By (1.7.8) we have

$$\underline{A} = \frac{1}{2}(\underline{E} - \underline{F}^{-1}) , \quad A_s = \frac{1}{2}(1 - v_s^{-2}) , \quad -\infty < A_s < \frac{1}{2} , \tag{2}$$

where $A_s = 0$ for the natural state, and the limits $(-\infty, 1/2)$ correspond to the infinite compression and stretching of the rod.

For a uniaxial state of stress, $A_2 = A_3$, $j_1' = A_1 + 2A_2$

$$\frac{1-2\nu}{2\mu}\sigma_1 = A_1(1-\nu) - \frac{3}{4}A_1^2 - A_1A_2 + 2\underline{\nu}A_2 + A_2^2 , \tag{3}$$

$$\frac{1-2\nu}{2\mu}\,\sigma_2 = \nu A_1 + \frac{1}{4}A_1^2 + A_2 - A_2^2 = 0\ . \tag{4}$$

The root of the quadratic equation (4) which is less than 1/2 is determined by the formula

$$A_2 = \frac{1}{2}\,(1-\Delta^{1/2})\ ,\qquad \Delta = A_1^2 + 4\nu A_1 + 1$$

and the value of Δ is nonnegative for the values of ν given by (5.2.16) and which are permitted by Signorini's theory. In view of (3), the principal force

$$t_1 = v_2^2\sigma_1 = (1-2A_2)^{-1}\sigma_1 = \Delta^{-1/2}\sigma_1$$

is given by the expression

$$t_1 = \frac{\mu}{1-2\nu}\,[A_1 - 1 - 2\nu + \Delta^{-1/2}(1+2\nu+A_1-A_1^2)]\ . \tag{5}$$

The breaking force $Q = t_1S_0$ corresponding to infinite elongation $A_1 = 1/2$ turns out to be finite (S_0 denotes the area of a cross-section in the natural state). It is given by

$$Q = \frac{\mu S_0}{2(1-2\nu)}\,(\sqrt{5+8\nu} - 1 - 4\nu)$$

and for all admissible values of ν

$$-\frac{5}{8} < \nu < \frac{1}{2}\ ,\qquad \frac{1}{3}\,\mu S_0 < Q < \frac{2}{3}\,\mu S_0\ . \tag{6}$$

The compressive force which reduces the length of the rod to zero ($A_1 \to -\infty$) is infinite.

2. For a Blatz and Ko material subjected to uniaxial tension ($\sigma_2 = \sigma_3 = 0$) we have by (5.6.5)

$$\sqrt{I_3} = v_1v_2^2 = v_2^{-2}\ ,\qquad v_2 = v_1^{-1/4}\ ,\qquad \sqrt{I_3} = \sqrt{v_1}\ ,\qquad \frac{\sigma_1}{\mu} = 1 - v_1^{-5/2}\ , \tag{7}$$

where σ_1 is a monotonically increasing function whose values are negative for compression ($0 < v_1 < 1$) and positive for tension ($v_1 > 1$). The behaviour of the principal force t_1

$$\frac{1}{\mu}\,t_1 = \frac{\sqrt{I_3}}{\mu v_1}\,\sigma_1 = v_1^{-1/2} - v_1^{-3}\ ,\qquad \frac{1}{\mu}\,\frac{\partial t_1}{\partial v_1} = -\frac{1}{2}\,v_1^{-3/2}(1-6v_1^{-5/2})\ , \tag{8}$$

is different. It increases monotonically from $-\infty$ to 0 for compression, and possesses a very flat maximum $t_1^\times \approx 0.58\mu$ in tension for $v_1 = v_1^\times \approx 2.1$. To any value $t_1 = t_1^0 < t_1^\times$ there are two corresponding values of $v_1 > 1$; for one of these $v_1 < v_1^\times$ and for the other $v_1 > v_1^\times$. This illustrates the effectiveness of the monotonicity criteria (5.10.13) not over the entire region of the values of v, but in the vicinity (in this case over quite a large region) of the natural configuration.

3. For a quasi-linear material in the state of uniaxial stretching we have by (5.5.5)

$$\sigma_1 = \frac{1}{v_2^2}\left[(\lambda + 2\mu)\delta_1 + 2\lambda\delta_2\right], \qquad \sigma_2 = \frac{1}{v_1 v_2}\left[\lambda\delta_1 + 2(\lambda + \mu)\delta_2\right] = 0$$

$$(\delta_s = v_s - 1)$$

and as is to be expected

$$\delta_2 = -\frac{\lambda}{2(\lambda + \mu)}\delta_1 = -\nu\delta_1, \tag{10}$$

so that

$$\sigma_1 = \frac{\mu}{v_2^2}\frac{3\lambda + 2\mu}{\lambda + \mu}\delta_1 = \frac{1}{v_2^2}E\delta_1, \qquad t_1 = E\delta_1, \qquad E = \frac{\mu(3\lambda + 2\mu)}{\lambda + \mu}, \tag{11}$$

as in the linear theory. For the uniaxial stretching of a prismatic rod, the principal axes of the strain measure remain the same, and therefore

$$\underline{O}^\times = \underline{E}, \qquad \overset{0}{\underline{\underline{\nabla R}}} = \overset{0}{\underline{\underline{\nabla R}}}^T = \underline{V}, \qquad \underline{\underline{\nabla r}} = \underline{\underline{\nabla r}}^T = \underline{V}^{-1},$$

$$\underline{P} = \sqrt{\frac{G}{g}}\,\underline{\underline{\nabla r}}^T \cdot \underline{T} = v_1 \underline{V}^{-1} \cdot \underline{i}_1\underline{i}_1 t_1 = \underline{i}_1\underline{i}_1 t_1;$$

in which, not unexpectedly, t_1 is the (11)-component of the Piola tensor. From (5.5.2) we have

$$\ni = \frac{1}{2}\lambda(\delta_1 + 2\delta_2)^2 + \mu(\delta_1^2 + 2\delta_2^2) = \delta_1^2\left[\frac{1}{2}\lambda(1 - 2\nu)^2 + \mu(1 + 2\nu^2)\right] = \frac{1}{2}\delta_1^2 E = \frac{1}{2}\frac{t_1^2}{E}.$$

Now, by virtue of (5.5.11) and (5.5.13) we have

$$\ni_\times = \ni + I_1((\underline{P}\cdot\underline{P}^T)^{1/2}) = \frac{t_1^2}{2E} + t_1, \qquad \overset{0}{\underline{\underline{\nabla R}}} = (\ni_\times)_{\underline{P}} = \underline{i}_1\underline{i}_1 \frac{\partial}{\partial t_1}\left(\frac{t_1^2}{2E} + t_1\right)$$

and, as expected,

$$\frac{\partial x^1}{\partial a^1} = \frac{t_1}{E} + 1 , \qquad x^1 = \left(\frac{t_1}{E} + 1\right) a^1 = (\delta_1 + 1) a^1 = v_1 a^1 .$$

This calculation has been carried out in order to illustrate for the simplest case the application of the principle of stationary complementary energy.

4. SIMPLE SHEAR

The problem of shear occupies a special place in non-linear theory: it provides an explanation of some phenomena predicted and experimentally discovered for a non-linearly elastic material (the Kelvin-Wertheim and Poynting effects); these phenomena cannot be explained within the framework of the linear theory.

A simple shear is prescribed by a linear transformation which converts the rectangular cross-section *ABCD* of a parallelepiped into the parallelogram *AB'C'D'*. The surface tractions are sought which produce deformation of this type.

The transformation of coordinates a^s of the natural reference configuration into those of x^s of the actual configuration and the inverse transformation are prescribed by the relationships

$$\underline{R} = \underline{i}_s x^s = \underline{r} + \underline{i}_1 s a^2 , \qquad \underline{r} = \underline{i}_s a^s = \underline{R} - \underline{i}_1 s x^2 , \qquad s = \operatorname{tg}\gamma = \frac{\partial x^1}{\partial a^1} = \text{const}$$

(the *X*- and *Y*-axes are directed along the sides *AD* and *AB* of the rectangle and the *Z*-axis is normal to the plane of shear). As has already been noted the cross-section a^3 = const. of the parallelepiped, the rectangle *ABCD*, is converted into a parallelogram with an acute vertex angle $\pi/2 - \gamma$. We now have

$$\begin{aligned} &\overset{0}{\underline{\underline{\nabla R}}} = \underline{\underline{E}} + \underline{i}_2 \underline{i}_1 s , \qquad \underline{\underline{\nabla r}} = \underline{\underline{E}} - \underline{i}_2 \underline{i}_1 s , \\ &\overset{0}{\underline{\underline{\nabla R}}}{}^T = \underline{\underline{E}} + \underline{i}_1 \underline{i}_2 s , \qquad \underline{\underline{\nabla r}}^T = \underline{\underline{E}} - \underline{i}_1 \underline{i}_2 s . \end{aligned} \tag{1}$$

The Finger and Almansi measures turn out to be equal to

$$\begin{aligned} &\underline{\underline{F}} = \overset{0}{\underline{\underline{\nabla R}}}{}^T \cdot \overset{0}{\underline{\underline{\nabla R}}} = \underline{\underline{E}} + (\underline{i}_1\underline{i}_2 + \underline{i}_2\underline{i}_1) s + \underline{i}_1\underline{i}_1 s^2 , \\ &\underline{\underline{g}} = \underline{\underline{\nabla r}} \cdot \underline{\underline{\nabla r}}^T = \underline{\underline{E}} - (\underline{i}_1\underline{i}_2 + \underline{i}_2\underline{i}_1) s + \underline{i}_2\underline{i}_2 s^2 , \end{aligned} \tag{2}$$

$$I_3(\underline{F}) = 1 \,, \qquad I_1(\underline{F}) = 3 + s^2 = I_2(\underline{g}) \,, \qquad I_1(\underline{g}) = I_2(\underline{F}) = 3 + s^2 \tag{3}$$

and by (4.3.4) the state equation is reduced to

$$\underline{T} = 2\left[\underline{E}\frac{\partial\ni}{\partial I_3} + \frac{\partial\ni}{\partial I_1}\underline{F} + \frac{\partial\ni}{\partial I_2}(I_1\underline{E} - \underline{F})\cdot\underline{F}\right] = 2\left[\left(\frac{\partial\ni}{\partial I_1} + 2\frac{\partial\ni}{\partial I_2} + \frac{\partial\ni}{\partial I_3}\right)\underline{E} + \right.$$

$$\left. + (\underline{i}_1\underline{i}_2 + \underline{i}_2\underline{i}_1)\left(\frac{\partial\ni}{\partial I_1} + \frac{\partial\ni}{\partial I_2}\right)s + \underline{i}_1\underline{i}_1\left(\frac{\partial\ni}{\partial I_1} + \frac{\partial\ni}{\partial I_2}\right)s^2 + \underline{i}_3\underline{i}_3\frac{\partial\ni}{\partial I_3}s^2\right] . \tag{4}$$

Setting

$$\ni(I_1, I_2, I_3) = \ni(3+s^2, 3+s^2, 1) = \hat{\ni}(s^2) \,,$$

we obtain

$$2\left(\frac{\partial\ni}{\partial I_1} + \frac{\partial\ni}{\partial I_2}\right) = 2\frac{\partial\hat{\ni}}{\partial s^2} = \hat{\mu}(s^2) \,. \tag{5}$$

Using the notation

$$t = 2\left(\frac{\partial\ni}{\partial I_1} + 2\frac{\partial\ni}{\partial I_2} + \frac{\partial\ni}{\partial I_3}\right) \tag{6}$$

we can represent the components of a stress tensor (its physical components) in the following Cartesian form

$$\begin{aligned} &t^{11} = t + 2s^2\frac{\partial\hat{\ni}}{\partial s^2} \,, \qquad t^{22} = t \,, \qquad t^{33} = t + 2\frac{\partial\ni}{\partial I_2}s^2 \,, \\ &t^{12} = \hat{\mu}(s^2)s \,, \qquad t^{23} = t^{31} = 0 \,. \end{aligned} \tag{7}$$

The expression for t^{12} makes it natural to call the quantity $\hat{\mu}(s^2)$ a shear modulus. By (4.7.12), $\hat{\mu}(0) = \mu$, which is the shear modulus of the linear theory. Note that $t(s^2)$ is an even function which by (4.3.22) is zero for $s = 0$. Note also the relationship

$$t^{11} - t^{22} = 2s^2\frac{\partial\hat{\ni}}{\partial s^2} = s^2\hat{\mu}(s^2) \tag{8}$$

which is universal in the sense that its form is the same for all materials. For $s \neq 0$ the condition $t^{12} = 0$ is necessary and sufficient for the equality

$t^{11} = t^{22}$ to hold.

For an elastic fluid $t^{12} = 0$ and the equality holds, but for a solid elastic body the presence of shear inevitably produces the tangential stress t^{12} and is associated with the inequality $t^{11} \neq t^{22}$ and this phenomenon cannot be explained by linear theory (the effect is proportional to s^2). The realization of a simple shear requires the application of normal stresses at all faces of the parallelepiped including that of t^{33} at the face normal to the plane of shear. These stresses are proportional to s^2, they are not predicted by linear theory since the terms t^{12} which are neglected by this theory are of order not less than s^3. The universal property of a deformed solid body expressed by Equation (8) has been noted by Poynting. This cannot be explained by linear theory.

The mean normal stress is also proportional to at least s^2

$$\frac{1}{3}\left(t^{11}+t^{22}+t^{33}\right) = t(s^2) + \frac{1}{3}\left(\hat{\mu}(s^2) + 2\,\frac{\partial \ni}{\partial I_2}\right)s^2 . \tag{9}$$

Kelvin foresaw that this expression must be non-zero.

The vector $\underline{N}$ of normal in the actual configuration is by (1.8.8) represented as

$$\underline{N} = (\underline{n}\cdot\underline{G}^{-1}\cdot\underline{n})^{-1/2}\underline{\underline{\nabla r}}\cdot\underline{n} = (1-2sn_1n_2+n_1^2s^2)^{-1/2}(\underline{n}-\underline{i}_2n_1s) , \tag{10}$$

since

$$\underline{G}^{-1} = \overset{0}{\underline{\underline{\nabla r}}}{}^{T}\cdot\underline{\underline{\nabla r}} = \underline{E} - (\underline{i}_1\underline{i}_2+\underline{i}_2\underline{i}_1)s + \underline{i}_1\underline{i}_1s^2 ,$$

$$\underline{n}\cdot\underline{G}^{-1}\cdot\underline{n} = 1 - 2sn_1n_2 + n_1^2s^2 .$$

The normal stress calculated on any face using (2.2.11) is

$$\sigma_{\underline{N}} = \underline{N}\cdot\underline{T}\cdot\underline{N} =$$

$$= t + 2(1-2sn_1n_2+n_1^2s^2)^{-1}\left(2n_1n_2s\,\frac{\partial\hat{\ni}}{\partial s^2} - n_1^2s^2\,\frac{\partial\hat{\ni}}{\partial s^2} + n_3^2s^2\,\frac{\partial \ni}{\partial I_2}\right) . \tag{11}$$

This is so in particular, on the faces $\underline{n} = \mp\underline{i}_1$ (in other words, on AB' and DC')

$$\sigma_{\underline{i}_1} = \frac{2}{1+s^2}\left[\frac{\partial \ni}{\partial I_1} + (2+s^2)\,\frac{\partial \ni}{\partial I_2} + (1+s^2)\,\frac{\partial \ni}{\partial I_3}\right] \tag{12}$$

and for $\underline{n} = \underline{i}_2$ and $\underline{n} = \underline{i}_3$ the normal stresses are of course equal to t^{22} and t^{33}. The tangential stress vector, $\underline{\tau}_{\underline{N}}$, is determined from (2.2.12)

$$\underline{\tau}_{\underline{N}} = \underline{N} \cdot \underline{T} - \underline{N}\sigma_{\underline{N}} = 2(1 - 2sn_1n_2 + s^2n_1^2)^{-3/2}\Big\{(1 - 2sn_1n_2 +$$

$$+ s^2n_1^2)\left[(n_1\underline{i}_2 + n_2\underline{i}_1)s\frac{\partial\hat{\ni}}{\partial s^2} + \underline{i}_3 s^2 n_3 \frac{\partial\ni}{\partial I_2}\right] -$$

$$- [n_1\underline{i}_1 + (n_2 - n_1 s)\underline{i}_2 + n_3\underline{i}_3]\left(2n_1n_2 s\frac{\partial\hat{\ni}}{\partial s^2} - n_1^2 s^2\frac{\partial\hat{\ni}}{\partial s^2} + n_3^2 s^2 \frac{\partial\ni}{\partial I_2}\right)\Big\} . \tag{13}$$

Of course no tangential stress is detected on the faces $\underline{n} = \underline{i}_3$. For $\underline{n} = \underline{i}_1$, $\underline{n} = \underline{i}_2$ we get

$$\underline{\tau}_{\underline{i}_1} = (1 + s^2)^{-3/2}(\underline{i}_2 + \underline{i}_1 s)s\hat{\mu}(s^2) , \qquad |\underline{\tau}_{\underline{i}_1}| = \frac{s\hat{\mu}(s^2)}{1 + s^2} , \qquad \underline{\tau}_{\underline{i}_2} = \underline{i}_1 s\hat{\mu}(s^2) . \tag{14}$$

One of the principal directions $\underline{e}_3 = \underline{i}_3$ of the tensor $\underline{T}$ corresponds to the principal stress $t^{33} = \sigma_3$; the other two principal stresses can be determined if we set the right-hand side of (13) equal to zero (for $n_3 = 0$, $\underline{\tau}_{\underline{N}} \cdot \underline{i}_1 = 0$, $\underline{\tau}_{\underline{N}} \cdot \underline{i}_2 = 0$). We then arrive at the equations

$$n_2 + n_1(n_1 + n_2 s)(n_1 s - 2n_2) = 0 , \qquad 1 + n_2(n_1 s - 2n_2) = 0 , \tag{15}$$

where n_1, n_2 denote projections of one of the principal directions $\underline{e}_1$ or $\underline{e}_2$ of the tensor $\underline{T}$ along the axes $\underline{i}_1$, $\underline{i}_2$

$$\underline{e}_1 : n_1 = \cos\alpha , \quad n_2 = \sin\alpha ; \qquad \underline{e}_2 : n_1 = -\sin\alpha , \quad n_2 = \cos\alpha .$$

By virtue of the second Equation (15) we find

$$\operatorname{ctg} 2\alpha = -\frac{1}{2}s \quad \text{or} \quad \sin 2\alpha = \frac{2\varepsilon}{\sqrt{s^2 + 4}} ,$$

$$\cos 2\alpha = -\frac{\varepsilon s}{\sqrt{s^2 + 4}} \qquad (\varepsilon = \pm 1) ,$$

the first Equation (15) is of course seen to hold. We may assume that $\pi/4 < \alpha < \pi/2$ for $\varepsilon = 1$ and $3\pi/4 < \alpha < \pi$ for $\varepsilon = -1$. We get

$$\varepsilon = 1 \,, \quad \underline{n} = \underline{e}_1 : \overset{1}{e}_1 = \cos\alpha = \left(\frac{\sqrt{s^2+4} - s}{2\sqrt{s^2+4}} \right)^{1/2} ,$$

$$\overset{2}{e}_1 = \sin\alpha = \left(\frac{\sqrt{s^2+4} + s}{2\sqrt{s^2+4}} \right)^{1/2} , \tag{16}$$

$$\varepsilon = -1 \,, \quad \underline{n} = \underline{e}_2 : \overset{1}{e}_2 = -\sin\alpha \,, \quad \overset{2}{e}_2 = \cos\alpha \,.$$

The principal stresses σ_1, σ_2 are found by substituting these expressions into Equation (11).

5. PURE SHEAR

In pure shear the stress is given by

$$\underline{T} = (\underline{i}_2\underline{i}_3 + \underline{i}_3\underline{i}_2)\tau \,, \qquad \tau = \text{const} > 0 \,, \tag{1}$$

and what is required is the transformation of the reference configuration into the actual configuration, in other words the principal values of the Finger measure are to be determined. This problem is associated with the inversion of the state equation and is therefore much more complex than that of a "simple shear".

We consider a cube in the actual configuration; the normal vectors to its faces form a trihedtron $\underline{e}_1$, $\underline{e}_2$, $\underline{e}_3$ of directions, and the state of stress within the cube is given by the tensor

$$\underline{T} = \tau(\underline{e}_2\underline{e}_2 - \underline{e}_3\underline{e}_3) \;; \tag{2}$$

the tensile stresses are directed along $\underline{e}_2$, the compressive stresses of the same absolute value - along $\underline{e}_3$. In the system of axes

$$\underline{i}_1 = \underline{e}_1 \,, \quad \underline{i}_2 = \frac{1}{\sqrt{2}}(\underline{e}_2 + \underline{e}_3) \,, \quad \underline{i}_3 = \frac{1}{\sqrt{2}}(\underline{e}_2 - \underline{e}_3) \tag{3}$$

($\underline{i}_2$, $\underline{i}_3$ are directed along the diagonals of the cross-section of the cube) the state of stress represents pure shear only (1).

The principal stresses and the principal invariants of $\underline{T}$ are

$$\sigma_1 = 0 \,, \quad \sigma_2 = \tau \,, \quad \sigma_3 = -\tau \;;$$

$$I_1(\underline{T}) = 0 \,, \quad I_2(\underline{T}) = -\tau^2 \,, \quad I_3(\underline{T}) = 0 \,.$$

The tensor $\underline{T}$ is equal to its own deviator, and by (I.13.5), (5.8.9). (5.8.11) we have

$$G_2 = 4\tau^2 , \qquad \chi = 0 , \qquad G_3 = 0 . \tag{4}$$

In the natural reference configuration, the cube had the form of a parallelepiped with the edges v_1^{-1}, v_2^{-1}, v_3^{-1}. In the plane of shear $\underline{e}_2$, $\underline{e}_3$ its cross-section is a rectangle with sides v_2^{-1}, v_3^{-1}, the angle 2α between its diagonals being given by the formulas

$$\cos\alpha = \frac{v_2}{\sqrt{v_2^2 + v_3^2}} , \qquad \cos 2\alpha = 2\cos^2\alpha - 1 = \frac{v_2^2 - v_3^2}{v_2^2 + v_3^2} .$$

From Chapter 1, Par. 4, the difference in the cosines of the angles 2α and $\pi/2$, in the reference and the actual configuration, respectively

$$\gamma = \frac{v_2^2 - v_3^2}{v_2^2 + v_3^2} , \qquad \frac{v_2^2}{v_3^2} = \frac{1+\gamma}{1-\gamma} \tag{5}$$

should be taken as a measure of the shear.

We now introduce the logarithmic strain measure $\underline{H}$ and its deviator. From formulas (Chapter 5, Par. 8)

$$\omega = -\psi , \qquad \sin\omega = -\sin\psi = -\frac{1}{\sqrt{3g_2}} \ln\frac{v_1^2}{v_2 v_3} , \qquad \cos\omega = \cos\psi = \frac{1}{\sqrt{g_2}} \ln\frac{v_2}{v_3} , \tag{6}$$

where g_2 is given by (5.8.4).

Assuming that the BE-criterion (5.13.9) is valid for the material considered and bearing in mind that the sign of the difference of two positive numbers is the same as that of the logarithm of their ratio, we get

$$(\sigma_1 - \sigma_2)\ln\frac{v_1}{v_2} = -\tau \ln\frac{v_1}{v_2} > 0 , \qquad (\sigma_2 - \sigma_3)\ln\frac{v_2}{v_3} = 2\tau \ln\frac{v_2}{v_3} > 0 ,$$

$$(\sigma_3 - \sigma_1)\ln\frac{v_3}{v_1} = -\tau \ln\frac{v_3}{v_1} > 0 ,$$

so that for such a material

$$v_2 > v_1 > v_3\,, \qquad 0 < \gamma < 1\,. \tag{7}$$

It also follows from the definition of a shearing state that the angle α cannot exceed 45°.

From (5.8.17) we have now

$$\frac{\partial \ni^\times}{\partial I_1(\underline{H})} = 0\,, \qquad \frac{\partial \ni^\times}{\partial \sqrt{g_2}} = \tau e^{I_1(\underline{H})} \cos\psi\,, \qquad \frac{\partial \ni^\times}{\partial \psi} = -\tau e^{I_1(\underline{H})} \sqrt{g_2} \sin\psi\,, \tag{8}$$

which implies that

$$\frac{\partial \ni^\times}{\partial \sqrt{g_2}} \sqrt{g_2} \sin\psi + \frac{\partial \ni^\times}{\partial \psi} \cos\psi = 0 \qquad \text{or} \qquad \frac{\partial \ni^\times}{\partial \ln \sqrt{g_2}} - \frac{\partial \ni^\times}{\partial \ln \cos\psi} = 0 \tag{9}$$

and a general solution to this first order partial differential equation is given by the function

$$\ni^\times = f(\sqrt{g_2} \cos\psi) = f\left(\ln \frac{v_2}{v_3}\right) = \varphi\left(\frac{v_2}{v_3}\right)\,. \tag{10}$$

At the same time, by (8) and (6)

$$d\ni^\times = \tau e^{I_1(\underline{H})} (\cos\psi d\sqrt{g_2} - \sqrt{g_2} \sin\psi\, d\psi) = \tau e^{I_1(\underline{H})} d\sqrt{g_2} \cos\psi =$$

$$= \tau e^{I_1(\underline{H})} d \ln \frac{v_2}{v_3} = \tau e^{I_1(\underline{H})} \frac{v_3}{v_2} d \frac{v_2}{v_3} = \varphi'\left(\frac{v_2}{v_3}\right) d \frac{v_2}{v_3}\,,$$

so that

$$\tau = e^{-I_1(\underline{H})} \frac{v_2}{v_3} \varphi'\left(\frac{v_2}{v_3}\right)\,. \tag{11}$$

The ratio v_2/v_3 can be expressed in terms of the shear measure. We get

$$\tau = \frac{1}{\sqrt{I_3}} (1-\gamma^2) \frac{d}{d\gamma} \varphi'\left(\sqrt{\frac{1+\gamma}{1-\gamma}}\right)\,. \tag{12}$$

The use of this relationship presupposes the knowledge of $\ni$ as a fucntion of $I_1(\underline{H})$, $\sqrt{g_2}$ and ψ whereas $\ni$ is commonly prescribed as a function either of invariants of the tensor $\underline{F}$ or of its principal values.

From the definition of the principal forces and the prescribed values of the principal stresses

$$\sigma_1 = \frac{v_1}{\sqrt{I_3}} t_1 = 0 \,, \qquad \sigma_2 + \sigma_3 = \frac{1}{\sqrt{I_3}} (v_2 t_2 + v_3 t_3) = 0 \,,$$

$$\sigma_2 + \sigma_3 = \frac{1}{\sqrt{I_3}} (v_2 t_2 - v_3 t_3) = 2\tau \tag{13}$$

and in view of (4.3.16)

$$\frac{\partial \text{Э}}{\partial v_1} = 0 \,, \qquad v_2 \frac{\partial \text{Э}}{\partial v_2} + v_3 \frac{\partial \text{Э}}{\partial v_3} = 0 \,, \qquad \tau = \frac{1}{2\sqrt{I_3}} \left(v_2 \frac{\partial \text{Э}}{\partial v_2} - v_3 \frac{\partial \text{Э}}{\partial v_3} \right) . \tag{14}$$

The first and the second relationships imply (cf. (10))

$$\text{Э} = \varphi\left(\frac{v_2}{v_3}\right) \tag{15}$$

and the third relationship together with (11) and (12) yields

$$\tau = \frac{1}{\sqrt{I_3}} \frac{v_2}{v_3} \varphi'\left(\frac{v_2}{v_3}\right) = \frac{1}{\sqrt{I_3}} (1 - \gamma^2) \frac{d}{d\gamma} \varphi\left(\sqrt{\frac{1+\gamma}{1-\gamma}} \right) . \tag{16}$$

When the Э-function is expressed in terms of the invariants $I_k(\underline{F})$ Equations (14) reduce to the form

$$(v_2^2 + v_3^2 - 2v_1^2) \frac{\partial \text{Э}}{\partial I_1} + [2v_2^2 v_3^2 - v_1^2 (v_2^2 + v_3^2)] \frac{\partial \text{Э}}{\partial I_2} = 0 \,, \tag{17}$$

$$I_1 \frac{\partial \text{Э}}{\partial I_1} + 2I_2 \frac{\partial \text{Э}}{\partial I_2} + 3I_3 \frac{\partial \text{Э}}{\partial I_3} = 0 \,, \tag{18}$$

$$\tau = \frac{1}{\sqrt{I_3}} (v_2^2 - v_3^2) \left(\frac{\partial \text{Э}}{\partial I_1} + v_1^2 \frac{\partial \text{Э}}{\partial I_2} \right) . \tag{19}$$

Calculations based on these formulas, that is to say representations of v_1, v_2 and v_3 in terms of τ, can only be solved completely for some especially "fortunate" prescriptions of the dependence of the function Э on the invariants In the general case, it is only possible to construct formulas describing the "second order effects" for sufficiently small values of τ. These formulas are comprehensively treated by Truesdell and Moon (C. Truesdell, H. Moon, 1974).

A Signorini material. The specific strain energy prescribed by (5.11.4) does not depend on I_1, and in the case of pure shear we have from (17)

$$v_1^2 = \frac{2v_2^2 v_3^2}{v_2^2 + v_3^2}, \qquad I_2 = 3v_2^2 v_3^2, \qquad I_3 = \frac{2v_2^4 v_3^4}{v_2^2 + v_3^2}. \tag{20}$$

Equation (18) is now reduced to the form

$$(\lambda + \mu)\left(\frac{I_2}{I_3}\right)^2 + 2(3\lambda + \mu)\frac{I_2}{I_3} - 3(9\lambda + 5\mu) = 0 .$$

This equation has the root $I_2/I_3 = 3$, its second root is negative and should therefore be discarded. From (10) we have

$$v_1 = 1, \qquad \frac{1}{v_2^2} + \frac{1}{v_3^2} = 2, \tag{21}$$

so that by (7) $v_2 > 1$ and $v_3 < 1$. In view of (19), (20) and (21) we now obtain

$$\frac{2\tau}{\mu} = \frac{1}{v_3^2} - \frac{1}{v_2^2}, \qquad v_2 = \left(1 - \frac{\tau}{\mu}\right)^{-1/2}, \qquad v_3 = \left(1 + \frac{\tau}{\mu}\right)^{-1/2} \tag{22}$$

and the law connecting the tangential stress τ and the shear for a simplified ($c = 0$) Signorini material yields the linear relationship

$$\tau = \mu\gamma, \tag{23}$$

here, however, the value of γ may be far from small.

The same dependence is also characteristic for a Blatz and Ko material. For this, in view of (5.6.5),

$$v_1^2 = \frac{1}{\sqrt{I_3}}, \qquad 1 - \frac{\tau}{\mu} = \frac{v_1^2}{v_2^2}, \qquad 1 + \frac{\tau}{\mu} = \frac{v_1^2}{v_3^2}, \qquad \tau = \mu\gamma . \tag{24}$$

For the problem of simple shear we have by (5.11.4), (4.3) and (4.5) for a Signorini material

$$\ni = \frac{1}{2}\mu s^2 + \frac{1}{8}(\lambda + \mu)s^4 + \mu, \qquad \hat{\mu}(s^2) = \mu + \frac{1}{2}(\lambda + \mu)s^2 \tag{25}$$

and by (4.7) the stresses are determined from the formulas

$$t^{11} = t^{33} = -[4\lambda - (\lambda + \mu)s^2]\,\frac{s^2}{8},$$

$$t^{22} = -[4(\lambda + 2\mu) + 3(\lambda + \mu)s^2]\,\frac{s^2}{8}, \qquad t^{12} = \hat{\mu}(s^2)s\,. \tag{26}$$

6. A QUASI-LINEAR MATERIAL. LAMÉ'S PROBLEM FOR A CYLINDER AND SPHERE

When the transformation of the reference configuration into the actual configuration is affine the stress tensor is constant and can be represented in the form of a state equation which is the same for all materials. The explicit specification of its coefficients or the representation of the specific stored energy in terms of the invariants of deformation is required for the determination of the quantitative relationships between deformations and stresses for a specific material.

So long as the transformation is different from affine, the equations of static equilibrium can no longer be isolated from the specification of material with the aid of its constitutive equation. The only possible approach consists in constructing solutions which approximate to those of the linear theory and which reduce to linear solutions when only terms which are linear with respect to the components of a gradient of the displacement vector are retained. This does not exclude the possibility of progress for certain materials under certain specific assumptions when either the equilibrium equations are expressed in terms of displacements or the application of the variational principles allow us to reduce the problem to a system of ordinary differential equations.

A "quasi-linear" material (Chapter 5, Par. 5) belongs to this selected group of "accessible" materials; of special value is the property of its constitutive equation to allow inversion with the aid of the principle of stationary complementary work. The problems associated with "quasi-linear" materials are discussed in Pars. 6-8.

1. The axisymmetric deformation of a circular hollow cylinder. Surface loading is produced by pressures p_0, p_1 distributed uniformly over the inner and outer surfaces ($r = r_0$, $r = r_1$) of the cylinder. The cylindrical coordinates r, φ, z in the reference configuration are chosen as material coordinates. For the usual loading conditions, the radial displacement must be assumed to be dependent only on r, and the axial displacement to be a linear function of z. The position of a point in the actual configuration is determined by the expression

$$\underline{R} = \underline{e}_r f(r) + \underline{k}\alpha z\,. \tag{1}$$

By the formulas III, Par. 7, and in particular by (III.7.17) we have

$$\underline{R}_1 = \underline{e}_r f'(r) , \quad \underline{R}_2 = \underline{e}_\varphi f(r) , \quad \underline{R}_3 = \alpha \underline{k} ;$$
$$\underline{r}_1 = \underline{r}^1 = \underline{e}_r , \quad \underline{r}_2 = r\underline{e}_\varphi , \quad \underline{r}^2 = \frac{\underline{e}_\varphi}{r} , \quad \underline{r}_3 = \underline{r}^3 = \underline{k} , \tag{2}$$

so that in view of Chapter 1, Par. 4

$$\overset{0}{\underline{\underline{\nabla R}}} = \underline{r}^s \underline{R}_s = \underline{e}_r \underline{e}_r f'(r) + \frac{1}{r} f(r) \underline{e}_\varphi \underline{e}_\varphi + \underline{\underline{kk}}\alpha = \overset{0}{\underline{\underline{\nabla R}}}^T ,$$
$$\underline{G} = \underline{F} = \underline{e}_r \underline{e}_r f'^2(r) + \underline{e}_\varphi \underline{e}_\varphi \frac{f^2(r)}{r^2} + \underline{\underline{kk}}\alpha^2 . \tag{3}$$

The tensors $\underline{G}$ and $\underline{F}$ are coaxial, and for a quasi-linear material the equations of equilibrium in the volume and on the surface are in view of (5.5.16) reduced to the form

$$\overset{0}{\underline{\nabla}}{}^2 \underline{u} = 0 , \quad \lambda \underline{e}_r \overset{0}{\underline{\nabla}} \cdot \underline{u} + 2\mu \underline{e}_r \cdot \overset{0}{\underline{\underline{\nabla u}}} = \begin{cases} -p_0 \underline{e}_r \left(\dfrac{dO}{do} \right)_0 , & r = r_0 , \\ -p_1 \underline{e}_r \left(\dfrac{dO}{do} \right)_1 , & r = r_1 , \end{cases} \tag{4}$$

because

$$\underline{n} = -\underline{e}_r , \quad \underline{f} = \underline{e}_r p_0 \quad \text{and} \quad r = r_0 ;$$
$$\underline{n} = \underline{e}_r , \quad \underline{f} = -\underline{e}_r p_1 \quad \text{and} \quad r = r_1 .$$

We shall need below the expressions for the physical components of the stress tensor $\underline{T}$ in terms of its contravariant components as well as the formulas which connect the latter with the components of the Piola tensor under transformation (1). We have

$$\underline{T} = \underline{R}_1 \underline{R}_1 t^{11} + \underline{R}_2 \underline{R}_2 t^{22} + \underline{R}_3 \underline{R}_3 t^{33} = \underline{e}_r \underline{e}_r f'^2 t^{11} +$$
$$+ \underline{e}_\varphi \underline{e}_\varphi f^2 t^{22} + \underline{\underline{kk}}\alpha^2 t^{33} = \sigma_R \underline{e}_R \underline{e}_R + \sigma_\Phi \underline{e}_\Phi \underline{e}_\Phi + \sigma_Z \underline{e}_Z \underline{e}_Z$$

(the non-diagonal elements are absent in the problems considered because of (3)). However, the directions of the unit vectors are preserved under the transformations (2), i.e. $\underline{e}_R = \underline{e}_r$, $\underline{e}_\Phi = \underline{e}_\varphi$, $\underline{e}_Z = \underline{k}$. Therefore

$$\sigma_R = f'^2 t^{11} , \quad \sigma_\Phi = f^2 t^{22} , \quad \sigma_Z = \alpha^2 t^{33} . \tag{5}$$

Now, by (2.6.3) and (5.5.5)

$$\underline{T} = \sqrt{\frac{g}{G}}\, \overset{0}{\underline{\nabla}}\underline{R}^{T} \cdot [\lambda(\overset{0}{\underline{\nabla}} \cdot \underline{R} - 3)\underline{E} + 2\mu(\overset{0}{\underline{\nabla}}\underline{R} - \underline{E})] =$$

$$= \frac{r}{ff'\alpha}\left\{\left[\lambda\left(f' + \frac{f}{r} + \alpha\right) - (3\lambda + 2\mu)\right]\left(f'\underline{e}_r\underline{e}_r + \frac{f}{r}\underline{e}_\varphi\underline{e}_\varphi + \alpha\underline{k}\underline{k}\right) + \right.$$

$$\left. + 2\mu\left(f'^2\underline{e}_r\underline{e}_r + \frac{f^2}{r^2}\underline{e}_\varphi\underline{e}_\varphi + \alpha^2\underline{k}\underline{k}\right)\right\} ,$$

so that

$$\sigma_R = \frac{r}{\alpha f}\left[\lambda\left(f' + \frac{f}{r} + \alpha\right) + 2\mu f' - (3\lambda + 2\mu)\right] ,$$

$$\sigma_\Phi = \frac{1}{f'\alpha}\left[\lambda\left(f' + \frac{f}{r} + \alpha\right) + 2\mu\frac{f}{r} - (3\lambda + 2\mu)\right] , \qquad (6)$$

$$\sigma_Z = \frac{r}{f'f}\left[\lambda\left(f' + \frac{f}{r} + \alpha\right) + 2\mu\alpha - (3\lambda + 2\mu)\right] .$$

But $\overset{0}{\underline{\nabla}}\underline{r} = \underline{E}$ and $\overset{0}{\underline{\nabla}}{}^2\underline{r} = \overset{0}{\underline{\nabla}} \cdot \overset{0}{\underline{\nabla}}\underline{r} = \underline{0}$, and therefore

$$\overset{0}{\underline{\nabla}}{}^2\underline{u} = \overset{0}{\underline{\nabla}}{}^2\underline{R} = \overset{0}{\underline{\nabla}}{}^2(\underline{e}_r f(r) + \underline{k}\alpha z) = \underline{e}_r\left(f' + \frac{f}{r}\right)' = \underline{0} ,$$

$$f' + \frac{f}{r} = 2C_1 , \qquad f = C_1 r + \frac{C_2}{r}$$

and the components of the stress tensor are given by

$$\sigma_R = \frac{1}{(C_1 + C_2/r^2)\alpha}\left[\lambda(2C_1 + \alpha) + 2\mu\left(C_1 - \frac{C_2}{r^2}\right) - (3\lambda + 2\mu)\right] ,$$

$$\sigma_\Phi = \frac{1}{(C_1 - C_2/r^2)\alpha}\left[\lambda(2C_1 + \alpha) + 2\mu\left(C_1 + \frac{C_2}{r^2}\right) - (3\lambda + 2\mu)\right] , \qquad (8)$$

$$\sigma_Z = \frac{r}{f'f}\ \lambda(2C_1 + \alpha) + 2\mu\alpha - (3\lambda + 2\mu) .$$

From the latter expression we determine the value of the longitudinal force Q which should be applied to the faces of a rod in order to produce the state of equilibrium considered

$$Q = 2\pi\int_{R_0}^{R_1} \sigma_Z R\,dR = 2\pi\int_{r_0}^{r_1} \sigma_Z f(r)f'(r)\,dr =$$

$$= \pi(r_1^2 - r_0^2)\,[\lambda(2C_1 + \alpha) + 2\mu\alpha - (3\lambda + 2\mu)] .$$

This is one of the equations required for the determination of the three constants C_1, C_2, α

$$\lambda(2C_1+\alpha) + 2\mu\alpha = \frac{Q}{\pi(r_1^2 - r_0^2)} + 3\lambda + 2\mu \ . \tag{9}$$

The other two follow from the boundary conditions (4) on the cylindrical surfaces. Bearing in mind that $\frac{dO}{do} = \alpha\,\frac{f(r)}{r}$ we reduce these equations to the form

$$\lambda(2C_1+\alpha) + 2\mu\left(C_1 - \frac{C_2}{r_0^2}\right) = -p_0\left(C_1 + \frac{C_2}{r_0^2}\right)\alpha + 3\lambda + 2\mu \ , \tag{10}$$

$$\lambda(2C_1+\alpha) + 2\mu\left(C_1 - \frac{C_2}{r_1^2}\right) = -p_1\left(C_1 + \frac{C_2}{r_1^2}\right)\alpha + 3\lambda + 2\mu \ . \tag{11}$$

Just as in linear theory the boundary condition on the face is satisfied in "the sense of Saint-Venant".

For a cyliner situated between two fixed smooth plates, $\alpha = 1$ (the length is fixed). Assuming the absence of internal pressure ($p_0 = 0$, $p_1 = p$) we get

$$R_0 = C_1 r_0 + \frac{C_2}{r_0} = \frac{\left(1 - \frac{p}{2\mu}\right) r_0}{1 + \frac{p}{2\mu}\,\frac{1+k-2\nu}{1-k}} \ , \qquad k = \frac{r_0^2}{r_1^2} \ , \qquad \nu = \frac{\lambda}{2(\lambda+\mu)} \ . \tag{12}$$

We have thus obtained an expression for the inner radius of a cylinder compressed by an external pressure. Retaining only the linear term of the small quantity $p/(2\mu)$ we arrive at the classical Lamé expression

$$R_0 = r_0\left(1 - \frac{p}{\mu}\,\frac{1-\nu}{1-k}\right) \ . \tag{13}$$

2. <u>The radially symmetric deformation of a hollow sphere</u>. Loading is produced by pressures p_0, p_1 uniformly distributed over the inner ($r = r_0$) and outer ($r = r_1$) surfaces of a sphere. The material coordinates are the spherical coordinates r, ϑ, λ of a particle in the natural reference configuration; the spherical coorinates in the actual configuration resulting from the radially symmetric deformation are designated by $R = f(r)$, $\theta = \vartheta$, $\Lambda = \lambda$. The vector bases in the reference and actual configurations are given by the formulas

$$\underline{r} = r\underline{e} \ : \ \underline{r}_1 = \underline{e}_r \ , \qquad \underline{r}_2 = r\underline{e}_\vartheta \ , \qquad \underline{r}_3 = r\underline{e}_\lambda \sin\vartheta \ ;$$

$$\underline{r}^1 = \underline{e}_r\,, \qquad \underline{r}^2 = \frac{\underline{e}_\vartheta}{r}\,, \qquad \underline{r}^3 = \frac{\underline{e}_\lambda}{r\sin\vartheta}\,;$$

$$\underline{R} = f(r)\underline{e}_r \;:\; \underline{R}_1 = f'(r)\underline{e}_r\,, \qquad \underline{R}_2 = f(r)\underline{e}_\vartheta\,, \qquad \underline{R}_3 = f(r)\underline{e}_\lambda \sin\vartheta\,;$$

$$\underline{R}^1 = f'(r)\underline{e}_r\,, \qquad \underline{R}^2 = \frac{\underline{e}_\vartheta}{f(r)}\,, \qquad \underline{R}^3 = \frac{\underline{e}_\lambda}{f(r)\sin\vartheta}\,,$$

so that

$$\underline{\underline{\overset{0}{\nabla}R}} = \underline{e}_r\underline{e}_r f'(r) + (\underline{e}_\vartheta\underline{e}_\vartheta + \underline{e}_\lambda\underline{e}_\lambda)\,\frac{f(r)}{r} = \underline{e}_r\underline{e}_r r\left(\frac{f}{r}\right)' + \underline{E}\,\frac{f}{r} = \underline{\underline{\overset{0}{\nabla}R}}^T\,,$$

$$\underline{G} = \underline{F} = \underline{e}_r\underline{e}_r\left(f'^2 - \frac{f^2}{r^2}\right) + \underline{E}\,\frac{f^2}{r^2}\,.$$

Recalling that $\underline{\underline{\overset{0}{\nabla}R}} = \underline{\underline{\overset{0}{\nabla}R}}^T$, $\overset{0}{\underline{\nabla}}{}^2\underline{R} = \overset{0}{\underline{\nabla}}\cdot\underline{\underline{\overset{0}{\nabla}R}}^T = \underline{\underline{\overset{00}{\nabla\nabla}}}\cdot\underline{R}$, we have

$$\overset{0}{\underline{\nabla}}\cdot\underline{R} = f' + 2\,\frac{f}{r}\,, \qquad \overset{0}{\underline{\nabla}}{}^2\underline{R} = \underline{e}_r\left(f' + 2\,\frac{f}{r}\right)'$$

and the equations (5.5.16) of equilibrium in the volume imply

$$f(r) = c_1 r + \frac{c_2}{r^2}\,. \tag{14}$$

On the surfaces $r = \text{const}$

$$\lambda\underline{e}_r\overset{0}{\underline{\nabla}}\cdot(\underline{R} - \underline{r}) + 2\mu\underline{e}_r\cdot\overset{0}{\underline{\nabla}}(\underline{R} - \underline{r}) = \left[(3\lambda + 2\mu)(c_1 - 1) - 4\mu\,\frac{c_2}{r^3}\right]\underline{e}_r\,,$$

$$\frac{dO}{do} = \frac{R^2}{r^2} = \left(c_1 + \frac{c_2}{r^3}\right)^2\,.$$

The constants c_1, c_2 are determined by the boundary conditions

$$\begin{aligned} r = r_0 &: (3\lambda + 2\mu)(c_1 - 1) - 4\mu\,\frac{c_2}{r_0^3} = -\left(c_1 + \frac{c_2}{r_0^3}\right)^2 p_0\,, \\ r = r_1 &: (3\lambda + 2\mu)(c_1 - 1) - 4\mu\,\frac{c_2}{r_1^3} = -\left(c_1 + \frac{c_2}{r_1^3}\right)^2 p_1 \end{aligned} \tag{15}$$

it can be seen that the equations for c_1, c_2 are nonlinear.

Consider the case of a spherical cavity in an unbounded elastic medium; one of the constants in (14) is determined by the condition that for $r \to \infty$ the radial displacement vanishes

$$(R-r)_{r\to\infty} = \left[(c_1-1)r + \frac{c_2}{r_2}\right]_{r\to\infty} = 0\,, \qquad c_1 = 1\,.$$

The constant c_2 is now determined as a root of the quadratic equation

$$\left(\frac{c_2}{r_0^3}\right)^2 - 2\left(\frac{2\mu}{p_0} - 1\right)\frac{c_2}{r_0^3} + 1 = 0\,, \tag{16}$$

its roots are both real for $p_0 < \mu$ and are determined from the formulas

$$\begin{aligned}\left(\frac{c_2}{r_0^3}\right)_1 &= \frac{2\mu}{p_0}\left(1 + \sqrt{1-\frac{p_0}{\mu}}\right) - 1 = \frac{4\mu}{p_0} - 2 - \frac{1}{4}\frac{p_0}{\mu} + \dots\,,\\ \left(\frac{c_2}{r_0^3}\right)_2 &= \frac{2\mu}{p_0}\left(1 - \sqrt{1-\frac{p_0}{\mu}}\right) - 1 = \frac{p_0}{4\mu} + \dots\,.\end{aligned} \tag{17}$$

The pressure distributed over the surface r = const is given by the expression (cf. (15))

$$\begin{aligned} p &= 4\mu\,\frac{c_2}{r_0^3}\,k\left(1 + \frac{c_2}{r_0^3}\,k\right)^{-2} = p_0\,\frac{k}{D}\,,\\ D &= 1 + \frac{1}{2}\,(k^2-1)\left(1 \pm \sqrt{1-\frac{p_0}{\mu}}\right) - \frac{p_0}{4\mu}\,(k-1)^2 \end{aligned} \tag{18}$$

in which $k = r_0^3/r^3$.

Taking the positive root would lead in the linear approximation ($p_0/\mu=0$) to the inadmissible result ($p = p_0\,\frac{r^3}{r_0^3}$); the negative root would yield the law $p = p_0\,\frac{r_0^3}{r^3}$ for decrease of pressure known from linear theory. The radial displacement of a particle of the body must therefore be determined from the expression

$$R - r = \frac{r_0^3}{r^2}\left[\frac{2\mu}{p_0}\left(1 - \sqrt{1-\frac{p_0}{\mu}}\right) - 1\right] = \frac{r_0^3}{r^2}\,\frac{p_0}{4\mu} + \dots \tag{19}$$

the first term in the right-hand side represents the solution of the linear theory.

7. THE CIRCULAR MEMBRANE

The equations of static equilibrium for the Piola tensor represented in terms of its physical components in cylindrical coordinates

$$\underline{P} = \hat{\sigma}_r \underline{e}_r \underline{e}_r + \hat{\sigma}_\varphi \underline{e}_\varphi \underline{e}_\varphi + \hat{\sigma}_z \underline{k}\underline{k} + \hat{\tau}_{r\varphi} \underline{e}_r \underline{e}_\varphi + \hat{\tau}_{\varphi r} \underline{e}_\varphi \underline{e}_r +$$

$$+ \hat{\tau}_{\varphi z} \underline{e}_\varphi \underline{k} + \hat{\tau}_{z\varphi} \underline{k}\underline{e}_\varphi + \hat{\tau}_{zr} \underline{k}\underline{e}_r + \hat{\tau}_{rz} \underline{e}_r \underline{k} , \tag{1}$$

are reduced by (III.7.28) to the form ($H_1 = 1,\ H_2 = r,\ H_3 = 1$)

$$\begin{aligned}
&\frac{\partial}{\partial r} r\hat{\sigma}_r - \hat{\sigma}_\varphi + \frac{\partial \hat{\tau}_{\varphi r}}{\partial \varphi} + r\frac{\partial \hat{\tau}_{zr}}{\partial z} + r\rho_0 k_r = 0 ,\\
&\frac{\partial}{\partial r} r\hat{\tau}_{r\varphi} + \hat{\tau}_{\varphi r} + \frac{\partial \hat{\sigma}_\varphi}{\partial \varphi} + r\frac{\partial \hat{\tau}_{z\varphi}}{\partial z} + r\rho_0 k_\varphi = 0 ,\\
&\frac{\partial}{\partial r} r\hat{\tau}_{rz} + \frac{\partial \hat{\tau}_{\varphi z}}{\partial \varphi} + r\frac{\partial \hat{\sigma}_z}{\partial z} + r\rho_0 k_z = 0 .
\end{aligned} \tag{2}$$

The notation $\hat{\sigma}_r,\ldots,\hat{\tau}_{rz}$ is used in order to distinguish these quantities from the physical components of the Cauchy stress tensor $\underline{T}$ and, obviously, in (1) and (2) $\hat{\tau}_{r\varphi} \neq \hat{\tau}_{\varphi r}$ etc.

On the surfaces r = const and z = const

$$\underline{e}_r \cdot \underline{P} = \hat{\sigma}_r \underline{e}_r + \hat{\tau}_{r\varphi} \underline{e}_\varphi + \hat{\tau}_{rz} \underline{k} = \underline{f}^{(r)} ,$$

$$\underline{k} \cdot \underline{P} = \hat{\tau}_{zr} \underline{e}_r + \hat{\tau}_{z\varphi} \underline{e}_\varphi + \hat{\sigma}_z \underline{k} = \underline{f}^{(k)} .$$

Koiter (W.T. Koiter, 1975) has considered the problem of a circular membrane of negligible thickness clamped along its boundary $r = a$ and we have chosen this as an example of the principle of stationary complementary energy. A possible static stress state is prescribed by the Piola tensor, its components σ_r, σ_φ, τ_{rz} depending only on r. By virtue of (1) and (2) and taking the weight of the membrane itself into consideration ($\rho_0 k_z = -\gamma$) we arrive at the relationships

$$\hat{\sigma}_\varphi = (r\hat{\sigma}_r)' , \qquad \hat{\tau}_{rz} = \frac{1}{2}\gamma r , \qquad \hat{\sigma}_z = 0 . \tag{3}$$

Of course, it would be more interesting to neglect the weight of the membrane and to take into account the pressure distributed over its surface.

We now have

$$\underline{P} = \hat{\sigma}_r \underline{e}_r \underline{e}_r + (r\hat{\sigma}_r)' \underline{e}_\varphi \underline{e}_\varphi + \frac{1}{2} \gamma r \underline{e}_r \underline{k} \ ,$$

$$\underline{P}^T = \hat{\sigma}_r \underline{e}_r \underline{e}_r + (\hat{\sigma}_r r)' \underline{e}_\varphi \underline{e}_\varphi + \frac{1}{2} \gamma r \underline{k} \underline{e}_r \ ,$$

so that

$$\begin{aligned} \underline{P} \cdot \underline{P}^T &= (\hat{\sigma}_r^2 + \frac{1}{4} \gamma^2 r^2) \underline{e}_r \underline{e}_r + (r\hat{\sigma}_r)'^2 \underline{e}_\varphi \underline{e}_\varphi \ , \\ (\underline{P} \cdot \underline{P}^T)^{1/2} &= (\hat{\sigma}_r^2 + \frac{1}{4} \gamma^2 r^2)^{1/2} \underline{e}_r \underline{e}_r + (r\hat{\sigma}_r)' \underline{e}_\varphi \underline{e}_\varphi \ . \end{aligned} \tag{4}$$

In view of (5.5.11) the expression for the specific additional work takes the form

$$Э_\times = \frac{1}{2E} \left\{ \hat{\sigma}^2 + (r\hat{\sigma}_r)'^2 + \frac{1}{4} \gamma^2 r^2 - 2\nu (r\hat{\sigma}_r)' \sqrt{\hat{\sigma}_r^2 + \frac{1}{4} \gamma^2 r^2} + \right.$$
$$\left. + 2E \left[\sqrt{\hat{\sigma}_r^2 + \frac{1}{4} \gamma^2 r^2} + (r\hat{\sigma}_r)' \right] \right\} , \tag{5}$$

where $E = 2\mu(1+\nu)$ is "Young's modulus".

From (5.5.12) we now determine the deformation gradient

$$\overset{0}{\underline{\nabla R}} = (Э_\times)_{\underline{P}} = \frac{1}{E} \underline{P} + \underline{e}_r \frac{\underline{e}_r \hat{\sigma}_r + \frac{1}{2} \underline{k} \gamma r}{\sqrt{\hat{\sigma}_r^2 + \frac{1}{4} \gamma^2 r^2}} \left(1 - \frac{\nu (r\hat{\sigma}_r)'}{E} \right) +$$
$$+ \underline{e}_\varphi \underline{e}_\varphi \left(1 - \frac{\nu}{E} \sqrt{\hat{\sigma}_r^2 + \frac{1}{4} \gamma^2 r^2} \right) . \tag{6}$$

An "analog of the Beltrami equation" (4.7.12) leads now to the differential equation for the stress

$$\overset{0}{\underline{\nabla}} \times \overset{0}{\underline{\nabla R}} = \left(\underline{e}_r \frac{\partial}{\partial r} + \frac{1}{r} \underline{e}_\varphi \frac{\partial}{\partial \varphi} \right) \times \overset{0}{\underline{\nabla R}} = \underline{0} \ .$$

On carrying out this fairly elaborate calculation we obtain

$$\frac{1}{E} \left(r^2 \hat{\sigma}_r'' + 3r\hat{\sigma}_r' - \frac{1}{2} \frac{\nu \gamma^2 r^2}{\sqrt{\hat{\sigma}_r^2 + \frac{1}{4} \gamma^2 r^2}} \right) + 1 - \frac{\hat{\sigma}_r}{\sqrt{\hat{\sigma}_r^2 + \frac{1}{4} \gamma^2 v^2}} = 0 \ . \tag{7}$$

Introducing, together with Koiter, the dimensionless quantities

$$\alpha = \left(\frac{\gamma^2 a^2}{E^2}\right)^{1/2} , \qquad \hat{\sigma}_r = E\alpha s , \qquad \rho = \frac{r}{a} , \qquad \hat{\sigma}_\varphi = E\alpha(\rho s)' ,$$

$$\tau^2 = \frac{1}{4}\gamma^2\rho^2 a^2 = \frac{1}{4}E^2\alpha^3\rho^2 \tag{8}$$

we may represent (7) in the form

$$1 - \frac{s}{\sqrt{s^2 + \frac{1}{4}\alpha\rho^2}} + \alpha\left(\rho^2 s'' + 3\rho s' - \frac{1}{2}\,\frac{\nu\alpha\rho^2}{\sqrt{s^2 + \frac{1}{4}\alpha\rho^2}}\right) = 0 . \tag{9}$$

Here we have denoted the derivatives with respect to ρ by primes. Since $\alpha \ll 1$ we may be contented with the simplified equation

$$\rho^2 s'' + 3\rho s' + \frac{\rho^2}{8s^2} = 0 , \tag{10}$$

which was obtained by Koiter by another route. This equation includes neither geometric (α) nor material (E,ν) parameters.

We now proceed to the determination of the position vector. From (6) we have

$$d\underline{R} = d\underline{r}\cdot\overset{0}{\underline{\underline{\nabla R}}} = (\underline{e}_r dr + r\underline{e}_\varphi d\varphi + \underline{k}dz)\cdot\overset{0}{\underline{\underline{\nabla R}}} = \frac{1}{2}\gamma r\underline{k}\left(\frac{1}{E} + \frac{1 - \frac{\nu}{E}(r\hat{\sigma}_r)'}{\sqrt{\hat{\sigma}_r^2 + \frac{1}{4}\gamma^2 r^2}}\right)dr +$$

$$+ \underline{e}_r\left(\frac{1}{E} + \frac{1 - \frac{\nu}{E}(r\hat{\sigma}_r)'}{\sqrt{\hat{\sigma}_r^2 + \frac{1}{4}\gamma^2 r^2}}\right)\hat{\sigma}_r dr + r\underline{e}_\varphi\left[\frac{(r\hat{\sigma}_r)'}{E} + 1 - \frac{\nu}{E}\sqrt{\hat{\sigma}_r^2 + \frac{1}{4}\gamma^2 r^2}\right]d\varphi . \tag{11}$$

The symbol w denotes the vertical component of $\underline{R}$ (the displacement of a membrane) which is zero for $r = a$

$$w = -\int_r^a \left(\frac{1}{E} + \frac{1 - \frac{\nu}{E}(r\hat{\sigma}_r)'}{\sqrt{\hat{\sigma}_r^2 + \frac{1}{4}\gamma^2 r^2}}\right)\frac{1}{2}\gamma r\,dr . \tag{12}$$

The expression for the position vector now takes the form

$$\underline{R} = \underline{k}w + \int_{M_0}^{M}\left\{\underline{e}_r\left[\frac{1}{E} + \frac{1 - \frac{\nu}{E}(r\hat{\sigma}_r)'}{\sqrt{\hat{\sigma}_r^2 + \frac{1}{4}\gamma^2 r^2}}\right]\hat{\sigma}_r dr +\right.$$

$$+ r\underline{e}_{\varphi}\left[\frac{(r\hat{\sigma}_r)'}{E} + 1 - \frac{\nu}{E}\sqrt{\hat{\sigma}_r^2 + \frac{1}{4}\gamma^2 r^2}\right] d\varphi\Big\} \tag{13}$$

and the condition of integrability in the form

$$\left(\frac{1}{E} + \frac{1 - \frac{\nu}{E}(r\hat{\sigma}_r)'}{\sqrt{\hat{\sigma}_r^2 + \frac{1}{4}\gamma^2 r^2}}\right)\hat{\sigma}_r = \left\{r\left[\frac{(r\hat{\sigma}_r)'}{E} + 1 - \frac{\nu}{E}\sqrt{\hat{\sigma}_r^2 + \frac{1}{4}\gamma^2 r^2}\right]\right\}' .$$

must hold.

This relationship is simply an alternative form of Equation (7). Equation (13) is now reduced to

$$\underline{R} = \underline{k} + \int_{r_0}^{r} \left[\left\{\left[\frac{(r\hat{\sigma}_r)'}{E} + 1 - \frac{\nu}{E}\sqrt{\hat{\sigma}_r^2 + \frac{1}{4}\gamma^2 r^2}\right] r\right\}' \underline{e}_r dr + \right.$$

$$\left. + \left[\frac{(r\hat{\sigma}_r)'}{E} + 1 - \frac{\nu}{E}\sqrt{\hat{\sigma}_r^2 + \frac{1}{4}\gamma^2 r^2}\right] r\underline{e}_{\varphi} d\varphi\right] ,$$

or

$$\underline{R} = \underline{k}\omega + \int_{r_0}^{r} d\left\{\underline{e}_r r\left[\frac{(r\hat{\sigma}_r)'}{E} + 1 - \frac{\nu}{E}\sqrt{\hat{\sigma}_r^2 + \frac{1}{4}\gamma^2 r^2}\right]\right\} = \underline{k}\omega + \underline{r} + \underline{e}_r u_r . \tag{14}$$

Here u_r denotes the radial displacement of particles of the membrane

$$u_r = r\left[\frac{(r\hat{\sigma}_r)'}{E} - \frac{\nu}{E}\sqrt{\hat{\sigma}_r^2 + \frac{1}{4}\gamma^2 r^2}\right]\Bigg|_{r_0}^{r} \tag{15}$$

For $r = a$ we obtain one of the boundary conditions for Equation (7)

$$(u_r)_{r=a} = 0 : \left[(r\hat{\sigma}_r)' - \nu\sqrt{\hat{\sigma}_r^2 + \frac{1}{4}\gamma^2 a^2}\right]_{r=a} = 0$$

or

$$\rho = 1 : s' + s - \nu\sqrt{s^2 + \frac{1}{4}\alpha} = 0 . \tag{16}$$

Returning to (15) it is natural to require (by symmetry) that the displacement u_r should vanish at the centre of the membrane. This leads to the condition that the quantity (15) in the square brackets must be finite at $r = 0$:

$$(r\hat{\sigma}_r)' - \nu\hat{\sigma}_r \quad \text{or} \quad \rho s' + (1-\nu)s \quad \text{is finite .} \tag{17}$$

But $\hat{\sigma}_r$ is also finite at the centre of the membrane and so (17) can be reduced to the requirement that $(r\hat{\sigma}_r)'_{r=0}$ is finite. Returning now to (7), multiplying by r and integrating with respect to r we arrive at the relationship

$$\frac{1}{2}r^2 - \int_0^r \frac{\hat{\sigma}_r r\,dr}{\sqrt{\sigma_r^2 + \frac{1}{4}\gamma^2 r^2}} + \frac{1}{E}r^3\hat{\sigma}_r' - \frac{1}{2}\frac{\nu\gamma^2}{E}\int_0^r \frac{r^3 dr}{\sqrt{\hat{\sigma}_r^2 + \frac{1}{4}\gamma^2 r^2}} = 0$$

since $(r^3\hat{\sigma}_r')_{r=0} = 0$ if $r\hat{\sigma}_r'$ is finite. For small values of r, since $\hat{\sigma}_r \neq 0$, this reduces to the form

$$\sigma_r' = \frac{1}{r^3}\left[-\frac{1}{8}E\gamma^2\int_0^r \frac{r^3 dr}{\hat{\sigma}_r^2} + \frac{1}{2}\nu\gamma^2\int_0^r \frac{r^3 dr}{\hat{\sigma}_r} - \frac{1}{16}\nu\gamma^4\int_0^r \frac{r^5 dr}{\hat{\sigma}_r^3}\right]_{r\to 0} = 0 \ ,$$

because the expression in the square brackets is at least of order ^{4}. For this reason the condition (17) expresses the boundary condition

$$r = 0 : \hat{\sigma}_r' = 0 \quad \text{or} \quad \rho = 0 : s' = 0 \ . \tag{18}$$

The displacement at the centre of the membrane is by (12) determined by the expression

$$-\frac{w}{a} = \sqrt{a}\left\{\frac{1}{4}\alpha + \frac{1}{2}\int_0^1 [1 - \nu\alpha(\rho s)'] \frac{\rho\,d\rho}{\sqrt{s^2 + \frac{1}{2}\alpha\rho^2}}\right\} . \tag{19}$$

Smallness of the parameter α allows us to substitute Equation (19) for the formula suggested by Koiter

$$-\frac{w}{a} = \frac{1}{2}\sqrt{a}\int_0^1 \frac{\rho\,d\rho}{s} \ . \tag{20}$$

Proceeding now from the boundary condition (16) to its approximate representation

$$\rho = 1 : s' + (1-\nu)s = 0 \tag{21}$$

the problem is reduced to the differential equation (10) under the conditions of (21) and (18). In his paper, Koiter represents the results of a numerical solution to this problem for several values of ν.

8. THE PLANE PROBLEM FOR A QUASI-LINEAR MATERIAL

1. Geometrical relations. The transformation of the reference configuration to the actual configuration is given by the relationships (the Greek indices take values 1 and 2)

$$x^\alpha = x^\alpha(a^1, a^2) , \qquad x^3 = c a^3 , \tag{1}$$

which determine plane deformation of a prismatic body. If the displacement of the body in the longitudinal direction is prohibited, then $c = 1$. The Cartesian coordinates a^1, a^2, a^3 of the reference configuration are chosen as material coordinates. The deformation gradient is

$$\overset{0}{\underline{\nabla R}} = \underline{i}_\alpha \underline{i}_\beta \frac{\partial x^\beta}{\partial a^\alpha} + \underline{i}_3 \underline{i}_3 c . \tag{2}$$

The Cauchy strain measure $\underline{G}$ and its components are given by the formulas

$$\underline{G} = \overset{0}{\underline{\nabla R}} \cdot \overset{0}{\underline{\nabla R}}^T = \underline{i}_\alpha \underline{i}_\beta \frac{\partial x^\gamma}{\partial a^\alpha} \frac{\partial x^\gamma}{\partial a^\beta} + \underline{i}_3 \underline{i}_3 c^2 , \qquad G_{11} = \left(\frac{\partial x^1}{\partial a^1}\right)^2 + \left(\frac{\partial x^2}{\partial a^1}\right)^2 ,$$

$$G_{22} = \left(\frac{\partial x^1}{\partial a^2}\right)^2 + \left(\frac{\partial x^2}{\partial a^2}\right)^2 , \qquad G_{12} = \frac{\partial x^1}{\partial a^1}\frac{\partial x^1}{\partial a^2} + \frac{\partial x^2}{\partial a^1}\frac{\partial x^2}{\partial a^2} , \tag{3}$$

$$G_{23} = G_{31} = 0 , \qquad G_{33} = c^2 .$$

The components of the stretch tensor $\underline{U} = \underline{G}^{1/2}$ are determined from the system of equations

$$\underline{U} \cdot \underline{U} = \underline{U}^2 = U_{\alpha\gamma} U_{\gamma\beta} \underline{i}_\alpha \underline{i}_\beta = G_{\alpha\beta} \underline{i}_\alpha \underline{i}_\beta + \underline{i}_3 \underline{i}_3 c^2 ,$$

$$U_{11}^2 + U_{12}^2 = G_{11} , \qquad U_{21}^2 + U_{22}^2 = G_{22} , \tag{4}$$

$$U_{12}(U_{11} + U_{22}) = G_{12} , \qquad U_{33} = c , \qquad U_{3\alpha} = 0 .$$

The invariants of the Cauchy measure turn out to be equal

$$I_1(\underline{G}) = G_{11} + G_{22} + c^2 = G_1 + G_2 + c^2 ,$$

$$I_2(\underline{G}) = G_1G_2 + c^2(G_1+G_2) = G_{11}G_{22} - G_{12}^2 + c^2(G_{11}+G_{22}) ,$$

$$I_3(\underline{G}) = \det\underline{G} = c^2(G_{11}G_{22} - G_{12}^2) ,$$

where G_1, G_2 and c^2 denote the principal values of $\underline{G}$. We make use of the notation

$$G = G_1G_2 = G_{11}G_{22} - G_{12}^2 = \left(\frac{\partial x^1}{\partial a^1}\frac{\partial x^2}{\partial a^2} - \frac{\partial x^1}{\partial a^2}\frac{\partial x^2}{\partial a^1}\right)^2 ,$$

$$\sqrt{G} = \frac{\partial x^1}{\partial a^1}\frac{\partial x^2}{\partial a^2} - \frac{\partial x^1}{\partial a^2}\frac{\partial x^2}{\partial a^1} , \qquad (6)$$

where $c\sqrt{G}$ denotes the determinant of the transformation (1). We now have

$$(U_{11}+U_{22})^2 = (\sqrt{G_1}+\sqrt{G_2})^2 = G_1 + G_2 + 2\sqrt{G_1G_2} =$$

$$= \left(\frac{\partial x^1}{\partial a^1} + \frac{\partial x^2}{\partial a^2}\right)^2 + \left(\frac{\partial x^2}{\partial a^1} - \frac{\partial x^1}{\partial a^2}\right)^2 .$$

Making use of the notation

$$q = \left[\left(\frac{\partial x^1}{\partial a^1} + \frac{\partial x^2}{\partial a^2}\right)^2 + \left(\frac{\partial x^2}{\partial a^1} - \frac{\partial x^1}{\partial a^2}\right)^2\right]^{1/2} , \qquad (7)$$

we have from Equations (4)

$$U_{11} + U_{22} = q , \qquad U_{12} = \frac{1}{q}\left(\frac{\partial x^1}{\partial a^1}\frac{\partial x^1}{\partial a^2} + \frac{\partial x^2}{\partial a^2}\frac{\partial x^2}{\partial a^1}\right) ,$$

$$U_{11}^2 - U_{22}^2 = q(U_{11}-U_{22}) = G_{11} - G_{22} , \qquad (8)$$

$$U_{11} = \frac{1}{2q}(G_{11}-G_{22}+q^2) = \frac{1}{q}\left[\frac{\partial x^1}{\partial a^1}\left(\frac{\partial x^1}{\partial a^1} + \frac{\partial x^2}{\partial a^2}\right) + \frac{\partial x^2}{\partial a^1}\left(\frac{\partial x^2}{\partial a^1} - \frac{\partial x^1}{\partial a^2}\right)\right] ,$$

$$U_{22} = \frac{1}{q}\left[\frac{\partial x^2}{\partial a^2}\left(\frac{\partial x^1}{\partial a^1} + \frac{\partial x^2}{\partial a^2}\right) + \frac{\partial x^1}{\partial a^2}\left(\frac{\partial x^1}{\partial a^2} - \frac{\partial x^2}{\partial a^1}\right)\right] .$$

These formulas show that it is reasonable to introduce the substitutions

$$\cos\chi = \frac{1}{q}\left(\frac{\partial x^1}{\partial a^1} + \frac{\partial x^2}{\partial a^2}\right), \qquad \sin\chi = \frac{1}{q}\left(\frac{\partial x^2}{\partial a^1} - \frac{\partial x^1}{\partial a^2}\right), \tag{9}$$

since these enable us to represent Equations (8) and the expression for the tensor $\underline{U}$ in the form

$$U_{11} = \frac{\partial x^1}{\partial a^1}\cos\chi + \frac{\partial x^2}{\partial a^1}\sin\chi ,$$

$$U_{22} = -\frac{\partial x^1}{\partial a^2}\sin\chi + \frac{\partial x^2}{\partial a^2}\cos\chi , \tag{10}$$

$$U_{12} = \frac{\partial x^1}{\partial a^2}\cos\chi + \frac{\partial x^2}{\partial a^2}\sin\chi = -\frac{\partial x^1}{\partial a^1}\sin\chi + \frac{\partial x^2}{\partial a^1}\cos\chi ,$$

$$\underline{U} = U_{\alpha\beta}\underline{i}_\alpha\underline{i}_\beta + c\underline{i}_3\underline{i}_3 = (\underline{i}_\alpha\underline{i}_\beta\cos\chi - \underline{i}_3\times\underline{i}_\alpha\underline{i}_\beta\sin\chi)\frac{\partial x^\alpha}{\partial a^\beta} + c\underline{i}_3\underline{i}_3 .$$

The angle χ is naturally connected with the orthogonal tensor $\underline{O}^\times$ of rotation accompanying the deformation. By (1.6.8) and (1.3.3) we have

$$\underline{O}^\times = \underline{U}\cdot\underline{\nabla r}^T = \left[(\underline{i}_\alpha\underline{i}_\beta\cos\chi - \underline{i}_3\times\underline{i}_\alpha\underline{i}_\beta\sin\chi)\frac{\partial x^\alpha}{\partial a^\beta} + c\underline{i}_3\underline{i}_3\right]\cdot\left(\underline{i}_\gamma\underline{i}_\delta\frac{\partial a^\gamma}{\partial x^\delta} + \frac{1}{c}\underline{i}_3\underline{i}_3\right) =$$

$$= (\underline{i}_\alpha\underline{i}_\delta\cos\chi - \underline{i}_3\times\underline{i}_\alpha\underline{i}_\delta\sin\chi)\frac{\partial x^\alpha}{\partial a^\beta}\frac{\partial a^\beta}{\partial x^\delta} + \underline{i}_3\underline{i}_3 =$$

$$= (\underline{i}_\alpha\underline{i}_\delta\cos\chi - \underline{i}_3\times\underline{i}_\alpha\underline{i}_\delta\sin\chi)\delta^\alpha_\delta + \underline{i}_3\underline{i}_3 ,$$

so that

$$\underline{O}^\times = \underline{E}_2\cos\chi - \underline{i}_3\times\underline{E}_2\sin\chi + \underline{i}_3\underline{i}_3 \qquad (\underline{E}_2 = \underline{i}_1\underline{i}_1 + \underline{i}_2\underline{i}_2) . \tag{11}$$

Subsequent analysis can be simplified by the introduction of complex notation (a bar above the symbol denotes transition to a conjugate number)

$$\zeta = a^1 + ia^2 , \qquad z = x^1 + ix^2 ,$$

so that

$$\frac{\partial}{\partial a^1} = \frac{\partial}{\partial \zeta} + \frac{\partial}{\partial \bar{\zeta}}, \qquad \frac{\partial}{\partial a^2} = i\left(\frac{\partial}{\partial \zeta} - \frac{\partial}{\partial \bar{\zeta}}\right);$$

$$2\frac{\partial}{\partial \zeta} = \frac{\partial}{\partial a^1} - i\frac{\partial}{\partial a^2}, \qquad 2\frac{\partial}{\partial \bar{\zeta}} = \frac{\partial}{\partial a^1} + i\frac{\partial}{\partial a^2}.$$

By virtue of (9) and (6) we get

$$2\frac{\partial z}{\partial \zeta} = \frac{\partial x^1}{\partial a^1} + \frac{\partial x^2}{\partial a^2} + i\left(\frac{\partial x^2}{\partial a^1} - \frac{\partial x^1}{\partial a^2}\right) = qe^{i\chi},$$

$$\sqrt{G} = \frac{\partial z}{\partial \zeta}\frac{\partial \bar{z}}{\partial \bar{\zeta}} - \frac{\partial z}{\partial \bar{\zeta}}\frac{\partial \bar{z}}{\partial \zeta}, \tag{12}$$

$$q = 2\left|\frac{\partial z}{\partial \zeta}\right| = 2\left(\frac{\partial z}{\partial \zeta}\frac{\partial \bar{z}}{\partial \bar{\zeta}}\right)^{1/2}, \qquad e^{i\chi} = \frac{\partial z}{\partial \zeta}\left|\frac{\partial z}{\partial \zeta}\right|^{-1}, \qquad e^{2i\chi} = \frac{\partial z}{\partial \zeta}\Big/\frac{\partial \bar{z}}{\partial \bar{\zeta}}. \tag{13}$$

2. The representation of the Piola tensor for a quasi-linear material. The equilibrium equations. By (5.5.5) and in view of (8) we have

$$\underline{P} = [\lambda(q + c - 3) - 2\mu]\underline{O}^{\times} + 2\mu\overset{0}{\underline{\nabla}}\underline{R} =$$

$$= [\lambda(q + c - 3) - 2\mu](\underline{E}_2\cos\chi - \underline{i}_3 \times \underline{E}_2 \sin\chi) + 2\mu\underline{i}_\alpha\underline{i}_\beta\frac{\partial x^\beta}{\partial a^\alpha} + \underline{i}_3\underline{i}_3 p^{33}; \tag{14}$$

$$p^{33} = \lambda(q + c - 3) + 2\mu(c - 1), \qquad p^{3\alpha} = p^{\alpha 3} = 0.$$

We now introduce representations for the components $p^{\alpha\beta}$ which are similar to the Kolosov-Muskhelishvili formulas of linear theory

$$p^{11} + ip^{12} = \psi(q)e^{i\chi} + 2\mu i\frac{\partial z}{\partial a^2},$$

$$p^{22} - ip^{21} = \psi(q)e^{i\chi} - 2\mu\frac{\partial z}{\partial a^1}, \tag{15}$$

where

$$\psi(q) = (\lambda + 2\mu)(q - 2) + 2\mu + \lambda(c - 1) = (\lambda + 2\mu)\left[q - \frac{1 - \nu(c - 1)}{1 - \nu}\right]$$

$$\left(\nu = \frac{\lambda}{2(\lambda + \mu)}\right). \tag{16}$$

In the absence of the body forces the equations of static equilibrium

$$\frac{\partial p^{11}}{\partial a^1} + \frac{\partial p^{21}}{\partial a^2} = 0 , \qquad \frac{\partial p^{12}}{\partial a^1} + \frac{\partial p^{22}}{\partial a^2} = 0 , \qquad \frac{\partial p^{33}}{\partial a^3} = 0$$

are now transformed to

$$\frac{\partial}{\partial a^1} (p^{11} + ip^{12}) + i \frac{\partial}{\partial a^2} (p^{22} - ip^{21}) = \left(\frac{\partial}{\partial a^1} + i \frac{\partial}{\partial a^2}\right)\psi(q)e^{i\chi} +$$

$$+ 2\mu i\left(\frac{\partial^2 z}{\partial a^1 \partial a^2} - \frac{\partial^2 z}{\partial a^2 \partial a^1}\right) = 0 ,$$

so that

$$\frac{\partial}{\partial \bar{\zeta}} \psi(q)e^{i\chi} = 0 , \qquad p^{33} = p^{33}(a^1, a^2) . \tag{17}$$

The first relation contains the main result and the expression $\psi(q)e^{i\chi}$ represents the function of the complex variable ζ defined below as

$$\psi(q)e^{i\chi} = \Phi'(\zeta) , \qquad e^{i\chi} = \frac{\Phi'(\zeta)}{|\Phi'(\zeta)|} . \tag{18}$$

It is therefore reasonable to call a quasi-linear material a "harmonic" material (John, 1960).

3. The boundary-value problem. The connection between the functions $z(\zeta,\bar{\zeta})$ and $\Phi'(\zeta)$ is given by the relationship

$$2 \frac{\partial z}{\partial \zeta} = \frac{\Phi'(\zeta)}{\lambda + 2\mu} + \frac{1 - \nu(c - 1)}{1 - \nu} \frac{\Phi'(\zeta)}{|\Phi'(\zeta)|} , \tag{19}$$

which follows from Equations (12), (16).

The equations of equilibrium on a lateral surface of a prismatic body may be written in the form

$$\underline{f}\, dO = \underline{n} \cdot \underline{P}\, do , \qquad f_1 \frac{dO}{do} = n_1 p^{11} + n_2 p^{21} , \qquad f_2 \frac{dO}{do} = n_1 p^{21} + n_2 p^{22} .$$

Here f_1 and f_2 denote projections of the contact force along the axes a^1, a^2; for the plane problem f_3 must be zero on the lateral surface. Introducing the complex vectors

$$f_1 + if_2 = f , \qquad n_1 + in_2 = n = \frac{da^2}{ds} - i \frac{da^1}{ds} = -i \frac{d\zeta}{ds} ,$$

where ds is the element of arc of contour l bounding the section in the reference configuration, and referring to (15) and (18), we obtain

$$if\,\frac{dO}{do} = icf\,\frac{dS}{ds} = \frac{d}{ds}\,\Phi'(\zeta) - 2\mu\,\frac{dz}{ds}\,. \tag{20}$$

We have denoted by dS the element of arc of the contour L which represents the result of the transformation into the actual configuration of the contour l of the reference configuration. Obviously,

$$dO = dS\,c\,da^3\,, \qquad do = ds\,da^3\,.$$

The resultant vector V of the contact forces on the arc M_0M is now given by

$$icV = ic\int_{S_0}^{S} f dS = \int_{M_0}^{M} \Phi'(\zeta)d\zeta - 2\mu\int_{M_0}^{M} dz =$$

$$= \Phi(\zeta) - \Phi(\zeta_0) - 2\mu\,[z(\zeta,\bar{\zeta}) - z(\zeta_0,\bar{\zeta}_0)]\,. \tag{21}$$

If the pressure p, which is uniformly distributed over the lateral surface, acts on l, then we have

$$f = -pN = pi\,\frac{dz}{dS}\,, \qquad icV = -cp(z - z_0)\,,$$

$$\Phi(\zeta) = (2\mu - cp)\,z(\zeta,\bar{\zeta}) + \text{const}\,. \tag{22}$$

The Cauchy stress tensor $\underline{T}$ is, from (2.7.3) and (6), represented by the expression

$$\underline{T} = \frac{1}{c}\sqrt{\frac{g}{G}}\,\overset{0}{\nabla}\underline{R}^T \cdot \underline{P} = \frac{1}{c}\sqrt{\frac{g}{G}}\left[\underline{i}_\alpha \underline{i}_\beta\,\frac{\partial x^\alpha}{\partial a^\gamma}\,p^{\gamma\beta} + \underline{i}_3\underline{i}_3 c p^{33}\right]. \tag{23}$$

The normal stress in the cross-section a^3 = const is given by

$$\sigma_3\sqrt{G} = p^{33}\,, \qquad \sigma_3 dO = p^{33}do = [\lambda(q - 2) + (\lambda + 2\mu)(c - 1)]\,do\,, \tag{24}$$

since the determinant $\sqrt{G}$ of transformation of the plane region a^3 = const into the plane region z^2 = const equals the ratio dO/do of the elementary areas. The longitudinal force in the body considered turns out to be

$$\iint_{\Omega} \sigma_3 dO = \iint_{\Omega_0} p^{33}do = \lambda\iint_{\Omega_0}(q - 2)\,do + (\lambda + 2\mu)(c - 1)\Omega_0\,, \tag{25}$$

where Ω_0 denotes the area of the cross-section in the reference configuration.

Returning to (23), we have

$$\sigma_1 + \sigma_2 = \frac{1}{c\sqrt{G}} \frac{\partial x^\alpha}{\partial a^\gamma} p^{\gamma\alpha}$$

and the calculation carried out using Equations (15) and (16) yields

$$c(\sigma_1 + \sigma_2) + 4\mu = \frac{q\psi(q)}{\sqrt{G}} . \tag{26}$$

This is the first of Kolosov-Muskhelishvili formula. The second formula yields the expression

$$\sigma_2 - \sigma_1 - 2i\tau_{12} = \underline{i}_2 \cdot \underline{T} \cdot \underline{i}_2 - \underline{i}_1 \cdot \underline{T} \cdot \underline{i}_1 - i(\underline{i}_1 \cdot \underline{T} \cdot \underline{i}_2 + \underline{i}_2 \cdot \underline{T} \cdot \underline{i}_1) =$$

$$= \frac{1}{c\sqrt{G}} \left[\frac{\partial x^2}{\partial a^\gamma} p^{\gamma 2} - \frac{\partial x^1}{\partial a^\gamma} p^{\gamma 1} - i \left(\frac{\partial x^1}{\partial a^\gamma} p^{\gamma 2} + \frac{\partial x^2}{\partial a^\gamma} p^{\gamma 1} \right) \right]$$

and calculation using Equations (15) leads to the relationship

$$c(\sigma_2 - \sigma_1 - 2i\tau_{12}) = - \frac{1}{\sqrt{G}} \psi(q) e^{i\chi} \left(\frac{\partial z}{\partial a^1} + i \frac{\partial z}{\partial a^2} \right) =$$

$$= - \frac{4}{\sqrt{G}} \frac{\psi(q)}{q} \frac{\partial z}{\partial \zeta} \frac{\partial z}{\partial \bar{\zeta}} . \tag{27}$$

Returning to the representation of the force $\underline{f}$ given by (20), we construct expressions for its normal, $f_{\underline{N}}$, and tangential, $f_{\underline{S}}$, components along the arc L. Denoting the unit vectors of the normal and tangent to this arc by $\underline{N}$ and $\underline{S}$, we have

$$f_{\underline{N}} = \underline{N} \cdot \underline{f} , \quad f_{\underline{S}} = \underline{S} \cdot \underline{f} ; \qquad N = -i \frac{dz}{dS} , \quad S = \frac{dz}{dS} . \tag{28}$$

The scalar product of two vectors is expressed in terms of their complex representations by the formula

$$\underline{a} \cdot \underline{b} = \frac{1}{2} (a\bar{b} + \bar{a}b) .$$

Substituting for $\underline{f}$ from (20), we have, from (28), the formulas

$$f_{\underline{N}} = \frac{i}{2} \left(f \frac{d\bar{z}}{dS} - \bar{f} \frac{dz}{dS} \right) = \frac{1}{2c} \left[\Phi'(\zeta) \frac{d\zeta}{dS} \frac{d\bar{z}}{dS} + \overline{\Phi'(\zeta)} \frac{d\bar{\zeta}}{dS} \frac{dz}{dS} - 4\mu \right] , \tag{29}$$

$$f_{\underset{-}{S}} = \frac{1}{2}\left(f\,\frac{d\bar{z}}{dS} + \bar{f}\,\frac{dz}{dS}\right) = -\frac{i}{2c}\left[\Phi'(\zeta)\,\frac{d\zeta}{dS}\,\frac{d\bar{z}}{dS} - \overline{\Phi'(\zeta)}\,\frac{d\bar{\zeta}}{dS}\,\frac{dz}{dS}\right], \tag{30}$$

$|dz| = dS,\ |d\zeta| = ds.$

Notes. 1. In the plane stress problem the deformation is considered of a thin plate (of the thickness $2h_0$) loaded on its lateral surface by forces parallel to its mean plane $a_3 = 0$ and distributed symmetrically with regard to this plane. The face surfaces $a_3 = \pm h_0$ are not loaded. This situation is realized approximately under a transformation of the reference configuration into the actual configuration given by the formulas

$$x^\alpha = x^\alpha(a^1,a^2)\,, \qquad x^3 = c(a^1,a^2)a^3 \qquad (|a_3| \le h_0) \tag{31}$$

provided that terms of the order h_0^2 are neglected and that only values of the components $p^{\alpha\beta}$ of the Piola tensors averaged over the thickness of the plate are considered. The mean values of $p^{\alpha 3}$ and $p^{3\alpha}$ turn out to be zero and the function $c(a^1,a^2)$ is determined from the condition that p^{33} equals zero.

In fact, we have by (31)

$$\overset{0}{\underset{--}{\nabla R}} = \underset{-s}{i}\underset{-t}{i}\,\frac{\partial x^t}{\partial a^s} = \underset{-\alpha}{i}\underset{-\beta}{i}\,\frac{\partial x^\beta}{\partial a^\alpha} + \underset{-\alpha}{i}\underset{-3}{i}\,\frac{\partial c}{\partial a^\alpha}\,a^3 + \underset{-3}{i}\underset{-3}{i}\,c\,,$$

so that

$$\underset{-}{G} = \overset{0}{\underset{--}{\nabla R}}\cdot\overset{0}{\underset{--}{\nabla R}}^T = \underset{-\alpha}{i}\underset{-\beta}{i}\left(\frac{\partial x^\gamma}{\partial a^\alpha}\,\frac{\partial x^\gamma}{\partial a^\beta} + a^{3^2}\,\frac{\partial c}{\partial a^\alpha}\,\frac{\partial c}{\partial a^\beta}\right) + a^3(\underset{-\alpha}{i}\underset{-3}{i} + \underset{-3}{i}\underset{-\alpha}{i})\,c\,\frac{\partial c}{\partial a^\alpha} + \underset{-3}{i}\underset{-3}{i}\,c^2$$

and the Cauchy measure averaged over the thickness of the plate

$$\underset{-}{\tilde{G}} = \frac{1}{2h_0}\int_{-h_0}^{h_0}\underset{-}{G}\,da^3 = \underset{-\alpha}{i}\underset{-\beta}{i}\left(\frac{\partial x^\gamma}{\partial a^\alpha}\,\frac{\partial x^\gamma}{\partial a^\beta} + \frac{1}{3}h_0^2\,\frac{\partial c}{\partial a^\alpha}\,\frac{\partial c}{\partial a^\beta}\right) + \underset{-3}{i}\underset{-3}{i}\,c^2\,,$$

since $\int_{-h_0}^{h_0} a^3 da^3 = 0$. Neglecting the term in h_0^2 this reduces to (3) and for the mean values of all quantities the relationships obtained earlier are preserved. (The sign ~ over their symbols is discarded.) In particular, $p^{\alpha 3} = p^{3\alpha} = 0$, and the requirement $p^{33} = 0$ leads by (4) to the following expression for

$$c - 1 = -\frac{\lambda}{\lambda + 2\mu}\,(q-2)\,. \tag{32}$$

Now the expression for $\psi(q)$ is in view of (16) reduced to the form

$$\psi(q) = (\lambda + 2\mu)(q - 2) - \frac{\lambda^2}{\lambda + 2\mu}(q - 2) + 2\mu = \frac{4\mu(\lambda + \mu)}{\lambda + 2\mu}\left(q - \frac{3\lambda + 2\mu}{2(\lambda + \mu)}\right) . \tag{33}$$

2. In the problem of plane deformation of a prismatic body when the displacement of its faces is prevented ($c = 1$), the representation for $\psi(q)$ takes the form

$$\psi(q) = (\lambda + 2\mu)\left(q - \frac{2(\lambda + \mu)}{\lambda + 2\mu}\right) \tag{34}$$

and it is easy to verify that the state equations for $p^{\alpha\beta}$ in the planar stress problem have the same form (15) as for the problem of plane deformation provided that we substitute λ by the quantity

$$\lambda' = \frac{2\mu\lambda}{\lambda + 2\mu} . \tag{35}$$

In fact, by (33) and (34) it is required that

$$(\lambda' + 2\mu)\left(q - \frac{2(\lambda' + \mu)}{\lambda' + 2\mu}\right) = \frac{4\mu(\lambda + \mu)}{\lambda + 2\mu}\left(q - \frac{3\lambda + 2\mu}{2(\lambda + \mu)}\right) ,$$

in other words

$$\lambda' + 2\mu = \frac{4\mu(\lambda + \mu)}{\lambda + 2\mu} , \qquad \lambda' + \mu = \mu\,\frac{3\lambda + 2\mu}{\lambda + 2\mu} .$$

Both relationships are satisfied if λ' is determined according to (35) (or by substitution of ν for $\nu/(1 + \nu)$). As is well known the same relationship between the constants in the law of elasticity also holds for linear plane problems.

3. For the transformation of the rotation $z = \zeta e^{i\theta}$, θ = const we have by (12)

$$\frac{\partial z}{\partial \bar{\zeta}} = 0 , \qquad 2\,\frac{\partial z}{\partial \zeta} = qe^{i\chi} = 2e^{i\theta} ,$$

$$q = 2 , \qquad \chi = \theta , \qquad \sqrt{G} = 1$$

and for $c = 1$, $\psi(q) = 2\mu$ by (16). As expected formulas (26) and (27) yield $\sigma_1 = \sigma_2 = \tau_{12} = 0$.

For $\psi(q) = 0$ and $q = (1 - \nu)^{-1}$ the same formulas imply $\sigma_1 = \sigma_2 = -2\mu$, $\tau_{12} = 0$, and the medium is found to be in a state of hydrostatic compression of intensity 2μ.

9. BENDING OF A STRIP INTO A RIGHT CYLINDRICAL FORM. DEFORMATION OF A HOLLOW CYLINDER

1. We now consider the deformation of the parallelepiped

$$-h \le a^1 \le h , \qquad -b \le a^2 \le b , \qquad -l \le a^3 \le l$$

into a cylindrical section

$$r_0 \le r \le r_1 , \qquad -\alpha < \varphi \le \alpha , \qquad -l \le a^3 \le l$$

with inner and outer radii r_0 and r_1 respectively and angle subtended at the centre equal to 2α. Longitudinal displacement of the faces $a^3 = \mp l$ is assumed to be prevented. The bending of the sheet into a section is due to the distribution of moments along the edges $\varphi = \pm\alpha$ of the sheet (along the edges $a^2 = \pm b$ in the reference configuration). The cylindrical surfaces $r = r_0$, $r = r_1$ (in the reference configuration, the sides $a^1 = \mp h$) are free from loading.

The transformation is described by the formulas

$$z = x^1 + ix^2 = C(a^1)e^{i\frac{\alpha}{b}a^2} = re^{i\varphi} , \qquad x^3 = a^3 , \tag{1}$$

so that $r = |z| = C(a^1)$, $\varphi = \alpha \dfrac{a^2}{b}$ - the values $a^2 = \pm b$ correspond to the azimuthal angle $\varphi = \alpha$. By (8.12) we have

$$2\frac{\partial z}{\partial \zeta} = \frac{\partial z}{\partial a^1} - i\frac{\partial z}{\partial a^2} = \left[C'(a^1) + \frac{\alpha}{b}C(a^1)\right]e^{i\frac{\alpha}{b}a^2} = qe^{i\chi} ,$$

which, together with (8.16) and (8.18), yields

$$q = C'(a^1) + \frac{\alpha}{b}C(a^1) , \qquad \chi = \frac{\alpha}{b}a^2 ,$$

$$\Phi'(\zeta) = (\lambda + 2\mu)\left[C'(a^1) + \frac{\alpha}{b}C(a^1) - (1-\nu)^{-1}\right]e^{i\frac{\alpha}{b}a^2}$$

and the differential equation specifying $C(a^1)$ is deduced from the condition

$$2\frac{\partial \Phi'(\zeta)}{\partial \bar{\zeta}} = \frac{\partial \Phi'(\zeta)}{\partial a^1} + \frac{\partial \Phi'(\zeta)}{\partial a^2} = 0 \; : \; C''(a^1) - \frac{\alpha^2}{b^2}C(a^1) = -\frac{\alpha}{b}(1-\nu)^{-1} . \tag{2}$$

The solution satisfying the boundary conditions

$$C(-h) = r_0 , \qquad C(h) = r_1 \tag{3}$$

is given by

$$C(a^1) = \frac{1}{2}\left[(r_1+r_0)\frac{\operatorname{ch}\frac{\alpha a^1}{b}}{\operatorname{ch}\gamma} + (r_1-r_0)\frac{\operatorname{sh}\frac{\alpha a^1}{b}}{\operatorname{sh}\gamma}\right] + \frac{b}{\alpha(1-\nu)}\left(1 - \frac{\operatorname{ch}\frac{\alpha a^1}{b}}{\operatorname{ch}\gamma}\right)$$

$$\left(\gamma = \frac{\alpha h}{b}\right) . \tag{4}$$

The unknown radii of the panel are determined from the condition that surfaces $a^1 = \pm h$ are not loaded. By (5.20), for $f = 0$ and $ds = \pm da^2$ we get

$$2\mu \frac{dz}{ds} = \Phi'(\zeta)\frac{d\zeta}{ds} : 2\mu \frac{\alpha}{b} C(\pm h) = (\lambda + 2\mu)[C'(\pm h) + \frac{\alpha}{b} C(\pm h)]$$

and in view of (4) we have

$$r_1 + r_0 = 2\frac{b}{\alpha} , \qquad r_1 - r_0 = 2\frac{b}{\alpha}\operatorname{th}\gamma . \tag{5}$$

The expressions for $C(a^1)$, $C'(a^1)$, q and $\psi(q)$ now become

$$C(a^1) = \frac{b}{\alpha \operatorname{ch}\gamma}\left(\operatorname{sh}\frac{\alpha a^1}{b} - \frac{\nu}{1-\nu}\operatorname{ch}\frac{\alpha a^1}{b}\right) ,$$

$$C'(a^1) = \frac{1}{\operatorname{ch}\gamma}\left(\operatorname{ch}\frac{\alpha a^1}{b} - \frac{\nu}{1-\nu}\operatorname{sh}\frac{\alpha a^1}{b}\right) ,$$

$$q = \frac{1}{\lambda + 2\mu}\left(\frac{2\mu}{\operatorname{ch}\gamma}\exp\frac{\alpha a^1}{b} + 2\lambda + 2\mu\right) , \tag{6}$$

$$\psi(q) = \frac{2\mu}{\operatorname{ch}\gamma}\exp\frac{\alpha a^1}{b} .$$

The components $f_{\underline{N}}$ and $f_{\underline{S}}$ of stress on the areas $a^2 = \pm b$ are given by formulas (8.29) and (8.30). On these areas,

$$ds = \mp da^1 , \qquad \frac{d\zeta}{ds}\frac{d\bar{z}}{ds} = C'(a^1)e^{-i\chi} , \qquad \Phi'(\zeta) = \psi(q)e^{i\chi} ,$$

$$\left|\frac{dz}{ds}\right| = \frac{dS}{ds} = C'(a^1) .$$

We now have

$$f_{\underline{N}}\frac{dS}{ds} = \psi(q) - 2\mu C'(a^1) = \frac{2\mu}{\operatorname{ch}\gamma}\left(\exp\frac{\alpha a^1}{b} - \operatorname{ch}\frac{\alpha a^1}{b} + \frac{\nu}{1-\nu}\operatorname{sh}\frac{\alpha a^1}{b}\right) =$$

$$= \frac{E}{1-\nu^2}\,\frac{1}{\operatorname{ch}\gamma}\,\operatorname{sh}\frac{\alpha a^1}{b}\,,$$

$$f_{\underline{S}} = 0 \qquad (E = 2\mu(1+\nu)) \tag{7}$$

in which tangential forces are absent on the edges $a^2 = \pm b$ and, as expected, the resultant vector of the normal stress is zero

$$\int_{-h}^{h} f_{\underline{N}}\,dS = \frac{E}{(1-\nu^2)\operatorname{ch}\gamma}\int_{-h}^{h}\operatorname{sh}\frac{\alpha a^1}{b}\,da^1 = 0\,.$$

The resultant moment of these stresses with respect to the axis of the form related to unit length along the a^3-axis is equal, for any cross-section $a^2 = \text{const}$, to

$$m^0 = \int_{-h}^{h} C(a^1) f_{\underline{N}}\,dS = \frac{E}{(1-\nu^2)\operatorname{ch}^2\gamma}\,\frac{b}{\alpha}\int_{-h}^{h}\left(\operatorname{sh}^2\frac{\alpha a^1}{b} - \frac{\nu}{1-\nu}\operatorname{sh}\frac{\alpha a^1}{b}\operatorname{ch}\frac{\alpha a^1}{b}\right)da^1 =$$

$$= \frac{Eh^2}{1-\nu^2}\,\frac{1}{\gamma^2}\left(\operatorname{th}\gamma - \frac{\gamma}{\operatorname{ch}^2\gamma}\right). \tag{8}$$

The angle subtended at the axis by the form, $\alpha = b\gamma/h$, may be determined from this equation in terms of the moment m^0.

Expanding the right-hand side of (8) in a power series in γ, we get

$$m^0 = \frac{2}{3}\,\frac{Eh^2}{1-\nu^2}\left(\gamma - \frac{4}{3}\gamma^3 + \ldots\right)$$

the first term determines the solution of the linear problem of the bending of a thin plate of thickness $2h$ and length $2l$ (along the a^3-axis) produced by the moments distributed along the edges $a^2 = \pm b$.

The longitudinal force (per unit length along the a^2-axis) is calculated from (8.25)

$$Q = \lambda\int_{-h}^{h}(q-2)\,da^1 = -\frac{2Eh\nu}{1-\nu^2}\left(1 - \frac{\operatorname{th}\gamma}{\gamma}\right). \tag{9}$$

2. Deformation of a hollow cylinder. The transformation producing planar deformation of a sufficiently long hollow cyliner is prescribed by the relationships

$$R = f(r)\ , \qquad \Phi = \varphi + \theta(r)\ , \qquad Z = z\ . \tag{10}$$

Here r, φ, z are the cylindrical coordinates of a particle in the natural reference configuration (its material coordinates), and R, Φ, Z its coordinates in the actual configuration. The functions $R(r)$ and $\theta(r)$ must satisfy the boundary conditions

$$f(r_0) = r_0\ , \qquad f(r_1) = r_1\ , \qquad \theta(r_0) = 0\ , \qquad \theta(r_1) = \alpha\ , \tag{11}$$

where r_0 and r_1 are the inner and outer radii of the cylinder.

This formulation of the problem was discussed in Chapter 4, Par. 13. The inner surface of the cylinder is clamped. The outer surface is held by a rigid band which constrains it along its whole length l. We consider the deformation produced by a rotation of the band about the axis of the cylinder through the angle α.

The position vector of a particle in the actual configuration may be specified by the complex coordinate

$$z = f(r)e^{i(\varphi+\theta(r))} = f(r)e^{i\theta(r)}\sqrt{\frac{\zeta}{\bar{\zeta}}}\ . \tag{12}$$

As in Par. 8, we denote the complex coordinate in the reference configuration by ζ and we have

$$r = \sqrt{\zeta\bar{\zeta}}\ , \qquad e^{i\varphi} = \sqrt{\frac{\zeta}{\bar{\zeta}}}\ , \qquad \frac{\partial r}{\partial \zeta} = \frac{1}{2}\sqrt{\frac{\bar{\zeta}}{\zeta}} = \frac{1}{2}e^{-i\varphi}\ ,$$

$$\frac{\partial}{\partial \zeta}e^{i\varphi} = \frac{\partial}{\partial \zeta}\sqrt{\frac{\zeta}{\bar{\zeta}}} = \frac{1}{2r}\ . \tag{13}$$

By virtue of (12) and (13) we find that

$$2\frac{\partial z}{\partial \zeta} = 2\frac{\partial z}{\partial r}\frac{\partial r}{\partial \zeta} + 2\frac{\partial z}{\partial e^{i\varphi}}\frac{\partial e^{i\varphi}}{\partial \zeta} =$$

$$= \left[f'(r) + \frac{1}{r}f(r) + i\theta'(r)f(r)\right]e^{i\theta(r)} = qe^{i(\theta(r)+\gamma(r))}\ . \tag{14}$$

Here the functions of r (but not of φ) are determined from the expressions

$$q = \left[\left(f'(r) + \frac{f(r)}{r}\right)^2 + f^2(r)\theta'^2(r)\right]^{1/2}\ ,$$

$$\cos\gamma = \frac{1}{q}\left[f'(r) + \frac{f(r)}{r}\right]\ , \qquad \sin\gamma = \frac{1}{q}f(r)\theta'(r)\ . \tag{15}$$

By (8.13) and (8.16) we have

$$\chi = \theta + \gamma , \qquad \psi(q) = (\lambda + 2\mu)\left[q - \frac{2(\lambda + \mu)}{\lambda + 2\mu}\right]$$

(it has already been assumed in (10) that $c = 1$). The function $\Phi'(\zeta)$ introduced by Equation (8.18) is now given by the expression

$$\Phi'(\zeta) = (\lambda + 2\mu)\left[q - \frac{2(\lambda + \mu)}{\lambda + 2\mu}\right]e^{i(\theta+\gamma)} . \tag{16}$$

This function of a complex variable sought for within the ring $r_0 \le r \le r_1$ and represented there by a Laurent series thus turns out to be dependent only upon r. For this reason this function is constant. We now adopt the following notation

$$\Phi'(\zeta) = (\lambda + 2\mu)\left[D - \frac{2(\lambda + \mu)}{\lambda + 2\mu}\right]e^{i\beta} , \qquad D = \text{const} , \qquad \beta = \chi = \text{const} . \tag{17}$$

The problem is thus reduced to the following system of ordinary differential equations

$$\left(f' + \frac{f}{r}\right)^2 + f^2\theta'^2 = D^2 \quad \text{or} \quad (rf)'^2 + r^2f^2\theta'^2 = r^2D^2 , \tag{18}$$

$$\cos\gamma = \frac{f' + \frac{f}{r}}{D} = \cos(\beta - \theta) , \qquad \sin\gamma = \frac{f\theta'}{D} = \sin(\beta - \theta) . \tag{19}$$

From the latter we find that

$$\frac{(fr)'}{fr} = \frac{\cos(\beta - \theta)}{\sin(\beta - \theta)}\,\theta' , \qquad rf\sin(\beta - \theta) = C , \tag{20}$$

where C denotes a new constant. Turning to (11), we have

$$r_0^2\sin\beta = C , \qquad r_1^2\sin(\beta - \alpha) = C , \qquad \frac{r_0^2}{r_1^2} = \frac{\sin(\beta - \alpha)}{\sin\beta} . \tag{21}$$

Eliminating rf from (18) with the aid of (20), substituting for C from (21) and rearranging we arrive at the differential equation

$$\frac{\theta'^2}{\sin^4(\beta - \theta)} = \frac{D^2r^2}{r_0^4\sin^2\beta} , \qquad \frac{d\theta}{\sin^2(\beta - \theta)} = \frac{Dr}{r_0^2|\sin\beta|} . \tag{22}$$

After determining the constants by Equations (11) the solution reduces to

$$\operatorname{ctg}(\beta-\theta) = \frac{r^2-r_0^2}{r_1^2-r_0^2}\operatorname{ctg}(\beta-\alpha) + \frac{r_1^2-r^2}{r_1^2-r_0^2}\operatorname{ctg}\beta \,. \tag{23}$$

This equation, together with the relationship

$$fr\sin(\beta-\theta) = r_0^2\sin\beta \tag{24}$$

and formula

$$\frac{r_0^2}{r_1^2} = \frac{\sin(\beta-\alpha)}{\sin\beta} \tag{25}$$

from which the constant β may be determined, yields the solution.

Note that the constant D determined above turns out to be

$$D = 2\frac{r_0^2}{r_1^2-r_0^2}\left[\operatorname{ctg}(\beta-\alpha) - \operatorname{ctg}\beta\right]\left|\sin\beta\right| \,. \tag{26}$$

The stresses are determined by Equations (8.26) and (8.27).*)

10. SECOND ORDER EFFECTS. THE BASIC EQUATIONS

The equation of state of a linearly elastic isotropic body

$$T = \lambda\vartheta\underline{E} + 2\mu\overset{0}{\underline{\varepsilon}} \qquad (\vartheta = I_1(\overset{0}{\underline{\varepsilon}}),\ \overset{0}{\underline{\varepsilon}} = \frac{1}{2}(\overset{0}{\underline{\underline{\nabla u}}} + \overset{0}{\underline{\underline{\nabla u}}}^T)) \tag{1}$$

is deduced on the assumption that the components of the gradient of the displacement vector $\underline{\underline{\nabla u}}$ are small and that the terms of second order and above may be neglected. As has already been mentioned in Chapter 6, Par. 6, the calculation of quantitative deviations from the predictions of linear theory is associated with the retention of quadratic terms - that is, with the consideration of "second order effects".

We have already noted in Chapter 4, Par. 14 that any attempt to take higher orders into account is usually associated with immensely cumbersome calcualtions. We do not consider here the construction of power series in $\overset{0}{\underline{\underline{\nabla u}}}$ or an investiga-

---

*) The suggestion that the transformation (10) be used for the solution of the problem is due to N.Kh. Sujunshkaliev. In his as yet unplublished paper alternative versions of the statement of the problem are considered.

tion of their convergence.

In the constitutive equation (4.3.4) of an isotropic elastic body, the strain measure $\underline{F}$ must be replaced by its representation in terms of $\overset{0}{\nabla}\underline{u}$

$$\underline{F} = \overset{0}{\nabla}\underline{R}^T \cdot \overset{0}{\nabla}\underline{R} = (\underline{E} + \overset{0}{\nabla}\underline{u}^T) \cdot (\underline{E} + \overset{0}{\nabla}\underline{u}) = \underline{E} + 2\overset{0}{\underline{\varepsilon}}(\underline{u}) + \overset{0}{\nabla}\underline{u}^T \cdot \overset{0}{\nabla}\underline{u} . \tag{2}$$

Then, in the accepted approximation,

$$\underline{F}^2 = \underline{E} + 4\overset{0}{\underline{\varepsilon}} + 4\overset{0}{\underline{\varepsilon}}{}^2 + 2\overset{0}{\nabla}\underline{u}^T \cdot \overset{0}{\nabla}\underline{u} ,$$

$$\begin{aligned} I_1(\underline{F}) &= 3 + 2\vartheta + 2\underline{\omega} \cdot \underline{\omega} + I_1(\overset{0}{\underline{\varepsilon}}{}^2) , \\ I_2(\underline{F}) &= 3 + 4\vartheta + 2\vartheta^2 + 4\underline{\omega} \cdot \underline{\omega} , \\ I_3(\underline{F}) &= 1 + 2\vartheta + 2\vartheta^2 + 2\underline{\omega} \cdot \underline{\omega} - I_1(\overset{0}{\underline{\varepsilon}}{}^2) \end{aligned} \tag{3}$$

here we have used formula (1.7.13). The constitutive equation is now reduced to the form

$$\underline{T} = 2I_3^{-1/2}[\underline{E}(\psi_0 + \psi_1 + \psi_2) + 2(\psi_1 + 2\psi_2)\overset{0}{\underline{\varepsilon}} + (\psi_1 + 2\psi_2)\overset{0}{\nabla}\underline{u}^T \cdot \overset{0}{\nabla}\underline{u} + 4\psi_2\overset{0}{\underline{\varepsilon}}{}^2] . \tag{4}$$

Here, by (4.3.5)

$$\begin{aligned} f &= \psi_0 + \psi_1 + \psi_2 = \ni_1 + (I_1 - 1)\ni_2 + I_3\ni_3 , \\ \varphi &= \psi_1 + 2\psi_2 = \ni_1 + (I_1 - 2)\ni_2 , \qquad \psi_2 = -\ni_2 , \end{aligned} \tag{5}$$

where $\ni_k = \partial\ni/\partial I_k$. These quantities are now determined to the required accuracy, f up to quadratic and φ to linear terms. The values in the natural reference configuration are designated by a zero superscript on the right. Using formulas (4.3.22), (4.7.12) and (4.7.13) we obtain

$$f^0 = (\ni_1 + 2\ni_2 + \ni_3)^0 = 0 , \qquad \varphi^0 = (\ni_1 + \ni_2)^0 = -(\ni_2 + \ni_3)^0 = \frac{1}{2}\mu ,$$

$$\begin{aligned} \frac{1}{4}\lambda &= \left[\left(\frac{\partial}{\partial I_1} + 2\frac{\partial}{\partial I_2} + \frac{\partial}{\partial I_3}\right)f\right]^0 = \\ &= [\ni_{11} + 2\ni_{22} + \ni_{33} + 2(2\ni_{12} + 2\ni_{23} + \ni_{13}) + \ni_2 + \ni_3]^0 . \end{aligned} \tag{6}$$

The quantities considered are expanded in a power series in $I_1 - 3$, $I_2 - 3$,

$I_3 - 1$. Making use of the expression for λ we get, to the required accuracy,

$$\varphi = \frac{1}{2}\mu + 2\vartheta\left(\frac{\partial\varphi}{\partial I_1} + 2\frac{\partial\varphi}{\partial I_2} + \frac{\partial\varphi}{\partial I_3}\right)^0 =$$

$$= \frac{1}{2}\mu + 2\vartheta(Э_{11} + 3Э_{12} + 2Э_{22} + Э_{31} + Э_{32} + Э_2)^0 =$$

$$= \frac{1}{2}\mu + \frac{1}{2}\lambda\vartheta - 2\vartheta(Э_{12} + Э_3 + Э_{13} + 3Э_{23} + Э_{33})^0 .$$

At this stage the stress tensor $\underline{T}$ can be represented by the expression

$$\underline{T} = 2I_3^{-1/2}\underline{E}f + 2\mu(1-\vartheta)\underline{\overset{0}{\varepsilon}} + 2\lambda\underline{\overset{0}{\varepsilon}}\vartheta -$$

$$- 8\underline{\overset{0}{\varepsilon}}\vartheta(Э_{12} + Э_3 + Э_{13} + 3Э_{23} + Э_{33})^0 - 8Э_2^0\underline{\overset{0}{\varepsilon}}^2 + \mu\underline{\overset{0}{\nabla u}}^T \cdot \underline{\overset{0}{\nabla u}} =$$

$$= 2I_3^{-1/2}\underline{E}f + 2(1-\vartheta)\underline{\overset{0}{\varepsilon}} + 2\underline{\overset{0}{\varepsilon}} \cdot \underline{T}^0 + 8Э_3^0\underline{\overset{0}{\varepsilon}}^2 + \mu\underline{\overset{0}{\nabla u}}^T \cdot \underline{\overset{0}{\nabla u}} -$$

$$- 8(Э_{12} + Э_3 + Э_{13} + 3Э_{23} + Э_{33})^0\underline{\overset{0}{\varepsilon}}\vartheta ,$$

where we have used the form given by (1) for the stress tensor in the reference configuration and formula (6) for μ. The quantity $I_3^{-1/2}$ is substituted for $1 - \vartheta$ and 1 respectively according to whether it is multiplied by a linear or a quadratic factor. All that now remains is to represent f in the form of a series up to quadratic terms

$$f = f^0 + 2(\vartheta + \underline{\omega}\cdot\underline{\omega})\left[\left(\frac{\partial}{\partial I_1} + 2\frac{\partial}{\partial I_2} + \frac{\partial}{\partial I_3}\right)f\right]^0 +$$

$$+ I_1(\underline{\overset{0}{\varepsilon}}^2)\left[\left(\frac{\partial}{\partial I_1} - \frac{\partial}{\partial I_3}\right)f\right]^0 + 2\vartheta^2\left[\left(\frac{\partial}{\partial I_2} + \frac{\partial}{\partial I_3}\right)f\right]^0 +$$

$$+ \frac{1}{2}\left[(I_1 - 3)^2\left(\frac{\partial^2 f}{\partial I_1^2}\right)^0 + \ldots + 2(I_3 - 1)(I_1 - 3)\left(\frac{\partial^2 f}{\partial I_3 \partial I_1}\right)^0\right] =$$

$$= \frac{\lambda}{2}(\vartheta + \underline{\omega}\cdot\underline{\omega}) + \frac{\lambda}{4}I_1(\underline{\overset{0}{\varepsilon}}^2) + 2(\vartheta^2 - I_1(\underline{\overset{0}{\varepsilon}}^2))\left[\left(\frac{\partial}{\partial I_2} + \frac{\partial}{\partial I_3}\right)f\right]^0 +$$

$$+ 2\vartheta^2\left[\left(\frac{\partial^2}{\partial I_1^2} + 4\frac{\partial^2}{\partial I_2^2} + 4\frac{\partial^2}{\partial I_1 \partial I_2} + \frac{\partial^2}{\partial I_3^2} + 4\frac{\partial^2}{\partial I_2 \partial I_3} + 2\frac{\partial^2}{\partial I_3 \partial I_1}\right)f\right]^0 =$$

$$= \frac{\lambda}{2}\vartheta + \frac{\lambda}{2}\left(\underline{\omega}\cdot\underline{\omega} + \frac{1}{2}I_1(\underline{\overset{0}{\varepsilon}}^2)\right) + 2(\vartheta^2 - I_1(\underline{\overset{0}{\varepsilon}}^2))\left[\left(\frac{\partial}{\partial I_2} + \frac{\partial}{\partial I_3}\right)f\right]^0 +$$

$$+ 2\vartheta^2\left[\left(\frac{\partial}{\partial I_1} + 2\frac{\partial}{\partial I_2} + \frac{\partial}{\partial I_3}\right)f\right]^0 .$$

The desired representation of the Cauchy stress tensor now takes the form

$$\underset{=}{T} = (1-\vartheta)\underset{=}{T}^0 + 2\overset{0}{\underset{=}{\varepsilon}}\cdot\underset{=}{T}^0 + \underset{=}{T}' ,$$

$$\underset{=}{T}' = \underset{=}{E}\left\{\lambda(\underset{=}{\omega}\cdot\underset{=}{\omega} + \frac{1}{2}I_1(\overset{0}{\underset{=}{\varepsilon}}{}^2)) + 4(\vartheta^2 - I_1(\overset{0}{\underset{=}{\varepsilon}}{}^2))\left[\left(\frac{\partial}{\partial I_2} + \frac{\partial}{\partial I_3}\right)f\right]^0 + A\vartheta^2\right\} +$$

$$+ 8\overset{0}{\ni}_3\overset{0}{\underset{=}{\varepsilon}}{}^2 + B\overset{0}{\underset{=}{\varepsilon}}\vartheta + \mu\overset{0}{\underset{=}{\nabla u}}{}^T\cdot\overset{0}{\underset{=}{\nabla u}} , \tag{7}$$

where

$$A = 4\left[\left(\frac{\partial}{\partial I_1} + 2\frac{\partial}{\partial I_2} + \frac{\partial}{\partial I_3}\right)^2 f\right]^0 , \qquad B = -8(\ni_{12} + \ni_3 + \ni_{13} + 3\ni_{23} + \ni_{33})^0 . \tag{8}$$

For example, for a Murnaghan type material, the value of f is given by (5.3.8) and the specific strain energy $\ni$ by (5.3.2). We now obtain

$$A = l , \qquad 4\left[\left(\frac{\partial}{\partial I_2} + \frac{\partial}{\partial I_3}\right)f\right]^0 = -m + \frac{n}{2} , \qquad B = -n + 2m , \qquad 8\overset{0}{\ni}_3 = n$$

and the stress tensor is given by the expression

$$\underset{=}{T} = (1-\vartheta)\underset{=}{T}^0 + 2\overset{0}{\underset{=}{\varepsilon}}\cdot\underset{=}{T}^0 + \underset{=}{T}' ,$$

$$\underset{=}{T}' = \underset{=}{E}\left[\lambda\,(\underset{=}{\omega}\cdot\underset{=}{\omega} + \frac{1}{2}I_1(\overset{0}{\underset{=}{\varepsilon}}{}^2)) - (m - \frac{n}{2})\,(\vartheta^2 - I_1(\overset{0}{\underset{=}{\varepsilon}}{}^2)) + l\vartheta^2\right] +$$

$$+ n\overset{0}{\underset{=}{\varepsilon}}{}^2 + (2m-n)\overset{0}{\underset{=}{\varepsilon}}\vartheta + \mu\overset{0}{\underset{=}{\nabla u}}{}^T\cdot\overset{0}{\underset{=}{\nabla u}} . \tag{9}$$

Proceeding to the Piola tensor, we may assume in the formula (2.7.2) that

$$I_3^{1/2} = 1 + \vartheta , \qquad \underset{=}{\nabla r}^T = \underset{=}{\nabla}(\underset{=}{R} - \underset{=}{u})^T = \underset{=}{E} - \overset{0}{\underset{=}{\nabla u}}{}^T ,$$

because the tensor $\underset{=}{T}^0$ in (8) is linear in $\overset{0}{\underset{=}{\nabla u}}$, $\underset{=}{\nabla u}^T \approx \overset{0}{\underset{=}{\nabla u}}{}^T$ and $I_3^{1/2} = 1 + \vartheta$ in the linear approximation. We now obtain

$$\underset{=}{P} = I_3^{1/2}\underset{=}{\nabla r}^T\cdot\underset{=}{T} = (1+\vartheta)(\underset{=}{E} - \overset{0}{\underset{=}{\nabla u}}{}^T)\cdot\underset{=}{T} = \underset{=}{T}^0 + \overset{0}{\underset{=}{\nabla u}}\cdot\underset{=}{T}^0 + \underset{=}{T}' . \tag{10}$$

The Piola and Cauchy tensors coincide only in the linear approximation. Second order terms must be taken into account in order to distinguish between the two. The equations of equilibrium in the volume and on its surface must be written as

$$\overset{0}{\underline{\nabla}} \cdot \underline{P} + \rho_0 \underline{k} = \underline{0} , \qquad \underline{n} \cdot \underline{P} = \underline{f} \frac{dO}{do} , \tag{11}$$

where, considering first order terms only, we obtain, from (1.8.7),

$$\frac{dO}{do} = \sqrt{\frac{G}{g}} (\underline{n} \cdot \underline{\nabla r}^T \cdot \underline{\nabla r} \cdot \underline{n})^{1/2} = (1 + \vartheta) [\underline{n} \cdot (\underline{E} + \overset{0}{\underline{\nabla u}}^T) \cdot (\underline{E} + \overset{0}{\underline{\nabla u}}) \cdot \underline{n}]^{1/2} =$$

$$= (1 + \vartheta) [1 + 2\underline{n} \cdot \overset{0}{\underline{\varepsilon}}(\underline{u}) \cdot \underline{n}]^{1/2} ,$$

so that

$$\frac{dO}{do} = 1 + \vartheta + \underline{n} \cdot \overset{0}{\underline{\varepsilon}}(\underline{u}) \cdot \underline{n} . \tag{12}$$

11. SECOND ORDER EFFECTS. THE CONSTRUCTION OF A SOLUTION

The starting point for this section is the equations (10.11). These are the equilibrium equations in a volume and on its surface expressed in terms of the Piola tensor determined by Equation (10.10). The surface forces are assumed to be "dead".

The displacement vector is represented as the sum of two vectors

$$\underline{u} = \underline{v} + \underline{w} , \tag{1}$$

where $\underline{v}$ denotes the solution to the linear problem

$$\overset{0}{\underline{\nabla}} \cdot \underline{T}^0 + \rho_0 \underline{k} = \underline{0} , \qquad \underline{n} \cdot \underline{T}^0 = \underline{f}^0 , \tag{2}$$

which is assumed to be known and the vector $\underline{w}$ represents a correction introduced in order to compensate terms of second degree with respect to the components of $\overset{0}{\underline{\nabla v}}$ which would otherwise enter (10.11). The tensor $\underline{T}^0$ is a linear operator in $\underline{u}$

$$\underline{T}^0(\underline{u}) = \underline{T}^0(\underline{v}) + \underline{T}^0(\underline{w})$$

and therefore

$$\overset{0}{\underline{\nabla u}} \cdot \underline{T}^0(\underline{u}) = (\overset{0}{\underline{\nabla v}} + \overset{0}{\underline{\nabla w}}) \cdot (\underline{T}^0(\underline{v}) + \underline{T}^0(\underline{w})) \approx \overset{0}{\underline{\nabla v}} \cdot \underline{T}^0(\underline{v}) , \qquad \underline{T}'(\underline{u}) \approx \underline{T}'(\underline{v}) \tag{3}$$

where terms of the third order and above have been discarded. Bearing (2) in mind, we can now transform equations (10.11) of equilibrium in the volume and on its surface to the following form

$$\overset{0}{\underline{\nabla}} \cdot \underline{T}^0(\underline{w}) + \rho_0 \underline{k}_\times = \underline{0} , \qquad \underline{n} \cdot \underline{T}^0(\underline{w}) = \underline{f}_\times , \tag{4}$$

here the vectors $\underline{k}_x$ and $\underline{f}_x$ play the role of the body and contact forces and represent operators of the second degree with respect to $\overset{0}{\nabla}\underline{v}$

$$\underline{k}_x = \overset{0}{\nabla} \cdot (\overset{0}{\nabla}\underline{v} \cdot \underline{T}^0(\underline{v}) + \underline{T}'(\underline{v})) \ , \qquad \underline{f}_x = -\underline{n} \cdot (\overset{0}{\nabla}\underline{v} \cdot \underline{T}^0(\underline{v}) + \underline{T}'(\underline{v})) \ . \tag{5}$$

As is well known, the necessary condition for the existence of a solution to the boundary value problem (4) requires that this system of forces be statically zero-equivalent. We have in fact

$$\iiint_v \rho_0 \underline{k}_x dv = \iiint_v \overset{0}{\nabla} \cdot (\overset{0}{\nabla}\underline{v} \cdot \underline{T}^0(\underline{v}) + \underline{T}'(\underline{v}))dv =$$
$$= \iint_o \underline{n} \cdot (\overset{0}{\nabla}\underline{v} \cdot \underline{T}^0(\underline{v}) + \underline{T}'(\underline{v}))do = - \iint_o \underline{f}_x do \ ,$$

which is the desired result. In view of (2.3.8), it only remains to express the condition of symmetry of the force tensor. From (2.1.16) and the Gauss-Ostrogradskii theorem we have

$$\underline{B} = \iiint_v \rho_0 \underline{k}_x \underline{r} dv - \iint_o \underline{n} \cdot (\overset{0}{\nabla}\underline{v} \cdot \underline{T}^0(\underline{v}) + \underline{T}'(\underline{v}))\underline{r} do =$$
$$= \iiint_v \rho_0 \underline{k}_x \underline{r} dv - \iiint_v [\overset{0}{\nabla} \cdot (\overset{0}{\nabla}\underline{v} \cdot \underline{T}^0(\underline{v}) + \underline{T}'(\underline{v}))]\underline{r} dv -$$
$$- \iiint_v \underline{r}^s \cdot (\overset{0}{\nabla}\underline{v} \cdot \underline{T}^0(\underline{v}) + \underline{T}'(\underline{v}))\underline{r}_s dv =$$
$$= - \iiint_v [\overset{0}{\nabla}\underline{v} \cdot \underline{T}^0(\underline{v}) + \underline{T}'(\underline{v})]^T dv \ .$$

We have used the following representation of the divergence of a tensor of the third rank

$$\overset{0}{\nabla} \cdot \underline{Q}\underline{r} = (\overset{0}{\nabla} \cdot \underline{Q})\underline{r} + \underline{r}^s \cdot \underline{Q}\underline{r}_s = (\overset{0}{\nabla} \cdot \underline{Q})\underline{r} + \underline{Q}^T \ .$$

Thus

$$\begin{aligned} \underline{B} &= - \iiint_v [\overset{0}{\nabla}\underline{v} \cdot \underline{T}^0(\underline{v}) + \underline{T}'(\underline{v})]^T dv \ , \\ \underline{B}^T &= - \iiint_v [\overset{0}{\nabla}\underline{v} \cdot \underline{T}^0(\underline{v}) + \underline{T}'(\underline{v})] dv \ . \end{aligned} \tag{6}$$

Denoting the linear tensor of deformation produced by the vector $\underline{v}$ of displacement by $\overset{0}{\underline{\varepsilon}}(\underline{v})$ and the vorticity tensor for the same vector by $\overset{0}{\underline{\Omega}}(\underline{v})$ and making use of the relationship $\overset{0}{\underline{\nabla}}\underline{v} = \overset{0}{\underline{\varepsilon}}(\underline{v}) - \overset{0}{\underline{\Omega}}(\underline{v})$ we may transform the force tensor as follows

$$\underline{B} = -\iiint\limits_{v} [\overset{0}{\underline{\varepsilon}}(\underline{v}) \cdot \underline{T}^{0}(\underline{v}) + \underline{T}^{0}(\underline{v}) \cdot \overset{0}{\underline{\Omega}}(\underline{v}) + \underline{T}'(\underline{v})]\,dv\,,$$

$$\underline{B}^{T} = -\iiint\limits_{v} [\overset{0}{\underline{\varepsilon}}(\underline{v}) \cdot \underline{T}^{0}(\underline{v}) - \overset{0}{\underline{\Omega}}(\underline{v}) \cdot \underline{T}^{0}(\underline{v}) + \underline{T}'(\underline{v})]\,dv\,, \tag{7}$$

this is so because tensors $\overset{0}{\underline{\varepsilon}}(\underline{v})$, $\underline{T}^{0}(\underline{v})$ and $\underline{T}'(\underline{v})$ from their definitions (10.1) and (10.7) are symmetric. The condition of symmetry of the force tensor now takes the form

$$\underline{B}^{T} = \underline{B} : -\iiint\limits_{v} (\underline{T}^{0} \cdot \overset{0}{\underline{\Omega}} + \overset{0}{\underline{\Omega}} \cdot \underline{T}^{0})\,dv = -\iiint\limits_{v} (\underline{T}^{0} \times \overset{0}{\underline{\omega}} + \overset{0}{\underline{\omega}} \times \underline{T}^{0})\,dv = \underline{0}\,, \tag{8}$$

where $\overset{0}{\underline{\omega}}$ denotes the vector associated with $\overset{0}{\underline{\Omega}}$. The vector associated with the skew-symmetric tensor under the integral sign is, by (I.14.17),

$$-\frac{1}{2}\underline{\epsilon} \cdot\cdot (\underline{T}^{0} \times \overset{0}{\underline{\omega}} + \overset{0}{\underline{\omega}} \times \underline{T}^{0}) = -\frac{1}{2}\underline{\epsilon} \cdot\cdot t^{0mn}\overset{0}{\omega}{}^{s}(\underline{r}_{m}\underline{r}_{n} \times \underline{r}_{s} - \underline{r}_{n} \times \underline{r}_{s}\underline{r}_{m}) =$$

$$= t^{0mn}\overset{0}{\omega}{}^{s}\underline{r}_{m} \times (\underline{r}_{n} \times \underline{r}_{s}) = t^{0mn}\overset{0}{\omega}{}^{s}(\underline{r}_{n}\underline{r}_{m} \cdot \underline{r}_{s} - \underline{r}_{s}\underline{r}_{m} \cdot \underline{r}_{n}) = \underline{T}^{0} \cdot \overset{0}{\underline{\omega}} - \overset{0}{\underline{\omega}} I_{1}(\underline{T}^{0})\,,$$

since $\underline{\epsilon} \cdot\cdot \underline{a}\underline{b} = \underline{b} \times \underline{a}$ and $t^{mn}\underline{r}_{m} \cdot \underline{r}_{n} = I_{1}(\underline{T})$. Condition (8) is now reduced to the form

$$\iiint\limits_{v} [\underline{T}^{0}(\underline{v}) \cdot \overset{0}{\underline{\omega}}(\underline{v}) - \overset{0}{\underline{\omega}}(\underline{v})\sigma]\,dv = \underline{0}\,, \qquad \sigma = I_{1}(\underline{T}^{0}(\underline{v})) \tag{9}$$

and this is a necessary condition for the existence of a solution to problem (2). This condition can be satisfied if we recall that the linear problem (2) for the vector $\underline{v}$ is solved except for some additive constant expressing the displacement $\underline{v} = \underline{v}_{0} + \overset{0}{\underline{\omega}}_{0} \times \underline{r}$ of a rigid body, and for the vector $\overset{0}{\underline{\omega}} = \frac{1}{2}\overset{0}{\underline{\nabla}} \times \underline{v}$ except for the constant term $\overset{0}{\underline{\omega}}_{0}$. For this reason, if we fix the value $\underline{\omega}'$ at some point, then we can assume

$$\overset{0}{\underline{\omega}}(\underline{v}) = \overset{0}{\underline{\omega}}{}'(\underline{v}) + \overset{0}{\underline{\omega}}_{0}$$

and represent (9) as follows

$$\overset{0}{\underline{\omega}}_{0} \cdot \iiint\limits_{v} (\underline{T}^{0}(\underline{v}) - \underline{E}\sigma)\,dv = -\iiint\limits_{v} [\overset{0}{\underline{\omega}}{}'(\underline{v}) \cdot \underline{T}^{0}(\underline{v}) - \overset{0}{\underline{\omega}}{}'(\underline{v})\sigma]\,dv = \underline{c}\,. \tag{10}$$

Vector $\underline{c}$ is calculated using solution to (2) which is assumed known. The tensor in the left-hand side (multiplied by $\overset{0}{\underline{\omega}}_0$) may be determined immediately by the prescribed forces $\rho_0\underline{k}$ and $\overset{0}{\underline{f}}$. In fact, we have, by (2.3.7),

$$\underline{B} = \iiint_v \rho_0 \underline{k}\underline{r}\, dv + \iint_o \overset{0}{\underline{f}}\underline{r}\, dv = \iiint_v \overset{0}{\underline{T}}(\underline{v})\, dv ,$$
$$I_1(\underline{B}) = \iiint_v I_1(\overset{0}{\underline{T}}(\underline{v}))\, dv = \iiint_v \rho_0 \underline{k}\cdot\underline{r}\, dv + \iint_o \overset{0}{\underline{f}}\cdot\underline{r}\, dv . \tag{11}$$

Equation (10) now takes the form

$$\overset{0}{\underline{\omega}}_0 \cdot [\underline{B} - \underline{E}I_1(\underline{B})] = \underline{c} \tag{12}$$

and is solvable provided that the tensor $\underline{B} + \underline{E}I_1(\underline{B})$ is non-singular

$$\det(\underline{B} - \underline{E}I_1(\underline{B})) \neq 0 . \tag{13}$$

This criterion can be verified by the specification of the external forces $\rho_0\underline{k}$, $\overset{0}{\underline{f}}$ and does not require the solution of the linear problem.

If condition (12) is not satisfied, the boundary value problem (2) has, in general, no solution.

The constant vector which is specified by the condition

$$\iiint_v (\underline{a}\times\underline{r}) \times \rho_0\underline{k}\, dv + \iint_o (\underline{a}\times\underline{r}) \times \overset{0}{\underline{f}}\, do = \underline{0} \tag{14}$$

in which the external forces are constant in magnitude and direction and whose resultant is zero will be called the axis of equilibrium, $\underline{a}$, of the body in the reference configuration. This condition asserts that the system of forces continues to be statically zero-equivalent when the body is rotated through 90 about $\underline{a}$. Equation (14) is now transformed to

$$\underline{a}\cdot \iiint_v \rho_0(\underline{k}\underline{r} - \underline{E}\underline{k}\cdot\underline{r})\, dv + \underline{a}\cdot \iint_o (\overset{0}{\underline{f}}\underline{r} - \underline{E}\overset{0}{\underline{f}}\cdot\underline{r})\, do = \underline{0}$$

or, in view of (11), to the homogeneous system of equations

$$\underline{a}\cdot[\underline{B} - \underline{E}I_1(\underline{B})] = \underline{0} \tag{15}$$

which may possess non-zero solutions provided that

$$\det(\underline{B} - \underline{E}I_1(\underline{B})) = 0 . \tag{16}$$

Representing the symmetric tensor $\underline{B}$ in its principal axes

$$\underline{B} = \sum_{s=1}^{3} b_s \underline{e}_s \underline{e}_s \ , \qquad I_1(\underline{B}) = b_1 + b_2 + b_3 \ ,$$

we may give Equations (15) in the following form

$$a_1(b_2 + b_3) = 0 \ , \qquad a_2(b_3 + b_1) = 0 \ , \qquad a_3(b_1 + b_2) = 0 \ . \tag{17}$$

Vector $\underline{a}$ remains arbitrary when $b_2 + b_3 = 0$, $b_3 + b_1 = 0$ and $b_1 + b_2 = 0$ or, which is the same thing, when $\underline{B} = \underline{0}$. If $b_2 + b_3 = 0$, $b_1 \neq -b_3$ and $b_1 \neq -b_2$ then $\underline{a} = \underline{e}_1$; if $\underline{a} \cdot \underline{e}_3 = 0$, then vector $\underline{a}$ lies in the plane $(\underline{e}_1, \underline{e}_2)$. Of course, the axis of equilibrium can also be oriented along the axes $\underline{e}_2$, $\underline{e}_3$ and lie in the planes $(\underline{e}_2, \underline{e}_3)$, $(\underline{e}_3, \underline{e}_1)$. In all these cases,

$$\det[\underline{B} - \underline{E} I_1(\underline{B})] = (b_1 + b_2)(b_2 + b_3)(b_3 + b_1) = 0 \ . \tag{18}$$

If condition (13) holds, then none of these multipliers is zero and by (16) $\underline{a} = \underline{0}$, i.e. the axis of equilibrium does not exist.

When Equation (16) holds, the non-homogeneous system (12) may have no solution at all or possesses more than one solution; in other words, when Equations (12) are incompatible the possibility that this system may become compatible under some specific circumstances can not be excluded. If the axis of equilibrium exists, then the loading does not determine the state of equilibrium uniquely because rotations about this axis transfer the body from one orientation to another which cannot be distinguished from the initial one under the same forces.

It has been shown that for $\underline{B} = \underline{0}$, when the axis of equilibrium may be any vector, the system (12) may prove to be incompatible for any orientation of the body. In this case nonlinearity cannot be taken into account by the introduction of a correction to a linear solution constructed using this approach.

A solution to the boundary value problem (4) for $\underline{w}$ is, as a rule, difficult to obtain because of the complexity of the expressions for the forces $\rho_0 \underline{k}_x$ and $\underline{f}_x$. However, the structure of these expressions facilitates the calculation of certain averaged quantities. This may be satisfactory when detailed knowledge of the state of stress is of minor importance.

We begin with the determination of the mean value of $\overset{0}{\underline{T}}(\underline{w})$. From (2.3.7) and (5) we have

$$\overset{0}{\underline{T}}(\underline{w}) = \frac{1}{v} \iiint_v \overset{0}{\underline{T}}(\underline{w})\, dv = \frac{1}{v} \underline{B} = \frac{1}{v} \iiint_v \overset{0}{\underline{\nabla}} \cdot [\overset{0}{\underline{\nabla}} \underline{v} \cdot \overset{0}{\underline{T}}(\underline{v}) + \underline{T}'(\underline{v})]\, \underline{r}\, dv \ -$$

$$- \iint\limits_{o} \underline{n} \cdot [\overset{0}{\underline{\nabla v}} \cdot \underline{T}^0(\underline{v}) + \underline{T}'(\underline{v})]\underline{r}\, do\Big\} = -\frac{1}{v} \iiint\limits_{v} [\underline{T}^0(\underline{v}) \cdot \overset{0}{\underline{\nabla v}}^T + \underline{T}'(\underline{v})]\, dv \quad (19)$$

in which we have used the Gauss-Ostrogradskii transformation and the rule for the calculation of the divergence of a product of a tensor and vector $\underline{r}$. The mean values of the first invariant of the stress tensor and of the volume extension (determined by this value) are equal to

$$\sigma_m(\underline{w}) = -\frac{1}{v} \iiint\limits_{v} [I_1(\underline{T}^0(\underline{v}) \cdot \overset{0}{\underline{\varepsilon}}(\underline{v})) + I_1(\underline{T}'(\underline{v}))]\, dv ,$$

$$\vartheta_m(\underline{w}) = \frac{\sigma_m(\underline{w})}{3\lambda + 2\mu} , \quad (20)$$

because the first invariant of a product of a symmetric tensor by a skew tensor is zero:

$$I_1(\underline{T}^0(\underline{v}) \cdot \overset{0}{\underline{\Omega}}(\underline{v})) = 0 .$$

The mean value of the linear strain tensor is also calculated from (19)

$$2\mu \underline{\varepsilon}_m(\underline{w}) = \underline{T}_m(\underline{w}) - \underline{E} \frac{\lambda}{3\lambda + 2\mu} \sigma_m(\underline{w}) . \quad (21)$$

These relationships can also be obtained by applying the duality theorem to the linear problem of determination $\underline{w}$ by the forces $\rho_0 \underline{k}_x$, $\underline{f}_x$.

12. THE DILATATION OF A BODY SUBJECTED TO DISTORTION

As follows from (2.3.7), the mean value $\underline{T}_m(\underline{u})$ of the stress tensor equals zero when external forces ($\underline{k} = \underline{0}$, $\underline{f} = \underline{0}$) are absent. Of course, the tensor $\underline{T}(\underline{u})$ itself may then be different from zero. For example, stresses arise in the presence of Volterra distortions or dislocations of more general nature.*)

---

*) The concept of Volterra distortion may be explained as follows: from a doubly-connected body (e.g. a torus) a thin section is removed and the resulting edges soldered together to reform the body (in this case a new torus). This procedure results in the edges being given a small progressive displacement $\underline{c}$ and a small rotation $\underline{b}$. It is this act of forming a new body from an intial one which Volterra has called distortion. In an elastic body subjected to distortion there arises a state of stress. This state can be calculated theoretically by description of vectors $\underline{c}$ and $\underline{b}$. The latter can be determined experimentally using measurements of the displacements and rotations of the edges of a "ring-shaped" body which has been treated as described above. The term "dislocation" is associated with physical phenomena similar to the breaking of the structure of a crystal lattice which also produces a state of stress in the absence of external forces.

In the approximation of the second order effects which is based upon the expression (10.7) for the tensor $\underline{T}(\underline{u})$, the condition for the vanishing of the mean value of this tensor is given by the relationship

$$\iiint_{V} \underline{T}(\underline{u})\,dV = \iiint_{v} (1+\vartheta)\underline{T}(\underline{u})\,dv =$$

$$= \iiint_{v} [\underline{T}^{0}(\underline{u}) + 2\overset{0}{\underline{\varepsilon}}(\underline{u})\cdot\underline{T}^{0}(\underline{u}) + \underline{T}'(\underline{u})]\,dv = \underline{0}\ . \qquad (1)$$

The mean value of the first invariant $\sigma_m(\underline{u})$ for a Murnaghan material is by (10.9) given by the formula

$$\sigma_m(\underline{u}) = \frac{1}{v}\iiint_{v} \Big[(3\lambda+2\mu)\vartheta(\underline{u}) + (2\lambda+3l-m+\frac{n}{2})\vartheta^2 +$$

$$+ (4\mu+3m-\frac{n}{2})I_1(\overset{0}{\underline{\varepsilon}}(\underline{u})^2) + (3\lambda+2\mu)(\overset{0}{\underline{\omega}}\cdot\overset{0}{\underline{\omega}} + \frac{1}{2}I_1(\overset{0}{\underline{\varepsilon}}{}^2))\Big]\,dv = 0\ . \qquad (2)$$

This gives the mean value of the first invariant of the linear strain tensor $\overset{0}{\underline{\varepsilon}}(\underline{u})$ in terms of quantities of the second order of magnitude. The same quantity may also be expressed in terms of the relative dilatation of a body. We have by (10.3)

$$\frac{V-v}{v} = D_m = \frac{1}{v}\iiint_{v}\Big(\frac{dV}{dv}-1\Big)dv = \frac{1}{v}\iiint_{v}(I_3^{1/2}-1)\,dv =$$

$$= \frac{1}{v}\iiint_{v}[\vartheta(\underline{u}) + \overset{0}{\underline{\omega}}\cdot\overset{0}{\underline{\omega}} + \frac{1}{2}\vartheta^2 - \frac{1}{2}I_1(\overset{0}{\underline{\varepsilon}}{}^2)]\,dv\ . \qquad (3)$$

Eliminating ϑ from these equations, we obtain the representation

$$D_m = -\frac{1}{3\lambda+2\mu}\iiint_{v}\Big[(\frac{1}{2}\lambda - \mu + 3l - m + \frac{n}{2})\vartheta^2 +$$

$$+ (3\lambda + 6\mu + 3m - \frac{n}{2})\,I_1(\overset{0}{\underline{\varepsilon}}{}^2)\Big]\,dv\ . \qquad (4)$$

It proves possible to express the Murnaghan constants entering this relationship in terms of the quantities (5.3.21), i.e. in terms of the second order variations of the bulk and shear moduli. We arrive at the formula obtained by Zener (Zener, 1942) from different considerations

$$D_m = -\frac{1}{\lambda+2/3\mu}\iiint_{v}\Big\{\frac{1}{2}[\lambda + k'_{xx}(0) - \frac{1}{6}\mu'_{xx}(0)]\vartheta^2 + [\mu + \mu'_{xx}(0)]I_1(\overset{0}{\underline{\varepsilon}}{}^2)\Big\}\,dv\ . \quad (5)$$

The invariants $\vartheta = I_1(\overset{0}{\varepsilon})$, $I_1(\overset{0}{\varepsilon}{}^2)$ entering this formula are determined from solution of the problem of the state of stress arising in a linearly elastic body subjected to distortion.

The more general results for anisotropic elastic media are cited in the paper of R.A. Toupin and R.S. Rivlin, 1960.

13. SECOND ORDER EFFECTS IN THE PROBLEM OF TORSION AND STRETCHING OF A ROD

1. The linear problem of torsion of a rod. The axis Oa^3 is directed along that of the rod and the axes Oa^1 and Oa^2 along the principal central axes of the cross-section a^3 = const. The length of the rod is designated by L, the area of the cross-section by S, and the polar moment of inertia by I_p. The radius vector of a point in the cross-section is designated by $\underline{\rho}$ so that

$$\underline{\rho} = \underline{i}_1 a^1 + \underline{i}_2 a^2 , \qquad \iint_S \underline{\rho}\, do = \underline{0} , \qquad \iint_S \underline{\rho} \cdot \underline{\rho}\, do = I_p . \tag{1}$$

The solution to the linear problem is presribed by the displacement vector

$$\underline{v} = \alpha [a^3 \underline{i}_3 \times \underline{\rho} + \underline{i}_3 \varphi(a^1, a^2)] \tag{2}$$

where we have omitted those terms which describe a rigid displacement. The angular distortion per unit length is given by α. The torsion function $\varphi(a^1, a^2)$ is a harmonic function in S given by the values of its normal derivative on the contour Γ of the cross-section

$$\overset{0}{\underline{\nabla}}{}^2 \varphi = \frac{\partial^2 \varphi}{\partial a^{1\,2}} + \frac{\partial^2 \varphi}{\partial a^{2\,2}} = 0 , \qquad \frac{\partial \varphi}{\partial n} + a^1 n_2 - a^2 n_1 = 0$$

$$\left(n_1 = \frac{da^2}{ds} , \quad n_2 = - \frac{da^1}{ds} \right) . \tag{3}$$

The tensor $\overset{0}{\underline{\underline{\nabla v}}}$ given by (2) is equal to

$$\overset{0}{\underline{\underline{\nabla v}}} = \alpha [a^3 (\underline{i}_1 \underline{i}_2 - \underline{i}_2 \underline{i}_1) + \underline{i}_3 (\underline{i}_3 \times \underline{\rho}) + \overset{0}{\underline{\nabla}} \varphi \underline{i}_3] \qquad \left(\overset{0}{\underline{\nabla}} \varphi = \underline{i}_1 \frac{\partial \varphi}{\partial a^1} + \underline{i}_2 \frac{\partial \varphi}{\partial a^2} \right) . \tag{4}$$

From here we deduce the expressions for a linear strain tensor $\overset{0}{\underline{\varepsilon}}(\underline{v})$, for its first invariant ϑ and for the vector $\overset{0}{\underline{\omega}}$ which describes a small rotation

$$\overset{0}{\underline{\varepsilon}}(\underline{v}) = \frac{1}{2} (\overset{0}{\underline{\underline{\nabla v}}} + \overset{0}{\underline{\underline{\nabla v}}}{}^T) = \frac{1}{2} \alpha [\underline{i}_3 (\underline{i}_3 \times \underline{\rho}) + (\underline{i}_3 \times \underline{\rho}) \underline{i}_3 + \underline{\nabla} \varphi \underline{i}_3 + \underline{i}_3 \underline{\nabla} \varphi] , \tag{5}$$

$$\overset{0}{\underline{\omega}}(\underline{v}) = \frac{1}{2}\overset{0}{\underline{\nabla}} \times \underline{v} = \frac{1}{2}\alpha(2a^3\underline{i}_3 - \underline{\rho} + \overset{0}{\underline{\nabla}}\varphi \times \underline{i}_3) , \qquad \vartheta = 0 . \tag{6}$$

The stress tensor $\overset{0}{\underline{T}}(\underline{v})$ turns out to be

$$\overset{0}{\underline{T}}(\underline{v}) = \lambda\vartheta\underline{E} + 2\mu\overset{0}{\underline{\varepsilon}}(\underline{v}) = \mu\alpha[\underline{i}_3(\underline{i}_3 \times \underline{\rho}) + (\underline{i}_3 \times \underline{\rho})\underline{i}_3 + \overset{0}{\underline{\nabla}}\varphi\underline{i}_3 + \underline{i}_3\overset{0}{\underline{\nabla}}\varphi] =$$

$$= \tau_{31}(\underline{i}_1\underline{i}_3 + \underline{i}_3\underline{i}_1) + \tau_{23}(\underline{i}_2\underline{i}_3 + \underline{i}_3\underline{i}_2) ,$$

$$\tau_{31} = \mu\alpha\left(\frac{\partial\varphi}{\partial a^1} - a^2\right) , \qquad \tau_{23} = \mu\alpha\left(\frac{\partial\varphi}{\partial a^2} + a^1\right) . \tag{7}$$

The constant α is given by the torsional moment m_z and the torsional C rigidity

$$m_z = \mu\alpha C , \qquad m_z = \iint_S (a^1\tau_{23} - a^2\tau_{31})\,do \tag{8}$$

and by (7) and (1) we have

$$C = I_p + \iint_S \left(a^1 \frac{\partial\varphi}{\partial a^2} - a^2 \frac{\partial\varphi}{\partial a^1}\right) do = I_p + \iint_S (\underline{i}_3 \times \underline{\rho}) \cdot \overset{0}{\underline{\nabla}}\varphi \, do . \tag{9}$$

An alternative representation of the torsional rigidity is given by the expression

$$C = I_p - \iint_S \overset{0}{\underline{\nabla}}\varphi \cdot \overset{0}{\underline{\nabla}}\varphi \, do . \tag{10}$$

We have

$$\iint_S \overset{0}{\underline{\nabla}}\varphi \, do = \underline{0} , \qquad \iint_S \underline{\rho} \cdot \overset{0}{\underline{\nabla}}\varphi \, do = 0 \tag{11}$$

the first of these relationships,together with (7), shows that the resultant of the stresses τ_{23}, τ_{31} is zero. This relationship follows easily from the boundary condition (3). The second relationship is somewhat more difficult to verify. We have

$$\iint_S \underline{\rho} \cdot \overset{0}{\underline{\nabla}}\varphi \, do = \iint_S \left(\frac{\partial}{\partial a^1} a^1\varphi + \frac{\partial}{\partial a^2} a^2\varphi\right) do - 2\iint_S \varphi \, do =$$

$$= \oint_\Gamma \varphi\underline{\rho} \cdot \underline{n} \, ds - 2\iint_S \varphi \, do = 0 ,$$

since Green's transformation applied to $\varphi(a^1,a^2)$ yields

$$\iint_S (\underline{\rho}\cdot\underline{\rho}\overset{0}{\nabla}{}^2\varphi - \varphi\overset{0}{\nabla}{}^2\underline{\rho}\cdot\underline{\rho})\,do = \oint_\Gamma \left(\underline{\rho}\cdot\underline{\rho}\frac{\partial\varphi}{\partial n} - \varphi\frac{\partial}{\partial n}\underline{\rho}\cdot\underline{\rho}\right)ds \ ,$$

$$4\iint_S \varphi\,do = 2\oint_\Gamma \varphi\underline{\rho}\cdot\underline{n}\,ds$$

where we have taken into account that φ is a harmonic function, and that the linear integral of $\underline{\rho}\cdot\underline{\rho}\ \partial\varphi/\partial n$ is zero from (3). Thus we have

$$\oint_\Gamma \underline{\rho}\cdot\underline{\rho}\frac{\partial\varphi}{\partial n}\,ds = \int_\Gamma (a^{12}+a^{22})(a^2 n_1 - a^1 n_2)\,ds = \iint_S 2(a^1 a^2 - a^2 a^1)\,do = 0 \ .$$

2. The second order effects. Since $I_1(\overset{0}{\underline{T}}) = 0$, the necessary condition for the existence of a correcting vector $\underline{w}$ (11.9) reduces to the form

$$\iint_S \overset{0}{\underline{T}}(\underline{v})\cdot\overset{0}{\underline{\omega}}(\underline{v})\,do = \mu\alpha^2\left(-\underline{i}_3\iint_S \underline{\rho}\cdot\overset{0}{\underline{\nabla}}\varphi\,do + a^3\iint_S(\overset{0}{\underline{\nabla}}\varphi + \underline{i}_3\times\underline{\rho})\,do\right) = \underline{0} \ . \qquad (12)$$

This condition holds in view of (11) and (1). The variation ΔL in the length of the rod is given by the mean value relative elongation

$$\Delta L = L\underline{i}_3\cdot\overset{0}{\underline{\varepsilon}}_m(\underline{v})\cdot\underline{i}_3$$

and by (11.20) and (11.21) this variation is represented as

$$\Delta L = \frac{1}{2\mu S}\int_0^L da^3\iint_S do\left\{\frac{\lambda}{3\lambda+2\mu}\left[I_1(\overset{0}{\underline{T}}\cdot\overset{0}{\underline{\nabla}}\underline{v}^T) + I_1(\underline{T}')\right] - \right.$$

$$\left. - \underline{i}_3\cdot(\overset{0}{\underline{T}}\cdot\overset{0}{\underline{\nabla}}\underline{v}^T + \underline{T}')\cdot\underline{i}_3\right\} \qquad (13)$$

it is sufficient here to know the first invariants and the (33) - components of the tensors $\overset{0}{\underline{T}}\cdot\overset{0}{\underline{\nabla}}\underline{v}^T$, $\underline{T}'$.

Calculation yields

$$I_1(\overset{0}{\underline{T}}\cdot\overset{0}{\underline{\nabla}}\underline{v}^T) = \overset{0}{\underline{T}}\cdot\cdot\overset{0}{\underline{\nabla}}\underline{v}^T = \mu\alpha^2[\underline{\rho}\cdot\underline{\rho} + 2(\underline{i}_3\times\underline{\rho})\cdot\overset{0}{\underline{\nabla}}\varphi + \overset{0}{\underline{\nabla}}\varphi\cdot\overset{0}{\underline{\nabla}}\varphi] \ ,$$

$$\underline{i}_3\cdot\overset{0}{\underline{T}}\cdot\overset{0}{\underline{\nabla}}\underline{v}^T\cdot\underline{i}_3 = \mu\alpha^2[\underline{\rho}\cdot\underline{\rho} + (\underline{i}_3\times\underline{\rho})\cdot\overset{0}{\underline{\nabla}}\varphi] \ ,$$

and, referring to representation (10.9) of $\underline{T}'$ we arrive at the expressions

$$I_1(\underline{T}') = \alpha^2\left[(3\lambda+2\mu)\left(a^{32} + \frac{1}{2}\underline{\rho}\cdot\underline{\rho} + \frac{1}{2}\overset{0}{\underline{\nabla}}\varphi\cdot\overset{0}{\underline{\nabla}}\varphi\right) + \right.$$

$$\left. + \frac{1}{2}\left(3m - \frac{n}{2}\right)(\underline{\rho}\cdot\underline{\rho} + 2(\underline{i}_3\times\underline{\rho})\cdot\overset{0}{\underline{\nabla}}\varphi + \overset{0}{\underline{\nabla}}\varphi\cdot\overset{0}{\underline{\nabla}}\varphi)\right] ,$$

$$\underline{i}_3 \cdot \underline{T}' \cdot \underline{i}_3 = \alpha^2 \Big[\lambda \, ({a^3}^2 + \frac{1}{2} \underline{\rho} \cdot \underline{\rho} + \frac{1}{2} \overset{0}{\underline{\nabla}}\varphi \cdot \overset{0}{\underline{\nabla}}\varphi) + $$

$$+ \frac{1}{2} m (\underline{\rho} \cdot \underline{\rho} + 2(\underline{i}_3 \times \underline{\rho}) \cdot \overset{0}{\underline{\nabla}}\varphi + \overset{0}{\underline{\nabla}}\varphi \cdot \overset{0}{\underline{\nabla}}\varphi) + \mu \overset{0}{\underline{\nabla}}\varphi \cdot \overset{0}{\underline{\nabla}}\varphi \Big] .$$

Substituting this into (13) and simplifying the resulting expression we get

$$\Delta L = \frac{\alpha^2}{2\mu S} \int_0^L da^3 \iint_S \Big[\Big(\frac{\lambda\mu}{3\lambda + 2\mu} + \frac{1}{2} \lambda \, \frac{3m - n/2}{3\lambda + 2\mu} - \frac{1}{2} m \Big) \times$$

$$\times (\underline{\rho} \cdot \underline{\rho} + 2(\underline{i}_3 \times \underline{\rho}) \cdot \overset{0}{\underline{\nabla}}\varphi + \overset{0}{\underline{\nabla}}\varphi \cdot \overset{0}{\underline{\nabla}}\varphi) - \mu (\underline{\rho} \cdot \underline{\rho} + (\underline{i}_3 \times \underline{\rho}) \cdot \overset{0}{\underline{\nabla}}\varphi + \overset{0}{\underline{\nabla}}\varphi \cdot \overset{0}{\underline{\nabla}}\varphi) \Big] do .$$

By (9) and (10)

$$\iint_S (\underline{\rho} \cdot \underline{\rho} + 2(\underline{i}_3 \times \underline{\rho}) \cdot \overset{0}{\underline{\nabla}}\varphi + \overset{0}{\underline{\nabla}}\varphi \cdot \overset{0}{\underline{\nabla}}\varphi) do = I_p + 2(C - I_p) + I_p - C = C ,$$

$$\iint_S (\underline{\rho} \cdot \underline{\rho} + (\underline{i}_3 \times \underline{\rho}) \cdot \overset{0}{\underline{\nabla}}\varphi) do = I_p + C - I_p = C , \qquad \iint_S \overset{0}{\underline{\nabla}}\varphi \cdot \overset{0}{\underline{\nabla}}\varphi \, do = I_p - C$$

and the required representation of the relative elongation is expressed by the formula*)

$$\frac{\Delta L}{L} = - \frac{\alpha^2 I_p}{2S} + \frac{\alpha^2 C}{2S} \, \frac{1}{3\lambda + 2\mu} \Big(\lambda - \frac{n\lambda}{4\mu} - m \Big) =$$

$$= - \frac{\alpha^2 I_p}{2S} + \frac{\alpha^2 C}{2S(1 + \nu)} \Big(\nu - \frac{n\nu}{4\mu} - (1 - 2\nu) \, \frac{m}{2\mu} \Big) . \qquad (14)$$

It is impossible to say in advance whether the rod is shortened or elongated, it depends on the material constants (for instance for many materials, $n < 0$, $m < 0$).

3. Stretching of a rod. When a prismatic rod is subjected to a longitudinal force R the displacement vector of points within the rod and the tensor of the stress produced in it can be determined by the well-known formulas

$$\underline{v} = -\nu e_3 \underline{\rho} + \underline{i}_3 e_3 a^3 , \qquad \underline{T}^0 = 2\mu(1 + \nu) e_3 \underline{i}_3 \underline{i}_3 ;$$

$$e_3 = \frac{S}{2(1 + \nu)\mu S} = \frac{\sigma_3}{2(1 + \nu)\mu} ,$$

*) This problem was considered by Rivlin (R.S. Rivlin, 1951). On rewriting this in his notation, a difference is detected between (14) and Rivlin's formula: in the latter, the first term in the brackets is 2ν and not ν. The calculation in the text has been checked many times and in a variety of ways.

so that

$$\overset{0}{\underline{\underline{\nabla v}}} = e_3(-\nu \underline{E}_2 + \underline{i}_3\underline{i}_3) = \overset{0}{\underline{\underline{\nabla v}}}^T = \overset{0}{\underline{\varepsilon}}(\underline{v}) , \qquad \overset{0}{\underline{\nabla}} \cdot \underline{v} = \vartheta = (1-2\nu)e_3 , \qquad \overset{0}{\underline{\omega}} = \underline{0} , \tag{15}$$

where $\underline{E}_2 = \underline{i}_1\underline{i}_1 + \underline{i}_2\underline{i}_2$. The quantities required for the calculation of second order effects are found to be

$$\underline{T}^0 \cdot \overset{0}{\underline{\underline{\nabla v}}}^T = 2\mu(1+\nu)e_3^2\underline{i}_3\underline{i}_3 , \qquad I_1(\overset{0}{\underline{\varepsilon}}{}^2) = (1+2\nu^2)e_3^2 ,$$
$$\overset{0}{\underline{\underline{\nabla v}}}^T \cdot \overset{0}{\underline{\underline{\nabla v}}} = e_3^2(\nu^2\underline{E}_2 + \underline{i}_3\underline{i}_3) . \tag{16}$$

By (10.9), the correcting stress tensor $\underline{T}'$ is represented by the expression

$$\underline{T}' = e_3^2\Big\{\underline{E}_2 [\tfrac{1}{2}\lambda(1+2\nu^2) + l(1-2\nu)^2 + (2m-n)\nu(1+\nu) + (n+\mu)\nu^2] +$$
$$+ \underline{i}_3\underline{i}_3[\tfrac{1}{2}\lambda(1+2\nu^2) + l(1-2\nu)^2 + (2m-n)(1-\nu^2) + n + \mu]\Big\} .$$

The necessary condition (11.9) for the existence of the correcting vector $\underline{w}$ certainly holds because $\overset{0}{\underline{\omega}} = \underline{0}$. By (11.21), the mean relative elongation superposed on e_3 turns out to be proportional to e_3^2 and depends only on the elastic constants of a material

$$\frac{\Delta L}{L} = -\frac{3}{2}e_3^2 - \frac{e_3^2}{E}[l(1-2\nu)^2 + 2m(1-\nu)(1+\nu)^2 + 3n\nu^2]$$
$$(E = 2\mu(1+\nu)) . \tag{17}$$

"The characteristic" of a material (the diagram e_3, σ_3 of stretching of a sample) is called soft for $\Delta L/L < 0$ and hard for $\Delta L/L > 0$.

4. Torsion of a stretching rod. In this section quantities referring to torsion will be indicated by the suffix "1", and those referring to stretching by "2". In particular, $\overset{0}{\underline{\omega}} = \overset{0}{\underline{\omega}}_1$, $\vartheta = \vartheta_2$.

The necessary condition (11.9) for the existence of vector $\underline{w}$ remains unfulfilled because

$$\iiint_v [\overset{0}{\underline{T}}_2 \cdot \overset{0}{\underline{\omega}}_1 - \overset{0}{\underline{\omega}}_1 I_1(\overset{0}{\underline{T}}_2)]\,dv = (1+\nu)\mu e_3\alpha \iiint_v (\underline{\rho} - \overset{0}{\underline{\nabla}}\varphi \times \underline{i}_3)\,dv = \underline{0} .$$

Further, by virtue of (4) and (5), we have

$$\overset{0}{\underline{\underline{\nabla v}}} = \overset{0}{\underline{\underline{\nabla v}}}_1 + \overset{0}{\underline{\underline{\nabla v}}}_2 = \alpha[a^3(\underline{i}_1\underline{i}_2 - \underline{i}_2\underline{i}_1) + \underline{i}_3\underline{i}_3 \times \underline{\rho} + \overset{0}{\underline{\nabla}}\varphi\underline{i}_3] + e_3(-\nu\underline{E}_2 + \underline{i}_3\underline{i}_3)$$

and therefore

$$\underline{T}^0 \cdot \overset{0}{\underline{\underline{\nabla v}}}{}^T = \underline{T}_1^0 \cdot \overset{0}{\underline{\underline{\nabla v}}}{}_1^T + \underline{T}_2^0 \cdot \overset{0}{\underline{\underline{\nabla v}}}{}_2^T + \underline{T}_1^0 \cdot \overset{0}{\underline{\underline{\nabla v}}}{}_2^T + \underline{T}_2^0 \cdot \overset{0}{\underline{\underline{\nabla v}}}{}_1^T =$$

$$= \underline{T}_1^0 \cdot \overset{0}{\underline{\underline{\nabla v}}}{}_1^T + \underline{T}_2^0 \cdot \overset{0}{\underline{\underline{\nabla v}}}{}^T + \mu\alpha e_3 [-\nu \underline{i}_3(\underline{i}_3 \times \underline{\rho}) + \underline{i}_3(\underline{i}_3 \times \underline{\rho}) + \overset{0}{\underline{\nabla}}\varphi \underline{i}_3] +$$

$$+ 2\mu\alpha e_3 (1+\nu) [\underline{i}_3(\underline{i}_3 \times \underline{\rho}) + \underline{i}_3 \overset{0}{\underline{\nabla}}\varphi] .$$

From here it follows that the interaction of torsional and stretching deformations does not influence the mean values $I_1(\underline{T}^0 \cdot \overset{0}{\underline{\underline{\nabla v}}}{}^T)$ and $\underline{i}_3 \cdot \underline{T}^0 \cdot \overset{0}{\underline{\underline{\nabla v}}}{}^T \cdot \underline{i}_3$. The effects produced by the two are additive. The interaction terms entering representation (10.9) of the correcting tensor form the tensor

$$\underline{T}' = \underline{E}\left(\frac{1}{2}\lambda + m - \frac{1}{2}n\right) I_1(\overset{0}{\underline{\varepsilon}}_1 \cdot \overset{0}{\underline{\varepsilon}}_2 + \overset{0}{\underline{\varepsilon}}_2 \cdot \overset{0}{\underline{\varepsilon}}_1) + n(\overset{0}{\underline{\varepsilon}}_1 \cdot \overset{0}{\underline{\varepsilon}}_2 + \overset{0}{\underline{\varepsilon}}_2 \cdot \overset{0}{\underline{\varepsilon}}_1) +$$

$$+ (2m - n)\overset{0}{\underline{\varepsilon}}_1 \vartheta_2 + \mu(\overset{0}{\underline{\underline{\nabla v}}}{}_1^T \cdot \overset{0}{\underline{\underline{\nabla v}}}{}_2 + \overset{0}{\underline{\underline{\nabla v}}}{}_2^T \cdot \overset{0}{\underline{\underline{\nabla v}}}{}_1)$$

and it is easily seen that the first invariant and the (33)-component of this tensor are equal to zero. The mean values of the second order effects for the problem of combined torsion and stretching are equal to the sum of these quantities calculated for each deformation taken separately. Of course, this situation represents an exception to the general rule that, for nonlinear problems, responses produced by separate factors are non-additive.

5. Torsion of a rod compressed over its lateral surface. The lateral surface of a rod is loaded by a pressure proportional to the axial coordinate: $\underline{f} = -qa^3 \underline{n}$. It can be verified immediately that the static equilibrium equations in the volume and on its surface

$$\overset{0}{\underline{\nabla}} \cdot \underline{T}^0 = \underline{0}\,, \qquad \underline{n} \cdot \underline{T}^0 = -qa^3 \underline{n}$$

will be satisfied if we assume that

$$\underline{T}^0 = -qa^3 \underline{E}_2 \,. \tag{18}$$

The Beltrami equations also hold because the tensor $\underline{T}^0$ depends linearly upon the coordinates.

The linear strain tensor calculated by the usual formula is found to be

$$\overset{0}{\underline{\varepsilon}} = \frac{1}{2\mu}\left[\underline{T}^0 - \frac{\nu}{1+\nu}\underline{E} I_1(\underline{T}^0)\right] = -\frac{qa^3}{2\mu}\left(\frac{1-\nu}{1+\nu}\underline{E} - \underline{i}_3\underline{i}_3 \frac{2\nu}{1+\nu}\right) . \tag{19}$$

Using this tensor we calculate the displacement vector (e.g. integrating using Cesaro's formula)

$$\underline{v} = -\frac{qa^3}{2\mu}\frac{1-\nu}{1+\nu}\underline{\rho} + \frac{q}{2\mu}\left(\frac{1-\nu}{1+\nu}\frac{1}{2}\underline{\rho}\cdot\underline{\rho} + \frac{\nu}{1+\nu}a^{3^2}\right)\underline{i}_3$$

and the vector for a small rotation

$$\overset{0}{\underline{\omega}} = -\frac{q}{2\mu}\frac{1-\nu}{1+\nu}\underline{i}_3 \times \underline{\rho} \,. \tag{20}$$

The necessary condition (6.11.9) for the existence of vector $\underline{w}$ is satisfied because

$$\iiint_v dv\,[\overset{0}{\underline{T}}\cdot\overset{0}{\underline{\omega}} - \overset{0}{\underline{\omega}}I_1(\overset{0}{\underline{T}})] =$$

$$= -\frac{q^2}{2\mu}\frac{1-\nu}{1+\nu}\int_{-L/2}^{L/2} a^3 da^3 \int_S do\,[\underline{E}_2\cdot(\underline{i}_3\times\underline{\rho}) - 2\underline{i}_3\times\underline{\rho}] = \underline{0}\,. \tag{21}$$

The origin has been placed at the centre of inertia of the rod

$$\int_{-L/2}^{L/2} a^3 da^3 = 0\,, \qquad \iint_S \underline{\rho}\,do = \underline{0}\,. \tag{22}$$

We now proceed to the consideration of the combined effects of torsion and compression over the lateral surface. The tensor in the left-hand side of (11.10) is now represented in the following form (the quantities referring to compression are indicated by the suffix "3", and those referring to torsion by "1")

$$\iiint_v [\overset{0}{\underline{T}}(\underline{v}_1+\underline{v}_3) - \underline{E}I_1(\overset{0}{\underline{T}}(\underline{v}_1+\underline{v}_3))]\,dv =$$

$$= \int_{-L/2}^{L/2} da^3 \iint_S do\,(\overset{0}{\underline{T}}(\underline{v}_1) - qa^3\underline{E}_2 + 2qa^3\underline{E}) = \underline{0}\,,$$

here we have also used Equations (22) and (7). We must now consider the right-hand side. Here we have

$$\iiint_v [\overset{0}{\underline{T}}(\underline{v}_1+\underline{v}_3)\cdot(\overset{0}{\underline{\omega}}_1+\overset{0}{\underline{\omega}}_3) - (\overset{0}{\underline{\omega}}_1+\overset{0}{\underline{\omega}}_3)I_1(\overset{0}{\underline{T}}(\underline{v}_1+\underline{v}_3))]\,dv =$$

$$= \iiint_v [\overset{0}{\underline{T}}(\underline{v}_1)\cdot\overset{0}{\underline{\omega}}_1 + \overset{0}{\underline{T}}(\underline{v}_3)\cdot\overset{0}{\underline{\omega}}_3 + \overset{0}{\underline{T}}(\underline{v}_1)\cdot\overset{0}{\underline{\omega}}_3 + \underline{T}(\underline{v}_3)\cdot\overset{0}{\underline{\omega}}_1 -$$

$$- \overset{0}{\underline{\omega}}_3 I_1(\overset{0}{\underline{T}}(\underline{v}_1)) - \overset{0}{\underline{\omega}}_1 I_1(\overset{0}{\underline{T}}(\underline{v}_3))]\,dv =$$

$$= \int_{-L/2}^{L/2} da^3 \iint_S do\,[\overset{0}{\underline{T}}(\underline{v}_1)\cdot\overset{0}{\underline{\omega}}_3 + \overset{0}{\underline{T}}(\underline{v}_3)\cdot\overset{0}{\underline{\omega}}_1 - \overset{0}{\underline{\omega}}_1 I_1(\overset{0}{\underline{T}}(\underline{v}_3))] = A\underline{i}_3\ ,$$

because the other terms vanish by (21) and (12) and there only remain those quantities stipulated by the mutual influence of deformations. Turning to Equations (6), (7), (18), (20) we obtain after some calculation

$$A = \frac{1}{2}\,qa\left(\frac{l^2}{3} - \frac{1-\nu}{1+\nu}\,\frac{C}{S}\right)\ .$$

This quantity is different from zero. We cannot satisfy the criterion of existence of a correcting vector $\underline{w}$ though this criterion is fulfilled for each deformation taken separately ($\alpha \neq 0$, $q = 0$), ($\alpha = 0$, $q \neq 0$). Any direction may be taken for the axis of equilibrium because the left-hand side of Equation (11.10) is identically zero.

This illustration of the impossibility of constructing the second order effects is oited in the monograph by Grioli (Grioli, 1962).

14. SECOND ORDER EFFECTS IN THE PLANE PROBLEM FOR A QUASI-LINEAR MATERIAL

The problem is considered under the assumption that in (8.1) $c = 1$. Instead of (8.18) we use

$$\psi(q)e^{i\chi} = \Phi'(\zeta) = 2\mu M^2(\zeta)\ . \tag{1}$$

We further assume that $q > 2(\lambda+\mu)/(\lambda+2\mu)$ so that $\psi(q) > 0$. If $\psi(q) = 0$, then $q = 2(\lambda+\mu)/(\lambda+2\mu)$, and by (8.26) and (8.27) we would have $\sigma_1 = \sigma_2 = -4\mu$ which means that the medium is in the state of hydrostatic compression of intensity 4μ. Under the above assumption function $M^2(\zeta)$ has no zeros within the plane domain $\mathcal{B}$ considered. It must also be recalled that for $q = 2$ $\psi(q) = 2\mu$ and $M^2(\zeta) = \exp i\chi$ and that there is no stress.

The basic relationship (8.19) connecting the unknown functions $z(\zeta,\bar{\zeta})$ and $M^2(\zeta)$ in the $\mathcal{B}$-domain is represented in the form

$$\frac{\partial z}{\partial \zeta} = \frac{\mu}{\lambda+2\mu}\,M^2(\zeta) + \frac{\lambda+\mu}{\lambda+2\mu}\,\frac{M(\zeta)}{\overline{M(\zeta)}}\ . \tag{2}$$

From this we get

$$z = \frac{\mu}{\lambda + 2\mu} \int^{\zeta} M^2(\zeta)\,d\zeta + \frac{\lambda + \mu}{\lambda + 2\mu} \left[\overline{M(\zeta)}^{-1} \int^{\zeta} M(\zeta)\,d\zeta - \overline{N(\zeta)} \right] , \tag{3}$$

so that

$$\frac{\partial z}{\partial \bar{\zeta}} = \frac{\lambda + \mu}{\lambda + 2\mu} \left[\int^{\zeta} M(\zeta) \frac{\partial}{\partial \bar{\zeta}} \overline{M(\zeta)}^{-1} d\zeta - \overline{N'(\zeta)} \right] . \tag{4}$$

The derivative of z along the arc of contour l in the b-region (in the reference configuration) is given by

$$\frac{dz}{ds} = \frac{\partial z}{\partial \zeta}\frac{d\zeta}{ds} + \frac{\partial z}{\partial \bar{\zeta}}\frac{d\bar{\zeta}}{ds} = i \left\{ n \left[\frac{\mu}{\lambda + 2\mu} M^2(\zeta) + \frac{\lambda + \mu}{\lambda + 2\mu} \frac{M(\zeta)}{\overline{M(\zeta)}} \right] - \right.$$

$$\left. - \bar{n} \frac{\lambda + \mu}{\lambda + 2\mu} \left[\int^{\zeta} \frac{\partial}{\partial \bar{\zeta}} \frac{M(\zeta)}{\overline{M(\zeta)}} d\zeta - \overline{N'(\zeta)} \right] \right. \tag{5}$$

and Equation (8.20) reduces to the relationship

$$\frac{\lambda + 2\mu}{2\mu(\lambda + \mu)} f \frac{dS}{ds} = n \left[M^2(\zeta) - \frac{M(\zeta)}{\overline{M(\zeta)}} \right] + \bar{n} \left[\int^{\zeta} \frac{\partial}{\partial \bar{\zeta}} \frac{M(\zeta)}{\overline{M(\zeta)}} d\zeta - \overline{N'(\zeta)} \right] . \tag{6}$$

We have thus arrived at the nonlinear boundary-value problem for the determination of two functions $M(\zeta)$ and $N(\zeta)$ which are analytic within the region b.

It can be verified immediately that the right-hand side of Equation (6) satisfies the conditions that the system of forces f distributed over the contour l of the domain is statically equivlent to zero. We make use of the relationships

$$\oint_l \varphi(\zeta,\bar{\zeta}) n\,ds = 2 \iint_b \frac{\partial \varphi}{\partial \bar{\zeta}} do , \qquad \oint_l \varphi(\zeta,\bar{\zeta}) \bar{n}\,ds = 2 \iint_b \frac{\partial \varphi}{\partial \zeta} do , \tag{7}$$

which are a form of the Gauss-Ostrogradskii transformations, e.g.

$$\oint_l \varphi(\zeta,\bar{\zeta}) n\,ds = \oint_l (n_1 + i n_2)\, \varphi(\zeta,\bar{\zeta})\,ds =$$

$$= \iint_b \left(\frac{\partial}{\partial a^1} + i \frac{\partial}{\partial a^2} \right) \varphi(\zeta,\bar{\zeta})\,ds = 2 \iint_b \frac{\partial \varphi}{\partial \bar{\zeta}} do .$$

From (5) the expression for the resultant V of the forces f is given by the function

$$\frac{\lambda + 2\mu}{2\mu(\lambda + \mu)} V = \frac{\lambda + 2\mu}{2\mu(\lambda + \mu)} \oint_L f dS =$$

$$= 2 \iint\limits_{b} \left[\frac{\partial M^2(\zeta)}{\partial \bar{\zeta}} - \frac{\partial}{\partial \bar{\zeta}} \frac{M(\zeta)}{\overline{M(\zeta)}} + \frac{\partial}{\partial \bar{\zeta}} \frac{M(\zeta)}{\overline{M(\zeta)}} - \frac{\partial \overline{N'(\zeta)}}{\partial \zeta} \right] do = 0 \ .$$

The expression for the moment of force is given by

$$xf_y - yf_x = \frac{1}{2} i (z\bar{f} - \bar{z}f)$$

and by (6) we have

$$\frac{\lambda + 2\mu}{2\mu(\lambda + \mu)} m^0 = \frac{i}{2} \oint\limits_{L} (z\bar{f} - \bar{z}f) dS =$$

$$= \frac{i}{2} \oint\limits_{l} \left\{ z\bar{n} \left[\overline{M^2(\zeta)} - \frac{\overline{M(\zeta)}}{M(\zeta)} \right] - \bar{z}n \left[M^2(\zeta) - \frac{M(\zeta)}{\overline{M(\zeta)}} \right] \right\} ds \ +$$

$$+ \frac{i}{2} \oint\limits_{l} \left\{ zn \left[\int^{\zeta} \frac{\partial}{\partial \zeta} \frac{\overline{M(\zeta)}}{M(\zeta)} d\bar{\zeta} - N'(\zeta) \right] - \bar{z}\bar{n} \left[\int^{\zeta} \frac{\partial}{\partial \bar{\zeta}} \frac{M(\zeta)}{\overline{M(\zeta)}} d\zeta - \overline{N'(\zeta)} \right] \right\} ds \ .$$

Eliminating the quantities

$$\frac{M(\zeta)}{\overline{M(\zeta)}} , \qquad \int^{\zeta} \frac{\partial}{\partial \bar{\zeta}} \frac{M(\zeta)}{\overline{M(\zeta)}} d\zeta - \overline{N'(\zeta)}$$

from this expression using (2) and (4) and applying transformation (7), we obtain

$$m^0 = i\mu \oint [z\bar{n}\overline{M^2(\zeta)} - \bar{z}n\, M^2(\zeta)] ds = 2i\mu \iint\limits_{b} \left[\frac{\partial z}{\partial \zeta} \overline{M^2(\zeta)} - \frac{\partial \bar{z}}{\partial \bar{\zeta}} M^2(\zeta) \right] do \ .$$

It can be seen that from (1) and (8.19) the integrand is zero as required.

We now proceed to the consideration of the second order effects. A function which has no zeros within the domain can be represented as an exponent in this domain, so that

$$M^2(\zeta) = \exp\Omega(\zeta) \ , \qquad \frac{M(\zeta)}{\overline{M(\zeta)}} = \exp \frac{1}{2} [\Omega(\zeta) - \overline{\Omega(\zeta)}] \ ,$$

$$\frac{\partial}{\partial \bar{\zeta}} \int^{\zeta} \frac{M(\zeta)}{\overline{M(\zeta)}} d\zeta = - \frac{1}{2} \overline{\Omega'(\zeta)} \int^{\zeta} \exp \frac{1}{2} [\Omega(\zeta) - \overline{\Omega(\zeta)}] d\zeta \ .$$

Introducing the Muskhelishvili potentials $\Phi(\zeta)$ and $\Psi(\zeta)$ which give a solution to the linear problem, we have

$$\frac{\lambda + 2\mu}{\mu(\lambda + \mu)} f^0 = n[\Phi(\zeta) + \overline{\Phi(\zeta)}] - \bar{n}[\zeta\overline{\Phi'(\zeta)} + \overline{\Psi(\zeta)}] \qquad (8)$$

(it is assumed here and below that the loading is "dead", in other words that $f\,\frac{dS}{ds} = f^0$, the force vector, is constant during deformation).

Equation (8) makes it possible to substitute boundary condition (6) for the following relationship

$$n(\Phi(\zeta) + \overline{\Phi(\zeta)}) - \bar{n}(\zeta\overline{\Phi'(\zeta)} + \overline{\Psi(\zeta)}) = 2n[\exp\Omega(\zeta) - \exp\frac{1}{2}(\Omega(\zeta) - \overline{\Omega(\zeta)})] - \bar{n}[\overline{\Omega'(\zeta)}\int^{\zeta}\exp\frac{1}{2}(\Omega(\zeta) - \overline{\Omega(\zeta)})d\zeta + \overline{N'(\zeta)}] \; . \tag{9}$$

The absolute values of the functions $\Phi(\zeta)$ and $\Psi(\zeta)$ which are used in the solution of the linear problem have the order of linear deformations. In order to take account of the second order effect it must be assumed that

$$\Omega(\zeta) = \Phi(\zeta) + \Phi_1(\zeta)\,, \qquad N'(\zeta) = \Psi(\zeta) + \Psi_1(\zeta) \tag{10}$$

the correcting terms $\Phi_1(\zeta)$ and $\Psi_1(\zeta)$ should be at least of the second order of magnitude. Discarding higher order terms, we obtain

$$\exp\Omega(\zeta) = 1 + \Phi(\zeta) + \Phi_1(\zeta) + \frac{1}{2}\Phi^2(\zeta)\,,$$

$$\exp\frac{1}{2}(\Omega(\zeta) - \overline{\Omega(\zeta)}) = 1 + \frac{1}{2}(\Phi(\zeta) - \overline{\Phi(\zeta)}) + \frac{1}{2}(\Phi_1(\zeta) - \overline{\Phi_1(\zeta)}) + \frac{1}{8}(\Phi(\zeta) - \overline{\Phi(\zeta)})^2\,,$$

$$2[\exp\Omega(\zeta) - \exp\frac{1}{2}(\Omega(\zeta) - \overline{\Omega(\zeta)})] = $$
$$= (\Phi(\zeta) + \overline{\Phi(\zeta)}) + (\overline{\Phi_1(\zeta)} + \overline{\Phi_1(\zeta)}) + \frac{3}{4}\Phi^2(\zeta) - \frac{1}{4}\overline{\Phi^2(\zeta)} + \frac{1}{2}\Phi(\zeta)\,\overline{\Phi(\zeta)}\,,$$

$$\overline{\Omega'(\zeta)}\int^{\zeta}\exp\frac{1}{2}(\Omega(\zeta) - \overline{\Omega(\zeta)})d\zeta = [\overline{\Phi'(\zeta)} + \overline{\Phi_1'(\zeta)}]\zeta + $$
$$+ \frac{1}{2}\overline{\Phi'(\zeta)}\int^{\zeta}[\Phi(\zeta) - \overline{\Phi(\zeta)}]d\zeta\;.$$

On substituting these expressions into (9), the terms linear with respect to $\Phi(\zeta)$, $\Psi(\zeta)$ cancel out. This serves to verify that the linear terms in (10) do actually provide a solution to the linear problem. To determine $\Phi_1(\zeta)$ and $\Psi_1(\zeta)$, we take the boundary condition

$$n(\Phi_1'(\zeta) + \overline{\Phi_1'(\zeta)} - \bar{n}\zeta\overline{\Phi_1'(\zeta)} - \bar{n}\overline{\Psi_1(\zeta)} = \frac{1}{2}n\left(-\frac{3}{2}\Phi^2(\zeta) + \frac{1}{2}\overline{\Phi^2(\zeta)} - \Phi(\zeta)\,\overline{\Phi(\zeta)}\right) + $$
$$+ \frac{1}{2}\bar{n}\left(\overline{\Phi'(\zeta)}\int^{\zeta}\Phi(\zeta)d\zeta - \zeta\overline{\Phi(\zeta)}\;\overline{\Phi'(\zeta)}\right)\,, \tag{11}$$

which corresponds to a linear problem with "surface tractions"

$$\frac{1}{2}\,n\left(-\,\frac{3}{2}\,\Phi^2(\zeta)\;+\frac{1}{2}\,\overline{\Phi^2(\zeta)}\;-\;\Phi(\zeta)\,\overline{\Phi(\zeta)}\right)+\frac{1}{2}\,\bar{n}\left(\overline{\Phi'(\zeta)}\int^{\zeta}\Phi(\zeta)d\zeta\;-\;\zeta\overline{\Phi(\zeta)}\;\overline{\Phi'(\zeta)}\right)=$$

$$=nP(\zeta,\bar{\zeta})\;+\;\bar{n}Q(\zeta,\bar{\zeta})\;, \tag{12}$$

known from solution (8). It is easily verified that

$$\frac{\partial P}{\partial\bar{\zeta}}+\frac{\partial Q}{\partial\zeta}=0\;,\qquad \frac{\partial\bar{P}}{\partial\zeta}+\frac{\partial\bar{Q}}{\partial\bar{\zeta}}=0\;.$$

These relationships together with (8) imply that the resultant of "forces" (12) is zero.

The expression for their resultant moment takes the form

$$m^0=\frac{1}{2i}\oint_l\,[\bar{\zeta}(nP+\bar{n}Q)\;-\;\zeta(\bar{n}\bar{P}+n\bar{Q})\,]\,ds\;=$$

$$=-i\iint_b(P-\bar{P})\,do\;=\;2i\iint_b(\Phi^2(\zeta)\;-\;\overline{\Phi^2(\zeta)})\,do$$

and the requirement that it should vanish leads to the relationship

$$\iint_b[\Phi^2(\zeta)\;-\;\overline{\Phi^2(\zeta)}\,]\,do\;=\;0\;. \tag{13}$$

The solution of the linear problem determines the function $\Phi(\zeta)$ except for an additive constant iC which is purely imaginary. Assuming that

$$\Phi(\zeta)\;=\;\Phi_{\times}(\zeta)\;+\;iC\;,\qquad \Phi^2(\zeta)\;=\;\Phi_{\times}^2(\zeta)\;+\;2iC\Phi_{\times}(\zeta)\;-\;C^2\;,$$

$$\Phi^2(\zeta)\;-\;\overline{\Phi^2(\zeta)}\;=\;\Phi_{\times}^2(\zeta)\;-\;\overline{\Phi_{\times}^2(\zeta)}\;+\;2iC(\Phi(\zeta)\;+\;\overline{\Phi(\zeta)})\;,$$

we can rewrite Equation (13) in the form

$$2C\iint_b[\Phi(\zeta)\;+\;\overline{\Phi(\zeta)}\,]\,do\;=\;i\iint_b[\Phi_{\times}^2(\zeta)\;-\;\overline{\Phi_{\times}^2(\zeta)}\,]\,do\;.$$

Constant C can be determined, provided that

$$2\iint_b[\Phi(\zeta)\;+\;\overline{\Phi(\zeta)}\,]\,do\;\neq\;0\;.$$

For the plane deformation problem

$$2(\Phi+\bar{\Phi}) = \sigma_1 + \sigma_2 = \frac{1}{\nu}\sigma_3 , \qquad \iint_b (\Phi+\bar{\Phi})\,do = \frac{1}{\nu} Q$$

where Q denotes a longitudinal force which is applied so as to prevent longitudinal displacement of the cylinder's faces (we recall here that c has been assumed equal to 1).

15. ON THE "PHYSICALLY NONLINEAR" THEORY OF ELASTICITY

It is well known that for certain materials (e.g. rocks, cast iron), the linear dependence between the stress tensor and the linear strain tensor $\underline{\varepsilon}$ is violated even for very small deformations. In the "physically nonlinear" theory it is proposed that this shortcoming be compensated by retaining second and higher order terms with respect to $\underline{\varepsilon}$ in the state equation.

This method of taking nonlinearity into account was suggested by Kauderer (1958) and has since been discussed in a number of papers. Here we compare it with formulations of the second order effects for which all terms quadratic with respect to the deformation gradient of the displacement vector $\underline{\underline{\nabla u}}$ are retained.

Confining ourselves to the consideration of an isotropic elastic material, we will begin with Equations (2.7.6) and (4.1.8)

$$\delta\ni = \sqrt{\frac{g}{G}}\,\ni_{\underline{C}} \cdot\cdot\, \delta\underline{C}^T = \underline{T}^{\vee} \cdot\cdot\, \delta\underline{C}^T , \tag{1}$$

where $\underline{C}$ denotes the Cauchy-Green strain tensor (1.7.9), and $\underline{T}^{\vee}$ - the energetic tensor (2.6.11) of stress

$$\underline{C} = \overset{0}{\underline{\varepsilon}}(\underline{u}) + \frac{1}{2}\overset{0}{\underline{\underline{\nabla u}}}\cdot\overset{0}{\underline{\underline{\nabla u}}}^T , \qquad \underline{T}^{\vee} = \sqrt{\frac{g}{G}}\,\ni_{\underline{C}} . \tag{2}$$

The Cauchy stress tensor is, by (4.3.26), connected with $\underline{T}^{\vee}$ by the relationship

$$\underline{T} = \overset{0}{\underline{\underline{\nabla R}}}^T\cdot\underline{T}^{\vee}\cdot\overset{0}{\underline{\underline{\nabla R}}} = (\underline{E}+\overset{0}{\underline{\underline{\nabla u}}}^T)\cdot\underline{T}^{\vee}\cdot(\underline{E}+\overset{0}{\underline{\underline{\nabla u}}}) =$$

$$= \underline{T}^{\vee} + \overset{0}{\underline{\underline{\nabla u}}}^T\cdot\underline{T}^{\vee} + \underline{T}^{\vee}\cdot\overset{0}{\underline{\underline{\nabla u}}} + \overset{0}{\underline{\underline{\nabla u}}}^T\cdot\underline{T}^{\vee}\cdot\overset{0}{\underline{\underline{\nabla u}}} . \tag{3}$$

Having assigned the specific stored energy $\ni$ in the form due to Murnaghan

$$\ni = \frac{1}{2}(\lambda+2\mu)I_1^2(\underline{C}) - 2\mu I_2(\underline{C}) + \frac{1}{3}(l+2m)I_1^3(\underline{C}) - 2mI_1(\underline{C})I_2(\underline{C}) + nI_3(\underline{C}) , \tag{4}$$

we obtain from (II.3.3) the relation

$$\underline{T}^{\vee} = \left[\frac{\partial \ni}{\partial I_1(\underline{C})} + I_1(\underline{C})\,\frac{\partial \ni}{\partial I_2(\underline{C})} + I_2(\underline{C})\,\frac{\partial \ni}{\partial I_3(\underline{C})}\right]\underline{E} -$$

$$- \left[\frac{\partial}{\partial I_2(\underline{C})} + I_1(\underline{C})\,\frac{\partial}{\partial I_3(\underline{C})}\right]\underline{C} + \frac{\partial}{\partial I_3(\underline{C})}\,\underline{C}^2 =$$

$$= \overset{0}{\underline{T}}(\underline{C}) + [lI_1^2(\underline{C}) - (2m-n)I_2(\underline{C})]\underline{E} + (2m-n)I_1(\underline{C})\underline{C} + n\underline{C}^2 , \tag{5}$$

where $\overset{0}{\underline{T}}(\underline{C})$ denotes the linear operator of the Hooke's law over $\underline{C}$

$$\overset{0}{\underline{T}}(\underline{C}_1) = \lambda I_1(\underline{C})\underline{E} + 2\mu\underline{C} = \overset{0}{\underline{T}}(\overset{0}{\underline{\varepsilon}}) + \frac{1}{2}\,\overset{0}{\underline{T}}(\overset{0}{\underline{\nabla u}}\cdot\overset{0}{\underline{\nabla u}}{}^{T}) =$$

$$= \lambda I_1(\overset{0}{\underline{\varepsilon}}) + 2\mu\overset{0}{\underline{\varepsilon}} + \frac{1}{2}\,\lambda I_1(\overset{0}{\underline{\nabla u}}\cdot\overset{0}{\underline{\nabla u}}{}^{T}) + \mu\overset{0}{\underline{\nabla u}}\cdot\overset{0}{\underline{\nabla u}}{}^{T} . \tag{6}$$

The symbol $\overset{0}{\underline{T}}(\overset{0}{\underline{\varepsilon}})$ of course, denotes the stress tensor of the linear theory of elasticity. Retaining now in (5) the terms of order less than the second with respect to $\overset{0}{\underline{\nabla u}}$, we get

$$\underline{T}^{\vee}(\underline{C}) = \overset{0}{\underline{T}}(\overset{0}{\underline{\varepsilon}}) + \frac{1}{2}\,\lambda I_1(\overset{0}{\underline{\nabla u}}\cdot\overset{0}{\underline{\nabla u}}{}^{T})\underline{E} + \mu\overset{0}{\underline{\nabla u}}\cdot\overset{0}{\underline{\nabla u}}{}^{T} +$$

$$+ [lI_1^2(\underline{\varepsilon}) - (2m-n)I_2(\underline{\varepsilon})]\underline{E} + (2m-n)I_1(\overset{0}{\underline{\varepsilon}})\overset{0}{\underline{\varepsilon}} + n\overset{0}{\underline{\varepsilon}}{}^2 . \tag{7}$$

By (5.3.6) and (3), an appropriate expression for the Cauchy stress tensor $\underline{T}$ may be rewritten as

$$\underline{T} = \overset{0}{\underline{T}}(\overset{0}{\underline{\varepsilon}}) - I_1(\underline{\varepsilon})\overset{0}{\underline{T}}(\underline{\varepsilon}) + \frac{1}{2}\,\lambda I_1(\overset{0}{\underline{\nabla u}}\cdot\overset{0}{\underline{\nabla u}}{}^{T}) + \mu\overset{0}{\underline{\nabla u}}\cdot\overset{0}{\underline{\nabla u}}{}^{T} +$$

$$+ \overset{0}{\underline{\nabla u}}{}^{T}\cdot\overset{0}{\underline{T}}(\underline{\varepsilon}) + \overset{0}{\underline{T}}(\underline{\varepsilon})\cdot\overset{0}{\underline{\nabla u}} + [lI_1^2(\underline{\varepsilon}) - (2m-n)I_2(\underline{\varepsilon})]\underline{E} +$$

$$+ (2m-n)I_1(\overset{0}{\underline{\varepsilon}})\overset{0}{\underline{\varepsilon}} + n\overset{0}{\underline{\varepsilon}}{}^2 . \tag{8}$$

It may certainly be transformed to the already known form

$$\underline{T} = \overset{0}{\underline{T}}(\overset{0}{\underline{\varepsilon}}) - I_1(\overset{0}{\underline{\varepsilon}})\overset{0}{\underline{T}}(\overset{0}{\underline{\varepsilon}}) + 2\overset{0}{\underline{\varepsilon}}\cdot\overset{0}{\underline{T}}(\overset{0}{\underline{\varepsilon}}) + \underline{T}' ,$$

$$\underline{T}' = \frac{1}{2}\,\overset{0}{\underline{T}}(\overset{0}{\underline{\nabla u}}{}^{T}\cdot\overset{0}{\underline{\nabla u}}) + [lI_1^2(\overset{0}{\underline{\varepsilon}}) - (2m-n)I_2(\overset{0}{\underline{\varepsilon}})]\underline{E} + (2m-n)I_1(\overset{0}{\underline{\varepsilon}})\overset{0}{\underline{\varepsilon}} + n\overset{0}{\underline{\varepsilon}}{}^2 \tag{9}$$

the notation has been somewhat changed in order to compare (8) or (9) with the representation for $\underline{T}$ in the "physically nonlinear" theory.

In this theory, the specific strain energy is assumed to be dependent only upon the linear strain tensor: $\ni(\underline{C})$ is substituted for $\ni(\underline{\varepsilon})$. For example, for a Murnaghan material it is assumed firstly that instead of (4)

$$\ni = \frac{1}{2}(\lambda + 2\mu)I_1^2(\overset{0}{\underline{\varepsilon}}) - 2\mu I_1(\overset{0}{\underline{\varepsilon}}) + \frac{1}{3}(l + 2m)I_1^3(\underline{\varepsilon}) - 2mI_1(\overset{0}{\underline{\varepsilon}})I_2(\overset{0}{\underline{\varepsilon}}) + nI_3(\overset{0}{\underline{\varepsilon}}) \,. \tag{10}$$

Secondly, when terms of second order in $\overset{0}{\underline{\varepsilon}}$ are taken into account, the relationships

$$\delta\ni = \underline{T} \cdot\cdot\, \delta\overset{0}{\underline{\varepsilon}}{}^{T} \,, \qquad \underline{T} = \ni_{\overset{0}{\underline{\varepsilon}}} \tag{11}$$

which are valid in linear theory, are used instead of (1). Thirdly, the tensor $\underline{T}$ calculated using (11), is neither the Cauchy stress tensor nor the stress tensor derived from an energy equation.

With these simplifications, the stress tensor of the "physically nonlinear" theory, denoted below as $\underline{\tilde{T}}$, is, by (10) and (11), represented by the expression

$$\underline{\tilde{T}} = \ni_{\overset{0}{\underline{\varepsilon}}} = \overset{0}{\underline{T}} + [lI_1^2(\underline{\varepsilon}) - (2m - n)I_2(\overset{0}{\underline{\varepsilon}})]\underline{E} + (2m - n)I_1(\overset{0}{\underline{\varepsilon}})\overset{0}{\underline{\varepsilon}} + n\overset{0}{\underline{\varepsilon}}{}^2 \,. \tag{12}$$

Those terms quadratic with respect to a gradient of the displacement vector turned out to have been lost while the other terms of the same order have been retained in this theory

$$\underline{T} - \underline{\tilde{T}} = \sqrt{\frac{g}{G}}\,\overset{0}{\underline{\nabla R}}{}^{T} \cdot \ni_{\underline{C}} \cdot \overset{0}{\underline{\nabla R}} - \ni_{\overset{0}{\underline{\varepsilon}}} = -I_1(\overset{0}{\underline{\varepsilon}})\overset{0}{\underline{T}}(\overset{0}{\underline{\varepsilon}}) + 2\overset{0}{\underline{\varepsilon}} \cdot \overset{0}{\underline{T}}(\overset{0}{\underline{\varepsilon}}) + \frac{1}{2}\overset{0}{\underline{T}}(\overset{0}{\underline{\nabla u}}{}^{T} \cdot \overset{0}{\underline{\nabla u}}) \,. \tag{13}$$

In the example illustrating simple shear given below, the components of the Cauchy stress tensor determined using Equations (4.7) by the specific strain energy (5.3.2) are represented by the formulas

$$t^{11} = \frac{1}{2}\left[(\lambda + 2\mu) + \frac{1}{2}ls^2 + m(1 + s^2) + 2\hat{\mu}(s^2)\right]s^2 \,,$$

$$t^{22} = \frac{1}{2}\left[(\lambda + 2\mu) + \frac{1}{2}ls^2 + m(1 + s^2)\right]s^2 \,,$$

$$t^{33} = \frac{1}{2}\left[\lambda + \frac{1}{2}\lambda s^2 + m - \frac{n}{2}\right]s^2 \,, \qquad t^{12} = \hat{\mu}(s^2)s \,, \qquad t^{23} = t^{31} = 0 \,.$$

Taking into account only the terms of second order or below with respect to we get

$$t^{11} = \frac{1}{2}(\lambda + 4\mu + m)s^2 \,, \qquad t^{22} = \frac{1}{2}(\lambda + 2\mu + m)s^2 \,, \qquad t^{33} = \frac{1}{2}(\lambda + m - \frac{n}{2}) \,.$$

We also have

$$\overset{0}{\underline{\underline{\nabla u}}}{}^T = \underline{i}_1\underline{i}_2 s , \qquad \overset{0}{\underline{\underline{\nabla u}}} = \underline{i}_2\underline{i}_1 s ,$$

$$\overset{0}{\underline{\varepsilon}} = \frac{1}{2} (\underline{i}_1\underline{i}_2 + \underline{i}_2\underline{i}_1) s , \qquad \overset{0}{\underline{\underline{\nabla u}}}{}^T \cdot \overset{0}{\underline{\underline{\nabla u}}} = \underline{i}_1\underline{i}_1 s^2 ,$$

$$I_1(\overset{0}{\underline{\varepsilon}}) = 0 , \qquad \overset{0}{\underline{T}} = \mu s (\underline{i}_1\underline{i}_2 + \underline{i}_2\underline{i}_1) , \qquad 2\overset{0}{\underline{\varepsilon}} \cdot \overset{0}{\underline{T}} = \mu s^2 (\underline{i}_1\underline{i}_1 + \underline{i}_2\underline{i}_2) ,$$

$$\frac{1}{2} \overset{0}{\underline{T}}(\overset{0}{\underline{\underline{\nabla u}}}{}^T \cdot \overset{0}{\underline{\underline{\nabla u}}}) = \frac{1}{2} \lambda s^2 \underline{E} + \underline{i}_1\underline{i}_1 \mu s^2$$

and by (13) we arrive at the relationships

$$t^{11} - \tilde{t}^{11} = \frac{1}{2} (\lambda + 4\mu) s^2 , \qquad t^{22} - \tilde{t}^{22} = \frac{1}{2} (\lambda + 2\mu) s^2 , \qquad t^{33} - \tilde{t}^{33} = \frac{1}{2} \lambda s^2 .$$

From here it follows that for $s \to 0$

$$\frac{t^{11}}{\tilde{t}^{12}} = 1 + \frac{\lambda + 4\mu}{m} , \qquad \frac{t^{22}}{\tilde{t}^{22}} = 1 + \frac{\lambda + 2\mu}{m} , \qquad \frac{t^{33}}{\tilde{t}^{33}} = 1 + \frac{\lambda}{m - n/2} . \tag{14}$$

For Hecla 17 steel we make use of the data presented in Chapter 5, Par. 11 and we have

$$\left| \frac{t^{11}}{\tilde{t}^{11}} \right| = 0{,}26 , \qquad \left| \frac{t^{22}}{\tilde{t}^{22}} \right| = 0{,}54 , \qquad \left| \frac{t^{33}}{\tilde{t}^{33}} \right| = 0{,}58 .$$

Seeger (A. Seeger, 1960) found experimentally that, for a particular variety of steel

$$\frac{\lambda}{\mu} = 1{,}39 , \qquad \frac{l}{\mu} = -1{,}6 , \qquad \frac{m}{\mu} = -10{,}1 , \qquad \frac{n}{\mu} = -22{,}7$$

and by (14) we get

$$\left| \frac{t^{11}}{\tilde{t}^{11}} \right| = 0{,}47 , \qquad \left| \frac{t^{22}}{\tilde{t}^{22}} \right| = 0{,}66 , \qquad \left| \frac{t^{33}}{\tilde{t}^{33}} \right| = 2{,}11 .$$

The inadmissibility of the "physically nonlinear" theory is a consequence both of these experimental results and of the erroneous premises of the theory itself cited above. It may also be added that the method of calculation of the physically nonlinear theory does not possess the advantage of simplicity as compared to the rules for the calculation of the second order effects described in Pars. 6-10.

CHAPTER 7

INCOMPRESSIBLE ELASTIC MATERIAL

1. AN ELASTIC MATERIAL WITH SUPERPOSED CONSTRAINTS

1. It has been assumed so far that the material is able to undergo arbitrary deformations under the action of suitably chosen forces. The deformation gradient $\overset{0}{\nabla}\underline{\underline{R}}$ and the corresponding strain measures have been regarded as being smooth arbitrary functions of material coordinates. It turns out, however, to be perfectly reasonable to examine models of materials which do not permit certain types of deformation whatever the applied forces, i.e. the so-called materials with superposed constraints. An incompressible fluid is a classical example of such a material.

The constraint consists of a condition imposed upon the deformation gradient

$$\gamma(\overset{0}{\nabla}\underline{\underline{R}}) = 0 \, . \tag{1}$$

This material property must satisfy the principle of material frame-indifference (Chapter 3, Par. 2), in other words, a scalar function γ should preserve its form under any orthogonal transformation

$$\forall \underline{O} \subset \mathcal{O} : \gamma(\overset{0}{\nabla}\underline{\underline{R}}) = \gamma(\overset{0}{\nabla}\underline{\underline{R}} \cdot \underline{O}) = 0 \, . \tag{2}$$

Substituting $\overset{0}{\nabla}\underline{\underline{R}}$ by its polar representation $\overset{0}{\nabla}\underline{\underline{R}} = \underline{U} \cdot \underline{O}^{\times}$ and assuming that $\underline{O} = \underline{O}^{\times T}$ we arrive at the relationship

$$\gamma(\overset{0}{\nabla}\underline{\underline{R}}) = \gamma(\underline{U} \cdot \underline{O}^{\times} \cdot \underline{O}^{\times T}) = \gamma(\underline{U}) = 0 \, , \tag{3}$$

which represents a solution to the functional equation (2). Retaining γ from (1), we obtain

$$\gamma(\overset{0}{\nabla}\underline{\underline{R}}) = \gamma(\underline{U}) = \gamma(\underline{G}^{1/2}) = \gamma(\underline{G}) \tag{4}$$

for all $\overset{0}{\nabla}\underline{\underline{R}}$. Hence, recalling the definition (II.2.7) of the derivative of a scalar by a tensor argument and referring to (II.3.5), we now have

$$\delta\gamma = \gamma(\overset{0}{\nabla}\underline{\underline{R}} + \delta\overset{0}{\nabla}\underline{\underline{R}}) - \gamma(\overset{0}{\nabla}\underline{\underline{R}}) = \gamma_{\overset{0}{\nabla}\underline{\underline{R}}} \cdot\cdot\, \delta\overset{0}{\nabla}\underline{\underline{R}}^{T} = 2\gamma_{\underline{G}} \cdot \overset{0}{\nabla}\underline{\underline{R}} \cdot\cdot\, \delta\overset{0}{\nabla}\underline{\underline{R}}^{T} = 0 \, . \tag{5}$$

2. From (2.7.5), the specific virtual work $\delta' a_{(e)}$ done by the external forces, which is equal to the variation of the specific strain energy Э, has been represented by a contraction

$$\delta' a_{(e)} = \delta\text{Э} = \text{Э}_{\overset{0}{\nabla}\underline{\underline{R}}} \cdot\cdot\, \delta\overset{0}{\nabla}\underline{\underline{R}}^{T} = \underline{P} \cdot\cdot\, \delta\overset{0}{\nabla}\underline{\underline{R}}^{T} = 2\text{Э}_{\underline{G}} \cdot \overset{0}{\nabla}\underline{\underline{R}} \cdot\cdot\, \delta\overset{0}{\nabla}\underline{\underline{R}}^{T} \tag{6}$$

making use of the fact that the variations $\delta\overset{0}{\nabla}\underline{\underline{R}}^{T}$ are arbitrary, we deduce from this equation a representation of the Piola tensor in the form of a derivative of $\text{Э}(\overset{0}{\nabla}\underline{\underline{R}})$ with respect to $\overset{0}{\nabla}\underline{\underline{R}}$. We now have to take into consideration that the variations $\delta\overset{0}{\nabla}\underline{\underline{R}}^{T}$ are constrained by condition (5). The introduction of the Lagrangian multiplier λ enables us to write (6) in the form

$$\delta\text{Э} = 2(\text{Э}_{\underline{G}} + \lambda\gamma_{\underline{G}}) \cdot \overset{0}{\nabla}\underline{\underline{R}} \cdot\cdot\, \delta\overset{0}{\nabla}\underline{\underline{R}}^{T} = \underline{P} \cdot\cdot\, \delta\overset{0}{\nabla}\underline{\underline{R}}^{T} \tag{7}$$

and again to regard the variations $\delta\overset{0}{\nabla}\underline{\underline{R}}^{T}$ as independent as a result of an appropriate choice of λ. This makes it possible to represent the Piola tensor for a constrained material in the form

$$\underline{P} = 2(\text{Э}_{\underline{G}} + \lambda\gamma_{\underline{G}}) \cdot \overset{0}{\nabla}\underline{\underline{R}} = \underline{P}_{E} + \underline{P}_{S}\,, \tag{8}$$

where the function Э should be prescribed with due regard to equation (4) of constraint. As in the classification of forces in general mechanics into applied (active) forces and reactions associated with constraints (reactive forces), the tensor is represented here as the sum of two terms, the stress $\underline{P}_{E}$ which is "determined" by Э, and the reactive stress $\underline{P}_{S}$

$$\underline{P}_{E} = 2\text{Э}_{\underline{G}} \cdot \overset{0}{\nabla}\underline{\underline{R}} = \text{Э}_{\overset{0}{\nabla}\underline{\underline{R}}}\,, \qquad \underline{P}_{S} = 2\lambda\gamma_{\underline{G}} \cdot \overset{0}{\nabla}\underline{\underline{R}} = \lambda\gamma_{\overset{0}{\nabla}\underline{\underline{R}}}\,. \tag{9}$$

In the same way the Cauchy stress tensor consists of a "determined" stress $\underline{T}_{E}$ and the "constraint stress" $\underline{T}_{S}$

$$\underline{T} = \underline{T}_{E} + \underline{T}_{S}\,, \qquad \underline{T}_{E} = \sqrt{\frac{g}{G}}\, \overset{0}{\nabla}\underline{\underline{R}}^{T} \cdot \underline{P}_{E} = 2\sqrt{\frac{g}{G}}\, \overset{0}{\nabla}\underline{\underline{R}}^{T} \cdot \text{Э}_{\underline{G}} \cdot \overset{0}{\nabla}\underline{\underline{R}}\,,$$

$$\underline{T}_{S} = 2\lambda\sqrt{\frac{g}{G}}\, \overset{0}{\nabla}\underline{\underline{R}}^{T} \cdot \gamma_{\underline{G}} \cdot \overset{0}{\nabla}\underline{\underline{R}}\,. \tag{10}$$

Generalization to the case of several constraints follows immediately

$$\gamma_{i}(\underline{G}) = 0 \qquad (i = 1,2,\ldots,s)\,. \tag{11}$$

Introducing the Lagrangian multipliers $\lambda_1,\lambda_2,\ldots,\lambda_3$, we obtain

$$\underline{P}_S = 2\sum_{i=1}^{s}\lambda_i\gamma_i(\underline{G})_{\underline{G}}\cdot\overset{0}{\underline{\nabla R}}\,,\qquad \underline{T}_S = 2\sqrt{\frac{g}{G}}\,\overset{0}{\underline{\nabla R}}^{T}\cdot\sum_{i=1}^{s}\lambda_i\gamma_i(\underline{G})_{\underline{G}}\cdot\overset{0}{\underline{\nabla R}}\,. \tag{12}$$

The constraints (11) are included in the equations to be analysed, this compensates for the introduction of the unknown functions $\lambda_1,\lambda_2,\ldots,\lambda_s$.

3. The specific magnitude of the constraints stresses is determined by a double contraction of the stress tensor with the rate of strain tensor $\underline{D}$; by (2.7.12), (1.10.10) and (1.13.10) this is

$$\underline{T}_S\cdot\cdot\,\underline{D} = 2\sqrt{\frac{g}{G}}\,\lambda\overset{0}{\underline{\nabla R}}^{T}\cdot\gamma_{\underline{G}}\cdot\overset{0}{\underline{\nabla R}}\cdot\cdot\,\underline{D} =$$

$$= 2\sqrt{\frac{g}{G}}\,\lambda\gamma_{\underline{G}}\cdot\cdot\,\overset{0}{\underline{\nabla R}}\cdot\underline{D}\cdot\overset{0}{\underline{\nabla R}}^{T} = \sqrt{\frac{g}{G}}\,\lambda\gamma_{\underline{G}}\cdot\cdot\,\dot{\underline{G}}\,.$$

We thus have

$$\underline{T}_S\cdot\cdot\,\underline{D} = \sqrt{\frac{g}{G}}\,\lambda\gamma_{\underline{G}}\cdot\cdot\,\dot{\underline{G}} = \sqrt{\frac{g}{G}}\,\lambda\dot{\gamma}(\underline{G}) = 0\,, \tag{13}$$

since $\dot{\gamma}(\underline{G}) = 0$ - the constraint equation must hold for all values of t. The specific magnitude of the constraints stresses turned out to be zero. In this way the concept of ideal constraints known from general mechanics is extended to continuum mechanics. The inverse statement is also proved since if Equation (13) holds for all motions permitted by the constraints (for all $\underline{D}$) then the tensor $\underline{T}_S$ has the structure (10). In fact,

$$\dot{\underline{G}} = 2\overset{0}{\underline{\nabla R}}\cdot\underline{D}\cdot\overset{0}{\underline{\nabla R}}^{T}\,,\qquad 2\underline{D} = \underline{\nabla r}\cdot\dot{\underline{G}}\cdot\underline{\nabla r}^{T}\,,$$

$$\underline{T}_S\cdot\cdot\,\underline{D} = \frac{1}{2}\,\underline{T}_S\cdot\cdot\,\underline{\nabla r}\cdot\dot{\underline{G}}\cdot\underline{\nabla r}^{T} = \frac{1}{2}\,\underline{\nabla r}^{T}\cdot\underline{T}_S\cdot\underline{\nabla r}\cdot\cdot\,\dot{\underline{G}}\,.$$

Moreover, from the constrain equation

$$\gamma(\underline{G}) = 0\,,\qquad \gamma_{\underline{G}}\cdot\cdot\,\dot{\underline{G}} = 0$$

and introducing the Lagrangian multiplier $-\lambda\sqrt{\frac{g}{G}}$, we obtain

$$\left(\underline{\nabla r}^{T}\cdot\underline{T}_S\cdot\underline{\nabla r} - 2\lambda\sqrt{\frac{g}{G}}\,\gamma_{\underline{G}}\right)\cdot\cdot\,\dot{\underline{G}} = 0\,,\qquad \underline{\nabla r}^{T}\cdot\underline{T}_S\cdot\underline{\nabla r} = 2\lambda\sqrt{\frac{g}{G}}\,\gamma_{\underline{G}}\,,$$

which again yields (10). Generalization of Equation (13) to continuum mechanics allows us to obtain the stress tensor without reference to the notion of the specific strain energy. In other words, it is admissible to consider an "elastic", rather than "hyperelastic" material.

4. As an example of superposition of constraints we will consider a material which contains a quite densely packed system of parallel inextensible fibres. By (1.4.2), the equation (4) of constraint may be represented as

$$\gamma(\underline{G}) = \overset{0}{\underline{e}} \cdot \underline{G} \cdot \overset{0}{\underline{e}} - 1 = 0 \,, \tag{14}$$

where $\overset{0}{\underline{e}}$ is a unit vector specifying direction of the fibres in the reference configuration. The material itself ("the matrix") is isotropic. It is easy to see that the vector $\overset{0}{\underline{e}}$ determines one of the principal directions of the strain measure $\underline{G}$ and that the eigenvalue corresponding to this direction is unity. This follows from the definition (I.9.1) of the principal directions of a tensor

$$\underline{G} \cdot \underline{e} = \lambda \underline{e} \,, \qquad \underline{e} \cdot \underline{G} \cdot \underline{e} = \lambda$$

and by (14) this relationship is satisfied by $\underline{e} = \overset{0}{\underline{e}}$ and $\lambda = 1$. In view of (I.9.7) and (I.9.8) we get

$$P_3(1) = -1 + I_1(\underline{G}) - I_2(\underline{G}) + I_3(\underline{G}) = 0 \,, \qquad I_2(\underline{G}) = I_1(\underline{G}) + I_3(\underline{G}) - 1 \tag{15}$$

and the specific strain energy $\ni$ may be thought of as a function of, say, I_1 and I_3

$$\ni(I_1, I_2, I_3) = \ni(I_1, I_1 + I_3 - 1, I_3) \,.$$

By (4.3.4), the determined stress turns out to be equal to

$$\underline{T}_E = 2\sqrt{\frac{g}{G}}\,(\psi_0 \underline{E} + \psi_1 \underline{F} + \psi_2 \underline{F}^2)_{I_2 = I_1 + I_3 - 1} \,,$$

where ψ_Γ is calculated by Equations (4.3.5) in which I_2 should be substituted for its value given by (15).

By (II.3.8) and (10) the tensor of the constraints stresses is represented by the expression

$$\gamma_{\underline{G}} = \overset{00}{\underline{\underline{ee}}} \,, \qquad \underline{T}_S = 2\sqrt{\frac{g}{G}}\,\lambda \overset{0}{\nabla}\underline{R}^T \cdot \overset{00}{\underline{\underline{ee}}} \cdot \overset{0}{\nabla}\underline{R} = 2\lambda\sqrt{\frac{g}{G}}\,\underline{\underline{ee}} \,,$$

where $\underline{e}$ denotes a unit vector prescribing the direction of the fibres in the actual configuration - see (1.4.17) and (14).

The state equation takes the form

$$\underline{T} = 2\sqrt{\frac{g}{G}}\,(\lambda\underline{\underline{ee}} + \psi_0\underline{E} + \psi_1\underline{F} + \psi_2\underline{F}^2)_{I_2=I_1+I_3-1} \,,$$

and the constraint equations can also be represented as

$$\overset{0}{\underline{e}}\cdot\underline{G}\cdot\overset{0}{\underline{e}} = \underline{e}\cdot\underline{\underline{\nabla r}}\cdot\underline{G}\cdot\underline{\underline{\nabla r}}^T\cdot\underline{e} = \underline{e}\cdot\underline{e} = 1\,, \qquad G_{sk}e^s e^k = 1\,.$$

Here e^k denotes the contravariant components of $\underline{e}$ in the vector basis of the actual configuration.

It has of course been implied here that the constraint (14) is restraining, in other words, the relationships given here are inapplicable if the material of the matrix is compressed in the direction of $\overset{0}{\underline{e}}$.

For a material which is inextensible in the three mutually orthogonal directions $\underline{e}_1$, $\underline{e}_2$, $\underline{e}_3$, all three eigenvalues of $\underline{G}$ equal 1, and $\underline{G} = \underline{E}$, i.e. strain is absent, and by (1.4.23) the "material" is an absolutely rigid body. The tensor of constraint stresses equals

$$\underline{T}_S = \lambda_1\underline{e}_1\underline{e}_1 + \lambda_2\underline{e}_2\underline{e}_2 + \lambda_3\underline{e}_3\underline{e}_3\,,$$

and therefore the stress tensor remains completely undetermined.

2. AN INCOMPRESSIBLE ELASTIC MATERIAL

This sort of material with "superposed constraints" is of major interest. Rubber may be treated to a good approximation as incompressible. It is the most important research material in the nonlinear theory of elasticity - "the most elastic of the elastic materials defying investigation by means of the linear theory of elasticity"*) (Truesdell).

Any elementary volume of an incompressible material is preserved in the course of deformation. By (1.8.1) and (1.4), the constrain equation has the form

$$\gamma(\underline{G}) = I_3(\underline{G}) - 1 = 0 \tag{1}$$

and in view of (II.3.2) and (1.12)

$$\gamma_{\underline{G}} = I_3(\underline{G})\underline{G}^{-1} = \underline{G}^{-1}\,, \qquad \underline{T}_S = 2\sqrt{\frac{g}{G}}\,\lambda\overset{0}{\underline{\underline{\nabla R}}}{}^T\cdot\underline{G}^{-1}\cdot\overset{0}{\underline{\underline{\nabla R}}} = 2\sqrt{\frac{g}{G}}\,\lambda\underline{E}\,. \tag{2}$$

In the representation of the specific strain energy of an incompressible material $I_3 = 1$, and for an isotropic material

*) Inverse translation from the Russian (Note by the translator).

$$\ni(I_1,I_2,I_3) = \ni(I_1,I_2) \ . \qquad (3)$$

Putting $2\sqrt{\frac{g}{G}}\,\lambda = -p$, we arrive at the state equation

$$\underline{T} = \underline{T}_S + \underline{T}_E = -p\underline{E} + 2\psi_1\underline{F} + 2\psi_2\underline{F}^2 \ , \qquad (4)$$

where, in view of (4.3.5),

$$\psi_1 = \left(\frac{\partial}{\partial I_1} + I_1\frac{\partial}{\partial I_2}\right)\ni(I_1,I_2) \ , \qquad \psi_2 = -\frac{\partial}{\partial I_2}\ni(I_1,I_2) \ . \qquad (5)$$

The value of p is of course not known in advance as a function of the material coordinates (and time in dynamic problems) determined from the equations of equilibrium (motion) with the additional condition of incompressibility. This situation is similar to that common for the hydrodynamics of an incompressible fluid where, as for an incompressible elastic medium, p denotes the pressure. The equation of continuity (1.14.13) plays the role of a constraint.

Denoting by v_1, v_2, v_3 the principal stretches, we represent the invariants of the strain measures $\underline{G}$ and $\underline{F}$ by the formulas

$$I_1 = v_1^2 + v_2^2 + v_3^2 \ , \qquad I_2 = v_1^2v_2^2 + v_2^2v_3^2 + v_3^2v_1^2 = v_1^{-2} + v_2^{-2} + v_3^{-2} \ , \qquad (6)$$

because for an incompressible material

$$I_3 = v_1^2v_2^2v_3^2 = 1 \ . \qquad (7)$$

Representation (4) of the stress tensor may also be written as follows

$$\underline{T} = -p\underline{E} + 2\frac{\partial\ni}{\partial I_1}\underline{F} - 2\frac{\partial\ni}{\partial I_2}\underline{F}^{-1} \qquad (8)$$

where we have used the Cayley-Hamilton transformation

$$\underline{F}^2 = I_1\underline{F} - I_2\underline{E} + \underline{F}^{-1}$$

and Equations (I.9.22). The sum $p + I_2\psi_2$ is again denoted by p.

From the constraint equation represented in the form $v_1v_2v_3 = 1$, there follows the relationship

$$\frac{\delta v_1}{v_1} + \frac{\delta v_2}{v_2} + \frac{\delta v_3}{v_3} = 0 \ . \qquad (9)$$

Making use of this formula and recalling the definition of the principal forces t_k we get

$$\delta \ni = \sum_{k=1}^{3} t_k \delta v_k = \sum_{k=1}^{3} \left(\frac{\partial \ni}{\partial v_k} - \frac{p}{v_k} \right) \delta v_k = 0 \ .$$

The principal forces and stresses σ_k are now represented by the formulas

$$t_k = -\frac{p}{v_k} + \frac{\partial \ni}{\partial v_k} , \qquad \sigma_k = -p + v_k \frac{\partial \ni}{\partial v_k} \ . \tag{10}$$

The equations of equilibrium in the volume and on its surface now take the form

$$\underline{\nabla} \cdot \underline{T} = \underline{\nabla} \cdot \underline{T}_S + \underline{\nabla} \cdot \underline{T}_E + \rho \underline{k} = -\underline{\nabla} p + \underline{\nabla} \cdot \underline{T}_E + \rho \underline{k} = \underline{0} \ ,$$

$$\underline{f} = -p\underline{N} + \underline{N} \cdot \underline{T}_E \ . \tag{11}$$

The variations $\delta \underline{R}$ of the position vector are not independent since by (1.5) and (2) they are connected by the relationship

$$\delta I_3 = 2\underline{G}^{-1} \cdot \overset{0}{\underline{\underline{\nabla R}}} \cdot\cdot \, \delta \overset{0}{\underline{\underline{\nabla R}}}^T = 2\underline{\underline{\nabla r}}^T \cdot\cdot \, \delta \overset{0}{\underline{\underline{\nabla R}}}^T = 2\underline{\underline{\nabla r}}^T \cdot\cdot \, (\underline{\nabla} \delta \underline{R})^T \cdot \overset{0}{\underline{\underline{\nabla R}}}^T = 2\underline{E} \cdot\cdot \, \underline{\nabla} \delta \underline{R}^T = 0 \tag{12}$$

where we have used the transformations (2.8.4) and the rules for contraction.

The derivation of the principle of stationary potential energy of a system is completely independent of the properties of the material. For this reason, this principle is also applicable in the form (4.16.8) to an incompressible medium. We have only to set $\rho = \rho_0$ and $\sqrt{g/G} = 1$. This principle now reduces to the relationship

$$\delta W_1 = \delta \left[\iiint\limits_V (\ni - \rho_0 \underline{k} \cdot \underline{R}) dV - \iint\limits_{O_1} \underline{f} \cdot \underline{R} \, dO \right] = 0 \ . \tag{13}$$

Here, however, we should take into account constraint (1) during variation. The functional subject to variation is represented in the form

$$W_1^{\times} = W_1 - \iiint\limits_V \frac{1}{2} p (I_3 - 1) dV \ , \tag{14}$$

where the variations $\delta \underline{R}$ are subjected to constraint (12), and by (III.3.10) we have

$$\delta \iiint_V \frac{1}{2} p(I_3 - 1)\,dV = \iiint_V p\underline{E} \cdot\cdot \underline{\nabla}\delta\underline{R}^T\,dV = \iiint_V [\underline{\nabla} \cdot (p\underline{E} \cdot \delta\underline{R}) - \underline{\nabla}p \cdot \delta\underline{R}]\,dV =$$

$$= \iint_{O_1} \underline{N}p \cdot \delta\underline{R}\,dO - \iiint_V \underline{\nabla}p \cdot \delta\underline{R}\,dV \tag{15}$$

(note that $\delta\underline{R} = \underline{0}$ on that part O_2 of the surface where the displacement is prescribed). For this reason

$$\delta W_1^{\times} = \iiint_V \delta\ni\,dV - \iiint_V (\rho_0\underline{k} - \underline{\nabla}p) \cdot \delta\underline{R}\,dV - \iint_{O_1} (\underline{N}p + \underline{f}) \cdot \delta\underline{R}\,dO = 0\ . \tag{16}$$

At the same time, in view of (II.3.6) and (4.3.4) we may write

$$\delta\ni = \ni_{\underline{\underline{\overset{0}{\nabla}R}}} \cdot\cdot\ \delta\underline{\underline{\overset{0}{\nabla}R}}^T = 2\underline{\underline{\overset{0}{\nabla}R}} \cdot \ni_{\underline{F}} \cdot\cdot\ \underline{\nabla}\delta\underline{R}^T \cdot \underline{\underline{\overset{0}{\nabla}R}}^T = 2\underline{\underline{\overset{0}{\nabla}R}}^T \cdot \underline{\underline{\overset{0}{\nabla}R}} \cdot \ni_{\underline{F}} \cdot\cdot\ \underline{\nabla}\delta\underline{R}^T =$$

$$= 2\underline{F} \cdot \ni_{\underline{F}} \cdot\cdot\ \underline{\nabla}\delta\underline{R}^T = \underline{\underline{T}}_E \cdot\cdot\ \underline{\nabla}\delta\underline{R}^T = \underline{\nabla} \cdot (\underline{\underline{T}}_E \cdot \delta\underline{R}) - (\underline{\nabla} \cdot \underline{\underline{T}}_E) \cdot \delta\underline{R}\ ,$$

$$\iiint_V \delta\ni\,dV = \iint_{O_1} \underline{N} \cdot \underline{\underline{T}}_E \cdot \delta\underline{R}\,dO - \iiint_V \underline{\nabla} \cdot \underline{\underline{T}}_E \cdot \delta\underline{R}\,dV$$

and substitution into (16) leads to the relationship

$$- \iiint_V (\underline{\nabla} \cdot \underline{\underline{T}}_E + \rho_0\underline{k} - \underline{\nabla}p)) \cdot \delta\underline{R}\,dV + \iint_{O_1} (\underline{N} \cdot \underline{\underline{T}}_E - \underline{N}p - \underline{f}) \cdot \delta\underline{R}\,dO = 0\ . \tag{17}$$

As expected, the equations of equilibrium, (11), follow.

In the reference configuration basis, we have by virtue of (15)

$$\delta \iiint_v \frac{1}{2} p(I_3 - 1)\,dv = \iint_{O_1} p\underline{n} \cdot \underline{\underline{\nabla r}}^T \cdot \delta\underline{R}\,do - \iiint_v (\underline{\underline{\nabla r}} \cdot \overset{0}{\underline{\nabla}}p) \cdot \delta\underline{R}\,dv$$

and at the same time

$$\iiint_v \delta\ni\,dv = \iiint_v \ni_{\underline{\underline{\overset{0}{\nabla}R}}} \cdot\cdot\ \delta\underline{\underline{\overset{0}{\nabla}R}}^T dv = \iiint_v \underline{\underline{P}}_E \cdot\cdot\ \delta\underline{\underline{\overset{0}{\nabla}R}}^T dv =$$

$$= \iiint_v [\overset{0}{\underline{\nabla}} \cdot (\underline{\underline{P}}_E \cdot \delta\underline{R}) - (\overset{0}{\underline{\nabla}} \cdot \underline{\underline{P}}_E) \cdot \delta\underline{R}]\,dv =$$

$$= \iint_{O_1} \underline{n} \cdot \underline{\underline{P}}_E \cdot \delta\underline{R}\,do - \iiint_v (\underline{\underline{\overset{0}{\nabla}P}}_E) \cdot \delta\underline{R}\,dv\ .$$

For this reason, we have in the case of "dead" loading ($\underline{f}\,dO = \underline{f}^0\,do$)

$$-\iiint\limits_{v} (\overset{0}{\underline{\nabla}} \cdot \underline{P}_E - \overset{0}{\underline{\nabla}} p + \rho_0 \underline{k}) \cdot \delta \underline{R}\, dv + \iint\limits_{O_1} (\underline{n} \cdot \underline{P}_E - p\underline{n} \cdot \underline{\nabla r}^{\mathrm{T}} - \underline{f}^0) \cdot \delta \underline{R}\, do = 0 \qquad (18)$$

and for the reference configuration, the equations of equilibrium in the volume and on its surface reduce to the form

$$\overset{0}{\underline{\nabla}} \cdot \underline{P}_E - \overset{0}{\underline{\nabla}} p + \rho_0 \underline{k} = \underline{0}\,, \qquad \underline{f}^0 = \underline{n} \cdot \underline{P}_E - p\underline{n} \cdot \underline{\nabla r}^{\mathrm{T}}\,. \qquad (19)$$

We may of course obtain these equations by direct transformation of Equations (11). In fact, the Piola tensor for an incompressible medium is by (2.6.2) represented as

$$\underline{P} = \underline{\nabla r}^{\mathrm{T}} \cdot \underline{T} = \underline{P}_E - \underline{\nabla r}^{\mathrm{T}} p \qquad (20)$$

and referring to the Piola identity (1.9.4) we get

$$\overset{0}{\underline{\nabla}} \cdot \underline{P} = \overset{0}{\underline{\nabla}} \cdot \underline{P}_E - \overset{0}{\underline{\nabla}} \cdot \underline{\nabla r}^{\mathrm{T}} p = \overset{0}{\underline{\nabla}} \cdot \underline{P}_E - \underline{r}^s \cdot \underline{\nabla r}^{\mathrm{T}} \frac{\partial p}{\partial q^s} = \overset{0}{\underline{\nabla}} \underline{P}_E - \overset{0}{\underline{\nabla}} p \cdot \underline{\nabla r}^{\mathrm{T}}\,,$$

which is the required result.

The expression for the Hamilton-Ostrogradskii principle (Chapter 4, Par. 19) in the case of an incompressible medium should be supplemented by the term

$$-\int_{t_0}^{t_1} dt \iiint\limits_{V} \frac{1}{2} p(I_3 - 1)\, dV\,,$$

which is furher transformed by (15). Using the derivation of Chapter 4, Par. 19, we arrive at the following form of the principle

$$\int_{t_0}^{t_1} dt \left[\delta L - \iiint\limits_{V} \underline{\nabla} p \cdot \delta \underline{R}\, dV + \iint\limits_{O_1} p\underline{N} \cdot \delta \underline{R}\, dO + \delta' A_{(e)} \right] = 0\,. \qquad (21)$$

Here L denotes the Langrangian defined as the difference between the kinetic and potential energies of a system

$$L = K - \Pi\,, \qquad K = \frac{1}{2} \iiint\limits_{V} \rho \underline{v} \cdot \underline{v}\, dV\,,$$

$$\Pi = \iiint\limits_{v} [\ni(I_1, I_2) - \rho_0 \underline{k} \cdot \underline{R}]\, dv - \iint\limits_{O_1} \underline{f}^0 \cdot \underline{R}\, do\,. \qquad (22)$$

In (21) the incremental work of non-potential surface tractions is represented by the term $\delta' A_{(e)}$

$$\delta' A_{(e)} = \iint\limits_{O_1} \underline{f} \cdot \delta \underline{R}\, dO \, . \tag{23}$$

Following (15), we may represent (21) also in the form

$$\int_{t_0}^{t_1} \left[\delta L + \iiint\limits_{V} p\underline{E} \cdot\cdot \underline{\nabla} \delta \underline{R}\, dV + \delta' A_{(e)} \right] = 0 \, . \tag{24}$$

3. SECOND ORDER EFFECTS IN AN INCOMPRESSIBLE ELASTIC BODY

If the quadratic terms with respect to $|\overset{0}{\underline{\nabla}}\underline{u}|$ are the only taken considered, then constraint (2.1) may by (6.10.3) be expressed by the relationship

$$\vartheta + \vartheta^2 + \overset{0}{\underline{\omega}} \cdot \overset{0}{\underline{\omega}} - \frac{1}{2} I_1(\overset{0}{\underline{\varepsilon}}{}^2) = 0 \tag{1}$$

and therefore

$$I_1(\underline{F}) - 3 = I_2(\underline{F}) - 3 = 2 I_1(\overset{0}{\underline{\varepsilon}}{}^2) + 2\vartheta^2 \, . \tag{2}$$

The invariants I_1, I_2 and consequently the coefficients ψ_Γ of the state equation differ from their values in the reference configuration by the second order terms.

After the elimination of the strain measures with the aid of the formulas of Chapter 6, Par. 10, the state equation (2.4) is reduced to the form

$$\underline{T} = -p\underline{E} + 4(\psi_1 + 2\psi_2)\overset{0}{\underline{\varepsilon}}(\underline{u}) + 4(\psi_1 + 2\psi_2)\overset{0}{\underline{\nabla}}\underline{u}^T \cdot \overset{0}{\underline{\nabla}}\underline{u} + 8\psi_2 \overset{0}{\underline{\varepsilon}}{}^2(\underline{u}) =$$

$$= -p\underline{E} + 4(\psi_1 + 2\psi_2)^0 \overset{0}{\underline{\varepsilon}}(\underline{u}) + 4(\psi_1 + 2\psi_2)^0 \overset{0}{\underline{\nabla}}\underline{u}^T \cdot \overset{0}{\underline{\nabla}}\underline{u} + 8\overset{0}{\psi}_2 \overset{0}{\underline{\varepsilon}}{}^2(\underline{u}) \, ,$$

because to the degree of accuracy required, there is no need to take into account the difference of the coefficients ψ_Γ from their values in the reference configuration. Turning to (6.10.6), we now have

$$(\psi_1 + 2\psi_2)^0 = \left(\frac{\partial \ni}{\partial I_1} + \frac{\partial \ni}{\partial I_2} \right)^0 = \frac{1}{2}\mu \, , \qquad \overset{0}{\psi}_2 = - \frac{\partial \ni^0}{\partial I_2} \tag{3}$$

and the state equation, taking all terms into account, reduces to the form

$$\underline{T} = -p\underline{E} + 2\mu \overset{0}{\underline{\varepsilon}}(\underline{u}) + 2\mu \overset{0}{\underline{\nabla}}\underline{u} \cdot \overset{0}{\underline{\nabla}}\underline{u}^T - 8\left(\frac{\partial \ni}{\partial I_2} \right)^0 \overset{0}{\underline{\varepsilon}}{}^2(\underline{u}) \, . \tag{4}$$

In the linear approximation (when the pressure equals p_0) we have

$$\underline{T} = \overset{0}{\underline{T}}(\underline{u}) = -p_0\underline{E} + 2\mu\overset{0}{\underline{\varepsilon}}(\underline{u}) \tag{5}$$

and this relationship should be taken into consideration along with the constraint

$$\vartheta = I_1(\overset{0}{\underline{\varepsilon}}) = \overset{0}{\underline{\nabla}}\cdot\underline{u} = 0 . \tag{6}$$

The equations of equilibrium in the volume and on its surface may by (2.11) be rewritten as

$$\mu\overset{0}{\underline{\nabla}}{}^2\underline{u} + \rho_0\underline{k} = \overset{0}{\underline{\nabla}}p , \qquad \underline{n}\cdot\overset{0}{\underline{T}}(\underline{u}) = -\underline{n}p_0 + 2\mu\underline{n}\cdot\overset{0}{\underline{\varepsilon}}(\underline{u}) = \underline{f} , \tag{7}$$

because

$$\overset{0}{\underline{\nabla}}\cdot\overset{0}{\underline{\varepsilon}} = \frac{1}{2}\overset{0}{\underline{\nabla}}\cdot(\overset{0}{\underline{\nabla}}\underline{u} + \overset{0}{\underline{\nabla}}\underline{u}^T) = \frac{1}{2}(\overset{0}{\underline{\nabla}}{}^2\underline{u} + \overset{00}{\underline{\nabla\nabla}}\cdot\underline{u}) = \frac{1}{2}\overset{0}{\underline{\nabla}}{}^2(\underline{u}) .$$

The bulk equation of motion and the equation of constraint again yield the linearized Navier-Stokes' equations for the slow time-independent motion of a viscous fluid, the difference arising in the boundary conditions.

By analogy with (6.11.1) we introduce the correcting vector $\underline{w}$ and represent, to the required accuracy, the stress tensor in the form

$$\underline{T} = \overset{0}{\underline{T}}(\underline{v}) + \overset{0}{\underline{T}}(\underline{w}) + \underline{T}'(\underline{v}) ,$$

$$\underline{T}'(\underline{v}) = 2\mu\overset{0}{\underline{\nabla}}\underline{v}^T\cdot\overset{0}{\underline{\nabla}}\underline{v} - 8\left(\frac{\partial\ni}{\partial I_2}\right)^{0}\overset{0}{\underline{\varepsilon}}{}^2(\underline{v}) , \tag{8}$$

where we have used the notation

$$\overset{0}{\underline{T}}(\underline{w}) = -p_1\underline{E} + 2\mu\overset{0}{\underline{\varepsilon}}(\underline{w}) , \qquad p = p_0 + p_1 . \tag{9}$$

It must also be noted that

$$\underline{N}\,dO = \underline{n}\cdot\underline{\nabla r}^T do = \underline{n}\cdot(\underline{E} - \underline{\nabla u}^T)do \approx \underline{n}\cdot(\underline{E} - \underline{\nabla u}^T)do ,$$

because in view of (1.3.9), $\underline{\nabla u}^T$ differs from $\overset{0}{\underline{\nabla}}\underline{u}^T$ by the second order term

$$\underline{\nabla u}^T = \overset{0}{\underline{\nabla}}\underline{u}^T\cdot\underline{\nabla r}^T = \overset{0}{\underline{\nabla}}\underline{u}^T\cdot(\underline{E} - \underline{\nabla u}^T) \approx \overset{0}{\underline{\nabla}}\underline{u}^T .$$

The equation of equilibrium on the boundary surface may now be written as follows

$$\underline{N}\cdot\underline{T}\,dO = (\underline{n}-\underline{n}\cdot\overset{0}{\underline{\underline{\nabla v}}}{}^{T})\cdot\overset{0}{\underline{T}}(\underline{v})\,do + \underline{n}\cdot\overset{0}{\underline{T}}(\underline{w})\,do + \underline{n}\cdot\underline{T}'(\underline{v})\,do =$$

$$= [\underline{n}\cdot\overset{0}{\underline{T}}(\underline{v}) + \underline{n}\cdot\overset{0}{\underline{T}}(\underline{w}) + \underline{n}\cdot\underline{T}'(\underline{v}) - \underline{n}\cdot\overset{0}{\underline{\underline{\nabla v}}}{}^{T}\cdot\overset{0}{\underline{T}}(\underline{v})]\,do = \underline{f}\,dO$$

and the boundary condition for the correcting vector $\underline{w}$ is, by (7), represented by the equation

$$\underline{n}\cdot\overset{0}{\underline{T}}(\underline{w}) = -\underline{n}p_1 + 2\mu\underline{n}\cdot\overset{0}{\underline{\varepsilon}}(\underline{w}) = \underline{f}_x\ ,$$

$$\underline{f}_x = -\underline{n}\cdot(-\overset{0}{\underline{\underline{\nabla v}}}{}^{T}\cdot\overset{0}{\underline{T}}(\underline{v}) + \underline{T}'(\underline{v}))\ . \tag{10}$$

Therefore

$$\iint\limits_{O}(\underline{n}\cdot\overset{0}{\underline{T}}(\underline{w}) - \underline{f}_x)\,do = \iiint\limits_{V}[\overset{0}{\underline{\nabla}}\cdot\overset{0}{\underline{T}}(\underline{w}) + \overset{0}{\underline{\nabla}}\cdot(-\overset{0}{\underline{\underline{\nabla v}}}{}^{T}\cdot\overset{0}{\underline{T}}(\underline{v}) + \underline{T}'(\underline{v}))]\,dv = \underline{0}$$

and the equation of equilibrium in the volume reduces to

$$\overset{0}{\underline{\nabla}}\cdot\underline{T}(\underline{w}) + \underline{k}_x = \underline{0}\ ,\qquad \underline{k}_x = \overset{0}{\underline{\nabla}}\cdot[-\overset{0}{\underline{\underline{\nabla v}}}{}^{T}\cdot\overset{0}{\underline{T}}(\underline{v}) + \underline{T}'(\underline{v})]\ . \tag{11}$$

To the same approximation, the constraint equation is represented as follows

$$\vartheta(\underline{w}) = \underline{\nabla}\cdot\underline{w} = \frac{1}{2}I_1(\overset{0}{\underline{\varepsilon}}{}^2(\underline{v})) - \overset{0}{\underline{\omega}}(\underline{v})\cdot\overset{0}{\underline{\omega}}(\underline{v})\ , \tag{12}$$

since $\vartheta(\underline{v}) = 0$ by (6). For this reason

$$\overset{0}{\underline{\nabla}}\cdot\overset{0}{\underline{\varepsilon}}(\underline{w}) = \frac{1}{2}\overset{0}{\underline{\nabla}}\cdot(\overset{0}{\underline{\underline{\nabla w}}} + \overset{0}{\underline{\underline{\nabla w}}}{}^{T}) = \frac{1}{2}(\overset{0}{\underline{\nabla}}{}^2\underline{w} + \overset{00}{\underline{\underline{\nabla\nabla}}}\cdot\underline{w}) =$$

$$= \frac{1}{2}\overset{0}{\underline{\nabla}}{}^2\underline{w} + \frac{1}{2}\overset{0}{\underline{\nabla}}\left[\frac{1}{2}I_1(\overset{0}{\underline{\varepsilon}}{}^2(\underline{v})) - \overset{0}{\underline{\omega}}(\underline{v})\cdot\overset{0}{\underline{\omega}}(\underline{v})\right]$$

and the equation (11) of equilibrium in the volume takes the form

$$-\overset{0}{\underline{\nabla}}p_1 + \mu\overset{0}{\underline{\nabla}}{}^2\underline{w} + \mu\overset{0}{\underline{\nabla}}\left[\frac{1}{2}I_1(\overset{0}{\underline{\varepsilon}}{}^2(\underline{v})) - \overset{0}{\underline{\omega}}(\underline{v})\cdot\overset{0}{\underline{\omega}}(\underline{v})\right] + \underline{k}_x = \underline{0}\ . \tag{13}$$

The correcting vector $\underline{w}$ and the scalar p_1 are determined by this equation, Equation (12) and by boundary condition (10). The resultant vector $\underline{k}_x$ of the body forces and the vector $\underline{f}_x$ of the surface tractions are equal to zero, and for the existence of the vector $\underline{w}$ it is necessary that the force tensor

$$\underline{B} = \iiint\limits_{V}\underline{k}_x\underline{r}\,dv + \iint\limits_{O}\underline{f}_x\underline{r}\,do = \iiint\limits_{V}\underline{Q}^{T}\,dv\ ,\qquad \underline{Q} = -\overset{0}{\underline{\underline{\nabla v}}}{}^{T}\cdot\overset{0}{\underline{T}}(\underline{v}) + \underline{T}'(\underline{v})\ . \tag{14}$$

be symmetric.

Having observed that $\underline{T}' = \underline{T}'^{T}$ and that the tensors $\overset{0}{\underline{T}}$ and $\overset{0}{\underline{\varepsilon}}$ are coaxial, we arrive at the relationship

$$\underline{B} - \underline{B}^{T} = \iiint\limits_{v} (\overset{0}{\underline{\Omega}} \cdot \overset{0}{\underline{T}} + \overset{0}{\underline{T}} \cdot \overset{0}{\underline{\Omega}}) = \iiint\limits_{v} [\overset{0}{\underline{\omega}}(\underline{v}) \times \overset{0}{\underline{T}}(\underline{v}) + \overset{0}{\underline{T}}(\underline{v}) \times \overset{0}{\underline{\omega}}(\underline{v})] = \underline{0} \ .$$

Like Equation (6.11.9), this is reducible to the form

$$\iiint\limits_{v} [\overset{0}{\underline{T}}(\underline{v}) \cdot \overset{0}{\underline{\omega}}(\underline{v}) - \overset{0}{\underline{\omega}}(\underline{v}) I_1(\overset{0}{\underline{T}}(\underline{v}))]\, dv = \underline{0} \ . \tag{15}$$

Making use of the equality

$$\overset{0}{\underline{T}}(\underline{v}) \cdot \overset{0}{\underline{\omega}}(\underline{v}) = -p_0 \overset{0}{\underline{\omega}}(\underline{v}) + 2\mu \overset{0}{\underline{\varepsilon}}(\underline{v}) \cdot \overset{0}{\underline{\omega}}(\underline{v}) \ ,$$

$$I_1(\overset{0}{\underline{T}}(\underline{v})) = -3p_0 + 2\mu I_1(\overset{0}{\underline{\varepsilon}}(\underline{v})) = -3p_0 \ ,$$

we may also rewrite this condition as follows

$$\iiint\limits_{v} [\mu \overset{0}{\underline{\varepsilon}}(\underline{v}) \cdot \overset{0}{\underline{\omega}}(\underline{v}) + p_0 \overset{0}{\underline{\omega}}(\underline{v})]\, dv = \underline{0} \ . \tag{16}$$

4. THE SPECIFIC STRAIN ENERGY FOR AN INCOMPRESSIBLE ELASTIC BODY

The most widely used prescription of $\ni$ is that given below which is in the form of the Mooney potential and contains two constants

$$\ni = C_1(I_1(\underline{G}) - 3) + C_2(I_2(\underline{G}) - 3) =$$

$$= C_1(v_1^2 + v_2^2 + v_3^2 - 3) + C_2(v_1^{-2} + v_2^{-2} + v_3^{-2} - 3) \ , \tag{1}$$

From (3.3), this expression can also be represented in the form

$$\ni = \frac{1}{4} \mu [(1+\beta)(I_1 - 3) + (1-\beta)(I_2 - 3)] \qquad (C_1 + C_2 = \frac{1}{2}\mu) \ . \tag{2}$$

Here μ denotes the shear modulus for small deformations, and β is a constant to be determined. For a simplified version of Mooney's potential one may choose the "Neo-Hookean" potential

$$\ni = C_1(I_1 - 3) \qquad (\beta = 1) \tag{3}$$

which is also called the Treloar potential after the author who obtained it from an analysis of a model of rubber regarded as a system of long molecular interlinking chains.

The necessary and sufficient conditions for the -function to be positive are given by the inequalities

$$C_1 \geq 0 , \quad C_2 \geq 0 \qquad (\mu > 0, -1 \leq \beta \leq 1) . \tag{4}$$

In fact, the sum of three positive numbers v_1^2, v_2^2, v_3^2 whose product equals 1 has a minimum when these numbers are equal and, consequently, each of them equals 1. We thus have

$$v_1^2 + v_2^2 + v_3^2 - 3 \geq 0 , \qquad v_1^{-2} + v_2^{-2} + v_3^{-2} - 3 \geq 0 ,$$

which implies that Inequalities (4) are sufficient for $\ni$ to be positive. These inequalities are also necessary. In fact, at least one constant, say C_1, is positive. Assuming that $C_2 < 0$, we may choose three positive numers v_1^2, v_2^2, v_3^2 so as to make their product unity so that inequality

$$\frac{C_1}{|C_2|} \leq \frac{I_2 - 3}{I_1 - 3} = \frac{v_1^{-2} + v_2^{-2} + v_3^{-2}}{v_1^2 + v_2^2 + v_3^2} , \qquad \ni < 0 .$$

The stress tensors of Cauchy and Piola determined by the Mooney potential are given by

$$\underline{T} = -p\underline{E} + 2[(C_1 + I_1 C_2)\underline{F} - C_2 \underline{F}^2] ,$$

$$\underline{P} = \underline{\nabla r}^T \cdot \underline{T} = \{-p\underline{G}^{-1} + 2[(C_1 + I_1 C_2)\underline{E} - C_2 \underline{G}]\} \cdot \overset{0}{\underline{\nabla R}} . \tag{5}$$

The mechanical properties of numerous rubberlike materials are, of course, very sensitive to their chemical composition and processing and so cannot be described by a unique Mooney potential. There is a great number of alternative descriptions which claim to fit the experimental data, within certain ranges of deformation and temperature, much better.

An example of such a potential is that of Rivlin and Saunders (R.S. Rivlin, D.W. Saunders, 1951) who discovered in the course of a number of careful experiments the dependence of the ratio $\frac{\partial \ni}{\partial I_1} / \frac{\partial \ni}{\partial I_2}$ on I_2. This observation has prompted the correction of Mooney's formula using the more general relationship

$$\ni = C_1(I_1 - 3) + f(I_2 - 3) \qquad (f > 0) . \tag{6}$$

G.M. Bartenev and T.N. Khasanovich (1960) have proposed the potential

$$\ni = 2\mu(v_1 + v_2 + v_3 - 3) = 2\mu I_1(\underline{U}) , \quad \underline{U} = \underline{G}^{1/2} , \tag{7}$$

which is based upon a statistical analysis of the behaviour of molecular chains with the orientating influence of the stresses taken into account. A generalized version of this potential, confirmed by a vast program of experimental investigations, has been put forward by K.F. Chernykh and I.M. Shubina (1976):

$$\ni = \mu[(1+\beta)(v_1 + v_2 + v_3 - 3) + (1-\beta)(v_1^{-1} + v_2^{-1} + v_3^{-1} - 3)] =$$

$$= \mu[(1+\beta)I_1(\underline{U}) + (1-\beta)I_1(\underline{U}^{-1})] - 6\mu . \tag{8}$$

It should be pointed out that this form, which is simpler than those given by the Treloar and Mooney potentials, makes the use of these potentials for solution of specific problems even more complex. Referring to (II.3.5), (II.3.10) and (7) we have

$$\underline{P}_E = \ni_{\overset{0}{\nabla}\underline{R}} = 2\ni_{\underline{G}} \cdot \overset{0}{\nabla}\underline{R} = 4\mu I_1(\underline{U})_{\underline{G}} \cdot \overset{0}{\nabla}R = 2\mu\underline{U}^{-1} \cdot \overset{0}{\nabla}\underline{R} = 2\mu\underline{O}^{\times} \qquad (\overset{0}{\nabla}\underline{R} = \underline{U} \cdot \underline{O}^{\times}) ,$$

so that

$$\underline{T}_E = 2\mu\overset{0}{\nabla}\underline{R}^T \cdot \underline{P}_E = 2\mu\underline{V} = 2\mu\underline{F}^{1/2} , \quad \underline{T} = -p\underline{E} + 2\mu\underline{F}^{1/2} , \tag{9}$$

whereas for the Treloar potential $\underline{T}_E = \mu\underline{F}$ from (5). This simplifies the use of this potential so long as the representation of $\underline{F}$ in its principal axes is not known.

5. THE PLANE DEFORMATION OF AN INCOMPRESSIBLE MATERIAL

1. The position vectors $\underline{r}$ and $\underline{R}$ in the reference and actual configurations are prescribed in accordance with (6.8.1)

$$\underline{r} = \underline{b} + \underline{i}_3 a^3 , \quad \underline{b} = a^1\underline{i}_1 + a^2\underline{i}_2 , \quad \underline{R} = \underline{B} + \underline{i}_3 c a^3 ,$$

$$\underline{B} = \underline{i}_1 x^1 + \underline{i}_2 x^2 , \quad x^3 = c a^3 , \tag{1}$$

where the parameter c is constant. We introduce the material coordinates q^1, q^2 (still general coordinates) and $q^3 = x^3$ so that

$$a^\alpha = a^\alpha(q^1,q^2)\,, \qquad x^\alpha = x^\alpha(q^1,q^2)\,; \qquad x^3 = q^3\,, \quad a^3 = \frac{q^3}{c} \tag{2}$$

and the Greek indices, as usual, take the values 1 and 2.

The vector fields $\underline{b}$ and $\underline{B}$ on the plane determine the bases

$$\underline{b}_\alpha = \underline{i}_\beta \frac{\partial a^\beta}{\partial q^\alpha}\,, \qquad \underline{B}_\alpha = \underline{i}_\beta \frac{\partial x^\beta}{\partial q^\alpha}\,, \qquad \underline{b}^\alpha = \varepsilon^{\alpha\beta}\underline{b}_\beta \times \underline{i}_3\,, \qquad \underline{B}^\alpha = \epsilon^{\alpha\beta}\underline{B}_\beta \times \underline{i}_3\,. \tag{3}$$

Here $\underline{\varepsilon}$ and $\underline{\epsilon}$ denote the plane tensor of Levi-Civita whose contravariant components are

$$\varepsilon^{\alpha\beta} = \frac{1}{\sqrt{b}}\,e^{\alpha\beta}\,, \qquad \epsilon^{\alpha\beta} = \frac{1}{\sqrt{B}}\,e^{\alpha\beta}\,; \qquad \varepsilon^{\alpha\beta} = \sqrt{b}\,e_{\alpha\beta}\,, \qquad \epsilon_{\alpha\beta} = \sqrt{B}\,e_{\alpha\beta}\,, \tag{4}$$

$$e_{\alpha\beta} = e^{\alpha\beta} = \begin{cases} 0\,, & \alpha = \beta\,, \\ 1\,, & \alpha = 1,\ \beta = 2\,, \\ -1\,, & \alpha = 2,\ \beta = 1\,, \end{cases}$$

and where $\sqrt{b}$ and $\sqrt{B}$ denote the Jacobians of the transformations

$$\sqrt{b} = \frac{\partial(a^1,a^2)}{\partial(q^1,q^2)} = (\underline{b}_1 \times \underline{b}_2)\cdot\underline{i}_3\,,$$
$$\sqrt{B} = \frac{\partial(x^1,x^2)}{\partial(q^1,q^2)} = \frac{\partial(x^1,x^2)}{\partial(a^1,a^2)}\sqrt{b} = (\underline{B}_1 \times \underline{B}_2)\cdot\underline{i}_3\,. \tag{5}$$

We determine the deformation gradients and the corresponding plane strain measures

$$\underline{\underline{\overset{0}{\nabla}B}} = \underline{b}^\alpha\underline{B}_\alpha\,, \qquad \underline{\underline{\overset{0}{\nabla}B}}^T = \underline{B}_\alpha\underline{b}^\alpha\,, \qquad \underline{\underline{G}}_\times = \underline{\underline{\overset{0}{\nabla}B}}\cdot\underline{\underline{\overset{0}{\nabla}B}}^T = B_{\alpha\beta}\underline{b}^\alpha\underline{b}^\beta\,,$$
$$\underline{\underline{F}}_\times = \underline{\underline{\overset{0}{\nabla}B}}^T\cdot\underline{\underline{\overset{0}{\nabla}B}} = b^{\alpha\beta}\underline{B}_\alpha\underline{B}_\beta\,, \qquad \underline{\underline{g}}_\times = \underline{\underline{F}}_\times^{-1} = b_{\alpha\beta}\underline{B}^\alpha\underline{B}^\beta\,. \tag{6}$$

Here $b^{\alpha\beta}$, $B^{\alpha\beta}$ are the contravariant components of the plane metric tensor $\underline{E}_2$, and $b_{\alpha\beta}$, $B_{\alpha\beta}$ denote its covariant components. The invariants of the strain measures are determined by the formulas

$$I_1(\underline{F}^\times) = \overset{0}{I}_1 = b^{\alpha\beta}B_{\alpha\beta} = \overset{0}{G}_1 + \overset{0}{G}_2\,, \qquad \overset{0}{I}_2 = \frac{B}{b} = \overset{0}{G}_1\overset{0}{G}_2\,, \qquad I_1(\underline{g}^\times) = \frac{\overset{0}{I}_1}{\overset{0}{I}_2}\,. \tag{7}$$

2. In the following transformation of a plane problem to complex variables, it proves useful to choose for material coordinates the complex coordinates

$q^1 = z,\ q^2 = \bar{z}$ of a point in the actual configuration.

Having agreed to set the complex number (vector) $c = c^1 + ic^2$ into correspondence with the vector $\underline{c} = c^1\underline{i}_1 + c^2\underline{i}_2$, we may represent the scalar product of two vectors in the form $\underline{c}\cdot\underline{d} = \operatorname{Re} c\bar{d}$. This leads to the relationships

$$\underline{b}_1 = \frac{\partial \underline{a}}{\partial x^1}, \qquad b_1 = \frac{\partial \zeta}{\partial x^1} = \left(\frac{\partial}{\partial z} + \frac{\partial}{\partial \bar{z}}\right)\zeta ,$$

$$\underline{b}_2 = \frac{\partial \underline{a}}{\partial x^2}, \qquad b_2 = \frac{\partial \zeta}{\partial x^2} = i\left(\frac{\partial}{\partial z} - \frac{\partial}{\partial \bar{z}}\right)\zeta ,$$

so that

$$b_{11} = \underline{b}_1\cdot\underline{b}_1 = \frac{\partial \zeta}{\partial x^1}\frac{\partial \bar{\zeta}}{\partial x^1}, \qquad b_{22} = \frac{\partial \zeta}{\partial x^2}\frac{\partial \bar{\zeta}}{\partial x^2}, \qquad b_{12} = \frac{1}{2}\left(\frac{\partial \zeta}{\partial x^1}\frac{\partial \bar{\zeta}}{\partial x^2} + \frac{\partial \bar{\zeta}}{\partial x^1}\frac{\partial \zeta}{\partial x^2}\right).$$

Using now the variables $z,\ \bar{z}$ we obtain the formulas used below

$$b_{11} + b_{22} = 2\left(\frac{\partial \zeta}{\partial z}\frac{\partial \bar{\zeta}}{\partial \bar{z}} + \frac{\partial \zeta}{\partial \bar{z}}\frac{\partial \bar{\zeta}}{\partial z}\right), \qquad b_{11} - b_{22} = 2\left(\frac{\partial \zeta}{\partial z}\frac{\partial \bar{\zeta}}{\partial z} + \frac{\partial \zeta}{\partial \bar{z}}\frac{\partial \bar{\zeta}}{\partial \bar{z}}\right),$$

$$b_{12} = i\left(\frac{\partial \zeta}{\partial z}\frac{\partial \bar{\zeta}}{\partial z} - \frac{\partial \zeta}{\partial \bar{z}}\frac{\partial \bar{\zeta}}{\partial \bar{z}}\right). \tag{8}$$

We will also need the following representation of the determinant

$$\frac{\partial(a^1,a^2)}{\partial(z,\bar{z})} = \frac{\partial \zeta}{\partial z}\frac{\partial \bar{\zeta}}{\partial \bar{z}} - \frac{\partial \zeta}{\partial \bar{z}}\frac{\partial \bar{\zeta}}{\partial z}$$

of the transformation $(z,\bar{z}) \to (a^1,a^2)$.

3. We will now turn to the consideration of the field of transformation (1). Here a superposed field which is perpendicular to a given plane vector field is considered. For the material coordinates we choose $q^1, q^2, q^3 = x^3$ and the vector bases are prescribed by the expressions:

$$\underline{r}_1 = \underline{b}_1, \quad \underline{r}_2 = \underline{b}_2, \quad \underline{r}_3 = \frac{\underline{i}_3}{c}; \qquad \underline{R}_1 = \underline{B}_1, \quad \underline{R}_2 = \underline{B}_2, \quad \underline{R}_3 = \underline{i}_3;$$

$$g = |g_{sk}| = \frac{b}{c^2}, \qquad G = |G_{sk}| = B. \tag{10}$$

The deformation gradients and the strain measures are equal to

$$\overset{0}{\underline{\underline{\nabla R}}} = \underline{r}^s\underline{R}_s = \underline{b}^\alpha\underline{B}_\alpha + \underline{r}^3\underline{R}_3 = \overset{0}{\underline{\underline{\nabla B}}} + c\underline{i}_3\underline{i}_3, \qquad \overset{0}{\underline{\underline{\nabla R}}}^{\mathrm{T}} = \overset{0}{\underline{\underline{\nabla B}}}^{\mathrm{T}} + c\underline{i}_3\underline{i}_3,$$

$$\underset{=}{G} = \underset{=}{G}^{\times} + c^2 \underline{i}_3 \underline{i}_3 , \qquad \underset{=}{F} = \underset{=}{F}^{\times} + c^2 \underline{i}_3 \underline{i}_3 . \tag{11}$$

These tensors have zero (α3)-components which is a fundamental property of the plane problem. The invariants of the strain measures turn out to be

$$I_1 = I_1^0 + c^2 , \qquad I_2 = I_2^0 + c^2 I_1^0 , \qquad I_3 = c^2 \frac{B}{b} = c^2 I_2^0 . \tag{12}$$

These formulas imply that the relationship between invariants $I_k(\underset{=}{G})$

$$I_2 = \frac{I_3}{c^2} + c^2 I_1 - c^4 \tag{13}$$

holds and that the specific strain energy may be treated as a function of the invariants I_1 and I_3

$$\ni\left(I_1 , c^2 I_1 - c^4 + \frac{I_3}{c^2} , I_3\right) = \ni^{\times}(I_1, I_3) , \tag{14}$$

so that

$$\frac{\partial \ni^{\times}}{\partial I_1} = \frac{\partial \ni}{\partial I_1} + c^2 \frac{\partial \ni}{\partial I_2} , \qquad \frac{\partial \ni^{\times}}{\partial I_3} = c^{-2} \frac{\partial \ni}{\partial I_2} + \frac{\partial \ni}{\partial I_3} . \tag{15}$$

Making use of (11), the state equation may be written in the form

$$\underset{=}{T} = 2 I_3^{-1/2} \left\{ \underset{=}{E} I_3 \frac{\partial \ni}{\partial I_3} + \left(\frac{\partial \ni}{\partial I_1} + I_1 \frac{\partial \ni}{\partial I_2} \right) \underset{=}{F}^{\times} - \frac{\partial \ni}{\partial I_2} \underset{=}{F}^{\times 2} + \right.$$

$$\left. + \left[\left(\frac{\partial \ni}{\partial I_1} + I_1 \frac{\partial \ni}{\partial I_2} \right) c^2 - \frac{\partial \ni}{\partial I_2} c^4 \right] \underline{i}_3 \underline{i}_3 \right\} . \tag{16}$$

From which we deduce the relationships

$$t^{\alpha 3} = 0 , \quad \alpha = 1,2 ; \qquad t^{33} = 2 I_3^{-1/2} \left\{ I_3 \frac{\partial \ni}{\partial I_3} + c^2 \left[\frac{\partial \ni}{\partial I_1} + (I_1 - c^2) \frac{\partial \ni}{\partial I_2} \right] \right\} \tag{17}$$

and determine the tensor

$$\underset{=}{T}^{\times} = t^{\alpha\beta} \underline{R}_{\alpha} \underline{R}_{\beta} = t^{\alpha\beta} \underline{B}_{\alpha} \underline{B}_{\beta} = 2 I_3^{-1/2} \left\{ \underset{=}{E}_2 I_3 \frac{\partial \ni}{\partial I_3} + \left(\frac{\partial \ni}{\partial I_1} + I_1 \frac{\partial \ni}{\partial I_2} \right) \underset{=}{F}^{\times} - \frac{\partial \ni}{\partial I_2} \underset{=}{F}^{\times 2} \right\} .$$

on a plane.

Substituting the plane tensor $\underset{=}{F}^{\times 2}$ by its representation by the Cayley--Hamilton theorem, we have

$$\underline{F}^{\times 2} = I_1^0 \underline{F}^{\times} - \underline{E}_2 I_2^0 ,$$

and obtain

$$\underline{T}^{\times} = 2I_3^{-1/2}\left[\underline{E}_2\left(I_3 \frac{\partial\ni}{\partial I_3} + I_2^0 \frac{\partial\ni}{\partial I_2}\right) + \left(\frac{\partial\ni}{\partial I_1} + I^2 \frac{\partial\ni}{\partial I_2}\right)\underline{F}^{\times}\right] .$$

By virtue of (12) and (15) this expression reduces to the form

$$\underline{T}^{\times} = 2I_3^{-1/2}\left(\underline{E}_2 I_3 \frac{\partial\ni^{\times}}{\partial I_3} + \frac{\partial\ni^{\times}}{\partial I_1}\underline{F}^{\times}\right) . \tag{18}$$

The stress tensor is given by

$$\underline{T} = \underline{T}^{\times} + \underline{i}_3\underline{i}_3 t^{33} = 2I_3^{-1/2}\left\{\underline{E}_2 I_3 \frac{\partial\ni^{\times}}{\partial I_3} + \frac{\partial\ni^{\times}}{\partial I_1}\underline{F}^{\times} + \right.$$

$$\left. + \underline{i}_3\underline{i}_3\left[I_3 \frac{\partial\ni}{\partial I_3} + c^2 \frac{\partial\ni}{\partial I_1} + c^2(I_1 - c^2)\frac{\partial\ni}{\partial I_2}\right]\right\} . \tag{19}$$

4. For an incompressible elastic body $\ni = \ni(I_1, I_2)$. The formulas (12), (14) and (15) have the form

$$I_1 = I_1^0 + c^2 , \qquad I_2^0 = c^{-2} , \qquad I_2 = c^{-2} + c^2 I_1 - c^4 = c^{-2} + c^2 I_1^0 , \tag{20}$$

$$\ni(I_1, I_2) = \ni(I_1, c^2 I_1 + c^{-2} - c^4) = \ni^{\times}(I_1) , \qquad \frac{d\ni^{\times}}{dI_1} = \frac{\partial\ni}{\partial I_1} + c^2 \frac{\partial\ni}{\partial I_2} , \tag{21}$$

the second Equation (15) vanishes since I_3 does not enter into the prescription of $\ni$ and the terms in $\partial\ni/\partial I_3$ and $\partial\ni^{\times}/\partial I_3$ should be omitted from (19). The "determined" stress $\underline{T}_E$ is given by the formula

$$\underline{T}_E = 2\left\{\frac{d\ni^{\times}}{dI_1}\underline{F}^{\times} + \underline{i}_3\underline{i}_3 c^2\left[\frac{\partial\ni}{\partial I_1} + (I_1 - c^2)\frac{\partial\ni}{\partial I_2}\right]\right\} \tag{22}$$

and the expression for the stress tensor $\underline{T}$ takes the form

$$\underline{T} = -p\underline{E} + \underline{T}_E = -p\underline{E}_2 - p\underline{i}_3\underline{i}_3 + \underline{T}_E = \underline{T}^{\times} + \underline{i}_3\underline{i}_3 t^{33} . \tag{23}$$

Here

$$\underline{T}^{\times} = t^{\alpha\beta}\underline{R}_{\alpha}\underline{R}_{\beta} = -p\underline{E}_2 + 2\frac{d\ni^{\times}}{dI_1}\underline{F}^{\times} ,$$

$$t^{33} = -p + 2c^2 \left[\frac{\partial \ni}{\partial I_1} + (I_1 - c^2) \frac{\partial \ni}{\partial I_2} \right] , \tag{24}$$

the stress tensor in the plane problem for an incompressible material, is represented as a sum of the plane tensor $\underline{T}^{\times}$ and the tensor $\underline{i}_3\underline{i}_3 t^{33}$. These tensors are independent of the third material coordinate $q^3 = x^3$.

In the absence of the body forces, the equilibrium equations are transformed to

$$\underline{\nabla} \cdot \underline{T} = \left(\underline{\nabla}^{\times} + \underline{i}_3\underline{i}_3 \frac{\partial}{\partial x^3} \right) \cdot \underline{T} =$$

$$= \left(\underline{B}^{\alpha} \frac{\partial}{\partial q^{\alpha}} + \underline{i}_3 \frac{\partial}{\partial x^3} \right) \cdot (\underline{T}^{\times} + \underline{i}_3\underline{i}_3 t^{33}) = \underline{\nabla}^{\times} \cdot \underline{T}^{\times} = \underline{0} , \tag{25}$$

because the derivatives with respect to x^3 are zero and $\underline{B}^{\alpha} \cdot \underline{i}_3 = 0$. The component t^{33} of the stress tensor is determined by Equation (24). The invariants I_1, I_2 entering this formula, and the constraint factor p (pressure) of constraint will be seen to be functions of the material coordinates q^1, q^2 and c from a solution to the problem of equilibrium for the tensor $\underline{T}^{\times}$. At this stage, the constant c is determined by prescription of the longitudinal force Q

$$Q = \iint_S t^{33} dO = \iint_S t^{33} \frac{\partial(x^1, x^2)}{\partial(a^1, a^2)} da^1 da^2 = \iint_S \sqrt{\frac{B}{b}}\, t^{33} da^1 da^2 . \tag{26}$$

5. In view of (2.4.7), the equation of static equilibrium is satsified identically by a representation of $\underline{T}^{\times}$ in terms of the stress function $U(q^1, q^2)$. Referring also to Equations (3), we obtain

$$\underline{T}^{\times} = t^{\alpha\beta} \underline{B}_{\alpha}\underline{B}_{\beta} = -\underline{i}_3 \times \underline{\nabla\nabla} U \times \underline{i}_3 = \underline{B}^{\gamma} \times \underline{i}_3\underline{B}^{\delta} \times \underline{i}_3 \nabla_{\gamma}\nabla_{\delta} U =$$

$$= \epsilon^{\gamma\alpha} (\underline{B}_{\alpha} \times \underline{i}_3) \times \underline{i}_3 \epsilon^{\delta\beta} (\underline{B}_{\beta} \times \underline{i}_3) \times \underline{i}_3 \nabla_{\gamma}\nabla_{\delta} U = \epsilon^{\gamma\alpha}\epsilon^{\delta\beta} \underline{B}_{\alpha}\underline{B}_{\beta} \nabla_{\gamma}\nabla_{\delta} U . \tag{27}$$

Introducing the notation

$$\underline{\nabla\nabla} U = \underline{M} = \underline{B}_{\alpha}\underline{B}_{\beta} m^{\alpha\beta} , \qquad \underline{T}^{\times} = -\underline{i}_3 \times \underline{M} \times \underline{i}_3 = \underline{B}_{\alpha} \times \underline{i}_3\underline{B}_{\beta} \times \underline{i}_3 m^{\alpha\beta} ,$$

it is easy to reverse this relationship by vector multiplication from the right and left by the vector $\underline{i}_3$. We then obtain

$$\underline{i}_3 \times \underline{T}^{\times} \times \underline{i}_3 = \underline{i}_3 \times (\underline{B}_{\alpha} \times \underline{i}_3)(\underline{B}_{\beta} \times \underline{i}_3) \times \underline{i}_3 m^{\alpha\beta} = -\underline{B}_{\alpha}\underline{B}_{\beta} m^{\alpha\beta} = -\underline{M} = \underline{\nabla\nabla} U .$$

Thus we have

$$\underline{\nabla\nabla}U = -\underline{i}_3 \times \underline{T}^{\times} \times \underline{i}_3 \, . \tag{28}$$

The unit vectors of tangent and normal to the curve Γ in the plane x^3 = const are denoted by $\underline{t}$ and $\underline{N}$; assuming that the vectors $\underline{N}$, $\underline{t}$ and $\underline{i}_3$ have the same orientation as the axes XYZ and denoting the element of an arclength of this curve by dS, we have

$$\underline{t} = \frac{\partial \underline{B}}{\partial S} \, , \qquad \underline{N} = \underline{t} \times \underline{i}_3 \, ,$$

$$\underline{N} \cdot \underline{T}^{\times} = -(\underline{t} \times \underline{i}_3) \cdot (\underline{i}_3 \times \underline{\nabla\nabla}U \times \underline{i}_3) = \underline{t} \cdot (\underline{\nabla\nabla}U \times \underline{i}_3)$$

and the resultant $\underline{P}$ of forces distributed along the arc Γ turns out to be

$$\underline{P} = \int_{M_0}^{M} \underline{t}\, dS(\underline{\nabla\nabla}U \times \underline{i}_3) = \int_{M_0}^{M} d\underline{B} \cdot \underline{\nabla\nabla}U \times \underline{i}_3 = \int_{M_0}^{M} d\underline{\nabla}U \times \underline{i}_3 = \underline{\nabla}U \times \underline{i}_3 \Big|_{M_0}^{M} \, ,$$

because $d\underline{B} \cdot \underline{\nabla\nabla}U = d\underline{\nabla}U$. The resultant moment of this distribution of forces is now determined by the expression

$$\underline{m}^0 = \int_{M_0}^{M} \underline{B} \times (d\underline{\nabla}U \times \underline{i}_3) = -\underline{i}_3 \int_{M_0}^{M} \underline{B} \cdot d\underline{\nabla}U$$

and the resultant moment $\underline{m}_Z^0$ with respect to the OZ-axis turns out to be equal to

$$m_Z^0 = -\int_{M_0}^{M} \underline{B} \cdot d\underline{\nabla}U = -\int_{M_0}^{M} d(\underline{B} \cdot \underline{\nabla}U) + \int_{M_0}^{M} d\underline{B} \cdot \underline{\nabla}U = (U - \underline{B} \cdot \underline{\nabla}U) \Big|_{M_0}^{M}$$

We thus see that to the extent of some additive terms,

$$\underline{P} = \underline{\nabla}U \times \underline{i}_3 \, , \qquad m_Z^0 = U - \underline{B} \cdot \underline{\nabla}U \, . \tag{29}$$

For any plane tensor $\underline{M}^{\times}$ we have

$$I_1(\underline{i}_3 \times \underline{M}^{\times} \times \underline{i}_3) = -I_1(\underline{M}^{\times}) \, .$$

Referring now to (28) and (24), we obtain for an incompressible medium

$$\underline{\nabla}^2 U = I_1(\underline{T}^{\times}) = -2p + 2\, \frac{d\ni^{\times}}{dI_1} I_1^0 \, , \qquad p = -\frac{1}{2}\left(\underline{\nabla}^2 U - 2I_1^0 \frac{d\ni^{\times}}{dI_1}\right) \, . \tag{30}$$

The tensor $\underline{T}^{\times}$ can now be represented by the expression

$$\underline{T}^{\times} = \frac{1}{2}\,\underline{E}_2\left(\underline{\nabla}^2 U - 2I_1^0\,\frac{d\ni^{\times}}{dI_1}\right) + 2\,\frac{d\ni^{\times}}{dI_1}\,\underline{F}^{\times} . \tag{31}$$

Observing that $\underline{i}_3 \times \underline{E} \times \underline{i}_3 = -\underline{E}_2$, we may reduce Equation (28) to the form

$$\underline{\nabla\nabla} U = \frac{1}{2}\,\underline{E}_2\left(\underline{\nabla}^2 U - 2\,\frac{d\ni}{dI_1}\,I_1^0\right) - 2\,\frac{d\ni^{\times}}{dI_1}\,\underline{i}_3 \times \underline{F}^{\times} \times \underline{i}_3 . \tag{32}$$

In the component representation

$$\underline{\nabla\nabla} U = \underline{B}^{\alpha}\underline{B}^{\beta}\nabla_{\alpha}\nabla_{\beta}U , \qquad \underline{E}_2 = \underline{B}^{\alpha}\underline{B}^{\beta}B_{\alpha\beta} , \qquad \underline{F}^{\times} = b^{\gamma\delta}\underline{B}_{\gamma}\underline{B}_{\delta} ,$$

$$-\underline{i}_3 \times \underline{F}^{\times} \times \underline{i}_3 = b^{\gamma\delta}\underline{B}_{\gamma} \times \underline{i}_3\underline{B}_{\delta} \times \underline{i}_3 = b^{\gamma\delta}\epsilon_{\gamma\alpha}\epsilon_{\delta\beta}\underline{B}^{\alpha}\underline{B}^{\beta} .$$

In the latter expression, in view of (3),

$$b^{\gamma\delta} = \underline{b}^{\gamma}\cdot\underline{b}^{\delta} = \varepsilon^{\gamma\mu}\varepsilon^{\delta\nu}(\underline{b}_{\mu} \times \underline{i}_3)\cdot(\underline{b}_{\nu} \times \underline{i}_3) = \varepsilon^{\gamma\mu}\varepsilon^{\delta\nu}b_{\mu\nu} ,$$

so that

$$b^{\gamma\delta}\epsilon_{\gamma\alpha}\epsilon_{\delta\beta} = b_{\mu\nu}\varepsilon^{\gamma\mu}\epsilon_{\gamma\alpha}\varepsilon^{\delta\nu}\epsilon_{\delta\beta} = \frac{B}{b}\,b_{\mu\nu}\varepsilon^{\gamma\mu}\varepsilon_{\gamma\alpha}\varepsilon^{\delta\nu}\varepsilon_{\delta\beta} =$$

$$= \frac{B}{b}\,b_{\mu\nu}\delta^{\mu}_{\alpha}\delta^{\nu}_{\beta} = \frac{B}{b}\,b_{\alpha\beta}$$

and now

$$-\underline{i}_3 \times \underline{F}^{\times} \times \underline{i}_3 = b^{\gamma\delta}\epsilon_{\gamma\alpha}\epsilon_{\delta\beta}\underline{B}^{\alpha}\underline{B}^{\beta} = \frac{B}{b}\,b_{\alpha\beta}\underline{B}^{\alpha}\underline{B}^{\beta} ;$$

lastly, in view of (12), we obtain that $B/b = c^{-2}$ for an incompressible medium. The component representation of Equation (32) now reduces to the form

$$\nabla_{\alpha}\nabla_{\beta}U = \frac{1}{2}\left(\underline{\nabla}^2 U - 2I_1^0\,\frac{d\ni^{\times}}{dI_1}\right)B_{\alpha\beta} + 2\,\frac{d\ni^{\times}}{dI_1}\,b^{\gamma\delta}\epsilon_{\gamma\alpha}\epsilon_{\delta\beta} =$$

$$= \frac{1}{2}\left(\underline{\nabla}^2 U - 2I_1^0\,\frac{d\ni^{\times}}{dI_1}\right)B_{\alpha\beta} + 2c^{-2}\,\frac{d\ni^{\times}}{dI_1}\,b_{\alpha\beta} . \tag{33}$$

6. We select $x^1 = q^1$, $x^2 = q^2$ as independent variables in system (33) of three ordinary differential equations. The function U and the coordinates

$a^\alpha(x^1, x^2)$ of a particle in its actual configuration are not known. We have

$$\underline{B}_\alpha = \underline{i}_\alpha\,, \qquad \underline{b}_\alpha = \frac{\partial \underline{a}}{\partial x^\alpha} = \underline{i}_\beta \frac{\partial a^\beta}{\partial x^\alpha}\,;$$

$$B_{\alpha\beta} = \begin{cases} 1, & \alpha = \beta, \\ 0, & \alpha \neq \beta, \end{cases} \qquad b_{\alpha\beta} = \frac{\partial a^\gamma}{\partial x^\alpha}\frac{\partial a^\gamma}{\partial x^\beta}\,, \qquad c^2 = \frac{b}{B} = b$$

here we have used Equation (10), the equality $B = 1$ and the incompressibility condition. By (7) we have

$$I_1^0 = b^{11} + b^{22} = \frac{1}{b}(b_{11} + b_{22}) = c^{-2}(b_{11} + b_{22}) \,. \tag{34}$$

Equations (33) now reduce to the form

$$\frac{1}{2}\left(\frac{\partial^2 U}{\partial x^{1\,2}} - \frac{\partial^2 U}{\partial x^{2\,2}}\right) = c^{-2}(b_{11} - b_{22})\frac{d\text{Э}^\times}{dI_1}\,, \qquad \frac{\partial^2 U}{\partial x^1 \partial x^2} = 2c^{-2}\frac{d\text{Э}^\times}{dI_1} b_{12}\,. \tag{35}$$

An additional equation expressing the condition of incompressibility is given by the relationship obtained from (5)

$$b = \left[\frac{\partial(a^1, a^2)}{\partial(x^1, x^2)}\right]^2 = b_{11}b_{22} - b_{12}^2 = c^2\,. \tag{36}$$

The expression (27) for the stress tensor yields the well-known formulas for its components

$$\sigma_{x^1} = \frac{\partial^2 U}{\partial x^{2\,2}}\,, \qquad \sigma_{x^2} = \frac{\partial^2 U}{\partial x^{1\,2}}\,, \qquad \tau_{x^1x^2} = -\frac{\partial^2 U}{\partial x^1 \partial x^2}\,, \tag{37}$$

these formulas allow us to write the "complex stresses" of Kolosov-Muskhelishvili in the following form instead of (35)

$$\sigma_{x^1} + \sigma_{x^2} = \underline{\nabla}^2 U = 4\frac{\partial^2 U}{\partial z \partial \bar{z}} = -2p + 4c^{-2}\frac{d\text{Э}^\times}{dI_1}\left(\frac{\partial \zeta}{\partial z}\frac{\partial \bar{\zeta}}{\partial \bar{z}} + \frac{\partial \zeta}{\partial \bar{z}}\frac{\partial \bar{\zeta}}{\partial z}\right), \tag{38}$$

$$\sigma_{x^2} - \sigma_{x^1} + 2i\tau_{x^1x^2} = \frac{\partial^2 U}{\partial x^{1\,2}} - \frac{\partial^2 U}{\partial x^{2\,2}} - 2i\frac{\partial^2 U}{\partial x^1 \partial x^2} = 4\frac{\partial^2 U}{\partial z^2} = 8c^{-2}\frac{d\text{Э}^\times}{dI_1}\frac{\partial \zeta}{\partial z}\frac{\partial \bar{\zeta}}{\partial z}\,, \tag{39}$$

here we have made use of Equations (8). The multiplying factor in this relationship is determined by the material and is, from (21), a function of its invariant I_1. Using (9), Equation (36) may also be given in the form

$$\frac{\partial \zeta}{\partial z}\frac{\partial \bar{\zeta}}{\partial \bar{z}} - \frac{\partial \zeta}{\partial \bar{z}}\frac{\partial \bar{\zeta}}{\partial z} = c \,, \tag{40}$$

and I_1^0 is given by the expression

$$I_1^0 = 2c^{-2}\left(\frac{\partial \zeta}{\partial z}\frac{\partial \bar{\zeta}}{\partial \bar{z}} + \frac{\partial \zeta}{\partial \bar{z}}\frac{\partial \bar{\zeta}}{\partial z}\right) = 2\left(2c^{-2}\frac{\partial \bar{\zeta}}{\partial z}\frac{\partial \zeta}{\partial \bar{z}} + c^{-1}\right) . \tag{41}$$

From (39) there follows the differential equation

$$\frac{\partial^2}{\partial \bar{z}^2}\frac{d\ni^{\times}}{dI_1}\frac{\partial \zeta}{\partial z}\frac{\partial \bar{\zeta}}{\partial z} = \frac{\partial^2}{\partial z^2}\frac{d\ni^{\times}}{dI_1}\frac{\partial \zeta}{\partial \bar{z}}\frac{\partial \bar{\zeta}}{\partial \bar{z}} \,, \tag{42}$$

which must be satisfied for any plane deformation $\zeta = \zeta(z,\bar{z},c)$ possible for a given material under longitudinal stretching c (if body forces are absent).

Representation (29) for the resultant vector of forces distributed along the arc of a contour Γ bounding the domain considered reduces to the form

$$P = P_1 + iP_2 = -2\frac{\partial U}{\partial \bar{z}} \tag{43}$$

which is the boundary condition when surface tractions are prescribed.

The coordinates a^1, a^2 of the reference configuration may be chosen as independent variables, in other words, ζ and $\bar{\zeta}$. From the relationships

$$\frac{\partial z}{\partial z} = 1 = \frac{\partial z}{\partial \zeta}\frac{\partial \zeta}{\partial z} + \frac{\partial z}{\partial \bar{\zeta}}\frac{\partial \bar{\zeta}}{\partial z} \,, \qquad \frac{\partial z}{\partial \bar{z}} = 0 = \frac{\partial z}{\partial \zeta}\frac{\partial \zeta}{\partial \bar{z}} + \frac{\partial z}{\partial \bar{\zeta}}\frac{\partial \bar{\zeta}}{\partial \bar{z}}$$

and from the condition of incompressibility, (40), we deduce

$$\frac{\partial z}{\partial \zeta} = \frac{1}{c}\frac{\partial \bar{\zeta}}{\partial \bar{z}} \,, \qquad \frac{\partial z}{\partial \bar{\zeta}} = -\frac{1}{c}\frac{\partial \zeta}{\partial \bar{z}} \,, \qquad \frac{\partial \bar{z}}{\partial \zeta} = -\frac{1}{c}\frac{\partial \bar{\zeta}}{\partial z} \,, \qquad \frac{\partial \bar{z}}{\partial \bar{\zeta}} = \frac{1}{c}\frac{\partial \zeta}{\partial z} \,,$$

so that by (21)

$$I_1^0 = 2\left(2\frac{\partial \bar{z}}{\partial \zeta}\frac{\partial z}{\partial \bar{\zeta}} + \frac{1}{c}\right) . \tag{44}$$

Assuming that the solution of Equation (42)

$$\zeta = f(z,\bar{z},c) \,, \tag{45}$$

is known, we now consider the transformation

$$z = \frac{1}{c} f(\zeta,\bar{\zeta},c) \ , \tag{46}$$

where f denotes the same function as in Equation (45). Then, by virtue of (44), the invariant I_1^0, and that means $d\ni^{\times}/dI_1$ too for the same material, will turn out to be given by exactly the same functions of ζ, $\bar{\zeta}$ and c as those used to express the dependence of the initial deformation (45) on z, $\bar{z}$ and c. The form invariance of the expression for $d\ni^{\times}/dI_1$ enables us to verify by an immediate (though rather elaborate) calculation that transformation (46) satisfies Equation (42) provided that the same property holds for (45). This is Adkins' "inverse theorem" (J.E. Adkins, 1958) which enables us to construct by a plane deformation together with a longitudinal stretching c, another plane deformation which is, for the same c, compatible with Equation (42).

Duality of the strain and stress fields for the plane problem. Relationships similar to those of Adkins' theorem hold for plane deformation of a compressible material

$$x^1 = x^1(a^1,a^2) \ , \qquad x^2 = x^2(a^1,a^2) \ , \qquad x^3 = ca^3 \ . \tag{47}$$

In the expressions for the Finger measure and for the stress tensor, the plane tensors are separated from the (33)-components

$$\underline{F} = \underline{F}^{\times} + \underline{i}_3\underline{i}_3c^2 \ , \qquad \underline{T} = \underline{T}^{\times} + \underline{i}_3\underline{i}_3t^{33} \qquad (F^{\alpha 3} = 0,\ t^{\alpha 3} = 0) \ . \tag{48}$$

Here, by virtue of (6),

$$\underline{F}^{\times} = \overset{0}{\underline{\nabla B}}{}^{T} \cdot \overset{0}{\underline{\nabla B}} \ , \qquad \underline{F}^{\times 2} = I_1^0 \underline{F}^{\times} - I_2^0 \underline{E}_2 \ , \qquad \overset{0}{\underline{\nabla B}} = \underline{i}^{\alpha}\underline{i}_{\beta}\frac{\partial x^{\beta}}{\partial a^{\alpha}} \ , \qquad \underline{B} = \underline{i}_{\beta}x^{\beta} \tag{49}$$

and by (12) and (16)

$$\underline{T}^{\times} = \frac{2}{c\sqrt{I_2^0}}\left[\underline{E}_2 I_2^0\left(\frac{\partial\ni}{\partial I_2} + c^2\frac{\partial\ni}{\partial I_3}\right) + \left(\frac{\partial\ni}{\partial I_1} + c^2\frac{\partial\ni}{\partial I_2}\right)\underline{F}^{\times}\right] \ ,$$

$$t^{33} = \frac{2c}{\sqrt{I_2^0}}\left(\frac{\partial\ni}{\partial I_1} + I_1^0\frac{\partial\ni}{\partial I_2} + I_2^0\frac{\partial\ni}{\partial I_3}\right) \ .$$

Here we have used the expressions

$$I_1 = I_1^0 + c^2 \ , \qquad I_2 = I_2^0 + c^2 I_1^0 \ ,$$

where I_α^0 denote the invariants of $\overset{\times}{\underline{F}}$. Now, on the basis of the relationships

$$\ni(I_1, I_2, I_3) = \ni(I_1^0 + c^2, I_2^0 + I_1^0 c^2, I_3 = I_2^0 c^2) ,$$

the following formulas are obtained

$$\frac{\partial \ni}{\partial I_1^0} = \frac{\partial \ni}{\partial I_1} + c^2 \frac{\partial \ni}{\partial I_2} , \qquad \frac{\partial \ni}{\partial I_2^0} = \frac{\partial \ni}{\partial I_2} + c^2 \frac{\partial \ni}{\partial I_3} ,$$

$$\frac{\partial \ni}{\partial c^2} = \frac{1}{2c} \frac{\partial \ni}{\partial c} = \frac{\partial \ni}{\partial I_1} + I_1^0 \frac{\partial \ni}{\partial I_2} + I_2^0 \frac{\partial \ni}{\partial I_3} .$$

This enables us to represent $\overset{\times}{\underline{T}}$ and t^{33} in the form

$$\overset{\times}{\underline{T}} = \frac{2}{c\sqrt{I_2^0}} \left(I_2^0 \frac{\partial \ni}{\partial I_2^0} \underline{E}_2 + \frac{\partial \ni}{\partial I_1^0} \overset{\times}{\underline{F}} \right) , \qquad t^{33} = \frac{1}{\sqrt{I_2^0}} \frac{\partial \ni}{\partial c} , \tag{50}$$

so that

$$I_1(\underline{T}) = \frac{2}{c\sqrt{I_2^0}} \left(I_1^0 \frac{\partial \ni}{\partial I_1^0} + 2I_2^0 \frac{\partial \ni}{\partial I_2^0} \right) + \frac{1}{\sqrt{I_2^0}} \frac{\partial \ni}{\partial c} . \tag{51}$$

Following Chapter 4, Par. 17, the specific complementary energy $\tilde{\ni}$ is determined by the expression

$$\tilde{\ni} = \underline{P} \cdot\cdot \overset{0}{\underline{\nabla R}}{}^T - \ni = \sqrt{I_3}\, \underline{\nabla r}^T \cdot \underline{T} \cdot\cdot \overset{0}{\underline{\nabla R}}{}^T - \ni = c\sqrt{I_2^0}\, I_1(\underline{T}) - \ni$$

and by (51)

$$\tilde{\ni} = 2\left(I_1^0 \frac{\partial \ni}{\partial I_1^0} + 2I_2^0 \frac{\partial \ni}{\partial I_2^0} \right) + c\frac{\partial \ni}{\partial c} - \ni . \tag{52}$$

Its plane component can also be separated from the expression for the Piola tensor

$$\underline{P} = c\sqrt{I_2^0}\, \underline{\nabla r}^T \cdot \underline{T} = \overset{\times}{\underline{P}} + \underline{i}_3\underline{i}_3 p^{33} = c\sqrt{I_2^0} \left(\underline{\nabla b}^T + \frac{1}{c}\, \underline{i}_3\underline{i}_3 \right) \cdot (\overset{\times}{\underline{T}} + \underline{i}_3\underline{i}_3 t^{33}) =$$

$$= 2\underline{\nabla b}^T \cdot \left(\underline{E}_2 I_2^0 \frac{\partial \ni}{\partial I_2^0} + I_1^0 \frac{\partial \ni}{\partial I_1^0} \overset{\times}{\underline{F}} \right) + \underline{i}_3\underline{i}_3 \frac{\partial \ni}{\partial c} ,$$

so that

$$\underline{P}^{\times} = 2\left(I_2^0 \frac{\partial \ni}{\partial I_2^0} \underline{\underline{\nabla b}}^{T} + I_1^0 \frac{\partial \ni}{\partial I_1^0} \underline{\underline{\overset{0}{\nabla} B}}\right) , \qquad p^{33} = \frac{\partial \ni}{\partial c} . \tag{53}$$

Turning now to the representations of the Piola tensor and the place gradient, (4.17.1) and (4.17.5), we have

$$\underline{P} = \ni_{\underline{\underline{\overset{0}{\nabla}R}}} , \qquad \underline{\underline{\overset{0}{\nabla}R}} = \underline{i}^{\alpha}\underline{i}_{\beta} \frac{\partial x^{\beta}}{\partial a^{\alpha}} + \underline{i}_3\underline{i}_3 c = \tilde{\ni}_{\underline{P}} \tag{54}$$

and from definition (II.2.7) of the derivative of a scalar by a tensor argument

$$\underline{P} = \underline{P}^{\times} + \underline{i}_3\underline{i}_3 p^{33} = \underline{i}_{\alpha}\underline{i}^{\beta} \frac{\partial \ni}{\partial \dfrac{\partial x^{\beta}}{\partial a^{\alpha}}} + \underline{i}_3\underline{i}_3 \frac{\partial \ni}{\partial c} = \ni_{\underline{\underline{\overset{0}{\nabla}B}}} + \underline{i}_3\underline{i}_3 \frac{\partial \ni}{\partial c} ,$$

$$\underline{\underline{\overset{0}{\nabla}R}} = \underline{\underline{\overset{0}{\nabla}B}} + c\underline{i}_3\underline{i}_3 = \underline{i}_{\alpha}\underline{i}^{\beta} \frac{\partial \tilde{\ni}}{\partial p_{\alpha}^{\cdot\beta}} + \underline{i}_3\underline{i}_3 \frac{\partial \tilde{\ni}}{\partial \dfrac{\partial \ni}{\partial c}} ,$$

so that the plane tensors $\underline{P}^{\times}$ and $\underline{\underline{\overset{0}{\nabla}B}}$ turn out to be

$$\underline{P}^{\times} = \ni_{\underline{\underline{\overset{0}{\nabla}B}}} , \qquad \underline{\underline{\overset{0}{\nabla}B}} = \tilde{\ni}_{\underline{P}^{\times}} . \tag{55}$$

The equilibrium equation in the absence of body forces and the condition for the existence of the vector $\underline{B}$ (the integrability condition) lead to the relationships

$$\underline{\overset{0}{\nabla}} \cdot \underline{P}^{\times} = \underline{\overset{0}{\nabla}} \cdot \ni_{\underline{\underline{\overset{0}{\nabla}B}}} = \underline{0} , \qquad \underline{\overset{0}{\nabla}} \times \underline{\underline{\overset{0}{\nabla}B}} = \underline{\overset{0}{\nabla}} \times \ni_{\underline{P}^{\times}} = \underline{0} . \tag{56}$$

The conditions (56) are satisfied if we introduce Piola's stress function tensor

$$\underline{\Phi} = \underline{i}_3\underline{i}^{\beta}\varphi_{\beta}(a^1,a^2) \qquad (\beta = 1,2) \tag{57}$$

and assume that

$$\underline{P}^{\times} = \ni_{\underline{\underline{\overset{0}{\nabla}B}}} = 2\mu\underline{\overset{0}{\nabla}} \times \underline{\Phi} = 2\mu e^{\alpha\gamma}\underline{i}_{\alpha}\underline{i}^{\beta} \frac{\partial \varphi_{\beta}}{\partial a^{\gamma}}$$

(where also $e^{12} = -e^{21} = 1$ and $e^{11} = e^{22} = 0$). We thus have

$$p_{\cdot\beta}^{\alpha} = 2\mu e^{\alpha\gamma} \frac{\partial \varphi_{\beta}}{\partial a^{\gamma}} , \qquad \underline{P}^{\times} = 2\mu e^{\alpha\gamma}\underline{i}_{\alpha}\underline{i}^{\beta} \frac{\partial \varphi_{\beta}}{\partial a^{\gamma}} . \tag{58}$$

Returning now to $(55)_2$, we have

$$\overset{0}{\underline{\nabla}}\underline{\underline{B}} = \underline{i}^{\alpha}\underline{i}_{\beta}\frac{\partial x^{\beta}}{\partial a^{\alpha}} = \underline{i}^{\alpha}\underline{i}_{\beta}\frac{\partial\tilde{\ni}}{\partial p^{\alpha}_{\cdot\beta}}\,, \qquad \frac{\partial x^{\beta}}{\partial a^{\alpha}} = \frac{\partial\ni}{\partial p^{\alpha}_{\cdot\beta}} \tag{59}$$

and by $(56)_2$

$$\underline{i}^{\gamma}\times\underline{i}^{\alpha}\underline{i}_{\beta}\frac{\partial^2 x^{\beta}}{\partial a^{\gamma}\partial a^{\alpha}} = \underline{i}_3\underline{i}_{\beta}e^{\gamma\alpha}\frac{\partial}{\partial a^{\gamma}}\frac{\partial\tilde{\ni}}{\partial p^{\alpha}_{\cdot\beta}} = 0\ .$$

The system of two equations

$$e^{\gamma\alpha}\frac{\partial}{\partial a^{\gamma}}\frac{\partial\tilde{\ni}}{\partial p^{\alpha}_{\cdot\beta}} = 0 \qquad (\beta = 1,2) \tag{60}$$

with the notation

$$\underline{\underline{Q}}^{\times} = \underline{i}^{\alpha}\underline{i}_{\beta}q^{\beta}_{\cdot\alpha} = \underline{i}_3\times\underline{i}^{\alpha}\underline{i}_{\beta}\frac{\partial\tilde{\ni}}{\partial p^{\alpha}_{\cdot\beta}} = \underline{i}_3\times\underline{i}^{\alpha}\underline{i}_{\beta}\frac{\partial x^{\beta}}{\partial a^{\alpha}} \tag{61}$$

are the equilibrium equations for the "Piola tensor" $\underline{\underline{Q}}$

$$\overset{0}{\underline{\nabla}}\cdot\underline{\underline{Q}}^{\times} = e^{\gamma\alpha}\underline{i}_{\beta}\frac{\partial}{\partial a^{\gamma}}\frac{\partial\tilde{\ni}}{\partial p^{\alpha}_{\cdot\beta}} = \underline{0}\ . \tag{62}$$

These formulas demonstrate the duality of the strain and stress fields. The plane displacement field in a material with specific strain energy $\ni$ is determined by Equations (47), and the state of stress in this material - by the formulas

$$\varphi_1 = \varphi_1(a^1,a^2)\ , \qquad \varphi_2 = \varphi_2(a^1,a^2)\ . \tag{63}$$

If we now introduce the "displacement field", (63), into a material with specific "strain" energy $\ni$, then, by (59), the stress field in this material will be determined by Equations (47).

The correspondence between the displacement and stress fields can be represented in a table

	$\ni$-material	$\tilde{\ni}$-material
The displacement field	x^1, x^2	φ_1,φ_2
The stress field	φ_1, φ_2	x^1,x^2

By (51) and (58) we have also

$$\underset{-}{P}^{\times} \cdot\cdot \overset{0}{\underset{--}{\nabla B}}{}^{T} = 2\mu \underset{-\alpha-}{i} \underset{}{i}^{\beta} e^{\alpha\gamma} \frac{\partial\varphi_\beta}{\partial a^\gamma} \cdot\cdot \underset{-\varepsilon-}{i} \underset{}{i}^{\delta} \frac{\partial x^\varepsilon}{\partial a^\delta} = 2\mu e^{\alpha\gamma} \frac{\partial\varphi_\beta}{\partial a^\gamma} \frac{\partial x^\beta}{\partial a^\alpha} ,$$

which implies the symmetric representations

$$\tilde{\ni} = 2\mu e^{\alpha\gamma} \frac{\partial\varphi_\beta}{\partial a^\gamma} \frac{\partial x^\beta}{\partial a^\alpha} - \ni , \qquad \ni = 2\mu e^{\alpha\gamma} \frac{\partial x^\beta}{\partial a^\alpha} \frac{\partial\varphi_\beta}{\partial a^\gamma} - \tilde{\ni} . \tag{64}$$

The simplest illustration is provided by the problem of the bending of a strip into a right cylindrical form considered in Chapter 6, Par. 9. In this case, by (58) and in view of Equations (6.9.6) and (6.8.14) the following representations hold

$$\varphi_1 = \psi(a^1)\cos\chi - x^2 , \qquad \varphi_2 = \psi(a^1)\sin\chi + x^1 .$$

We also have

$$\frac{\partial\varphi_1}{\partial a^1} + \frac{\partial\varphi_2}{\partial a^2} = -\frac{\partial x^2}{\partial a^1} + \frac{\partial x^1}{\partial a^2} , \qquad \frac{\partial\varphi_2}{\partial a^1} - \frac{\partial\varphi_1}{\partial a^2} = \frac{\partial x^1}{\partial a^1} + \frac{\partial x^2}{\partial a^2} ,$$

where

$$\psi(a^1) = \frac{b}{\alpha} \frac{\exp \dfrac{\alpha a^1}{b}}{\operatorname{ch}\gamma} , \qquad \chi = \frac{\alpha a^2}{b} , \qquad \operatorname{ch}\gamma = \frac{h\alpha}{2b} .$$

6. SECOND ORDER EFFECTS IN THE PROBLEM OF THE PLANE DEFORMATION OF AN INCOMPRESSIBLE MATERIAL

1. The complex representation of the displacement vector $\underset{-}{D}$ is given by the expression

$$D(z,\bar{z}) = z - \zeta(z,\bar{z}) \tag{1}$$

and in Equations (5.39) and (5.40) the substitutions should be made

$$\frac{\partial\zeta}{\partial z} = 1 - \frac{\partial D}{\partial z} , \qquad \frac{\partial\zeta}{\partial\bar{z}} = -\frac{\partial D}{\partial\bar{z}} , \qquad \frac{\partial\bar{\zeta}}{\partial\bar{z}} = 1 - \frac{\partial\bar{D}}{\partial\bar{z}} , \qquad \frac{\partial\bar{\zeta}}{\partial z} = -\frac{\partial\bar{D}}{\partial z} .$$

We further assume that $c = 1$, i.e. that the faces of the deformed cylindrical body are not displaced longitudinally ($x_3 = a_3$). We then have

$$\frac{\partial^2 U}{\partial z^2} = 2\frac{d\partial^\times}{dI_1}\left(\frac{\partial D}{\partial z} - 1\right)\frac{\partial \bar{D}}{\partial z}, \qquad \frac{\partial D}{\partial z} + \frac{\partial \bar{D}}{\partial \bar{z}} - \left(\frac{\partial D}{\partial z}\frac{\partial \bar{D}}{\partial \bar{z}} - \frac{\partial D}{\partial \bar{z}}\frac{\partial \bar{D}}{\partial z}\right) = 0 , \tag{2}$$

and from (5.41) the invariant I_1^0 is represented as

$$I_1^0 = 2 + 4\frac{\partial \bar{D}}{\partial z}\frac{\partial D}{\partial \bar{z}} . \tag{3}$$

In order to determine the second order effects we assume that

$$U = U_0 + U_1 , \qquad D = D_0 + D_1 , \tag{4}$$

retaining in the subsequent relations those terms which are quadratic in D_0 and linear in D_1. From (3) and (5.21) we obtain, to the required approximation,

$$\frac{d\partial^\times}{dI_1} = \frac{\partial \partial}{\partial I_1} + \frac{\partial \partial}{\partial I_2} = \frac{1}{2}\mu + \kappa\frac{\partial D_0}{\partial \bar{z}}\frac{\partial \bar{D}_0}{\partial z} , \qquad \kappa = 2\left(\frac{d^2\partial^\times}{dI_1^2}\right)_{I_1^0=2}$$

and from (2) the equations for the first and the second approximations reduce to the forms

$$\frac{\partial^2 U_0}{\partial z^2} = -\mu\frac{\partial \bar{D}_0}{\partial z} , \qquad \frac{\partial D_0}{\partial z} + \frac{\partial \bar{D}_0}{\partial \bar{z}} = 0 , \tag{5}$$

$$\frac{\partial^2 U_1}{\partial z^2} = -\mu\frac{\partial \bar{D}_1}{\partial z} + \mu\frac{\partial D_0}{\partial z}\frac{\partial \bar{D}_0}{\partial z} , \qquad \frac{\partial D_1}{\partial z} + \frac{\partial \bar{D}_1}{\partial \bar{z}} = \frac{\partial D_0}{\partial z}\frac{\partial \bar{D}_0}{\partial \bar{z}} - \frac{\partial D_0}{\partial \bar{z}}\frac{\partial \bar{D}_0}{\partial z} . \tag{6}$$

These equations have the same form for all incompressible materials - the value of the constant parameter κ need not be known for the determination of U_1 and D_1.

Equation (5) is the equation of a plane problem in the linear elasticity theory of an incompressible material (of Poisson ratio 1/2). In this theory, the unknowns U_0 and D_0 are expressed through two analytic functions - the Muskhelishvili potentials

$$\begin{aligned} U_0 &= \frac{1}{2}\left[\bar{z}\varphi_0(z) + z\overline{\varphi_0(z)} + \chi_0(z) + \overline{\chi_0(z)}\right] , \\ 2\mu D_0 &= \varphi_0(z) - z\overline{\varphi_0'(z)} + \overline{\psi_0(z)} , \end{aligned} \tag{7}$$

where $\psi_0(z) = \chi_0'(z)$. These expressions are easily shown to be compatible with (5). The potentials $\varphi_0(z)$ and $\psi_0(z)$ are easily determined from the solution of a linear boundary value problem.

Equation $(6)_1$ can be integrated as follows

$$\frac{\partial U_1}{\partial z} + \mu\bar{D}_1 = \mu \int^{z} \frac{\partial D_0}{\partial z} \frac{\partial \bar{D}_0}{\partial z} dz + \overline{\varphi_1(z)} \ . \tag{8}$$

Making also use of $(5)_2$ we have

$$\frac{\partial^2 U_1}{\partial\bar{z}\partial z} + \mu \frac{\partial \bar{D}_1}{\partial \bar{z}} = \mu \int^{z} \left(\frac{\partial^2 D_0}{\partial\bar{z}\partial z} \frac{\partial \bar{D}_0}{\partial z} + \frac{\partial D_0}{\partial z} \frac{\partial}{\partial z} \frac{\partial \bar{D}_0}{\partial \bar{z}} \right) dz + \overline{\varphi_1'(z)} =$$

$$= -\mu \int^{z} \frac{\partial^2 \bar{D}_0}{\partial \bar{z}^2} \frac{\partial \bar{D}_0}{\partial z} dz - \frac{1}{2}\mu \left(\frac{\partial D_0}{\partial z} \right)^2 + \overline{\varphi_1'(z)} \ .$$

Further, by virtue of $(6)_2$ and $(5)_2$ and because the value of $\partial^2 U/\partial z\partial\bar{z}$ is real

$$2 \frac{\partial^2 U_1}{\partial z\partial\bar{z}} = -\mu \left[\int^{z} \frac{\partial^2 D_0}{\partial \bar{z}^2} \frac{\partial \bar{D}_0}{\partial z} dz + \int^{\bar{z}} \frac{\partial^2 D_0}{\partial z^2} \frac{\partial D_0}{\partial \bar{z}} d\bar{z} \right] -$$

$$- \frac{1}{2}\mu \left[\left(\frac{\partial D_0}{\partial z} \right)^2 + \left(\frac{\partial \bar{D}_0}{\partial \bar{z}} \right)^2 \right] + \varphi_1'(z) + \overline{\varphi_1'(z)} - \mu \frac{\partial D_0}{\partial z} \frac{\partial \bar{D}_0}{\partial \bar{z}} + \mu \frac{\partial D_0}{\partial \bar{z}} \frac{\partial \bar{D}_0}{\partial z} =$$

$$= -\mu \left[\int^{z} \frac{\partial^2 \bar{D}_0}{\partial \bar{z}^2} \frac{\partial \bar{D}_0}{\partial z} dz + \int^{\bar{z}} \frac{\partial^2 D_0}{\partial z^2} \frac{\partial D_0}{\partial \bar{z}} d\bar{z} \right] + \mu \frac{\partial D_0}{\partial \bar{z}} \frac{\partial \bar{D}_0}{\partial z} + \varphi_1'(z) + \overline{\varphi_1'(z)} \ .$$

After integrating with the aid of $(5)_2$ and the relationships of Equation $(7)_2$

$$\frac{\partial^3 D_0}{\partial\bar{z}\partial z^2} = 0 \ , \qquad \frac{\partial^3 \bar{D}_0}{\partial z\partial\bar{z}^2} = 0 \ ,$$

we get

$$\frac{\partial^2 U_1}{\partial z\partial\bar{z}} = \frac{1}{2} [\varphi_1'(z) + \overline{\varphi_1'(z)}] + \frac{1}{2}\mu \left[\frac{\partial^2 D_0}{\partial\bar{z}\partial z} \bar{D}_0 + \frac{\partial^2 \bar{D}_0}{\partial\bar{z}\partial z} D_0 + \frac{\partial D_0}{\partial \bar{z}} \frac{\partial \bar{D}_0}{\partial z} \right] \tag{9}$$

and integrating again with respect to z we obtain the required representations

$$\frac{\partial U_1}{\partial \bar{z}} = \frac{1}{2} [\varphi_1(z) + z\overline{\varphi_1'(z)} + \overline{\psi_1(z)}] + \frac{1}{2}\mu\Gamma(z,\bar{z}) \ , \tag{10}$$

$$\mu D_1 = \frac{1}{2} [\varphi_1(z) - z\overline{\varphi_1'(z)} - \overline{\psi_1(z)}] + \mu \int^{\bar{z}} \frac{\partial \bar{D}_0}{\partial \bar{z}} \frac{\partial D_0}{\partial \bar{z}} d\bar{z} - \frac{\mu}{2}\Gamma(z,\bar{z}) \ , \tag{11}$$

$$\Gamma(z,\bar{z}) = \frac{\partial D_0}{\partial \bar{z}} \bar{D}_0 + \frac{\partial \bar{D}_0}{\partial \bar{z}} D_0 + \int^z \left(\frac{\partial D_0}{\partial z}\right)^2 dz \ . \tag{12}$$

2. The potentials $\varphi_1(z)$ and $\psi_1(z)$ are determined by the solution of a linear boundary value problem to second order approximation. The procedure has been worked out in great detail, and the associated calculations are tedious and complex.

In the first approximation, as well as for the second order effects, the stresses are determined by (5.38) and (5.39) with the aid of the expressions

$$\frac{1}{2}(\sigma_{x^1} + \sigma_{x^2})^0 = 2\frac{\partial^2 U_0}{\partial z \partial \bar{z}} = \varphi_0'(z) + \overline{\varphi_0'(z)} \ ,$$

$$\frac{1}{2}(\sigma_{x^2} - \sigma_{x^1} + 2i\tau_{x^1x^2})^0 = 2\frac{\partial^2 U_0}{\partial z^2} = -2\mu\frac{\partial \bar{D}_0}{\partial z} = \bar{z}\varphi_0''(z) + \varphi_0'(z) \ , \tag{13}$$

$$\frac{1}{2}(\sigma_{x^1} + \sigma_{x^2})^1 = 2\frac{\partial^2 U_1}{\partial z \partial \bar{z}} = \varphi_1'(z) + \overline{\varphi_1'(z)} + \mu\left(\frac{\partial^2 D_0}{\partial z \partial \bar{z}}\bar{D}_0 + \frac{\partial^2 \bar{D}_0}{\partial z \partial \bar{z}} D_0 + \frac{\partial D_0}{\partial \bar{z}}\frac{\partial \bar{D}_0}{\partial z}\right) , \tag{14}$$

$$\frac{1}{2}(\sigma_{x^2} - \sigma_{x^1} + 2i\tau_{x^1x^2})^1 = 2\frac{\partial^2 U_1}{\partial z^2} = \bar{z}\varphi_1''(z) + \psi_1'(z) + \mu\overline{\frac{\partial \Gamma(z,\bar{z})}{\partial z}} \ .$$

By (5.38) or (5.30) we also get

$$p = p^0 + p^1 = \left(-2\frac{\partial^2 U_0}{\partial z \partial \bar{z}} + \mu\right) + \left[-2\frac{\partial^2 U_1}{\partial z \partial \bar{z}} + 2(\mu + \kappa)\frac{\partial D_0}{\partial \bar{z}}\frac{\partial \bar{D}_0}{\partial z}\right] . \tag{15}$$

3. It is well known that the boundary value problem for the state of stress in the domain whose boundary is loaded can be reduced to the functional equation

$$\frac{1}{2}[(\sigma_{x^1} + \sigma_{x^2})n - (\sigma_{x^2} - \sigma_{x^1} - 2i\tau_{x^1x^2})\bar{n}] =$$

$$= [\Phi(z) + \overline{\Phi(z)}]n - [z\overline{\Phi'(z)} + \overline{\Psi(z)}]\bar{n} = -f \ . \tag{16}$$

Here $\Phi(z) = \varphi'(z)$, $\Psi(z) = \psi'(z)$ denote the "second" Muskhelishvili potentials, n denotes the outward normal to the boundary of the domain, and f is the surface traction which is statically equivalent to zero. We first represent the terms in (16) by sums such as those of (3), i.e. of linear terms and those associated with second order effects. We then assume that the solution to the linear problem is given by the boundary condition

$$[\Phi_0(z) + \overline{\Phi_0(z)}]n - [z\overline{\Phi_0'(z)} + \overline{\Psi_0(z)}]\bar{n} = -f \ . \tag{17}$$

The second order effects are now determined from the boundary condition

$$\frac{1}{2}(\sigma_{x^1} + \sigma_{x^2})^1 n - \frac{1}{2}(\sigma_{x^2} - \sigma_{x^1} + 2i\tau_{x^1x^2})^1\bar{n} = 0 \ , \tag{18}$$

in which the stresses σ_{x^1} etc. are given by Equations (14). We thus arrive at the functional equation

$$[\Phi_1(z) + \overline{\Phi_1(z)}]n - [z\overline{\Phi_1'(z)} + \overline{\Psi_1(z)}]\bar{n} = nP + \bar{n}Q \ , \tag{19}$$

where, from (14), P and Q are prescribed as follows

$$\begin{aligned} P &= -\mu\left(\frac{\partial^2 D_0}{\partial z\partial\bar{z}}\bar{D}_0 + \frac{\partial^2\bar{D}_0}{\partial z\partial\bar{z}}D_0 + \frac{\partial D_0}{\partial\bar{z}}\frac{\partial\bar{D}_0}{\partial z}\right) \ , \\ Q &= \mu\left(\frac{\partial^2 D_0}{\partial\bar{z}^2}\bar{D}_0 + \frac{\partial^2\bar{D}_0}{\partial\bar{z}^2}D_0 + 2\frac{\partial D_0}{\partial\bar{z}}\frac{\partial\bar{D}_0}{\partial\bar{z}} + 2\int^z\frac{\partial D_0}{\partial z}\frac{\partial^2 D_0}{\partial z\partial\bar{z}}dz\right) \end{aligned} \tag{20}$$

and we have also used Equation (10).

The system of "forces" in the right-hand side of (19) should be statically zero-equivalent. As it has already been pointed out in Chapter 6, Par. 14, for its resultant to be zero it is sufficient that one of the equalities

$$\frac{\partial P}{\partial\bar{z}} + \frac{\partial Q}{\partial z} = 0 \ , \qquad \frac{\partial\bar{P}}{\partial z} + \frac{\partial\bar{Q}}{\partial\bar{z}} = 0 \tag{21}$$

be satisfied. If this condition holds, then the resulting moment

$$P - \bar{P} = 0 \ , \tag{22}$$

will also be zero because, by (20), the value of P is real.

The verification of Equations (21) is cumbersome. After elimination, differentiation and subsequent simplification with the use of (5), we arrive at the condition

$$\frac{\partial^2 D_0}{\partial z\partial\bar{z}} + \frac{\partial^2\bar{D}_0}{\partial\bar{z}^2} = 0 \ ,$$

which can be seen to hold from $(7)_2$. The necessary condition of the solvability of the boundary value problem (19) is thus satisfied.

7. THE AFFINE TRANSFORMATION OF A REFERENCE CONFIGURATION IN AN INCOMPRESSIBLE ELASTIC BODY

1. The linear state of stress produced by uniaxial stretching. The principal values of the Finger strain measure and its invariants are

$$v_1^2 = v^2, \quad v_2^2 = v_3^2 = v^{-1}, \quad I_1 = v^2 + 2v^{-1}, \quad I_2 = 2v + v^{-2}, \quad I_3 = 1 \tag{1}$$

and from (2.8), the expressions for the principal stresses σ_s take the form

$$\sigma_2 = \sigma_3 = -p + 2\left(\frac{\partial \ni}{\partial I_1} v^{-1} - v \frac{\partial \ni}{\partial I_2}\right) = 0 ,$$

$$\sigma_1 = -p + 2\left(\frac{\partial \ni}{\partial I_2} v^2 - \frac{\partial \ni}{\partial I_2} v^{-2}\right) . \tag{2}$$

The elimination of the pressure p from these relationships leads to the relationship for pure extension $\sigma_1 = \sigma_1(v)$ and to the representation of the principal force $t_1 = t_1(v)$ by the formulas

$$\sigma_1 = 2\left[\frac{\partial \ni}{\partial I_1} (v^2 - v^{-1}) - \frac{\partial \ni}{\partial I_2} (v^{-2} - v)\right] ,$$

$$t_1 = 2(v - v^{-2})\left(\frac{\partial \ni}{\partial I_1} + v^{-1} \frac{\partial \ni}{\partial I_2}\right) . \tag{3}$$

From (2) it follows that for any prescribed value of v we may require the choice of p to be restricted bv the condition $\sigma_2 = \sigma_3 = 0$. This leads to the conclusion that to any initially prescribed value of v there is a corresponding unique value of stress σ_1.

In accordance with the BE-criterion (4.12.12)

$$\frac{\sigma_1 - \sigma_2}{v_1^2 - v_2^2} = \frac{\sigma_1 - \sigma_3}{v_1^2 - v_3^2} = \frac{\sigma_1}{v - \dfrac{1}{\sqrt{v}}} > 0 ,$$

$$\operatorname{sgn} \sigma_1 = \operatorname{sgn}\left(v - \frac{1}{\sqrt{v}}\right) = \operatorname{sgn}(\sqrt{v} - 1) ,$$

where $\sqrt{v} - 1 > 0$ when the rod is lengthening and $\sqrt{v} - 1 < 0$ when it is shortening. In other words, when the rod is stretched ($\sigma_1 > 0$), its length increases, and when it is compressed ($\sigma_1 < 0$), its length decreases. From the BE-criterion alone we would not be able to assert that this "trivial" conclusion could be extended to an isotropic compressible material.

2. Simple shear. For this type of deformation (Chapter 6, Par. 4) to be realized in a compressible material, it is necessary to apply normal stresses to each face of a deformable parallelepiped, including a stress t^{33} perpendicular to the plane of shear. In an incompressible material, a simple shear is possible without loading these faces. In fact, assuming

$$p = 2\frac{\partial \ni}{\partial I_1} + 2(2+s^2)\frac{\partial \ni}{\partial I_2}, \tag{4}$$

we get

$$t = 2\frac{\partial \ni}{\partial I_1} + 4\frac{\partial \ni}{\partial I_2} - p = -2s^2\frac{\partial \ni}{\partial I_2}$$

and in accordance with (6.4.7) the formulas determining the normal stresses in an incompressible material take the form

$$t^{11} = 2\frac{\partial \ni}{\partial I_1}s^2, \qquad t^{22} = -2\frac{\partial \ni}{\partial I_2}s^2, \qquad t^{33} = 0$$

and the normal stresses on the planes a^3 = const are seen to be superfluous. The same formulas combined with the empirical criteria (5.11.3) (acceptable for rubber) show that the inequality $t^{22} < 0$ (compression holds on the planes a^2 = = const which are displaced along the axis OX, whereas $t^{11} > 0$ (stretching) holds on the planes a^1 = const.

Similarly it can be shown that simple shear can be realized in the absence of normal loading on the faces a^1 = const or a^2 = const.

3. Plane stress. The planes normal to the principal axis $\underline{e}_3$ of the stress tensor are load-free and for this case

$$p = 2\left(\frac{\partial \ni}{\partial I_1}v_3^2 - \frac{\partial \ni}{\partial I_2}v_3^{-2}\right),$$

$$t_3 = 0, \qquad t_1 = 2\left(v_1^2 - \frac{1}{v_1^2v_2^2}\right)\left(\frac{\partial \ni}{\partial I_1} + v_2^2\frac{\partial \ni}{\partial I_2}\right),$$

$$t_2 = 2\left(v_2^2 - \frac{1}{v_1^2v_3^2}\right)\left(\frac{\partial \ni}{\partial I_1} + v_1^2\frac{\partial \ni}{\partial I_2}\right);$$

$$I_1 = v_1^2 + v_2^2 + (v_1v_2)^{-2}, \qquad I_2 = v_1^{-2} + v_2^{-2} + v_1^2v_2^2 .$$

These formulas show that the function $\ni(I_1, I_2)$ may be determined by a series of experiments on the state of plane stress.

The experiments of Rivlin and Saunders (R.S. Rivlin and D.W. Saunders, 1951) on slabs of zero-hysteresis rubber consisted of measuring the stresses associated with specific deformations in which one of the principal invariants was held constant and the other varied over a fairly wide range, the linear dimensions being increased up to four times. The program assumed values of I_1 and I_2 such that $5 \le I_1 \le 11$ and $5 \le I_2 \le 30$. It has been found that the stresses are only slightly sensitive to variation of I_1 and depend considerably upon I_2. Measurements were taken of extension, contraction and pure shear and for pure shear combined with a superposed extension. On the basis of these experiments Rivlin and Saunders proposed that the Mooney potential, (4.1), be replaced by the modified expression (4.6).

8. THE UNIVERSAL DEFORMATIONS OF AN INCOMPRESSIBLE MATERIAL

In his definitive papers of 1948-54, Rivlin gave rigorous 'universal' ('general') solutions to certain problems of the continuum mechanics of an incompressible medium, i.e. solutions which did not depend upon a particular prescription of the specific strain energy $\ni(I_1, I_2)$.

In Chapter 4, Par. 15 we gave Shield's proof of Ericksen's theorem (1955) on the non-existence of universal solutions to the equilibrium equations of the non-linear theory of a compressible elastic medium for non-affine transformations of a reference configuration into an actual configuration. Somewhat earlier (in 1954) Ericksen had demonstrated, in a complex and difficult analysis, the existence of four classes of (non-linear) transformations which permit universal, rigorous solutions for an incompressible elastic body (these solutions include, of course, the problems considered by Rivlin). One more class of transformations was later discovered by Shield and Klingbeil (W.W. Klingbeil, R.T. Shield, 1966). These solutions seem to exhaust all the possibilities in the construction of universal solutions to the equilibrium equations of an incompressible elastic body.

We shall consider here solutions to the equilibrium equations of an incompressible elastic body in the absence of body forces

$$\underline{\nabla} \cdot \underline{T} = -\underline{\nabla} p + \underline{\nabla} \cdot \underline{T}_E =$$

$$= -\underline{\nabla} p + 2\left(\underline{F} \cdot \underline{\nabla} \frac{\partial \ni}{\partial I_1} - \underline{F}^{-1} \cdot \underline{\nabla} \frac{\partial \ni}{\partial I_2} + \frac{\partial \ni}{\partial I_1} \underline{\nabla} \cdot \underline{F} - \frac{\partial \ni}{\partial I_2} \underline{\nabla} \cdot \underline{F}^{-1}\right) = \underline{0} \,, \tag{1}$$

these solutions must preserve their form irrespective of the specific prescription of $\ni(I_1, I_2)$ and they should in any case satisfy the conditions for the existence of the position vector $\underline{R}(q^1,q^2,q^3)$, i.e. the Ricci tensor associated with the Finger (or Almansi) strain measure must be zero. The contact forces which produce this equilibrium state can be found from this state whenever it obtains. There is no question of solving some boundary value problem which is formulated in advance.

Equations (1) can be rewritten in a more detailed form as

$$\underline{\nabla}\cdot\underline{T} = -\underline{\nabla}p + 2\Big[\Big(\frac{\partial^2\ni}{\partial I_1^2}\underline{\nabla}I_1 + \frac{\partial^2\ni}{\partial I_2\partial I_1}\underline{\nabla}I_2\Big)\cdot\underline{F} + \frac{\partial\ni}{\partial I_1}\underline{\nabla}\cdot\underline{F} - $$

$$- \Big(\frac{\partial^2\ni}{\partial I_1\partial I_2}\underline{\nabla}I_1 + \frac{\partial^2\ni}{\partial I_2^2}\underline{\nabla}I_2\Big)\cdot\underline{F}^{-1} - \frac{\partial\ni}{\partial I_2}\underline{\nabla}\cdot\underline{F}^{-1}\Big] = \underline{0} \;. \tag{2}$$

In order to eliminate the unknown scalar p we use the equation

$$\underline{\nabla}\times\underline{\nabla}\cdot\underline{T} = \underline{0} \;. \tag{3}$$

and take into account the relationship $\underline{\nabla}\times\underline{\nabla}p = \underline{0}$. The calculation of the other terms (each of which is a product of a scalar and vector quantity) is based upon the rule

$$\underline{\nabla}\times\varphi\underline{a} = (\underline{\nabla}\varphi)\times\underline{a} + \varphi\underline{\nabla}\times\underline{a} \;.$$

For example

$$\underline{\nabla}\times\frac{\partial^2\ni}{\partial I_1^2}\underline{\nabla}I_1\cdot\underline{F} = \underline{\nabla}\frac{\partial^2\ni}{\partial I_1^2}\times\underline{\nabla}I_1\cdot\underline{F} + \frac{\partial^2\ni}{\partial I_1^2}\underline{\nabla}\times(\underline{\nabla}I_1\cdot\underline{F}) =$$

$$= \Big(\frac{\partial^3\ni}{\partial I_1^3}\underline{\nabla}I_1 + \frac{\partial^2\ni}{\partial I_2\partial I_1^2}\underline{\nabla}I_2\Big)\times\underline{\nabla}I_1\cdot\underline{F} + \frac{\partial^2\ni}{\partial I_1^2}\underline{\nabla}\times(\underline{\nabla}I_1\cdot\underline{F}) \;.$$

Equation (3) may now be rewritten as follows

$$\frac{\partial^3\ni}{\partial I_1^3}\underline{\nabla}I_1\times(\underline{\nabla}I_1\cdot\underline{F}) - \frac{\partial^3\ni}{\partial I_2^3}\underline{\nabla}I_2\times(\underline{\nabla}I_2\cdot\underline{F}^{-1}) +$$

$$+ \frac{\partial^3\ni}{\partial I_1^2\partial I_2}[\underline{\nabla}I_2\times(\underline{\nabla}I_1\cdot\underline{F}) + \underline{\nabla}I_1\times(\underline{\nabla}I_2\cdot\underline{F}) - \underline{\nabla}I_1\times(\underline{\nabla}I_1\cdot\underline{F}^{-1})] +$$

$$+ \frac{\partial^3\ni}{\partial I_1\partial I_2^2}[\underline{\nabla}I_2\times(\underline{\nabla}I_2\cdot\underline{F}) - \underline{\nabla}I_2\times(\underline{\nabla}I_1\cdot\underline{F}^{-1}) - \underline{\nabla}I_1\times(\underline{\nabla}I_2\cdot\underline{F}^{-1})] +$$

$$+ \frac{\partial^2\ni}{\partial I_1^2}(\underline{\nabla}\times(\underline{\nabla}I_1\cdot\underline{F}) + \underline{\nabla}I_1\times\underline{\nabla}\cdot\underline{F}) -$$

$$- \frac{\partial^2 \ni}{\partial I_2^2} (\underline{\nabla} \times (\underline{\nabla I}_2 \cdot \underline{F}^{-1}) + \underline{\nabla I}_2 \times \underline{\nabla} \cdot \underline{F}^{-1}) +$$

$$+ \frac{\partial^2 \ni}{\partial I_1 \partial I_2} (\underline{\nabla} \times (\underline{\nabla I}_2 \cdot \underline{F}) + \underline{\nabla I}_2 \times \underline{\nabla} \cdot \underline{F} - \underline{\nabla} \times (\underline{\nabla I}_1 \cdot \underline{F}^{-1}) - \underline{\nabla I}_1 \times \underline{\nabla} \cdot \underline{F}^{-1}) +$$

$$+ \frac{\partial \ni}{\partial I_1} \underline{\nabla} \times \underline{\nabla} \cdot \underline{F} - \frac{\partial \ni}{\partial I_2} \underline{\nabla} \times \underline{\nabla} \cdot \underline{F}^{-1} = \underline{0} \,. \tag{4}$$

For universal solutions which do not depend on the form of the expression for the specific strain energy as a function of invariants, this relationship will be satisfied only provided that the coefficients of the derivatives of $\ni$ are all equal to zero. We arrive then at the following system of equations

$$\underline{\nabla I}_1 \times (\underline{\nabla I}_1 \cdot \underline{F}) = \underline{0} \,, \qquad \underline{\nabla I}_2 \times (\underline{\nabla I}_2 \cdot \underline{F}^{-1}) = \underline{0} \,, \tag{5}$$

$$\begin{aligned} &\underline{\nabla I}_2 \times (\underline{\nabla I}_1 \cdot \underline{F}) + \underline{\nabla I}_1 \times (\underline{\nabla I}_2 \cdot \underline{F}) - \underline{\nabla I}_1 \times (\underline{\nabla I}_1 \cdot \underline{F}^{-1}) = \underline{0} \,, \\ &\underline{\nabla I}_1 \times (\underline{\nabla I}_2 \cdot \underline{F}^{-1}) + \underline{\nabla I}_2 \times (\underline{\nabla I}_1 \cdot \underline{F}^{-1}) - \underline{\nabla I}_2 \times (\underline{\nabla I}_2 \cdot \underline{F}) = \underline{0} \,, \end{aligned} \tag{6}$$

$$\underline{\nabla} \times (\underline{\nabla I}_1 \cdot \underline{F}) + \underline{\nabla I}_1 \times \underline{\nabla} \cdot \underline{F} = \underline{0} \,, \qquad \underline{\nabla} \times (\underline{\nabla I}_2 \cdot \underline{F}^{-1}) + \underline{\nabla I}_2 \times \underline{\nabla} \cdot \underline{F}^{-1} = \underline{0} \,, \tag{7}$$

$$\underline{\nabla} \times (\underline{\nabla I}_2 \cdot \underline{F}) + \underline{\nabla I}_2 \times \underline{\nabla} \cdot \underline{F} - \underline{\nabla} \times (\underline{\nabla I}_1 \cdot \underline{F}^{-1}) - \underline{\nabla I}_1 \times \underline{\nabla} \cdot \underline{F}^{-1} = \underline{0} \,, \tag{8}$$

$$\underline{\nabla} \times \underline{\nabla} \cdot \underline{F} = \underline{0} \,, \qquad \underline{\nabla} \times \underline{\nabla} \cdot \underline{F}^{-1} = \underline{0} \,. \tag{9}$$

This system is supplemented by the condition of incompressibility

$$I_3(\underline{F}) = 1$$

and by the six equations which indicate the vanishing of the components of the Ricci tensor (III.10.21). These components are expressed in terms of the components g^{sk}, g_{sk} of the unknown tensors $\underline{F}$ and $\underline{F}^{-1} = g$. It is also essential that the inequalities which express the positiveness of the tensor $\underline{F}$ should be fulfilled, i.e. that its eigenvalues $c_s = v_s^2$ which are connected by the equality $c_1 c_2 c_3 = 1$ should be positive.

This system of $9 \times 3 + 1 + 6 = 34$ equations is very much overspecified. It is satisfied by a tensor $\underline{F}$ which remains constant within the volume of the body and determines, as is to be exptected, a linear transformation of the reference configuration into the actual configuration. It turns out, however, that this solution does not exhaust the set of all possible transformations, i.e. the totality of tensors $\underline{F}$ satisfying all the cited requirements.

9. THE EXACT GENERAL SOLUTIONS

An investigation of the systems of equations of Par. 8 and the derivation of representations of the tensor $\underline{F}$ is carried out in Par. 21 of this chapter. We cite here only the results obtained.

We shall use the notation a^1, a^2, a^3 and x^1, x^2, x^3 for the Cartesian coordinates in the reference and the actual configurations, r, φ, z and R, Φ, Z for the cylindrical coordinates, and r, ϑ, λ, R, θ, Λ for the spherical ccordinates. The rules of differentiation of the unit vectors of the cylindrical ($\underline{e}_r$, $\underline{e}_\varphi$, $\underline{k}$ and $\underline{e}_R$, $\underline{e}_\Phi$, $\underline{k}$) and spherical ($\underline{e}_r$, $\underline{e}_\vartheta$, $\underline{e}_\lambda$ and $\underline{e}_R$, $\underline{e}_\theta$, $\underline{e}_\Lambda$) coordinate systems are given in III, Par. 7 (see Equations (III.7.18)-(III.7.20)).

The coordinates (Cartesian, cylindrical or spherical) of the reference configuration are chosen for the material coordinates. The Ericksen sets prescribe the position coordinates in the actual configuration as functions of the material coordinates, in other words, they represent the transformations $\underline{R} = \underline{R}(\underline{r})$ satisfying the condition of incompressibility $\det \overset{0}{\underline{\nabla}}\underline{R} = 1$.

1. <u>The bending of a plate into a right cylindrical form</u>. The set of transformations

$$R = \sqrt{2Aa^1}\,, \qquad \Phi = Ba^2 + Ca^3\,, \qquad Z = Da^2 + Ea^3\,,$$

$$\underline{R} = \underline{e}_R\sqrt{2Aa^1} + \underline{k}(Da^2 + Ea^3) \tag{1}$$

sets the correspondence between a plane sheet ($a^1_0 \le a^1 < a^1_0 + h$, $-b \le a^2 \le b$, $-l \le a^3 \le l$) and a cylindrical section. The constants are connected by the condition of incompressibility

$$A(BE - CD) = 1\,. \tag{2}$$

The vector bases of the actual state are represented by the formulas

$$\underline{R}_1 = \frac{\partial \underline{R}}{\partial a^1} = \frac{A}{R}\underline{e}_R\,,$$

$$\underline{R}_2 = \frac{\partial \underline{R}}{\partial a^2} = R\frac{\partial \underline{e}_R}{\partial a^2} + \frac{\partial Z}{\partial a^2}\underline{k} = R\frac{\partial \underline{e}_R}{\partial \Phi}\frac{\partial \Phi}{\partial a^2} + \underline{k}D = R\underline{e}_\Phi B + \underline{k}D\,,$$

$$\underline{R}_3 = R\underline{e}_\Phi C + \underline{k}E\,;$$

$$\underline{R}^1 = \frac{R}{A}\underline{e}_R\,, \qquad \underline{R}^2 = A\left(\frac{E}{R}\underline{e}_\Phi - \underline{k}C\right), \tag{3}$$

$$\underline{R}^3 = A\left(-\frac{D}{R}\,\underline{e}_\Phi + \underline{k}B\right) .$$

The expressions for the deformation gradients and for the Finger and Almansi measures are reduced to the form

$$\overset{0}{\underline{\underline{\nabla R}}} = \underline{i}^s\underline{R}_s = \frac{A}{R}\,\underline{i}_1\underline{e}_R + R(B\underline{i}_2 + C\underline{i}_3)\underline{e}_\Phi + (D\underline{i}_2 + E\underline{i}_3)\underline{k} ,$$

$$\underline{\underline{\nabla r}} = \frac{R}{A}\,\underline{e}_R\underline{i}_1 + \frac{A}{R}\,\underline{e}_\Phi(E\underline{i}_2 - D\underline{i}_3) + A\underline{k}(-C\underline{i}_2 + B\underline{i}_3) , \tag{4}$$

$$\underline{F} = \overset{0}{\underline{\underline{\nabla R}}}^{T} \cdot \overset{0}{\underline{\underline{\nabla R}}} = \frac{A^2}{R^2}\,\underline{e}_R\underline{e}_R + R^2(B^2 + C^2)\underline{e}_\Phi\underline{e}_\Phi + (D^2 + E^2)\underline{\underline{kk}} +$$

$$+ R(BD + EC)(\underline{e}_\Phi\underline{k} + \underline{k}\underline{e}_\Phi) , \tag{5}$$

$$\underline{g} = \underline{F}^{-1} = \underline{\underline{\nabla r}} \cdot \underline{\underline{\nabla r}}^{T} = \frac{R^2}{A^2}\,\underline{e}_R\underline{e}_R + \frac{A^2}{R^2}\,(E^2 + D^2)\underline{e}_\Phi\underline{e}_\Phi +$$

$$+ A^2(C^2 + B^2)\underline{\underline{kk}} - \frac{A^2}{R}\,(BD + AC)(\underline{e}_\Phi\underline{k} + \underline{k}\underline{e}_\Phi) . \tag{6}$$

The components of these tensors, i.e. their invariants too, depend on the single coordinate R of the actual configuration, that is on the coordinate a^1 of the reference configuration. This is a general property of universal solutions. It predetermines the simplicity of the equations of static equilibrium, their solution is reduced to the evaluation of the function p which enables us to calculate the components of a stress tensor with the aid of the state equation.

The invariants $I_1(\underline{F})$, $I_2(\underline{F}) = I_1(\underline{g})$ and the divergences $\underline{\nabla} \cdot \underline{F}$ and $\underline{\nabla} \cdot \underline{g}$ are given by

$$I_1(\underline{F}) = \frac{A^2}{R^2} + R^2(B^2 + C^2) + (D^2 + E^2) ,$$

$$I_2(\underline{F}) = \frac{R^2}{A^2} + \frac{A^2}{R^2}\,(E^2 + D^2) + A^2(C^2 + B^2) , \tag{7}$$

$$\underline{\nabla} \cdot \underline{F} = -\left[\frac{A^2}{R^2} + R^2(B^2 + C^2)\right]\underline{e}_R , \qquad \underline{\nabla} \cdot \underline{g} = \left[\frac{3R}{A^2} - \frac{A^2}{R^3}\,(E^2 + D^2)\right]\underline{e}_R . \tag{8}$$

The function $\ni(I_1, I_2)$ depends, of course, only upon R. This enables us to write down the equations of static equilibrium in the following form

$$\underline{\nabla}p = 2\underline{\nabla} \cdot \underline{T}_E = 2\underline{\nabla} \cdot \left(\frac{\partial \ni}{\partial I_1}\,\underline{F} - \frac{\partial \ni}{\partial I_2}\,\underline{F}^{-1}\right) =$$

$$= 2\left[\left(\frac{\partial \ni}{\partial I_1}\underline{\nabla}\cdot\underline{F} - \frac{\partial \ni}{\partial I_2}\underline{\nabla}\cdot\underline{F}^{-1}\right) + \left(\underline{\nabla}\frac{\partial \ni}{\partial I_1}\right)\cdot\underline{F} - \left(\underline{\nabla}\frac{\partial \ni}{\partial I_2}\right)\cdot\underline{F}^{-1}\right] =$$

$$= 2\underline{e}_R\left\{-\frac{\partial \ni}{\partial I_1}\left[\frac{A^2}{R^3} + R(B^2+C^2)\right] - \frac{\partial \ni}{\partial I_2}\left[\frac{3R}{A^2} - \frac{A^2}{R^3}(E^2+D^2)\right] +\right.$$

$$\left. + \frac{A^2}{R^2}\frac{d}{dR}\frac{\partial \ni}{\partial I_1} - \frac{R^2}{A^2}\frac{d}{dR}\frac{\partial \ni}{\partial I_2}\right\} .$$

Taking this together with the equality

$$\underline{\nabla}\ni = \underline{e}_R\left(\frac{\partial \ni}{\partial I_1}\frac{\partial I_1}{\partial R} + \frac{\partial \ni}{\partial I_2}\frac{\partial I_2}{\partial R}\right) =$$

$$= 2\left\{\frac{\partial \ni}{\partial I_1}\left[-\frac{A^2}{R^3} + R(B^2+C^2)\right] + \frac{\partial \ni}{\partial I_2}\left[\frac{R}{A^2} - \frac{A^2}{R^3}(E^2+D^2)\right]\right\}\underline{e}_R ,$$

we arrive at the relationship

$$\underline{\nabla}(p+\ni) = 2\underline{e}_R\left[-\frac{2A^2}{R^3}\frac{\partial \ni}{\partial I_1} - \frac{2R}{A^2}\frac{\partial \ni}{\partial I_2} + \frac{A^2}{R^2}\frac{d}{dR}\frac{\partial \ni}{\partial I_1} - \frac{R^2}{A^2}\frac{d}{dR}\frac{\partial \ni}{\partial I_2}\right] =$$

$$= 2\underline{\nabla}\left(\frac{A^2}{R^2}\frac{\partial \ni}{\partial I_1} - \frac{R^2}{A^2}\frac{\partial \ni}{\partial I_2}\right) .$$

which determines p to the extent of an additive constant

$$p = p_0 + 2\left(\frac{A^2}{R^2}\frac{\partial \ni}{\partial I_1} - \frac{R^2}{A^2}\frac{\partial \ni}{\partial I_2}\right) - \ni .$$

The expression for the stress tensor $\underline{T}$ is given by the state equation

$$\underline{T} = -p\underline{E} + 2\left(\frac{\partial \ni}{\partial I_1}\underline{F} - \frac{\partial \ni}{\partial I_2}\underline{F}^{-1}\right) . \tag{9}$$

By (5) and (6) we arrive at the following representations of its components

$$\sigma_R = \ni - p_0 , \qquad \sigma_\Phi = \ni - p_0 + R\frac{\partial \ni}{\partial R} = \frac{d}{dR}R\sigma_R ,$$

$$\sigma_Z = \sigma_R + 2\left[\frac{\partial \ni}{\partial I_1}\left(D^2 + E^2 - \frac{A^2}{R^2}\right) + \frac{\partial \ni}{\partial I_2}\left(\frac{R^2}{A^2} - A^2C^2 - A^2B^2\right)\right] , \tag{10}$$

$$\tau_{Z\Phi} = 2(BD+EC)\left(\frac{\partial \ni}{\partial I_1}R + \frac{\partial \ni}{\partial I_2}\frac{A^2}{R}\right) , \qquad \tau_{R\Phi} = 0 , \quad \tau_{ZR} = 0 .$$

The solution is seen to be "universal" and is represented uniquely in terms of the strain energy.

It now remains to determine which distribution of the surface tractions produces this stress state. On the "basic" cylindrical surfaces $R = R_0 = \sqrt{2Aa_0^1}$, $R = R_1 = \sqrt{2Aa_1^1}$ of a panel only uniform distribution of the normal forces is admissible

$$[\ni(I_1(R_0), I_2(R_0)) - p_0]\underline{e}_R = f_0\underline{e}_R\ , \qquad [\ni(I_1(R_1), I_2(R_1)) - p_0]\underline{e}_R = f_1\underline{e}_R\ .$$

This imposes the following constraint

$$\ni(I_1(R_0), I_2(R_0)) - \ni(I_1(R_1), I_2(R_1)) = f_0 - f_1$$

on the constants of solution (1), and determines the parameter p_0. The surface tractions which must be distributed over the remaining surfaces of the panel (which are the images of the rectangular faces $a^2 = \pm b$, $a^3 = \pm l$ of the plate) depend only upon R. These tractions are determined by the solution itself, and the prescription of the values of their resultant and moment for any of these surfaces provides a system of equations for the constants $A,\ldots,E$.

2. <u>The flattening of a cylindrical section</u>. The set of transformations

$$\underline{R} = \frac{1}{2}Ar^2\underline{i}_1 + (B\varphi + Cz)\underline{i}_2 + (D\varphi + Ez)\underline{i}_3$$

$$\left(r = (a^{1\,2} + a^{2\,2})^{1/2}\ , \quad \varphi = \operatorname{arctg}\frac{a^2}{a^1}\right) \tag{11}$$

determines the unbending of a cylindrical panel $r_0 \le r \le r_1$, $-\varphi_0 \le \varphi \le \varphi_0$, $-l \le z \le l$ into a plane plate. The condition of incompressibility preserves the form (2).

The calculation of no. 1 is reproduced with a slight modification. The vector bases of the actual configuration and the deformation gradients are equal to

$$\underline{R}_1 = Ar\underline{i}_1\ , \qquad \underline{R}_2 = B\underline{i}_2 + D\underline{i}_3\ , \qquad \underline{R}_3 = C\underline{i}_2 + E\underline{i}_3\ ,$$

$$\underline{R}^1 = \frac{1}{Ar}\underline{i}_1\ , \qquad \underline{R}^2 = A(E\underline{i}_2 - C\underline{i}_3)\ , \qquad \underline{R}^3 = A(-D\underline{i}_2 + B\underline{i}_3)\ ,$$

$$\overset{0}{\underline{\nabla}}\underline{R} = \underline{r}^s\underline{R}_s = Ar\underline{e}_r\underline{i}_1 + \frac{\underline{e}_\varphi}{r}(B\underline{i}_2 + D\underline{i}_3) + \underline{k}(C\underline{i}_2 + E\underline{i}_3)\ ,$$

$$\underline{\nabla}\underline{r} = \underline{R}^s\underline{r}_s = \frac{1}{Ar}\underline{i}_1\underline{e}_r + A(E\underline{i}_2 - C\underline{i}_3)r\underline{e}_\varphi + A(-D\underline{i}_2 + B\underline{i}_3)\underline{k}\ ,$$

so that

$$\underline{F} = 2Ax^1\underline{i}_1\underline{i}_1 + \left(\frac{A}{2x^1}B^2 + C^2\right)\underline{i}_2\underline{i}_2 + \left(\frac{A}{2x^1}D^2 + E^2\right)\underline{i}_3\underline{i}_3 + $$

$$+ \left(\frac{A}{2x^1}BD + EC\right)(\underline{i}_2\underline{i}_3 + \underline{i}_3\underline{i}_2) \ ,$$

$$\underline{g} = \frac{1}{2Ax^1}\underline{i}_1\underline{i}_1 + (2Ax^1E^2 + A^2D^2)\underline{i}_2\underline{i}_2 + (2Ax^1C^2 + A^2B^2)\underline{i}_3\underline{i}_3 + $$

$$+ (-2Ax^1EC - A^2BD)(\underline{i}_2\underline{i}_3 + \underline{i}_3\underline{i}_2) \ ,$$

$$I_1(\underline{F}) = 2Ax^1 + \frac{A}{2x^1}(B^2 + D^2) + C^2 + E^2 \ ,$$

$$I_2(\underline{F}) = \frac{1}{2Ax^1} + 2Ax^1(E^2 + C^2) + A^2(D^2 + B^2) \ , \qquad (12)$$

$$\underline{\nabla}\cdot\underline{F} = 2A\underline{i}_1 \ , \qquad \underline{\nabla}\cdot\underline{g} = -\frac{1}{2Ax^1}\underline{i}_1 \ .$$

Moreover, we have

$$\underline{\nabla}p = \underline{i}_1\frac{\partial}{\partial x^1}\left(4Ax^1\frac{\partial\ni}{\partial I_1} - \frac{1}{Ax^1}\frac{\partial\ni}{\partial I_2}\right) \ , \qquad p = p_0 + 4Ax^1\frac{\partial\ni}{\partial I_1} - \frac{1}{Ax^1}\frac{\partial\ni}{\partial I_2}$$

and from the state equation we find that

$$\sigma_1 = -p_0 \ ,$$

$$\sigma_2 = -p_0 + 2\frac{\partial\ni}{\partial I_1}\left(-2Ax^1 + \frac{A}{2x^1}B^2 + C^2\right) + 2\frac{\partial\ni}{\partial I_2}\left(\frac{1}{2Ax^1} - 2AE^2x^1 - A^2D^2\right) \ ,$$

$$\sigma_3 = -p_0 + 2\frac{\partial\ni}{\partial I_1}\left(-2Ax^1\frac{\partial\ni}{\partial I_1} + \frac{A}{2x^1}D^2 + E^2\right) + $$

$$+ 2\frac{\partial\ni}{\partial I_2}\left(\frac{1}{2Ax^1} - 2AC^2x^1 - A^2B^2\right) \ , \qquad (13)$$

$$\tau_{23} = 2\frac{\partial\ni}{\partial I_1}\left(\frac{A}{2x^1}BD + EC\right) + 2\frac{\partial\ni}{\partial I_2}(2Ax^1EC + A^2BD) \ .$$

On the surface of the sheet x^1 = const the pressure $\sigma_1 = -p_0$ is assumed known. The constants $A,\dots,E$ are determined by the values of the forces and moments on the lateral surfaces of the sheet.

3. The transformation of a right cylindrical form. The form is bounded by coaxial cylindrical surfaces. The transformation is prescribed by the relationships

$$R^2 = Ar^2 + B\,, \qquad \Phi = C\varphi + Dz\,, \qquad Z = E\varphi + Fz\,, \qquad \underline{R} = R\underline{e}_R + Z\underline{k} \tag{14}$$

which hold together with the incompressibility condition

$$A(CF - DE) = 1\,. \tag{15}$$

The calculation yields the following representations of the tensors $\underline{F}$, $\underline{g}$, their invariants and divergences

$$\underline{F} = A^2 \frac{r^2}{R^2}\underline{e}_R\underline{e}_R + R^2\left(\frac{C^2}{r^2} + D^2\right)\underline{e}_\Phi\underline{e}_\Phi + \left(\frac{E^2}{r^2} + F^2\right)\underline{kk} + R\left(\frac{CE}{R^2} + DF\right)(\underline{e}_\Phi\underline{k} + \underline{k}\underline{e}_\Phi)\,,$$

$$\underline{g} = \frac{R^2}{r^2A^2}\underline{e}_R\underline{e}_R + \frac{A^2}{R^2}(r^2F^2 + E^2)\underline{e}_\Phi\underline{e}_\Phi + A^2(C^2 + r^2D^2)\underline{kk} - \frac{A^2}{R^2}(DFr^2 + EC)(\underline{e}_\Phi\underline{k} + \underline{k}\underline{e}_\Phi)\,,$$

$$I_1(\underline{F}) = A^2\frac{r^2}{R^2} + \frac{R^2}{r^2}C^2 + R^2D^2 + \frac{E^2}{r^2} + F^2\,,$$

$$I_2(\underline{F}) = \frac{R^2}{r^2A^2} + \frac{A^2r^2}{R^2}F^2 + \frac{A^2}{R^2}E^2 + A^2C^2 + r^2A^2D^2\,,$$

$$\underline{\nabla}\cdot\underline{F} = \left[A^2\frac{d}{dR}\frac{r^2}{R^2} + A^2\frac{r^2}{R^3} - R\left(\frac{C^2}{r^2} + D^2\right)\right]\underline{e}_R\,,$$

$$\underline{\nabla}\cdot\underline{g} = \left[\frac{1}{A^2}\frac{d}{dR}\frac{R^2}{r^2} + \frac{R^2}{A^2r^2} - \frac{A^2}{R^3}(r^2F^2 + E^2)\right]\underline{e}_R\,. \tag{16}$$

The pressure p is determined by the equilibrium equation and is

$$p = p_0 + 2\left(A^2\frac{r^2}{R^2}\frac{\partial\partial}{\partial I_1} - \frac{R^2}{A^2r^2}\frac{\partial\partial}{\partial I_2}\right) + 2\int_{R_0}^{R}\frac{dR}{R}\left[\frac{\partial\partial}{\partial I_1}\left(A^2\frac{r^2}{R^2} - \frac{R^2}{r^2}C^2 - R^2D^2\right) - \right.$$

$$\left. - \frac{\partial\partial}{\partial I_2}\left(\frac{R^2}{A^2r^2} - A^2F^2\frac{r^2}{R^2} - A^2\frac{E^2}{R^2}\right)\right]\,. \tag{17}$$

The components of the stress tensor are determined by the formulas

$$\sigma_R = -p_0 - 2\int_{R_0}^{R} \frac{dR}{R}\left[\frac{\partial \ni}{\partial I_1}\left(A^2\frac{r^2}{R^2} - \frac{R^2}{r^2}C^2 - R^2D^2\right) - \right.$$

$$\left. - \frac{\partial \ni}{\partial I_2}\left(\frac{R^2}{A^2r^2} - A^2F^2\frac{r^2}{R^2} - A^2\frac{E^2}{R^2}\right)\right] ,$$

$$\sigma_\Phi = \frac{d}{dR}R\sigma_R , \tag{18}$$

$$\sigma_Z = \sigma_R - 2\left[\frac{\partial \ni}{\partial I_1}\left(A^2\frac{r^2}{R^2} - \frac{E^2}{r^2} - F^2\right) - \frac{\partial \ni}{\partial I_2}\left(\frac{R^2}{A^2r^2} - A^2C^2 - A^2D^2r^2\right)\right] ,$$

$$\tau_{Z\Phi} = 2\left[\frac{\partial \ni}{\partial I_1}R\left(\frac{CE}{R^2} + DF\right) + \frac{\partial \ni}{\partial I_2}\frac{A^2}{R^2}(DFr^2 + EC)\right] .$$

Here we have also uniform distribution of the normal contact forces over the inner ($R = R_0$) and outer ($R = R_1$) cylindrical surfaces.

The constant B in Equation (14) can be expressed in terms of the inner and outer radii r_0 and r_1 of the reference configuration

$$B = R_0^2 - Ar_0^2 = R_1^2 - Ar_1^2 , \qquad R_0^2 - R_1^2 = A(r_0^2 - r_1^2) . \tag{19}$$

For a hollow cylinder the requirement that the displacements be single--valued implies that

$$C = 1 , \qquad E = 0 . \tag{20}$$

The continuity condition (15) is now reduced to the form

$$AF = 1 . \tag{21}$$

4. <u>The axisymmetric transformation of a hollow sphere</u>. This transformation is given by the relationships

$$R^3 = \varepsilon r^3 + A , \qquad \Theta = \vartheta , \qquad \Lambda = \lambda \qquad (\varepsilon = \pm 1) . \tag{22}$$

For this transformation

$$\underline{F} = \left(\frac{r^4}{R^4} - \frac{R^2}{r^2}\right)\underline{e}_R\underline{e}_R + \frac{R^2}{r^2}\underline{E} , \qquad \underline{g} = \left(\frac{R^4}{r^4} - \frac{r^2}{R^2}\right)\underline{e}_R\underline{e}_R + \frac{r^2}{R^2}\underline{E} ,$$

$$\underline{\nabla} \cdot \underline{F} = \underline{e}_R\left[\frac{d}{dR}\frac{r^4}{R^4} + \frac{2}{R}\left(\frac{r^4}{R^4} - \frac{R^2}{r^2}\right)\right] ,$$

$$\underline{\nabla} \cdot \underline{g} = \underline{e}_R \left[\frac{d}{dR} \frac{R^4}{r^4} + \frac{2}{R} \left(\frac{R^4}{r^4} - \frac{r^2}{R^2} \right) \right] ,$$

$$I_1(\underline{F}) = \frac{r^4}{R^4} + 2\,\frac{R^2}{r^2} , \qquad I_2(\underline{F}) = \frac{R^4}{r^4} + 2\,\frac{r^2}{R^2} .$$

The pressure p and the components of the stress tensor are

$$p = p_0 + 2\left(\frac{r^4}{R^4} \frac{\partial \text{Э}}{\partial I_1} - \frac{R^4}{r^4} \frac{\partial \text{Э}}{\partial I_2} \right) + 4 \int_{R_0}^{R} \frac{r^6 - R^6}{R^4 r^2} \left(\frac{\partial \text{Э}}{\partial I_1} + \frac{R^2}{r^2} \frac{\partial \text{Э}}{\partial I_2} \right) \frac{dR}{R} , \tag{23}$$

$$\sigma_R = -p_0 + 4 \int_{R_0}^{R} \frac{R^6 - r^6}{R^4 r^2} \left(\frac{\partial \text{Э}}{\partial I_1} + \frac{R^2}{r^2} \frac{\partial \text{Э}}{\partial I_2} \right) \frac{dR}{R} , \qquad \sigma_\theta = \sigma_\Lambda = \frac{1}{2R} \frac{d}{dR} R^2 \sigma_R . \tag{24}$$

The normal forces uniformly distributed over the inner ($R = R_0$) and outer ($R = R_1$) surfaces are given by the expressions

$$f_0 = -p_0 , \qquad f_1 = f_0 + 4 \int_{R_0}^{R_1} \frac{R^6 - r^6}{R^4 r^2} \left(\frac{\partial \text{Э}}{\partial I_1} + \frac{R^2}{r^2} \frac{\partial \text{Э}}{\partial I_2} \right) \frac{dR}{R} .$$

The constant A can be determined by the formula

$$A = R_0^3 - \varepsilon r_0^3 = R_1^3 - \varepsilon r_1^3 . \tag{25}$$

5. The transformation of a right cylindrical form (Shield, Klingbeil, 1966) into a cylindrical form of different curvature is given by the expressions

$$R = Ar , \qquad \Phi = B \ln r + C\varphi , \qquad Z = Dz ,$$

$$\underline{R} = R\underline{e}_R + \underline{k}Z ; \qquad A^2 CD = 1 . \tag{26}$$

For this transformation

$$\underline{R}_1 = A(\underline{e}_R + B\underline{e}_\Phi) , \qquad \underline{R}_2 = RC\underline{e}_\Phi , \qquad \underline{R}_3 = D\underline{k} ;$$

$$\underline{R}^1 = \frac{\underline{e}_R}{A} , \qquad \underline{R}^2 = \frac{1}{CR} (\underline{e}_\Phi - \underline{e}_R) , \qquad \underline{R}^3 = \frac{\underline{k}}{D} ;$$

$$\overset{0}{\underline{\underline{\nabla R}}} = A\underline{e}_r(\underline{e}_R + B\underline{e}_\Phi) + AC\underline{e}_\varphi\underline{e}_\Phi + D\underline{\underline{kk}} ,$$

$$\underline{\underline{\nabla r}} = \frac{1}{A}\, \underline{e}_R\underline{e}_r + \frac{1}{CA} (\underline{e}_\Phi - B\underline{e}_R)\underline{e}_\varphi + \frac{\underline{\underline{kk}}}{D}$$

and the representations for the strain measures are reduced to the form

$$\underline{F} = A^2 [\underline{e}_R\underline{e}_R + (B^2 + C^2)\underline{e}_\Phi\underline{e}_\Phi + B(\underline{e}_R\underline{e}_\Phi + \underline{e}_\Phi\underline{e}_R)] + D^2\underline{kk} ,$$

$$\underline{g} = \frac{1}{A^2}\left[\underline{e}_R\underline{e}_R\left(1 + \frac{B^2}{C^2}\right) + \frac{1}{C^2}\underline{e}_\Phi\underline{e}_\Phi - B(\underline{e}_R\underline{e}_\Phi + \underline{e}_\Phi\underline{e}_R)\right] + \frac{\underline{kk}}{D^2} . \quad (27)$$

The invariants and the physical components of these tensors are constant

$$I_1(\underline{F}) = A^2(1 + B^2 + C^2) + D^2 , \qquad I_1(\underline{g}) = \frac{1}{A^2C^2}(B^2 + C^2) + \frac{1}{D^2} ,$$

but this is of course not the case for the tensors $\underline{F}$ and $\underline{F}^{-1}$ themselves. The specific strain energy is constant which is also true for the coefficients of the constitutive equation. The divergences $\underline{\nabla} \cdot \underline{F}$ and $\underline{\nabla} \cdot \underline{g}$ are equal to

$$\underline{\nabla} \cdot \underline{F} = \frac{A^2}{R^2}[(1 - B^2 - C^2)\underline{e}_R + 2B\underline{e}_\Phi] , \qquad \underline{\nabla} \cdot \underline{g} = -\frac{1}{C^2A^4}\underline{\nabla} \cdot \underline{F}$$

and the pressure gradient, determined by the equilibrium equations, is represented by the expression

$$\underline{\nabla}p = \frac{2}{R}\left(A^2\frac{\partial \ni}{\partial I_1} + \frac{1}{A^2C^2}\frac{\partial \ni}{\partial I_2}\right)[(1 - B^2 - C^2)\underline{e}_R + 2B\underline{e}_\Phi] = \frac{\partial p}{\partial R}\underline{e}_R + \frac{1}{R}\frac{\partial p}{\partial \Phi}\underline{e}_\Phi .$$

We now obtain

$$p = 2\left(A^2\frac{\partial \ni}{\partial I_1} + \frac{1}{A^2C^2}\frac{\partial \ni}{\partial I_2}\right)\left[(1 - B^2 - C^2)\ln\frac{R}{R_0} + 2B\Phi\right] + p_0 ,$$

$$\sigma_R = -p_0 + 2\left[A^2\frac{\partial \ni}{\partial I_1} - \frac{1}{A^2C^2}(B^2 + C^2)\frac{\partial \ni}{\partial I_2}\right] ,$$

$$\sigma_\Phi = -p + 2\left[A^2(B^2 + C^2)\frac{\partial \ni}{\partial I_1} - \frac{1}{A^2C^2}\frac{\partial \ni}{\partial I_2}\right] ,$$

$$\sigma_Z = -p + 2\left(D^2\frac{\partial \ni}{\partial I_1} - \frac{1}{D^2}\frac{\partial \ni}{\partial I_2}\right) , \qquad \tau_{R\Phi} = B\left(A^2\frac{\partial \ni}{\partial I_1} + \frac{1}{A^2}\frac{\partial \ni}{\partial I_2}\right) . \quad (28)$$

In contrast with transformations 1-4, transformation (26) requires the application to the surfaces $R = R_0$ and $R = R_1$ of normal forces which depend upon Φ together with constant tangential forces.

In his investigations, Ericksen did not consider tensors $\underline{F}$ with constant invariants.

6. Note. The "arbitrariness" peculiar to the solutions 1-4 is associated with the constants of these solutions. This restricts the number of the boundary value problems which can be examined using these solutions. The "major surfaces" may be required to be either free of loads or subjected only to uniformly distributed normal stresses. On the remaining surfaces, one has to be content with fulfillment of the boundary conditions on integrating ("in the sense of Saint--Venant").

The value of these solutions consists not only in their "universal" character, e.g. in their representations which are valid for all isotropic incompressible materials and therefore suitable for any material provided that the dependence of its specific strain energy on the invariants of the strain measure is known. These solutions are attractive for their simplicity, their descriptive character and the surprising nature of the results they yield. They compel one to reject certain commonly-held concepts of linear theory and make it unnecessary to use linear theory to make construction of phenomena which it cannot in fact explain.

As has been mentioned earlier, the solutions of the problems considered below were originally given in the definitive papers of Rivlin. The distinguished analysis of Ericksen is of rather negative value in that it demonstrates that it is pointless to look for some new "universal" solutions which are different from those contained in 1-4 supplemented by solution 5.

In the subsequent sections the applications of the solutions 1-4 to some particular problem will be considered.

10. TORSION, STRETCHING AND CHANGE OF DIAMETER OF A CIRCULAR CYLINDER

In the formulas (9.14) it is assumed that

$$B = 0\,, \qquad C = 1\,, \qquad E = 0\,; \qquad AF = 1\,;$$

$$R = \sqrt{Ar}\,, \qquad \Phi = \varphi + Dz\,, \qquad Z = Fz\,. \tag{1}$$

The constant F determines the change of length of a cylinder ($L = Fl$) and $\sqrt{A}a = R_0$ denotes its radius in the actual configuration whereas a denotes its value in the reference configuration. The twist angle per unit length ($L\alpha = Dl$) is given by $\alpha = D/F$.

If the surface of the cylinder is free of load ($p_0 = 0$), then the formulas (9.17) for the components of stress are transformed to

$$\sigma_R = -2\alpha^2F^2 \int_{R_0}^{R} \frac{\partial \ni}{\partial I_1} R\,dR\,, \qquad \sigma_\Phi = \frac{d}{dR} R\sigma_R\,,$$

$$\sigma_Z = \sigma_R - 2\alpha^2FR^2 \frac{\partial \ni}{\partial I_2} + \frac{F^3-1}{R\alpha^2F^3}\frac{\partial \ni}{\partial R}\,, \qquad \tau_{Z\Phi} = \frac{1}{\alpha}\frac{\partial \ni}{\partial R}\,, \tag{2}$$

this because for the case under consideration

$$\frac{\partial \ni}{\partial R} = 2R\alpha^2F\left(\frac{\partial \ni}{\partial I_1}F + \frac{\partial \ni}{\partial I_2}\right) .$$

The constants α and F are determined by the prescription of the torque and the longitudinal force m_Z

$$m_Z = 2\pi \int_0^{R_0} R^2\tau_{Z\Phi}dR = \frac{2\pi}{\alpha}\int_0^{R_0} \frac{\partial \ni}{\partial R} R^2 dR = \frac{2\pi}{\alpha}\left[R_0^2\ni(R_0) - \int_0^{R_0} R\ni\,dR\right]\,, \tag{3}$$

$$Q = 2\pi \int_0^{R_0} \sigma_Z R\,dR = 2\pi\frac{F^3-1}{\alpha^2F^3}[\ni(R_0) - \ni(0)] -$$

$$- 2\pi\alpha^2F^2 \int_0^{R_0}\left(\frac{\partial \ni}{\partial I_1} + \frac{2}{F}\frac{\partial \ni}{\partial I_2}\right)R^3dR\,. \tag{4}$$

Here, the double integral has been transformed to a single integral. The values of $\ni(I_1,I_2)$ on the surface and on the axis of the cylinder are given by $\ni(\ _0)$ and $\ni(0)$.

If the length of the cylinder remains constant, then $F = 1$ and

$$Q = -2\pi\alpha^2 \int_0^{R_0} R^3\left(\frac{\partial \ni}{\partial I_1} + 2\frac{\partial \ni}{\partial I_2}\right)dR = -2\pi\alpha^2\int_0^{a} r^3\left(\frac{\partial \ni}{\partial I_1} + 2\frac{\partial \ni}{\partial I_2}\right)dr\,. \tag{5}$$

According to the empirical criterion (5.11.3), the cylinder is then compressed.

In the absence of the longitudinal force (faces of the cylinder are free of load)

$$\frac{F^3-1}{\alpha^2F^3}[\ni(R_0) - \ni(0)] = \frac{F^3-1}{\alpha^2F^3}\int_0^{R_0}\frac{\partial \ni}{\partial R}dR =$$

$$= 2\frac{F^3-1}{F}\int_0^{R_0}\left(\frac{\partial \ni}{\partial I_1} + \frac{1}{F}\frac{\partial \ni}{\partial I_2}\right)R\,dR = \alpha^2F^2\int_0^{R_0}\left(\frac{\partial \ni}{\partial I_1} + \frac{2}{F}\frac{\partial \ni}{\partial I_2}\right)R^3\,dR\,, \tag{6}$$

which means that $F - 1$ is of order α^2. Taking the quantities of this order only into account, we have, referring to (3.3)

$$\left[2\,\frac{F^3-1}{F}\int_0^{R_0}\left(\frac{\partial \ni}{\partial I_1}+\frac{1}{F}\,\frac{\partial \ni}{\partial I_2}\right)R\,dR\right]_{F=1} = 3(F-1)\left(\frac{\partial \ni}{\partial I_1}+\frac{\partial \ni}{\partial I_2}\right)^0 a^2 =$$

$$= 3(F-1)\,\frac{\mu}{2}\,a^2$$

The substitution of I_1 and I_2 by $I_1 = I_2 = 3$ in the reference configuration is indicated by a zero subscript. To that degree of accuracy, we have by (6)

$$F - 1 = \frac{1}{12}\,a^2\alpha^2\left(1+\frac{2}{\mu}\left(\frac{\partial \ni}{\partial I_2}\right)^0\right)\,. \tag{7}$$

This corresponds to the change in the length of a wire under torsion (the Poynting effect), this effect cannot be explained by linear theory since Equation (7) contains a term which is beyond the scope of this theory.

For Mooney material, formulas (4.1), (3) and (4) lead in the case of free faces to the relationships

$$m_Z = 2I_p\left(C_1+\frac{C_2}{F}\right)\alpha\,,\qquad \frac{F^3-1}{F^2}\,\frac{C_1+C_2/F}{C_1+2C_2/F} = \frac{(\alpha a)^2}{4}\qquad \left(I_p = \frac{\pi a^4}{2}\right)\,. \tag{8}$$

11. LAMÉ'S PROBLEM FOR A HOLLOW CYLINDER

Here we use the same solution (9.14) in which we now set

$$C = 1\,,\ D = 0\,,\ E = 0\,,\ R^2 = Ar^2 + B\,,\ \Phi = \varphi\,,\ Z = Fz\,;\qquad AF = 1\,. \tag{1}$$

The constant B is given in terms of the radii (r_0, r_1) and (R_0, R_1) in the reference and actual configurations of the cylinder (r_0 being the inner, r_1 the outer radius; $r_1^2 = r_0^2(1+\delta)$, $\delta > 0$)

$$B = R^2 - Ar^2 = R_0^2 - Ar_0^2 = R_1^2 - Ar_1^2\,,$$

$$R_1^2 - R_0^2 = A(r_1^2 - r_0^2) = A\delta r_0^2\,. \tag{2}$$

For $A > 0$ we have $R_1 > R_0$ where R_1 denotes the outer radius and R_0 the inner radius in the actual configuration. The case $A < 0$ considered in Par. 12 represents the problem of an everted cylinder.

From (9.17) and (1), it can be seen that the tangential stress $\tau_{Z\Phi}$ is absent, and that the normal stresses are equal to

$$\sigma_R = -p_0 - 2\int_{R_0}^{R}\left[\frac{\partial \ni}{\partial I_1}\left(A^2\frac{r^2}{R^2}-\frac{R^2}{r^2}\right)-\frac{\partial \ni}{\partial I_2}\left(\frac{R^2}{A^2r^2}-\frac{r^2}{R^2}\right)\right]\frac{dR}{R},$$

$$\sigma_\Phi = \frac{d}{dR}R\sigma_R\,, \qquad \sigma_Z = \sigma_R + 2\left(A^{-2}-A^2\frac{r^2}{R^2}\right)\left(\frac{\partial \ni}{\partial I_1}+\frac{R^2}{r^2}\frac{\partial \ni}{\partial I_2}\right).$$

Here

$$I_1(\underline{F}) = A^2\frac{r^2}{R^2}+\frac{R^2}{r^2}+A^{-2}\,, \qquad I_2(\underline{F}) = \frac{R^2}{A^2r^2}+\frac{r^2}{R^2}+A^2$$

and in view of the relationship $RdR = Ardr$ we get

$$\frac{dI_1}{dR} = 2\left(\frac{r^2}{R^2}-\frac{R^2}{A^2r^2}\right)\left(\frac{RA}{r^2}-\frac{A^2}{r}\right) = \frac{AB}{r^2}\frac{2}{R}\left(\frac{r^2}{R^2}-\frac{R^2}{A^2r^2}\right),$$

$$\frac{dI_2}{dR} = A^{-2}\frac{dI_1}{dR}\,, \qquad \frac{d\ni}{dR} = \frac{B}{Ar^2}\frac{2}{R}\left(\frac{r^2}{R^2}-\frac{R^2}{A^2r^2}\right)\left(A^2\frac{\partial \ni}{\partial I_1}+\frac{\partial \ni}{\partial I_2}\right). \tag{3}$$

This also enables us to represent σ_R in the form

$$\sigma_R = -2\int_{R_0}^{R}\frac{dR}{R}\left(\frac{r^2}{R^2}-\frac{R^2}{A^2r^2}\right)\left(A^2\frac{\partial \ni}{\partial I_1}+\frac{\partial \ni}{\partial I_2}\right) - p_0 = -\frac{A}{B}\int_{R_0}^{R} r^2\frac{d\ni}{dR}\,dR - p_0\,, \tag{4}$$

where $p_0 = -\sigma_R(R_0)$ denotes the pressure on the inner surface of the cylinder. The pressure on the outer surface is given by $p_1 = -\sigma_R(R_1)$ so that

$$\sigma_R(R_0) - \sigma_R(R_1) = -\frac{A}{B}\int_{R_0}^{R_1} r^2\frac{d\ni}{dR}\,dR =$$

$$= -2\int_{R_0}^{R_1}\frac{dR}{R}\left(\frac{r^2}{R^2}-\frac{R^2}{A^2r^2}\right)\left(A^2\frac{\partial \ni}{\partial I_1}+\frac{\partial \ni}{\partial I_2}\right). \tag{5}$$

The longitudinal force is determined from the expression

$$Q = 2\pi\int_{R_0}^{R_1} R\sigma_z dR =$$

$$= 2\pi A \left[\int_{r_0}^{r_1} \sigma_R r\, dr + 2 \int_{r_0}^{r_1} \left(A^{-2} - A^2 \frac{r^2}{R^2}\right)\left(\frac{\partial \ni}{\partial I_1} + \frac{R^2}{r^2} \frac{\partial \ni}{\partial I_2}\right) r\, dr \right] .$$

After eliminating σ_R with the aid of (4) and (5) and after transforming the double integral to a single integral this reduces to the form

$$Q = 2\pi A \left\{ \frac{1}{2} [r_0^2 \sigma_R(R_0) - r_1^2 \sigma_R(R_1)] + \frac{A}{2B} \int_{r_0}^{r_1} \frac{d\ni}{dr} r^4\, dr + \right.$$

$$\left. + 2 \int_{r_0}^{r_1} \left(A^{-2} - A^2 \frac{r^2}{r^2}\right)\left(\frac{\partial \ni}{\partial I_1} + \frac{R^2}{r^2} \frac{\partial \ni}{\partial I_2}\right) r\, dr \right\} . \tag{6}$$

Substituting for R^2 from (2) we may rewrite Equation (5) as follows

$$\sigma_R(R_0) - \sigma_R(R_1) = 2 \frac{B}{A^2} \int_{r_0}^{r_1} \left(A^2 \frac{\partial \ni}{\partial I_1} + \frac{\partial \ni}{\partial I_2}\right)\left(1 + \frac{r^2}{B + Ar^2}\right) \frac{dr}{r^3} . \tag{7}$$

Equations (6) and (7) determine the unknown constants A, B.

For the case of fixed faces, when the cylinder is placed between two rigid plates, $F = A = 1$, and Equation (7) takes the form

$$\sigma_R(R_0) - \sigma_R(R_1) = B \int_{r_0}^{r_1} \left(1 + \frac{r^2}{B + r^2}\right) \hat{\mu}(I_1, I_2) \frac{dr}{r^3} , \qquad \hat{\mu} = 2\left(\frac{\partial \ni}{\partial I_1} + \frac{\partial \ni}{\partial I_2}\right) , \tag{8}$$

where $\hat{\mu}(3,3) = \mu$ denotes the shear modulus in the reference configuration. For free faces, we have $Q = 0$.

12. THE EVERTED CYLINDER

It is assumed that in this state normal stresses on the surface of a cylinder ($\sigma_R(R_0) = \sigma_R(R_1) = 0$) are absent, and that this is also the case for the longitudinal force ($Q = 0$). Denoting $-A = \beta > 0$, we have by (11.5)

$$\int_{R_0}^{R_1} \frac{dR}{R} \left(\frac{r^2}{R^2} - \frac{R^2}{\beta^2 r^2}\right)\left(\beta^2 \frac{\partial \ni}{\partial I_1} + \frac{\partial \ni}{\partial I_2}\right) = 0 . \tag{1}$$

After eliminating $d\ni/dr$ in (11.6) with the aid of (11.3), condition $Q = 0$ is reduced to the form

$$\int_{R_0}^{R_1} r^2\left(\frac{r^2}{R^2} - \frac{R^2}{\beta^2 r^2}\right)\left(\beta^2 \frac{\partial \ni}{\partial I_1} + \frac{\partial \ni}{\partial I_2}\right) \frac{dR}{R} =$$

$$= \frac{2}{\beta} \int_{R_0}^{R_1} \left(\beta^{-2} - \beta^2 \frac{r^2}{R^2}\right)\left(\frac{\partial \ni}{\partial I_1} + \frac{R^2}{r^2} \frac{\partial \ni}{\partial I_2}\right) R\,dR \ . \tag{2}$$

Introducing the independent variable

$$\xi = \frac{R^2}{r^2} \tag{3}$$

we get

$$\frac{1}{2}\, d\xi = \frac{R}{r^2}\, dR - \frac{R^2}{r^3}\, dr = \frac{R\,dR}{r^2}\left(1 + \frac{\xi}{\beta}\right) ,$$

$$R\,dR = \frac{1}{2}\, r^2 \frac{\beta\, d\xi}{\xi + \beta} , \qquad \frac{dR}{R} = \frac{1}{2} \frac{\beta\, d\xi}{\xi(\xi + \beta)} \ .$$

Equation (1) now takes the form

$$\int_{x_1^2}^{x_0^2} \left(\beta^2 \frac{\partial \ni}{\partial I_1} + \frac{\partial \ni}{\partial I_2}\right) (\beta - \xi)\, \frac{d\xi}{\xi^2} = 0 \ . \tag{4}$$

Here $x_0 = R_0/r_0$ and $x_1 = R_1/r_1$, and by (11.2) we have

$$R_0^2 + \beta r_0^2 = R_1^2 + \beta r_1^2 ,$$

$$x_0^2 + \beta = (1 + \delta)(x_1^2 + \beta) ,$$

$$x_0^2 = (1 + \delta) x_1^2 + \beta\delta , \tag{5}$$

where $1 + \delta = r_1^2/r_2^2$ and $\delta > 0$, so that $x_0^2 > (1+\delta)x_1^2$ and $x_0^2 > x_1^2$. With this notation, Equation (2) becomes

$$\int_{x_1^2}^{x_0^2} \frac{\beta - \xi}{\beta + \xi}\left(\beta^2 \frac{\partial \ni}{\partial I_1} + \frac{\partial \ni}{\partial I_2}\right) \frac{d\xi}{\xi^2} = \frac{2}{\beta} \int_{x_1^2}^{x_0^2} \frac{\xi - \beta^4}{(\beta + \xi)^2}\left(\frac{\partial \ni}{\partial I_1} + \xi \frac{\partial \ni}{\partial I_2}\right) \frac{d\xi}{\xi} \ . \tag{6}$$

The problem thus reduces to the determination of three unknown quantities x_0^2, x_1^2 and β from the system of equations (4), (5) and (6). Without specifying the function $\ni(I_1, I_2)$, it is not possible to investigate this system. Consider the case of the Mooney potential (4.1). Discarding the factor $\frac{1}{3}$ in (4) and integrating we arrive at the equation

$$\frac{\beta}{x_0^2} - \ln\frac{\beta}{x_0^2} = \frac{\beta}{x_1^2} - \ln\frac{\beta}{x_1^2}, \qquad f\left(\frac{\beta}{x_0^2}\right) = f\left(\frac{\beta}{x_1^2}\right). \tag{7}$$

The function $f(\lambda) = \lambda - \ln\lambda$, determined for positive λ, decreases monotonically (remaining positive) from ∞ to 1 for $0 \leq \lambda \leq 1$, and increases monotonically from 1 to ∞ for $1 \leq \lambda \leq \infty$. Hence, for each value of $\lambda_0 = \beta/x_0^2 < 1$, Equation (7) yields a unique value of $\lambda_1 = \beta/x_1^2 > 1$.

For the "Neo-Hookean" material (4.3), Equation (6) leads to the equation

$$\left(1 + \frac{\beta}{x_0^2}\right)\left(1 + \frac{\beta}{x_1^2}\right) = 2(1 + \beta^{-3}). \tag{8}$$

Equations (7) and (8) together with (5) determine the unknown quantities β and the outer and the inner radii of the everted cylinder x_0^2 and x_1^2.

We also note here the formulas for the stress, σ_ϕ, evaluated for $R = R_0$ and $R = R_1$. On these surfaces, from (9.18) and (11.4), $\sigma_R = 0$ and for Mooney material

$$(\sigma_\phi)_{R=R_0} = \frac{2}{\beta}\left(\frac{x_0^2}{\beta} - \frac{\beta}{x_0^2}\right)(C_1\beta^2 + C_2),$$

$$(\sigma_\phi)_{R=R_1} = \frac{2}{\beta}\left(\frac{x_1^2}{\beta} - \frac{\beta}{x_1^2}\right)(C_1\beta^2 + C_2). \tag{9}$$

As expected, the stress σ_ϕ is that of tension on the outer, and of compression on the inner surface.

13. LAMÉ'S PROBLEM FOR A HOLLOW SPHERE

The transformation of the reference configuration into the actual configuration is determined by Equations (9.22). By (9.25) we have

$$x_0^3 - \varepsilon = (1 + \delta)(x_1^3 - \varepsilon), \qquad 1 + \delta = \frac{r_1^3}{r_0^3}, \qquad \delta > 0, \tag{1}$$

where r_0 and r_1 denote the inner and outer radii of a sphere in its reference configuration. In the Lamé problem $\varepsilon = 1$ and q_0 and q_1 denote the internal and external pressures. Here

$$R_1^3 - R_0^3 = r_1^3 - r_0^3 > 0 , \qquad R_1 > R_0 \tag{2}$$

and in the actual configuration, R_1 is the outer and R_0 the inner radius of the sphere. For $\varepsilon = -1$

$$R_1^3 - R_0^3 = r_0^3 - r_1^3 , \qquad R_1 < R_0 \tag{3}$$

this problem being related to an everted sphere.

1. Lamé's problem. By (9.24) we have

$$q_0 - q_1 = 4 \int_{R_0}^{R_1} \frac{R^6 - r^6}{R^4 r^2} \left(\frac{\partial \ni}{\partial I_1} + \frac{R^2}{r^2} \frac{\partial \ni}{\partial I_2} \right) \frac{dR}{R} .$$

Introducing the independent variable

$$\xi = \frac{R^3}{r^3} , \qquad \frac{1}{3} \frac{d\xi}{\xi(1 - \xi)} = \frac{dR}{R} , \tag{4}$$

we arrive at the equation

$$q_0 - q_1 = \int_{x_1^3}^{x_0^3} (\xi + 1)\xi^{-7/3} \left(\frac{\partial \ni}{\partial I_1} + \xi^{2/3} \frac{\partial \ni}{\partial I_2} \right) d\xi , \tag{5}$$

which determines, together with (1), the unknowns x_0^3 and x_1^3 for $\varepsilon = 1$. Making use of the expressions

$$I_1 = \xi^{-4/3} + 2\xi^{2/3} , \qquad I_2 = \xi^{4/3} + 2\xi^{-2/3} ,$$

for the invariants it is easy to verify that Equation (5) can also be rewritten in the form

$$q_0 - q_1 = \int_{x_1^3}^{x_0^3} \frac{d\ni}{d\xi} \frac{d\xi}{\xi - 1} . \tag{6}$$

2. Eversion. Now R_0 denotes the outer, and R_1 the inner radius of a sphere, $q_0 = 0$ and $q_1 = 0$. The unknowns x_0^3 and x_1^3 given by (1) for $\varepsilon = -1$, are determined from the equation

$$\int_{x_1^3}^{x_0^3} \frac{d\ni}{d\xi}\,\frac{d\xi}{\xi+1} = 0 \;. \tag{7}$$

The process of eversion may be thought of as being carried out extruding the material through a small hole in the wall which is then filled.

14. BENDING OF A PLANE SHEET INTO A RIGHT CYLINDRICAL FORM

Assuming in (9.1) that $C = 0$ and $D = 0$ and making use of (9.2) and $E = (AB)^{-1}$, we have

$$R^2 = 2Aa^1 \;, \qquad \Phi = Ba^2 \;, \qquad Z = \frac{a^3}{AB} = Ea^3 \;. \tag{1}$$

The surfaces $R_0 = \sqrt{2Aa_0^1}$ and $R_1 = \sqrt{2Aa_1^1}$ of the form are free of load, and by (9.10)

$$\ni(I_1(R_0), I_2(R_0)) = \ni(I_1(R_1), I_2(R_1)) \;,$$

$$I_1(R_0) = I_1(R_1), I_2(R_0) = I_2(R_1) \;. \tag{2}$$

The expressions (9.7) for the invariants take the form

$$I_1 = \frac{A^2}{R^2} + R^2B^2 + (AB)^{-2} \;, \qquad I_2 = \frac{R^2}{A^2} + \frac{1}{R^2B^2} + A^2B^2 \tag{3}$$

and Equations (2) are satisfied by

$$B^2 = \frac{\alpha^2}{b^2} = \frac{1}{4}\,[a_0^1(a_0^1 + h)]^{-1} \;. \tag{4}$$

Here 2α denotes the angle subtended at the axis by the section ($2\alpha = \Phi$ for $a^2 = b$). The resultant stress σ_Φ is given by the expression

$$\int_{-lE}^{lE} dZ \int_{R_0}^{R_1} \frac{d}{dr} R\sigma_R dR = 2lER\sigma_R \Big|_{R_0}^{R_1} = 0$$

The resultant moment of these stresses about the axis of the section can, from (2) and (9.10), be represented by the expression

$$m_Z = 2El \int_{R_0}^{R} \left[R(\ni - p_0) + R^2 \frac{d\ni}{dR} \right] dR =$$

$$= 2El \left[-\frac{1}{2} p_0 (R_1^2 - R_0^2) + \int_{R_0}^{R_1} R \frac{d}{dR} \ni R \, dR \right] =$$

$$= 2El \left[-\frac{1}{2} p_0 (R_1^2 - R_0^2) + R^2 \ni \Big|_{R_0}^{R_1} - \int_{R_0}^{R_1} R \ni \, dR \right] =$$

$$= 2El \left[\frac{1}{2} (R_1^2 - R_0^2) \ni (I_1(R_0), I_2(R_0)) - \int_{R_0}^{R_1} R \ni \, dR \right] .$$

Substituting R_1 and R_0 and taking Equation (4) into account, we arrive at the following expression for the moment

$$m_Z = 2l \sqrt{a_0^1 (a_0^1 + h)} \left[h \ni (I_1(R_0), I_2(R_0)) - \int_{a_0^1}^{a_0^1 + h} \ni (I_1, I_2) da^1 \right] . \tag{5}$$

An alternative representation is given by the equality

$$m_Z = 2lE \int_{R_0}^{R_1} R \frac{d}{dR} R \sigma_R dR = 2lE \left[R^2 \sigma_R \Big|_{R_0}^{R_1} - \int_{R_0}^{R_1} R \sigma_R dR \right] = -2lE \int_{R_0}^{R_1} R \sigma_R dR . \tag{6}$$

This enables us to represent the expression for the longitudinal force in the following form (see (9.10))

$$Q = 2\alpha \int_{R_0}^{R_1} R \sigma_Z dR = -\frac{\alpha}{lE} m_Z + 4\alpha \int_{R_0}^{R_1} R \left[\frac{\partial \ni}{\partial I_1} \left(E^2 - \frac{A^2}{R^2} \right) + \frac{\partial \ni}{\partial I_2} \left(\frac{R^2}{A^2} - \frac{1}{E^2} \right) \right] dR \tag{7}$$

and the condition that it be zero is given by the equation determining E. We thus obtain

$$h \ni (I_1(R_0), I_2(R_0)) - \int_{a_0^1}^{a_0^1 + h} \ni (I_1, I_2) da^1 =$$

$$= 4E^2 \int_{a_0^1}^{a_0^1 + h} \left[1 - \frac{\sqrt{a_0^1 (a_0^1 + h)}}{a^1 E^3} \right] \left[\frac{\partial \ni}{\partial I_1} + \frac{a^1}{E \sqrt{a_0^1 (a_0^1 + h)}} \frac{\partial \ni}{\partial I_2} \right] da^1 \tag{8}$$

and the invariants I_1 and I_2 should be expressed through the variable a^1

$$I_1 = \frac{\sqrt{a_0^1(a_0^1+h)}}{Ea^1} + \frac{a^1}{E\sqrt{a_0^1(a_0^1+h)}} + E^2 ,$$
$$I_2 = \frac{Ea^1}{\sqrt{a_0^1(a_0^1+h)}} + \frac{E\sqrt{a_0^1(a_0^1+h)}}{a^1} + E^{-2} . \tag{9}$$

The choice of the constant a_0^1 depends upon the width $2b$ of the form and the angle 2α

$$\alpha = lB = \frac{b}{2}\,[a_0^1(a_0^1+h)]^{-1/2} , \qquad a_0^1 = \frac{h}{2}\left(\sqrt{1+\frac{b^2}{\alpha^2h^2}} - 1\right) . \tag{10}$$

The above formulas give the stresses within the panel, its outer and inner radii R_1 and R_0 and the length $2lE$ as long as we are given the material, the dimensions h, $2b$ and $2l$ of the sheet and the angle subtended by the section at the axis.

15. THE EXACT GENERAL SOLUTIONS IN THE CASE OF BODY FORCES

1. For potential body forces, the equilibrium equation has the form

$$\underline{\nabla}\cdot\underline{T} = \underline{\nabla}\pi , \qquad \rho\underline{k} = -\underline{\nabla}\pi \tag{1}$$

and its solution is given as the sum of a partial solution $\underline{E}\pi$ and a "general" solution $\underline{\tilde{T}}$ of a homogeneous equation. For an incompressible medium

$$\underline{T} = \underline{E}\pi + \underline{T}_E , \qquad \underline{\nabla}\cdot\underline{\tilde{T}} = -\underline{\nabla}\tilde{p} + \underline{\nabla}\cdot\underline{T}_E , \qquad \underline{\tilde{T}} = -\tilde{p}\underline{E} + 2\left(\underline{F}\frac{\partial\Im}{\partial I_1} - \underline{g}\frac{\partial\Im}{\partial I_2}\right) \tag{2}$$

and therefore

$$\underline{T} = \underline{E}(\pi-\tilde{p}) + 2\left(\underline{F}\frac{\partial\Im}{\partial I_1} - \underline{g}\frac{\partial\Im}{\partial I_2}\right) . \tag{3}$$

The first step in the formulation of universal solutions is the elimination of p from the equilibrium equation - see (7.8.3). Hence, these solutions determine the transformation of the reference configuration into the actual configuration, in other words, determine the position vector $\underline{R}$. This is a function of the material coordinates and certain constants but is independent of $\tilde{p}$. Hence, these solutions also hold for the stress tensor (3). Differences in the solutions arise at the stage at which the surface tractions which give rise to the state under consideration

$$\underline{f} = \underline{N} \cdot \underline{T} = \underline{N}\pi - \underline{N}\tilde{p} + 2\underline{N} \cdot \left(\underline{F} \frac{\partial \ni}{\partial I_1} - \underline{g} \frac{\partial \ni}{\partial I_2} \right) \tag{4}$$

are determined. In the universal solutions, the quantities $\tilde{p}$, $\underline{F}$ and $\underline{g}$ depend upon one coordinate only (say q^1), on the major coordinate surfaces $q^1 = \text{const}$ these solutions have constant values. In order to avoid some unrealistic surface tractions (which would depend on q^2 and q^3), it is therefore necessary that the potential π be a function of q^1 only. This sets limitations on applicability of universal solutions to problems which require body forces to be taken into account.

2. The rotating cylinder. The cylinder is assumed to rotate with constant angular velocity ω about its axis. In the system of axes associated with the rotating cylinder, the equation of motion takes the form of an equilibrium equation which includes "centrifugal forces" with the potential

$$\pi = -\frac{1}{2}\rho\omega^2 R^2 . \tag{5}$$

In the universal solution (9.14) we set $B = 0$, $C = 1$, $D = 0$, $E = 0$ and $F = A^{-1}$. From (9.16) we have

$$\begin{gathered} \underline{F} = A[\underline{E} + (A^{-3} - 1)\underline{kk}] , \qquad \underline{g} = A^{-1}[\underline{E} + (A^3 - 1)\underline{kk}] , \\ I_1 = 2A + A^{-2} , \qquad I_2 = 2A^{-1} + A^2 , \end{gathered} \tag{6}$$

so that $\ni(I_1, I_2) = \ni(A) = \text{const}$. By virtue of (9.17) and (3),

$$\tilde{p} = 2\left(A \frac{\partial \ni}{\partial I_1} - A^{-1} \frac{\partial \ni}{\partial I_2} \right) ,$$

$$\underline{T} = -\frac{1}{2}(A\rho\omega^2 r^2 + q_0)\underline{E} + 2(A^{-3} - 1)A\left(\frac{\partial \ni}{\partial I_1} + A \frac{\partial \ni}{\partial I_2} \right)\underline{kk} .$$

The surface $R = R_1$ of the cylinder (its radius in the reference configuration is denoted by r_1) is free of load. Therefore

$$\sigma_R = \frac{1}{2} A\rho\omega^2 (r_1^2 - r^2) \; ,$$

$$\sigma_Z = \frac{1}{2} A\rho\omega^2 (r_1^2 - r^2) + 2(A^{-3} - 1)A\left(\frac{\partial \ni}{\partial I_1} + A\frac{\partial \ni}{\partial I_2}\right) . \tag{7}$$

The constant A is determined from the condition of the absence of longitudinal force. Bearing in mind that $RdR = Ardr$, we have

$$Q = 2\pi A \int_0^{r_1} \sigma_Z r\, dr = 0 \; , \qquad \frac{1}{8}\rho\omega^2 r_1^2 = (1 - A^{-3})\left(\frac{\partial \ni}{\partial I_1} + A\frac{\partial \ni}{\partial I_2}\right) . \tag{8}$$

Here, from the empirical criterion (5.11.3), $\frac{\partial \ni}{\partial I_1} > 0$, $\frac{\partial \ni}{\partial I_2} > 0$. For this reason $A > 1$, that is the cylinder is thickened and shortened.

3. The rotating hollow cylinder. This problem is substantially more complicated. As for Lamé's problem (Par. 11), we are seeking the two-parameter solution (11.1). Repeating the calculation of Par. 11, we arrive at the following formulas for normal stresses

$$\sigma_R = \frac{1}{2}\rho\omega^2 (R_0^2 - R^2) - \frac{A}{B}\int_{R_0}^{R_1} r^2 \frac{\partial \ni}{\partial R}\, dR \; ,$$

$$\sigma_\Phi = \frac{d}{dR} R\sigma_R + \rho\omega^2 R^2 \; , \tag{9}$$

$$\sigma_Z = \sigma_R + 2\left(A^{-2} - A^2\frac{r^2}{R^2}\right)\left(\frac{\partial \ni}{\partial I_1} + \frac{R^2}{r^2}\frac{\partial \ni}{\partial I_2}\right) ,$$

so that $\sigma_R = 0$ on the inner surface of a cylinder. In these formulas

$$\frac{\partial \sigma_R}{\partial R} = -\rho\omega^2 R - \frac{A}{B} r^2 \frac{\partial \ni}{\partial R} \; , \qquad \frac{\partial \ni}{\partial R} = \frac{B}{Ar^2}\frac{2}{R}\left(\frac{r^2}{R^2} - \frac{R^2}{A^2 r^2}\right)\left(A^2\frac{\partial \ni}{\partial I_1} + \frac{\partial \ni}{\partial I_2}\right) .$$

The constants A and B are determined by the condition of the absence of normal stress on the outer surface $R = R_1$ of the cylinder as well as of absence of the longitudinal force Q. We arrive at the two equations

$$\frac{1}{2}\rho\omega^2 (R_1^2 - R_0^2) = 2\int_{R_1}^{R_0} \frac{dR}{R}\left(\frac{r^2}{R^2} - \frac{R^2}{A^2 r^2}\right)\left(A^2\frac{\partial \ni}{\partial I_1} + \frac{\partial \ni}{\partial I_2}\right) , \tag{10}$$

$$\int_{R_0}^{R_1} \sigma_R R\,dR + 2\int_{R_0}^{R_1} R\,dR\left(A^{-2} - A^2\frac{r^2}{R^2}\right)\left(\frac{\partial \ni}{\partial I_1} + \frac{R^2}{r^2}\frac{\partial \ni}{\partial I_2}\right) = 0 . \tag{11}$$

Introducing the variable of integration (12.3), these equations become

$$\frac{1}{4}\rho\omega^2 r_0^2\delta = \int_{x_1^2}^{x_0^2} (\xi + A)\left(\frac{\partial \ni}{\partial I_1} + A^{-2}\frac{\partial \ni}{\partial I_2}\right)\frac{d\xi}{\xi^2} ,$$

$$x_0^2 = \frac{R_0^2}{r_0^2} , \qquad x_1^2 = \frac{x_0^2 + A\delta}{1+\delta} , \tag{12}$$

$$\frac{1}{4}\rho\omega^2 r_0^2\delta\,\frac{2+\delta}{x_0^2 - A} = \int_{x_1^2}^{x_0^2} \frac{\xi + A}{\xi - A}\left(\frac{\partial \ni}{\partial I_1} + A^{-2}\frac{\partial \ni}{\partial I_2}\right)\frac{d\xi}{\xi^2} +$$

$$+ \frac{2}{A^3}\int_{x_1^2}^{x_0^2} (A^4 - \xi)\left(\frac{\partial \ni}{\partial I_1} + \xi\frac{\partial \ni}{\partial I_2}\right)\frac{d\xi}{\xi(\xi - A)^2} . \tag{13}$$

According to the empirical criterion, the right-hand side of Equation (12) is positive so that $x_0^2 > x_1^2$. By (12), this implies the inequalities

$$x_0^2(1+\delta) > x_0^2 + A\delta , \qquad x_0^2 > A , \qquad x_1^2 > A .$$

It is to be expected that under rotation, a hollow cylinder (and also a solid cylinder), its diameters (inner and outer) will be increased and its length reduced, in other words, $A > 1$.

16. THE UNIVERSAL SOLUTIONS OF THE EQUATIONS OF MOTION

1. We now consider the equation of motion of an incompressible elastic medium in the absence of body forces

$$-\underline{\nabla}p + \underline{\nabla}\cdot\underline{\mathrm{T}}_E = \rho\underline{\mathrm{b}} . \tag{1}$$

This one is compared with the "quasibalanced motion" defined by the equation of equilibrium in the pressure field $\tilde{p}(\underline{\mathrm{R}},t)$

$$-\underline{\nabla}\tilde{p} + \underline{\nabla}\cdot\underline{\mathrm{T}}_E = \underline{0} . \tag{2}$$

The position vector $\underline{\mathrm{R}}$ determined by this equation (we confine ourselves here to the universal solutions) includes among its parameters the fixed value of t.

From Equations (1) and (2) it follows that

$$\underline{\nabla}(\tilde{p} - p) = \rho \underline{b} ,$$

this condition can be satisfied if we postulate the existence of the acceleration potential

$$\underline{b} = \underline{\nabla}\zeta , \qquad \rho\zeta = \tilde{p} - p . \tag{3}$$

Here ζ denotes a single-valued function of coordinates which exists by virtue of Thomson's theorem (1.14.16) for those motions whose velocity circulation is uniform

$$\oint_{\Gamma} \dot{\underline{R}} \cdot d\underline{R} = \text{const} . \tag{4}$$

The quasibalanced motion is dynamically possible if and only if there can be found a single-valued acceleration potential, i.e. if the position vector satisfies condition (4). The stress tensor is then given by the relationship

$$\underline{T} = \rho\zeta\underline{E} + \bar{\underline{T}} , \tag{5}$$

where $\bar{\underline{T}}$ is a general representation of the stress tensor corresponding to equilibrium at the instant t.

2. On the acceleration potential. Equation (4) is the necessary and sufficient condition for the existence of the acceleration potential. The sufficient condition is provided by the vanishing of the velocity circulation which means that the motion is potential and irrotational:

$$\underline{\nabla} \times \underline{v} = \underline{0} , \qquad \underline{v} = \underline{\nabla}\chi = \underline{R}^s \frac{\partial}{\partial q^s} \chi , \qquad \underline{\nabla} \cdot \underline{v} = 0 , \qquad \underline{\nabla}^2 \chi = 0 .$$

Here χ denotes the velocity potential from which the acceleration potential ζ may be calculated

$$\underline{b} = \underline{\nabla}\zeta = \dot{\underline{v}} = \left(\underline{R}^s \frac{\partial\chi}{\partial q^s}\right)^{\cdot} = \underline{R}^s \frac{\partial^2\chi}{\partial t \partial q^s} + \dot{\underline{R}}^s \frac{\partial\chi}{\partial q^s} =$$

$$= \underline{\nabla} \frac{\partial\chi}{\partial t} - \underline{\nabla}\underline{v} \cdot \underline{R} \frac{\partial\chi}{\partial q^s} = \underline{\nabla} \frac{\partial\chi}{\partial t} - \underline{\nabla}\underline{v} \cdot \underline{v} = \underline{\nabla}\left(\frac{\partial\chi}{\partial t} - \frac{1}{2} v^2 \right) .$$

Here we have made use of Equation (1.10.5). We thus see that whenever the velocity potential exists, the acceleration potential is determined by the expression

$$\zeta = \frac{\partial\chi}{\partial t} - \frac{1}{2}v^2 \tag{6}$$

to the extent of some additive function of time.

3. *Examples*. a) The affine transformation. For this case

$$\underline{R} = \underline{c}(t) + \underline{r}\cdot\underline{\Lambda}(t) , \qquad \det\underline{\Lambda} \neq 0 .$$

$$\underline{v} = \dot{\underline{c}} + \underline{r}\cdot\dot{\underline{\Lambda}}(t) = \dot{\underline{c}} + (\underline{R}-\underline{c})\underline{\Lambda}^{-1}\cdot\dot{\underline{\Lambda}}$$

and denoting by $\underline{\omega}$ the vector associated with the tensor $\underline{\Lambda}^{-1}\cdot\dot{\underline{\Lambda}}$ we have by (I.11.1)

$$\underline{\nabla}\times\underline{v} = \underline{\nabla}\times\underline{R}\cdot\underline{\Lambda}^{-1}\cdot\dot{\underline{\Lambda}} = \underline{R}^s\times\underline{R}_s\cdot(\underline{\Lambda}^{-1}\cdot\dot{\underline{\Lambda}})_{mn}\underline{R}^m\underline{R}^n =$$

$$= (\underline{R}^s\times\underline{R}^n)(\underline{\Lambda}^{-1}\cdot\dot{\underline{\Lambda}})_{sn} = -2\underline{\omega}$$

and the sufficient condition for the existence of the potential ζ is provided by the requirement that the tensor $\underline{\Lambda}^{-1}\cdot\dot{\underline{\Lambda}}$ be symmetric

$$\underline{\omega} = \underline{0} : \underline{\Lambda}^{-1}\cdot\dot{\underline{\Lambda}} = (\underline{\Lambda}^{-1}\cdot\dot{\underline{\Lambda}})^T , \qquad \dot{\underline{\Lambda}}^T = \underline{\Lambda}^{-1}\cdot\dot{\underline{\Lambda}}\cdot\underline{\Lambda}^T . \tag{7}$$

For example, this condition is violated for the deformation of a simple shear

$$\underline{\Lambda} = \underline{E} + \underline{i}_2\underline{i}_1 s , \qquad \underline{\Lambda}^{-1} = \underline{E} - \underline{i}_2\underline{i}_1 s ,$$

$$\underline{\Lambda}^{-1}\cdot\dot{\underline{\Lambda}} = (\underline{E}-\underline{i}_2\underline{i}_1 s)\cdot\underline{i}_2\underline{i}_1\dot{s} = \underline{i}_2\underline{i}_1\dot{s} , \qquad 2\underline{\omega} = \underline{i}_3\dot{s} \neq \underline{0} .$$

The necessary and sufficient conditions are given by the relationships

$$\underline{b} = \ddot{\underline{c}} + \underline{r}\cdot\ddot{\underline{\Lambda}}(t) = \ddot{\underline{c}} + (\underline{R}-\underline{c})\underline{\Lambda}^{-1}\cdot\ddot{\underline{\Lambda}}(t) , \qquad \ddot{\Lambda}^T = \underline{\Lambda}^{-1}\cdot\ddot{\underline{\Lambda}}\cdot\underline{\Lambda}^T \tag{8}$$

and the acceleration potential is represented by

$$\zeta = \frac{1}{2}\underline{R}\cdot\underline{\Lambda}^{-1}\cdot\ddot{\underline{\Lambda}}\cdot\underline{R} + (\ddot{\underline{c}} - \underline{c}\cdot\underline{\Lambda}^{-1}\cdot\ddot{\underline{\Lambda}})\cdot\underline{R} . \tag{9}$$

b) Transformation of a hollow cylinder into a hollow cylinder, while preserving its axial dimension. In Equations (11.4)

$$R^2 = r^2 + A , \qquad \Phi = \varphi , \qquad Z = z , \qquad \underline{R} = \underline{e}_R\sqrt{r^2 + A} + \underline{k}z$$

and bearing in mind that $\dot{\underline{e}}_R = \underline{e}_\Phi \dot{\Phi} = \underline{0}$, we have

$$\underline{v} = \frac{\dot{A}}{2R}\,\underline{e}_R = \frac{1}{2}\dot{A}\underline{\nabla}\ln R\,, \qquad \chi = \frac{1}{2}\dot{A}\ln R\,, \qquad \frac{\partial\chi}{\partial t} = \frac{1}{2}\ddot{A}\ln R + \frac{1}{4}\frac{\dot{A}^2}{R^2}$$

and by virtue of (4)

$$\zeta = \frac{1}{2}\left(\ddot{A}\ln R + \frac{1}{4}\frac{\dot{A}^2}{R^2}\right). \tag{10}$$

At the same time

$$A = R_0^2 - r_0^2\,, \qquad \dot{A} = 2R_0\dot{R}_0\,, \qquad \ddot{A} = 2(R_0\ddot{R}_0 + \dot{R}_0^2)$$

and therefore

$$\underline{v} = \frac{R_0}{R}\dot{R}_0\underline{e}_R\,, \qquad \chi = \frac{1}{2}R_0\dot{R}_0\ln R^2\,, \qquad \zeta = \frac{1}{2}\left[(R_0\ddot{R}_0 + \dot{R}_0^2\ln R^2) + \frac{R_0^2}{R^2}\dot{R}_0^2\right]. \tag{11}$$

c. <u>The transformation of a hollow cylinder into a hollow cylinder when its axial dimension is not preserved</u>. By (11.1) we have

$$R^2 = Ar^2 + B\,, \qquad \Phi = \varphi\,, \qquad Z = A^{-1}z\,; \qquad \underline{R} = R\underline{e}_R + \underline{k}\frac{z}{A}\,,$$

$$\underline{v} = \dot{R}\underline{e}_R + \underline{k}\dot{Z} = \frac{1}{2}\left[\frac{\dot{A}}{A}R + \frac{1}{R}\left(\dot{B} - \frac{\dot{A}}{A}B\right)\right]\underline{e}_R - \frac{\dot{A}}{A}Z\underline{k}\,,$$

$$\chi = \left[\frac{1}{2}\frac{\dot{A}}{A}R^2 + \left(\dot{B} - \frac{\dot{A}}{A}\right)\ln R + \frac{1}{2}\frac{\dot{A}}{A}Z^2\right];$$

and the acceleration potential determined by (4) is

$$\zeta = \frac{1}{4}\left(\frac{\ddot{A}}{A} - \frac{\dot{A}^2}{2A^2}\right)R^2 + \frac{1}{8R^2}\left(\dot{B} - \frac{\dot{A}}{A}B\right)^2 +$$
$$+ \frac{1}{2}\left[\frac{\dot{A}}{A}\left(\frac{\dot{A}}{A}B - \dot{B}\right) + \left(\ddot{B} - \frac{\ddot{A}}{A}\dot{B}\right)\right]\ln R + \left(\frac{\dot{A}^2}{A^2} - \frac{1}{2}\frac{\ddot{A}}{A}\right)Z^2\,. \tag{12}$$

d) <u>The central-symmetric transformation of a hollow sphere</u>. By (9.22) we have

$$R^3 = r^3 + A\,, \qquad \Theta = \vartheta\,, \qquad \Lambda = \lambda\,;$$
$$\underline{R} = R\underline{e}_R\,, \qquad \underline{v} = \dot{R}\underline{e}_R = \frac{\dot{A}}{3R^2}\underline{e}_R\,, \qquad \chi = -\frac{\dot{A}}{3R} \tag{13}$$

and in view of (4)

$$\zeta = \frac{\dot{A}^2}{18R^4} - \frac{1}{3}\frac{\ddot{A}}{R} = -\left(\ddot{R}R + \frac{3}{2}\dot{R}^2\right) . \tag{14}$$

17. THE LAGRANGIAN DIFFERENTIAL EQUATIONS FOR THE PARAMETERS IN UNIVERSAL SOLUTIONS

Equations (16.2) define the stress tensor as a function of the material coordinates and the parameters A, B, C etc. of the position vector $\underline{R}$. These parameters are represented below by μ^α and their first and second time derivatives. The differential equations for determining these parameters are governed by the boundary conditions just as the finite relationships between them, which are found from Pars. 10-15. It seems, however, that the procedure for constructing these equations based upon the Hamilton-Ostrogradskii principle leads more directly to the desired result.

The assignment of the position vector

$$\underline{R} = \underline{R}(q^1, q^2, q^3 ; \mu^\alpha) \tag{1}$$

implies the following representations for its variation, the velocity vector and the kinetic energy

$$\delta\underline{R} = \frac{\partial \underline{R}}{\partial \mu^\alpha}\delta\mu^\alpha , \qquad \underline{v} = \frac{\partial \underline{R}}{\partial \mu^\alpha}\dot{\mu}^\alpha ,$$

$$K = \frac{1}{2}\dot{\mu}^\alpha\dot{\mu}^\beta \iiint_V \frac{\partial \underline{R}}{\partial \mu^\alpha}\cdot\frac{\partial \underline{R}}{\partial \mu^\alpha} dV . \tag{2}$$

Recalling also the expression for the variation of the Lagrangian

$$\delta L = -E_\alpha(L)\delta\mu^\alpha , \qquad E_\alpha(L) = \frac{d}{dt}\frac{\partial L}{\partial \dot{\mu}^\alpha} - \frac{\partial L}{\partial \mu^\alpha} \tag{3}$$

and following (2.24), we may now put the expression for Hamilton's principle in the following form

$$\int_{t_0}^{t_1} dt\left[E_\alpha(L) - \iiint_V \tilde{p}\underline{\nabla}\cdot\frac{\partial \underline{R}}{\partial \mu^\alpha} dV - \iint_{O_1} \underline{f}\cdot\frac{\partial \underline{R}}{\partial \mu^\alpha} dO\right]\delta\mu^\alpha = 0 . \tag{4}$$

For an incompressible medium we have, from (1.10.18), that

$$\dot{I}_3 = I_3\underline{\nabla}\cdot\underline{v} = \underline{\nabla}\cdot\underline{v} = \dot{\mu}^\alpha\underline{\nabla}\cdot\frac{\partial\underline{R}}{\partial\mu^\alpha} = 0\ , \qquad \underline{\nabla}\cdot\frac{\partial\underline{R}}{\partial\mu^\alpha} = 0$$

and from (4) that

$$\iiint\limits_V \tilde{p}\underline{\nabla}\cdot\frac{\partial\underline{R}}{\partial\mu^\alpha}\,\delta\mu^\alpha dV = 0\ . \tag{5}$$

There is no need to include this in the expression for Hamilton's principle when the parameters μ^α are independent. In this case, neither is it necessary to introduce a multiplying factor. The expression (4) then takes the form

$$\int_{t_0}^{t_1} dt\left[E_\alpha(L) - \iint\limits_{O_1}\underline{f}\cdot\frac{\partial\underline{R}}{\partial\mu^\alpha}\,dO\right]\delta\mu^\alpha = 0\ . \tag{6}$$

Note also that the acceleration vector and the condition for the existence of the acceleration potential can also be represented by the expressions

$$\begin{aligned} \underline{b} &= \frac{\partial\underline{R}}{\partial\mu^\alpha}\ddot{\mu}^\alpha + \frac{\partial^2\underline{R}}{\partial\mu^\alpha\partial\mu^\beta}\dot{\mu}^\alpha\dot{\mu}^\beta\ , \\ \underline{\nabla}\times\left(\frac{\partial\underline{R}}{\partial\mu^\alpha}\ddot{\mu}^\alpha + \frac{\partial^2\underline{R}}{\partial\mu^\alpha\partial\mu^\beta}\dot{\mu}^\alpha\dot{\mu}^\beta\right) &= \underline{0}\ . \end{aligned} \tag{7}$$

18. RADIAL VIBRATIONS OF A CYLINDRICAL TUBE

The tube is assumed to be confined between two smooth plates which do not allow axial displacements but at the same time do not prevent radial displacements of the faces.

Following notation of Par. 11, we have for $A = 1$

$$B = R^2 - r^2 = R_0^2 - r_0^2 = R_1^2 - r_1^2\ , \qquad \underline{R} = R\underline{e}_R = \sqrt{B + r^2}\,\underline{e}_R + \underline{k}z\ . \tag{1}$$

The parameter B represents a function of time to be determined and for any t, $B > -r_0^2$, and in the reference configuration $B = 0$. The velocity vector and the kinetic energy of a tube are

$$\underline{v} = \dot{\underline{R}} = \frac{1}{2R}\dot{B}\underline{e}_R\ ,$$

$$K = \frac{1}{4}\pi l\rho\dot{B}^2\ln\frac{R_1}{R_0} = \frac{1}{8}\pi l\rho\dot{B}^2\ln\frac{B + r_1^2}{B + r_0^2}\ . \tag{2}$$

The strain energy of deformation of a tube is determined by the expression*)

$$\Pi(B) = 2\pi l \int_{r_0}^{r_1} Э(I_1, I_2)\, r\, dr\,, \qquad I_1 = I_2 = \frac{R^2}{r^2} + \frac{r^2}{R^2} + 1 = 3 + \frac{B^2}{(r^2 + B)\, r^2}\,, \tag{3}$$

so that $\Pi(0) = 0$. The derivative $\Pi'(B)$ turns out to be

$$\Pi'(B) = \pi l B \int_{r_0}^{r_1} \hat{\mu}(I_1, I_2) \frac{B + 2r^2}{(B + r^2)^2} \frac{dr}{r}\,, \qquad \hat{\mu} = 2\left(\frac{\partial Э}{\partial I_1} + \frac{\partial Э}{\partial I_2}\right)\,, \tag{4}$$

where $\hat{\mu}(3,3) = \mu$ denotes the shear modulus in the reference configuration. We have obviously

$$\Pi'(0) = 0\,;$$

$$\Pi'(B) < 0\,, \qquad -r_0^2 < B < 0\,;$$

$$\Pi'(B) > 0\,, \qquad 0 < B < \infty\,,$$

so that the positive function $\Pi(B)$ is monotonic and decreasing from ∞ to 0 for negative values of the argument and increasing from 0 to ∞ for positive values.

The universal solution considered allows the surface pressure to be distributed over the inner and outer surfaces of the tube. In (17.4) we should then set

$$f_0 = q_0(t)\,\underline{e}_R\,, \qquad f_1 = -q_1(t)\,\underline{e}_R\,;$$

$$\frac{\partial \underline{R}}{\partial B} = \frac{1}{2R}\,\underline{e}_R\,, \qquad \iint_{O_1} \underline{f} \cdot \frac{\partial \underline{R}}{\partial B}\, dO = \pi l (q_0 - q_1)$$

and the Lagrange equation for the "generalized coordinate" B is reduced to the form

$$\frac{1}{4}\rho\left[\ddot{B} \ln \frac{B + r_1^2}{B + r_0^2} + \frac{1}{2}\dot{B}^2\left(\frac{1}{R_1^2} - \frac{1}{R_0^2}\right)\right] + B \int_{r_0}^{r_1} \frac{B + 2r^2}{(B + r^2)^2}\, \hat{\mu}(I_1, I_2) \frac{dr}{r} = q_0 - q_1\,. \tag{6}$$

Its first integral - the energy integral - is given by the expression

*) Recall that Э denotes the strain energy per unit volume in the reference configuration.

$$K + \Pi = \pi l \int_0^t [q_0(t) - q_1(t)]\dot{B}\,dt + 2h \,. \tag{7}$$

The constant of integration $h = 0$ if the initial values of B and $\dot{B}$ are zero. For constant values of q_0 and q_1 and $h = 0$, Equation (7) determines the motion of a tube which has been fixed in its reference configuration and suddenly been subjected to a constant pressure load at the instant $t = 0$.

We consider below an example of the free vibrations ($q_0 = q_1 = 0$) of a very thin tube made of material which has Mooney potential (4.1)

$$\frac{r_1^2}{r_0^2} = 1 + \delta \,, \qquad \delta \ll 1 \,; \qquad \hat{\mu} = 2(C_1 + C_2) \,.$$

A calculation carried out up to terms of first order with respect to δ enables us to represent the energy integral in the form

$$\frac{\dot{x}^2}{1+x} + \nu^2 \frac{x^2}{1+x} = 2h \,, \qquad \nu^2 = \frac{4\hat{\mu}}{\rho r_0^2} \,; \qquad x = \frac{B}{r_0^2} \,.$$

In the vibrational motion, the range of possible values of x is restricted by the roots x_0 and x_1 of the equation

$$-\nu^2 x^2 + 2hx + 2h = 0 \,, \qquad x_0 = \frac{h}{\nu^2}\left(1 - \sqrt{1 + \frac{2\nu^2}{h}}\right) \,,$$

$$x_1 = \frac{h}{\nu^2}\left(1 + \sqrt{1 + \frac{2\nu^2}{h}}\right) \,.$$

The period of vibrations (which is not dependent on the initial conditions) is

$$\tau = \frac{2\pi}{\nu} = \frac{\pi r_0 \sqrt{\rho}}{\sqrt{\hat{\mu}}} \,.$$

The relationships (16.5) and (16.10) and the formulas of Par. 11 enable us to construct the expressions for the stress.

The vibrations of a cylindrical tube were originally examined by Knowles (Knowles, 1960). His solution has been reproduced and augmented by Truesdell and Wang in their book (1974). The starting point of the analysis given in these papers was the representation of the stress σ_R. In view of Equations (16.5), (16.10) and (11.4) this can be set out as follows

$$\sigma_R = \frac{1}{4}\rho\left(\ddot{B}\ln R^2 + \frac{1}{2}\frac{\dot{B}^2}{R^2}\right) + B\int_{r_0}^{r_1} \hat{\mu}(I_1,I_2)\,\frac{B+2r^2}{(B+r^2)^2}\,\frac{dr}{r} - p_0 \,. \tag{8}$$

The stresses σ_R on the inner and outer surfaces $r = r_0$ and $r = r_1$ of the tube were denoted by $-q_0$ and $-q_1$ respectively. Equation (8) thus leads to the relationship

$$\frac{1}{4}\rho\left[\ddot{B}\ln\frac{B+r_1^2}{B+r_0^2} + \frac{1}{2}\dot{B}^2\left(\frac{1}{R_1^2} - \frac{1}{R_0^2}\right)\right] + B\int_{r_0}^{r_1} \hat{\mu}(I_1,I_2)\,\frac{B+2r^2}{(B+r^2)^2}\,\frac{dr}{r} = q_0 - q_1 \,,$$

which, of course, again yields Equation (6). It seems then that the calculation based on the Lagrange equation has lead us more quickly to the desired result.

19. RADIAL VIBRATIONS OF A HOLLOW SPHERE

The mapping which carries the reference configuration into the actual configuration is prescribed by Equations (16.13). The inner and outer radii $R_0(t)$ and $R_1(t)$ of the sphere are connected with the unknown function $A(t)$ via the relationships

$$A = R^3 - r^3 = R_0^3 - r_0^3 = R_1^3 - r_1^3 \,, \qquad -r_1^3 < A < \infty \,. \tag{1}$$

The expression for the kinetic energy has the form

$$K = \frac{1}{2}\rho\iiint_V v^2 dV = \frac{1}{18}\rho\dot{A}^2 4\pi\int_{R_0}^{R_1}\frac{dR}{R^2} = \frac{2\pi}{9}\rho\dot{A}^2\left(\frac{1}{R_0} - \frac{1}{R_1}\right) \,. \tag{2}$$

The invariants $I_1(\underline{F})$ and $I_2(\underline{F})$ are, by Par. 9, determined from the formulas

$$I_1 = \frac{r^4}{R^4} + 2\,\frac{R^2}{r^2} \,, \qquad I_2 = \frac{R^4}{r^4} + 2\,\frac{r^2}{R^2} \,. \tag{3}$$

The functions R and $\frac{\partial R}{\partial A}$ below must be substituted by their expressions

$$R = (A + r^3)^{1/3} \,, \qquad \frac{\partial R}{\partial A} = \frac{1}{3R^2} \tag{4}$$

and the potential energy of deformation is represented by

$$\Pi(A) = 4\pi\int_{r_0}^{r_1} \ni(I_1,I_2)\,r^2 dr \,.$$

This is a positive function which is zero in the reference configuration for $A = 0$. Here $\Pi(A) \to \infty$ for $A \to -r_0^3$ and also for $A \to \infty$. The expression for its derivative is reduced to

$$\Pi'(A) = 4\pi \int_{r_0}^{r_1} \left(\frac{\partial \ni}{\partial I_1} \frac{\partial I_1}{\partial A} + \frac{\partial \ni}{\partial I_2} \frac{\partial I_2}{\partial A} \right) r^2 dr =$$

$$= \frac{16\pi}{3} A \int_{r_0}^{r_1} \left(\frac{\partial \ni}{\partial I_1} + \frac{R^2}{r^2} \frac{\partial \ni}{\partial I_2} \right) (R^3 + r^3) \frac{dr}{R^7} , \tag{5}$$

from this, assuming the validity of the empirical criteria (5.11.3), we conclude that $\Pi(A)$ is monotonically decreasing for negative values of the argument and increasing for positive values thus having a single minimum (equal to zero) at the point $A = 0$.

The equation of motion is now

$$\rho \left[\ddot{A} \frac{R_1 - R_0}{R_1 R_0} - \frac{1}{6} \dot{A}^2 \frac{R_1^4 - R_0^4}{R_0^4 R_1^4} \right] + 12A \int_{r_0}^{r_1} \frac{dr}{R^7} (R^3 + r^3) \left(\frac{\partial \ni}{\partial I_1} + \frac{R^2}{r^2} \frac{\partial \ni}{\partial I_2} \right) = 3(q_0 - q_1) . \tag{6}$$

This can, of course, be deduced from an equation similar to (18.8) where, according to (9.24), we should set

$$\sigma_R = -\rho\zeta + 4A \int_{r_0}^{r_1} \ni(I_1, I_2) r^2 dr = 6 \int_0^t [q_0(t) - q_1(t)] \dot{A}\, dt + h , \tag{7}$$

and define ζ from Equation (16.10).

The energy integral is now reduced to the form

$$\rho \dot{A}^2 \left(\frac{1}{R_0} - \frac{1}{R_1} \right) + 18 \int_{r_0}^{r_1} \ni(I_1, I_2) r^2 dr = 6 \int_0^t [q_0(t) - q_1(t)] \dot{A}\, dt + h . \tag{8}$$

This problem has been considered by Guo Zhong-Heng and Solecki [7.14] and their solution has been reproduced by Truesdell and Wang who took Equation (19.7) as their starting point. For a very thin spherical shell $r_1^3 = r_0^3(1 + \delta)$, $\delta \ll 1$ and retaining in (8) only terms linear with respect to δ, we have

$$\frac{1}{R_0} - \frac{1}{R_1} = \frac{\delta}{3r_0} (x + 1)^{-4/3} , \qquad \int_{r_0}^{r_1} \ni(I_1, I_2) r^2 dr = \frac{r_0^2 \delta}{3} \ni(I_1^0, I_2^0) , \qquad x = \frac{A}{r_0^3} ,$$

where

$$I_1^0 = \frac{r_0^4}{R_0^4} + 2\frac{R_0^2}{r_0^2} = (1+x)^{-4/3} + 2(1+x)^{2/3} ,$$

$$I_2^0 = \frac{R_0^4}{r_0^4} + 2\frac{r_0^2}{R_0^2} = (1+x)^{4/3} + 2(1+x)^{-2/3} .$$

Considering free vibrations only, Equation (8) reduces to the form

$$\rho r_0^2 \dot{x}^2 (1+x)^{-4/3} + 18\Im(I_1^0, I_2^0) = \frac{3h}{\delta r_0^3} = E = \text{const} ,$$

where $-1 < x < \infty$, and the above properties of $\Pi(A)$ allow us to assert that for any $E > 0$ the equation

$$E - 18\Im(I_1^0, I_2^0) = 0$$

possesses two and only two real roots $x_0 < 0$ and $x_1 > 0$. The motion represents vibration within these limits $x_0 \le x \le x_1$ with the period

$$\tau = 2r_0\sqrt{\rho}\int_{x_0}^{x_1} \frac{dx}{(1+x)^{2/3}\sqrt{E - \Im(I_1^0, I_2^0)}} . \tag{9}$$

For example, for a Neo-Hookean material we have

$$\tau = \frac{2r_0\sqrt{\rho}}{\sqrt{C_1}}\int_{x_0}^{x_1} \left\{\frac{E}{C_1}(1+x)^{4/3} - [1 + 2(1+x)^2]\right\}^{-1/2} dx . \tag{10}$$

20. THE ANTIPLANE DEFORMATION OF AN INCOMPRESSIBLE MATERIAL

Deformation of a cylindrical body is termed "antiplane" if the displacements of its particles are parallel to some fixed generator and do not depend on the axial coordinate.

The Cartesian coordinates a^s of the reference configuration serve as its material coordinates. It is assumed that the antiplane deformation is carried out in a prestressed cylinder so that in the actual configuration the Cartesian coordinates are given by the formulas

$$x^1 = \lambda^{-1/2}a^1 , \quad x^2 = \lambda^{-1/2}a^2 , \quad x^3 = \lambda a^3 + u(a^1, a^2) , \tag{1}$$

where λ is a constant and $u(a^1,a^2)$ is a function to be determined and which is twice continuously differentiable within the hollow of the cylinder.

The deformation gradient $\overset{0}{\nabla}\underline{R}$, the strain measure $\underline{G}$ and its invariants are

$$\overset{0}{\nabla}\underline{R} = (\underline{i}_1\underline{i}_1 + \underline{i}_2\underline{i}_2)\lambda^{-1/2} + \underline{i}_3\underline{i}_3\lambda + \overset{0}{\nabla}u\underline{i}_3 \,, \tag{2}$$

$$\underline{G} = \overset{0}{\nabla}\underline{R}\cdot\overset{0}{\nabla}\underline{R}^T = (\underline{i}_1\underline{i}_1 + \underline{i}_2\underline{i}_2)\lambda^{-1} + \lambda^2\underline{i}_3\underline{i}_3 + \lambda(\overset{0}{\nabla}u\underline{i}_3 + \underline{i}_3\overset{0}{\nabla}u) + \overset{0}{\nabla}u\overset{0}{\nabla}u \,, \tag{3}$$

$$I_1(\underline{G}) = 2\lambda^{-1} + \lambda^2 + \overset{0}{\nabla}u\cdot\overset{0}{\nabla}u \,,$$

$$I_2(\underline{G}) = \lambda^{-2} + 2\lambda + \lambda^{-1}\overset{0}{\nabla}u\cdot\overset{0}{\nabla}u \,, \qquad I_3(\underline{G}) = 1 \,, \tag{4}$$

and the representation of $\nabla\underline{r}$ has the form

$$\nabla\underline{r} = (\overset{0}{\nabla}\underline{R})^{-1} = (\underline{i}_1\underline{i}_1 + \underline{i}_2\underline{i}_2)\lambda^{1/2} + \underline{i}_3\underline{i}_3\lambda^{-1} - \lambda^{-1/2}\overset{0}{\nabla}u\underline{i}_3 \,. \tag{5}$$

By (2.4) and (2.5) the Cauchy stress tensor $\underline{T}$ and the Piola tensor $\underline{P}$ (which is determined by $\underline{T}$) are

$$\underline{T} = 2\left[\frac{\partial\ni}{\partial I_1}\underline{F} + \frac{\partial\ni}{\partial I_2}(I_1\underline{E} - \underline{F})\cdot\underline{F}\right] + p\underline{E} \,,$$

$$\underline{P} = \nabla\underline{r}^T\cdot\underline{T} = 2\left[\frac{\partial\ni}{\partial I_1}\underline{E} + \frac{\partial\ni}{\partial I_2}(I_1\underline{E} - \underline{G})\right]\cdot\overset{0}{\nabla}\underline{R} - p\nabla\underline{r}^T \,. \tag{6}$$

The three equilibrium equations

$$\overset{0}{\nabla}\cdot\underline{P} = 0 \tag{7}$$

include two unknown functions p and $u(a^1,a^2)$. Hence "antiplane" deformation is only feasible for a given function $\ni(I_1,I_2)$ of the specific strain energy if it satisfies some additional condition. In particular, as we shall see below, this deformation can be realized in a Mooney-Rivlin material (4.1). The general case has been examined by Knowles (J.K. Knowles, 1976).

In view of (2)-(5), the Piola tensor (6) for a Mooney-Rivlin material takes the form

$$\underline{P} = -\lambda^{-1/2}(q\underline{E} + 2C_2\overset{0}{\nabla}u\overset{0}{\nabla}u) + 2(C_1 + \lambda^{-1}C_2)\overset{0}{\nabla}u\underline{i}_3 +$$

$$+ (-2\lambda^{-1/2}C_2 + \lambda^{-1/2}p)\underline{i}_3\overset{0}{\nabla}u + (2\lambda C_1 + 4C_2 - \lambda^{-1}p)\underline{i}_3\underline{i}_3 \,, \tag{8}$$

where

$$q = \lambda p - 2C_1 - 2(\lambda^2 + \lambda^{-1} + \overset{0}{\underline{\nabla}}u \cdot \overset{0}{\underline{\nabla}}u)C_2 \; . \tag{9}$$

On the surface of the cylinder

$$\underline{f} = \underline{n} \cdot \underline{P} = -\lambda^{-1/2}(q\underline{n} + 2C_2\underline{n} \cdot \overset{0}{\underline{\nabla}}u\overset{0}{\underline{\nabla}}u) + 2(C_1 + \lambda^{-1}C_2)\underline{n} \cdot \overset{0}{\underline{\nabla}}u\underline{i}_3 \tag{10}$$

and from (9) it follows that if the surface traction $\underline{f}$ does not depend on a^3, $p = p(a^1, a^2)$. Under this condition, the equilibrium equations (7) reduce to the form

$$\overset{0}{\underline{\nabla}}q + 2C_2\overset{0}{\underline{\nabla}} \cdot (\overset{0}{\underline{\nabla}}u\overset{0}{\underline{\nabla}}u) = \underline{0} \; , \tag{11}$$

$$\overset{0}{\underline{\nabla}}{}^2 u = 0 \; . \tag{12}$$

However, in view of (11),

$$\overset{0}{\underline{\nabla}} \cdot (\overset{0}{\underline{\nabla}}u\overset{0}{\underline{\nabla}}u) = \overset{0}{\underline{\nabla}}{}^2 u\overset{0}{\underline{\nabla}}u + \overset{0}{\underline{\nabla}}u \cdot \overset{00}{\underline{\nabla\nabla}}u = \overset{0}{\underline{\nabla}}u \cdot \overset{00}{\underline{\nabla\nabla}}u$$

and Equations (12) and (13) take the form

$$\overset{0}{\underline{\nabla}}q + 2C_2\overset{0}{\underline{\nabla}}u \cdot \overset{00}{\underline{\nabla\nabla}}u = \underline{0} \; . \tag{13}$$

Equations (12) and (13) are compatible, since

$$\overset{0}{\underline{\nabla}} \times (\overset{0}{\underline{\nabla}}u \cdot \overset{00}{\underline{\nabla\nabla}}u) = (\overset{0}{\underline{\nabla}} \times \overset{0}{\underline{\nabla}}u) \cdot \overset{00}{\underline{\nabla\nabla}}u - \overset{0}{\underline{\nabla}}u \times \overset{0}{\underline{\nabla}} \cdot \overset{00}{\underline{\nabla\nabla}}u =$$

$$= (\overset{0}{\underline{\nabla}} \times \overset{0}{\underline{\nabla}}u) \cdot \overset{00}{\underline{\nabla\nabla}}u - \overset{0}{\underline{\nabla}}u \times \overset{00}{\underline{\nabla\nabla}}{}^2 u = \underline{0} \; ,$$

because $\overset{0}{\underline{\nabla}} \times \overset{0}{\underline{\nabla}}u = \underline{0}$ and $\overset{0}{\underline{\nabla}}{}^2 u = 0$. Hence, $\overset{0}{\underline{\nabla}} \times \overset{0}{\underline{\nabla}}q = \underline{0}$ which is the desired result.

Example. A tube made of an incompressible Mooney-Rivlin material and bounded by coaxial cylindrical surfaces of radii r_1 and r_2 $(r_1 > r_2)$ is confined between two rigid bands. The outer band (of inner radius r_1) is given a longitudinal displacement c. The inner band (of outer radius r_2) is kept fixed, i.e. we have in mind a rubber bush between two metal cylinders.

The longitudinal displacement of the material particles of a tube is, by (12), an axisymmetric harmonic function which takes the value c for $r = r_1$ and equals zero for $r = r_2$. It is given by

$$u(r) = c \ln \frac{r}{r_2} \Big/ \ln \frac{r_1}{r_2} . \tag{14}$$

From this we get

$$\overset{0}{\underline{\nabla}} u = \frac{c}{\ln \frac{r_1}{r_2}} \frac{\underline{e}_r}{r} , \qquad \overset{00}{\underline{\nabla}\underline{\nabla}} u = \frac{c}{r^2 \ln \frac{r_1}{r_2}} (-\underline{e}_r\underline{e}_r + \underline{e}_\varphi\underline{e}_\varphi)$$

and by (13)

$$\overset{0}{\underline{\nabla}} q = 2C_2 \frac{c^2 \underline{e}_r}{r^3 \left(\ln \frac{r_1}{r_2}\right)^2} , \qquad q = q_0 - C_2 \frac{c^2}{r^2 \left(\ln \frac{r_1}{r_2}\right)^2} ,$$

where q_0 is the constant of integration.

We will confine ourselves below to the case in which there is not initial tension ($\lambda = 1$). Then from (9) and (8) we have

$$p = q + 2(C_1 + 2C_2) + 2C_2 \overset{0}{\underline{\nabla}} u \cdot \overset{0}{\underline{\nabla}} u , \tag{15}$$

$$\underline{P} = -\left[q_0 \underline{E}_2 + C_2 \frac{c^2}{r^2 \left(\ln \frac{r_1}{r_2}\right)^2} (\underline{e}_r\underline{e}_r - \underline{e}_\varphi\underline{e}_\varphi)\right] + 2(C_1 + C_2)(\underline{e}_r\underline{i}_3 + \underline{i}_3\underline{e}_r) \frac{c}{\ln \frac{r_1}{r_2}} \frac{1}{r} +$$

$$+ \left(q_0 + C_2 \frac{c^2}{r^2 \left(\ln \frac{r_1}{r_2}\right)^2}\right)\left(\frac{c}{r \ln \frac{r_1}{r_2}} \underline{i}_3\underline{e}_r - \underline{i}_3\underline{i}_3\right) . \tag{16}$$

The surface traction on the oriented area $\underline{N}\, dO = \underline{\nabla}\underline{r}^T \cdot \underline{n}\, do$ is represented by the expressions

$$\underline{f}\, dO = \underline{N} \cdot \underline{T}\, dO = \underline{n} \cdot \underline{P}\, dO$$

and on the cylindrical surfaces of the tube ($\underline{n} = \pm\underline{e}_r$) is

$$\underline{f}\, dO\Big|_{r=r_i} = \pm \left\{\left[-q_0 + C_2 \frac{c^2}{r_i^2 \left(\ln \frac{r_1}{r_2}\right)^2}\right] \underline{e}_r + 2(C_1 + C_2) \frac{c}{r_i \ln \frac{r_1}{r_2}} \underline{i}_3\right\} r_i\, d\varphi da^3 ,$$

where $i = 1,2$, the positive sign refers to the outer and the negative sign to the inner surface.

The resultants of these distributions of the surface tractions per unit length in the longitudinal direction are now determined from the expressions

$$\left(\iint \underline{f}\, dO\right)\Bigg|_{r=r_i} = \pm 4\pi (C_1 + C_2) \frac{c}{\ln \frac{r_1}{r_2}} \underline{i}_3 , \tag{17}$$

where the positive sign corresponds to the force applied to the outer band and produces the displacement c and the negative sign specifies the reaction of the inner band. Both here and below, when carrying out the calculation, we have borne in mind that

$$\int_0^{2\pi} \underline{e}_r d\varphi = \int_0^{2\pi} d\underline{e}_\varphi = 0 .$$

On the faces $\underline{n} = \pm \underline{i}_3$ of the tube $dO = r\, dr d\varphi$, so that

$$\underline{f}\, dO = \pm \left\{ \left[2(C_1 + C_2) \frac{c}{r \ln \frac{r_1}{r_2}} + \left(q_0 + \frac{C_2 c^2}{r^2 \left(\ln \frac{r_1}{r_2} \right)^2} \right) \frac{c}{r \ln \frac{r_1}{r_2}} \right] \underline{e}_r + \right.$$

$$\left. + \left[q_0 + C_2 \frac{c^2}{r^2 \left(\ln \frac{r_1}{r_2} \right)^2} \right] \underline{i}_3 \right\} r\, dr d\varphi$$

and the longitudinal forces are given by the formulas

$$Q = \pi \left[q_0 (r_1^2 - r_2^2) + 2C_2 \frac{c^2}{\ln \frac{r_1}{r_2}} \right] . \tag{18}$$

The constant q_0 is determined from the condition that these forces are actually absent

$$q_0 = -2C_2 \frac{c^2}{(r_1^2 - r_2^2) \ln \frac{r_1}{r_2}} . \tag{19}$$

The Cauchy stress tensor is now represented by the expression

$$\underline{T} = q_0(-\underline{E}_2 + \underline{i}_3\underline{i}_3) + [C_2(\underline{e}\,\underline{e} - \underline{e}_r\underline{e}_r) - (2C_1 + C_2)\underline{e}_3\underline{e}_3]\,\frac{c^2}{r^2\left(\ln\dfrac{r_1}{r_2}\right)^2} +$$

$$+\, 2(C_1 + C_2)(\underline{e}_r\underline{i}_3 + \underline{i}_3\underline{e}_r)\,\frac{c}{r\ln\dfrac{r_1}{r_2}}\,. \qquad (20)$$

The problem of the "antiplane" deformation of the region bounded inside and out by two smooth contours which allow conformal mapping into a circular ring is likewise reduced to a boundary-value problem of the theory of analytic functions.

21. THE CONSTRUCTION OF ERICKSEN'S UNIVERSAL SOLUTION

1. The structure of tensors $\underline{F}$ and $\underline{F}^{-1}$ satisfying Equations (8.5)-(8.9). Equations (8.5) enable us to establish the relationships

$$\underline{F}\cdot\nabla\underline{I}_1 = A\nabla\underline{I}_1\,,\quad \underline{F}^{-1}\cdot\nabla\underline{I}_1 = A^{-1}\nabla\underline{I}_1\,;\qquad \underline{F}^{-1}\cdot\nabla\underline{I}_2 = B^{-1}\nabla\underline{I}_2\,,\quad \underline{F}\cdot\nabla\underline{I}_2 = B\nabla\underline{I}_2\,. \qquad (1)$$

where A and B are scalar quantities. Substitution of these expressions into the two Equations (8.6) leads to the single relationship

$$(B - A)\nabla\underline{I}_1 \times \nabla\underline{I}_2 = \underline{0} \qquad \left(\text{or } \left(\frac{1}{A} - \frac{1}{B}\right)\nabla\underline{I}_1 \times \nabla\underline{I}_2 = \underline{0}\right)\,. \qquad (2)$$

If the invariants I_1 and I_2 are not constants, then $\nabla\underline{I}_1 \neq \underline{0}$ and $\nabla\underline{I}_2 \neq \underline{0}$, and Equation (2) implies that vectors $\nabla\underline{I}_1$ and $\nabla\underline{I}_2$ are parallel

$$\frac{\nabla\underline{I}_1}{|\nabla\underline{I}_1|} = \frac{\nabla\underline{I}_2}{|\nabla\underline{I}_2|} = \underline{e}_1\,, \qquad (3)$$

where either $A \neq B$, or $A = B$. Assuming the former, we obtain

$$\underline{F}\cdot\underline{e}_1 = A\underline{e}_1\,,\qquad \underline{F}\cdot\underline{e}_1 = B\underline{e}_1 \qquad (4)$$

and two distinct eigenvalues are associated with the direction $\underline{e}_1$. For this reason the quantity $A = B = c_1$ represents the eigenvalue of the tensor $\underline{F}$, and $\underline{e}_1$ denotes the corresponding eigenvector. As can be seen from (3), the vectors may be parallel (if Equation (4) holds) but equally may be not so. In the latter case, c_1 represents the multiple eigenvalue (corresponding to the distinct eigenvectors $\nabla\underline{I}_1/|\nabla\underline{I}_1|$ and $\nabla\underline{I}_2/|\nabla\underline{I}_2|$). Turning now to the representations of the invariants

$$I_1 = c_1 + c_2 + c_3\,,\qquad I_2 = c_1c_2 + c_2c_3 + c_3c_1\,,\qquad I_3 = c_1c_2c_3 = 1\,, \qquad (5)$$

we must conclude that the quantities c_1, c_2 and c_3 are functionally dependent. When vectors $\underline{\nabla}I_1$ and $\underline{\nabla}I_2$ are parallel this follows from (3), and for the case of a multiple eigenvalue this is a consequence of the vanishing of the Jacobian

$$D\frac{(I_1,I_2,I_3)}{(c_1,c_2,c_3)} = (c_1^2-c_2^2)c_3 + (c_2^2-c_3^2)c_1 + (c_3^2-c_1^2)c_3 = 0 \ .$$

This allows us to regard I_1 and I_2 as functions of a single scalar quantity

$$I_1 = I_1(\Phi) \ , \qquad I_2 = I_2(\Phi) \ ; \qquad \underline{\nabla}I_1 = I_1'\underline{\nabla}\Phi \ , \qquad \underline{\nabla}I_2 = I_2'\underline{\nabla}\Phi \tag{6}$$

the primes being associated with differentiation by Φ. The roots of a characteristic equation are functions of its invariants and therefore $c_k = c_k(\Phi)$ where $k = 1,2,3$. Equations (1), (3) now take the form

$$\underline{F}\cdot\underline{\nabla}\Phi = c_1\underline{\nabla}\Phi \ , \qquad \underline{e}_1 = \frac{\underline{\nabla}\Phi}{|\underline{\nabla}\Phi|} = \frac{\underline{\nabla}\Phi}{(\underline{\nabla}\Phi\cdot\underline{\nabla}\Phi)^{1/2}} \ . \tag{7}$$

The first term in (8.7) therefore turns out to be zero; we have in fact

$$\underline{\nabla}\times(\underline{\nabla}I_1\cdot\underline{F}) = \underline{\nabla}\times I_1'\underline{\nabla}\Phi\cdot\underline{F} = \underline{\nabla}\times I_1'c_1\underline{\nabla}\Phi = \underline{\nabla}(I_1'c_1)\times\underline{\nabla}\Phi + I_1'c_1\underline{\nabla}\times\underline{\nabla}\Phi =$$

$$= (I_1'c_1)'\underline{\nabla}\Phi\times\underline{\nabla}\Phi + I_1'c_1\underline{\nabla}\times\underline{\nabla}\Phi \equiv \underline{0}$$

and Equations (8.7) take the form

$$I_1'\underline{\nabla}\Phi\times\underline{\nabla}\cdot\underline{F} = \underline{0} \ , \qquad I_2'\underline{\nabla}\Phi\times\underline{\nabla}\cdot\underline{F}^{-1} = \underline{0} \ , \tag{8}$$

so that Equations (8.8) are also satisfied. From (8) it follows that the vectors $\underline{\nabla}\Phi$ and $\underline{\nabla}\cdot\underline{F}$ are parallel

$$\underline{\nabla}\cdot\underline{F} = \alpha^{+}\underline{\nabla}\Phi \ , \qquad \underline{\nabla}\cdot\underline{F}^{-1} = \alpha^{-1}\underline{\nabla}\Phi \ . \tag{9}$$

In order to satisfy the remaining Equations (8.9) it is sufficient to assume that $\alpha^{+} = \alpha^{+}(\Phi)$ and $\alpha^{-} = \alpha^{-}(\Phi)$. In fact, we would then have e.g.

$$\underline{\nabla}\times\underline{\nabla}\cdot\underline{F} = \underline{\nabla}\times\alpha^{+}\underline{\nabla}\Phi = \underline{\nabla}\alpha^{+}(\Phi)\times\underline{\nabla}\Phi = \alpha^{+'}(\Phi)\underline{\nabla}\Phi\times\underline{\nabla}\Phi = \underline{0} \ .$$

All conditions (8.5)-(8.8) are thus fulfilled.

2. <u>Representation of dyadics of eigenvectors of the tensors $\underline{F}$ and $\underline{F}^{-1}$ in terms of Φ.</u> From the decompositions

$$\underline{F} = c_1\underline{e}_1\underline{e}_1 + c_2\underline{e}_2\underline{e}_2 + c_3\underline{e}_3\underline{e}_3 , \qquad \underline{F}^{-1} = c_1^{-1}\underline{e}_1\underline{e}_1 + c_2^{-2}\underline{e}_2\underline{e}_2 + c_3^{-1}\underline{e}_3\underline{e}_3$$

we eliminate the dyad $\underline{e}_3\underline{e}_3 = \underline{E} - \underline{e}_1\underline{e}_1 - \underline{e}_2\underline{e}_2$ and obtain

$$\underline{F} = (c_1 - c_3)\underline{e}_1\underline{e}_1 + (c_2 - c_3)\underline{e}_2\underline{e}_2 + c_3\underline{E} ,$$

$$\underline{F}^{-1} = \left(\frac{1}{c_1} - \frac{1}{c_3}\right)\underline{e}_1\underline{e}_1 + \left(\frac{1}{c_2} - \frac{1}{c_3}\right)\underline{e}_2\underline{e}_2 + \frac{1}{c_3}\underline{E} , \tag{10}$$

where by (7.7)

$$\underline{e}_1\underline{e}_1 = \frac{\underline{\nabla}\Phi\underline{\nabla}\Phi}{\underline{\nabla}\Phi \cdot \underline{\nabla}\Phi} . \tag{11}$$

In view of (10) we have

$$\underline{\nabla} \cdot \underline{F} = \alpha^{+}(\Phi)\underline{\nabla}\Phi = \underline{\nabla}\Phi \cdot [(c_1' - c_3')\underline{e}_1\underline{e}_1 + (c_2' - c_3')\underline{e}_2\underline{e}_2 + c_3'\underline{E}] +$$

$$+ (c_1 - c_3)\underline{\nabla} \cdot \underline{e}_1\underline{e}_1 + (c_2 - c_3)\underline{\nabla} \cdot \underline{e}_2\underline{e}_2 = \underline{\nabla}\Phi c_1' + (c_1 - c_3)\underline{\nabla} \cdot \underline{e}_1\underline{e}_1 +$$

$$+ (c_2 - c_3)\underline{\nabla} \cdot \underline{e}_2\underline{e}_2 ,$$

this is because by (11) $\underline{\nabla}\Phi \cdot \underline{e}_1\underline{e}_1 = \underline{\nabla}\Phi$, and by (7) $\underline{\nabla}\Phi \cdot \underline{e}_2 = 0$.

In order to determine the unknown divergences the equations were obtained

$$(c_1 - c_3)\underline{\nabla} \cdot \underline{e}_1\underline{e}_1 + (c_2 - c_3)\underline{\nabla} \cdot \underline{e}_2\underline{e}_2 = (\alpha^{+} - c_1')\underline{\nabla}\Phi ,$$

$$(c_1^{-1} - c_2^{-1})\underline{\nabla} \cdot \underline{e}_1\underline{e}_1 + (c_2^{-1} - c_3^{-1})\underline{\nabla} \cdot \underline{e}_2\underline{e}_2 = \left(\alpha^{-} - \left(\frac{1}{c_1}\right)'\right)\underline{\nabla}\Phi$$

with the determinant

$$\Delta = (c_1 - c_2)(c_2 - c_3)(c_3 - c_1) .$$

In the absence of multiple roots for the characteristic equation we get

$$\underline{\nabla} \cdot \underline{e}_1\underline{e}_1 = \underline{\nabla}\Phi\psi_1(\Phi) , \qquad \underline{\nabla} \cdot \underline{e}_2\underline{e}_2 = \underline{\nabla}\Phi\psi_2(\Phi) , \tag{12}$$

where $\psi_1(\Phi)$ and $\psi_2(\Phi)$ are rational functions of c_1, c_2, c_3, c_1', α^{+} and α^{-}. For

$$c_2 = c_3 \neq c_1 , \quad \underline{\nabla} \cdot \underline{e}_1\underline{e}_1 = \psi_1(\Phi)\underline{\nabla}\Phi ; \qquad c_3 = c_1 \neq c_2 , \quad \underline{\nabla} \cdot \underline{e}_2\underline{e}_2 = \psi_2(\Phi)\underline{\nabla}\Phi , \tag{13}$$

where in the first case $\underline{\nabla} \cdot \underline{e}_2\underline{e}_2$ cannot be determined and in the second case one cannot determine $\underline{\nabla} \cdot \underline{e}_1\underline{e}_1$.

Recalling the representation (III.3.5) for the divergence of a vector dyad, we may by (7) represent (12) in the form

$$[\underline{\nabla} \cdot \underline{e}_1 - (\underline{\nabla}\Phi \cdot \underline{\nabla}\Phi)^{1/2}\psi_1(\Phi)]\,\underline{e}_1 + \underline{e}_1 \cdot \underline{\nabla}\underline{e}_1 = \underline{0} . \tag{14}$$

But vectors $\underline{e}_1$ and $\underline{e}_1 \cdot \underline{\nabla}\underline{e}_1$ are mutually perpendicular since by (III.3.2)

$$(\underline{e}_1 \cdot \underline{\nabla}\underline{e}_1) \cdot \underline{e}_1 = \underline{e}_1 \cdot (\underline{\nabla}\underline{e}_1 \cdot \underline{e}_1) = \frac{1}{2}\, \underline{e}_1 \cdot \underline{\nabla}(\underline{e}_1 \cdot \underline{e}_1) = 0$$

and relation (14) is possible only if

$$\underline{\nabla} \cdot \underline{e}_1 = (\underline{\nabla}\Phi \cdot \underline{\nabla}\Phi)^{1/2}\psi_1(\Phi) , \qquad \underline{e}_1 \cdot \underline{\nabla}\underline{e}_1 = \underline{0} . \tag{15}$$

From (7) we have

$$\underline{\nabla}\underline{e}_1 = \underline{\nabla}\, \frac{\underline{\nabla}\Phi}{(\underline{\nabla}\Phi \cdot \underline{\nabla}\Phi)^{1/2}} = \frac{\underline{\nabla}\underline{\nabla}\Phi}{(\underline{\nabla}\Phi \cdot \underline{\nabla}\Phi)^{1/2}} - \frac{\underline{\nabla}\underline{\nabla}\Phi \cdot \underline{\nabla}\Phi}{(\underline{\nabla}\Phi \cdot \underline{\nabla}\Phi)^{3/2}}\, \underline{\nabla}\Phi ,$$

since $\underline{\nabla}(\underline{\nabla}\Phi \cdot \underline{\nabla}\Phi) = 2\underline{\nabla}\underline{\nabla}\Phi \cdot \underline{\nabla}\Phi$. The second relationship (15) turns out to be equivalent to the equality

$$\underline{\nabla}\Phi \cdot \underline{\nabla}\underline{\nabla}\Phi = \frac{1}{2}\, \underline{\nabla}(\underline{\nabla}\Phi \cdot \underline{\nabla}\Phi) = (\underline{\nabla}\Phi \cdot \underline{\nabla}\Phi)^{-1}(\underline{\nabla}\Phi \cdot \underline{\nabla}\underline{\nabla}\Phi \cdot \underline{\nabla}\Phi)\underline{\nabla}\Phi , \tag{16}$$

in other words, the vectors $\underline{\nabla}(\underline{\nabla}\Phi \cdot \underline{\nabla}\Phi)$ and $\underline{\nabla}\Phi$ are parallel

$$\underline{\nabla}(\underline{\nabla}\Phi \cdot \underline{\nabla}\Phi) \times \underline{\nabla}\Phi = \underline{0} . \tag{17}$$

Recalling the implicit functions' theorem, which asserts that the necessary and sufficient conditions for the existence of a functional relationship between two functions $f_1(q^1,q^2,q^3)$ and $f_2(q^1,q^2,q^3)$ of three variables amount to the vanishing of their Jacobians calculated with respect to any pair of variables, we have

$$\frac{\mathcal{D}(f_1,f_2)}{\mathcal{D}(q^1,q^2)} = 0 , \qquad \frac{\mathcal{D}(f_1,f_2)}{\mathcal{D}(q^2,q^3)} = 0 , \qquad \frac{\mathcal{D}(f_1,f_2)}{\mathcal{D}(q^3,q^1)} = 0 .$$

In other words, the gradients $\underline{\nabla}f_1$ and $\underline{\nabla}f_2$ of these functions are parallel vectors. Returning to (17), we arrive at the statement which is of fundamental importance for what follows. The quantity $\underline{\nabla}\Phi \cdot \underline{\nabla}\Phi$ is a function of Φ. In other words, representation (12) of divergence of the dyad $\underline{\nabla} \cdot \underline{e}_1\underline{e}_1$ is possible only for those functions $\Phi(q^1, q^2, q^3)$ which satisfy the condition formulated. In view of (15) we arrive at the relationship

$$\underline{\nabla} \cdot \underline{e}_1 = \varphi(\Phi)\psi_1(\Phi) , \qquad \underline{\nabla}\Phi \cdot \underline{\nabla}\Phi = \varphi(\Phi) . \tag{18}$$

The second relationship (12) is now rewritten as follows

$$\underline{\nabla} \cdot \underline{e}_2\underline{e}_2 = (\underline{\nabla} \cdot \underline{e}_2)\underline{e}_2 + \underline{e}_2 \cdot \underline{\underline{\nabla e}}_2 = \varphi(\Phi)\psi_2(\Phi)\underline{e}_1 .$$

But $\underline{e}_2 \cdot \underline{e}_1 = 0$ and $\underline{e}_2 \cdot \underline{\underline{\nabla e}}_2 \cdot \underline{e}_2 = \frac{1}{2}\underline{e}_2 \cdot (\underline{e}_2 \cdot \underline{e}_2) = 0$ and therefore

$$\underline{\nabla} \cdot \underline{e}_2 = 0 \qquad (c_2 \neq c_3) ; \tag{19}$$

this equality fails if $c_2 = c_3$. We will also need the representation for the contraction $\underline{\underline{\nabla e}}_1 \cdot\cdot \underline{\underline{\nabla e}}_1$. We calculate divergence of the vector $\underline{e}_1 \cdot \underline{\underline{\nabla e}}_1$ which is zero in view of (15). By (III.3.9) we have

$$\underline{\nabla} \cdot (\underline{e}_1 \cdot \underline{\underline{\nabla e}}_1) = \underline{\nabla} \cdot (\underline{\underline{\nabla e}}_1^T \cdot \underline{e}_1) = (\underline{\nabla} \cdot \underline{\underline{\nabla e}}_1^T) \cdot \underline{e}_1 + \underline{\underline{\nabla e}}_1^T \cdot\cdot \underline{\underline{\nabla e}}_1^T = 0 ,$$

$$\underline{\underline{\nabla e}}_1^T \cdot\cdot \underline{\underline{\nabla e}}_1^T = \underline{\underline{\nabla e}}_1 \cdot\cdot \underline{\underline{\nabla e}}_1 = -(\underline{\nabla} \cdot \underline{\underline{\nabla e}}_1^T) \cdot \underline{e}_1 .$$

The right-hand side of this equation can now be expressed in terms of Φ using the above relationships, i.e. from (18) and (III.6.16) we have

$$\underline{\nabla} \cdot \underline{\underline{\nabla e}}_1^T = \underline{\underline{\nabla\nabla}} \cdot \underline{e}_1 = \underline{\nabla}\psi_1(\Phi)\varphi(\Phi) = [\psi_1(\Phi)\varphi(\Phi)]'\underline{\nabla}\Phi , \qquad (\underline{\nabla} \cdot \underline{\underline{\nabla e}}_1^T) \cdot \underline{e}_1 = (\psi_1\varphi)'\varphi(\Phi) .$$

We have thus proved the relationship

$$\underline{\underline{\nabla e}}_1 \cdot\cdot \underline{\underline{\nabla e}}_1 = -\varphi(\Phi)[\psi_1(\Phi)\varphi(\Phi)]' \tag{20}$$

it is essential here that this contraction is also a function of Φ.

3. The surfaces Φ = const. It has been shown that the vector $\underline{n} = \underline{e}_1$ of the normal to the surfaces Φ = const is characterized by the properties (18) and (20) in that its divergence $\underline{\nabla} \cdot \underline{n}$ and the double contraction $\underline{\underline{\nabla n}} \cdot\cdot \underline{\underline{\nabla n}}$ of its gradient are constant on these surfaces.

The necessary relationships of the theory of surfaces are given in (III, Par. 11). From (III.11.11) and (III.11.12) we have

$$\underline{\underline{\nabla n}} = \underline{\rho}^{\alpha} n_{\alpha} = -\underline{\rho}^{\alpha}\underline{\rho}_{\beta} b^{\beta}_{\alpha} , \qquad \underline{\nabla} \cdot \underline{n} = -b^{\alpha}_{\alpha} = -2H ,$$

$$\underline{\underline{\nabla n}} \cdot\cdot \underline{\underline{\nabla n}} = \underline{\rho}^{\alpha}\underline{\rho}_{\beta} b^{\beta}_{\alpha} \cdot\cdot \underline{\rho}^{\gamma}\underline{\rho}_{\delta} b^{\delta}_{\gamma} = b^{\beta}_{\alpha} b^{\alpha}_{\beta} = K , \tag{21}$$

where H denotes the mean curvature and K the Gaussian curvature of the surface. We thus see that on the surface Φ = const both the mean and Gaussian curvatures are constant. Such surfaces can only be spheres, circular cylinders or planes. Moving along the normal $\underline{n}$, that is setting $d\underline{n} = d\underline{R}$, we have

$$d\underline{n} = d\underline{R} \cdot \underline{\underline{\nabla n}} = d\underline{R} , \qquad \underline{\underline{\nabla n}} = \underline{E} ,$$

in other words, for this displacement $\underline{n} = \underline{e}_1$ - the normal is a constant vector and the surfaces Φ = const are either concentric spheres, coaxial cylinders or parallel planes. Denoting by $\underline{R}$ the position vector on these surfaces and making use of the notation of Par. 9, we can represent these families of surfaces by the following relationships:

$$\underline{R} = R\underline{e}_R , \qquad \underline{R} = R\underline{e}_R + \underline{k}Z , \qquad \underline{n} \cdot \underline{R} = \text{const} . \tag{22}$$

4. This analysis reveals the structure of the Finger and Almansi strain measures $\underline{F}$ and $\underline{F}^{-1} = \underline{g}$. The eigenvalues $c_k(\underline{F})$ of the tensor $\underline{F}$ (c_k^{-1} of the Almansi tensor) take constant values on the above surfaces (21). The eigenvector $\underline{e}_1$ is a unit vector normal to these surfaces and the direction $\underline{e}_2$ lying in the tangential plane at $c_2 = c_3$ satisfies condition (19). Of course, $c_1 c_2 c_3 = 1$ and tensors $\underline{F}$ and $\underline{g}$ are positive.

For the tensors $\underline{F}$ (or $\underline{g}$) which are defined by these conditions, the conditions of integrability must hold, i.e. six components of the Ricci tensor should be zero.

The following cases may be considered

$$\text{a. } c_1 \neq c_2 \neq c_3 , \quad \underline{F} = (c_1 - c_3)\underline{e}_1\underline{e}_1 + (c_2 - c_3)\underline{e}_2\underline{e}_2 + c_3\underline{E} , \quad c_1 = (c_2 c_3)^{-1} , \tag{23}$$

$$\text{b. } c_2 = c_3 \neq c_1 , \quad \underline{F} = (c_1 - c_3)\underline{e}_1\underline{e}_1 + c_3\underline{E} , \quad c_1 = c_2^{-2} ,$$

$$\text{c. } c_1 = c_3 \neq c_2 , \quad \underline{F} = (c_2 - c_3)\underline{e}_2\underline{e}_2 + c_3\underline{E} , \quad c_2 = c_3^{-2} . \tag{24}$$

The latter case is not considered in what follows because Ericksen has shown that it corresponds either to a constant tensor $\underline{F}$ (a linear transformation) or to tensors which do not satisfy the integrability conditions. The case when both invariants $I_1(\underline{F})$ and $I_2(\underline{F})$ are constant (Shield and Klingbeil) has been excluded from the present analysis.

5. For the three families of surfaces given above, the vector bases $\underline{R}_s$ and $\underline{R}^s$ of their actual configurations and the non-zero components of their unit tensors $\underline{E}$ are specified by the following formulas

α) Planes: $\underline{R}_s = \underline{i}_s = \underline{R}^s$ $(s = 1,2,3)$;

$$G_{11} = G_{22} = G_{33} = G^{11} = G^{22} = G^{33} = 1 \,, \tag{25α}$$

where $\underline{i}_s$ denote the unit vectors of a Cartesian coordinate system and $\underline{i}_1 = \underline{e}_1$.

β) Cylinders:

$$\underline{R}_1 = \underline{R}^1 = \underline{e}_R = \underline{e}_1 \,, \qquad \underline{R}_2 = R\underline{e}_\Phi \,, \qquad \underline{R}_3 = \underline{R}^3 = \underline{k} \,, \qquad \underline{R}^2 = \frac{1}{R}\,\underline{e}_\Phi \,;$$

$$G_{11} = G^{11} = 1 \,, \qquad G_{33} = G^{33} = 1 \,, \qquad G_{22} = R^2 \,, \qquad G^{22} = R^{-2} \,, \tag{25β}$$

γ) Spheres:

$$\underline{R}_1 = \underline{R}^1 = \underline{e}_R = \underline{e}_1 \,, \qquad \underline{R}_2 = R\underline{e}_\theta \,, \qquad \underline{R}_3 = R\sin\theta\,\underline{e}_\Lambda \,, \qquad \underline{R}^2 = R^{-1}\underline{e}_\theta \,,$$

$$\underline{R}^3 = (R\sin\theta)^{-1}\underline{e}_\Lambda \,; \tag{25γ}$$

$$G_{11} = G^{11} = 1 \,, \; G_{22} = R^2 \,, \; G_{33} = R^2\sin^2\theta \,, \; G^{22} = R^{-2} \,, \; G^{33} = (R\sin\theta)^{-2} \,.$$

6. The vectors $\underline{e}_2$ and $\underline{e}_3$. Denoting the co- and contravariant components of the unit vector $\underline{e}_2$ by e_s, e^s (introduced to avoid notation of the type e_{2s}, e_2^s) we have

$$\underline{e}_2 = \underline{R}_s e^s = \underline{R}^s e_s \,, \qquad G_{sk}e^s e^k = G^{sk}e_s e_k = 1$$

and because $\underline{e}_2 \cdot \underline{e}_1 = 0$ and $e^1 = e_1 = 0$, we can assume that for the above three cases

$$\text{α)}\; e_2 = e^2 = \cos\psi \,, \qquad e_3 = e^3 = \sin\psi \,;$$

$$\text{β)}\; e_2 = R\cos\psi \,, \qquad e_3 = e^3 = \sin\psi \,, \qquad e^2 = R^{-1}\cos\psi \,;$$

$$\text{γ)}\; e_2 = R\cos\psi \,, \quad e_3 = R\sin\theta\sin\psi \,, \quad e^2 = R^{-1}\cos\psi \,, \quad e^3 = (R\sin\theta)^{-1}\sin\psi$$

and decompositions for the vectors $\underline{e}_2$, $\underline{e}_3 = \underline{e}_1 \times \underline{e}_2$ are reduced to the form

$$\text{α)}\; \underline{e}_2 = \underline{i}_2\cos\psi + \underline{i}_3\sin\psi \,, \qquad \underline{e}_3 = -\underline{i}_2\sin\psi + \underline{i}_3\cos\psi \,, \tag{26α}$$

$$\beta)\ \underline{e}_2 = \underline{R}^2 R\cos\psi + \underline{R}^3\sin\psi = \underline{R}_2 R^{-1}\cos\psi + \underline{R}_3\sin\psi\ ,$$

$$\underline{e}_3 = -\underline{R}^2 R\sin\psi + \underline{R}^3\cos\psi = -\underline{R}_2 R^{-1}\sin\psi + \underline{R}_3\cos\psi\ , \tag{26β}$$

$$\gamma)\ \underline{e}_2 = \underline{R}^2 R\cos\psi + \underline{R}^3 R\sin\theta\sin\psi = \underline{R}_2 R^{-1}\cos\psi + \underline{R}_3\frac{\sin\psi}{R\sin\theta}\ ,$$

$$\underline{e}_3 = -\underline{R}^2 R\sin\psi + \underline{R}^3 R\sin\theta\cos\psi = -\underline{R}_2 R^{-1}\sin\psi + \underline{R}_3\frac{\cos\psi}{R\sin\theta}\ . \tag{26γ}$$

Below, up to no. 10, we will be dealing with the case of distinct roots $c_1 \neq c_2 \neq c_3$. The vector $\underline{e}_2$ is subject to condition (19) and is, moreover, a unit vector. Therefore

$$\underline{\nabla}\cdot\underline{e}_2 = 0\ ,$$

$$\underline{\nabla e}_2\cdot\underline{e}_2 - \underline{e}_2\underline{\nabla}\cdot\underline{e}_2 = \underline{e}_2\times(\underline{\nabla}\times\underline{e}_2) = e^s\underline{R}_s\times(\underline{R}^m\times\underline{R}^k)\nabla_m e_k = \underline{R}^m e_k(\nabla_m e_k - \nabla_k e_m) = \underline{0}\ .$$

The function ψ which enters representations (26) should be selected in accordance with these differential relationships. We thus arrive at the following systems of equations

$$\alpha)\ q^1 = x^1\ ,\quad q^2 = x^2\ ,\quad q^3 = x^3\ :\ \cos\psi\frac{\partial\psi}{\partial x^2} + \sin\psi\frac{\partial\psi}{\partial x^3} = 0\ ,$$

$$-\sin\psi\frac{\partial\psi}{\partial x^2} + \cos\psi\frac{\partial\psi}{\partial x^3} = 0\ ,$$

from here it follows that

$$\frac{\partial\psi}{\partial x^2} = \frac{\partial\psi}{\partial x^3} = 0\ ,\qquad \psi = \psi(x^1)\ . \tag{27α}$$

$$\beta)\ q^1 = R\ ,\quad q^2 = \Phi\ ,\quad q^3 = Z\ :\ \cos\psi\frac{\partial\psi}{\partial\Phi} + \sin\psi\frac{\partial\psi}{\partial Z} = 0\ ,$$

$$-\sin\psi\frac{\partial\psi}{\partial\Phi} + \cos\psi\frac{\partial\psi}{\partial Z} = 0\ ,$$

$$\frac{\partial\psi}{\partial\Phi} = 0\ ,\quad \frac{\partial\psi}{\partial Z} = 0\ ,\quad \psi = \psi(R)\ , \tag{27β}$$

$$\gamma)\ q^1 = R\ ,\quad q^2 = \theta\ ,\quad q^3 = \Lambda\ :\ \cos\psi\sin\theta\frac{\partial\psi}{\partial\theta} + \sin\psi\left(\cos\theta + \frac{\partial\psi}{\partial\Lambda}\right) = 0\ ,$$

$$-\sin\psi\sin\theta\frac{\partial\psi}{\partial\theta} + \cos\psi\left(\cos\theta + \frac{\partial\psi}{\partial\Lambda}\right) = 0\ ,$$

$$\frac{\partial\psi}{\partial\theta} = 0\ ,\quad \frac{\partial\psi}{\partial\Lambda} = -\cos\theta\ ;\qquad \frac{\partial^2\psi}{\partial\Lambda\partial\theta} = 0\ ,\quad \frac{\partial^2\psi}{\partial\theta\partial\Lambda} = \sin\theta$$

the latter system having no solution. In other words, the conditions $c_1 \neq c_2 \neq c_3$ exclude the possibility of the existence of universal solutions for a family of concentric spheres. We will therefore consider below the cases α) and β). For these, ψ is a function of the coordinate q^1 only (x^1 and R) which is constant on the surface Φ = const (i.e. a plane or a cylinder).

7. Almansi's measure. Its decomposition is written as follows

$$\underline{g} = c_1^{-1}\underline{e}_1\underline{e}_1 + c^{-2}\underline{e}_2\underline{e}_2 + c^{-3}\underline{e}_3\underline{e}_3 = g_{sk}\underline{R}^s\underline{R}^k , \qquad g_{sk} = \underline{r}_s \cdot \underline{r}_k .$$

From here we deduce

$$g_{sk} = c_1^{-1}\underline{e}_1 \cdot \underline{R}_s\underline{R}_k \cdot \underline{e}_1 + c_2^{-1}\underline{e}_2 \cdot \underline{R}_s\underline{R}_k \cdot \underline{e}_2 + c_3^{-1}\underline{e}_3 \cdot \underline{R}_s\underline{R}_k \cdot \underline{e}_3$$

or in a more detailed form

$$g_{11} = c_1^{-1}, \; g_{12} = 0, \; g_{13} = 0 ; \qquad g_{22} = c_2^{-1}(\underline{e}_2 \cdot \underline{R}_2)^2 + c_3^{-1}(\underline{e}_3 \cdot \underline{R}_2)^2 ;$$

$$g_{33} = c_2^{-1}(\underline{e}_2 \cdot \underline{R}_3)^2 + c_3^{-1}(\underline{e}_3 \cdot \underline{R}_3)^2 ; \qquad g_{23} = c_2^{-1}\underline{e}_2 \cdot \underline{R}_2\underline{e}_2 \cdot \underline{R}_3 + c_3^{-1}\underline{e}_3 \cdot \underline{R}_2\underline{e}_3 \cdot \underline{R}_3 ,$$

so that

$$g = |g_{sk}| = g_{11}(g_{22}g_{33} - g_{23}^2) ; \qquad \alpha): g = 1 , \; \beta): g = R^2 . \tag{28}$$

From (26) we arrive at the following representations of the matrices of the covariant components of the Almansi measure (or, which is the same thing, of the covariant components of $\underline{E}$ in the reference configuration)

$$\alpha) \begin{Vmatrix} g_{11} = c_1^{-1} & 0 & 0 \\ 0 & g_{22} = c_2^{-1}\cos^2\psi + c_3^{-1}\sin^2\psi & g_{23} = (c_2^{-1} - c_3^{-1})\sin\psi\cos\psi \\ 0 & g_{23} = (c_2^{-1} - c_3^{-1})\sin\psi\cos\psi & g_{33} = c_2^{-1}\sin^2\psi + c_3^{-1}\cos^2\psi \end{Vmatrix} , \tag{29α}$$

$$\beta) \begin{Vmatrix} g_{11} = c_1^{-1} & 0 & 0 \\ 0 & g_{22} = R^2(c_2^{-1}\cos^2\psi + c_3^{-1}\sin^2\psi) & g_{23} = R(c_2^{-1} - c_3^{-1})\sin\psi\cos\psi \\ 0 & g_{23} = R(c_2^{-1} - c_3^{-1})\sin\psi\cos\psi & g_{33} = c_2^{-1}\sin^2\psi + c_3^{-1}\cos^2\psi \end{Vmatrix} . \tag{29β}$$

The elements of these matrices are functions of the sole argument q^1 (α): $q^1 = x_1$; β): $q^1 = R$. These functions are specified by the conditions of integrability of the systems of equations for coordinates in the reference configuration visualized as functions of its position coordinates in the actual configuration.

8. The Ricci equations. The expressions for the components (III.10.15) and (III.10.21) of the Ricci tensor are considerably simplified. Two of these components, namely R_{1223} and R_{2331} are identically zero. The requirement that R_{2323} be zero yields the condition

$$g'_{22}g'_{33} - g'^2_{23} = 0$$

(the prime denoting differentiation by q^1). Making use of this, we may reduce the remaining three conditions to the form

$$R_{1212} = 0 : g''_{22} - \frac{1}{2} g'_{22}S = 0 ;$$

$$R_{3131} = 0 : g''_{33} - \frac{1}{2} g'_{33}S = 0 ;$$

$$R_{1231} = 0 : g''_{23} - \frac{1}{2} g'_{23}S = 0 ,$$

$$S = g^{11}g'_{11} + g^{22}g'_{22} + g^{33}g'_{33} + 2g^{23}g'_{23} .$$

Substituting the elements g^{sk} of the inverse matrix for their values in the expression for S, we obtain

$$S = (\ln g_{11})' + [\ln(g_{22}g_{33} - g^2_{23})]' = (\ln g)' , \qquad g = g_{11}(g_{22}g_{33} - g^2_{23}) ,$$

where we have α) $g = 1$, β) $g = R^2$. The integrability conditions may now be written

$$\left(\ln \frac{g'_{22}}{\sqrt{g}}\right)' = 0 , \quad \left(\ln \frac{g'_{33}}{\sqrt{g}}\right)' = 0 , \quad \left(\ln \frac{g'_{23}}{\sqrt{g}}\right)' = 0 , \quad g'_{22}g'_{33} - g'^2_{23} = 0 . \tag{30}$$

From these we get

$$\alpha)\ g'_{22} = 2AE^2 , \quad g'_{33} = 2AC^2 , \quad g'_{23} = -2ACE ; \qquad (g'_{22}g'_{33} - g'^2_{23}) = 0 ,$$

$$g_{22} = 2AE^2x^1 + A^2D^2 , \quad g_{33} = 2AC^2x^1 + A^2B^2 , \quad g_{23} = -(2ACEx^1 + A^2BD) \tag{31α}$$

the notation for the constants of integration being the same as that of (9.12). By (28) we have

$$g_{11}(g_{22}g_{33}-g_{23}^2) = 2A^3(EB-CD)^2x^1g_{11} = 1$$

and applying the condition of (9.2) to the constants we get

$$g_{11} = \frac{1}{2Ax^1} . \tag{32α}$$

It now remains to show that these constants must be chosen so as to ensure that $\underline{g}$ is positive. The following inequalities must hold

$$g_{11} > 0 , \quad g_{22} > 0 , \quad g_{22}g_{33} - g_{23}^2 > 0 . \tag{33α}$$

These conditions are fulfilled when $A > 0$ and when x^1 is nonnegative. The signs of constants B, C, D and E which themselves enter as squares and $(EB-CD)^2$ are indeterminate. These signs are prescribed so as to make this difference positive.

β) By (30) and because $\sqrt{g} = R$ we have

$$g_{22} = a_1R^2 + a_2 , \quad g_{23} = a_3R^2 + a_4 , \quad g_{33} = a_5R^2 + a_6 ; \qquad a_1a_5 = a_3^2 . \tag{31β}$$

The criteria for $\underline{g}$ to be positively definite reduce to the form

$$g_{22} = a_1R^2 + a_2 > 0 , \quad g_{22}g_{33} - g_{23}^2 = (a_1a_6 + a_2a_5 - 2a_3a_4)R^2 + a_2a_6 - a_4^2 > 0 . \tag{33β}$$

β_1) For the first version of the choice of coefficients a_k consistent with (9.6) and with the requirement (33α)

$$a_1 = a_3 = a_5 = 0 , \quad a_2 = g_{22} = A^2(E^2+D^2) > 0 ,$$
$$a_4 = g_{23} = -A^2(BD+EC) , \quad a_6 = g_{33} = A^2(C^2+B^2) > 0 . \tag{33β₁}$$

We recall here that g_{sk} are defined for the basis $\underline{R}_s$. For this, $R\underline{e}_\Phi = \underline{R}_s$ so that Equations ($31\beta_1$) are in complete agreement with (9.6).

The component g_{11} is determined by the expression for the determinant

$$g = R^2 = g_{11}(g_{22}g_{33} - g_{23}^2) = g_{11}A^4(BE-CD)^2$$

and by (9.2)

$$g_{11} = \frac{R^2}{A^2} \, . \tag{32β_1}$$

β_2) The second version of the choice of the coefficients a_k satisfying Inequality (33β) and the condition $a_1a_5 = a_3^2$ is more complicated. We have

$$g_{22} = A[AE^2 + F^2(R^2 - B)] \, , \qquad g_{33} = A[AC^2 + D^2(R^2 - B)] \, ,$$

$$g_{23} = -A[AEC + FD(R^2 - B)] \, .$$

Now, in view of the normalizing condition (9.16), we get

$$g = R^2 = g_{11}A^3(R^2 - B)(FC - ED)^2 = g_{11}A(R^2 - B) \, ,$$

so that

$$g_{11} = \frac{R^2}{A(R^2 - B)} > 0 \qquad (B < R^2) \, .$$

This analysis does not reveal any other relationships satisfying the conditions (33β) for positive definiteness.

9. The position vector in the reference configuration. The vector basis $\underline{r}_s$ of the reference configuration may be found from the already known values of the covariant components $g_{sk} = \underline{r}_s \cdot \underline{r}_k$ of the metric tensor $\underline{E}$ of this configuration by virtue of the integrable system (1.18.12). The position vector $\underline{r}$ is subsequently determined by squaring, using (1.18.13). The resulting formulas must, of course, be solutions of equations (9.1), (9.11) and (9.14) for coordinates in the reference configuration corresponding to the cases α, β_1 and β_2.

The coefficient of the system of six equations

$$\frac{\partial \underline{r}_s}{\partial q^t} = \overset{0}{\left\{ {q \atop st} \right\}} \underline{r}_q = g^{qr} [\overset{0}{st},r] \underline{r}_q =$$

$$= g^{11}\underline{r}_1 [\overset{0}{st},1] + \underline{r}_2(g^{22}[\overset{0}{st},2] + g^{23}[\overset{0}{st},3]) + \underline{r}_3(g^{23}[\overset{0}{st},2] + g^{33}[\overset{0}{st},3]) \, ,$$

depends on the coordinate q^1 (α)$q^1 = x^1$, β) $q^1 = R$). This system may be written in the following expanded form

$$\frac{\partial \underline{r}_1}{\partial q^1} = \frac{1}{2}\frac{g'_{11}}{g_{11}}\underline{r}_1 \, ,$$

$$\frac{\partial \underline{r}_1}{\partial q^2} = \frac{\partial \underline{r}_2}{\partial q^1} = \frac{1}{2}\frac{g_{11}}{g}[(g'_{22}g_{33} - g'_{23}g_{23})\underline{r}_2 + (-g'_{22}g_{23} + g'_{23}g_{23})\underline{r}_3] \, ,$$

$$\frac{\partial \underline{r}_1}{\partial q^3} = \frac{\partial \underline{r}_3}{\partial q^1} = \frac{1}{2}\,\frac{g_{11}}{g}\,[(-g'_{33}g_{23} + g'_{23}g_{23})\underline{r}_2 + (g'_{33}g_{22} - g'_{23}g_{23})\underline{r}_3]\ ,$$

$$\frac{\partial \underline{r}_2}{\partial q^2} = -\frac{1}{2}\,\frac{g'_{22}}{g_{11}}\,\underline{r}_1\ ,\qquad \frac{\partial \underline{r}_2}{\partial q^3} = \frac{\partial \underline{r}_3}{\partial q^2} = -\frac{1}{2}\,\frac{g'_{23}}{g_{11}}\,\underline{r}_1\ ,\qquad \frac{\partial \underline{r}_3}{\partial q^3} = -\frac{1}{2}\,\frac{g'_{33}}{g_{11}}\,\underline{r}_1\ . \tag{34}$$

The integration of this system for the above cases is technically difficult and would require considerable space and so it cannot be carried out here. This procedure would of course lead to the formulas anticipated.

10. <u>Concentric spheres (solutions of the fourth class)</u>. Here $c_2 = c_3$ and $c_1 = c_3^{-2}$, and the Almansi measure is represented by the expression

$$\underline{\underline{g}} = (c_3^{-2} - c_3)\underline{e}_1\underline{e}_1 + c_3\underline{\underline{E}} = g_{sk}\underline{R}^s\underline{R}^k\ . \tag{35}$$

For a spherical coordinate system we have

$$\underline{R} = R\underline{e}_R\ ,\qquad \underline{R}_1 = \underline{e}_R = \underline{e}_1\ ,\qquad \underline{R}_2 = R\underline{e}_\theta\ ,\qquad \underline{R}_3 = R\underline{e}_\Lambda \sin\theta$$

and by (I.4.10) the covariant components of the Almansi tensor are seen to be

$$g_{sk} = \underline{R}_s \cdot \underline{\underline{g}} \cdot \underline{R}_k = (c_3^{-2} - c_3)\underline{R}_s \cdot \underline{e}_R\underline{e}_R \cdot \underline{R}_k + c_3\underline{R}_s \cdot \underline{R}_k\ ;$$

$$g_{11} = c_3^{-2}\ ,\qquad g_{22} = c_3R^2\ ,\qquad g_{33} = c_3R^2\sin^2\theta\ ,\qquad g_{12} = g_{23} = g_{31} = 0\ .$$

The Almansi measure is a diagonal tensor

$$\underline{\underline{g}} = c_3^{-2}\underline{R}^1\underline{R}^1 + c_3R^2\underline{R}^2\underline{R}^2 + c_3R^2\sin^2\theta\underline{R}^2\underline{R}^2\ . \tag{36}$$

The function $c_3(R)$ should satisfy the six conditions expressing the vanishing of the components of the Ricci tensor. The components R_{1223}, R_{1231} and R_{2331} are identically zero. The calculation of components R_{1212} and R_{1313} leads to the same differential equation ($z = c_3R^2$)

$$z'' - \frac{1}{2}R^{-4}z^2\left(\frac{R^4}{z^2}\right)' - \frac{1}{2}\,\frac{z'^2}{z} = 0\ .$$

Its first integral (which is easily determined) is given by

$$z'\sqrt{z} = CR^2\ ,\qquad z'^2c_3 = C^2R^2\ . \tag{37}$$

The condition $R_{2323} = 0$ now takes the form

$$4R^4 = zz'^2 \quad \text{or} \quad 4R^2 = c_3 z'^2$$

and it is possible to satisfy the conditions of integrability assuming $C = 4$ in (37). Introducing the new independent variable $x = R^3$ we transform this equation to

$$\left(\frac{dz}{dx}\right)^2 z = \frac{4}{9} \, .$$

Integration shows that

$$z^3 = (x - A)^2 \quad \text{or} \quad c_3(R) = R^{-2}(R^3 - A)^{2/3} \, . \tag{38}$$

Assuming that the deformation is radially symmetric we have

$$\underline{r} = r(R)\underline{e}_r \, , \quad \underline{r}_1 = r'(R)\underline{e}_r \, , \quad \underline{r}_2 = r\underline{e}_\vartheta \, , \quad \underline{r}_3 = r\underline{e}_\lambda \sin\vartheta \, ,$$

so that by (38)

$$g_{11} = \underline{r}_1 \cdot \underline{r}_1 = r'^2(R) = R^4(R^3 - A)^{-4/3} \, , \qquad g_{22} = \underline{r}_2 \cdot \underline{r}_2 = r^2 = (R^3 - A)^{2/3} \, ,$$

$$g_{33} = \underline{r}_3 \cdot \underline{r}_3 = r^2 \sin^2\vartheta = (R^3 - A)^{2/3} \sin^2\vartheta \, , \qquad g_{12} = g_{23} = g_{31} = 0 \, .$$

The first relationship implies that

$$r = \pm \int^R \frac{R^2 dR}{(R^3 - A)^{2/3}} = \pm (R^3 - A)^{1/3} \, ,$$

which also satisfies the second and the third equalities (for $\theta = \pm\vartheta$). The reference configuration is thus transformed into the actual configuration by Equations (9.22). The assumption that the deformation is radially symmetric is justified since, when analysed, any solution of the system (i.18.12) differs from that obtained by a rigid displacement.

CHAPTER 8

SMALL DEFORMATION OF AN INITIALLY LOADED BODY

1. EQUILIBRIUM UNDER CHANGING STATE OF STRESS

The problem of variation of the state of stress has been considered in Chapter 4, Par. 5. We have, in the usual way, been dealing with the reference v- and the actual V-configurations. The state of stress is specified either by the Piola tensor $\underline{P}$ or by the Cauchy tensor $\underline{T}$. The equilibrium equations for internal and boundary points of a body are represented by the relationships

$$\text{in } v:\quad \overset{0}{\underline{\nabla}}\cdot\underline{P}+\rho_0\underline{k}=\underline{0}\,,\qquad \text{on } o:\quad \underline{n}\cdot\underline{P}=\underline{f}\,\frac{dO}{do}\,, \tag{1}$$

$$\text{in } V:\quad \underline{\nabla}\cdot\underline{T}+\rho\underline{k}=\underline{0}\,,\qquad \text{on } O:\quad \underline{N}\cdot\underline{T}=\underline{f}\,. \tag{2}$$

The particles of a medium which are in a state of equilibrium under the action of body forces $\underline{k}$ and surface tractions $\underline{f}$ are subject in the actual V-configuration to virtual displacements $\eta\underline{w}(q^1,q^2,q^3)$ where η denotes a small parameter. Quantities in the new $\overset{\times}{V}$-configuration are denoted by a cross $(\ )^{\times}$ superscript and in particular, $\overset{\times}{\underline{R}}=\underline{R}+\eta\underline{w}$ denotes the position vector. By definition, the virtual displacements are consistent with the constraints imposed and must be subject to restrictions formulated in advance. For example, for an incompressible medium we have by (1.10.18)

$$\underline{\nabla}\cdot\underline{w}=0\,. \tag{3}$$

On that part of the body's surface (O_2) where the displacement ($\delta\underline{R}=\underline{0}$) is prescribed, it is necessary that

$$\underline{w}=\underline{0}\,. \tag{4}$$

Whenever surface tractions $\underline{f}$ are prescribed on the entire surface O of the body the vector $\underline{w}$ should be specified so that the resultant moment of the forces in the $\overset{\times}{V}$-configuration is zero provided that this configuration is itself in equilibrium. The latter condition will be formulated below (see (21)). Finally, one has sometimes to be content with a consideration of virtual displacements of some special form.

The quantity Φ in the V-configuration (which is denoted as $\overset{\times}{\Phi}$ in the $\overset{\times}{V}$-configuration) is defined in Chapter 1, Par. 10 as

$$\Phi^{\times} = \Phi + \eta\dot{\Phi} . \tag{5}$$

The term $\eta\dot{\Phi}$ arises due to a variation in the geometry of the V-configuration which is accompanied by the superposition of the field of virtual displacements upon that of the displacement vector $\underline{u} = \underline{R} - \underline{r}$ from the v- into the V-configuration. We do not exclude the possibility of the insertion into $\Phi^{\times}$ of quantities of order η not associated with the V-configuration.

The body force $\underline{k}\,dm$ per elementary volume of the V-configuration is by (5) assumed to be

$$\underline{k}^{\times} dm^{\times} = (\underline{k} + \eta\dot{\underline{k}})\,dm = \underline{k}\,dm + \eta\dot{\underline{k}}\,dm , \qquad \underline{k}^{\times} + \eta\dot{\underline{k}} , \tag{6}$$

in the $V^{\times}$-configuration, this because $dm^{\times} = dm$ by the law of conservation of mass. For example, in the gravitational field $\underline{k} = -\underline{i}_3 g$, $\dot{\underline{k}} = \underline{0}$, and in the centrifugal force field (2.1.6) $\dot{\underline{k}} = -\underline{\omega} \times (\underline{\omega} \times \underline{w})$ provided that the angular velocity remains constant. In what follows, we will define $\underline{k}^{\times}$ by the expression (6) though $\dot{\underline{k}}$ does not necessarily coincide with the convected derivative of $\underline{k}$ because body forces might arise which were absent in the V-configuration.

Similarly, we consider the surface tractions

$$(\underline{f}\,dO)^{\times} = \underline{f}\,dO + \eta(\underline{f}\,dO)^{\boldsymbol{\cdot}} = \underline{N}\cdot\underline{T}\,dO + \eta(\underline{N}\cdot\underline{T}\,dO)^{\boldsymbol{\cdot}} =$$

$$= \underline{f}\,dO + \eta\underline{N}\,dO\cdot(\underline{T}^{\boldsymbol{\cdot}} + \underline{T}\underline{\nabla}\cdot\underline{w} - \underline{\nabla}\underline{w}^{T}\cdot\underline{T}) = \underline{f}\,dO + \eta\underline{N}\cdot\underline{\theta}\,dO$$

here we have used Equation (1.10.22) and the definition (4.5.8) of a tensor $\underline{\theta}$. We arrive at the representation

$$(\underline{f}\,dO)^{\times} = \underline{f}\,dO + \eta\underline{N}\cdot\underline{\theta}\,dO + \lambda\underline{f}_{\times}dO \tag{7}$$

the term $\eta\underline{f}_{\times}$ takes account of the surface tractions which were absent in the V-configuration.

Here we will for the most part be considering "dead" loading by the body and surface forces (see (4.6.8))

$$\dot{\underline{k}} = \underline{0} : \underline{k}^{\times} = \underline{k} ; \qquad (\underline{f}\,dO)^{\times} = \underline{f}\,dO = \overset{0}{\underline{f}}\,do , \quad \underline{f}_{\times} = \underline{0} , \quad \underline{N}\cdot\underline{\theta} = \underline{0} . \tag{8}$$

We will also deal with the case of a surface loading represented by a uniformly distributed pressure $\underline{f} = -p\underline{N}$. By (1.10.19) we have

$$(\underline{f}\,dO)^{\times} = \underline{f}\,dO - \eta p(\underline{N}\,dO)^{\times} = \underline{f}\,dO - \eta p\underline{N}\,dO\cdot(\underline{E}\underline{\nabla}\cdot\underline{w} - \underline{\nabla}\underline{w}^{T})$$

and for $\underline{f}_x = \underline{0}$

$$\underline{N} \cdot \underline{\theta} = -p\underline{N} \cdot (\underline{E\nabla} \cdot \underline{w} - \underline{\nabla w}^T) \ . \tag{9}$$

The tensor $\underline{\theta}$ introduced by Equations (4.5.9) and (4.5.8)

$$\underline{\theta} = \underline{T}^{\cdot} + \underline{T\nabla} \cdot \underline{w} - \underline{\nabla w}^T \cdot \underline{T} \ , \tag{10}$$

is connected with the convected derivative $\dot{\underline{P}}$ of the Piola tensor by the formula

$$\dot{\underline{P}} = \sqrt{\frac{G}{g}}\, \underline{\nabla r}^T \cdot \underline{\theta} \ , \qquad \underline{\theta} = \sqrt{\frac{g}{G}}\, \underline{\overset{0}{\nabla}R}^T \cdot \dot{\underline{P}} \tag{11}$$

and in view of (5)

$$\underline{P}^{\times} - \underline{P} = \underline{P}(\underline{\overset{0}{\nabla}R}^{\times}) - \underline{P}(\underline{\overset{0}{\nabla}R}) = \underline{P}(\underline{\overset{0}{\nabla}R} + \eta\underline{\overset{0}{\nabla}w}) - \underline{P}(\underline{\overset{0}{\nabla}R}) = \eta\dot{\underline{P}}(\underline{\overset{0}{\nabla}w}) \ .$$

Recalling the definition of the derivative with respect to a tensor argument

$$\dot{\underline{P}}(\underline{\overset{0}{\nabla}w}) = \underline{P}_{\underline{\overset{0}{\nabla}R}} \cdot\cdot \underline{\overset{0}{\nabla}w}^T \tag{12}$$

and the representation (4.1.4) of the Piola tensor in terms of the specific strain energy, we obtain

$$\underline{P}_{\underline{\overset{0}{\nabla}R}} = \ni_{\underline{\overset{0}{\nabla}R\overset{0}{\nabla}R}} = \frac{\partial^2 \ni}{\partial\overset{0}{\nabla}_p \chi_m \partial\overset{0}{\nabla}_q \chi_n}\, \underline{r}_p\underline{r}_m\underline{r}_q\underline{r}_n = K^{pmqn}\, \underline{r}_p\underline{r}_m\underline{r}_q\underline{r}_n \ . \tag{13}$$

In this formula, χ_m denote the covariant components of $\underline{R}$ in the v-basis

$$\underline{R} = \underline{r}^m \chi_m \ , \qquad \underline{\overset{0}{\nabla}R} = \underline{r}^p\underline{r}^m \overset{0}{\nabla}_p \chi_m \ , \tag{14}$$

and K^{pmqn} are the components of the tensor $\underline{P}_{\underline{\overset{0}{\nabla}R}}$ of elasticities which are symmetric with respect to the pairs of indices (pm) and (qn). This type of symmetry implies the property of duality

$$\dot{\underline{P}}(\underline{\overset{0}{\nabla}w}_1) \cdot\cdot \underline{\overset{0}{\nabla}w}_2^T = \dot{\underline{P}}(\underline{\overset{0}{\nabla}w}_2) \cdot\cdot \underline{\overset{0}{\nabla}w}_1^T \ . \tag{15}$$

We have, in fact

$$\dot{\underline{P}}(\overset{0}{\nabla}\underline{w}_1)\cdot\cdot\overset{0}{\nabla}\underline{w}_2^T = \underline{P}_{\overset{0}{\nabla}\underline{R}}\cdot\cdot\overset{0}{\nabla}\underline{w}_1^T\cdot\cdot\overset{0}{\nabla}\underline{w}_2^T = K^{pmqn}\overset{0}{\nabla}_p w_{1m}\overset{0}{\nabla}_q w_{2n} =$$

$$= K^{qnpm}\overset{0}{\nabla}_q w_{1n}\overset{0}{\nabla}_p w_{2m} = K^{pmqn}\overset{0}{\nabla}_p w_{2m}\overset{0}{\nabla}_q w_{1n} = \dot{\underline{P}}(\overset{0}{\nabla}\underline{w}_2)\cdot\cdot\overset{0}{\nabla}\underline{w}_1^T .$$

Expressing $\dot{\underline{P}}$ in terms of $\underline{\theta}$ from (11) and using (1.3.9) we have from (15) the expression

$$\dot{\underline{P}}(\overset{0}{\nabla}\underline{w}_1)\cdot\cdot\overset{0}{\nabla}\underline{w}_2^T = \sqrt{\frac{G}{g}}\,\nabla\underline{r}^T\cdot\underline{\theta}(\nabla\underline{w}_1)\cdot\cdot\overset{0}{\nabla}\underline{w}_2^T =$$

$$= \sqrt{\frac{G}{g}}\,\underline{\theta}(\nabla\underline{w}_1)\cdot\cdot\overset{0}{\nabla}\underline{w}^T\cdot\nabla\underline{r}^T = \sqrt{\frac{G}{g}}\,\underline{\theta}(\nabla\underline{w}_1)\cdot\cdot\nabla\underline{w}_2^T$$

which allows us to extend the duality property to the tensor $\underline{\theta}$

$$\underline{\theta}(\nabla\underline{w}_1)\cdot\cdot\nabla\underline{w}_2^T = \underline{\theta}(\nabla\underline{w}_2)\cdot\cdot\nabla\underline{w}_1^T . \tag{16}$$

By (4.6.1) and (4.6.2) the static equilibrium equations in the volume of the $\overset{\times}{V}$-configuration have the form

$$\overset{0}{\nabla}\cdot\dot{\underline{P}} + \rho_0\dot{\underline{k}} = \underline{0} , \qquad \nabla\cdot\underline{\theta} + \rho\dot{\underline{k}} = \underline{0} . \tag{17}$$

On the surface $\overset{\times}{O}$ we have by (11) and (1.8.9)

$$\underline{N}\cdot\underline{\theta}\,dO = \sqrt{\frac{G}{g}}\,\underline{n}\cdot\nabla\underline{r}^T\cdot\underline{\theta}\,dO = \underline{n}\cdot\dot{\underline{P}}\,do \tag{18}$$

and for "dead" surface tractions, in view of (7),

$$\underline{N}\cdot\underline{\theta} = -\underline{f}_\times , \qquad \underline{n}\cdot\dot{\underline{P}} = -\underline{f}_\times\frac{dO}{do} . \tag{19}$$

This representation does not exclude the forces which were absent in the V-configuration.

The force tensor (2.1.6) in the $\overset{\times}{V}$-configuration is by (6), (7) and (8) represented for the "dead" body and surface forces by the expression

$$\overset{\times}{\underline{B}} = \underline{B} + \eta\dot{\underline{B}} = \iiint_V \rho\underline{k}(\underline{R}+\eta\underline{w})dV + \iint_O \underline{f}(\underline{R}+\eta\underline{w})dO =$$

$$= \underline{B} + \eta\left[\iiint_V \rho\underline{k}\underline{w}\,dV + \iint_O \underline{f}\underline{w}\,dO\right] .$$

Here by (2.3.1) and (2.3.5)

$$\iint_O \underline{\underline{f}}\underline{w}\, dO = \iint_O \underline{N} \cdot \underline{\underline{T}}\underline{w}\, dO = \iiint_V (\underline{\nabla} \cdot \underline{\underline{T}})\underline{w}\, dV + \iiint_V \underline{R}^s \cdot \underline{\underline{T}} \frac{\partial \underline{w}}{\partial q^s}\, dV =$$

$$= - \iiint_V \rho \underline{\underline{k}}\underline{w}\, dV + \iiint_V \underline{\underline{T}} \cdot \underline{\nabla}\underline{w}\, dV ,$$

so that

$$\dot{\underline{B}} = \iiint_V \underline{\underline{T}} \cdot \underline{\nabla}\underline{w}\, dV , \qquad \dot{\underline{B}}^T = \iiint_V \underline{\nabla}\underline{w}^T \cdot \underline{\underline{T}}\, dV ,$$

$$\dot{\underline{B}} - \dot{\underline{B}}^T = \iiint_V (\underline{\underline{T}} \cdot \underline{\nabla}\underline{w} - \underline{\nabla}\underline{w}^T \cdot \underline{\underline{T}})dV \tag{20}$$

and for the equilibrated V-configuration

$$\iiint_V (\underline{\underline{T}} \cdot \underline{\nabla}\underline{w} - \underline{\nabla}\underline{w}^T \cdot \underline{\underline{T}})dV = \underline{0} . \tag{21}$$

From this it follows that when the "dead" surface tractions are prescribed over the entire surface O of a body, the superposed field of virtual displacement $\eta\underline{w}$ must satisfy Equation (21).

For the "dead" forces together with a uniform pressure applied over the entire surface ($\underline{f} = -p\underline{N}$) we have

$$\dot{\underline{B}} = \iiint_V \rho \underline{\underline{k}}\underline{w}\, dV + \iint_O \underline{\underline{f}}\underline{w}\, dO - p \iint_O \underline{N} \cdot (\underline{\underline{E}}\underline{\nabla} \cdot \underline{w} - \underline{\nabla}\underline{w}^T)\underline{R}\, dO =$$

$$= \iiint_V \rho \underline{\underline{k}}\underline{w}\, dV + \iint_O \underline{N} \cdot \underline{\underline{T}}\underline{w}\, dO - p \iint \underline{\nabla} \cdot (\underline{\underline{E}}\underline{\nabla} \cdot \underline{w} - \underline{\nabla}\underline{w}^T)\underline{R}\, dO .$$

Taking into account that

$$\underline{\nabla} \cdot (\underline{\underline{E}}\underline{\nabla} \cdot \underline{w} - \underline{\nabla}\underline{w}^T)\underline{R} = (\underline{\nabla}\underline{\nabla} \cdot \underline{w} - \underline{\nabla} \cdot \underline{\nabla}\underline{w}^T)\underline{R} + \underline{\underline{E}}\underline{\nabla} \cdot \underline{w} - \underline{\nabla}\underline{w} ,$$

$$\underline{\nabla}\underline{\nabla} \cdot \underline{w} - \underline{\nabla} \cdot \underline{\nabla}\underline{w}^T = \underline{0} ,$$

$$\iint_O \underline{N} \cdot \underline{\underline{T}}\underline{w}\, dO = \iiint_V (\underline{\nabla} \cdot \underline{\underline{T}})\underline{w}\, dV + \iiint_V \underline{\underline{T}} \cdot \underline{\nabla}\underline{w}\, dV ,$$

we arrive at the relationship

$$\dot{\underline{B}} = \iiint_V [(\underline{\underline{T}} + p\underline{\underline{E}}) \cdot \underline{\nabla}\underline{w} - \underline{\underline{E}}\underline{\nabla} \cdot \underline{w}]\, dV ,$$

$$\dot{\underline{B}} - \dot{\underline{B}}^{T} = \iiint\limits_{V} [(\underline{T} \cdot \underline{\underline{\nabla w}} - \underline{\underline{\nabla w}}^{T} \cdot \underline{T}) + p(\underline{\underline{\nabla w}} - \underline{\underline{\nabla w}}^{T})]\, dV \ . \tag{22}$$

For the case when a surface pressure is the only applied load, we have $\underline{T} = -p\underline{E}$ and as may be expected

$$\dot{\underline{B}} - \dot{\underline{B}}^{T} = \underline{0} \tag{23}$$

i.e. the requirement that the $V^{\times}$-configuration be itself equilibrated does not impose any restrictions upon the prescribed field of virtual displacements.

2. THE STRAIN ENERGY AND THE DEFINING EQUATION IN A $V^{\times}$-CONFIGURATION

The specific strain energy $\ni(\overset{0}{\underline{\underline{\nabla R}}})$ is represented by (II.4.23) and (1.12) in the form of a Maclaurin series to the second power of η i.e. by

$$\ni(\overset{0}{\underline{\underline{\nabla R}}}{}^{\times}) = \ni(\overset{0}{\underline{\underline{\nabla R}}} + \eta\overset{0}{\underline{\underline{\nabla w}}}) = \ni(\overset{0}{\underline{\underline{\nabla R}}}) + \eta \ni_{\overset{0}{\underline{\underline{\nabla R}}}} \cdot\cdot\, \overset{0}{\underline{\underline{\nabla w}}}{}^{T} +$$

$$+ \frac{1}{2}\eta^{2} \overset{0}{\underline{\underline{\nabla w}}}{}^{T} \cdot\cdot \ni_{\overset{0}{\underline{\underline{\nabla R}}}\overset{0}{\underline{\underline{\nabla R}}}} \cdot\cdot\, \overset{0}{\underline{\underline{\nabla w}}}{}^{T} = \ni(\overset{0}{\underline{\underline{\nabla R}}}) + \eta \underline{P} \cdot\cdot\, \overset{0}{\underline{\underline{\nabla w}}}{}^{T} + \frac{1}{2}\eta^{2} \overset{0}{\underline{\underline{\nabla w}}}{}^{T} \cdot\cdot\, \dot{\underline{P}} \ . \tag{1}$$

The strain energy of a system in its V-configuration under "dead" body and surface forces ($W(\underline{R})$) has been defined by the formula (4.16.8)

$$W(\underline{R}) = \iiint\limits_{v} \ni\, dv - \iiint\limits_{v} \rho_{0}\underline{k} \cdot \underline{R}\, dv - \iint\limits_{o_1} \underline{f}^{0} \cdot \underline{R}\, do \ . \tag{2}$$

In the $V^{\times}$-configuration this expression has the form

$$W(\underline{R} + \eta\underline{w}) = W(\underline{R}) + \eta W_{1}(\underline{R}) + \eta^{2} W_{2}(\underline{R}) \ . \tag{3}$$

Here by (1)

$$W_{1} = \iiint\limits_{v} \underline{P} \cdot\cdot\, \overset{0}{\underline{\underline{\nabla w}}}{}^{T} dv - \iiint\limits_{v} \rho_{0}\underline{k} \cdot \underline{w}\, dv - \iint\limits_{o_1} \underline{f}^{0} \cdot \underline{w}\, do \ , \tag{4}$$

$$W_{2} = \frac{1}{2}\iiint\limits_{v} \overset{0}{\underline{\underline{\nabla w}}}{}^{T} \cdot\cdot \ni_{\overset{0}{\underline{\underline{\nabla R}}}\overset{0}{\underline{\underline{\nabla R}}}} \cdot\cdot\, \overset{0}{\underline{\underline{\nabla w}}}{}^{T} dv = \frac{1}{2}\iiint\limits_{v} \overset{0}{\underline{\underline{\nabla w}}}{}^{T} \cdot\cdot\, \underline{P}_{\overset{0}{\underline{\underline{\nabla R}}}} \cdot\cdot\, \overset{0}{\underline{\underline{\nabla w}}}{}^{T} dv =$$

$$= \frac{1}{2}\iiint\limits_{v} \dot{\underline{P}} \cdot\cdot\, \overset{0}{\underline{\underline{\nabla w}}}{}^{T} dv = \frac{1}{2}\iiint\limits_{v} \sqrt{\frac{G}{g}}\, \underline{\underline{\nabla r}}^{T} \cdot \underline{\Theta} \cdot\cdot\, \overset{0}{\underline{\underline{\nabla w}}}{}^{T} dv = \frac{1}{2}\iiint\limits_{V} \underline{\Theta} \cdot\cdot\, \underline{\underline{\nabla w}}^{T} dV \ . \tag{5}$$

The quantity W_1 obviously represents the variation in the stored energy (2) of a system which, according to the stationary value principle (Par. 16), is zero provided that the V-configuration is itself in equilibrium

$$W_1 = 0 \,. \tag{6}$$

An alternative representation of Equations (5) is given by

$$W_2 = \iiint\limits_{v} \Phi(\overset{0}{\nabla}\underline{\underline{w}})\,dv \,, \qquad W_2 = \iiint\limits_{V} \Psi(\nabla\underline{\underline{w}})\,dV \,. \tag{7}$$

Here Φ denotes the qudratic form of the variables $\overset{0}{\nabla}\underline{\underline{w}}$, and Ψ that of the variables $\nabla\underline{\underline{w}}$

$$\Phi = \frac{1}{2}\,\dot{\underline{P}} \cdot\cdot \overset{0}{\nabla}\underline{\underline{w}}^T = \frac{1}{2}\,\overset{0}{\nabla}\underline{\underline{w}}^T \cdot\cdot \ni_{\overset{0}{\nabla}R\overset{0}{\nabla}R} \cdot\cdot \overset{0}{\nabla}\underline{\underline{w}}^T \,, \qquad \Psi = \frac{1}{2}\,\underline{\boldsymbol{\theta}} \cdot\cdot \nabla\underline{\underline{w}}^T \,, \tag{8}$$

the quantities $\dot{\underline{P}}$ and $\underline{\boldsymbol{\theta}}$ are linear functions of these variables. Therefore

$$\dot{\underline{P}}(\overset{0}{\nabla}\underline{\underline{w}} + \delta\overset{0}{\nabla}\underline{\underline{w}}) = \dot{\underline{P}}(\overset{0}{\nabla}\underline{\underline{w}}) + \dot{\underline{P}}(\delta\overset{0}{\nabla}\underline{\underline{w}}) \,,$$

$$\underline{\boldsymbol{\theta}}(\nabla\underline{\underline{w}} + \delta\nabla\underline{\underline{w}}) = \underline{\boldsymbol{\theta}}(\nabla\underline{\underline{w}}) + \underline{\boldsymbol{\theta}}(\delta\nabla\underline{\underline{w}}) \,.$$

The operations $\delta\overset{0}{\underline{\nabla}}$ and $\delta\underline{\nabla}$ are interchangeable whenever the variation is carried out in the $V^{\times}$-configuration. We thus obtain

$$\delta\Phi = \frac{1}{2}\,\delta(\dot{\underline{P}} \cdot\cdot \overset{0}{\nabla}\underline{\underline{w}}^T) = \frac{1}{2}\,[\dot{\underline{P}}(\overset{0}{\nabla}\underline{\underline{w}}) \cdot\cdot \overset{0}{\underline{\nabla}}\delta\underline{w}^T + \dot{\underline{P}}(\underline{\overset{0}{\nabla}}\delta\underline{w}) \cdot\cdot \overset{0}{\nabla}\underline{\underline{w}}^T] \,,$$

$$\delta\Psi = \frac{1}{2}\,\delta(\underline{\boldsymbol{\theta}} \cdot\cdot \nabla\underline{\underline{w}}^T) = \frac{1}{2}\,[\underline{\boldsymbol{\theta}}(\nabla\underline{\underline{w}}) \cdot\cdot \underline{\nabla}\delta\underline{w}^T + \underline{\boldsymbol{\theta}}(\underline{\nabla}\delta\underline{w}) \cdot\cdot \nabla\underline{\underline{w}}^T] \,.$$

By the duality property (1.15) and (1.16), assuming $\underline{w}_1 = \underline{w}$, $\underline{w}_2 = \delta\underline{w}$, we get

$$\delta\Phi = \dot{\underline{P}} \cdot\cdot \delta\overset{0}{\nabla}\underline{\underline{w}}^T \,, \qquad \delta\Psi = \underline{\boldsymbol{\theta}}(\nabla\underline{\underline{w}}) \cdot\cdot \delta\nabla\underline{\underline{w}} \,. \tag{9}$$

Recalling the definition (II.2.7) for the derivative of a scalar with respect to a tensor argument we arrive at the relationships

$$\dot{\underline{P}} = \Phi_{\overset{0}{\nabla}\underline{\underline{w}}} \,, \qquad \underline{\boldsymbol{\theta}} = \Psi_{\nabla\underline{\underline{w}}} \,, \tag{10}$$

which are analogous to the state equations of an isotropic elastic material for the case of the superposition of a small deformation upon some finite deformation of a non-linearly elastic isotropic body. The formulas (10), however, follow from the known properties of quadratic forms, that is from the Euler's theorem for the derivative of a homogeneous function.

The explicit expression for the quadratic form $\Psi(\nabla w)$ follows immediately from the representation (4.5.8) of the tensor θ. Bearing in mind that the first invariant of the product of symmetric and skew-symmetric tensors is zero, we have the relationships

$$\varepsilon \cdot\cdot \nabla w^T = \varepsilon \cdot\cdot (\varepsilon + \Omega^T) = \varepsilon \cdot\cdot \varepsilon = I_1(\varepsilon^2) \ ,$$

$$F \cdot \varepsilon \cdot F \cdot\cdot \nabla w^T = F \cdot \varepsilon \cdot F \cdot\cdot \varepsilon = I_1(F \cdot \varepsilon \cdot F \cdot \varepsilon) = I_1((F \cdot \varepsilon)^2)$$

and so on and we get

$$\Psi = \frac{1}{2}\theta \cdot\cdot \nabla w^T = \frac{1}{2} T \cdot\cdot \nabla w \cdot \nabla w^T + 2\sqrt{\frac{g}{G}}\{-\psi_0 I_1(\varepsilon^2) +$$
$$+ \psi_2 I_1((F \cdot \varepsilon)^2) + [\vartheta_{00} I_1^2(\varepsilon) + \vartheta_{11} I_1^2(F \cdot \varepsilon) + \vartheta_{22} I_1^2(F^2 \cdot \varepsilon) +$$
$$+ 2\vartheta_{01} I_1(\varepsilon) I_1(F \cdot \varepsilon) + 2\vartheta_{02} I_1(\varepsilon) I_1(F^2 \cdot \varepsilon) + 2\vartheta_{12} I_1(F \cdot \varepsilon) I_1(F^2 \cdot \varepsilon)]\} \ . \qquad (11)$$

The quantity W_2 may be termed the potential energy in the $V^\times$-configuration for the "dead" forces in the V-configuration. Recalling that the symbol ε in (11) denotes a linear strain tensor for the vector w in the R_s-basis we have

$$\varepsilon = \frac{1}{2}(\nabla w + \nabla w^T) \ . \qquad (12)$$

W_2 may also contain the term

$$-\iiint_V \rho \dot{k} w \, dV - \iint_{O_1} f_\times \cdot w \, dO \ , \qquad (13)$$

where $\dot{k}$, $f_\times$ denote the body and surface forces which were absent in the V-configuration.

In the vector basis of the proper vectors e_s of tensors F and T the invariants $I_1(F^N \cdot \varepsilon)$ are represented through the diagonal components ε_s of the tensor ε in this basis

$$I_1(F^N \cdot \varepsilon) = \sum_{smk} v_s^{2N} e_s e_s \cdot\cdot \varepsilon_{mk} e_m e_k = \sum_{s=1}^{3} v_s^{2N} \varepsilon_s \qquad (\varepsilon_{ss} = \varepsilon_s) \ .$$

The quadratic term in the square brackets (11) may be rewritten as follows

$$\sum_N \sum_\Gamma \vartheta_{N\Gamma} I_1(\underline{F}^N \cdot \underline{\varepsilon}) I_1(\underline{F}^\Gamma \cdot \underline{\varepsilon}) = \sum_{N,\Gamma=0}^{2} \sum_{s,k=1}^{3} \vartheta_{N\Gamma} v_s^{2N} v_k^{2\Gamma} \varepsilon_s \varepsilon_k =$$

$$= \sum_{s=1}^{3} \sum_{k=1}^{3} \alpha_{sk} \varepsilon_s \varepsilon_k \ .$$

We have

$$\alpha_{sk} = \alpha_{ks} = \sum_{N=0}^{2} \sum_{\Gamma=0}^{2} \vartheta_{N\Gamma} v_s^{2N} v_k^{2\Gamma} = \vartheta_{00} + \vartheta_{11} v_s^2 v_k^2 + \vartheta_{22} v_s^4 v_k^4 +$$

$$+ \vartheta_{01}(v_s^2 + v_k^2) + \vartheta_{02}(v_s^4 + v_k^4) + \vartheta_{12}(v_s^2 v_k^4 + v_s^4 v_k^2) \ . \tag{14}$$

Recalling the definition of the quantities

$$\vartheta_{00} = \frac{\partial \ni}{\partial I_3} I_3 + \frac{\partial^2 \ni}{\partial I_3^2} I_3^2 \ , \qquad \vartheta_{11} = \frac{\partial \ni}{\partial I_2} + \frac{\partial^2 \ni}{\partial I_1^2} + 2I_1 \frac{\partial^2 \ni}{\partial I_1 \partial I_2} + I_1^2 \frac{\partial^2 \ni}{\partial I_2^2} \ ,$$

$$\vartheta_{22} = \frac{\partial^2 \ni}{\partial I_2^2} \ ,$$

$$\vartheta_{01} = I_3 \frac{\partial^2 \ni}{\partial I_1 \partial I_3} + I_1 I_3 \frac{\partial^2 \ni}{\partial I_2 \partial I_3} \ , \qquad \vartheta_{02} = -I_3 \frac{\partial^2 \ni}{\partial I_2 \partial I_3} \ ,$$

$$\vartheta_{12} = -\frac{\partial^2 \ni}{\partial I_1 \partial I_2} - I_1 \frac{\partial^2 \ni}{\partial I_2^2} \ ,$$

we get

$$\alpha_{sk} = I_3 \frac{\partial \ni}{\partial I_3} + v_s^2 v_k^2 \frac{\partial \ni}{\partial I_2} + v_s^2 v_k^2 \left\{ \frac{\partial^2 \ni}{\partial I_1^2} + \frac{\partial^2 \ni}{\partial I_1 \partial I_2} [(I_1 - v_k^2) + (I_1 - v_s^2)] + \right.$$

$$+ \frac{\partial^2 \ni}{\partial I_2^2} (I_1 - v_s^2)(I_1 - v_k^2) + \frac{\partial^2 \ni}{\partial I_2 \partial I_2} \left[\frac{I_3}{v_s^2} (I_1 - v_k^2) + \frac{I_3}{v_k^2} (I_1 - v_s^2) \right] +$$

$$\left. + \frac{\partial^2 \ni}{\partial I_3 \partial I_1} \left(\frac{I_3}{v_s^2} + \frac{I_3}{v_k^2} \right) + \frac{\partial^2 \ni}{\partial I_3^2} \frac{I_3}{v_s^2} \frac{I_3}{v_k^2} \right\} \ . \tag{15}$$

Continuing the transformation, we introduce the derivatives of $\ni$ with respect to v_s^2

$$\frac{\partial \ni}{\partial v_s^2} = \frac{\partial \ni}{\partial I_1} + (I_1 - v_s^2)\frac{\partial \ni}{\partial I_2} + \frac{I_3}{v_s^2}\frac{\partial \ni}{\partial I_3} .$$

These derivatives depend explicitly on the invariants as well as on v_t^2 for $t \neq s$. Therefore

$$\frac{\partial}{\partial v_s^2}\frac{\partial \ni}{\partial v_s^2} = \frac{\partial^2 \ni}{(\partial v_s^2)^2} = \frac{\partial^2 \ni}{\partial I_1^2} + 2(I_1 - v_s^2)\frac{\partial^2 \ni}{\partial I_1 \partial I_2} + (I_1 - v_s^2)^2\frac{\partial^2 \ni}{\partial I_2^2} +$$

$$+ 2(I_1 - v_s^2)\frac{I_3}{v_s^2}\frac{\partial^2 \ni}{\partial I_2 \partial I_3} + 2\frac{I_3}{v_s^2}\frac{\partial^2 \ni}{\partial I_3 \partial I_1} + \left(\frac{I_3}{v_s^2}\right)^2\frac{\partial^2 \ni}{\partial I_3^2} ,$$

so that

$$\alpha_{ss} = I_3\frac{\partial \ni}{\partial I_3} + v_s^4\frac{\partial \ni}{\partial I_2} + v_s^4\frac{\partial^2 \ni}{(\partial v_s^2)^2} . \tag{16}$$

Calculating the mixed second derivatives it should be borne in mind that

$$\frac{\partial}{\partial v_k^2}(I_1 - v_s^2) = 1 , \qquad \frac{\partial}{\partial v_k^2}\frac{I_3}{v_s^2} = \frac{I_3}{v_s^2 v_k^2} \qquad (k \neq s)$$

and therefore

$$\alpha_{sk} = v_s^2 v_k^2 \frac{\partial^2 \ni}{\partial v_k^2 \partial v_s^2} \qquad (s \neq k) . \tag{17}$$

The quadratic form (11) can now be represented as

$$\Psi = \frac{1}{2}\Big\{\underline{T} \cdot\cdot \underline{\nabla w} \cdot \underline{\nabla w}^T - 4\sqrt{\frac{g}{G}}\left[I_3\frac{\partial \ni}{\partial I_3} I_1(\underline{\varepsilon}^2) + \frac{\partial \ni}{\partial I_2} I_1((\underline{F} \cdot \underline{\varepsilon})^2)\right] +$$

$$+ 4\sqrt{\frac{g}{G}}\sum_{s=1}^{3}\sum_{k=1}^{3}\alpha_{sk}\varepsilon_s\varepsilon_k\Big\} , \tag{18}$$

where α_{sk} are determined by Equations (14), (15) or (16) and (17). An alternative representation follows from the relation

$$\sigma_s = 2\sqrt{\frac{g}{G}}\, v_s^2\left[\frac{\partial \ni}{\partial I_1} + (I_1 - v_s^2)\frac{\partial \ni}{\partial I_2} + \frac{I_3}{v_s^2}\frac{\partial \ni}{\partial I_3}\right] .$$

This yields

$$\alpha_{sk} = v_s^2 v_k^2 \frac{\partial^2 \ni}{\partial v_s^2 \partial v_k^2} = \frac{\sqrt{I_3}}{2}\left(v_s^2 \frac{\partial \sigma_k}{\partial v_s^2} + \frac{1}{2}\sigma_k\right) = \frac{\sqrt{I_3}}{2}\left(v_k^2 \frac{\partial \sigma_s}{\partial v_k^2} + \frac{1}{2}\sigma_s\right), \tag{19}$$

$$\alpha_{ss} = \frac{\sqrt{I_3}}{2}\left[v_s^2 \frac{\partial \sigma_s}{\partial v_s^2} + \frac{3}{2}\sigma_s - \hat{\mu}_s(v_1^2, v_2^2, v_3^2)\right]. \tag{20}$$

The identity of the representations (19) is easily verified. In (20) we have introduced the notation

$$\hat{\mu}_s(v_1^2, v_2^2, v_3^2) = \frac{2}{\sqrt{I_3}} v_s^2 \left(\frac{\partial \ni}{\partial I_1} + I_1 \frac{\partial \ni}{\partial I_2} - 2v_s^2 \frac{\partial \ni}{\partial I_2}\right). \tag{21}$$

In the natural configuration, $\hat{\mu}(1,1,1) = \mu$ by virtue of (4.7.12).

3. THE STATIONARY VALUE PRINCIPLE IN A $V^{\times}$-CONFIGURATION

From (2.7) and (2.13) we may represent the second variation of the stored energy as a functional of $\underline{w}$ having the form

$$W_2 = \iiint\limits_V \Psi \, dV - \iiint\limits_V \rho \dot{\underline{k}} \cdot \underline{w} \, dV - \iint\limits_{O_1} \underline{f}_{\times} \cdot \underline{w} \, dO . \tag{1}$$

Its variation (caused by a variation of $\underline{w}$) is, by (2.9), given by the expression

$$\delta W_2 = \iiint\limits_V \underline{\Theta} \cdot\cdot\, \delta \underline{\nabla} \underline{w}^T \, dV - \iiint\limits_V \rho \dot{\underline{k}} \cdot \delta \underline{w} \, dV - \iint\limits_{O_1} \underline{f}_{\times} \cdot \delta \underline{w} \, dO$$

and this is, by (III.3.10), transformed to

$$\delta W_2 = - \iiint\limits_V (\underline{\nabla} \cdot \underline{\Theta} + \rho \dot{\underline{k}}) \cdot \delta \underline{w} \, dV + \iiint\limits_V \underline{\nabla} \cdot (\underline{\Theta} \cdot \delta \underline{w}) dV -$$

$$- \iint\limits_{O_1} \underline{f}_{\times} \cdot \delta \underline{w} \, dO = - \iiint\limits_V (\underline{\nabla} \cdot \underline{\Theta} + \rho \dot{\underline{k}}) \cdot \delta \underline{w} \, dV +$$

$$+ \iint\limits_{O_1} (\underline{N} \cdot \underline{\Theta} - \underline{f}_{\times}) \cdot \delta \underline{w} \, dO + \iint\limits_{O_2} \underline{N} \cdot \underline{\Theta} \cdot \delta \underline{w} \, dO . \tag{2}$$

The displacement $\underline{w}$ is described in the region O_2 in the V-configuration and, by (1.4), in O_2 in the $V^{\times}$-configuration $\delta \underline{w} = \underline{0}$. Bearing (1.17) in mind we arrive at the principle of stationary potential energy O_2 in the $V^{\times}$-configuration

$$\delta W_2 = 0 \ . \tag{3}$$

Of course, the inverse is also true: (2) embodys the Equations (1.17) in V and on O_1 and the requirement of the coincidence of $\underline{w}$ and its value (1.4) on O_2.

The homogeneous boundary value problem for the linear system (1.17) of differential equations may possess non-trivial (non-zero) solutions $\underline{w}$ for some selected values of parameters of loading in the V-configuration, these parameters entering the representation for $\underline{\theta}$. The equilibrium in the V-configuration is for this case called neutral equilibrium, and the corresponding values of the loading parameters are termed critical (or bifurcational). These statements are associated with the problem of stability of equilibrium and a detailed account of this will be given in Pars. 10-25 of this chapter.

In the systems (2.10) of equations (which are linear with respect to $\overset{0}{\underline{\underline{\nabla w}}}$, $\underline{\underline{\nabla w}}$), the quadratic forms Φ and Ψ represent generating functions of transformations $\overset{0}{\underline{\underline{\nabla w}}} \to \dot{\underline{P}}$, $\underline{\underline{\nabla w}} \to \underline{\theta}$.

The inverse transformations $\dot{\underline{P}} \to \overset{0}{\underline{\underline{\nabla w}}}$, $\underline{\theta} \to \underline{\underline{\nabla w}}$ are generated by the quadratic forms of $\dot{\underline{P}}$ and $\underline{\theta}$

$$\Phi_\times(\dot{\underline{P}}) = \dot{\underline{P}} \cdot\cdot \overset{0}{\underline{\underline{\nabla w}}}{}^T(\dot{\underline{P}}) - \Phi(\overset{0}{\underline{\nabla}}(\underline{w}(\dot{\underline{P}}))) \ ,$$

$$\Psi_\times(\underline{\theta}) = \underline{\theta} \cdot\cdot \underline{\underline{\nabla w}}{}^T(\underline{\theta}) - \Psi(\underline{\underline{\nabla w}}(\underline{\theta})) \ . \tag{4}$$

In these, $\overset{0}{\underline{\underline{\nabla w}}}(\dot{\underline{P}})$ and $\underline{\underline{\nabla w}}(\underline{\theta})$ denote solutions of the linear systems (2.10) and which, from the Legendre transformation, may be written in the form

$$\overset{0}{\underline{\underline{\nabla w}}}(\dot{\underline{P}}) = (\Phi_\times)_{\dot{\underline{P}}} \ , \qquad \underline{\underline{\nabla w}}(\underline{\theta}) = (\Psi_\times)_{\underline{\theta}} \ . \tag{5}$$

These relationships imply the principle of stationary complementary energy, i.e. that the variation of the functional

$$W_{2\times} = \iiint\limits_V \Psi_\times(\theta)dV - \iint\limits_{O_2} \underline{w} \cdot \underline{f}_\times \, dO \tag{6}$$

is zero for all states of stress satisfying Equations (1.17) within the body and on that part O_1 of the boundary where the forces are prescribed. For these states we have

$$\text{in } V\text{: } \underline{\nabla} \cdot \delta\underline{\theta} = \underline{0} \ ; \qquad \text{on } O_2\text{: } \underline{N} \cdot \delta\underline{\theta} = \delta\underline{f}_\times \ ; \qquad \text{on } O_1\text{: } \delta\underline{f}_\times = \underline{N} \cdot \delta\underline{\theta} = \underline{0} \ . \tag{7}$$

In fact, by (4) and (2.10)

$$\delta W_{2\times} = \iiint_V (\delta\underline{\theta} \cdot\cdot \underline{\nabla w}^T(\underline{\theta}) + \underline{\theta} \cdot\cdot \delta\underline{\nabla w}^T - \underline{\theta} \cdot\cdot \delta\underline{\nabla w}^T)dV - \iint_{O_2} \underline{w} \cdot \delta\underline{f}_{\times}\, dO =$$

$$= \iiint_V [\underline{\nabla} \cdot (\delta\underline{\theta} \cdot \underline{w}) - (\underline{\nabla} \cdot \delta\underline{\theta}) \cdot \underline{w}]\, dV - \iint_{O_2} \underline{w} \cdot \delta\underline{f}_{\times}\, dO =$$

$$= - \iiint_V (\underline{\nabla} \cdot \delta\underline{\theta}) \cdot \underline{w}\, dV + \iint_{O_2} (\underline{N} \cdot \delta\underline{\theta} - \delta\underline{f}_{\times}) \cdot \underline{w}\, dO + \iint_{O_1} \underline{N} \cdot \delta\underline{\theta} \cdot \underline{w}\, dO = 0 ,$$

which follows from the definition (7) of the set of possible static states.

4. SMALL DEFORMATION OF A HYDROSTATIC STRESS STATE

The state of stress in the V-configuration is produced by a pressure p uniformly distributed on the surface of a body, so that

$$\underline{k} = \underline{0} , \qquad \underline{f} = -p\underline{N} . \tag{1}$$

The equilibrium equations within the body and on its surface in this configuration

$$\underline{\nabla} \cdot \underline{T} = \underline{0} , \qquad \underline{N} \cdot \underline{T} = -p\underline{N}$$

are satisfied by a spherical stress tensor

$$\underline{T} = -p\underline{E} . \tag{2}$$

The Finger measure $\underline{F}$ is also a spherical tensor whose invariants are

$$\underline{F} = v^2\underline{E} , \quad I_1(\underline{F}) = 3v^2 , \quad I_2(\underline{F}) = 3v^4 , \quad I_3(\underline{F}) = v^6$$

and the state equation (4.3.13) is given in the form

$$-p = 2v^{-3}(\psi_0 + \psi_1 v^2 + \psi_2 v^4) = 2v^{-1}\left(\frac{\partial \ni}{\partial I_1} + 2v^2 \frac{\partial \ni}{\partial I_2} + v^4 \frac{\partial \ni}{\partial I_3}\right) =$$

$$= 2v^{-1}\left[\frac{\partial \ni}{\partial I_1} + v^2 \frac{\partial \ni}{\partial I_2} + v^2\left(\frac{\partial \ni}{\partial I_2} + v^2 \frac{\partial \ni}{\partial I_3}\right)\right] \tag{3}$$

or, alternatively, by

$$\hat{\mu}(v) = 2v^1\left(\frac{\partial \ni}{\partial I_1} + v^2 \frac{\partial \ni}{\partial I_2}\right) = -\left[p + 2v\left(\frac{\partial \ni}{\partial I_2} + v^2 \frac{\partial \ni}{\partial I_3}\right)\right] . \tag{4}$$

Here $\hat{\mu}(1) = \mu$ denotes the shear modulus in the undistorted reference configuration. From the BE-criterion used in the form of (5.13.11), $\hat{\mu}(v) > 0$ for a strongly elliptic material.

The potential energy in the V-configuration is by (4.16.14) represented by the expression

$$W(\underline{R}) = \iiint\limits_{v} \left(\ni + p\sqrt{\frac{G}{g}}\right) dv = \iiint\limits_{v} \ni\, dv + p \iiint\limits_{V} dV = \iiint\limits_{v} \ni\, dv + pV\ .$$

Its representation in the $V^{\times}$-configuration may by (2.3) and (2.7) and (1.12.6) be expressed in the form

$$W(\underline{R} + \eta\underline{w}) = W(\underline{R}) + \eta W_1 + \eta^2 W_2 = W(\underline{R}) + \eta\left[\iiint\limits_{V} \underline{T} \cdot\cdot \underline{\underline{\nabla w}}^T dV + p \iiint\limits_{V} \underline{\nabla} \cdot \underline{w}\, dV\right] +$$

$$+ \frac{1}{2}\eta^2 \left\{\iiint\limits_{V} \Psi\, dV + p \iiint\limits_{V} [(\underline{\nabla} \cdot \underline{w})^2 - \underline{\underline{\nabla w}} \cdot\cdot \underline{\underline{\nabla w}}]\, dV\right\}\ . \tag{5}$$

To construct the expression for Ψ from Equation (2.11) we make use of the formulas

$$I_1(\underline{F} \cdot \underline{\varepsilon}) = v^2 I_1(\underline{\varepsilon})\ , \quad I_1(\underline{F}^2 \cdot \underline{\varepsilon}) = v^4 I_1(\underline{\varepsilon})\ , \quad I_1(\underline{F}^2 \cdot \underline{\varepsilon}^2) = v^4 I_1(\underline{\varepsilon}^2)\ ,$$

$$[-\psi_0 I_1(\underline{\varepsilon}^2) + \psi_2 I_1(\underline{F}^2 \cdot \underline{\varepsilon}^2)] = -v^4\left(\frac{\partial\ni}{\partial I_2} + v^2 \frac{\partial\ni}{\partial I_3}\right) I_1(\underline{\varepsilon}^2) =$$

$$= \frac{1}{2} v^3 [p + \hat{\mu}(v)] I_1(\underline{\varepsilon}^2)\ ,$$

$$\vartheta_{00} = v^6 \frac{\partial\ni}{\partial I_3} + v^{12} \frac{\partial^2\ni}{\partial I_3^2}\ , \quad \vartheta_{11} = \frac{\partial\ni}{\partial I_2} + \frac{\partial^2\ni}{\partial I_1^2} + 6v^2 \frac{\partial^2\ni}{\partial I_1 \partial I_2} + 9v^4 \frac{\partial^2\ni}{\partial I_2^2}\ ,$$

$$\vartheta_{22} = \frac{\partial^2\ni}{\partial I_2^2}\ , \quad \vartheta_{01} = v^6 \frac{\partial^2\ni}{\partial I_1 \partial I_3} + 3v^8 \frac{\partial^2\ni}{\partial I_2 \partial I_3}\ , \quad \vartheta_{02} = -v^6 \frac{\partial^2\ni}{\partial I_2 \partial I_3}\ ,$$

$$\vartheta_{12} = -\frac{\partial^2\ni}{\partial I_1 \partial I_2} - 3v^2 \frac{\partial^2\ni}{\partial I_2^2}\ ,$$

obtaining

$$\Psi = -\frac{1}{2} p\underline{\underline{\nabla w}} \cdot\cdot \underline{\underline{\nabla w}}^T + [\mu(v) + p] I_1(\underline{\varepsilon}^2) + 2v\left\{\left[\frac{\partial^2\ni}{\partial I_1^2} + 4v^4 \frac{\partial^2\ni}{\partial I_2^2} + v^8 \frac{\partial^2\ni}{\partial I_3^2} +\right.\right.$$

$$\left.\left. + 4v^2 \frac{\partial^2\ni}{\partial I_1 \partial I_2} + 4v^6 \frac{\partial^2\ni}{\partial I_2 \partial I_3} + 2v^4 \frac{\partial^2\ni}{\partial I_3 \partial I_1}\right] + \frac{\partial\ni}{\partial I_2} + 2v^4 \frac{\partial\ni}{\partial I_3}\right\} I_1^2(\underline{\varepsilon})\ . \tag{6}$$

Regarding the specific strain stored energy as a function of v^2

$$\ni(I_1,I_2,I_3) = \ni(3v^2,3v^4,v^6) = \hat{\ni}(v^2) \ , \tag{7}$$

we have

$$\frac{d\hat{\ni}}{dv^2} = 3\left(\frac{\partial\ni}{\partial I_1} + 2v^2\,\frac{\partial\ni}{\partial I_2} + v^4\,\frac{\partial\ni}{\partial I_3}\right) ,$$

$$\frac{d^2\hat{\ni}}{(dv^2)^2} = 9\left[\frac{\partial^2\ni}{\partial I_1^2} + 4v^4\,\frac{\partial^2\ni}{\partial I_2^2} + v^8\,\frac{\partial^2\ni}{\partial I_3^2} + 4v^2\,\frac{\partial^2\ni}{\partial I_1\partial I_2} + 4v^6\,\frac{\partial^2\ni}{\partial I_2\partial I_3} + 2v^4\,\frac{\partial^2\ni}{\partial I_3\partial I_1}\right] + $$
$$+ 6\left(\frac{\partial\ni}{\partial I_2} + v^2\,\frac{\partial\ni}{\partial I_3}\right)$$

and this allows us to substitute (6) by the expression

$$\Psi = -\frac{1}{2}\,p\underline{\underline{\nabla w}}\cdot\cdot\underline{\underline{\nabla w}}^{\mathrm{T}} + [\hat{\mu}(v) + p]\,I_1(\underline{\varepsilon}^2) + $$
$$+ \frac{2}{9}\,v\left(\frac{d^2\hat{\ni}}{(dv^2)^2} + 3\left(\frac{\partial\ni}{\partial I_2} + v^2\,\frac{\partial\ni}{\partial I_3}\right)\right)I_1^2(\underline{\varepsilon}) \ . \tag{8}$$

Bearing also in mind the relationships

$$-\frac{1}{2}\,p\underline{\underline{\nabla w}}\cdot\cdot\underline{\underline{\nabla w}}^{\mathrm{T}} + \frac{1}{2}\,p(\underline{\nabla}\cdot\underline{w})^2 - \frac{1}{2}\,p\underline{\underline{\nabla w}}\cdot\cdot\underline{\underline{\nabla w}} = \frac{1}{2}\,pI_1^2(\underline{\varepsilon}) - $$
$$-\frac{p}{2}\,\underline{\underline{\nabla w}}\cdot\cdot(\underline{\underline{\nabla w}} + \underline{\underline{\nabla w}}^{\mathrm{T}}) = \frac{1}{2}\,pI_1^2(\underline{\varepsilon}) - p\underline{\underline{\nabla w}}\cdot\cdot\underline{\varepsilon} = \frac{1}{2}\,pI_1^2(\underline{\varepsilon}) - pI_1(\underline{\varepsilon}^2) \ ,$$

the second variation of potential energy may be expressed as follows

$$W_2 = \iiint\limits_V \Psi\, dV \ ,$$
$$\Psi = \hat{\mu}(v)I_1(\underline{\varepsilon}^2) + \frac{1}{2}\left[p + \frac{4}{9}\,v\,\frac{d^2\hat{\ni}}{(dv^2)^2} + \frac{4}{3}\,v\left(\frac{\partial\ni}{\partial I_2} + v^2\,\frac{\partial\ni}{\partial I_3}\right)\right]I_1^2(\underline{\varepsilon}) \ . \tag{9}$$

The quantity $\hat{\Psi}$ can be interpreted as the specific strain energy in the $V^{\times}$-configuration (referred to unit volume in the V-configuration). Comparison with the known expression

$$\frac{1}{2}\,\lambda I_1^2(\underline{\varepsilon}) + \mu I_1(\underline{\varepsilon}^2)$$

for the specific strain energy of a linearly elastic body makes it natural to introduce a "reduced Lamé modulus" $\hat{\lambda}(v)$ together with the modulus $\hat{\mu}(v)$ determined above

$$\hat{\lambda}(v) = p + \frac{4}{9} v \frac{d^2\hat{\ni}}{(dv^2)^2} + \frac{4}{3} v\left(\frac{\partial\ni}{\partial I_2} + v^2 \frac{\partial\ni}{\partial I_3}\right) = \frac{1}{3} p - \frac{2}{3}\hat{\mu}(v) + \frac{4}{9} v \frac{d^2\hat{\ni}}{(dv^2)^2} . \tag{10}$$

We thus obtain

$$\Psi = \frac{1}{2}\hat{\lambda}(v) I_1^2(\underline{\varepsilon}) + \hat{\mu}(v) I_1(\underline{\varepsilon}^2) . \tag{11}$$

A "reduced bulk modulus" is defined by the formula

$$\hat{k}(v) = \hat{\lambda}(v) + \frac{2}{3}\hat{\mu}(v) = \frac{1}{3} p + \frac{4}{9} v \frac{d^2\hat{\ni}}{(dv^2)^2} . \tag{12}$$

Introducing the variable $\tau = \sqrt{I_3} = v^3$, which is the ratio of elementary volume in the state of hydrostatic compression and in the natural reference configuration, we have by (3)

$$\frac{d\hat{\ni}}{d\tau} = \frac{d\hat{\ni}}{dv^2} \frac{2}{3} v^{-1} = \frac{\partial\ni}{\partial I_1}\frac{dI_1}{d\tau} + \frac{\partial\ni}{\partial I_2}\frac{dI_2}{d\tau} + \frac{\partial\ni}{\partial I_3}\frac{dI_3}{d\tau} =$$

$$= 2v^{-1}\left(\frac{\partial\ni}{\partial I_1} + 2v^2\frac{\partial\ni}{\partial I_2} + v^4\frac{\partial\ni}{\partial I_3}\right) = -p .$$

Therefore

$$\frac{d^2\hat{\ni}}{d\tau^2} = \frac{4}{9}\frac{v}{\tau}\frac{d^2\ni}{(dv^2)^2} - \frac{1}{3\tau}\frac{d\ni}{d\tau} = \frac{4}{9}\frac{v}{\tau}\frac{d^2\ni}{(dv^2)^2} + \frac{1}{3\tau} p .$$

The expression for $\hat{k}(v)$ is thus reduced to

$$\hat{k}(v) = \tau \frac{d^2\hat{\ni}}{d\tau^2} . \tag{13}$$

The expression

$$\frac{1}{2}(\lambda + 2\mu) I_1^2(\underline{\varepsilon}) - 2\mu I_2(\underline{\varepsilon}) = \frac{1}{2} k I_1^2(\underline{\varepsilon}) + 2\mu\left(\frac{1}{3} I_1^2(\underline{\varepsilon}) - I_2(\underline{\varepsilon})\right) =$$

$$= \frac{1}{2} k I_1^2(\underline{\varepsilon}) - 2\mu I_2(\operatorname{dev}\underline{\varepsilon})$$

for the specific strain energy of a linearly elastic medium is identical in form to the expression

$$\Psi = \frac{1}{2}\hat{k}(v)I_1^2(\underline{\varepsilon}) - 2\hat{\mu}(v)I_2(\operatorname{dev}\underline{\varepsilon}) \tag{14}$$

which is valid for a hydrostatically compressed medium (see (11)).

But from either (I.9.23) or (I.13.3) $I_2(\operatorname{dev}\underline{\varepsilon}) < 0$ and $\Psi > 0$ for all values of v which satisfy the natural condition

$$\hat{k}(v) = \tau\frac{d^2\hat{\ni}}{d\tau^2} > 0\ , \qquad \frac{d^2\hat{\ni}}{d\tau^2} > 0\ , \tag{15}$$

which means that the rate of increase of $\hat{\ni}$ increases with increase of deformation.

By (8), (10) and (11) the expression for Ψ reduces to

$$\Psi = -\frac{1}{2}p\underline{\underline{\nabla w}}\cdot\cdot\underline{\underline{\nabla w}}^T + pI_1(\underline{\varepsilon}^2) - \frac{1}{2}pI_1^2(\underline{\varepsilon}) + \frac{1}{2}\hat{\lambda}(v)I_1^2(\underline{\varepsilon}) + \mu I_1(\underline{\varepsilon}^2)$$

and by (2.10) we have

$$\underline{\theta} = \Psi_{\underline{\underline{\nabla w}}} = \hat{\lambda}(v)I_1(\underline{\varepsilon})\underline{E} + 2\hat{\mu}(v)\underline{\varepsilon} - p(\underline{\underline{E\nabla}}\cdot\underline{w} - \underline{\underline{\nabla w}}^T)\ . \tag{16}$$

Here we have made use of the formulas

$$I_1(\underline{\varepsilon}^2)_{\underline{\underline{\nabla w}}} = 2\underline{\varepsilon} = \underline{\underline{\nabla w}} + \underline{\underline{\nabla w}}^T\ , \quad I_1^2(\underline{\varepsilon})_{\underline{\underline{\nabla w}}} = \underline{E}I_1(\underline{\varepsilon})\ , \quad (\underline{\underline{\nabla w}}\cdot\cdot\underline{\underline{\nabla w}}^T)_{\underline{\underline{\nabla w}}} = 2\underline{\underline{\nabla w}}\ .$$

The equilibrium equations (1.17) and (1.9) in the volume and on its surface in the $V^\times$-configuration

$$\underline{\nabla}\cdot\underline{\theta} = \underline{0}\ , \qquad \underline{N}\cdot\underline{\theta} = -p\underline{N}\cdot(\underline{\underline{E\nabla}}\cdot\underline{w} - \underline{\underline{\nabla w}}^T)$$

are transformed to

$$\underline{\nabla}\cdot[\hat{\lambda}(v)I_1(\underline{\varepsilon})\underline{E} + 2\hat{\mu}(v)\underline{\varepsilon} - p(\underline{\underline{E\nabla}}\cdot\underline{w} - \underline{\underline{\nabla w}}^T)] = (\hat{\lambda}+\hat{\mu})\underline{\underline{\nabla\nabla}}\cdot\underline{w} + \hat{\mu}\underline{\nabla}^2\underline{w}\ , \tag{17}$$

$$\underline{N}\cdot\underline{\theta} = \underline{N}\cdot(\hat{\lambda}\underline{E}I_1(\underline{\varepsilon}) + 2\hat{\mu}\underline{\varepsilon}) - p\underline{N}\cdot(\underline{\underline{E\nabla}}\cdot\underline{w} - \underline{\underline{\nabla w}}^T) = -p\underline{N}(\underline{\underline{E\nabla}}\cdot\underline{w} - \underline{\underline{\nabla w}}^T)\ . \tag{18}$$

Here we have taken account of the fact that $\underline{\nabla}\cdot(\underline{\underline{E\nabla}}\cdot\underline{w} - \underline{\underline{\nabla w}}^T) = \underline{0}$ and have thus arrived at the equations of equilibrium of the linear theory of elasticity in terms of displacements where λ, μ are substituted by the reduced moduli $\hat{\lambda}$, $\hat{\mu}$, and the displacement vector $\underline{u}$ is substituted by $\underline{w}$

$$(\hat{\lambda}+\hat{\mu})\underline{\nabla\nabla}\cdot\underline{w}+\hat{\mu}\nabla^2\underline{w}=\underline{0}\,,\qquad \underline{N}\cdot(2\underline{E\nabla}\cdot\underline{w}+2\mu\underline{\varepsilon})=\underline{0}\,.$$

Body and surface forces are absent, and by Kirchhoff's theorem, which asserts uniqueness of the equilibrium state, we conclude that the Equations (17) possess no solutions different from those for a rigid displacement of a body provided that

$$\hat{\lambda}+\frac{2}{3}\hat{\mu}=\hat{k}>0\,,\qquad \hat{\mu}>0\,,$$

these inequalities being consistent with the restrictions formulated above. Under these conditions neutral equilibrium states do not exist for a hydrostatically compressed medium.

5. A SMALL DEFORMATION SUPERPOSED ON A UNIFORM STATE OF STRESS

For the affine transformation of the undistorted reference v-configuration into the actual V-configuration we have by (4.15.1) and (4.15.2)

$$\underline{R}=\underline{r}\cdot\underline{\Lambda}\,,\quad \underline{F}=\underline{\Lambda}^T\cdot\underline{\Lambda}\,,\quad \underline{T}=2(\det\underline{\Lambda})^{-1}(\psi_0\underline{E}+\psi_1\underline{\Lambda}^T\cdot\underline{\Lambda}+\psi_2(\underline{\Lambda}^T\cdot\underline{\Lambda})^2) \tag{1}$$

and the coefficients of a linear transformation over the vector $\underline{w}$ in Equation (4.5.8) are constant, so that

$$\underline{\theta}=\underline{T}\cdot\underline{\nabla w}+4\sqrt{\frac{g}{G}}\left[-\psi_0\underline{\varepsilon}(\underline{w})+\psi_2\underline{F}\cdot\underline{\varepsilon}(\underline{w})\cdot\underline{F}+\sum_{\Gamma=0}^{2}\sum_{N=0}^{2}\vartheta_{N\Gamma}\underline{F}^{\Gamma}\underline{F}^{N}\cdot\cdot\,\underline{\varepsilon}(\underline{w})\right]. \tag{2}$$

We will confine ourselves here to consideration of the uniaxial tension of a rod (Chapter 6, Par. 2). The axis of the rod is directed along the axis OZ, and the Finger measure and its square is expressed by

$$\underline{F}=(\underline{i}_1\underline{i}_1+\underline{i}_2\underline{i}_2)\alpha^2v^2+\underline{i}_3\underline{i}_3v^2=\underline{E}\alpha^2v^2+v^2(1-\alpha^2)\underline{i}_3\underline{i}_3\,,$$

$$\underline{F}^2=\underline{E}\alpha^4v^4+v^4(1-\alpha^4)\underline{i}_3\underline{i}_3\,. \tag{3}$$

The principal stresses σ_1 and σ_2 are equal to zero

$$\psi_0+\alpha^2v^2\psi_1+\alpha^4v^4\psi_2=2v\left[\frac{1}{v^2}\frac{\partial\Im}{\partial I_1}+(1+\alpha^2)\frac{\partial\Im}{\partial I_2}+v^2\alpha^2\frac{\partial\Im}{\partial I_3}\right]=0\,, \tag{4}$$

and the stress σ_3 is represented by the expression

$$\sigma_3 = 2v\left(\frac{1}{v^2\alpha^2}\frac{\partial \ni}{\partial I_1} + 2\frac{\partial \ni}{\partial I_2} + v^2\alpha^2\frac{\partial \ni}{\partial I_3}\right), \tag{5}$$

which, by (4), can be expressed in the alternative forms

$$\sigma_3 = \frac{2(1-\alpha^2)}{\alpha^2 v}\left(\frac{\partial \ni}{\partial I_1} + v^2\alpha^2\frac{\partial \ni}{\partial I_2}\right) = -\frac{2v}{\alpha^2}(1-\alpha^2)\left(\frac{\partial \ni}{\partial I_2} + \alpha^2 v^2\frac{\partial \ni}{\partial I_3}\right). \tag{6}$$

By (2) tensor $\underline{\theta}$ is seen to be equal to

$$\begin{aligned}\underline{\theta} = {} & \underline{i}_3\underline{i}_k\sigma_3\frac{\partial w_k}{\partial x^3} - \frac{4v}{\alpha^2}\left\{\alpha^4\left(\frac{\partial \ni}{\partial I_2} + v^2\frac{\partial \ni}{\partial I_3}\right)\underline{\varepsilon}(\underline{w}) + \right.\\ & + \frac{\partial \ni}{\partial I_2}(1-\alpha^2)\left[\left.(1-\alpha^2)\underline{i}_3\underline{i}_3\varepsilon_{33} + \alpha^2\sum_{s=1}^{3}(\underline{i}_3\underline{i}_s + \underline{i}_s\underline{i}_3)\varepsilon_{s3}\right]\right\} + \\ & + \frac{4}{v^3\alpha^2}\left\{\vartheta_{00}\underline{E}\underline{\nabla}\cdot\underline{w} + \vartheta_{11}v^2\underline{F}[\alpha^2\underline{\nabla}\cdot\underline{w} + (1-\alpha^2)\varepsilon_{33}] + \right.\\ & + \vartheta_{22}v^4\underline{F}^2[\alpha^4\underline{\nabla}\cdot\underline{w} + (1-\alpha^4)\varepsilon_{33}] + \vartheta_{01}v^2\underline{E}[\alpha^2\underline{\nabla}\cdot\underline{w} + (1-\alpha^2)\varepsilon_{33}] + \\ & + \vartheta_{10}\underline{F}\underline{\nabla}\cdot\underline{w} + \vartheta_{02}v^4\underline{E}[\alpha^4\underline{\nabla}\cdot\underline{w} + (1-\alpha^4)\varepsilon_{33}] + \vartheta_{20}\underline{F}^2\underline{\nabla}\cdot\underline{w} + \\ & \left. + \vartheta_{12}v^4\underline{F}[\alpha^4\underline{\nabla}\cdot\underline{w} + (1-\alpha^4)\varepsilon_{33}] + \vartheta_{21}v^2\underline{F}^2[\alpha^2\underline{\nabla}\cdot\underline{w} + (1-\alpha^2)\varepsilon_{33}]\right\}. \end{aligned} \tag{7}$$

Following the *BE*-criterion (5.13.11) and making use of (6), we infer that $\alpha > 1$ in the case of compression ($\sigma_3 < 0$) and $\alpha < 1$ for tension ($\sigma_3 > 0$).

6. TORSIONAL DEFORMATION SUPERPOSED ON A UNIAXIAL STATE OF STRESS

The displacement vector $\underline{w}$ is prescribed by its projections

$$w_1 = -\gamma yz, \quad w_2 = \gamma zx, \quad w_3 = \gamma\varphi(x,y) \tag{1}$$

along the principal central axes of inertia of the cross-section of a prismatic rod in its reference v-configuration. These axes preserve their direction in the V-configuration because the v-configuration is transformed into the former by a similarity transformation

$$x = \alpha v a^1, \quad y = \alpha v a^2 \tag{2}$$

(here a^1 and a^2 denote coordinates in the v-configuration, and x and y are the coordinates in the V-configuration of a point in the cross-section). The area of the cross-section, its polar moment of inertia and the torsional rigidity in the V-configuration are connected with the same quantities in the reference configuration by the formulas

$$S = \alpha^2 v^2 S_0 , \qquad I_p = \alpha^4 v^4 I_p^0 , \qquad C = \alpha^4 v^4 C^0 . \tag{3}$$

We also have the relationships

$$\iint_O x\, dO = 0 , \qquad \iint_O y\, dO = 0 , \qquad \iint_O xy\, dO = 0 , \tag{4}$$

which express the fact that the axes OX and OY are the principal central axes of inertia.

The components ε_{23} and ε_{31} of a linear strain tensor over $\underline{w}$ are the only ones which are different from zero and they are

$$\varepsilon_{23} = \frac{1}{2}\gamma\left(\frac{\partial\varphi}{\partial y} + x\right) , \qquad \varepsilon_{31} = \frac{1}{2}\left(\frac{\partial\varphi}{\partial x} - y\right) . \tag{5}$$

In the formulas of Par. 5

$$I_1(\underline{\varepsilon}) = 0 , \qquad \varepsilon_{33} = 0 , \qquad \underline{\varepsilon} = (\underline{i}_2\underline{i}_3 + \underline{i}_3\underline{i}_2)\varepsilon_{23} + (\underline{i}_3\underline{i}_1 + \underline{i}_1\underline{i}_3)\varepsilon_{31} \tag{6}$$

and after some calculation using (5.7)

$$\underline{\theta} = \gamma\sigma_3\underline{i}_3(-\underline{i}_1 y + \underline{i}_2 x) - 4v\left(\frac{\partial\ni}{\partial I_2} + \alpha^2 v^2 \frac{\partial\ni}{\partial I_3}\right)[(\underline{i}_3\underline{i}_1 + \underline{i}_1\underline{i}_3)\varepsilon_{31} + (\underline{i}_2\underline{i}_3 + \underline{i}_3\underline{i}_2)\varepsilon_{23}]$$

and by (5.6)

$$\underline{\theta} = \gamma\sigma_3\left\{\underline{i}_3(-\underline{i}_1 y + \underline{i}_2 x) + \frac{\alpha^2}{1-\alpha^2}\left[(\underline{i}_3\underline{i}_1 + \underline{i}_1\underline{i}_3)\left(\frac{\partial\varphi}{\partial x} - y\right) + \right.\right.$$

$$\left.\left. + (\underline{i}_2\underline{i}_3 + \underline{i}_3\underline{i}_2)\left(\frac{\partial\varphi}{\partial y} + x\right)\right]\right\} . \tag{7}$$

The equations of equilibrium within the body and on the sides of a rod

$$\underline{\nabla}\cdot\underline{\theta} = \underline{0} , \qquad \underline{N}\cdot\underline{\theta} = (N_1\underline{i}_1 + N_2\underline{i}_2)\cdot\underline{\theta} = \underline{0}$$

are reduced to the well-known equations of Saint-Venant of the linear theory of torsion

$$\underline{\nabla}^2\varphi = 0 \, , \qquad \frac{\partial\varphi}{\partial N} = yN_1 - xN_2 \, . \tag{8}$$

The stress vector in the cross-section is determined by the expression

$$\underline{f} = \underline{i}_3 \cdot \underline{\theta} = \gamma\sigma_3\left\{(-\underline{i}_1 y + \underline{i}_2 x) + \frac{\alpha^2}{1-\alpha^2}\left[\underline{i}_2\left(\frac{\partial\varphi}{\partial y} + x\right) + \underline{i}_1\left(\frac{\partial\varphi}{\partial x} - y\right)\right]\right\} \tag{9}$$

and it is easily seen that the resultant of these stresses is zero. We have in fact, by (8) for example, that

$$\iint_O \left(\frac{\partial\varphi}{\partial x} - y\right)dO = \iint_O \left[\frac{\partial}{\partial x}\, x\left(\frac{\partial\varphi}{\partial x} - y\right) + \frac{\partial}{\partial y}\, x\left(\frac{\partial\varphi}{\partial y} + x\right)\right]dO =$$

$$= \oint_\Gamma x\left(\frac{\partial\varphi}{\partial N} - yN_1 + xN_2\right)dl = 0$$

and by (4) and (9)

$$\iint_O \underline{f}\, dO = \underline{0} \, . \tag{10}$$

which is the desired result.

The resultant moment of the stresses is equal to

$$\iint_O (\underline{i}_1 x + \underline{i}_2 y + \underline{i}_3 z) \times \underline{f}\, dO =$$

$$= \gamma\sigma_3\underline{i}_3 \iint_O \left\{x^2 + y^2 + \frac{\alpha^2}{1-\alpha^2}\left[x\left(\frac{\partial\varphi}{\partial y} + x\right) - y\left(\frac{\partial\varphi}{\partial x} - y\right)\right]\right\}dO + \underline{i}_3 z \times \iint_O \underline{f}\, dO \, .$$

From (10) the second term vanishes and, from (3) and (6.13.9), we get

$$m_z = \gamma\sigma_3\alpha^4 v^4\left(\frac{\alpha^2}{1-\alpha^2}\, C^0 + I_p^0\right) . \tag{11}$$

The torque may be zero for $\alpha > 1$ in the V-configuration when the rod is compressed (see (5.6)). In this case, by (11)

$$\frac{1}{\alpha^2} = 1 - \frac{C^0}{I_p^0} \, . \tag{12}$$

It is known that $C^0 \leq I_p^0$ where the strict equality holds only for a circular cross-section and for that of a circular ring. Therefore the inequality

$$0 < \frac{1}{\alpha^2} = 1 - \frac{C^0}{I_p^0} < 1 \tag{13}$$

allows us to determine geometric characteristics of a rod whose cross-section differs from that of a circle (or a ring) when the uniaxial state of stress is neutral for a suitable value of the compressing stress $|\sigma_3|$. The case when a torsional deformation arises without any torque being applied to the face of the rod is not excluded.

The critical (bifurcational) value of the compressing force is determined by (5.6) and (3)

$$Q = |\sigma_3|\alpha^2 v^2 S_0 = 2S_0 v(\alpha^2 - 1)\left(\frac{\partial \ni}{\partial I_1} + \alpha^2 v^2 \frac{\partial \ni}{\partial I_2}\right), \tag{14}$$

where v and α are calculated by (12) and (5.4). The latter condition vanishes if the material is incompressible. It is then replaced by the condition

$$\sqrt{I_3} = \alpha^2 v^3 = 1 , \qquad v^3 = \alpha^{-2} = 1 - \frac{C^0}{I_p^0}$$

and the critical compressing force is represented by the relationship

$$Q = \frac{2S_0 C^0}{I_p^0 - C^0}\left[\left(1 - \frac{C^0}{I_p^0}\right)^{1/3} \frac{\partial \ni}{\partial I_1} + \frac{\partial \ni}{\partial I_2}\right] . \tag{15}$$

For an incompressible medium we have

$$I_1 = v^2 + 2v^{-1} , \qquad I_2 = v^{-2} + 2v , \qquad \ni(I_1, I_2) = \hat{\ni}(v) ,$$

$$\frac{d\hat{\ni}(v)}{dv} = 2(1 - v^{-3})\left(\frac{\partial \ni}{\partial I_1} v + \frac{\partial \ni}{\partial I_2}\right)$$

which allows us to represent Q by the formula

$$Q = S_0 \left|\frac{d\hat{\ni}}{dv}\right| . \tag{16}$$

The displacement vector from the V- into the $\overset{\times}{V}$-configuration is, for this problem, given by the Equations (1). Therefore we cannot reject in advance the existence of other neutral forms of equilibrium (e.g., some buckled forms) corresponding to some value of the compressing force which might be lower than that prescribed by (14).

7. PLANE WAVES IN A UNIFORMLY STRESSED ELASTIC MEDIUM

A plane wave propagating in the direction $\underline{N}$ is determined by the displacement vector

$$\underline{w}(t) = \underline{w}_0(t)\exp ik\underline{N}\cdot\underline{R}\,, \qquad \underline{w}_0(t) = \underline{w}_0 e^{i\omega t}\,. \tag{1}$$

Here k denotes the wave number, ω the frequency, and $c = \omega/k$ the velocity of the wave. The equation of motion of a particle of a prestressed unbounded elastic medium is represented in the form (1.17) where the body force, absent in the V-configuration, is treated in the $\overset{\times}{V}$-configuration as the "inertia" (this terminology can of course be avoided)

$$\rho\dot{\underline{k}} = -\rho\ddot{\underline{w}}(t) = -\rho\ddot{\underline{w}}_0(t)\exp ik\underline{N}\cdot\underline{R} = \rho\omega^2\underline{w}(t)\,. \tag{2}$$

The equation of motion now takes the form

$$\underline{\nabla}\cdot\underline{\theta} + \rho\omega^2\underline{w}(t) = \underline{0}\,. \tag{3}$$

In expression (5.2) for the tensor $\underline{\theta}$ the gradient $\underline{\underline{\nabla w}}$ is substituted by the expression

$$\underline{\underline{\nabla w}} = \underline{R}^s\underline{w}_0(t)\exp ik\underline{N}\cdot\underline{R}ik\frac{\partial}{\partial q^s}\underline{N}\cdot\underline{R} = ik\underline{R}^s\underline{w}(t)\underline{N}\cdot\underline{R}_s = ik\underline{\underline{Nw}}(t)\,. \tag{4}$$

Here the operator nabla has been replaced by the vector $ik\underline{N}$, whence

$$\underline{\varepsilon} = \frac{1}{2}ik(\underline{\underline{Nw}} + \underline{\underline{wN}})\,, \qquad \underline{F}^N\cdot\underline{\varepsilon} = \frac{1}{2}ik\,[\underline{F}^N\cdot\underline{\underline{Nw}} + \underline{F}^N\cdot\underline{\underline{wN}}]\,,$$

$$I_1(\underline{F}^N\cdot\underline{\varepsilon}) = ik\underline{N}\cdot\underline{F}^N\cdot\underline{w}$$

etc. We thus obtain

$$\underline{\theta} = ik\left\{\underline{T}\cdot\underline{\underline{Nw}} + 2\sqrt{\frac{g}{G}}\left[-\psi_0(\underline{\underline{Nw}} + \underline{\underline{wN}}) + \psi_2\underline{F}\cdot(\underline{\underline{Nw}} + \underline{\underline{wN}})\cdot\underline{F} + \right.\right.$$

$$\left.\left. + 2\sum_{\Gamma=0}^{2}\sum_{N=0}^{2}\vartheta_{N\Gamma}\underline{F}^{\Gamma}\underline{N}\cdot\underline{F}^N\cdot\underline{N}\right]\right\} \tag{5}$$

and moreover

$$\underline{\nabla} \cdot \underline{\theta} = -k^2 \Big\{ \underline{N} \cdot \underline{T} \cdot \underline{NE} + 2\sqrt{\frac{g}{G}} \Big[-\psi_0 (\underline{E} + \underline{NN}) + \psi_2 (\underline{N} \cdot \underline{F} \cdot \underline{NF} + \underline{N} \cdot \underline{FF} \cdot \underline{N}) + \\ + 2 \sum_{N=0}^{2} \sum_{\Gamma=0}^{2} \vartheta_{N\Gamma} \underline{N} \cdot \underline{F}^{N} \underline{F}^{\Gamma} \cdot \underline{N} \Big] \Big\} \cdot \underline{w} \, .$$

Substituting for $\underline{T}$ in this formula from (4.3.4) we arrive at the representation

$$\underline{\nabla} \cdot \underline{\theta} = -4k^2 \sqrt{\frac{g}{G}} \Big[\frac{1}{2} (\psi_1 \underline{N} \cdot \underline{F} \cdot \underline{N} + \psi_2 \underline{N} \cdot \underline{F}^2 \cdot \underline{N}) \underline{E} + \\ + \frac{1}{2} \psi_2 (\underline{N} \cdot \underline{F} \cdot \underline{NF} + \underline{N} \cdot \underline{FF} \cdot \underline{N}) + (\vartheta_{00} - \frac{1}{2} \psi_0) \underline{NN} + \vartheta_{11} \underline{N} \cdot \underline{FF} \cdot \underline{N} + \\ + \vartheta_{22} \underline{N} \cdot \underline{F}^2 \underline{F}^2 \cdot \underline{N} + \vartheta_{01} (\underline{NF} \cdot \underline{N} + \underline{F} \cdot \underline{NN}) + \vartheta_{02} (\underline{N} \cdot \underline{F}^2 \underline{N} + \underline{NF}^2 \cdot \underline{N}) + \\ + \vartheta_{12} (\underline{N} \cdot \underline{FF}^2 \cdot \underline{N} + \underline{N} \cdot \underline{F}^2 \underline{F} \cdot \underline{N}) \Big] \cdot \underline{w} \qquad (6)$$

in the form of a product

$$\underline{\nabla} \cdot \underline{\theta} = -4k^2 \sqrt{\frac{g}{G}} \, \underline{Q} \cdot \underline{w} \qquad (7)$$

with respect to $\underline{w}$ of a tensor which differs from the acoustical tensor (4.11.14) only by a multiplying factor. The equation (3) of motion may now be put into the form

$$\Big(\underline{Q} - \frac{1}{4k^2} \sqrt{\frac{G}{g}} \rho \omega^2 \underline{E}\Big) \cdot \underline{w} = \underline{0} \, , \qquad \Big(\underline{Q} - \frac{1}{4} \rho_0 c^2 \underline{E}\Big) \cdot \underline{w} = \underline{0} \qquad (8)$$

which is the characteristic equation of an acoustical tensor whose eigenvalues are $1/4 \, \rho_0 c^2$. These will be shown in Par. 11 to be positive, that is the velocities c along any direction $\underline{N}$ are real for the deformed states in the V-configuration so long as the material remains strongly elliptic.

8. PLANE WAVES IN A HYDROSTATICALLY STRESSED ELASTIC MEDIUM

The simplest procedure is to transform (4.17) using Equation (7.4) without any reference to the general representation (7.6)

$$\underline{\nabla} \cdot \underline{\theta} = (\hat{\lambda} + \hat{\mu}) \underline{\nabla\nabla} \cdot \underline{w} + \hat{\mu} \underline{\nabla}^2 \underline{w} = -k [(\hat{\lambda} + \hat{\mu}) \underline{NN} + \hat{\mu} \underline{E}] \cdot \underline{w} \, . \qquad (1)$$

Equation (7.3) is reduced to

$$[(\hat{\lambda} + \hat{\mu}) \underline{NN} + \hat{\mu} \underline{E} - \rho c^2 \underline{E}] \cdot \underline{w} = \underline{0} \, , \qquad \underline{E} = \underline{NN} + \underline{t}_1 \underline{t}_1 + \underline{t}_2 \underline{t}_2 \, , \qquad (2)$$

where the vectors $\underline{t}_1$, $\underline{t}_2$ and $\underline{N}$ constitute an orthonormal trihedron in which the vectors $\underline{t}_1$ and $\underline{t}_2$ lie in the plane perpendicular to $\underline{N}$. Equation (2) may now be rewritten in the form

$$[(\hat{\lambda} + 2\hat{\mu} - \rho c^2)\underline{NN} + (\hat{\mu} - \rho c^2)(\underline{t}_1\underline{t}_1 + \underline{t}_2\underline{t}_2)] \cdot \underline{w} = 0 .$$

For the longitudinal waves, $\underline{t}_s \cdot \underline{w} = 0$, and the velocity $c_{\parallel}$ of their propagation turns out to be

$$c_{\parallel} = \sqrt{\frac{\hat{\lambda} + 2\hat{\mu}}{\rho}} = v^{3/2}\sqrt{\frac{\hat{\lambda} + 2\hat{\mu}}{\rho_0}} . \tag{3}$$

For the transverse waves polarised perpendicularly to the direction of their propagation, $\underline{N} \cdot \underline{w} = 0$. Their velocity $c_{\perp}$ is

$$c_{\perp} = \sqrt{\frac{\hat{\mu}}{\rho}} = v^{3/2}\sqrt{\frac{\hat{\mu}}{\rho_0}} . \tag{4}$$

For an unstressed linearly elastic medium $v = 1$, $\hat{\lambda} = \lambda$ and $\hat{\mu} = \mu$. We thus arrive at the well-known expressions for the longitudinal and transverse velocities

$$c_{\parallel} = \sqrt{\frac{\lambda + 2\mu}{\rho_0}} , \qquad c_{\perp} = \sqrt{\frac{\mu}{\rho_0}} . \tag{5}$$

9. PRINCIPAL WAVES

The eigenvalues of an acoustical tensor generally depend on the direction $\underline{N}$ of the wave and calculation of their velocities is associated with the solution of a cubic equation whose coefficients are functions of $\underline{N}$. The problem is considerably simplified when we consider those waves which have been termed principal waves by Truesdell: these waves propagate along the principal axes $\underline{e}_1$, $\underline{e}_2$ and $\underline{e}_3$ of the stress tensor (or the Finger measure) in the V-configuration.

For the first principal wave $\underline{e}_1 = \underline{N}$

$$\underline{F} = v_1^2\underline{NN} + v_2^2\underline{e}_2\underline{e}_2 + v_3^2\underline{e}_3\underline{e}_3 ,$$

$$N_1 = \underline{N} \cdot \underline{e}_1 = 1 , \qquad N_2 = \underline{N} \cdot \underline{e}_2 = 0 , \qquad N_3 = \underline{N} \cdot \underline{e}_3 = 0 . \tag{1}$$

The expression (4.11.14) for the acoustical tensor is reduced to

$$\underset{\approx}{Q} = \left[\left(\vartheta_{00} - \frac{1}{2}\psi_0\right) + v_1^4\left(\vartheta_{11} + \frac{1}{2}\psi_2\right) + v_1^8\vartheta_{22} + 2\vartheta_{01}v_1^2 + 2\vartheta_{02}v_1^4 + 2\vartheta_{12}v_1^6\right]\underline{\underline{NN}} +$$

$$+ \frac{1}{2}v_1^2(\psi_1 + \psi_2 v_1^2)\underline{\underline{E}} + \frac{1}{2}\psi_2(v_1^2\underline{\underline{NN}} + v_2^2\underline{e}_2\underline{e}_2 + v_3^2\underline{e}_3\underline{e}_3) . \tag{2}$$

The velocities of waves polarised along $\underline{e}_2$ and $\underline{e}_3$, are, by (7.8), determined by the formulas

$$\frac{1}{2v_1^2}\rho_0 c_{12}^2 = \psi_1 + \psi_2(v_1^2 + v_2^2) , \qquad \frac{1}{2v_1^2}\rho_0 c_{13}^2 = \psi_1 + \psi_2(v_1^2 + v_3^2) . \tag{3}$$

We have by (4.3.4)

$$\psi_1 + \psi_2(v_1^2 + v_2^2) = \frac{1}{2}\sqrt{\frac{G}{g}}\,\frac{\sigma_1 - \sigma_2}{v_1^2 - v_2^2} , \qquad \psi_1 + \psi_2(v_1^2 + v_3^2) = \frac{1}{2}\sqrt{\frac{G}{g}}\,\frac{\sigma_1 - \sigma_3}{v_1^2 - v_3^2}$$

and the equations (3) take the form

$$\frac{\rho c_{12}^2}{v_1^2} = \frac{\sigma_1 - \sigma_2}{v_1^2 - v_2^2} , \qquad \frac{\rho c_{13}^2}{v_1^2} = \frac{\sigma_1 - \sigma_3}{v_1^2 - v_3^2} . \tag{4}$$

The square of the velocity of the longitudinal waves is, from (2), represented by the expression

$$\frac{1}{4}\rho_0 c_{11}^2 = \vartheta_{00} - \frac{1}{2}\psi_0 + v_1^4\left(\vartheta_{11} + \frac{1}{2}\psi_2\right) + v_1^8\vartheta_{22} + 2\vartheta_{01}v_1^2 + 2\vartheta_{02}v_1^4 +$$

$$+ 2\vartheta_{12}v_1^6 + \frac{1}{2}v_1^2(\psi_1 + 2\psi_2) . \tag{5}$$

This quantity is also expressed in terms of the principal stress σ_1

$$\sigma_1 = \frac{2}{\sqrt{I_3}}(\psi_0 + \psi_1 v^2 + \psi_2 v^4) \tag{6}$$

and recalling definition (4.5.4) it is easy to verify that the right-hand side of (3) can be represented as

$$\frac{1}{4}\sqrt{I_3}\,v_1\frac{\partial\sigma_1}{\partial v_1} .$$

We thus arrive at the expression

$$\rho c_{11}^2 = v_1\frac{\partial\sigma_1}{\partial v_1} . \tag{7}$$

This calculation may be avoided if we recall the representations (4.12.11) of the components of the acoustical tensor for the case when the vector $\underline{N}$ of the normal is defined by (1).

As may be seen from (4.12.12) and (4.12.13) the expressions obtained for the velocities are real for a strongly elliptic material.

From Chapter 5, Par. 3 it can be seen that for a Murnaghan material, provided that the following approximations

$$I_1 = 3 + 2\vartheta\ , \qquad I_2 = 3 + 4\vartheta\ , \qquad I_3 = 1 + 2\vartheta\ ,$$

$$v_s^2 = 1 + 2\delta_s \qquad (s = 1,2,3)\ ,$$

may be regarded as reasonable, the formulas for velocities of the transverse waves reduce to

$$\frac{\rho_0}{\mu} c_{1\alpha} = 1 + 2\delta_1 + (\delta_1 + \delta_\alpha)\left(2 + \frac{n}{2\mu}\right) + \frac{1}{\mu}\vartheta\left(\lambda + m - \frac{n}{2}\right) \qquad (\alpha = 2,3)\ . \tag{8}$$

The coefficients

$$\vartheta_{00} = \frac{1}{8} n I_3\ , \qquad \vartheta_{11} = \frac{1}{4}\left[\lambda + l(I_1 - 3) + m(1 - I_1) - \frac{n}{2}\right]\ , \qquad \vartheta_{12} = \frac{1}{4} m \tag{9}$$

prove to be non-zero and the velocity of the longitudinal waves is, to the same approximation, given by the formula

$$\frac{\rho_0}{\mu} c_{11}^2 = \frac{\lambda}{\mu} + 2 + \left(\frac{4\lambda}{\mu} + \frac{4m}{\mu} + 10\right)\delta_1 + \left(\frac{\lambda}{\mu} + 2\frac{l}{\mu}\right)\vartheta\ . \tag{10}$$

For a quasi-linear material, the acoustical tensor is given by Equation (4.12.17). Comparing this with Equation (4.12.11) and assuming $\underline{N} = \underline{e}_1$ we arrive at the relationships

$$v_1 \frac{\partial \sigma_1}{\partial v_1} = (\lambda + 2\mu) v_1^2\ , \qquad \frac{\sigma_2 - \sigma_1}{v_2^2 - v_1^2} = \frac{3\lambda + 2\mu}{v_1 + v_2}\,[(1 - 2q)(v_1 + v_2) + q v_3 - 1]\ . \tag{11}$$

By virtue of (7), the velocity of the principal longitudinal wave ($\underline{N} = \underline{e}_1$) is determined from the formula

$$c_{11} = v_1 \sqrt{\frac{\lambda + 2\mu}{\rho}}\ ,$$

whereas the velocities of waves polarised along $\underline{e}_2$ and $\underline{e}_3$ turn out to be equal to (see (4))

$$c_{12} = v_1 \sqrt{\frac{3\lambda + 2\mu}{\rho(v_1 + v_2)}} \, [(1 - 2q)(v_1 + v_2) + qv_3 - 1]^{1/2} \, ,$$

$$c_{13} = v_1 \sqrt{\frac{3\lambda + 2\mu}{\rho(v_1 + v_3)}} \, [(1 - 2q)(v_1 + v_3) + qv_2 - 1]^{1/2} \, . \tag{12}$$

10. NEUTRAL EQUILIBRIUM AND STABILITY

We will confine ourselves to the consideration of a one-parameter family S_p of equilibrium states in the V-configuration in which p is the parameter describing the load which transforms the unstrained natural v-configuration into the V-configuration.

It will be recalled that a field of virtual displacements $\eta \underline{w}(q^1, q^2, q^3)$ which, from (1.4), vanishes on the region O_2 of the surface of V, is superposed upon the displacement field $\underline{u} = \underline{R} - \underline{r}$ in the S_p-state. The second variation $\eta^2 W_2^{(p)}$ of the stored energy of a system is by (2.5), (2.7) represented as a functional of $\underline{\underline{\nabla w}}$

$$W_2^p = \frac{1}{2} \iiint_V \underline{\theta} \cdot\cdot \underline{\underline{\nabla w}}^T dV = \iiint_V \Psi(\underline{\underline{\nabla w}}) dV \, , \tag{1}$$

where Ψ is a quadratic in $\underline{\underline{\nabla w}}$, and by (2.4)

$$\underline{\theta} = \Psi_{\underline{\underline{\nabla w}}} \, . \tag{2}$$

In view of the relationship

$$\underline{\theta} \cdot\cdot \underline{\underline{\nabla w}}^T = \underline{\nabla} \cdot (\underline{\theta} \cdot \underline{w}) - (\underline{\nabla} \cdot \underline{\theta}) \cdot \underline{w} \, , \tag{3}$$

we can represent expression (1) in the form

$$W_2^p = -\frac{1}{2} \iiint_V (\underline{\nabla} \cdot \underline{\theta}) \cdot \underline{w} \, dV + \frac{1}{2} \iint_1 \underline{N} \cdot \underline{\theta} \cdot \underline{w} \, dO \, . \tag{4}$$

Definitions. 1. The equilibrium state S_p will be termed stable if

$$W_2^{(p)} > 0 \tag{5}$$

for all non-zero vectors $\underline{w}$ which satisfy Equation (1.4) and is unstable if

$$W_2^{(p)} \leq 0 \tag{6}$$

for some vector $\underline{w}$ satisfying Equation (1.4).

2. The equilibrium state S_{p_0} will be called neutral if for some $\underline{w} = \underline{w}_0$ not equal to zero ($\underline{w}_0 \not\equiv \underline{0}$) and satisfying Equation (1.4), the equations of neutral equilibrium*)

$$\text{in } V\text{: } \underline{\nabla} \cdot \underline{\theta}(\underline{w}_0) = \underline{0} \ ; \qquad \text{on } O_1\text{: } \underline{N} \cdot \underline{\theta} = \underline{0} \ ; \qquad \text{on } O_2\text{: } \underline{w}_0 = \underline{0} \ . \tag{7}$$

are satisfied for some $p = p_0$.

From these definitions it follows that the "neutral equilibrium" is unstable. Indeed, from (4) we infer that $W_2^{p_0} = 0$.

According to the principle of stationary potential energy of Par. 3, the Equations (7) of neutral equilibrium formally amount to the existence of the forms S_{p_0} of equilibrium which are close to the form under consideration.

Let S_p and $p \geq p_0$ be a family of equilibrium states, and let $p_0 > 0$ be such that for some $\varepsilon > 0$ the states S_p, $p \in (p_0 - \varepsilon, p_0)$ are stable, and the states S_p, $p \in (p_0, p_0 + \varepsilon)$ are unstable. It is natural to assume that for this case

$$W_2^{p_0}(w) \geq 0 \tag{8}$$

for all admissible $\underline{w}$, but for some $\underline{w}_0 \not\equiv \underline{0}$

$$W_2^{p_0}(\underline{w}_0) = 0 \ . \tag{9}$$

In which case S_{p_0} is the state of neutral equilibrium. In fact, assuming

$$\underline{w} = \underline{w}_0 + \eta \underline{v} \ ,$$

we have by (4)

$$W_2^{p_0}(\underline{w}) = W_2^{p_0}(\underline{w}_0) + \frac{1}{2}\eta\left(-\iiint_V [\underline{\nabla} \cdot \underline{\theta}(\underline{w}_0)] \cdot \underline{v}\, dV + \iint_{O_1} \underline{N} \cdot \underline{\theta}(\underline{w}_0) \cdot \underline{v}\, dO + o(\eta)\right) \ , \tag{10}$$

where $\eta^{-1} o(\eta)|_{\eta \to 0} \to 0$. From this it follows that the coefficient in the neighbourhood of η is zero for all $\underline{v}$, otherwise, by (9), Inequality (8) would not hold for some value of η.

This shows that the value $p = p_0$ is the value of the loading parameter at the point of bifurcation (at the critical value) and that the homogeneous

---

*) In Equation (1.17) it is assumed that $\underline{\dot{k}} = \underline{0}$ and $\underline{f}_x = \underline{0}$.

boundary value problem (7) possesses a non-trivial solution $\underline{w}_0$ which defines the neutral state (which is by definition unstable) S_{p_0} of equilibrium of the body in its V-configuration.

Note. Definition (5) simplifies, but at the same time restricts, the not on of stability. For instance by this definition, the equilibrium of a heavy sphere fixed at the vertex (0,0) of a parabola $y = ax^2$ is stable for $a > 0$ and unstable for $a < 0$. However, when applied to the parabola $y = ax^4$, this definition yields no information about the actual stability or instability of the equilibrium state at its vertex.

11. NEUTRAL EQUILIBRIUM OF A QUASI-LINEAR MATERIAL

The state equation of a semi-linear material is given in terms of the Piola tensor in the form (5.5.5)

$$\underline{P} = (\lambda s_1 - 2\mu)\underline{O}^{\times} + 2\mu\overset{0}{\underline{\underline{\nabla R}}} . \tag{1}$$

Turning to the Equations (1.10.7), (1.11.5) and (1.11.11) we obtain the following expression for its convected derivative

$$\dot{\underline{P}} = \lambda\dot{s}_1\underline{O}^{\times} + (\lambda s_1 - 2\mu)\dot{\underline{O}}^{\times} + 2\mu\overset{0}{\underline{\underline{\nabla R}}}\cdot\underline{\underline{\nabla w}} = [(\lambda s_1 - 2\mu)\underline{O}^{\times} + 2\mu\overset{0}{\underline{\underline{\nabla R}}}]\cdot\underline{\underline{\nabla w}} +$$

$$+ \lambda\underline{O}^{\times}\underline{V}\cdot\cdot\underline{\underline{\nabla w}} - 2(\lambda s_1 - 2\mu)\underline{\underline{\nabla r}}^{T}\cdot\sum_{s=1}^{3}\sum_{k=1}^{3}\frac{v_s v_k}{v_s + v_k}\underline{e}_k\underline{e}_s\underline{e}_k\cdot\underline{\varepsilon}\cdot\underline{e}_s , \tag{2}$$

where the first term can also be written as $\underline{P}\cdot\underline{\underline{\nabla w}}$.

We will confine ourselves below to the case of a linear transformation of the undistorted configuration into the actual configuration. This transformation will be assumed to preserve the principal directions

$$x^s = v_s a^s , \quad \underline{R} = \sum_{s=1}^{3}\underline{i}_s v_s a^s , \quad \underline{R}_s = \underline{i}_s v_s , \quad \overset{0}{\underline{\underline{\nabla R}}} = \overset{0}{\underline{\underline{\nabla R}}}^{T} = \underline{V} ,$$
$$\underline{\underline{\nabla r}}^{T} = \underline{V}^{-1} , \quad \underline{e}_s = \overset{0}{\underline{e}}_s = \underline{i}_s , \quad \underline{O}^{\times} = \underline{E} . \tag{3}$$

The Piola tensor is symmetric in this case, its principal directions constitute a trihedron $\underline{i}_s$

$$\underline{P} = \sum_{s=1}^{3}\underline{i}_s\underline{i}_s p_s = (\lambda s_1 - 2\mu)\underline{E} + 2\mu\underline{V} , \qquad p_s = \lambda s_1 - 2\mu + 2\mu v_s \tag{4}$$

and the following relationships hold

$$\frac{\lambda s_1 - 2\mu}{v_s + v_k} = \frac{1}{2}\,\frac{p_s + p_k}{v_s + v_k} - \mu \,. \qquad (5)$$

Now, also making use of the transformations

$$\underline{\underline{\overset{0}{\nabla}R}} \cdot \underline{\underline{\nabla w}} = \underline{\underline{\overset{0}{\nabla}w}}\,, \qquad \underline{\underline{\nabla w}} = \underline{V}^{-1} \cdot \underline{\underline{\overset{0}{\nabla}w}}\,,$$

$$2\underline{e}_k \cdot \underline{\varepsilon} \cdot \underline{e}_s = \underline{i}_k \cdot \underline{\underline{\nabla w}} \cdot \underline{i}_s + \underline{i}_k \cdot \underline{\underline{\nabla w}}^T \cdot \underline{i}_s = v_k^{-1}\underline{\underline{\overset{0}{\nabla}w}} \cdot\cdot\, \underline{i}_s\underline{i}_k + v_s^{-1}\underline{\underline{\overset{0}{\nabla}w}}^T \cdot\cdot\, \underline{i}_k\underline{i}_s\,,$$

$$\underline{\underline{\nabla r}}^T \cdot \underline{e}_k = v_k^{-1}\underline{i}_k\,, \qquad \underline{V} \cdot \underline{\underline{\nabla w}} = \underline{\underline{\overset{0}{\nabla}w}}\,,$$

we can rewrite (2) in the form

$$\dot{\underline{P}} = (\lambda s_1 - 2\mu)\underline{V}^{-1} \cdot \underline{\underline{\overset{0}{\nabla}w}} + 2\mu\underline{\underline{\overset{0}{\nabla}w}} + \lambda\underline{\underline{E}}\overset{0}{\nabla} \cdot \underline{w} -$$

$$- \sum_{s=1}^{3}\sum_{k=1}^{3} \frac{\lambda s_1 - 2\mu}{v_s + v_k}\,\underline{i}_k\underline{i}_s\left(\frac{v_s}{v_k}\,\underline{\underline{\overset{0}{\nabla}w}} \cdot\cdot\, \underline{i}_s\underline{i}_k + \underline{\underline{\overset{0}{\nabla}w}} \cdot\cdot\, \underline{i}_k\underline{i}_s\right).$$

Bearing in mind also that

$$\underline{V}^{-1} \cdot \underline{\underline{\overset{0}{\nabla}w}} = \sum_{s=1}^{3}\sum_{k=1}^{3} \frac{\underline{i}_k\underline{i}_s}{v_k}\,\frac{\partial w_s}{\partial a^k}\,, \qquad \underline{\underline{\overset{0}{\nabla}w}} \cdot\cdot\, \underline{i}_s\underline{i}_k = \frac{\partial w_s}{\partial a^k}\,, \qquad \underline{\underline{\overset{0}{\nabla}w}} \cdot\cdot\, \underline{i}_k\underline{i}_s = \frac{\partial w_k}{\partial a^s}$$

and appealing to (5), we obtain

$$\dot{\underline{P}} = 2\mu\underline{\underline{\overset{0}{\nabla}w}} + \lambda\underline{\underline{E}}\overset{0}{\nabla} \cdot \underline{w} +$$

$$+ \sum_{s=1}^{3}\sum_{k=1}^{3}(\lambda s_1 - 2\mu)\left[\frac{1}{v_k}\left(1 - \frac{v_s}{v_s + v_k}\right)\frac{\partial w_s}{\partial a^k} - \frac{1}{v_s + v_k}\,\frac{\partial w_k}{\partial a^s}\right]\underline{i}_k\underline{i}_s =$$

$$= 2\mu\underline{\underline{\overset{0}{\nabla}w}} + \lambda\underline{\underline{E}}\overset{0}{\nabla} \cdot \underline{w} + \sum_{s=1}^{3}\sum_{k=1}^{3}\frac{\lambda s_1 - 2\mu}{v_s + v_k}\,\frac{\partial w_s}{\partial a^k}\,(\underline{i}_k\underline{i}_s - \underline{i}_s\underline{i}_k) =$$

$$= 2\mu\underline{\underline{\overset{0}{\nabla}w}} + \lambda\underline{\underline{E}}\overset{0}{\nabla} \cdot \underline{w} + \frac{1}{2}\sum_{s=1}^{3}\sum_{k=1}^{3}\frac{p_s + p_k}{v_s + v_k}\,\frac{\partial w_s}{\partial a^k}\,(\underline{i}_k\underline{i}_s - \underline{i}_s\underline{i}_k) - \mu\underline{\underline{\overset{0}{\nabla}w}} + \mu\underline{\underline{\overset{0}{\nabla}w}}^T =$$

$$= \lambda\underline{\underline{E}}\overset{0}{\nabla} \cdot \underline{w} + 2\mu\overset{0}{\underline{\underline{\varepsilon}}}(\underline{w}) + \frac{1}{2}\left[\frac{p_1 + p_2}{v_1 + v_2}\left(\frac{\partial w_2}{\partial a^1} - \frac{\partial w_1}{\partial a^2}\right)(\underline{i}_1\underline{i}_2 - \underline{i}_2\underline{i}_1) + \right.$$

$$\left. + \frac{p_2 + p_3}{v_2 + v_3}\left(\frac{\partial w_3}{\partial a^2} - \frac{\partial w_2}{\partial a^3}\right)(\underline{i}_2\underline{i}_3 - \underline{i}_3\underline{i}_2) + \frac{p_3 + p_1}{v_3 + v_1}\left(\frac{\partial w_1}{\partial a^3} - \frac{\partial w_3}{\partial a^1}\right)(\underline{i}_3\underline{i}_1 - \underline{i}_1\underline{i}_3)\right].$$

If we now introduce a diagonal tensor $\underline{A}$ with components

$$A_1 = \frac{1}{\mu}\,\frac{p_2+p_3}{v_2+v_3}\,, \qquad A_2 = \frac{1}{\mu}\,\frac{p_3+p_1}{v_3+v_1}\,, \qquad A_3 = \frac{1}{\mu}\,\frac{p_1+p_2}{v_1+v_2} \tag{6}$$

and the vorticity vector

$$\overset{0}{\underline{\omega}} = \frac{1}{2}\,\overset{0}{\underline{\nabla}} \times \underline{w}\,, \qquad \overset{0}{\omega}_1 = \frac{1}{2}\left(\frac{\partial w_3}{\partial a^2} - \frac{\partial w_2}{\partial a^3}\right),$$

$$\overset{0}{\omega}_2 = \frac{1}{2}\left(\frac{\partial w_1}{\partial a^3} - \frac{\partial w_3}{\partial a^1}\right), \qquad \overset{0}{\omega}_3 = \frac{1}{2}\left(\frac{\partial w_2}{\partial a^1} - \frac{\partial w_1}{\partial a^2}\right). \tag{7}$$

$\dot{\underline{P}}$ may be written in the following form

$$\dot{\underline{P}} = \lambda \underline{\underline{E}}\,\overset{0}{\nabla} \cdot \underline{w} + 2\mu \overset{0}{\underline{\varepsilon}}(\underline{w}) +$$

$$+ \mu\,[A_1\overset{0}{\omega}_1(\underline{i}_2\underline{i}_3 - \underline{i}_3\underline{i}_2) + A_2\overset{0}{\omega}_2(\underline{i}_3\underline{i}_1 - \underline{i}_1\underline{i}_3) + A_3\overset{0}{\omega}_3(\underline{i}_1\underline{i}_2 - \underline{i}_2\underline{i}_1)] \tag{8}$$

or in the invariant form

$$\dot{\underline{P}} = \lambda \underline{\underline{E}}\,\overset{0}{\nabla} \cdot \underline{w} + 2\mu \overset{0}{\underline{\varepsilon}}(\underline{w}) - \mu \underline{E} \times \underline{A} \cdot \overset{0}{\underline{\omega}}\,, \tag{9}$$

which is independent of the coordinate system.

By (2.8) the quadratic form Φ may now be given by the expression

$$\Phi = \frac{1}{2}\,\dot{\underline{P}} \cdot\cdot\, \underline{\underline{\overset{0}{\nabla} w}}^{T} = \frac{1}{2}\,\lambda(\overset{0}{\underline{\nabla}} \cdot \underline{w})^2 + \mu I_1(\overset{0}{\underline{\varepsilon}}{}^2) + \mu(A_1\overset{0}{\omega}{}_1^2 + A_2\overset{0}{\omega}{}_2^2 + A_3\overset{0}{\omega}{}_3^2) =$$

$$= \frac{1}{2}\,\lambda(\overset{0}{\underline{\nabla}} \cdot \underline{w})^2 + \mu I_1(\overset{0}{\underline{\varepsilon}}{}^2) + \mu \underline{A} \cdot\cdot\, \underline{\underline{\overset{0}{\omega}\overset{0}{\omega}}}\,. \tag{10}$$

The equations of neutral equilibrium within the body and on its surface are as follows

$$\overset{0}{\underline{\nabla}} \cdot \dot{\underline{P}} = (\lambda + \mu)\,\underline{\underline{\overset{0}{\nabla}\overset{0}{\nabla}}} \cdot \underline{w} + \mu \overset{0}{\underline{\nabla}}{}^2 \underline{w} - \mu \overset{0}{\underline{\nabla}} \times (\underline{A} \cdot \overset{0}{\underline{\omega}}) = \underline{0}\,, \tag{11}$$

$$\lambda \underline{\underline{n\overset{0}{\nabla}}} \cdot \underline{w} + 2\mu \underline{n} \cdot \overset{0}{\underline{\varepsilon}}(\underline{w}) - \mu \underline{n} \times \underline{A} \cdot \overset{0}{\underline{\omega}} = \underline{0}\,. \tag{12}$$

Equations (11) in the form

$$\left(\frac{\lambda}{\mu} + 1\right)\frac{\partial \overset{0}{\underline{\nabla}} \cdot \underline{w}}{\partial a^1} + \overset{0}{\underline{\nabla}}{}^2 w_1 + A_2\,\frac{\partial \overset{0}{w}_2}{\partial a^3} - A_3\,\frac{\partial \overset{0}{w}_3}{\partial a^2} = 0 \quad \text{etc.} \tag{13}$$

were obtained by Southwell (R.V. Southwell, 1913) from other considerations. A generalization to a non-homogeneous state of stress in the V-configuration has been suggested by Biezeno and Hencky (C.B. Biezeno, H. Hencky, 1928-1929) and is reproduced in the book by C.B. Biezeno and R. Grammel, Technische Dynamik, Bd. I. - Springer-Verlag, Berlin, Göttingen, Heidelberg, 1953.

Note that the expressions (8) and (9) for the tensor $\dot{\underline{P}}$ differ from that for the stress tensor $\underline{T}$ of the linear theory of elasticity in the terms given by the curl of the vector $\underline{w}$. The quadratic form Φ also differs from the specific strain stored energy of a linearly elastic body in a similar fashion. Equations (11) and (12) of neutral equilibrium are converted into the homogeneous equations of linear theory when $\overset{\circ}{\underline{\nabla}} \times \underline{w} = \underline{0}$ in exactly the same way.

From Kirchhoff's theorem, which asserts the uniqueness of the solutions of the equations of equilibrium of the linear theory of elasticity, we may conclude that the equilibrium state in the V-configuration is stable with respect to irrotational virtual displacements. A non-trivial solution to the equations of neutral equilibrium may exist (with the same values of Lamé's moduli as in the linear theory) only if $\overset{\circ}{\underline{\nabla}} \times \underline{w} \neq \underline{0}$.

After making the substitution

$$\underline{\overset{\circ\circ}{\nabla\nabla}} \cdot \underline{w} - \overset{\circ}{\underline{\nabla}} \times (\overset{\circ}{\underline{\nabla}} \times \underline{w}) = \overset{\circ}{\underline{\nabla}}{}^2 \underline{w}$$

Equations (11) may be put into the form

$$\underline{\overset{\circ\circ}{\nabla\nabla}} \cdot \underline{w} - \overset{\circ}{\underline{\nabla}} \times \underline{B} \cdot (\overset{\circ}{\underline{\nabla}} \times \underline{w}) = \underline{0} \, , \qquad \underline{B} = \frac{\mu}{\lambda + 2\mu} (\underline{E} + \frac{1}{2} \underline{A}) \, . \tag{14}$$

The general solution to this system of equations may be represented as the vector sum*)

$$\underline{w} = \underline{w}' + \underline{w}'' \, . \tag{15}$$

Here $\underline{w}'$ can be represented as the gradient of a biharmonic scalar φ

$$\underline{w}' = (\underline{B} \cdot\cdot \underline{\overset{\circ\circ}{\nabla\nabla}} + B_1 B_2 B_3 \underline{B}^{-1} \cdot\cdot \underline{\overset{\circ\circ}{\nabla\nabla}}) \overset{\circ}{\underline{\nabla}} \varphi - \underline{B} \cdot \underline{\overset{\circ\circ}{\nabla\nabla}}{}^2 \varphi \, , \qquad \overset{\circ}{\underline{\nabla}}{}^4 \varphi = 0 \, , \tag{16}$$

and $\underline{w}''$ expressed by means of a solenoidal vector $\underline{q}$

$$\underline{w}'' = \underline{B} \cdot \overset{\circ}{\underline{\nabla}}{}^2 \underline{q} - \overset{\circ}{\underline{\nabla}} (\underline{B} \cdot \overset{\circ}{\underline{\nabla}}) \cdot \underline{q} \, , \qquad \overset{\circ}{\underline{\nabla}} \cdot \underline{q} = 0 \, , \qquad \underline{B}^{-1} \cdot\cdot \underline{\overset{\circ\circ\circ}{\nabla\nabla\nabla}}{}^2 \underline{q} = 0 \, . \tag{17}$$

---

*) A derivation of these relationships is given in Chapter IX, Par. 7.10 of the author's book "Theory of Elasticity" (in Russian), Nauka, Moscow, 1970. In Par. 7.13 of the same chapter these solutions have been applied to the problem of the axisymmetric neutral equilibrium of a hollow sphere compressed by an external pressure distributed uniformly over its outer surface. See also [8.6].

In Cartesian axes, the components of these vectors are given by the formulas

$$w_1' = [(B_1 + B_2 B_3)\partial_1^2 + (B_2 + B_3 B_1)\partial_2^2 + (B_3 + B_1 B_2)\partial_3^2 - B_1 \overset{0}{\nabla}{}^2]\partial_1 \varphi \qquad \left(\partial_s = \frac{\partial}{\partial a^s}\right),$$

$$w_1'' = B_1 \overset{0}{\nabla}{}^2 q_1 - \partial_1 (B_1 \partial_1 q_1 + B_2 \partial_2 q_2 + B_3 \partial_3 q_3)$$

etc.

12. APPLICATION TO THE PROBLEM OF THE STABILITY OF A COMPRESSED ROD

The state of stress in the V-configuration is characterised for this problem by the Piola tensor

$$\underline{P} = \sqrt{\frac{G}{g}}\, \underline{\nabla r}^{\mathrm{T}} \cdot \underline{T} = v_1^2 v_3 \underline{V}^{-1} \cdot \underline{i}_3 \underline{i}_3 \sigma_3 = v_1^2 \sigma_3 \underline{i}_3 \underline{i}_3, \quad p_1 = p_2 = 0, \quad p_3 = v_1^2 \sigma_3$$

and by (11.4) the components $v_1 = v_2$, v_3 of the tensor $\underline{V}$ are determined from the equations

$$2(\lambda + \mu) v_1 + \lambda v_3 = 3\lambda + 2\mu, \qquad 2\lambda v_1 + (\lambda + 2\mu) v_3 = 3\lambda + 2\mu + p_3 .$$

From these we find

$$v_1 = v_2 = 1 - \nu \frac{p_3}{E} = 1 - \nu \delta_3, \quad v_3 = 1 + \frac{p_3}{E} = 1 + \delta_3$$

$$\left(\nu = \frac{\lambda}{2(\lambda + \mu)}, \quad E = 2\mu(1 + \nu)\right),$$

where

$$\delta_3 = \frac{Q}{ES_0}, \qquad Q = \sigma_3 S = \sigma_3 v_1^2 S_0 = p_3 S_0 . \tag{1}$$

Here Q denotes a longitudinal force, δ_3 a stretch in the V-configuration and S and S_0 are the areas of the cross-sections in this and in the reference configuration, respectively.

By (11.6) we obtain

$$A_1 = A_2 = \frac{2(1 + \nu)\delta_3}{2 + (1 - \nu)\delta_3}, \qquad A_3 = 0 . \tag{2}$$

The expression (11.10) for the quadratic form Φ is now transformed to

$$\Phi = \frac{1}{2} \lambda I_1^2 (\underline{\overset{0}{\varepsilon}}(\underline{w})) + \mu I_1 (\underline{\overset{0}{\varepsilon}}{}^2 (\underline{w})) + \frac{2\mu(1 + \nu)\delta_3}{2 + (1 - \nu)\delta_3} (\overset{0}{\omega}{}_1^2 + \overset{0}{\omega}{}_2^2) . \tag{3}$$

The problem of the stability of a compressed circular cylinder has been considered by Sensenig (C.B. Sensenig, 1964) his investigation was based on the exact equations (11.14). The solution is of course very cumbersome, a value of the parameter at the bifurcation point is determined by a complicated equation and is expressed in terms of Bessel functions.*)

To estimate the critical load, it is natural to use some *a priori* assignment of the vector $\underline{w}$ together with the condition that the second variation of the functional (10.1) has a stationary value

$$W_2(\underline{w}) = \frac{1}{2}\iiint\limits_V \underline{\underline{\Theta}} \cdot\cdot \underline{\underline{\nabla w}}^T dV = \frac{1}{2}\iiint\limits_V \sqrt{\frac{g}{G}}\, \overset{0}{\underline{\underline{\nabla R}}}^T \cdot \underline{\dot{P}} \cdot\cdot \underline{\underline{\nabla w}}^T dV =$$

$$= \frac{1}{2}\iiint\limits_v \underline{\dot{P}} \cdot\cdot \overset{0}{\underline{\underline{\nabla w}}}^T dv = \iiint\limits_v \Phi\, dv \,. \tag{4}$$

It might well be foreseen that the critical load parameter so determined will exceed its exact value because the *a priori* specification of the vector $\underline{w}$ is equivalent to the superposition of constraints restricting the rod's deformation. It is therefore necessary to treat the value of the parameter found in this way with caution.

Let us assume for example that

$$w_1 = 0\,, \qquad w_2 = f(a^3)\,, \qquad w_3 = -a^2 f'(a^3)\,. \tag{5}$$

This assumption amounts to a lateral displacement (along the axis OY) of the axis of the rod combined with rotation of its cross-section (which preserves its shape) around the axis OX. We have then

$$\overset{0}{\varepsilon}_{11} = 0\,, \quad \overset{0}{\varepsilon}_{22} = 0\,, \quad \overset{0}{\varepsilon}_{33} = -a^2 f'(a^3)\,, \quad \overset{0}{\varepsilon}_{12} = \overset{0}{\varepsilon}_{23} = \overset{0}{\varepsilon}_{31} = 0\,,$$

$$\overset{0}{\omega}_1 = \frac{1}{2}\left(\frac{\partial w_3}{\partial a^2} - \frac{\partial w_2}{\partial a^3}\right) = -f'(a^3)\,, \qquad \overset{0}{\omega}_2 = \overset{0}{\omega}_3 = 0$$

and by (11.10), (4) and (3)

$$W_2 = \frac{1}{2}(\lambda + 2\mu)\iiint\limits_v [f''(a^3)a^2]^2 dv + \frac{2\mu(1+\nu)\delta_3}{2 + (1-\nu)\delta_3}\iiint\limits_v f'^2(a^3)dv =$$

*) This solution is reproduced in Chapter IX, Par. 7.11 of the author's "Theory of Elasticity". The representations (11.16), (11.17) have been used in this text.

$$= \int_0^l da^3 \left[\frac{1}{2} (\lambda + 2\mu) I_x g'^2 + \frac{Q}{2 + (1 - \nu)\delta_3} g^2 \right] . \tag{6}$$

In which

$$f'(a^3) = g(a^3) , \qquad I_x = \iint_{S_0} (a^2)^2 da^1 da^2 , \qquad Q = 2\mu(1 + \nu) S_0 \delta_3 .$$

The condition that the functional (6) has a stationary value is reduced to the form

$$\int_0^l \left[-(\lambda + 2\mu) I_x g'' + \frac{2Q}{2 + (1 - \nu)\delta_3} g \right] \delta g da^3 + g'(l)\delta g(l) - g'(0)\delta g(0) = 0$$

and for a cantilevered rod (for which $\delta g(0) = 0$ at the lower end and $\delta g(1)$ is arbitrary at the upper end) the homogeneous boundary value problem is thus reduced to the differential equation and the boundary condition

$$g'' + \frac{E}{\lambda + 2\mu} \frac{|Q|}{EI_x} g = 0 , \qquad g(0) = 0 , \qquad g'(l) = 0 , \tag{7}$$

where in the denominator of A_1 the small term $(1 - \nu)\delta_3 = (1 - \nu)Q/(ES_0)$ vanishes. A critical value of the compressing force is now given by the expression

$$|Q| = \frac{\pi^2 EI_x}{4l^2} \frac{\lambda + 2\mu}{E} = Q_E \frac{1 - \nu}{(1 - 2\nu)(1 + \nu)} = Q_E \left(1 - \frac{2\nu^2}{1 - \nu} \right)^{-1} > Q_E , \quad Q_E = \frac{\pi^2 EI_x}{4l^2} \tag{8}$$

this exceeds the Eulerian value for all $\nu \in (-1, 1/2)$ except $\nu = 0$. The same result is given by Pearson [8.11].

Following Pearson, we will prescribe the field of the vector $\underline{w}$ by the relationships

$$w_1 = -\nu a^1 a^2 f''(a^3) , \qquad w_2 = \frac{1}{2} \nu (a^{1^2} - a^{2^2}) f'' + f , \qquad w_3 = -a^2 f' \tag{9}$$

representing the displacements in the Saint-Venant bending problem by a moment m_x, the constant $m_x/(EI_x)$ being substituted in these formulas for the curvature $f''(a^3)$ of the rod. We now have

$$\overset{0}{\varepsilon}_{11} = -\nu a^2 f''(a^3) , \qquad \overset{0}{\varepsilon}_{22} = -\nu a^2 f'' , \qquad \overset{0}{\varepsilon}_{33} = -a^2 f'' ,$$

$$\overset{0}{\varepsilon}_{12} = 0 , \qquad \overset{0}{\varepsilon}_{23} = \frac{1}{4} \nu (a^{1^2} - a^{2^2}) f''' , \qquad \overset{0}{\varepsilon}_{31} = -\frac{1}{2} \nu a^1 a^2 f''' ,$$

$$\overset{0}{\omega}_1 = -\frac{1}{4}\nu(a^{1^2} - a^{2^2})f''' - f' , \qquad \overset{0}{\omega}_2 = -\frac{1}{2}\nu a^1 a^2 f''' , \qquad \overset{0}{\omega}_3 = 0$$

and expression (3) for the quadratic form is reduced to

$$\Phi = \mu(1+\nu)a^{2^2}f''^2 + \mu A f'^2 + \frac{1}{8}\nu^2\mu\left(1 + \frac{1}{2}A\right)(a^{1^2} + a^{2^2})^2 f'''^2 +$$

$$+ \frac{1}{2}\mu\nu A(a^{1^2} - a^{2^2})f'f''' . \tag{10}$$

If in this relationship the first and the second terms are retained and as before the parameter A is replaced by its approximate value $(1+\nu)\delta_3$ we would arrive at the Eulerian value for critical longitudinal force.

It would be natural here to neglect the third term in (10). This introduces a term proportional to

$$\iint_{S_0} (a^{1^2} + a^{2^2})^2 da^1 da^2 ,$$

into the representation of the functional W_2, the latter term being small for a thin rod. It would be less acceptable, however, to disregard the latter term in (10). If it is retained we obtain for W_2 an expression proportional to

$$\frac{1}{2}\int_0^l da^3\left[g'^2 - \frac{|Q|}{EI_x}g^2 - \frac{1}{2}\nu\frac{|\sigma_3|}{E}\left(\frac{I_y}{I_x} - 1\right)gg''\right] =$$

$$= \frac{1}{2}\int_0^l da^3\left\{\left[1 + \frac{1}{2}\nu\frac{|\sigma_3|}{E}\left(\frac{I_y}{I_x} - 1\right)\right]g'^2 - \frac{|Q|}{EI_x}g^2\right\} - \frac{1}{2}\nu\frac{|\sigma_3|}{E}\left(\frac{I_y}{I_x} - 1\right)gg'\Bigg|_0^l$$

$$\left(g = f',\ \sigma_3 = \frac{Q}{S_0}\right) ,$$

where it is to be assumed that $I_y/I_x > 1$ since the specification (9) of the vector $\underline{w}$ presupposes bending in the YZ-plane. The corresponding boundary value problem for a cantilevered rod

$$g''\left[1 + \frac{1}{2}\nu\frac{|\sigma_3|}{E}\left(\frac{I_y}{I_x} - 1\right)\right] + \frac{|Q|}{EI_x}g = 0 , \quad g(0) = 0 , \quad g'(l) = 0$$

leads to a critical value for the compressing force which is greater than the Eulerian value

$$Q = \left[1 + \frac{1}{2}\nu\frac{|\sigma_3|}{E}\left(\frac{I_y}{I_x} - 1\right)\right]Q_E .$$

13. THE SECURE LOADING. HOLDEN'S ESTIMATES

The purpose of the problem is to determine those loads which ensure that the integrand, a quadratic in Ψ, and also its integral (10.1) are positive. By definition (10.5), if this condition holds, it is impossible for the equilibrium state in the V-configuration to become unstable.

It is assumed that over the entire boundary surface of a body a "dead" loading is prescribed and that condition (1.21) holds, in other words, that, after superposition of the field of virtual displacements, the resultant moment of the external forces remains zero. This statement of the problem is due to Holden (J.T. Holden, 1964) and it has been developed by Beatty (M.F. Beatty, 1967-1971).

Following Holden, we resolve the vector $\underline{w}$ into the sum

$$\underline{w} = \underline{w}' + \underline{w}'' , \tag{1}$$

where the mean value of the vorticity of the first term throughout the volume is zero

$$\frac{1}{2}\iiint_V \underline{\nabla} \times \underline{w}' dV = \iiint_V \underline{\omega}' dV = \underline{0} \quad \text{or} \quad \iiint_V (\underline{\underline{\nabla w}}' - \underline{\underline{\nabla w}}'^{\mathrm{T}}) dV = \underline{0} . \tag{2}$$

The term $\underline{w}''$ represents a small rotation of a rigid body

$$\underline{\underline{\varepsilon}}(\underline{w}'') = \frac{1}{2}(\underline{\underline{\nabla w}}'' + \underline{\underline{\nabla w}}'^{\mathrm{T}}) = \underline{0} \quad \text{or} \quad \frac{\partial w''_s}{\partial x^t} = -\frac{\partial w''_t}{\partial x^s} . \tag{3}$$

We have for this decomposition

$$\underline{\underline{T}} \cdot \underline{\underline{\nabla w}} - \underline{\underline{\nabla w}}^{\mathrm{T}} \cdot \underline{\underline{T}} = \sum_{s=1}^{3} \sum_{t=1}^{3} \underline{e}_s \underline{e}_t \left(\sigma_s \frac{\partial w_t}{\partial x^s} - \sigma_t \frac{\partial w_s}{\partial x^t} \right) =$$

$$= \sum_s \sum_t \underline{e}_s \underline{e}_t \left[\left(\sigma_s \frac{\partial w'_t}{\partial x^s} - \sigma_t \frac{\partial w'_s}{\partial x^t} \right) + (\sigma_s + \sigma_t) \frac{\partial w''_t}{\partial x^s} \right] ,$$

where $\underline{e}_s$ denotes a trihedron of the principal axes of the stress tensor, w_t the components of $\underline{w}$ along those axes, and σ_s the principal stresses.

We will consider below the equilibrated V-configurations in a uniform stress field. Tensors $\underline{T}$ and $\underline{F}$ will be assumed constant within the volume. Equation (1.21) takes the form

$$\iiint_V (\underline{\underline{T}} \cdot \underline{\underline{\nabla w}} - \underline{\underline{\nabla w}}^{\mathrm{T}} \cdot \underline{\underline{T}}) dV = \sum_s \sum_t \underline{e}_s \underline{e}_t \left[(\sigma_s - \sigma_t) \iiint_V \frac{\partial w'_t}{\partial x^s} dV + V(\sigma_s + \sigma_t) \frac{\partial w''_t}{\partial x^s} \right] = \underline{0} , \tag{4}$$

because by (2) and (3)

$$\iiint_V \left(\frac{\partial w'_t}{\partial x^s} - \frac{\partial w'_s}{\partial x^t}\right) dV = 0 \ , \qquad \frac{\partial w''_t}{\partial x^s} = \text{const.} \tag{5}$$

For $\sigma_s + \sigma_t \neq 0$ we obtain Holden's formulas

$$\frac{\partial w''_s}{\partial x^t} = \frac{1}{V}\,\frac{\sigma_s - \sigma_t}{\sigma_s + \sigma_t} \iiint_V \frac{\partial w'_s}{\partial x^t}\, dV \qquad (s,t = 1,2,3) \ , \tag{6}$$

and for $\sigma_s + \sigma_t = 0$

$$\iiint_V \frac{\partial w'_t}{\partial x^s}\, dV = 0 \ . \tag{7}$$

The Equation (1.21) implies the relationship

$$\iiint_V (\underline{T} \cdot \underline{\underline{\nabla w}} - \underline{\underline{\nabla w}}^T \cdot \underline{T}) \cdot\cdot\, \underline{\underline{\nabla w''}}^T dV = \iiint_V (\underline{T} \cdot\cdot\, \underline{\underline{\nabla w}} \cdot \underline{\underline{\nabla w''}}^T - \underline{T} \cdot\cdot\, \underline{\underline{\nabla w''}}^T \cdot \underline{\underline{\nabla w}}^T)\, dV =$$

$$= 2 \iiint_V (\underline{T} \cdot\cdot\, \underline{\underline{\nabla w'}} \cdot \underline{\underline{\nabla w''}}^T + \underline{T} \cdot\cdot\, \underline{\underline{\nabla w''}} \cdot \underline{\underline{\nabla w''}}^T)\, dV = 0 \ . \tag{8}$$

Here we have made use of Equations (2) and (3). We now turn to the transformation of the first term of the quadratic form (2.11). Bearing (8) in mind, we have

$$\frac{1}{2} \iiint_V \underline{T} \cdot\cdot\, \underline{\underline{\nabla w}} \cdot \underline{\underline{\nabla w}}^T dV = \frac{1}{2} \iiint_V \underline{T} \cdot\cdot\, \underline{\underline{\nabla w'}} \cdot \underline{\underline{\nabla w'}}^T dV +$$

$$+ \iiint_V \underline{T} \cdot\cdot\, \underline{\underline{\nabla w'}} \cdot \underline{\underline{\nabla w''}}^T dV + \frac{1}{2} \iiint_V \underline{T} \cdot\cdot\, \underline{\underline{\nabla w''}} \cdot \underline{\underline{\nabla w''}}^T dV =$$

$$= \frac{1}{2} \left(\iiint_V \underline{T} \cdot\cdot\, \underline{\underline{\nabla w'}} \cdot \underline{\underline{\nabla w'}}^T dV + \iiint_V \underline{T} \cdot\cdot\, \underline{\underline{\nabla w'}} \cdot \underline{\underline{\nabla w''}}^T dV \right) \ . \tag{9}$$

These terms are evaluated below. Turning to (6) we get

$$\iiint_V \underline{T} \cdot\cdot\, \underline{\underline{\nabla w'}} \cdot \underline{\underline{\nabla w''}}^T dV = \sum_t \sum_s \sigma_t \frac{\partial w''_s}{\partial x^t} \iiint_V \frac{\partial w'_s}{\partial x^t}\, dV =$$

$$= (\sigma_1 - \sigma_2) \frac{\partial w''_2}{\partial x^1} \iiint_V \frac{\partial w'_2}{\partial x^1}\, dV + (\sigma_2 - \sigma_3) \frac{\partial w''_3}{\partial x^2} \iiint_V \frac{\partial w'_3}{\partial x^2}\, dV +$$

$$+ (\sigma_3 - \sigma_1) \frac{\partial w''_1}{\partial x^3} \iiint_V \frac{\partial w'_1}{\partial x^3} dV = -\frac{1}{V}\left[\frac{(\sigma_1 - \sigma_2)^2}{\sigma_1 + \sigma_2}\left(\iiint_V \frac{\partial w'_1}{\partial x^2} dV\right)^2 + \right.$$

$$\left. + \frac{(\sigma_2 - \sigma_3)^2}{\sigma_2 + \sigma_3}\left(\iiint \frac{\partial w'_2}{\partial x^3} dV\right)^2 + \frac{(\sigma_3 - \sigma_1)^2}{\sigma_3 + \sigma_1}\left(\iiint \frac{\partial w'_3}{\partial x^1} dV\right)^2\right] . \qquad (10)$$

The nonnegative quantity σ is given by

$$0 \le \sigma = \text{Max}\left\{\frac{(\sigma_1 - \sigma_2)^2}{\sigma_1 + \sigma_2}, \frac{(\sigma_2 - \sigma_3)^2}{\sigma_2 + \sigma_3}, \frac{(\sigma_3 - \sigma_1)^2}{\sigma_3 + \sigma_1}, 0\right\} \qquad (11)$$

and use the Schwartz inequality

$$\iiint_V \left(\frac{\partial w'_s}{\partial x^t}\right)^2 dV \ge \frac{1}{V}\left(\iiint_V \frac{\partial w'_s}{\partial x^t} dV\right)^2 .$$

By (10) and (11) we arrive at the relationship

$$\iiint_V \underline{T} \cdot\cdot \underline{\underline{\nabla w}}' \cdot \underline{\underline{\nabla w}}''^{T} dV \ge -\sigma \iiint_V \left[\left(\frac{\partial w'_1}{\partial x^2}\right)^2 + \left(\frac{\partial w'_2}{\partial x^3}\right)^2 + \left(\frac{\partial w'_3}{\partial x^1}\right)^2\right] dV . \qquad (12)$$

In the calculation, which is based on Equation (6), we may substitute

$$\iiint_V \frac{\partial w'_t}{\partial x^s} dV \quad \text{for} \quad \iiint_V \frac{\partial w'_s}{\partial x^t} dV$$

and Inequality (12) may thus be expressed in the following form

$$\iiint \underline{T} \cdot\cdot \underline{\underline{\nabla w}}' \cdot \underline{\underline{\nabla w}}''^{T} dV \ge -\frac{1}{2}\sigma \iiint \left[\left(\frac{\partial w'_1}{\partial x^2}\right)^2 + \left(\frac{\partial w'_2}{\partial x^1}\right)^2 + \right.$$

$$\left. + \left(\frac{\partial w'_2}{\partial x^3}\right)^2 + \left(\frac{\partial w'_3}{\partial x^2}\right)^2 + \left(\frac{\partial w'_3}{\partial x^1}\right)^2 + \left(\frac{\partial w'_1}{\partial x^3}\right)^2\right] dV . \qquad (13)$$

We now turn to the first term in (9), Let $\sigma_1 \ge \sigma_2 \ge \sigma_3$, then

$$\underline{T} \cdot\cdot \underline{\underline{\nabla w}}' \cdot \underline{\underline{\nabla w}}'^{T} = \sum_s \sum_q \sigma_s \left(\frac{\partial w'_q}{\partial x^s}\right)^2 = \sigma_1 \sum_q \left(\frac{\partial w'_q}{\partial x^1}\right)^2 + \sigma_2 \sum_q \left(\frac{\partial w'_q}{\partial x^2}\right)^2 + \sigma_3 \sum_q \left(\frac{\partial w'_q}{\partial x^3}\right)^2 \ge$$

$$\ge \sigma_3 \sum_{q=1}^{3} \left[\left(\frac{\partial w'_q}{\partial x^1}\right)^2 + \left(\frac{\partial w'_q}{\partial x^2}\right)^2 + \left(\frac{\partial w'_q}{\partial x^3}\right)^2\right] \ge$$

$$\geq \sigma_3 \left[\left(\frac{\partial w_3'}{\partial x^1} \right)^2 + \left(\frac{\partial w_1'}{\partial x^2} \right)^2 + \left(\frac{\partial w_2'}{\partial x^3} \right)^2 \right]$$

and by virtue of the above observation,

$$\iiint_V \underline{T} \cdot\cdot \underline{\underline{\nabla w}}' \cdot \underline{\underline{\nabla w}}'^{T} dV \geq \frac{1}{2} \sigma_3 \iiint_V \left[\left(\frac{\partial w_1'}{\partial x^3} \right)^2 + \left(\frac{\partial w_3'}{\partial x^1} \right)^2 + \left(\frac{\partial w_2'}{\partial x^1} \right)^2 + \right.$$

$$\left. + \left(\frac{\partial w_1'}{\partial x^2} \right)^2 + \left(\frac{\partial w_3'}{\partial x^2} \right)^2 + \left(\frac{\partial w_2'}{\partial x^3} \right)^2 \right] dV \ . \tag{14}$$

Bearing in mind the identity

$$\underline{\underline{\nabla w}} \cdot\cdot \underline{\underline{\nabla w}}^{T} = I_1(\underline{\varepsilon}^2) + 2\underline{\omega} \cdot \underline{\omega} \ ,$$

which follows immediately from (1.7.10), we have

$$I_1(\underline{\varepsilon}^2) + \underline{\omega} \cdot \underline{\omega} = \underline{\underline{\nabla w}} \cdot\cdot \underline{\underline{\nabla w}}^{T} - \underline{\omega} \cdot \underline{\omega} = \sum_s \sum_q \left[\frac{3}{4} \left(\frac{\partial w_q}{\partial x^s} \right)^2 + \frac{1}{4} \frac{\partial w_q}{\partial x^s} \frac{\partial w_s}{\partial x^q} \right] =$$

$$= \varepsilon_{11}^2 + \varepsilon_{22}^2 + \varepsilon_{33}^2 + \varepsilon_{12}^2 + \varepsilon_{23}^2 + \varepsilon_{31}^2 +$$

$$+ \frac{1}{2} \left[\left(\frac{\partial w_1}{\partial x^3} \right)^2 + \left(\frac{\partial w_3}{\partial x^1} \right)^2 + \left(\frac{\partial w_2}{\partial x^1} \right)^2 + \left(\frac{\partial w_1}{\partial x^2} \right)^2 + \left(\frac{\partial w_3}{\partial x^2} \right)^2 + \left(\frac{\partial w_2}{\partial x^3} \right)^2 \right]$$

and therefore

$$\frac{1}{2} \iiint \left[\left(\frac{\partial w_1'}{\partial x^3} \right)^2 + \left(\frac{\partial w_3'}{\partial x^1} \right)^2 + \left(\frac{\partial w_2'}{\partial x^1} \right)^2 + \left(\frac{\partial w_1'}{\partial x^2} \right)^2 + \left(\frac{\partial w_3'}{\partial x^2} \right)^2 + \left(\frac{\partial w_2'}{\partial x^3} \right)^2 \right] dV \leq$$

$$\leq \iiint [I_1(\underline{\varepsilon}^2) + \underline{\omega} \cdot \underline{\omega}] \, dV \ . \tag{15}$$

14. KORN'S INEQUALITY

In treating the problem of the existence of solutions to the equations of the linear theory of elasticity, Korn (A. Korn, 1908) made use of the inequality

$$K \iiint_V I_1(\underline{\varepsilon}^2) dV \geq \iiint_V \underline{\omega} \cdot \underline{\omega} \, dV \qquad \left(\underline{\omega} = \frac{1}{2} \underline{\nabla} \times \underline{w},\ \underline{\varepsilon} = \frac{1}{2} (\underline{\underline{\nabla w}} + \underline{\underline{\nabla w}}^{T}) \right) \ , \tag{1}$$

which holds provided that the mean value of the vorticity of the displacement vector is zero

$$\iiint_V \underline{\omega}\, dV = \underline{0} \,. \tag{2}$$

The Korn constant K depends only upon the geometry of the volume V, its surface being not necessarily smooth, i.e. it may possess edges and corners. An exact value of K is known only for a sphere [8.14]. Bernstein and Toupin have suggested a lower limit for K having specified the vector $\underline{w}$ in the form

$$\underline{w} = \underline{e} \cdot \underline{Re} \times \underline{R} \,, \tag{3}$$

where $\underline{e}$ denotes some constant unit vector. For such a prescription,

$$\underline{\nabla w} = \underline{ee} \times \underline{R} - \underline{e} \cdot \underline{RE} \times \underline{e} \,, \qquad \underline{\nabla w}^T = \underline{e} \times \underline{Re} + \underline{e} \cdot \underline{RE} \times \underline{e} \,,$$

$$\underline{\nabla w} \cdot\cdot \underline{\nabla w} = \underline{\nabla w}^T \cdot\cdot \underline{\nabla w}^T = (\underline{e} \cdot \underline{R})^2 \underline{E} \times \underline{e} \cdot\cdot \underline{E} \times \underline{e} = -2(\underline{e} \cdot \underline{R})^2 \,,$$

$$\underline{\nabla w} \cdot\cdot \underline{\nabla w}^T = \underline{\nabla w}^T \cdot\cdot \underline{\nabla w} = |\underline{e} \times \underline{R}|^2 - (\underline{e} \cdot \underline{R})^2 \underline{E} \times \underline{e} \cdot\cdot \underline{E} \times \underline{e} =$$

$$= |\underline{e} \times \underline{R}|^2 + 2(\underline{e} \cdot \underline{R})^2 = R^2 + (\underline{e} \cdot \underline{R})^2$$

and further

$$(\underline{\nabla w} + \underline{\nabla w}^T) \cdot\cdot (\underline{\nabla w} + \underline{\nabla w}^T) = 2\,[R^2 - (\underline{e} \cdot \underline{R})^2] \,,$$

$$(\underline{\nabla w}^T - \underline{\nabla w}) \cdot\cdot (\underline{\nabla w} - \underline{\nabla w}^T) = 2\,[R^2 + 3(\underline{e} \cdot \underline{R})^2] = 2\,[R^2 - (\underline{e} \cdot \underline{R})^2] + 8(\underline{e} \cdot \underline{R})^2 \,.$$

Vector (3) also satisfies Equation (2) and in fact, we have

$$\underline{\nabla w}^T - \underline{\nabla w} = 2\underline{E} \times \underline{ee} \cdot \underline{R} \,, \qquad \frac{1}{2} \iiint_V (\underline{\nabla w}^T - \underline{\nabla w})\, dV = \underline{E} \times \underline{e} \iiint_V \underline{e} \cdot \underline{R}\, dV = \underline{0} \,,$$

provided that the origin is placed at the centre of mass of the volume.

Now, representing Korn's inequality in the form

$$K \iiint (\underline{\nabla w} + \underline{\nabla w}^T) \cdot\cdot (\underline{\nabla w} + \underline{\nabla w}^T)\, dV \geq \iiint_V (\underline{\nabla w}^T - \underline{\nabla w}) \cdot\cdot (\underline{\nabla w} - \underline{\nabla w}^T)\, dV \,,$$

we arrive at the relationship

$$K \geq 1 + \frac{4 \iiint_V (\underline{e} \cdot \underline{R})^2 dV}{\iiint_V R^2 dV - \iiint_V (\underline{e} \cdot \underline{R})^2 dV} = 1 + \frac{4\underline{e} \cdot \underline{\theta} \cdot \underline{e}}{I_1(\underline{\theta}) - \underline{e} \cdot \underline{\theta} \cdot \underline{e}} \,. \tag{4}$$

Here $\underline{\theta}$ denotes the tensor of the principal moments of inertia of the volume calculated with respect to the coordinate planes

$$\underline{\theta} = \sum_s \sum_t \theta_{st} \underline{i}_s \underline{i}_t , \qquad \theta_{st} = \iiint_V x^s x^t dV .$$

The ratio of the quadratic forms in the right-hand side of (4) reaches a maximum at the same time as does the quadratic form in the numerator. The latter maximum is attained for the direction $\underline{e}$ corresponding to the maximum eigenvalue θ_3 of the tensor $\underline{\theta}$. We thus have

$$K \geq 1 + \frac{4\theta_3}{\theta_1 + \theta_2} \qquad (I(\underline{\theta}) = \theta_1 + \theta_2 + \theta_3) . \tag{5}$$

The denominator reaches a maximum for $\theta_1 = \theta_2$ so that

$$K \geq 1 + \frac{2\theta_3}{\theta_1}$$

and this value is a minimum for $\theta_3 = \theta_1$. Thus

$$K \geq 3 , \tag{6}$$

which inequality determines the lower limit for the Korn constant.

For a circular cylinder of radius r and length l the principal moments of inertia with respect to the longitudinal and transverse axes are

$$\frac{\pi}{4} l r^4 , \qquad \frac{\pi}{12} l^3 r^2 ,$$

so that

$$\left(\frac{l}{r}\right)^2 > 3 , \quad \theta_3 = \frac{\pi}{12} l^3 r^2 , \quad \theta_1 + \theta_2 = \frac{\pi}{2} l r^4 ;$$

$$\left(\frac{l}{r}\right)^2 < 3 , \quad \theta_3 = \frac{\pi}{4} l r^4 , \quad \theta_1 + \theta_2 = \frac{\pi}{4} l r^4 + \frac{\pi}{12} l^3 r^2$$

and by (6)

$$K \geq \begin{cases} 1 + \dfrac{2}{3}\left(\dfrac{l}{r}\right)^2 & \text{and} \quad \left(\dfrac{l}{r}\right)^2 > 3 , \\[2ex] 1 + \dfrac{4}{1 + \dfrac{1}{3}\left(\dfrac{l}{r}\right)^2} & \text{and} \quad \left(\dfrac{l}{r}\right)^2 < 3 . \end{cases} \tag{7}$$

Note that for a circular disk ($l \to 0$) this formula shows that $K \geq 5$ and according to Payne and Weinberger, this is the best possible value for a plane domain.

The advantages of form (3) for the specification of $\underline{w}$ also lies in the fact that the quantities entering Inequality (5) are expressed directly in terms of the geometrical properties of the volume.

We note also that Korn's inequality (1) may equally be represented in the form

$$(K+1) \iiint_V I_1(\underline{\varepsilon}^2)\,dV \geq \iiint_V [\underline{\omega} \cdot \underline{\omega} + I_1(\underline{\varepsilon}^2)]\,dV \,. \tag{8}$$

15. THE INEQUALITIES OF HOLDEN AND BEATTY

Summing up the Inequalities (13.13) and (13.14) and making use of (13.9) we arrive at the inequality

$$\iiint \underline{T} \cdot\cdot \underline{\underline{\nabla w}} \cdot \underline{\underline{\nabla w}}^T dV \geq (\sigma_3 - \sigma)A \,,$$

$$A = \frac{1}{2} \iiint_V \left[\left(\frac{\partial w_1'}{\partial x^2}\right)^2 + \left(\frac{\partial w_2'}{\partial x^1}\right)^2 + \left(\frac{\partial w_2'}{\partial x^3}\right)^2 + \left(\frac{\partial w_3'}{\partial x^2}\right)^2 + \left(\frac{\partial w_3'}{\partial x^1}\right)^2 + \left(\frac{\partial w_1'}{\partial x^3}\right)^2\right] dV \,. \tag{1}$$

At the same time, bearing (14.8) in mind, the Inequality (13.15) may be strengthened if we rewrite it in the form

$$(K+1) \iiint_V I_1(\underline{\varepsilon}^2)\,dV \geq A \,. \tag{2}$$

Assuming the natural state (it must be recalled that σ_3 denotes the least principal stress)

$$\sigma_3 - \sigma \leq 0 \tag{3}$$

and comparing Inequalities (1) and (2) we deduce the strengthened inequality (Holden)

$$\iiint_V \underline{T} \cdot\cdot \underline{\underline{\nabla w}} \cdot \underline{\underline{\nabla w}}^T dV \geq (\sigma_3 - \sigma)(K+1) \iint_V I_1(\underline{\varepsilon}^2)\,dV \tag{4}$$

here the right-hand side of (1) has been substituted by a smaller quantity (by virtue of (3)).

We now proceed to an evaluation of the second group of terms in (2.18). We have

$$I_1((\underline{F}\cdot\underline{\varepsilon})^2) = \sum_s\sum_t v_s^4\varepsilon_{st}\varepsilon_{ts} \le V^4\sum_s\sum_t \varepsilon_{st}\varepsilon_{ts} = V^4 I_2(\underline{\varepsilon}^2) ,$$

where

$$V^2 = \mathrm{Max}(v_1^2, v_2^2, v_3^2) . \tag{5}$$

We have thus arrived at the inequality

$$I_3\frac{\partial\Im}{\partial I_3} I_1(\underline{\varepsilon}^2) + \frac{\partial\Im}{\partial I_2} I_1((\underline{F}\cdot\underline{\varepsilon})^2) \le \left(I_3\frac{\partial\Im}{\partial I_3} + V^4\left|\frac{\partial\Im}{\partial I_2}\right|\right) I_1(\underline{\varepsilon}^2) . \tag{6}$$

With the notation

$$C(\varepsilon_1,\varepsilon_2,\varepsilon_3) = 4\sqrt{\frac{g}{G}}\sum_s\sum_k \alpha_{sk}\varepsilon_s\varepsilon_k \tag{7}$$

we now obtain by (2.18), (4) and (6)

$$2\Psi = \underline{T}\cdot\cdot\underline{\nabla w}\cdot\underline{\nabla w}^{\mathrm{T}} - 4\sqrt{\frac{g}{G}}\left[I_3\frac{\partial\Im}{\partial I_3} I_1(\underline{\varepsilon}^2) + \frac{\partial\Im}{\partial I_2} I_1((\underline{F}\cdot\underline{\varepsilon})^2)\right] + C(\varepsilon_1,\varepsilon_2,\varepsilon_3) \ge$$

$$\ge \left[(K+1)(\sigma_3-\sigma) - 4\sqrt{\frac{g}{G}}\left(I_3\frac{\partial\Im}{\partial I_3} + V^4\left|\frac{\partial\Im}{\partial I_2}\right|\right)\right] I_1(\underline{\varepsilon}^2) + C(\varepsilon_1,\varepsilon_2,\varepsilon_3) =$$

$$= BI_1(\underline{\varepsilon}^2) + C(\varepsilon_1,\varepsilon_2,\varepsilon_3) . \tag{8}$$

Beatty's constant, B, is given by

$$B = (K+1)(\sigma_3-\sigma) - 4\sqrt{\frac{g}{G}}\left(I_3\frac{\partial\Im}{\partial I_3} + V^4\left|\frac{\partial\Im}{\partial I_2}\right|\right) =$$

$$= (K+1)(\sigma_3-\sigma) + 2(\hat{\mu}_V-\sigma_V) + 4\sqrt{\frac{g}{G}}\,V^4\left(\frac{\partial\Im}{\partial I_2} - \left|\frac{\partial\Im}{\partial I_2}\right|\right) , \tag{9}$$

see Equation (2.20), and the subscript V indicates the subscript given in Equation (5).

The sufficient condition

$$BI_1(\varepsilon^2) + C(\varepsilon_1,\varepsilon_2,\varepsilon_3) > 0 \tag{10}$$

for the atability of the equilibrium state has been substitúted by Beatty by the two conditions

$$B \geq 0 \ , \qquad C(\varepsilon_1, \varepsilon_2, \varepsilon_3) > 0 \tag{11}$$

considered separately.

This substitution obviously guarantees that Inequality (10) holds. The safe load is determined below by the condition

$$B = 0 \ ,$$

in order to simplify the calculation associated with the determination of the sign of the quadratic form (7) it is proposed that this sign be tested for positiveness only under condition (12) which is assumed to hold for the material considered, i.e. for the usual form of the specific strain energy $\text{Э}(I_1, I_2, I_3)$. It is obvious from the definitions (2.15) of coefficients α_{sk} of the form $C(\varepsilon_1, \varepsilon_2, \varepsilon_3)$ that its sign depends only on this prescription and may be positive for one material and negative for another. In the latter case, the substitution of the single Inequality (10) for the two Inequalities (11) ceases to be acceptable.

Note. By virtue of (2.22) and (2.23), the quadratic form (7) can be represented as

$$C(\varepsilon_1, \varepsilon_2, \varepsilon_3) = I_3 \frac{\partial \text{Э}}{\partial I_3} I_1(\underline{\varepsilon}^2) + \frac{\partial \text{Э}}{\partial I_2} \sum_{s=1}^{3} v_s^4 \varepsilon_s^2 + \sum_{s=1}^{3} \sum_{k=1}^{3} \frac{\partial^2 \text{Э}}{\partial v_s^2 \partial v_k^2} v_s^2 v_k^2 \varepsilon_s \varepsilon_k \ , \tag{13}$$

$$\begin{aligned} C(\varepsilon_1, \varepsilon_2, \varepsilon_3) &= I_3 \frac{\partial \text{Э}}{\partial I_3} I_1(\underline{\varepsilon}^2) + \frac{\partial \text{Э}}{\partial I_2} \sum_{s=1}^{3} v_s^4 \varepsilon_s^2 + \\ &\quad + \frac{1}{4} \sum_{s=1}^{3} \sum_{k=1}^{3} \frac{\partial^2 \text{Э}}{\partial v_s \partial v_k} v_s \varepsilon_s v_k \varepsilon_k - \frac{1}{2} \sum_{s=1}^{3} \frac{\partial^2 \text{Э}}{\partial v_s^2} v_s^2 \varepsilon_s^2 = \\ &= I_3 \frac{\partial \text{Э}}{\partial I_3} I_1(\underline{\varepsilon}^2) + \frac{\partial \text{Э}}{\partial I_2} \sum_{s=1}^{3} v_s^4 \varepsilon_s^2 + \frac{1}{4} \sum_{s=1}^{3} \sum_{k=1}^{3} \frac{\partial^2 \text{Э}}{\partial v_s \partial v_k} v_s \varepsilon_s v_k \varepsilon_k - \\ &\quad - \frac{1}{4} \sqrt{\frac{G}{g}} \sum_{k=1}^{3} \sigma_s \varepsilon_s^2 \ , \end{aligned} \tag{14}$$

and this allows us to write (8) in one of the following forms

$$\begin{aligned} 2\Psi &= \underline{T} \cdot\cdot \underline{\nabla w} \cdot \underline{\nabla w}^{\mathrm{T}} + 4 \sqrt{\frac{g}{G}} \frac{\partial \text{Э}}{\partial I_2} \left[\sum_{s=1}^{3} v_s^4 \varepsilon_s^2 - I_1((\underline{F} \cdot \underline{\varepsilon})^2) \right] + \\ &\quad + 4 \sqrt{\frac{g}{G}} \sum_{s=1}^{3} \sum_{k=1}^{3} \frac{\partial^2 \text{Э}}{\partial v_s^2 \partial v_k^2} v_s^2 v_k^2 \varepsilon_s \varepsilon_k \ , \end{aligned} \tag{15}$$

$$2\Psi = \underline{T} \cdot\cdot \underline{\nabla w} \cdot \underline{\nabla w}^T + 4\sqrt{\frac{g}{G}}\, \frac{\partial \ni}{\partial I_2} \left[\sum_{s=1}^{3} v_s^4 \varepsilon_s^2 - I_1((\underline{F} \cdot \underline{\varepsilon})^2) \right] -$$

$$- \sum_{s=1}^{3} \sigma_s \varepsilon_s^2 + \sqrt{\frac{g}{G}} \sum_{s=1}^{3} \sum_{k=1}^{3} \frac{\partial^2 \ni}{\partial v_s \partial v_k} v_s \varepsilon_s v_k \varepsilon_k ; \tag{16}$$

in these formulas,

$$\sum_{s=1}^{3} [v_s^4 \varepsilon_s^2 - I_1((\underline{F} \cdot \underline{\varepsilon})^2)] = \sum_{s=1}^{3} v_s^4 \varepsilon_s^2 - \sum_{s=1}^{3} \sum_{t=1}^{3} v_s^2 v_t^2 \varepsilon_{st} \varepsilon_{ts} =$$

$$= -2(v_1^2 v_2^2 \varepsilon_{12}^2 + v_2^2 v_3^2 \varepsilon_{23}^2 + v_3^2 v_1^2 \varepsilon_{31}^2) < 0 . \tag{17}$$

Provided that a material satisfies the monotonicity criterion (5.10.13), inequality

$$\sum_{s=1}^{3} \sum_{k=1}^{3} \frac{\partial^2}{\partial v_s \partial v_k} v_s \varepsilon_s v_k \varepsilon_k \geq 0 \tag{18}$$

holds.

The sufficient condition for stability is now expressed by any of the inequalities

$$(K+1)(\sigma_3 - \sigma) I_1(\underline{\varepsilon}^2) - 4\sqrt{\frac{g}{G}} \sum_{s=1}^{3} \sum_{k=1}^{3} \left[\frac{\partial^2 \ni}{\partial v_s^2 \partial v_k^2} v_s^2 v_k^2 \varepsilon_s \varepsilon_k - \right.$$

$$\left. - 2 \frac{\partial \ni}{\partial I_2} (v_1^2 v_2^2 \varepsilon_{12}^2 + v_2^2 v_3^2 \varepsilon_{23}^2 + v_3^2 v_1^2 \varepsilon_{31}^2) \right] > 0 , \tag{19}$$

$$(K+1)(\sigma_3 - \sigma) I_1(\underline{\varepsilon}^2) - 8\sqrt{\frac{g}{G}} \frac{\partial \ni}{\partial I_2} (v_1^2 v_2^2 \varepsilon_{12}^2 + v_2^2 v_3^2 \varepsilon_{23}^2 + v_3^2 v_1^2 \varepsilon_{31}^2) - \sum_{s=1}^{3} \sigma_s \varepsilon_s^2 > 0 , \tag{20}$$

where in the latter Inequality (18) has been taken into account.

When $\partial \ni / \partial I_2 > 0$, both Inequalities (19) and (20) are violated. It is sufficient to assume that $\varepsilon_s = 0$ for all $s = 1,2,3$ and that $\varepsilon_{sk} \neq 0$ for some ε_{sk} $(s \neq k)$. We are therefore forced to abandon the attempt to evaluate Inequality (10) directly and instead to require initially that B be nonnegative and that it be subsequently verified that $C(\varepsilon_1, \varepsilon_2, \varepsilon_3)$ is positive.

16. THE COMPRESSED ROD (EULER'S COLUMN). MURNAGHAN'S MATERIAL

The principal stresses arising within a rod compressed by a force Q are

$$\sigma_1 = 0 , \quad \sigma_2 = 0 , \quad \sigma_3 = -\frac{Q}{S} \tag{1}$$

and by (13.11)

$$\sigma = 0 . \tag{2}$$

Using the notation of Chapter 6, Par. 2 we have by (15.5) and (2.21)

$$v_1 = v_2 = \alpha v , \quad v_3 = v , \quad \alpha > 1 , \quad V = v_1 ; \qquad \sqrt{I_3} = \alpha^2 v^3 , \tag{3}$$

$$\hat{\mu}_1 = 2v^{-1}\left(\frac{\partial \ni}{\partial I_1} + v^2 \frac{\partial \ni}{\partial I_2}\right) = -2\alpha^2 v\left(\frac{\partial \ni}{\partial I_2} + v^2 \frac{\partial \ni}{\partial I_3}\right) . \tag{4}$$

From here and from Equation (5.6) which represents the stress σ_3 we can calculate the values of α, v and $\hat{\mu}_1$.

We assume that for this material at hand

$$\frac{\partial \ni}{\partial I_2} > 0 \tag{5}$$

and the Beatty criterion leads, by (15.9) and (15.11), to the inequality

$$Q \leq \frac{2\hat{\mu}_1 S}{K+1} . \tag{6}$$

If the V-configuration is the natural configuration and S_0 denotes the cross-sectional area in this configuration, then

$$\hat{\mu}_1 = \mu , \quad S = S_0 , \quad Q \leq \frac{2\mu S_0}{K+1} . \tag{7}$$

For a rod of circular cross-section with the ratio $l/r > \sqrt{3}$, the safe load is by (4.7) given by

$$Q = \frac{\mu S_0}{1 + \frac{1}{3}\left(\frac{l}{r}\right)^2} . \tag{8}$$

It is interesting to compare this expression with the Eulerian value

$$Q_E = \frac{\pi^2 E S_0 r^2}{4l^2} \tag{9}$$

for a rod whose ends are prevented from rotating and a field $\underline{w}$ of virtual displacements is superposed ("clamped ends"). For such a construction Equation (1.21) holds, the latter being effectively used in the derivation of Holden's inequality (15.4). From (8) and (9) we have

$$\frac{Q}{Q_E} = \frac{6}{\pi^2(1+\nu)} \tag{10}$$

this ratio being approximately equal to 0.6-0.4 for $0 < \nu < 1/2$.

a) *Murnaghan's material.* By (5.3.2) we have

$$\frac{\partial \ni}{\partial I_1} = \frac{1}{4}\left[\lambda(I_1 - 3) + 2\mu(I_1 - 1) + \frac{1}{2}\,l(I_1 - 3)^2 + m(I_1^2 - 2I_1 - I_2) + \frac{n}{2}\right],$$

$$\frac{\partial \ni}{\partial I_2} = -\frac{1}{4}\left[2\mu + m(I_1 - 3) + \frac{1}{2}\,n\right], \qquad \frac{\partial \ni}{\partial I_3} = \frac{1}{8}\,n,$$

and the second derivatives

$$\frac{\partial^2 \ni}{\partial I_1^2} = \frac{1}{4}\left[\lambda + 2\mu + l(I_1 - 3) + 2m(I_1 - 1)\right],$$
$$\frac{\partial^2 \ni}{\partial I_1 \partial I_2} = -\frac{1}{4}\,m. \tag{12}$$

are different from zero.

In the following calculation we confine ourselves to terms which are linear with respect to the stretches-longitudinal ($-\delta$) and transverse (δ_1) ($\delta_1 > 0$, $\delta > 0$) of a compressed rod. To this approximation

$$v_1 = v_2 = 1 + \delta_1, \quad v = 1 - \delta; \qquad v_1^2 = v_2^2 = 1 + 2\delta_1, \quad v^2 = 1 - 2\delta,$$

$$I_1(\underline{F}) = I_1 - 3 = 2(2\delta_1 - \delta), \quad I_2 - 3 = 4(2\delta_1 - \delta), \quad I_3 - 1 = 2(2\delta_1 - \delta). \tag{13}$$

From the condition

$$\sigma_1 = 2\sqrt{\frac{g}{G}}\left[I_3\,\frac{\partial \ni}{\partial I_3} + \left(\frac{\partial \ni}{\partial I_1} + I_1\,\frac{\partial \ni}{\partial I_2}\right)v_1^2 - \frac{\partial \ni}{\partial I_2}\,v_1^4\right] = 0$$

and formulas (11)-(13) it follows that

$$(2\delta_1 - \delta)\lambda + 2\mu\delta_1 = 0, \qquad \delta_1 = \frac{\lambda}{2(\lambda+\mu)} = \nu, \tag{14}$$

where ν denotes the Poisson ratio of linear theory. To this approximation

$$\sigma_3 = -2\mu(1+\nu)\delta \ . \tag{15}$$

Turning to (4), we obtain, to the same approximation,

$$\hat{\mu}_1 = \mu(1+\tau\delta) \ , \qquad \tau = 1 + 2\nu - (1-2\nu)\frac{m}{\mu} + \frac{n}{2\mu} \ . \tag{16}$$

The constant τ may turn out to be negative in which case the value of the safe load given by (6) and which is valid only for $\hat{\mu}_1 > 0$, retains its value over the range of contractions $\delta < |\tau|^{-1}$.

By virtue of (6), (15) and (16), the safe compression δ and the safe longitudinal force Q are determined by the relationships

$$[(K+1)(1+\nu) - \tau]\delta < 1 \ , \qquad Q = S|\sigma_3| = \frac{2\mu S}{K+1-\dfrac{\tau}{1+\nu}} \ . \tag{17}$$

In the latter formula

$$S = S_0 v_1^2 = S_0(1+2\delta_1) = S_0(1+2\nu\delta) = S_0\{1 + 2\nu[(K+1)(1+\nu) - \tau]^{-1}\} \ ,$$

however, this correction plainly lacks practical significance and so we may put $S = S_0$.

We now proceed to the determination of the sign of the quadratic form $(15.11)_2$. The calculation is carried out using (2.15) and (13) and yields

$$C(\varepsilon_1,\varepsilon_2,\varepsilon_3) = \frac{n}{8}[I_3 I_1^2(\underline{\varepsilon}) - I_1^2(\underline{F}\cdot\underline{\varepsilon})] + \frac{1}{4}[\lambda I_1^2(\underline{F}\cdot\underline{\varepsilon}) + l(I_1-3)I_1^2(\underline{F}\cdot\underline{\varepsilon})] +$$
$$+ \frac{m}{2}[I_1(\underline{F}\cdot\underline{\varepsilon})I_1(\underline{F}^2\cdot\underline{\varepsilon}) - \frac{1}{2}(I_1-1)I_1^2(\underline{F}\cdot\underline{\varepsilon})] \ . \tag{18}$$

For a majority of materials

$$l < 0 \ , \quad m < 0 \ , \quad n < 0 \ , \quad \lambda = \frac{2\mu\nu}{1-2\nu} > 0 \qquad (0 < \nu < \tfrac{1}{2})$$

and at the same time

$$v^2 \le \frac{I_1(\underline{F}\cdot\underline{\varepsilon})}{I_1(\underline{\varepsilon})} \le v_1^2 \ , \qquad v^4 \le \frac{I_1(\underline{F}^2\cdot\underline{\varepsilon})}{I_1(\underline{\varepsilon})} \ .$$

By (13 and (14) we have

$$I_3 - 1 = 2(2\delta_1 - \delta) = -2(1-2\nu)\delta < 0\,, \quad I_1 - 3 = 2(2\delta_1 - \delta) = -2(1-2\nu)\delta < 0$$

and returning to (18),

$$C_{\min} = \frac{\lambda}{4} I_1^2(\underline{\varepsilon}^2)v^4 + \frac{l}{4}(I_1 - 3)I_1^2(\underline{\varepsilon})v^4 +$$

$$+ \frac{m}{2} I_1^2(\underline{\varepsilon})\left[v_1^6 - v^4 \frac{1}{2}(I_1 - 1)\right] + \frac{n}{8} I_1^2(\underline{\varepsilon})(I_3 - v^4) =$$

$$= \frac{1}{4}\lambda v^4 I_1^2(\underline{\varepsilon})\left\{1 - \frac{\delta}{\lambda}\,[2l(1-2\nu) - 2m(5+4\nu) - n(1+2\nu)]\right\}\,. \tag{19}$$

The value of the safe load given by (17) turns out to be acceptable for the value δ of the longitudinal compression determined from the inequality

$$\delta \le \lambda \left|\, [2l(1-2\nu) - 2m(5+4\nu) - n(1+2\nu)]\,\right|\,. \tag{20}$$

From the data for steel (see e.g. Chapter 5, Par. 11), $\tau/(1+\nu) \approx 1$, and for sufficiently long rods, when $(l/r)^2 \approx 90$ and $K + 1 = 62$, the correction in Equation (17) becomes meaningless when compared to the "engineering" criterion (7). From the same data we have

$$-\frac{n}{2\mu} > 2 - 2m(1-2\nu)\delta\,, \tag{21}$$

which confirms the validity of assumption (5) for sufficiently small values of δ.

b) *A compressible Mooney-Rivlin material.* The specific strain energy of this material (also called a Hadamard material) is prescribed by the expression

$$\ni = f(I_3) + C_1 I_1 + C_2 I_2\,, \qquad C_1 > 0\,,\ C_2 > 0\,. \tag{22}$$

In Equation (6) we should now set, (see (2.21)),

$$\hat{\mu}_1 = 2v^{-1}(C_1 + C_2 v^2)\,, \qquad |\sigma_3| \le \frac{4(C_1 v^{-1} + C_2 v)}{K+1}\,. \tag{23}$$

By (2.15)

$$\begin{aligned} C(\varepsilon_1, \varepsilon_2, \varepsilon_3) &= I_3\,[I_3 f'(I_3)]'\, I_1^2(\underline{\varepsilon}) + C_2 I_1^2(\underline{F}\cdot\underline{\varepsilon})\,, \\ C_{\min} &= [I_3(I_3 f')' + C_2 v^4]\, I_1^2(\underline{\varepsilon}) \end{aligned} \tag{24}$$

and the value of the critical compressive stress given by (23) is acceptable provided that $C_{\min} \ge 0$

$$I_3 [I_3 f'(I_3)]' \ge -C_2 v^4 , \qquad (25)$$

where the values of v and v_1 are determined from the equations

$$\sigma_1 = 0 , \qquad I_3 f_3'(I_3) + v_1^2 [C_1 + C_2 (v_1^2 + v^2)] = 0 ,$$

$$|\sigma_3| = -\frac{2}{v_1^2 v} [I_3 f_3'(I_3) + v^2 (C_1 + 2C_2 v_1^2)] ,$$

so that the value of the specific strain energy of a rod should satisfy the inequality

$$I_3 f'(I_3) \le -v^2 (C_1 + 2C_2 v_1^2) \qquad (26)$$

together with that of (25).

17. THE SAFE LOAD. QUASILINEAR MATERIAL

It is assumed that the specific strain energy is specified as a function of the invariants $I_k(\underline{V})$ of the tensor $\underline{V} = \underline{F}^{1/2}$ which are designated below by I_k' $(k = 1,2,3)$

$$\ni = \ni(I_1', I_2', I_3') . \qquad (1)$$

The following relationships are seen to hold

$$I_1'^2 - 2I_2' = I_1 , \qquad -I_1' I_3' + 2I_2'^2 = I_2 , \qquad I_3'^2 = I_3 , \qquad (2)$$

from these it is easy to obtain the table of derivatives

$$2D \frac{\partial I_1'}{\partial I_1} = I_2' , \qquad 2D \frac{\partial I_2'}{\partial I_1} = I_3' , \qquad \frac{\partial I_3'}{\partial I_1} = 0 ,$$

$$2 \frac{\partial I_1'}{\partial I_2} = 1 , \qquad 2D \frac{\partial I_2'}{\partial I_2} = I_1' , \qquad \frac{\partial I_3'}{\partial I_2} = 0 ,$$

$$2 \frac{\partial I_1'}{\partial I_3} = \frac{I_1'}{I_3'} , \qquad 2D \frac{\partial I_2'}{\partial I_3} = \frac{\partial I_1'^2}{I_3'} , \qquad 2 \frac{\partial I_3'}{\partial I_3} = \frac{1}{I_3'} , \qquad (3)$$

where

$$D = I_1'I_2' - I_3' \, . \tag{4}$$

Turning to Equations (4.3.5) we get

$$\begin{aligned}
\psi_0 &= I_3 \frac{\partial \ni}{\partial I_3} = \frac{I_3'I_1'}{2D}\left(\frac{\partial \ni}{\partial I_1'} + I_1' \frac{\partial \ni}{\partial I_2'}\right) + \frac{1}{2} I_3' \frac{\partial \ni}{\partial I_3'} \, , \\
\psi_1 &= \frac{1}{2D}\,(I_1'^2 - I_2')\left(\frac{\partial \ni}{\partial I_1'} + I_1' \frac{\partial \ni}{\partial I_2'}\right) - \frac{1}{2}\frac{\partial \ni}{\partial I_2'} \, , \\
\psi_2 &= -\frac{1}{2D}\left(\frac{\partial \ni}{\partial I_1'} + I_1' \frac{\partial \ni}{\partial I_2'}\right)
\end{aligned} \tag{5}$$

and representation (4.3.4) of the Cauchy stress tensor takes the form

$$\begin{aligned}
\underline{T} &= \frac{1}{D}\sqrt{\frac{g}{G}}\,[I_3'I_1'\underline{E} + (I_1'^2 - I_2')\underline{F} - \underline{F}^2]\left(\frac{\partial \ni}{\partial I_1'} + I_1' \frac{\partial \ni}{\partial I_2'}\right) + \\
&\quad + \sqrt{\frac{g}{G}}\left(I_3' \frac{\partial \ni}{\partial I_3'}\underline{E} - \frac{\partial \ni}{\partial I_2'}\underline{F}\right) \, .
\end{aligned} \tag{6}$$

Eliminating $\underline{F}^2$ in this formula by the Cayley-Hamilton identity

$$\underline{F}^2 = \underline{V}^4 = (I_1'^2 - I_2')\underline{V}^2 - D\underline{V} + I_1'I_3'\underline{E} \, ,$$

we obtain

$$\underline{T} = \sqrt{\frac{g}{G}}\left[I_3' \frac{\partial \ni}{\partial I_3'}\underline{E} + \left(\frac{\partial \ni}{\partial I_1'} + I_1' \frac{\partial \ni}{\partial I_2'}\right)\underline{V} - \frac{\partial \ni}{\partial I_2'}\underline{V}^2\right] = \sqrt{\frac{g}{G}}\,(\psi_0'\underline{E} + \psi_1'\underline{V} + \psi_2'\underline{V}^2) \, . \tag{7}$$

The expressions for ψ_Γ' again yield the formulas (4.3.5)

$$\psi_0' = I_3' \frac{\partial \ni}{\partial I_3'} \, , \qquad \psi_1' = \frac{\partial \ni}{\partial I_1'} + I_1' \frac{\partial \ni}{\partial I_2'} \, , \qquad \psi_2' = -\frac{\partial \ni}{\partial I_2'} \, , \tag{8}$$

but in representation (7) which is similar to that of (4.3.4), the factor 2 is absent.

For a quasilinear material we have by (5.5.4)

$$\ni = \frac{1}{2}\,(\lambda + 2\mu)I_1'^2 - (3\lambda + 2\mu)I_1' - 2\mu I_2' + \frac{1}{2}\,(9\lambda + 6\mu) \tag{9}$$

and by (8) and (7)

$$\psi_0' = 0\,, \qquad \psi_1' = \lambda(I_1' - 3) - 2\mu\;; \qquad \frac{\partial \ni}{\partial I_2'} = -2\mu\;;$$

$$\underline{T} = \sqrt{\frac{g}{G}}\,\{[\lambda(I_1' - 3) - 2\mu]\,\underline{V} + 2\mu\underline{V}^2\}\,, \tag{10}$$

where the representation of $\underline{T}$ is, of course, consistent with (5.5.5).

Applying the calculation of (4.5.2) to (7) we have

$$\dot{\underline{T}} = -\underline{\underline{T\nabla}}\cdot\underline{w} + \sqrt{\frac{g}{G}}\,(\psi_1'\dot{\underline{V}} + \psi_2'\dot{\underline{F}}) + \sqrt{\frac{g}{G}}\,(\dot{\psi}_0'\underline{E} + \dot{\psi}_1'\underline{V} + \dot{\psi}_2'\underline{V}^2) =$$

$$= -\underline{\underline{T\nabla}}\cdot\underline{w} + \sqrt{\frac{g}{G}}\left[\psi_1'\left(\underline{\underline{\nabla w}}^{\mathrm{T}}\cdot\underline{V} + \underline{V}\cdot\underline{\underline{\nabla w}} - 2\sum_s\sum_k \frac{v_s v_k}{v_s + v_k}\,\underline{e}_k\cdot\underline{\varepsilon}\cdot\underline{e}_s\underline{e}_k\underline{e}_s\right) + \right.$$

$$\left. + \psi_2'(\underline{\underline{\nabla w}}^{\mathrm{T}}\cdot\underline{V}^2 + \underline{V}^2\cdot\underline{\underline{\nabla w}})\right] + 2\sqrt{\frac{g}{G}}\sum_{\Gamma=0}^{2}\sum_{N=0}^{2}\vartheta_{N\Gamma}'\underline{V}^{\Gamma}\underline{V}^{N}\cdot\cdot\,\underline{\varepsilon}(\underline{w})\,.$$

We have made use here of expressions (1.11.10) and (1.10.11) for the tensors $\dot{\underline{V}}$ and $\underline{F} = (\underline{V}^2)^{\cdot}$ and the quantities $\vartheta_{N\Gamma}'$ are expressed in terms of ψ_Γ' by Equations (4.5.5). The resulting expression is now transformed to

$$\dot{\underline{T}} = -\underline{\underline{T\nabla}}\cdot\underline{w} + \underline{\underline{\nabla w}}^{\mathrm{T}}\cdot\underline{T} + \underline{T}\cdot\underline{\underline{\nabla w}} - 2\sqrt{\frac{g}{G}}\left(\psi_0'\underline{\varepsilon}(\underline{w}) + \psi_1'\sum_s\sum_k \frac{v_s v_k}{v_s + v_k}\,\underline{e}_k\cdot\underline{\varepsilon}(\underline{w})\cdot\underline{e}_s\underline{e}_k\underline{e}_s\right) +$$

$$+ 2\sqrt{\frac{g}{G}}\sum_{\Gamma=0}^{2}\sum_{N=0}^{2}\vartheta_{N\Gamma}'V^{\Gamma}V^{N}\cdot\cdot\,\underline{\varepsilon}(\underline{w})$$

and by (4.5.8) we arrive at the following representation for the tensor $\underline{\theta}$

$$\underline{\theta} = \underline{T}\cdot\underline{\underline{\nabla w}} - 2\sqrt{\frac{g}{G}}\left[\psi_0'\underline{\varepsilon}(\underline{w}) + \psi_1'\sum_s\sum_k \frac{v_s v_k}{v_s + v_k}\,\underline{e}_k\cdot\underline{\varepsilon}(\underline{w})\cdot\underline{e}_s\underline{e}_k\underline{e}_s\right] +$$

$$+ 2\sqrt{\frac{g}{G}}\sum_{\Gamma=0}^{2}\sum_{N=0}^{2}\vartheta_{N\Gamma}'\underline{V}^{\Gamma}\underline{V}^{N}\cdot\cdot\,\underline{\varepsilon}(\underline{w})\,. \tag{11}$$

The quadratic form $\underline{\theta}\cdot\cdot\,\underline{\underline{\nabla w}}^{\mathrm{T}}$, whose positiveness guarantees that the loading is safe now takes the form

$$\underline{\theta}\cdot\cdot\,\underline{\underline{\nabla w}}^{\mathrm{T}} = \underline{\underline{\underline{T\nabla w}}}\cdot\cdot\,\underline{\underline{\nabla w}}^{\mathrm{T}} - 2\sqrt{\frac{g}{G}}\left[\psi_0' I_1(\underline{\varepsilon}^2) + \psi_1'\sum_s\sum_k \frac{v_s v_k}{v_s + v_k}\,(\underline{e}_k\cdot\underline{\varepsilon}\cdot\underline{e}_s)^2\right] +$$

$$+ 2\sqrt{\frac{g}{G}}\sum_{N=0}^{2}\sum_{\Gamma=0}^{2}\vartheta_{N\Gamma}' I_1(\underline{V}^{N}\cdot\underline{\varepsilon})I_1(\underline{V}^{\Gamma}\cdot\underline{\varepsilon})\,. \tag{12}$$

The first term is evaluated using formula (15.4), and application of the criterion (15.11) leads to the inequality

$$H \iiint I_1(\underline{\varepsilon}^2)dV > 2\sqrt{\frac{g}{G}}\left[\psi_0' \iiint_V I_1(\underline{\varepsilon}^2)dV + \right.$$

$$\left. + \psi_1' \sum_s \sum_k \frac{v_s v_k}{v_s + v_k} \iiint_V (\underline{e}_k \cdot \underline{\varepsilon} \cdot \underline{e}_s)^2 dV \right] . \tag{13}$$

Introducing here, instead of (15.4), the notation $\tilde{v}$ for the largest eigenvalue of the tensor $\underline{V}$, we get

$$\frac{v_s v_k}{v_s + v_k} \le \frac{1}{4}(v_s + v_k) \le \frac{1}{2}\tilde{v} ,$$

$$\sum_{s=1}^{3} \sum_{k=1}^{3} \frac{v_s v_k}{v_s + v_k} (\underline{e}_k \cdot \underline{\varepsilon} \cdot \underline{e}_s)^2 \le \frac{1}{2}\tilde{v} I_1(\underline{\varepsilon}^2) \tag{14}$$

and on reinforcing the Inequality (13) we obtain

$$(K+1)(\sigma - \sigma_3) \le \sqrt{\frac{g}{G}} (|\psi_1'|\tilde{v} - \psi_0') . \tag{15}$$

For a quasi-linear material, this reduces to the form (see (10))

$$(K+1)(\sigma - \sigma_3) \le \sqrt{\frac{g}{G}} \tilde{v} |\lambda(I_1' - 3) - 2\mu| \tag{16}$$

of course, there still remains the problem of the verification of the sign of the quadratic form

$$C(\varepsilon_1, \varepsilon_2, \varepsilon_3) = 2\sqrt{\frac{g}{G}} [\vartheta_{11}' I_1^2(\underline{V} \cdot \underline{\varepsilon}) + \vartheta_{22}' I_1^2(\underline{F} \cdot \underline{\varepsilon}) + 2\vartheta_{12}' I_1(\underline{V} \cdot \underline{\varepsilon}) I_1(\underline{F} \cdot \underline{\varepsilon})] \tag{17}$$

(because $\vartheta_{0\Gamma}' = 0$, $\Gamma = 0,1,2$).

For a rod compressed by a longitudinal force, we have, in the notation of Chapter 8, Par. 16

$$\psi_1' = \lambda(2\delta_1 - \delta) - 2\mu , \qquad |\psi_1'| = 2\mu(1 + \nu\delta) ,$$

$$\sqrt{\frac{G}{g}} = (1 + \nu\delta)^2 (1 - \delta) , \qquad \tilde{v} = 1 + \nu\delta$$

and the safe load is given by the expression

$$Q = -\sigma_3 S \le \frac{2\mu(1 + \nu\delta)^2 S_0}{(1 - \delta)(K + 1)} . \tag{18}$$

18. AN INCOMPRESSIBLE ELASTIC BODY

For an incompressible elastic body, the specific strain energy is a function of the invariants I_1 and I_2

$$\ni = \ni(I_1, I_2) \tag{1}$$

and Equations (2.3) specify the "defined quantities" (designated by the index E).

The strain energy is now represented as follows

$$W_E(\underline{R} + \eta\underline{w}) - W_E(\underline{R}) = \eta W_{1E} + \eta^2 W_{2E} \,. \tag{2}$$

Here

$$W_{1E} = \iiint_V \underline{T}_E \cdot\cdot \underline{\nabla w}^T dV - \iiint_V \rho_0 \underline{k} \cdot \underline{w}\, dV - \iint_{O_1} \underline{f}^0 \cdot \underline{w}\, dO \,, \tag{3}$$

$$W_{2E} = \iiint_V \Psi_E dV = \frac{1}{2} \iiint_V \underline{\theta}_E \cdot\cdot \underline{\nabla w}^T dV \tag{4}$$

and by Equation (2.11)

$$\Psi_E = \frac{1}{2} \underline{T}_E \cdot\cdot \underline{\nabla w} \cdot \underline{\nabla w}^T + 2\,[\psi_2 I_1((\underline{F} \cdot \underline{\varepsilon})^2) + \vartheta_{11} I_1^2(\underline{F} \cdot \underline{\varepsilon}) +$$

$$+ \vartheta_{22} I_1^2(\underline{F}^2 \cdot \underline{\varepsilon}) + 2\vartheta_{12} I_1(\underline{F} \cdot \underline{\varepsilon}) I_1(\underline{F}^2 \cdot \underline{\varepsilon})] \,. \tag{5}$$

Here $\underline{T}_E$ denotes the defined stress

$$\underline{T}_E = 2(\psi_1 \underline{F} + \psi_2 \underline{F}^2) \,, \tag{6}$$

and ψ_Γ and $\vartheta_{N\Gamma}$ $(N, \Gamma = 1,2)$ are given by the formulas (4.3.5) and (4.5.5). The tensor $\underline{\theta}_E$ given by (2.10) is linear in $\underline{\nabla w}$

$$\underline{\theta}_E = (\Psi_E)_{\underline{\nabla w}} = \underline{T}_E \cdot \underline{\nabla w} + 4\{\psi_2 \underline{F} \cdot \underline{\varepsilon} \cdot \underline{F} + \vartheta_{11} I_1(\underline{F} \cdot \underline{\varepsilon})\underline{F} +$$

$$+ \vartheta_{22} I_1(\underline{F}^2 \cdot \underline{\varepsilon})\underline{F}^2 + \vartheta_{12}\,[I_1(\underline{F}^2 \cdot \underline{\varepsilon})\underline{F} + I_1(\underline{F} \cdot \underline{\varepsilon})\underline{F}^2]\} \,. \tag{7}$$

The constraint

$$\sqrt{I_3(\underline{R} + \eta\underline{w})} - \sqrt{I_3(\underline{R})} = 0$$

where the terms quadratic in η are taken into consideration, is represented by the relationship (see (1.12.5))

$$\eta I_1(\underline{\varepsilon}) + \frac{1}{2}\eta^2 [I_1^2(\underline{\varepsilon}) - \underline{\underline{\nabla w}}^T \cdot\cdot \underline{\underline{\nabla w}}^T] = 0 . \tag{8}$$

Multiplying this by $p + \frac{1}{2}\eta p'$ and integrating throughout the volume we obtain

$$\eta \iiint_V p\underline{\nabla} \cdot \underline{w}\, dV + \frac{1}{2}\eta^2 \iiint_V (p'\underline{\nabla} \cdot \underline{w} - p\underline{\underline{\nabla w}}^T \cdot\cdot \underline{\underline{\nabla w}}^T) dV =$$

$$= \eta\left(\iint_O p\underline{N} \cdot \underline{w}\, dO - \iiint_V \underline{w} \cdot \underline{\nabla} p\, dV\right) + \frac{1}{2}\eta^2\left[\iint_O p'\underline{N} \cdot \underline{w}\, dO - \right.$$

$$\left. - \iiint_V (\underline{w} \cdot \underline{\nabla} p' + p\underline{\underline{\nabla w}}^T \cdot\cdot \underline{\underline{\nabla w}}^T) dO\right] = 0 . \tag{9}$$

The term

$$\iiint_V pI_1^2(\underline{\varepsilon}) dV = \iiint_V \underline{\nabla} \cdot (p\underline{\underline{w\nabla}} \cdot \underline{w}) dV - \iiint_V \underline{w} \cdot \underline{\nabla}(p\underline{\nabla} \cdot \underline{w}) dV =$$

$$= \iint_O p\underline{N} \cdot \underline{\underline{w\nabla}} \cdot \underline{w}\, dO - \iiint_V \underline{w} \cdot \underline{\nabla}(p \cdot \underline{\underline{\nabla w}}) dV$$

is not included in (9) because $\underline{\nabla} \cdot \underline{w} = 0$ up to terms of order η^2 and the former would contribute to neither the equilibrium equations nor the boundary conditions.

Subtracting (9) from (3) and (4), we arrive at the representations

$$W_1 = W_{1E} - \iiint_V p\underline{\nabla} \cdot \underline{w}\, dV = \iiint_V (\underline{\underline{T}}_E \cdot\cdot \underline{\underline{\nabla w}}^T - p\underline{\nabla} \cdot \underline{w} - \rho_0 \underline{k} \cdot \underline{w}) dV - \iint_{O_1} \underline{f}^0 \cdot \underline{w}\, dO =$$

$$= \iiint_V (\underline{\underline{T}}_E \cdot\cdot \underline{\underline{\nabla w}}^T + \underline{w} \cdot \underline{\nabla} p - \rho_0 \underline{k} \cdot \underline{w}) dV - \iint_{O_1} (\underline{f}^0 + p\underline{N}) \cdot \underline{w}\, dO , \tag{10}$$

$$2W_2 = 2W_{1E} - \iiint_V (p'\underline{\nabla} \cdot \underline{w} + p_0 \underline{\underline{\nabla w}}^T \cdot\cdot \underline{\underline{\nabla w}}^T) dV =$$

$$= \iiint_V [-p'\underline{\nabla} \cdot \underline{w} + (\underline{\underline{\theta}}_E + p\underline{\underline{\nabla w}}^T) \cdot\cdot \underline{\underline{\nabla w}}^T] dV . \tag{11}$$

In accordance with the principle of stationary potential energy we have $W_1 = 0$. This also follows immediately from the transformed expression for (10)

$$\iiint_V \underline{\underline{T}}_E \cdot\cdot \underline{\underline{\nabla w}}^T dV = \iiint_V [\underline{\nabla} \cdot (\underline{\underline{T}}_E \cdot \underline{w}) - (\underline{\nabla} \cdot \underline{\underline{T}}_E) \cdot \underline{w}]\, dV =$$

$$= \iint_{O_1} \underline{N} \cdot \underline{\underline{T}}_E \cdot \underline{w}\, dO - \iiint_V (\underline{\nabla} \cdot \underline{\underline{T}}_E) \cdot \underline{w}\, dV$$

and taking the equilibrium equations in the V-configuration into account

$$W_1 = - \iiint_V (\underline{\nabla} \cdot \underline{\underline{T}}_E - \underline{\nabla} p + \rho_0 \underline{k}) \cdot \underline{w}\, dV + \iiint_{O_1} [\underline{N} \cdot (\underline{\underline{T}}_E - \underline{\underline{E}} p) - \underline{f}^0] \cdot \underline{w}\, dO \ .$$

By the definition of Par. 10 this equilibrium is stable if the second order variation of the potential energy is positive. From (11) we obtain

$$\iiint_V (\underline{\underline{\theta}}_E + p \underline{\underline{\nabla w}}^T) \cdot\cdot \underline{\underline{\nabla w}}^T dV > 0 \ , \tag{12}$$

this because the sign is not influenced by the introduction of the term $p' \underline{\nabla} \cdot \underline{w}$ (which equals zero) into the integrand.

This term should, however, be retained when formulating the equations of neutral equilibrium. We have

$$\iiint_V \{-p' \underline{\nabla} \cdot \underline{w} + (\underline{\underline{\theta}}_E + p \underline{\underline{\nabla w}}^T) \cdot\cdot \underline{\underline{\nabla w}}^T\} dV =$$

$$= \iiint_V \{-\underline{\nabla} \cdot p' \underline{w} + \underline{w} \cdot \underline{\nabla} p' + \underline{\nabla} \cdot [(\underline{\underline{\theta}}_E + p \underline{\underline{\nabla w}}^T) \cdot \underline{w}] - [\underline{\nabla} \cdot (\underline{\underline{\theta}}_E + p \underline{w}^T)] \cdot \underline{w}\} dV =$$

$$= - \iiint_V [\underline{\nabla} \cdot (\underline{\underline{\theta}}_E + p \underline{\underline{\nabla w}}^T) - \underline{\nabla} p'] \cdot \underline{w}\, dV + \iint_{O_1} \underline{N} \cdot [(\underline{\underline{\theta}}_E + p \underline{\underline{\nabla w}}^T) - \underline{\underline{E}} p'] \cdot \underline{w}\, dO \ .$$

The appearance of the Lagrange multiplier p' enables us to write down the equilibrium equations for interior and boundary points of the volume

$$\text{in } V\text{: } \underline{\nabla} \cdot \underline{\underline{\theta}}_E + \underline{\nabla} \cdot p \underline{\underline{\nabla w}}^T - \underline{\nabla} p' = \underline{0} \ , \tag{13}$$

$$\text{on } O_1\text{: } \underline{N} \cdot (\underline{\underline{\theta}}_E + p \underline{\underline{\nabla w}}^T) - \underline{N} p' = \underline{0} \ , \tag{14}$$

these relationships incorporate into the constraints

$$\underline{\nabla} \cdot \underline{w} = 0 \ . \tag{15}$$

In these equations function p is assumed to be known from the equilibrium equations in the V-configuration augmented by the constraint

$I_3 - 1 = 0$.

19. THE SAFE LOADING OF AN INCOMPRESSIBLE ELASTIC BODY

As before in Par. 13, "dead" loading is assumed over the entire surface of the body, and the field of virtual displacements is assumed to satisfy condition (1.21) and of course the incompressibility condition (18.15). The tensor which enters criterion (18.12) is transformed by virtue of (18.7) to the form

$$\underline{\theta}_E + p\underline{\nabla w}^T = \underline{T}_E \cdot \underline{\nabla w} + p\underline{\nabla w}^T + 4\psi_2 \underline{F} \cdot \underline{\varepsilon} \cdot \underline{F} + 4 \sum_{N=1}^{2} \sum_{\Gamma=1}^{2} \vartheta_{N\Gamma} I_1(\underline{F}^\Gamma \cdot \underline{\varepsilon})\underline{F}^N . \tag{1}$$

Substituting $\underline{T}_E$ by $\underline{T} + p\underline{E}$ we get

$$\underline{T}_E \cdot \underline{\nabla w} + p\underline{\nabla w}^T = (\underline{T} + p\underline{E}) \cdot \underline{\nabla w} + p\underline{\nabla w}^T = \underline{T} \cdot \underline{\nabla w} + 2p\underline{\varepsilon} \tag{2}$$

and furthermore

$$(\underline{T}_E \cdot \underline{\nabla w} + p\underline{\nabla w}^T) \cdot\cdot \underline{\nabla w}^T = \underline{T} \cdot\cdot \underline{\nabla w} \cdot \underline{\nabla w}^T + 2pI_1(\underline{\varepsilon}^2) ,$$

since $\underline{\varepsilon} \cdot\cdot \underline{\nabla w}^T = \underline{\varepsilon} \cdot\cdot \underline{\varepsilon}$. We also have, referring to (15.5) and (15.6)

$$\underline{F} \cdot \underline{\varepsilon} \cdot \underline{F} \cdot\cdot \underline{\nabla w}^T = \underline{F} \cdot \underline{\varepsilon} \cdot \underline{F} \cdot\cdot \underline{\varepsilon} = I_1((\underline{F} \cdot \underline{\varepsilon})^2) ,$$

$$\psi_2 I_1((\underline{F} \cdot \underline{\varepsilon})^2) \geq -|\psi_2| V^4 I_1(\underline{\varepsilon}^2) , \tag{3}$$

$$\sum_{N=1}^{2} \sum_{\Gamma=1}^{2} \vartheta_{N\Gamma} I_1(\underline{F}^\Gamma \cdot \underline{\varepsilon})\underline{F} \cdot\cdot \underline{\nabla w}^T = \sum_{N=1}^{2} \sum_{\Gamma=1}^{2} \vartheta_{N\Gamma} I_1(\underline{F}^\Gamma \cdot \underline{\varepsilon}) I_1(\underline{F}^N \cdot \underline{\varepsilon}) = \sum_{s=1}^{3} \sum_{k=1}^{3} \alpha_{sk}\varepsilon_s\varepsilon_k , \tag{4}$$

where just as in (2.14)

$$\alpha_{sk} = \sum_{N=1}^{2} \sum_{\Gamma=1}^{2} \vartheta_{N\Gamma} v_s^{2N} v_k^{2\Gamma} = \vartheta_{11} v_s^2 v_k^2 + \vartheta_{12} v_s^2 v_k^2 (v_s^2 + v_k^2) +$$

$$+ \vartheta_{22} v_s^4 v_k^4 = v_s^2 v_k^2 \left\{ \frac{\partial \ni}{\partial I_2} + \frac{\partial^2 \ni}{\partial I_1^2} + \frac{\partial^2 \ni}{\partial I_1 \partial I_2} [(I_1 - v_s^2) + (I_1 - v_k^2)] + \right.$$

$$\left. + \frac{\partial^2 \ni}{\partial I_2^2} (I_1 - v_s^2)(I_1 - v_k^2) \right\} , \tag{5}$$

or

$$\alpha_{sk} = v_s^2 v_k^2 \frac{\partial^2 \ni}{\partial v_s^2 \partial v_k^2} \quad \text{and} \quad s \neq k \ ;$$

$$\alpha_{ss} = v_s^4 \left(\frac{\partial \ni}{\partial I_2} + \frac{\partial^2 \ni}{(\partial v_s^2)^2} \right) \qquad (s,k = 1,2,3) \ . \tag{6}$$

The Beatty criterion is given in the same way as in (15.9). We thus arrive at the expressions

$$B = H + 2p - 4v_v^4 \left| \frac{\partial \ni}{\partial I_2} \right| \ , \quad B = H + 2\left[\hat{\mu}_v - \sigma_v + 2V^4 \left(\frac{\partial \ni}{\partial I_2} - \left| \frac{\partial \ni}{\partial I_2} \right| \right) \right] \ , \tag{7}$$

where now, just as in (2.21), we have

$$\hat{\mu}_s = \sigma_s + p - 2v_s^4 \frac{\partial \ni}{\partial I_2} = 2\left[\left(\frac{\partial \ni}{\partial I_1} + I_1 \frac{\partial \ni}{\partial I_2} \right) v_s^2 - 2v_s^4 \frac{\partial \ni}{\partial I_2} \right] \ . \tag{8}$$

The conditions for safe loading retain the form of Inequality (15.11) or, in their more complete version, of Inequality (15.10).

20. THE STABILITY OF AN UNDEFORMED STATE OF AN INCOMPRESSIBLE MATERIAL

The material in its undistorted V-configuration is assumed to be subjected to uniform tension or compression. In this state the tensors $\underline{T}$ and $\underline{F}$ are spherical, i.e.

$$\underline{T} = \sigma \underline{E} \ , \qquad \underline{F} = v^2 \underline{E} = \underline{E} \ ,$$

because for an incompressible material, $I_3 = v^6 = 1$. The state equation is prescribed by the relationship

$$\sigma = -p + \sigma_E \ , \quad \sigma_E = 2\left(\frac{\partial \ni}{\partial I_1} + I_1 \frac{\partial \ni}{\partial I_2} - \frac{\partial \ni}{\partial I_2} \right) = 2\left(\frac{\partial \ni}{\partial I_1} + 2 \frac{\partial \ni}{\partial I_2} \right) = \mu + 2 \frac{\partial \ni}{\partial I_2} \ .$$

Here we have by (19.8)

$$\mu = 2\left(\frac{\partial \ni}{\partial I_1} + \frac{\partial \ni}{\partial I_2} \right) \ , \tag{1}$$

so that

$$p = -\sigma + \mu + 2 \frac{\partial \ni}{\partial I_2} \ .$$

There is no need to consider the double summation in (18.1) because we now have by (18.5)

$$I_1(\underline{F}^N \cdot \underline{\varepsilon}) = I_1(\underline{\varepsilon}) = \underline{\nabla} \cdot \underline{w} = 0 \ .$$

We now obtain

$$\underline{\theta}_E + p\underline{\nabla w}^T = \left(\mu + 2\frac{\partial \ni}{\partial I_2}\right)(\underline{\nabla w} + \underline{\nabla w}^T) - \sigma\underline{\nabla w}^T - 4\frac{\partial \ni}{\partial I_2}\underline{\varepsilon} = 2\mu\underline{\varepsilon} - \sigma\underline{\nabla w}^T \ ,$$

so that

$$(\underline{\theta}_E + p\underline{\nabla w}^T) \cdot\cdot \underline{\nabla w}^T = 2\mu I_1(\underline{\varepsilon}^2) - \sigma\underline{\nabla w}^T \cdot\cdot \underline{\nabla w}^T \ .$$

In order to transform the second term we use the identity

$$\underline{\nabla w}^T \cdot\cdot \underline{\nabla w}^T = (\underline{\varepsilon} + \underline{\Omega}) \cdot\cdot (\underline{\varepsilon} + \underline{\Omega}) = I_1(\underline{\varepsilon}^2) + I_1(\underline{\Omega}^2) = I_1(\underline{\varepsilon}^2) - 2\underline{\omega} \cdot \underline{\omega} \qquad (I_1(\underline{\Omega} \cdot \underline{\varepsilon}) = 0)$$

and the stability condition (18.12) takes the form

$$(2\mu - \sigma)\iiint_V I_1(\underline{\varepsilon}^2)\,dV + 2\sigma\iiint_V \underline{\omega} \cdot \underline{\omega}\,dV > 0 \ ,$$

so that stability holds for $\sigma > 0$ (tension) provided that

$$2\mu > \sigma \ . \tag{3}$$

Holden's estimate is inapplicable if $\sigma > 0$ because Inequality (15.3) does not hold for this case.

For the compressed state $\sigma < 0$, and by (19.7), (1) and (2) we have

$$B = H + 2p - 4\left|\frac{\partial \ni}{\partial I_2}\right| = (K+1)\sigma - 2\sigma + 2\mu - 4\left(\left|\frac{\partial \ni}{\partial I_2}\right| - \frac{\partial \ni}{\partial I_2}\right) =$$

$$= (K-1)\sigma + 2\mu - 4\left(\left|\frac{\partial \ni}{\partial I_2}\right| - \frac{\partial \ni}{\partial I_2}\right)$$

and the sufficient condition for safe loading is expressed by the inequality

$$B > 0: \ 2\mu > -(K-1)\sigma + 4\left(\left|\frac{\partial \ni}{\partial I_2}\right| - \frac{\partial \ni}{\partial I_2}\right) =$$

$$= (K-1)|\sigma| + \begin{cases} 0 & \text{and} \quad \dfrac{\partial \ni}{\partial I_2} > 0 \ , \\ 8\left|\dfrac{\partial \ni}{\partial I_2}\right| & \text{and} \quad \dfrac{\partial \ni}{\partial I_2} < 0 \ . \end{cases} \tag{4}$$

In particular, for a two-constant Mooney material we have by (7.4.1) and (7.4.4) that

$$2\mu > (K-1)\,|\sigma| \; . \tag{5}$$

21. INCOMPRESSIBLE MATERIAL. A COMPRESSED ROD

The compressive (longitudinal) stress will be denoted by $\sigma_3 < 0$; $\sigma_1 = \sigma_2 = 0$ and $\sigma = 0$ by (13.11). The eigenvalues of the Finger measure are denoted as before by $v_1^2 = v_2^2 = v^{-1}$, $v_3 = v^2$ and in accordance with the BE-inequality (5.13.10)

$$\frac{\sigma_3}{v^2 - 1/v} > 0 \; , \qquad v^3 - 1 < 0 \; , \qquad v < 1 \; . \tag{1}$$

The stress components are given by (7.2.4) and (7.2.5) and are

$$\sigma_1 = -p + \frac{2}{v}\left[\frac{\partial \ni}{\partial I_1} + (v^2 + v^{-1})\frac{\partial \ni}{\partial I_2}\right] = 0 \; ,$$

$$\sigma_3 = -p + 2v^2\left(\frac{\partial \ni}{\partial I_1} + 2v^{-1}\frac{\partial \ni}{\partial I_2}\right) = 2(v - v^2)\left(v\frac{\partial \ni}{\partial I_1} + \frac{\partial \ni}{\partial I_2}\right) \tag{2}$$

and the Beatty constant (19.7) takes the form

$$B = 2(K+1)(v - v^{-2})\left(v\frac{\partial \ni}{\partial I_1} + \frac{\partial \ni}{\partial I_2}\right) + 4v^{-1}\left[\frac{\partial \ni}{\partial I_1} + (v^2 + v^{-1})\frac{\partial \ni}{\partial I_2}\right] -$$

$$- 4v^{-2}\left|\frac{\partial \ni}{\partial I_2}\right| = 2\frac{\partial \ni}{\partial I_1}\left[(K+1)v^2 - (K-1)v^{-1}\right] +$$

$$+ 2\frac{\partial \ni}{\partial I_2}\left[(K+3)v - (K+1)v^{-2}\right] - 4v^{-2}\left(\left|\frac{\partial \ni}{\partial I_2}\right| - \frac{\partial \ni}{\partial I_2}\right) .$$

For a two-constant Mooney material the criterion for a safe loading reduces to

$$(C_1 v + C_2)\left[(K+1)v^3 - (K-1)\right] + 2C_2(v^3 - 1) > 0 \; .$$

For $C_2 = 0$ we arrive at the inequalities

$$1 > v^3 > 1 - \frac{2}{K+1} \tag{3}$$

and by (1) and (7.4.9)

$$\sigma_3 = 2C_1 v^{-1}(v^3 - 1) > -\frac{4C_1 v^{-1}}{K+1}, \qquad |\sigma_3| < \frac{4C_1 v^{-1}}{K+1},$$
$$Q = S|\sigma_3| < \frac{2\mu S_0 v^{-2}}{K+1}, \tag{4}$$

where S_0 is the cross-sectional area in the natural configuration ($S = S_0 v_1^2 = S_0 v^{-1}$).

We have thus arrived for $v \approx 1$ at the same "engineering" formula as (16.7).

22. CONVEXITY WITH RESPECT TO A GRADIENT. HADAMARD'S CONDITION

A differentiable function $\ni(\overset{0}{\underline{\underline{\nabla R}}})$ is called "convex with respect to a gradient" $\overset{0}{\underline{\underline{\nabla R}}}$ within a certain range of its values if for any tensor $\underline{Q}$ the inequality

$$\left[\frac{d^2}{d\eta^2}\ni(\overset{0}{\underline{\underline{\nabla R}}}\cdot(\underline{E}+\eta\underline{Q}))\right]_{\eta=0} \geq 0 \tag{1}$$

holds.

By the principle of material frame-indifference (Chapter 3, Par. 2) we may substitute the argument

$$\overset{0}{\underline{\underline{\nabla R}}}{}^{\times} = \overset{0}{\underline{\underline{\nabla R}}}\cdot(\underline{E}+\eta\underline{Q})$$

in this inequality by the argument

$$\overset{0}{\underline{\underline{\nabla R}}}{}^{\times}\cdot\underline{O} = \overset{0}{\underline{\underline{\nabla R}}}\cdot(\underline{E}+\eta\underline{Q})\cdot\underline{O},$$

where $\underline{O}$ denotes an orthogonal tensor. For an isotropic material, the substitution of $\overset{0}{\underline{\underline{\nabla R}}}{}^{\times}$ by $\underline{O}\cdot\overset{0}{\underline{\underline{\nabla R}}}{}^{\times}$ is equally acceptable.

In particular the orthogonal tensor $\underline{O}$ may be set equal to $\exp(-\eta\underline{\Omega})$ where $\underline{\Omega}$ denotes a skew tensor. This will enable us to write Inequality (1) in the following form

$$\left[\frac{d^2}{d\eta^2}\ni(\overset{0}{\underline{\underline{\nabla R}}}\cdot(\underline{E}+\eta\underline{Q})\cdot\exp(-\eta\underline{\Omega}))\right]_{\eta=0} \geq 0\,. \tag{2}$$

The tensor $\underline{Q}$ is now expressed by its decomposition into symmetric and skew parts

$$\underline{Q} = \underline{S} + \underline{\Omega}\,, \qquad \underline{S} = \frac{1}{2}\,(\underline{Q} + \underline{Q}^T)\,, \qquad \underline{\Omega} = \frac{1}{2}\,(\underline{Q} - \underline{Q}^T)\,,$$

and the expression for the specific strain energy takes the form

$$\begin{aligned}
\ni(\overset{0}{\nabla}\underline{\underline{R}} \cdot (\underline{E} + \eta\underline{Q})) &= \ni(\overset{0}{\nabla}\underline{\underline{R}} \cdot (\underline{E} + \eta\underline{S} + \eta\underline{\Omega}) \cdot \exp(-\eta\underline{\Omega})) = \\
&= \ni(\overset{0}{\nabla}\underline{\underline{R}} \cdot (\underline{E} + \eta\underline{S} + \eta\underline{Q}) \cdot (\underline{E} - \eta\underline{\Omega} + \frac{1}{2}\eta^2\underline{\Omega}^2 + \ldots)) = \\
&= \ni(\overset{0}{\nabla}\underline{\underline{R}} \cdot [\underline{E} + \eta\underline{S} - \frac{1}{2}\eta^2(\underline{\Omega}^2 + 2\underline{S} \cdot \underline{\Omega}) + \ldots]) = \ni(\overset{0}{\nabla}\underline{\underline{R}}) + \\
&\quad + \ni_{\overset{0}{\nabla}\underline{\underline{R}}} \cdot\cdot [\eta\underline{S} - \frac{1}{2}\eta^2(\underline{\Omega}^2 + 2\underline{S} \cdot \underline{\Omega})] \cdot \overset{0}{\nabla}\underline{\underline{R}}^T + \\
&\quad + \frac{1}{2}\eta^2\underline{S} \cdot \overset{0}{\nabla}\underline{\underline{R}}^T \cdot\cdot \ni_{\overset{0}{\nabla}\underline{\underline{R}}\overset{0}{\nabla}\underline{\underline{R}}} \cdot\cdot \underline{S} \cdot \overset{0}{\nabla}\underline{\underline{R}}^T + \ldots = \ni(\overset{0}{\nabla}\underline{\underline{R}} \cdot (\underline{E} + \eta\underline{S})) - \\
&\quad - \frac{1}{2}\eta^2 \ni_{\overset{0}{\nabla}\underline{\underline{R}}} \cdot\cdot (\underline{\Omega}^2 - 2\underline{\Omega} \cdot \underline{S}) \cdot \overset{0}{\nabla}\underline{\underline{R}}^T + \ldots = \ni(\overset{0}{\nabla}\underline{\underline{R}} \cdot (\underline{E} + \eta\underline{S})) - \\
&\quad - \frac{1}{2}\eta^2 \sqrt{\frac{G}{g}}\,\underline{T} \cdot\cdot (\underline{\Omega}^2 - 2\underline{S} \cdot \underline{\Omega})\,.
\end{aligned}$$

In this formula we have separated out the terms

$$\ni(\overset{0}{\nabla}\underline{\underline{R}} \cdot (\underline{E} + \eta\underline{S})) = \ni(\overset{0}{\nabla}\underline{\underline{R}}) + \eta\ni_{\overset{0}{\nabla}\underline{\underline{R}}} \cdot\cdot \underline{S} \cdot \overset{0}{\nabla}\underline{\underline{R}}^T + \frac{1}{2}\eta^2\underline{S} \cdot \overset{0}{\nabla}\underline{\underline{R}}^T \cdot\cdot \ni_{\overset{0}{\nabla}\underline{\underline{R}}\overset{0}{\nabla}\underline{\underline{R}}} \cdot\cdot \underline{S} \cdot \overset{0}{\nabla}\underline{\underline{R}}^T + \ldots\,,$$

made the substitution

$$\ni_{\overset{0}{\nabla}\underline{\underline{R}}} \cdot\cdot (\underline{\Omega}^2 - 2\underline{S} \cdot \underline{\Omega}) \cdot \overset{0}{\nabla}\underline{\underline{R}}^T = \overset{0}{\nabla}\underline{\underline{R}}^T \cdot \underline{P} \cdot\cdot (\underline{\Omega}^2 - 2\underline{\Omega} \cdot \underline{S}) = \sqrt{\frac{G}{g}}\,\underline{T} \cdot\cdot (\underline{\Omega}^2 - 2\underline{\Omega} \cdot \underline{S})$$

and omitted terms of second order and above in η^2.

We have thus arrived at the relationship

$$\left[\frac{d^2}{d\eta^2} \ni(\overset{0}{\nabla}\underline{\underline{R}} \cdot (\underline{E} + \eta\underline{Q}))\right]_{\eta=0} = \left[\frac{d^2}{d\eta^2} \ni(\overset{0}{\nabla}\underline{\underline{R}} \cdot (\underline{E} + \eta\underline{S}))\right]_{\eta=0} - \sqrt{\frac{G}{g}}\,\underline{T} \cdot\cdot (\underline{\Omega}^2 - 2\underline{\Omega} \cdot \underline{S})\,, \qquad (3)$$

which enables us to distinguish that part of the function which is convex with respect to the gradient and which corresponds to the symmetric part $\underline{S}$ of the tensor $\underline{Q}$.

In the specific case of a symmetric tensor $\underline{Q}$ when $\underline{\Omega} = \underline{0}$, the property of convexity with respect to a gradient proves to be equivalent to the monotonicity criterion (5.9.4) of Coleman and Noll.

For $\underline{S} = \underline{0}$, $\underline{Q} = \underline{\Omega}$ the convexity criterion (1) leads to the inequality (cf.

(3) and (I.11.4))

$$\underline{T} \cdot\cdot \underline{\Omega}^2 = I_1(\underline{T} \cdot \underline{\Omega}^2) = \underline{\omega} \cdot \underline{T} \cdot \underline{\omega} - I_1(\underline{T})\underline{\omega} \cdot \underline{\omega} \leq 0 \ . \tag{4}$$

Assuming that $\underline{\omega}$ is the unit vector $\underline{e}_1$ of one of the principal directions of the stress tensor, we have

$$\sigma_1 - (\sigma_1 + \sigma_2 + \sigma_3) \leq 0 \ , \qquad \sigma_2 + \sigma_3 \geq 0$$

which inequalities are already known from Chapter 5, Par. 9.

The transformation carried out below is designed to eliminate the skew tensor $\underline{\Omega}$ from (3). We thus have

$$\underline{T} \cdot\cdot \underline{Q} \cdot \underline{Q}^T = \underline{T} \cdot\cdot (\underline{S} + \underline{\Omega}) \cdot (\underline{S} - \underline{\Omega}) = \underline{T} \cdot\cdot \underline{S}^2 - \underline{T} \cdot\cdot \underline{S} \cdot \underline{\Omega} + \underline{T} \cdot\cdot \underline{\Omega} \cdot \underline{S} -$$

$$- \underline{T} \cdot\cdot \underline{\Omega}^2 = \underline{T} \cdot\cdot \underline{S}^2 - \underline{T} \cdot\cdot (\underline{\Omega}^2 - 2\underline{\Omega} \cdot \underline{S}) \ .$$

Returning to (3), we obtain

$$\left[\frac{d^2}{d\eta^2} \ni(\overset{0}{\underline{\nabla R}} \cdot (\underline{E} + \eta \underline{Q}))\right]_{\eta=0} = \sqrt{\frac{G}{g}} \, (\underline{T} \cdot\cdot \underline{Q} \cdot \underline{Q}^T - \underline{T} \cdot\cdot \underline{S}^2) + \left[\frac{d^2}{d\eta^2} \ni(\overset{0}{\underline{\nabla R}} \cdot (\underline{E} + \eta \underline{S}))\right]_{\eta=0} . \tag{5}$$

This equality enables us to represent the specific value of the second order variation of the specific strain energy in a different way. In fact, assuming that

$$\underline{Q} = \underline{\nabla w} \ , \qquad \underline{S} = \underline{\varepsilon}$$

and recalling (2.8), we have

$$\frac{d^2}{d\eta^2} \ni(\overset{0}{\underline{\nabla R}} \cdot (\underline{E} + \eta \underline{\nabla w})) = \underline{\nabla w}^T \cdot \overset{0}{\underline{\nabla R}}{}^T \cdot \ni_{\underline{\overset{0}{\nabla R}\overset{0}{\nabla R}}} \cdot\cdot \underline{\nabla w}^T \cdot \overset{0}{\underline{\nabla R}}{}^T \ ,$$

$$2\Phi = \sqrt{\frac{G}{g}} \, (\underline{T} \cdot\cdot \underline{\nabla w} \cdot \underline{\nabla w}^T - \underline{T} \cdot\cdot \underline{\varepsilon}^2) + \left[\frac{d^2}{d\eta^2} \ni(\overset{0}{\underline{\nabla R}} \cdot (\underline{E} + \eta \underline{\varepsilon}))\right]_{\eta=0} . \tag{6}$$

The expression for twice the second of the specific strain energy now takes the form

$$2W_2 = \iiint\limits_{v} \Phi \, dv = \iiint\limits_{V} \underline{T} \cdot\cdot \underline{\nabla w} \cdot \underline{\nabla w}^T dV - \iiint\limits_{V} \underline{T} \cdot\cdot \underline{\varepsilon}^2 dV +$$

$$+\left[\iiint_{v} \frac{d^2}{d\eta^2} \ni(\overset{0}{\underline{\underline{\nabla R}}} \cdot (\underline{E} + \eta \underline{\varepsilon}))dv\right]_{\eta=0}, \tag{7}$$

where, according to the monotonicity criterion,

$$\iiint_{v} \frac{d^2}{d\eta^2} \ni(\overset{0}{\underline{\underline{\nabla R}}} \cdot (\underline{E} + \eta \underline{\varepsilon}))dv > 0 \,. \tag{8}$$

Substituting $\underline{Q}$ by the dyad $\underline{\underline{ab}}$ in the definition of convexity with respect to a gradient we obtain Hadamard's inequality (J. Hadamard, 1903)

$$A(\underline{a},\underline{b}) = \left[\frac{d^2}{d\eta^2} \ni(\overset{0}{\underline{\underline{\nabla R}}} \cdot (\underline{E} + \eta \underline{\underline{ab}}))\right]_{\eta=0} \geq 0 \,. \tag{9}$$

Alternatively, bearing in mind the relationship

$$\ni(\underline{\underline{\nabla R}} \cdot (\underline{E} + \eta \underline{\underline{ab}})) = \ni(\overset{0}{\underline{\underline{\nabla R}}}) + \eta \ni_{\overset{0}{\underline{\underline{\nabla R}}}} \cdot\cdot\, \underline{\underline{ba}} \cdot \overset{0}{\underline{\underline{\nabla R}}}{}^{T} + \frac{1}{2}\eta^2 \underline{\underline{ba}} \cdot \overset{0}{\underline{\underline{\nabla R}}}{}^{T} \cdot\cdot \ni_{\overset{0}{\nabla R}\overset{0}{\nabla R}} \cdot\cdot\, \underline{\underline{ba}} \cdot \overset{0}{\underline{\underline{\nabla R}}}{}^{T},$$

this inequality may be written as

$$\underline{\underline{ba}} \cdot \overset{0}{\underline{\underline{\nabla R}}}{}^{T} \cdot\cdot \ni_{\overset{0}{\nabla R}\overset{0}{\nabla R}} \cdot\cdot\, \underline{\underline{ba}} \cdot \overset{0}{\underline{\underline{\nabla R}}}{}^{T} \geq 0 \,. \tag{10}$$

This form is connected by (4.11.5), with the definition of the ellipticity of the equilibrium equations and with that of strong ellipticity when (9) is strictly satisfied.

In the component representation

$$\ni_{\overset{0}{\nabla R}\overset{0}{\nabla R}} = \underline{r}_s \underline{r}_t \underline{r}_p \underline{r}_q \frac{\partial^2 \ni}{\partial \overset{0}{\underline{\nabla}}_s \chi_t \, \partial \overset{0}{\underline{\nabla}}_p \chi_q} = A^{stpq} \underline{r}_s \underline{r}_t \underline{r}_p \underline{r}_q \,, \tag{11}$$

and Hadamard's inequality may be rewritten in the form

$$A(\underline{a},\underline{b}) = A^{stpq} a_s b^0_t a_p b^0_q \geq 0 \qquad (a_s = \underline{a} \cdot \underline{R}_s,\; b^0_t = \underline{b} \cdot \underline{r}_t) \,. \tag{12}$$

By (4.9.12) we have

$$\ni_{\overset{0}{\nabla R}\overset{0}{\nabla R}} = \underline{P}_{\overset{0}{\nabla R}} = \underline{\underline{\underline{P\nabla r}}}^{T} - \underline{r}_s \underline{P}^{T} \underline{R}^{s} + \sqrt{\frac{G}{g}}\, \underline{\underline{\nabla r}}^{T} \cdot \underline{T}_F \cdot\cdot (\underline{C}_{II} + \underline{C}_{III}) \cdot\cdot \underline{\underline{\nabla R}}^{T} \cdot \underline{C}_{II} \,, \tag{13}$$

where, carrying out the calculation of Chapter 4, Par. 10 again we see that the terms $\underline{\underline{\underline{P\nabla r}}}^{T}$ and $\underline{r}_s \underline{P}^{T} \underline{R}^{s}$ cancel out in this expression and we obtain

$$A(\underline{a},\underline{b}) = \sqrt{\frac{G}{g}}\, \underline{\underline{ba}} \cdot \underline{\underline{\overset{0}{\nabla}R}}^T \cdot\cdot \underline{\underline{\nabla r}}^T \cdot \underline{\underline{T}}_F \cdot\cdot (\underline{C}_{II} + \underline{C}_{III}) \cdot\cdot \underline{\underline{\overset{0}{\nabla}R}}^T \cdot \underline{C}_{II} \cdot\cdot \underline{\underline{ba}} \cdot \underline{\underline{\overset{0}{\nabla}R}}^T =$$

$$= \sqrt{\frac{G}{g}}\, \underline{\underline{ba}} \cdot\cdot \underline{\underline{T}}_F \cdot\cdot (\underline{C}_{II} + C_{III}) \cdot\cdot \underline{\underline{\overset{0}{\nabla}R}}^T \cdot\cdot \underline{\underline{\overset{0}{\nabla}R}} \cdot \underline{\underline{ab}} =$$

$$= \sqrt{\frac{G}{g}}\, \underline{b} \cdot [\underline{a} \cdot (\underline{\underline{T}}_F + \underline{\underline{T}}_F \cdot\cdot \underline{C}_{II}) \cdot \underline{F} \cdot \underline{a}] \cdot \underline{b} = \sqrt{\frac{G}{g}}\, \underline{b} \cdot \underline{\underline{Q}} \cdot \underline{b} \,. \tag{14}$$

Here we have introduced, by (4.11.13), the acoustical tensor

$$\underline{\underline{Q}} = \underline{a} \cdot (\underline{\underline{T}}_F + \underline{\underline{T}}_F \cdot\cdot \underline{C}_{II}) \cdot \underline{F} \cdot \underline{a} \,, \tag{15}$$

and the satisfying of Hadamard's inequality, as givenbby (10) or (12), guarantees the ellipticity of the material while its strong ellipticity is ensured by the strengthened form

$$A(\underline{a},\underline{b}) > 0 \,. \tag{16}$$

For a strongly elliptic material, all the eigenvalues of its acoustical tensor are positive for all values of $\underline{a} \neq \underline{0}$.

If the function $f(x)$ is convex for all $x > 0$ and a material with the specific strain energy $\ni$ iw strongly elliptic, then the same will hold for a material with specific strain energy $f(\ni)$. In fact, using the notation

$$\varphi(\eta) = \ni(\underline{\underline{\overset{0}{\nabla}R}} + \eta\underline{\underline{\overset{0}{\nabla}R}} \cdot \underline{\underline{ab}}) \,, \qquad F(\eta) = f(\varphi(\eta))$$

we get

$$F'(\eta) = \frac{df}{d\varphi}\, \varphi'(\eta) \,, \qquad F''(\eta) = \frac{d^2 f}{d\varphi^2}\, [\varphi'(\eta)]^2 + \frac{df}{d\varphi}\, \varphi''(\eta) \,,$$

so that

$$[F''(\eta)]_{\eta=0} = \frac{d^2 f}{d\varphi^2} \left(\ni_{\underline{\underline{\overset{0}{\nabla}R}}} \cdot\cdot \underline{\underline{ba}} \cdot \underline{\underline{\overset{0}{\nabla}R}}^T \right)^2 + \frac{df}{d\varphi} \left(\underline{\underline{ba}} \cdot \underline{\underline{\overset{0}{\nabla}R}}^T \cdot\cdot \ni_{\underline{\underline{\overset{0}{\nabla}R\overset{0}{\nabla}R}}} \cdot\cdot \underline{\underline{ba}} \cdot \underline{\underline{\overset{0}{\nabla}R}}^T \right) > 0 \,, \tag{17}$$

because $f''_\varphi > 0$, $f'_\varphi > 0$ from the definition of a convex function.

The following example*) illustrates how much less restrictive Hadamard's inequality is than are the convexity criteria given in the form (1). Let us consider a material with the specific strain energy given by (5.6.17), i.e.

$$\ni = \frac{1}{2}\,\mu \left[I_1 + \frac{1}{\alpha}\,(I_3^{-\alpha} - 1) - 3 \right] \,, \tag{18}$$

*) Communicated to the author by E.L. Gurvich.

this material being strongly elliptic for arbitrary deformations.

By (1.9.1) and (1.9.12) we now have

$$\partial_{\overset{0}{\nabla}\underline{\underline{R}}} = \mu(\overset{0}{\nabla}\underline{\underline{R}} - I_3^{-\alpha}\underline{\underline{\nabla r}}^{T}) ,$$

$$\partial_{\overset{0}{\nabla}\underline{\underline{R}}\overset{0}{\nabla}\underline{\underline{R}}} = \mu I_3^{-\alpha}(I_3^{\alpha}\underline{\underline{C}}_{3-II} + 2\alpha\underline{\underline{\nabla r}}^{T}\underline{\underline{\nabla r}}^{T} + \underline{r}_m \underline{R}^{t}\underline{r}_t\underline{R}^{m}) ,$$

and by the definition (1) of convexity it is necessary to consider the inequality

$$\left[\frac{d^2}{d\eta^2}(\partial(\overset{0}{\nabla}\underline{\underline{R}} + \eta\overset{0}{\nabla}\underline{\underline{R}}\cdot\underline{\underline{Q}}))\right]_{\eta=0} = \overset{0}{\nabla}\underline{\underline{R}}^{T}\cdot\underline{\underline{Q}}^{T}\cdot\cdot\,\partial_{\overset{0}{\nabla}\underline{\underline{R}}\overset{0}{\nabla}\underline{\underline{R}}}\cdot\cdot\,\overset{0}{\nabla}\underline{\underline{R}}^{T}\cdot\cdot\,\underline{\underline{Q}}^{T} =$$

$$= \mu I_3^{-\alpha}[I_3^{\alpha}\underline{\underline{Q}}^{T}\cdot\overset{0}{\nabla}\underline{\underline{R}}^{T}\cdot\cdot\,\overset{0}{\nabla}\underline{\underline{R}}\cdot\underline{\underline{Q}} + 2\alpha\underline{\underline{Q}}^{T}\cdot\cdot\,\underline{\underline{EE}}\cdot\cdot\,\underline{\underline{Q}}^{T} + \underline{\underline{Q}}^{T}\cdot\cdot\,\underline{R}_m\underline{R}^{t}\underline{\underline{Q}}^{T}\cdot\cdot\,\underline{R}_t\underline{R}^{m}] =$$

$$= \mu I_3^{-\alpha}\,[I_3^{\alpha}I_1(\underline{\underline{F}}\cdot\underline{\underline{Q}}\cdot\underline{\underline{Q}}^{T}) + 2\alpha I_1^2(\underline{\underline{Q}}) + I_1(\underline{\underline{Q}}^2)] \geq 0 .$$

The problem is reduced to the determination of the sign of the quadratic form

$$E(\underline{\underline{Q}}) = I_3^{\alpha}I_1(\underline{\underline{F}}\cdot\underline{\underline{Q}}\cdot\underline{\underline{Q}}^{T}) + 2\alpha I_1^2(\underline{\underline{Q}}) + I_1(\underline{\underline{Q}}^2) ,$$

of nine variables which, in the proper basis of the Finger measure, can be represented as a sum of quadratic forms

$$E(Q) = E_1(q_{11},q_{22},q_{33}) + E_2(q_{23},q_{32}) + E_3(q_{31},q_{13}) + E_4(q_{12},q_{21}) .$$

Using the notation $\lambda_1 = (v_1^{\alpha+1}v_2^{\alpha}v_3^{\alpha})^2$, $\lambda_2 = (v_1^{\alpha}v_2^{\alpha+1}v_3^{\alpha})^2$, $\lambda_3 = (v_1^{\alpha}v_2^{\alpha}v_3^{\alpha+1})^2$ we may represent the matrices of the coefficients of these forms as follows

$$\|E_1\| = \begin{Vmatrix} \lambda_1+2\alpha+1 & 2\alpha & 2\alpha \\ 2\alpha & \lambda_2+2\alpha+1 & 2\alpha \\ 2\alpha & 2\alpha & \lambda_3+2\alpha+1 \end{Vmatrix} , \quad \|E_2\| = \begin{Vmatrix} \lambda_2 & 1 \\ 1 & \lambda_3 \end{Vmatrix}$$

etc. The matrix of the form $E(\underline{\underline{Q}})$ turns out to be a diagonal block matrix

$$\begin{Vmatrix} \|E_{11}\| & 0 & 0 & 0 \\ 0 & \|E_{22}\| & 0 & 0 \\ 0 & 0 & \|E_{33}\| & 0 \\ 0 & 0 & 0 & \|E_{44}\| \end{Vmatrix} .$$

Confining ourselves to the case $\alpha = \lambda/2\mu \geq 0$ which implies that $0 \leq \nu \leq 0.5$, we see that $\|E_1\|$ is a positive "Sylvesterian" matrix, and the matrices $\|E_2\|$, $\|E_3\|$, $\|E_4\|$ will be positive provided that $\lambda_2\lambda_3 > 0$, $\lambda_3\lambda_1 > 0$, $\lambda_1\lambda_2 > 0$, these conditions are reducible to the inequality

$$I_3 > v_s^{-(\alpha+1)} , \qquad v_s = \min(v_1, v_2, v_3) , \tag{19}$$

while the condition of strong ellipticity imposes no restrictions upon the eigenvalues of the Finger measure $\underline{F}$.

23. HADAMARD'S CONDITION AND STABILITY

We confine ourselves as before to a consideration of the affine transformations of the natural reference v-configuration into the actual equilibrium V-configuration. We recall here that the latter configuration is stable if its strain energy is a minimum, i.e. if its second order variation is positive for all $\underline{w} \not\equiv \underline{0}$ such that $\underline{w} = \underline{0}$ at that part O_2 of the boundary where the displacement vector is prescribed.

By (2.5) and (10.5) this condition can be represented as

$$\iiint_v \overset{0}{\underline{\underline{\nabla w}}}{}^T \cdot\cdot\, \ni_{\underline{\underline{\overset{0}{\nabla} R \overset{0}{\nabla} R}}} \cdot\cdot\, \overset{0}{\underline{\underline{\nabla w}}}{}^T dv > 0 \tag{1}$$

for all $\underline{w}$ satisfying the above requirements.

We will show below that stability implies that Hadamard's inequality is satisfied

$$\underline{\underline{ba}} \cdot\cdot\, \ni_{\underline{\underline{\overset{0}{\nabla} R \overset{0}{\nabla} R}}} \cdot\cdot\, \underline{\underline{ba}} \geq 0 \tag{2}$$

for all pairs of vectors $\underline{a}$, $\underline{b}$. In other words, if this condition is violated, then the V-configuration is unstable. In proceeding to an analysis of the stability of an equilibrium configuration, it must be verified that within the range of stretches v_1, v_2, v_3 considered, Hadamard's inequality is satisfied.

Introducing the scalar $\varphi(q^1, q^2, q^3)$, we assume that

$$\underline{w} = \varphi \underline{b}\,, \qquad \overset{0}{\underline{\underline{\nabla w}}} = (\overset{0}{\nabla}\varphi)\underline{b}\,, \qquad (\overset{0}{\underline{\underline{\nabla w}}})^T = \underline{\underline{b\overset{0}{\nabla}\varphi}}\,, \tag{3}$$

and that the restrictions imposed upon the choice of $\underline{w}$ are satisfied provided that $\varphi \not\equiv 0$ and $\varphi = 0$ over the entire boundary O. This will ensure that this condition is satisfied on O_2. On the remaining part O_1 of the boundary the vector $\underline{w}$ may be arbitrary, in particular, it may also be assumed to be zero.

Inequality (1) now takes the form

$$\iiint_v \underline{\nabla\varphi} \cdot \underline{\underline{S}} \cdot \underline{\nabla\varphi}\, dv > 0\,, \qquad \underline{\underline{S}} = A^{qrts}\overset{0}{b}_r \overset{0}{b}_s \underline{r}_q \underline{r}_t = s^{qt}\underline{r}_q\underline{r}_t\,. \tag{4}$$

The expressions (4) and (2) allow the following alternative representations

$$\iiint_v s^{qt}\frac{\partial\varphi}{\partial q^q}\frac{\partial\varphi}{\partial q^t}\,dv > 0\,, \qquad s^{qt}\overset{0}{a}_q\overset{0}{a}_t > 0 \tag{5}$$

and it is necessary to verify that for the first (integral) inequality to hold, matrix $\| s^{qt}\|$ should be nonnegative. This condition is "not obvious" because the positiveness of an integral does not necessarily imply the same property for an integrand. In order to prove this assertion the appropriate function $\varphi(q^1,q^2,q^3)$ which satisfies the above conditions should be chosen.

A constant (independent of coordinates q^1, q^2, q^3) tensor $\underline{\underline{S}} = \underline{\underline{S}}^T$ may be reduced to its principal axes

$$\underline{\underline{S}} = \lambda_1\underline{i}_1\underline{i}_1 + \lambda_2\underline{i}_2\underline{i}_2 + \lambda_3\underline{i}_3\underline{i}_3\,. \tag{6}$$

Here λ_k denote the eigenvalues of the tensor and $\underline{i}_k$ its principal directions, these are for all points of the body. These directions may be taken as the axes of the Cartesian coordinate system x^1, x^2, x^3

$$\underline{\nabla\varphi}\cdot\underline{\underline{S}}\cdot\underline{\nabla\varphi} = s^{qt}\frac{\partial\varphi}{\partial q^q}\frac{\partial\varphi}{\partial q^t} = \sum_{k=1}^{3}\lambda_k\underline{\nabla\varphi}\cdot\underline{i}_k\underline{i}_k\cdot\underline{\nabla\varphi} = \sum_{k=1}^{3}\lambda_k\left(\frac{\partial\varphi}{\partial x^k}\right)^2\,. \tag{7}$$

The Inequalities (5) now take the form

$$\sum_{k=1}^{3}\lambda_k\iiint_v\left(\frac{\partial\varphi}{\partial x^k}\right)^2 dx^1dx^2dx^3 > 0\,, \qquad \sum_{k=1}^{3}\lambda_k\overset{0}{a}{}_k^2 \ge 0 \qquad (\overset{0}{a}_k = \underline{i}_k\cdot\underline{a}) \tag{8}$$

and it remains only to show that the first inequality implies the nonnegativeness of λ_k.

To this end we assume that the point x_0^1, x_0^2, x_0^3 is a vertex of the parallelepiped K_h with edges h_1, h_2, h_3 located entirely within the body. The coordinates of the points contained satisfy the inequalities

$$x_0^k \le x^k \le x_0^k + h_k \ ;$$

we set

$$\varphi(x^1,x^2,x^3) = \begin{cases} 0 & \text{outside } K_h \\ f\left(\frac{x^1 - x_0^1}{h_1}\right) f\left(\frac{x^2 - x_0^2}{h_2}\right) f\left(\frac{x^3 - x_0^3}{h}\right) & \text{within } K_h \end{cases} . \tag{10}$$

The function $f(t)$ is defined here for $0 \le t \le 1$ and vanishes, together with its derivatives up to some sufficiently high order, for $t = 0$ and $t = 1$. The integrals (8) are now transformed to

$$\iiint\limits_v \left(\frac{\partial\varphi}{\partial x^1}\right)^2 dx^1 dx^2 dx^3 =$$

$$= \int_{x_0^1}^{x^1} \frac{1}{h_1^2}\left[f'\left(\frac{x^1 - x_0^1}{h_1}\right)\right]^2 dx^1 \cdot \int_{x_0^2}^{x^2} \left[f\left(\frac{x^2 - x_0^2}{h_2}\right)\right]^2 dx^2 \int_{x_0^3}^{x^3} \left[f\left(\frac{x^3 - x_0^3}{h_3}\right)\right]^2 dx^3 .$$

Using the notation

$$\int_0^1 f'^2(t)dt = \alpha , \qquad \int_0^1 f^2(t)dt = \beta \qquad (\alpha > 0,\ \beta > 0)$$

we now obtain

$$\iiint\limits_v \left(\frac{\partial\varphi}{\partial x^k}\right)^2 dx^1 dx^2 dx^3 = \alpha\beta^2 \frac{h_1 h_2 h_3}{h_k^2} .$$

Substituting this into (8), we arrive at the inequality

$$\sum_{k=1}^{3} \lambda_k \iiint\limits_v \left(\frac{\partial\varphi}{\partial x^k}\right)^2 dx^1 dx^2 dx^3 = \alpha\beta^2 h_1 h_2 h_3 \left(\frac{\lambda_1}{h_1^2} + \frac{\lambda_2}{h_2^2} + \frac{\lambda_3}{h_3^2}\right) > 0 ,$$

which implies that

$$\lambda_k \ge 0 \qquad (k = 1,2,3) ,$$

here all three inequalities cannot be satisfied simultaneously. This completes the proof of the theorem:
stability of equilibrium ⇒ Hadamard's condition.

The inverse statement:

strong ellipticity ⇒ stability of the affine equilibrium configuration can be proved for the first boundary value problem when the displacement vector is prescribed over the entire surface of a body, i.e. when the superposed field of virtual displacements is subject to the condition that

$$\text{on } O\colon\ \underline{w} = \underline{0}\ . \tag{1}$$

This condition may be regarded as attainable if it is assumed that the body, in its natural configuration, is constrained by an absolutely rigid band and that displacements of the material particles adjacent to the inner surface of the body are prevented. It must also be assumed that there is some device which enables the dimensions of the band to be varied in the course of the affine transformation.

We introduce the vector

$$\underline{W}(q^1,q^2,q^3) = \begin{cases} \underline{w}(q^1,q^2,q^3)\ , & (q^1,q^2,q^3) \in v\ , \\ 0\ , & (q^1,q^2,q^3) \bar{\in} v\ , \end{cases} \tag{2}$$

which from (1) is continuous and piecewise differentiable in E_3. This allows us to use instead of (23.1) the expression

$$\iiint\limits_{v} \underline{\underline{\overset{0}{\nabla}\underline{W}}}^{T} \cdot\cdot\, \underline{\underline{\underline{\underline{\partial_{\overset{0}{\nabla}R\overset{0}{\nabla}R}}}}} \cdot\cdot\, \underline{\underline{\overset{0}{\nabla}\underline{W}}}^{T} dv = \iiint\limits_{E_3} \underline{\underline{\overset{0}{\nabla}\underline{W}}}^{T} \cdot\cdot\, \underline{\underline{\underline{\underline{\partial_{\overset{0}{\nabla}R\overset{0}{\nabla}R}}}}} \cdot\cdot\, \underline{\underline{\overset{0}{\nabla}\underline{W}}}^{T} dv\ , \tag{3}$$

where $\underline{\underline{\underline{\underline{\partial_{\overset{0}{\nabla}R\overset{0}{\nabla}R}}}}}$ is a tensor of the fourth rank which is dependent of the coordinates.

The vector $\underline{W}$ defined in E_3 will be represented by its Fourier transformation

$$\underline{W}(q^1,q^2,q^3) = (2\pi)^{-3/2} \iiint\limits_{-\infty}^{\infty} \exp(i\underline{r}\cdot\underline{\omega})\,\underline{C}(\omega_1,\omega_2,\omega_3)\,d\omega_1 d\omega_2 d\omega_3\ , \tag{4}$$

where $\underline{C}$ denotes the Fourier transform of $\underline{W}$

$$\underline{C} = (2\pi)^{-3/2} \iint\limits_{E_3} \exp(-i\underline{r}\cdot\underline{\omega})\,\underline{W}(q^1,q^2,q^3)\,dv\ . \tag{5}$$

Denoting conjugate values by a superimposed dot and going to Cartesian coordinates a^q, we have

$$\iiint\limits_{E_3} \underline{\overset{0}{\nabla}\mathrm{W}}^T \cdot\cdot \, \exists_{\underline{\overset{0}{\nabla}\mathrm{R}\overset{0}{\nabla}\mathrm{R}}} \cdot\cdot \, \underline{\overset{0}{\nabla}\mathrm{W}}^T dv = \sum_{qrts} A^{qrts} \iiint\limits_{E_3} \frac{\partial W_r}{\partial a^q} \frac{\partial W_s}{\partial a^t} dv =$$

$$= \sum_{qrts} A^{qrts} \iiint\limits_{-\infty}^{\infty} (-i\omega_q)\bar{C}_r(-i\omega)(i\omega_t)C_s(i\omega)\, d\omega_1 d\omega_2 d\omega_3 \,, \tag{6}$$

where the latter equality follows from Plancherel's theorem and from the differentiation theorem which asserts that the transformation of a derivative with respect to a^m is equal to the product of $i\omega_m$ and the transformation of the function itself.

Equation (3) may now be rewritten in the form

$$\iiint\limits_{E_3} \underline{\overset{0}{\nabla}\mathrm{W}}^T \cdot\cdot \, \exists_{\underline{\overset{0}{\nabla}\mathrm{R}\overset{0}{\nabla}\mathrm{R}}} \cdot\cdot \, \underline{\overset{0}{\nabla}\mathrm{W}}^T dv = \iiint\limits_{-\infty}^{\infty} \sum_{rs} \bar{C}_r C_s \sum_{qt} A^{qrts} \omega_q \omega_t \, d\omega_1 d\omega_2 d\omega_3 =$$

$$= \iiint\limits_{-\infty}^{\infty} \sum_{rs} |C|^2 \bar{c}_r c_s \sum_{qt} A^{qrts} \omega^2 e_q e_t \, d\omega_1 d\omega_2 d\omega_3 \,, \tag{7}$$

where we have introduced the vectors $\underline{c}$ and $\underline{\omega}$

$$c_s = \frac{C_s}{|C|} \,, \qquad e_q = \frac{\omega_q}{\omega} \,.$$

For a strongly elliptic material the quadratic form of the components of the unit vector $\underline{e}$

$$\sum_{qt} A^{qrts} e_q e_t$$

is positive definite. Its least eigenvalue will be denoted by λ_{rs}.

Using the symbol $\varepsilon > 0$ for the least value of λ_{rr} and setting λ_{rs} equal to zero for $r \neq s$, we obtain

$$\sum_{qt} A^{qrts} e_q e_t > \varepsilon \delta_{rs} \,,$$

$$\sum_r \sum_s \bar{c}_r c_s \sum_{q,t} A^{qrts} e_q e_t > \varepsilon \sum_r \sum_s \delta_{rs} \bar{c}_r c_s = \frac{\varepsilon}{3} \sum_s \bar{c}_s c_s = \varepsilon$$

and returning to (3) and (7)

$$\iiint\limits_{v} \underline{\overset{0}{\nabla}\mathrm{W}}^T \cdot\cdot \, \exists_{\underline{\overset{0}{\nabla}\mathrm{R}\overset{0}{\nabla}\mathrm{R}}} \cdot\cdot \, \underline{\overset{0}{\nabla}\mathrm{W}}^T dv > \varepsilon \iiint\limits_{-\infty}^{\infty} \sum_{j,k} \omega_j^2 |C_k|^2 d\omega_1 d\omega_2 d\omega_3 =$$

$$= \varepsilon \iiint_{E_3} \sum_{j,k} \left(\frac{\partial W_k}{\partial a^j}\right)^2 dv = \varepsilon \iiint_{v} \sum_{j,k} \left(\frac{\partial w_k}{\partial a^j}\right)^2 dv > 0 . \tag{8}$$

We have thus proved that the equilibrium affine deformation of a strongly elliptic material is stable for the first boundary value problem.

Strong ellipticity represents a property of material which is exhibited within a certain range of its deformation. For a quasi-linear material for example, this range is determined by Inequalities (5.5.20), for a Blatz and Ko material by Inequalities (5.6.16) and the hypothetical material (5.6.17) is strongly elliptic for all deformations.

Whenever the parameters of affine deformation in our thought experiment fall outside the range for strong ellipticity, the V-configuration which has hitherto been stable ceases to be so. The new equilibrated $\tilde{V}$-configuration, provided that it exists, is not affine, the value of its potential energy is less than that corresponding to the V-configuration. The process of transition from the V- to the $\tilde{V}$-configuration must be accompanied by the displacement of the particles within the fixed volume which they occupy. The material "has nowhere to go" and so we must expect the emergence, in a stable $\tilde{V}$-configuration (possessing the minimum value of energy), of folds and points and lines of discontinuity of the derivatives.

25. EXAMPLE. A DEFORMABLE DISK IN A RIGID BAND

A disk of unit radius is confined in its reference natural configuration by a cylindrical band applied by some device which allows its inner radius to vary. The material of the band (a hollow cylinder with plates forming its faces) is absolutely rigid. It is assumed that material particles of the disk are fixed with respect to the confining surfaces of the band. In particular, the component of the stretch tensor along the direction perpendicular to the disk's plane is equal to unity, i.e. $v_3 = 1$.

The material of the disk is assumed to be quasi-linear. By (5.5.20) it remains strongly elliptic provided that

$$L_0 = v_1 + v_2 - \frac{1}{1-\nu} > 0 ,$$

$$L_1 = (1-\nu)v_1 + \nu v_2 - 2\nu > 0 , \tag{1}$$

$$L_2 = \nu v_1 + (1-\nu)v_2 - 2\nu > 0 .$$

The domain of strong ellipticity, i.e. of stable affine deformation, is bounded in the quadrant $v_1 > 0$, $v_2 > 0$ of the v_1, v_2 plane by those line segments whose ends are defined by the following coordinates

$$\left[(0\,;\,2)\,,\left(\frac{\nu}{1-\nu}\,;\,1\right)\right] \qquad \text{on the line } L_1 = 0\,,$$

$$\left[(2\,;\,0)\,,\left(1\,;\,\frac{\nu}{1-\nu}\right)\right] \qquad \text{on the line } L_2 = 0\,, \tag{2}$$

$$\left[\left(\frac{\nu}{1-\nu}\,;\,1\right),\left(1\,;\,\frac{\nu}{1-\nu}\right)\right] \qquad \text{on the line } L_0 = 0\,,$$

the domain lying above these lines. The point nearest the origin lies on the bisector of the angle between the axes v_1 and v_2 and its coordinates are given by

$$v_1 = v_2 = \frac{1}{2(1-\nu)}\,, \tag{3}$$

and for

$$v_s \le \frac{1}{2(1-\nu)} \qquad (s = 1,2) \tag{4}$$

the material ceases to be strongly elliptic.

The material coordinates in the reference configuration are the polar coordinates r, φ. The transformation

$$R = kr\,, \qquad \Phi = \varphi\,, \qquad Z = z\,, \qquad \underline{R} = kr\underline{e}_r + \underline{i}_3 z \tag{5}$$

determines an affine deformation.

For

$$k > k_0 = \frac{1}{2(1-\nu)} \qquad \left(\text{for } \nu = 0\text{: } k_0 = \frac{1}{2};\ \text{ for } \nu = \frac{1}{2} : k_0 = 1\right) \tag{6}$$

the material is strongly elliptic, and the affine deformation is stable. For $k < k_0$ this deformation is unstable.

Under the constraints imposed the deformation may be non-affine, in particular, if a rotation through $\psi(r)$, which equals zero on the circular boundary of the disk, is superposed upon deformation (5), i.e. we have

$$\psi(1) = 0\,. \tag{7}$$

Instead of (5), we now have

$$R = kr\,, \qquad \Phi = \varphi + \psi(r)\,, \qquad Z = z \quad \text{and} \quad \underline{R} = kr\underline{e}_R + \underline{i}_3 z\,, \tag{8}$$

where

$$\underline{e}_R = \underline{e}_r \cos\psi + \underline{e}_\varphi \sin\psi ,$$

$$\frac{\partial \underline{e}_R}{\partial \Phi} = \underline{e}_\Phi = \frac{\partial \underline{e}_R}{\partial \varphi} = -\underline{e}_r \sin\psi + \underline{e}_\varphi \cos\psi ,$$

$$\frac{\partial \underline{e}_R}{\partial r} = \psi'(r)\underline{e}_\Phi . \tag{9}$$

The base vectors $\underline{R}_s$ in this configuration and the deformation gradient are given by the formulas

$$\underline{R}_1 = \underline{R}_r = k(\underline{e}_R + r\psi'\underline{e}_\Phi) , \qquad \underline{R}_2 = \underline{R}_\varphi = kr\underline{e}_\Phi , \qquad \underline{R}_3 = \underline{R}_z = \underline{i}_3 ,$$

$$\overset{0}{\underline{\underline{\nabla R}}} = k\,[\underline{e}_r(\underline{e}_R + r\psi'\underline{e}_\Phi) + \underline{e}_\varphi\underline{e}_\Phi] + \underline{i}_3\underline{i}_3 .$$

The Cauchy strain measure and its invariants are equal to

$$\underline{G} = \overset{0}{\underline{\underline{\nabla R}}} \cdot \overset{0}{\underline{\underline{\nabla R}}}{}^T = k^2\,[\underline{e}_r\underline{e}_r(1 + r^2\psi'^2) + (\underline{e}_r\underline{e}_\varphi + \underline{e}_\varphi\underline{e}_r)r\psi' + \underline{e}_\varphi\underline{e}_\varphi] + \underline{i}_3\underline{i}_3 ,$$

$$I_1(\underline{G}) = k^2(2 + r^2\psi'^2) + 1 = v_1^2 + v_2^2 + 1 ,$$

$$I_2(\underline{G}) = k^4 + k^2(2 + r^2\psi'^2) = v_1^2v_2^2 + v_1^2 + v_2^2 ,$$

$$I_3(\underline{G}) = v_1^2v_2^2 .$$

From here we obtain

$$v_1^2 + v_2^2 = k^2(2 + r^2\psi'^2) , \quad v_1v_2 = k^2 , \quad v_1 + v_2 = kA , \quad A = (4 + r^2\psi'^2)^{1/2} \tag{10}$$

and the invariants $I_k(\underline{G}^{1/2}) = I_k'$ are given by

$$I_1' = v_1 + v_2 + 1 = kA + 1 , \quad I_2' = v_1v_2 + v_1 + v_2 = k^2 + kA , \quad I_3' = k^2 .$$

The specific strain energy (17.9) of a quasi-linear material is reduced after some obvious transformations to the form

$$Э(A) = \frac{1}{2}(\lambda + 2\mu)\left[k^2A^2 - \frac{2}{1-\nu}kA\right] + 2\lambda - 2\mu(1 - k^2) . \tag{11}$$

The parameter A is defined by (10) and for the affine deformation, is equal to 2. The minimum value of the energy coincides for this case with that of the potential energy

$$\mathfrak{Z} = 2\pi \int_0^1 r\,dr\,\exists(A) \;, \tag{12}$$

because the work produced by surface tractions for a virtual displacement from the equilibrium configuration (8) is zero.

By (11), the functional $\exists(A)$ is a minimum for

$$A = \frac{1}{k(1-\nu)} \;. \tag{13}$$

If the affine deformation is unstable, i.e. if

$$k < k_0 = \frac{1}{2(\;\;-\nu)} \;,$$

then there exists some non-affine deformation (8) for which (13) holds and by (10)

$$r^2\psi'^2 = \frac{1-4k^2(1-\nu)^2}{k^2(1-\nu)^2} \;.$$

By virtue of (7) we have

$$\psi' = -\frac{\sqrt{1-4k^2(1-\nu)^2}}{k(1-\nu)}\,\frac{1}{r} \;,$$

$$\psi = \frac{\sqrt{1-4k^2(1-\nu)^2}}{k(1-\nu)}\ln\frac{1}{r} \;. \tag{14}$$

The angle ψ thus increases without limit as we approach the centre of the disk.

The centre is a singular point of the stress field, a real material would not preserve on its continuity in its neighbourhood and a crack must be expected to form.

26. PLANE WAVES IN A UNIFORMLY STRESSED INCOMPRESSIBLE ELASTIC MEDIUM

Here we return to the analysis of Pars. 7-9 with the assumption that an elastic medium is incompressible. The equations of motion (18.13) take the following form in the $V^{\times}$-configuration

$$\underline{\nabla}\cdot\underline{\theta}_E + \underline{\nabla}\cdot p\underline{\underline{\nabla w}}^{\mathrm{T}} - \underline{\nabla}p' + \rho\omega^2\underline{w} = \underline{0} \;, \tag{1}$$

where the field of virtual displacements satisfies the continuity equation

$$\underline{\nabla} \cdot \underline{w} = 0 \ . \tag{2}$$

The state of stress in the V-configuration is assumed to be uniform, i.e. that tensors $\underline{T}$ and $\underline{F}$ are constant

$$\underline{T} = -p\underline{E} + 2[(\ni_1 + I_1 \ni_2)\underline{F} - \ni_2 \underline{F}^2] \ ,$$

$$\underline{F} = v_1^2 \underline{e}_1 \underline{e}_1 + v_2^2 \underline{e}_2 \underline{e}_2 + v_3^2 \underline{e}_3 \underline{e}_3 \ . \tag{3}$$

Here $\underline{e}_s$ denote the principal directions of the Finger measure, and v_s^2 its constant principal values where

$$I_3(\underline{F}) = v_1^2 v_2^2 v_3^2 = 1 \ . \tag{4}$$

A pressure p is also constant and therefore

$$\underline{\nabla} \cdot p\underline{\nabla w}^T = p\underline{\nabla} \cdot \underline{\nabla w}^T = p\underline{\nabla\nabla} \cdot \underline{w} = \underline{0} \ . \tag{5}$$

Since now $\ni = \ni(I_1, I_2)$, an acoustical tensor $\underline{\underline{Q}}$ in the equation of motion (7.7)

$$\underline{\nabla} \cdot \underline{\theta}_E = -4k^2 \underline{\underline{Q}} \cdot \underline{w} \tag{6}$$

is given by the expression

$$\underline{\underline{Q}} = \frac{1}{2}(\psi_1 \underline{N} \cdot \underline{F} \cdot \underline{N} + \psi_2 \underline{N} \cdot \underline{F}^2 \cdot \underline{N})\underline{E} + \frac{1}{2}\psi_2(\underline{N} \cdot \underline{F} \cdot \underline{NF} + \underline{N} \cdot \underline{FF} \cdot \underline{N}) +$$

$$+ \vartheta_{11} \underline{N} \cdot \underline{FF} \cdot \underline{N} + \vartheta_{12}(\underline{N} \cdot \underline{FF}^2 \cdot \underline{N} + \underline{N} \cdot \underline{F}^2 \underline{F} \cdot \underline{N}) + \vartheta_{22} \underline{N} \cdot \underline{F}^2 \underline{F}^2 \cdot \underline{N} \ . \tag{7}$$

Describing the plane wave by (7.1) and making use of (7.4), we obtain by (2)

$$\underline{\nabla} \cdot \underline{w} = ik\underline{N} \cdot \underline{w} \ , \qquad \underline{N} \cdot \underline{w} = 0 \tag{8}$$

as expected longitudinal waves are absent in an incompressible medium. From (1) and (5) we have

$$\underline{N} \cdot \underline{\nabla} p' = -4k^2 \underline{N} \cdot \underline{\underline{Q}} \cdot \underline{w} \ , \qquad \underline{NN} \cdot \underline{\nabla} p' = -4k^2 \underline{NN} \cdot \underline{\underline{Q}} \cdot \underline{w} \ .$$

Moreover,

$$\underline{N}\cdot\underline{\nabla}p' = ik\underline{N}\cdot\underline{N}p' = ikp' , \qquad \underline{\underline{NN}}\cdot\underline{\nabla}p' = ik\underline{N}p' = \underline{\nabla}p$$

and therefore

$$\underline{\nabla}p' = -4k^2\underline{\underline{NN}}\cdot\underline{\underline{Q}}\cdot\underline{w} . \tag{9}$$

The equation of motion (1) now reduces to the form

$$\underline{\underline{Q}}^{\times}\cdot\underline{w} = \frac{1}{4}\rho c^2\underline{w} , \qquad \underline{\underline{Q}}^{\times} = \underline{\underline{Q}} - \underline{\underline{NN}}\cdot\underline{\underline{Q}} \qquad (c^2 = \omega^2/k^2) . \tag{10}$$

Here we should term the tensor $\underline{\underline{Q}}^{\times}$ the acoustical tensor of an incompressible elastic medium. From (7) it may be put into the form

$$\begin{aligned}\underline{\underline{Q}}^{\times} = {} & \frac{1}{2}(\psi_1\underline{N}\cdot\underline{F}\cdot\underline{N}+\psi_2\underline{N}\cdot\underline{F}^2\cdot\underline{N})(\underline{E}-\underline{\underline{NN}}) + \\ & + \frac{1}{2}\psi_2[\underline{N}\cdot\underline{F}\cdot\underline{\underline{NF}}+\underline{N}\cdot\underline{\underline{FF}}\cdot\underline{N}-2\underline{N}\cdot\underline{F}\cdot\underline{\underline{NNN}}\cdot\underline{F}] + \\ & + \vartheta_{11}(\underline{N}\cdot\underline{\underline{FF}}\cdot\underline{N}-\underline{N}\cdot\underline{F}\cdot\underline{\underline{NNN}}\cdot\underline{F}) + \\ & + \vartheta_{12}[(\underline{N}\cdot\underline{F}-\underline{\underline{NN}}\cdot\underline{F}\cdot\underline{N})\underline{F}^2\cdot\underline{N}+(\underline{N}\cdot\underline{F}^2-\underline{\underline{NN}}\cdot\underline{F}^2\cdot\underline{N})\underline{F}\cdot\underline{N}] + \\ & + \vartheta_{22}(\underline{N}\cdot\underline{F}^2-\underline{\underline{NN}}\cdot\underline{F}^2\cdot\underline{N})\underline{F}^2\cdot\underline{N} .\end{aligned} \tag{11}$$

Substituting ψ_Γ and $\vartheta_{N\Gamma}$ for their values

$$\psi_1 = \ni_1 + I_1\ni_2 , \qquad \psi_2 = -\ni_2 , \qquad \vartheta_{11} = \ni_{11} + 2I_1\ni_{12} + I_1^2\ni_{22} + \ni_2 ,$$

$$\vartheta_{12} = -\ni_{12} - I_1\ni_{22} , \qquad \vartheta_{22} = \ni_{22} ; \qquad \ni_i = \frac{\partial\ni}{\partial I_i} , \qquad \ni_{ik} = \frac{\partial^2\ni}{\partial I_i\partial I_k} \qquad (k = 1,2)$$

we arrive at the expressions

$$\begin{aligned}\underline{\underline{Q}}^{\times} = {} & \frac{1}{2}\ni_1\underline{N}\cdot\underline{F}\cdot\underline{N}(\underline{E}-\underline{\underline{NN}}) + \\ & + \frac{1}{2}\ni_2[(I_1\underline{N}\cdot\underline{F}\cdot\underline{N}-\underline{N}\cdot\underline{F}^2\cdot\underline{N})(\underline{E}-\underline{\underline{NN}}) - \underline{N}\cdot\underline{F}\cdot\underline{\underline{NF}} + \underline{N}\cdot\underline{\underline{FF}}\cdot\underline{N}] + \\ & + \ni_{11}(\underline{N}\cdot\underline{\underline{FF}}\cdot\underline{N}-\underline{N}\cdot\underline{F}\cdot\underline{\underline{NNN}}\cdot\underline{F}) + \ni_{22}[\underline{N}\cdot\underline{F}^2\underline{F}^2\cdot\underline{N}-\underline{N}\cdot\underline{F}^2\cdot\underline{\underline{NN}}\cdot\underline{F}^2 + \\ & + I_1^2(\underline{N}\cdot\underline{\underline{FF}}\cdot\underline{N}-\underline{N}\cdot\underline{F}\cdot\underline{\underline{NNF}}\cdot\underline{N}) - I_1(\underline{N}\cdot\underline{F}^2\underline{F}^2\cdot\underline{\underline{NNF}}\cdot\underline{N})] +\end{aligned}$$

$$+ \ni_{12}[2I_1(\underline{N}\cdot\underline{FF}\cdot\underline{N}-\underline{N}\cdot\underline{F}\cdot\underline{NNF}\cdot\underline{N}) - (\underline{N}\cdot\underline{FF}^2\cdot\underline{N}-\underline{N}\cdot\underline{F}\cdot\underline{NNF}^2\cdot\underline{N}) -$$

$$- (\underline{N}\cdot\underline{F}^2\underline{F}^2\cdot\underline{N}-\underline{N}\cdot\underline{F}^2\cdot\underline{NNN}\cdot\underline{F})] \, . \tag{12}$$

We have of course $\underline{\underline{Q}}^{\times}\cdot\underline{N} = \underline{N}\cdot\underline{\underline{Q}}^{\times} = \underline{0}$ because all terms in $\underline{\underline{Q}}^{\times}$ are zero. In the orthonormal trihedron $\underline{N}$, $\underline{t}_1$, $\underline{t}_2$ the tensor $\underline{\underline{Q}}^{\times}$ therefore turns out to be a plane tensor

$$\underline{\underline{Q}}^{\times} = \lambda_{11}\underline{t}_1\underline{t}_1 + \lambda_{12}(\underline{t}_1\underline{t}_2+\underline{t}_2\underline{t}_1) + \lambda_{22}\underline{t}_2\underline{t}_2 \, , \tag{13}$$

where

$$\lambda_{sk} = \frac{1}{2}[\ni_1\underline{N}\cdot\underline{F}\cdot\underline{N}+\ni_2(I_1\underline{N}\cdot\underline{F}\cdot\underline{N}-\underline{N}\cdot\underline{F}^2\cdot\underline{N})]\delta_{sk} +$$

$$+ \frac{1}{2}\ni_2(\underline{N}\cdot\underline{F}\cdot\underline{t}_s\underline{N}\cdot\underline{F}\cdot\underline{t}_k-\underline{N}\cdot\underline{F}\cdot\underline{Nt}_s\cdot\underline{F}\cdot\underline{t}_k) + \ni_{11}\underline{N}\cdot\underline{F}\cdot\underline{t}_s\underline{N}\cdot\underline{F}\cdot\underline{t}_k +$$

$$+ \ni_{22}[\underline{N}\cdot\underline{F}^2\cdot\underline{t}_s\underline{N}\cdot\underline{F}^2\cdot\underline{t}_k + I_1^2\underline{N}\cdot\underline{F}\cdot\underline{t}_s\underline{N}\cdot\underline{F}\cdot\underline{t}_k -$$

$$- I_1(\underline{N}\cdot\underline{F}^2\cdot\underline{t}_s\underline{N}\cdot\underline{F}\cdot\underline{t}_k+\underline{N}\cdot\underline{F}\cdot\underline{t}_s\underline{N}\cdot\underline{F}^2\cdot\underline{t}_k)] +$$

$$+ \ni_{12}[2I_1\underline{N}\cdot\underline{F}\cdot\underline{t}_s\underline{N}\cdot\underline{F}\cdot\underline{t}_k-\underline{N}\cdot\underline{F}\cdot\underline{t}_s\underline{N}\cdot\underline{F}^2\cdot\underline{t}_k-\underline{N}\cdot\underline{F}^2\cdot\underline{t}_k\underline{N}\cdot\underline{F}\cdot\underline{t}_s] \, ; \tag{14}$$

$s,k = 1,2$, and δ_{sk} denotes the Kronecker delta.

We now proceed to the analysis of some special cases.

1. The principal waves. The normal $\underline{N}$ points along the principal direction of the Finger measure and we may assume $\underline{t}_1 = \underline{e}_2$, $\underline{t}_2 = \underline{e}_3$ because the vectors $\underline{t}_1$, $\underline{t}_2$ are determined up to some rotation around $\underline{N}$. We thus obtain

$$\lambda_{11} = \frac{1}{2}v_1^2(\ni_1+\ni_2v_3^2) \, , \qquad \lambda_{22} = \frac{1}{2}v_1^2(\ni_1+\ni_2v_2^2) \, , \qquad \lambda_{12} = 0 \, ,$$

so that

$$\underline{\underline{Q}}^{\times} = \frac{1}{2}v_1^2[(\ni_1+\ni_2v_3^2)\underline{e}_2\underline{e}_2 + (\ni_1+\ni_2v_2^2)\underline{e}_3\underline{e}_3] \, , \tag{15}$$

and the squares of the velocities of the principal transverse waves propagating along $\underline{e}_1$ and polarised in the planes $(\underline{e}_1,\underline{e}_2)$ and $(\underline{e}_1,\underline{e}_3)$ are given by the formulas

$$\frac{1}{2}\rho\frac{c_{12}^2}{v_1^2} = \ni_1 + \ni_2v_3^2 \, , \qquad \frac{1}{2}\rho\frac{c_{13}^2}{v_1^2} = \ni_1 + \ni_2v_2^2 \, . \tag{16}$$

Directing $\underline{N}$ along $\underline{e}_2$ and $\underline{e}_3$, we get likewise

$$\frac{1}{2}\rho\frac{c_{23}^2}{v_2^2} = \ni_1 + \ni_2 v_1^2 , \qquad \frac{1}{2}\rho\frac{c_{21}^2}{v_2^2} = \ni_1 + \ni_2 v_3^2 , \tag{17}$$

$$\frac{1}{2}\rho\frac{c_{31}^2}{v_3^2} = \ni_1 + \ni_2 v_2^2 , \qquad \frac{1}{2}\rho\frac{c_{32}^2}{v_3^2} = \ni_1 + \ni_2 v_1^2 . \tag{18}$$

Referring to the state equation (3) we have

$$\frac{1}{2}\frac{\sigma_1 - \sigma_2}{v_1^2 - v_2^2} = \ni_1 + \ni_2 v_3^2 , \qquad \frac{1}{2}\frac{\sigma_2 - \sigma_3}{v_2^2 - v_3^2} = \ni_1 + \ni_2 v_1^2 , \qquad \frac{1}{2}\frac{\sigma_3 - \sigma_1}{v_3^2 - v_1^2} = \ni_1 + \ni_2 v_2^2$$

and this enables us to express the squares of velocities of the principal waves in the form

$$\frac{\rho c_{12}^2}{v_1^2} = \frac{\rho c_{21}^2}{v_2^2} = \frac{\sigma_1 - \sigma_2}{v_1^2 - v_2^2} , \qquad \frac{\rho c_{23}^2}{v_2^2} = \frac{\rho c_{32}^2}{v_3^2} = \frac{\sigma_2 - \sigma_3}{v_2^2 - v_3^2} , \qquad \frac{\rho c_{31}^2}{v_3^2} = \frac{\rho c_{13}^2}{v_1^2} = \frac{\sigma_3 - \sigma_1}{v_3^2 - v_1^2} . \tag{19}$$

Whenever the $\mathcal{BE}$-criteria hold, i.e. as long as the material continues to be strongly elliptic, the velocities are real and the plane waves propagate along the principal directions of the Finger measure. These conditions are necessary for a medium to be stable. They are, however, not complete because the possibility that the velocities would not be real for propagation along directions not coincident with the principal directions cannot be excluded.

If any of the $\mathcal{BE}$-criteria is violated, then an incompressible medium is unstable in a state of uniform stress.

2. The plane problem. Following Rivlin and Sawyers (1973), we will consider the waves propagating in the $(\underline{e}_1,\underline{e}_2)$ plane and which are linearly polarised in this plane so that

$$\underline{w}\cdot\underline{N} = 0 , \quad \underline{w}\cdot\underline{e}_3 = 0 , \quad \underline{N}\cdot\underline{e}_3 = 0 , \quad \underline{N} = \underline{e}_1 N_1 + \underline{e}_2 N_2 , \tag{20}$$

where, as before, $\underline{e}_s$ denotes the principal directions of the Finger measure. Here, with the usual notation,

$$\underline{e}_3 = \underline{t}_1 , \quad \underline{t}_2 = \underline{N}\times\underline{t}_1 = -N_1\underline{e}_2 + N_2\underline{e}_1 , \tag{21}$$

and the Equation (10) takes the form

$$[\lambda_{11}\underline{e}_3\underline{e}_3 + \lambda_{12}(\underline{e}_3\underline{t}_2 + \underline{e}_2\underline{t}_3) + \lambda_{22}\underline{t}_2\underline{t}_2]\cdot w_2\underline{t}_2 =$$

$$= (\lambda_{12}\underline{e}_3 + \lambda_{22}\underline{t}_2)w_2 = \frac{1}{4}\rho c_{12}^2 w_2\underline{t}_2 , \tag{22}$$

where c_{12} denotes the wave velocity in the plane $(\underline{e}_1,\underline{e}_2)$. For $w_2 \neq 0$ we obtain

$$\lambda_{12} = 0 , \qquad \lambda_{22} = \frac{1}{4}\rho c_{12}^2 \tag{23}$$

and by (14)

$$\lambda_{22} = \frac{1}{2}\partial_1\underline{N}\cdot\underline{F}\cdot\underline{N} + \frac{1}{2}\partial_2[I_1\underline{N}\cdot\underline{F}\cdot\underline{N} - \underline{N}\cdot\underline{F}^2\cdot\underline{N} + (\underline{N}\cdot\underline{F}\cdot\underline{t}_2)^2 - \underline{N}\cdot\underline{F}\cdot\underline{N}\underline{t}_2\cdot\underline{N}\cdot\underline{t}_2] +$$

$$+ \partial_{11}(\underline{N}\cdot\underline{F}\cdot\underline{t}_2)^2 + \partial_{22}(I_1\underline{N}\cdot\underline{F}\cdot\underline{t}_2 - \underline{N}\cdot\underline{F}^2\cdot\underline{t}_2)^2 + 2\partial_{12}(I_1\underline{N}\cdot\underline{F}\cdot\underline{t}_2 - \underline{N}\cdot\underline{F}^2\cdot\underline{t}_2) .$$

Substituting $\underline{N}$ and $\underline{t}_2$ by their values given by (20) and (21), we get

$$\underline{N}\cdot\underline{F}\cdot\underline{N} = N_1^2 v_1^2 + N_2 v_2^2 , \qquad \underline{N}\cdot\underline{F}\cdot\underline{t}_2 = N_1 N_2(v_1^2 - v_2^2) ,$$

$$\underline{t}_2\cdot\underline{F}\cdot\underline{t}_2 = N_1 v_2^2 + N_2^2 v_1^2$$

etc. We arrive at the formulas

$$\frac{1}{2}\rho c_{12}^2 = (\partial_1 + v_3^2\partial_2)\left[N_1^2 v_1^2 + N_2^2 v_2^2 + 2N_1^2 N_2^2(v_1^2 - v_2^2)^2 \frac{\partial_{11} + 2\partial_{12}v_3^2 + \partial_{22}v_3^4}{\partial_1 + \partial_2 v_3^2}\right] , \tag{24}$$

derived along different lines in the paper by Rivlin and Sawyers (loc. cit.).

It is assumed that $\partial_1 + v_3^2\partial_2 > 0$ - otherwise the uniform state of stress would be unstable with respect to the propagation of the principal waves. The necessary condition of stability for the type of waves considered here therefore amounts to a requirement that the value

$$\Phi_3 = N_1^2 v_1^2 + N_2^2 v_2^2 + 2N_1^2 N_2^2(v_1^2 - v_2^2)^2 B_3 ,$$

$$B_3 = \frac{\partial_{11} + 2\partial_{12}v_3^2 + \partial_{22}v_3^4}{\partial_1 + \partial_2 v_3^2} \tag{25}$$

be positive.

The sign of Φ_3 cen be determined from a consideration of those waves for which

$$N_1^2 = \frac{v_2}{v_1 + v_2}, \qquad N_2^2 = \frac{v_1}{v_1 + v_2} \tag{26}$$

satisfy the conditions $N_1^2 + N_2^2 = 1$, $0 \le N_1^2 \le 1$ and $0 \le N_2^2 \le 1$. We get

$$\Phi_3 = v_1 v_2 [1 + 2(v_1 - v_2)^2 B_3] , \tag{27}$$

so that

$$\Phi_3 \gtrless 0 \quad \text{for} \quad B_3 = \frac{\partial_{11} + 2v_3^2 \partial_{12} + \partial_{22} v_3^4}{\partial_1 + \partial_2 v_3^2} \gtrless -\frac{1}{2}(v_1 - v_2)^{-2} . \tag{28}$$

For the planes $(\underline{e}_2, \underline{e}_3)$ and $(\underline{e}_3, \underline{e}_1)$ we have in the same way

$$\Phi_1 \gtrless 0 \quad \text{for} \quad B_1 = \frac{\partial_{11} + 2v_1^2 \partial_{12} + \partial_{22} v_1^4}{\partial_1 + \partial_2 v_1^2} \gtrless -\frac{1}{2}(v_2 - v_3)^{-2} ,$$

$$\Phi_2 \gtrless 0 \quad \text{for} \quad B_2 = \frac{\partial_{11} + 2v_2^2 \partial_{12} + \partial_{22} v_2^4}{\partial_1 + \partial_2 v_2^2} \gtrless -\frac{1}{2}(v_3 - v_1)^{-2} . \tag{29}$$

The necessary condition of stability requires that these three Inequalities should hold with the upper sign (>). Instability arises with the sign (<) for any of these Inequalities.

3. <u>A Mooney-Rivlin material.</u> For this material

$$\partial = C_1(I_1 - 3) + C_2(I_2 - 3) , \qquad \partial_1 = C_1 > 0 , \qquad \partial_2 = C_2 \ge 0 . \tag{30}$$

Here $\partial_{11} = \partial_{22} = \partial_{12} = 0$, and omitting some cumbersome elementary transformations, we arrive by (14) at the formulas

$$2\lambda_{11} = \partial_1 \underline{N} \cdot \underline{F} \cdot \underline{N} + \partial_2 \left[\left(\frac{t_1^1}{v_1}\right)^2 + \left(\frac{t_1^2}{v_2}\right)^2 + \left(\frac{t_1^3}{v_3}\right)^2 \right] , \tag{31}$$

$$2\lambda_{22} = \partial_1 \underline{N} \cdot \underline{F} \cdot \underline{N} + \partial_2 \left[\left(\frac{t_2^1}{v_1}\right)^2 + \left(\frac{t_2^2}{v_2}\right)^2 + \left(\frac{t_2^3}{v_3}\right)^2 \right] , \tag{32}$$

$$2\lambda_{12} = \partial_2 \left[\frac{t_1^1 t_2^1}{v_1^2} + \frac{t_1^2 t_2^2}{v_2^2} + \frac{t_1^3 t_2^3}{v_3^2} \right] , \tag{33}$$

where

$$t_1^k = \underline{t}_1 \cdot \underline{e}_k\,, \qquad t_2^k = \underline{t}_2 \cdot \underline{e}_k \qquad (k = 1,2,3)$$

denote components of the vectors $\underline{t}_1$, $\underline{t}_2$ along the principal directions of the measure $\underline{F}$.

The eigenvalues of an acoustical tensor are given by the roots of the quadratic equation

$$\sigma^2 - (\lambda_{11} + \lambda_{22})\sigma + \lambda_{11}\lambda_{22} - \lambda_{12}^2 = 0 \tag{34}$$

and because $\lambda_{11} > 0$ and $\lambda_{22} > 0$, we arrive by (30) at the inequality

$$\lambda_{11}\lambda_{22} - \lambda_{12}^2 > 0\,. \tag{35}$$

This equation is by (31)-(33) transformed to

$$\ni_1^2(\underline{N} \cdot \underline{F} \cdot \underline{N})^2 + \ni_1\ni_2\underline{N} \cdot \underline{F} \cdot \underline{N}\left[\frac{1 - N_1^2}{v_1^2} + \frac{1 - N_2^2}{v_2^2} + \frac{1 - N_3^2}{v_3^2}\right] +$$

$$+ \ni_2^2\left\{\frac{1}{v_1^2 v_2^2}\,[(t_1^1 t_2^2)^2 + (t_1^2 t_2^1)^2 - 2t_1^1 t_2^1 t_1^2 t_2^2] +\right.$$

$$+ \frac{1}{v_2^2 v_3^2}\,[(t_1^2 t_2^3)^2 + (t_1^3 t_2^2)^2 - 2t_1^2 t_2^2 t_1^3 t_2^3] +$$

$$\left.+ \frac{1}{v_3^2 v_1^2}\,[(t_1^3 t_2^1)^2 + (t_1^1 t_2^3)^2 - 2t_1^3 t_2^3 t_1^1 t_2^1]\right\},$$

and the expression within the braces can be rewritten as follows (see (4))

$$\frac{1}{v_1^2 v_2^2}\,(t_1^1 t_2^2 - t_1^2 t_2^1)^2 + \frac{1}{v_2^2 v_3^2}\,(t_1^2 t_2^3 - t_1^3 t_2^2)^2 + \frac{1}{v_3^2 v_1^2}\,(t_1^3 t_2^1 - t_1^1 t_2^3) =$$

$$= v_3^2 N_3^2 + v_1^2 N_1^2 + v_2^2 N_2^2 = \underline{N} \cdot \underline{F} \cdot \underline{N}\,.$$

We have thus arrived at the inequality

$$\lambda_{11}\lambda_{22} - \lambda_{12}^2 = \underline{N} \cdot \underline{F} \cdot \underline{N}\left[\ni_1^2\underline{N} \cdot \underline{F} \cdot \underline{N} + \ni_1\ni_2\left(\frac{1 - N_1^2}{v_1^2} + \frac{1 - N_2^2}{v_2^2} + \frac{1 - N_3^2}{v_3^2}\right) + \ni_2^2\right] > 0 \tag{36}$$

which asserts the stability of a Mooney-Rivlin material.

Rivlin and Sawyers (1977) solved the same problem rather differently, the equation of propagation of the plane waves being written in a different form from that of (10) and (11). These authors established in their paper stability criteria not only for a Mooney-Rivlin material, but also for materials whose potential energy has the form $\ni = \ni(I_1), \ni(I_2)$. However, the general case of an arbitrary incompressible material proved to be beyond their analysis, neither were they able to obtain any algebraic stability criteria in terms of $\ni_i$, $\ni_{ik}$ for any $\underset{\sim}{N}$. The attempt of the same authors in their 1978 papers to obtain more definite results was also unsucessful.

27. HADAMARD'S CRITERION FOR A UNIFORMLY STRESSED INCOMPRESSIBLE ELASTIC MEDIUM

The criteria that the squares of the eigenvalues of an acoustical tensor should be positive

$$\lambda_{11} > 0 , \qquad \lambda_{11}\lambda_{22} - \lambda_{12}^2 > 0$$

include not only the initial specification of the quantities v_1, v_2, v_3 describing the deformation, and the functions $\ni_i$, $\ni_{ik}$ of these quantities defined by the choice of material, but also the components N_1, N_2, N_3 of the vector $\underset{\sim}{N}$ normal to the wave and of the vectors $\underset{\sim}{t}_1$ and $\underset{\sim}{t}_2$ perpendicular to $\underset{\sim}{N}$. These quantities are related to each other by the six equations

$$\underset{\sim}{N} \cdot \underset{\sim}{N} = 1 , \qquad \underset{\sim}{t}_\alpha \cdot \underset{\sim}{t}_\beta = \delta_{\alpha\beta} , \qquad \underset{\sim}{N} \cdot \underset{\sim}{t}_\alpha = 0 \qquad (\alpha,\beta = 1,2) ,$$

any attempt to eliminate these parameters from the above inequalities, in other words, to find some criteria defined by the initial data only, is obviously fraught with unsurmountable difficulties. A glance at Equations (26.14) is sufficient for this to be fully appreciated. The inequalities given by Rivlin and Sawyers in their papers (1977, 1978) are written differently, though they also contain N_1, N_2, N_3. These inequalities are, however, equally cumbersome.

1. In what follows, the problem will be considered in a different way. In contrast to Par. 26, we will define the derivatives of $\ni$ with respect to v_1, v_2, v_3 by $\ni_k$ and $\ni_{ks}$

$$\ni_k = \frac{\partial \ni}{\partial v_k} , \qquad \ni_{ks} = \frac{\partial^2 \ni}{\partial v_k \partial v_s} ,$$

and the components of an acoustical tensor in the orthonormal basis $\underset{\sim}{e}_1$, $\underset{\sim}{e}_2$, $\underset{\sim}{e}_3$ of principal directions of the Finger measure will be determined from Equations (4.12.9) and (4.12.10). In these Equations, the principal stresses σ_k are substituted by their expressions in terms of $\ni_k$ given by (4.3.12) and (4.3.16)

$$\sigma_1 = \frac{v_1}{\sqrt{I_3}} \ni_1 , \qquad \frac{\partial \sigma_1}{\partial v_1} = \frac{v_1}{\sqrt{I_3}} \frac{\partial^2 \ni}{\partial v_1^2} ,$$

$$\frac{\partial \sigma_1}{\partial v_2} = \frac{\partial}{\partial v_2} \frac{1}{v_2 v_3} \ni_1 = \frac{v_1}{\sqrt{I_3}} \left(\ni_{12} - \frac{\ni_1}{v_2} \right)$$

etc. We thus obtain

$$\sqrt{I_3} Q^{11} = \ni_{11} v_1^2 N_1^2 + \frac{v_1 \ni_1 - v_2 \ni_2}{v_1^2 - v_2^2} v_2^2 N_2^2 + \frac{v_1 \ni_1 - v_3 \ni_3}{v_1^2 - v_3^2} v_3^2 N_3^2 . \tag{1}$$

The expression for Q^{12} is transformed to

$$Q^{12} = N_1 N_2 \left(v_2 \frac{\partial \sigma_1}{\partial v_2} + v_1^2 \frac{\sigma_1 - \sigma_2}{v_1^2 - v_2^2} \right) = N_1 N_2 \left(v_1 v_2 \ni_{12} - v_1 \ni_1 + v_1^2 \frac{v_1 \ni_1 - v_2 \ni_2}{v_1^2 - v_2^2} \right)$$

and after simplification we have

$$\sqrt{I_3} Q^{12} = v_1 v_2 N_1 N_2 \left(\ni_{12} + \frac{\ni_1 v_2 - \ni_2 v_1}{v_1^2 - v_2^2} \right) . \tag{2}$$

Using the notation

$$A_{12} = A_{21} = \frac{v_1 \ni_1 - v_2 \ni_2}{v_1^2 - v_2^2} , \qquad A_{23} = A_{32} = \frac{v_2 \ni_2 - v_3 \ni_3}{v_2^2 - v_3^2} ,$$
$$A_{31} = A_{13} = \frac{v_3 \ni_3 - v_1 \ni_1}{v_3^2 - v_1^2} , \tag{3}$$

$$B_{12} = B_{21} = \frac{\ni_1 v_2 - \ni_2 v_1}{v_1^2 - v_2^2} , \qquad B_{23} = B_{32} = \frac{\ni_2 v_3 - \ni_3 v_2}{v_2^2 - v_3^2} ,$$
$$B_{31} = B_{13} = \frac{\ni_3 v_1 - \ni_1 v_3}{v_3^2 - v_1^2} \tag{4}$$

we may represent an acoustical tensor by the expression

$$\sqrt{I_3} \underline{Q} = \sum_{i=1}^{3} \sum_{k=1}^{3} \ni_{ik} N_i N_k v_i v_k \underline{e}_i \underline{e}_k + (A_{12} v_2^2 N_2^2 + A_{13} v_3^2 N_3^2) \underline{e}_1 \underline{e}_1 +$$

$$+ (A_{23} v_3^2 N_3^2 + A_{21} v_1^2 N_1^2) \underline{e}_2 \underline{e}_3 + (A_{31} v_1^2 N_1^2 + A_{32} v_2^2 N_2^2) \underline{e}_3 \underline{e}_3 +$$

$$+ B_{12}N_1N_2v_1v_2(\underline{e}_1\underline{e}_2 + \underline{e}_2\underline{e}_1) + B_{23}N_2N_3v_2v_3(\underline{e}_2\underline{e}_3 + \underline{e}_3\underline{e}_2) +$$

$$+ B_{31}N_3N_1v_3v_1(\underline{e}_3\underline{e}_1 + \underline{e}_1\underline{e}_3) \ . \tag{5}$$

We also note the relationships

$$A_{12} + B_{12} = \frac{\partial_1 - \partial_2}{v_1 - v_2}\ , \qquad A_{23} + B_{23} = \frac{\partial_2 - \partial_3}{v_2 - v_3}\ , \qquad A_{31} + B_{31} = \frac{\partial_3 - \partial_1}{v_2 - v_1}\ , \tag{6}$$

$$A_{12} - B_{12} = \frac{\partial_1 + \partial_2}{v_1 + v_2}\ , \qquad A_{23} - B_{23} = \frac{\partial_2 + \partial_3}{v_2 + v_3}\ , \qquad A_{31} - B_{31} = \frac{\partial_3 + \partial_1}{v_3 + v_1}\ . \tag{7}$$

2. From (22.14) and using the notation

$$N_i v_i = a_i \qquad (i = 1,2,3) \tag{8}$$

we may rewrite Hadamard's inequality (22.12) in the following way

$$A(\underline{a},\underline{b}) = \sum_{i=1}^{3}\sum_{k=1}^{3} \partial_{ik}a_i a_k b_i b_k + A_{12}(a_2^2b_1^2 + a_1^2b_2^2) + A_{23}(a_3^2b_2^2 + a_2^2b_3^2) +$$

$$+ A_{31}(a_1^2b_3^2 + a_3^2b_1^2) + 2B_{12}a_1b_2a_2b_1 + 2B_{23}a_2b_3a_3b_2 + 2B_{31}a_3b_1a_1b_3 \geq 0\ . \tag{9}$$

Assuming in particular that $a_1 = a_2 = 0$, $a_3 = 1$, $b_1 = 1$ and $b_2 = b_3 = 0$, we arrive at the inequality

$$A_{31} \geq 0\ .$$

This and two similar conditions

$$A_{12} \geq 0\ , \qquad A_{23} \geq 0\ , \qquad A_{31} \geq 0 \tag{10}$$

represent the already known $\mathcal{BE}$-criteria (5.13.10) expressing the necessary conditions for ellipticity of the system of the equilibrium equations for the material considered.

Introducing the quantities

$$a_i b_i = x_i \qquad (i = 1,2,3)\ , \tag{11}$$

we may replace (9) by the inequality

$$A(\underline{a},\underline{b}) = \sum_{i=1}^{3}\sum_{k=1}^{3} \partial_{ik}x_i x_k + 2B_{12}x_1x_2 + 2B_{23}x_2x_3 + 2B_{31}x_3x_1 +$$

$$+ A_{12}\left[x_1^2\left(\frac{b_2}{b_1}\right)^2 + x_2^2\left(\frac{b_1}{b_2}\right)^2\right] + A_{23}\left[x_2^2\left(\frac{b_3}{b_2}\right)^2 + x_3^2\left(\frac{b_2}{b_3}\right)^2\right] +$$

$$+ A_{31}\left[x_3^2\left(\frac{b_1}{b_3}\right)^2 + x_1^2\left(\frac{b_3}{b_1}\right)^2\right] \geq 0 \ , \qquad (12)$$

which holds, see (10), provided that we substitute the expression

$$x_1^2\left(\frac{b_2}{b_1}\right)^2 + x_2^2\left(\frac{b_1}{b_2}\right)^2 = x_1^2 z + \frac{x_2^2}{z} = f(z) \ , \qquad z = \left(\frac{b_2}{b_1}\right)^2$$

and similar quantities by their minimum values (for fixed values of x_s). But $f(z)$ reaches a minimum for

$$f'(z) = x_1^2 - \frac{x_2^2}{z^2} = 0 \ , \qquad z = \left(\frac{b_2}{b_1}\right)^2 = \left|\frac{x_2}{x_1}\right| \ ,$$

this minimum being equal to

$$f\left(\left|\frac{x_2}{x_1}\right|\right) = x_1^2\left|\frac{x_2}{x_1}\right| + x_2^2\left|\frac{x_1}{x_2}\right| = 2|x_1 x_2| \ .$$

Inequality (12) is now reduced to the form

$$A(x) = Э_{11}x_1^2 + Э_{22}x_2^2 + Э_{33}x_3^2 + 2(Э_{12} + B_{12})x_1x_2 + 2(Э_{23} + B_{23})x_2x_3 +$$

$$+ 2(Э_{31} + B_{31})x_3x_1 + 2A_{12}|x_1x_2| + 2A_{23}|x_2x_3| + 2A_{31}|x_3x_1| \geq 0 \ . \qquad (13)$$

It expresses not only the sufficient but also the necessary condition of validity of Inequality (9) because the minimum value of the expression

$$A_{12}(a_1^2b_2^2 + a_2^2b_1^2) + A_{23}(a_2^2b_3^2 + a_3^2b_2^2) + A_{31}(a_3^2b_1^2 + a_1^2b_3^2)$$

in fact falls within the region prescribed by the variables a_s, b_s.

3. The transition from Inequality (9) to that of (13) is the decisive step in the whole process, the form in six variables being substituted by that in three. However, this form is still fairly complicated since it contains not only the variables x_1, x_2, x_3 themselves, but also their absolute values. In its general form the problem thus remains cumbersome but may, however, be simplified if we confine our analysis to the case of an incompressible medium.

With this assumption the argument $\overset{0}{\underline{\nabla R}}$ of the expression for the specific strain energy $\Im(\overset{0}{\underline{\nabla R}})$ is subject to the condition $\det \overset{0}{\underline{\nabla R}} = \sqrt{I_3} = 1$. This condition of incompressibility must hold for all states which are comparable to the actual state, that is for the value of the argument

$$\overset{0}{\underline{\nabla R}} \cdot (\underline{E} + \eta \underline{Nb}) \ ,$$

which is contained in the definition of Hadamard's inequality (22.8). We thus have

$$\det \overset{0}{\underline{\nabla R}} \cdot (\underline{E} + \eta \underline{Nb}) = \det \overset{0}{\underline{\nabla R}} \cdot \det(\underline{E} + \eta \underline{Nb}) = \det(\underline{E} + \eta \underline{Nb}) =$$

$$= 1 + \eta \underline{N} \cdot \underline{b} = 1 \ , \qquad \underline{N} \cdot \underline{b} = 0 \ ,$$

since $\det(\underline{E} + \eta \underline{Nb}) = 1 + \eta \underline{N} \cdot \underline{b}$.

Now, by virtue of (8) and (11) we arrive at the constraint

$$\underline{N} \cdot \underline{b} = \sum_{i=1}^{3} N_i b_i = \sum_{i=1}^{3} \frac{a_i b_i}{v_i} = \sum_{i=1}^{3} \frac{x_i}{v_i} = 0 \ , \tag{14}$$

which is imposed on the variables x_i in an incompressible medium. Because $v_i > 0$, this constraint is not satisfied if the variables x_1, x_2, x_3 are of the same sign and so two of the eight octants vanish from the analysis. The possible combinations of signs for the remaining octants are exhausted by the formulas

$$\begin{aligned} &1.\ x_1 > 0,\ x_2 < 0,\ x_3 < 0,\quad 1'.\ x_1 < 0,\ x_2 > 0,\ x_3 > 0, \\ &2.\ x_2 > 0,\ x_3 < 0,\ x_1 < 0,\quad 2'.\ x_2 < 0,\ x_3 > 0,\ x_1 > 0, \\ &3.\ x_3 > 0,\ x_1 < 0,\ x_2 < 0,\quad 3'.\ x_3 < 0,\ x_1 > 0,\ x_2 > 0. \end{aligned} \tag{15}$$

Using the notation

$$|x_i| = v_i \xi_i \qquad (i = 1,2,3) \tag{16}$$

we obtain only three different combinations of solutions of the Equation (14)

$$\begin{aligned} &(1,1')\ x_1 = \pm v_1 \xi_1,\ x_2 = \mp v_2 \xi_2,\ x_3 = \mp v_3 \xi_3; \qquad \xi_1 = \xi_2 + \xi_3, \\ &(2,2')\ x_2 = \pm v_2 \xi_2,\ x_3 = \mp v_3 \xi_3,\ x_1 = \mp v_1 \xi_1; \qquad \xi_2 = \xi_3 + \xi_1, \\ &(3,3')\ x_3 = \pm v_3 \xi_3,\ x_1 = \mp v_1 \xi_1,\ x_2 = \mp v_2 \xi_2; \qquad \xi_3 = \xi_1 + \xi_2. \end{aligned} \tag{17}$$

The upper signs correspond to the cases 1, 2, 3 and the lower to those of 1', 2', 3'.

4. The problem is reduced to the determination of the conditions which would ensure the nonnegativeness of the three forms of two nonnegative variables obtained by successive elimination of x_1, x_2, x_3 from expression (13)

$$
\begin{aligned}
f^{(1)}(\xi_2,\xi_3) &= f^{(1)}_{22}\xi_2^2 + 2f^{(1)}_{23}\xi_2\xi_3 + f^{(1)}_{33}\xi_3^2 ,\\
f^{(2)}(\xi_3,\xi_1) &= f^{(1)}_{33}\xi_3^2 + 2f^{(2)}_{31}\xi_3\xi_1 + f^{(2)}_{11}\xi_1^2 ,\\
f^{(3)}(\xi_1,\xi_2) &= f^{(3)}_{11}\xi_1^2 + 2f^{(3)}_{12}\xi_1\xi_2 + f^{(3)}_{22}\xi_2^2 .
\end{aligned}
\tag{18}
$$

By (3), (4) and (6), (7) the coefficients of these forms turn out to be

$$
\begin{aligned}
f^{(1)}_{22} &= f^{(2)}_{11} = \ni_{11}v_1^2 + \ni_{22}v_2^2 - 2\left(\ni_{12} - \frac{\ni_1+\ni_2}{v_1+v_2}\right)v_1v_2 ,\\
f^{(2)}_{33} &= f^{(3)}_{22} = \ni_{22}v_2^2 + \ni_{33}v_3^2 - 2\left(\ni_{23} - \frac{\ni_2+\ni_3}{v_2+v_3}\right)v_2v_3 ,\\
f^{(3)}_{11} &= f^{(1)}_{33} = \ni_{33}v_3^2 + \ni_{11}v_1^2 - 2\left(\ni_{31} - \frac{\ni_3+\ni_1}{v_3+v_1}\right)v_3v_1 ,\\
f^{(1)}_{23} &= \ni_{11}v_1^2 + \left(\ni_{23} + \frac{\ni_2-\ni_3}{v_2-v_3}\right)v_2v_3 -\\
&\quad - \left(\ni_{31} - \frac{\ni_3+\ni_1}{v_3+v_1}\right)v_3v_1 - \left(\ni_{12} - \frac{\ni_1+\ni_2}{v_1+v_2}\right)v_1v_2 ,\\
f^{(2)}_{31} &= \ni_{22}v_2^2 + \left(\ni_{31} + \frac{\ni_3-\ni_1}{v_3-v_1}\right)v_3v_1 -\\
&\quad - \left(\ni_{12} - \frac{\ni_1+\ni_2}{v_1+v_2}\right)v_1v_2 - \left(\ni_{23} - \frac{\ni_2+\ni_3}{v_2+v_3}\right)v_2v_3 ,\\
f^{(3)}_{12} &= \ni_{33}v_3^2 + \left(\ni_{12} + \frac{\ni_1-\ni_2}{v_1-v_2}\right)v_1v_2 -\\
&\quad - \left(\ni_{23} - \frac{\ni_2+\ni_3}{v_2+v_3}\right)v_2v_3 - \left(\ni_{31} - \frac{\ni_3+\ni_1}{v_3+v_1}\right)v_3v_1 .
\end{aligned}
\tag{19}
$$

The necessary and sufficient conditions for the nonnegativeness of the quadratic form of two nonnegative variabels z_1, z_2

$$f = a_{11}z_1^2 + 2a_{12}z_1z_2 + a_{22}z_2^2 \tag{20}$$

are expressed by the inequalities

$$a_{11} \geq 0 , \qquad a_{22} \geq 0$$

and

$$\text{if } a_{12} \leq 0 , \quad \text{then } a_{11}a_{22} - a_{12}^2 \geq 0 . \tag{21}$$

In fact the conditions $a_{11} \geq 0$, $a_{22} \geq 0$ ensure the nonnegativeness of the form when $z_2 = 0$, $z_1 \neq 0$ or $z_1 = 0$, $z_2 \neq 0$, respectively. If $a_{12} \geq 0$, then the polynomial

$$a_{11} + 2a_{12}u + a_{22}u^2$$

has no positive roots, and for $a_{11}a_{22} - a_{12}^2 \geq 0$ it has no real roots. The criteria (21) can be written in the unique form

$$a_{11} \geq 0 , \qquad a_{22} \geq 0 , \qquad \sqrt{a_{11}a_{22}} + a_{12} \geq 0 . \tag{22}$$

The algebraic criteria of ellipticity (or that Hadamard's inequality holds or the squares of the velocities of propagation of the plane waves are positive) are reduced to nine inequalities, i.e. the three inequalities

$$f_{11}^{(2)} = f_{22}^{(1)} \geq 0 , \qquad f_{22}^{(3)} = f_{33}^{(2)} \geq 0 , \qquad f_{33}^{(1)} = f_{11}^{(3)} \geq 0 , \tag{23}$$

the three inequalities

$$\sqrt{f_{22}^{(1)}f_{33}^{(1)}} + f_{23}^{(1)} \geq 0 , \qquad \sqrt{f_{33}^{(2)}f_{11}^{(2)}} + f_{31}^{(2)} \geq 0 , \qquad \sqrt{f_{11}^{(3)}f_{22}^{(3)}} + f_{12}^{(3)} \geq 0 \tag{24}$$

and the three $\mathcal{BE}$-criteria (10). These formulas are rather cumbersome but calculations based on them are not difficult as long as the functional dependence of $\ni(v_1,v_2,v_3)$ or $\ni(I_1,I_2)$ is specified.

5. For a Mooney-Rivlin material (7.4.1) all the inequalities hold. For example,

$$\frac{1}{2} f_{22}^{(1)} = C_1(v_1 + v_2)^2 + C_2\left(\frac{1}{v_1} + \frac{1}{v_2}\right)^2 ,$$

$$\frac{1}{2} f_{33}^{(1)} = C_1 (v_3 + v_1)^2 + C_2 \left(\frac{1}{v_3} + \frac{1}{v_1} \right)^2 ,$$

$$\frac{1}{2} f_{23}^{(1)} = C_1 v^2 + \frac{C_2}{v_2^2} + C_1 \left(\frac{1}{v_1} + \frac{1}{v_2} + \frac{1}{v_3} \right) + C_2 (v_1 + v_2 + v_3)$$

and $C_1 > 0$, $C_2 > 0$. The material is elliptic for arbitrary deformations.

As a second example, let us consider the material whose specific strain energy is prescribed by

$$\ni = \ni(I_1) = \ni(v_1^2 + v_2^2 + v_3^2) .$$

Here

$$\ni_1 = 2v_1 \ni' , \qquad \ni_{11} = 2\ni' + 4v_1^2 \ni'' , \qquad \ni_{12} = 4v_1 v_2 \ni''$$

etc., and $\ni' > 0$ from (10). By (19) we obtain

$$\frac{1}{2} f_{22}^{(1)} = (v_1 + v_2)^2 [\ni' + 2\ni'' (v_1 - v_2)^2] ,$$

$$\frac{1}{2} f_{33}^{(1)} = (v_3 + v_1)^2 [\ni' + 2\ni'' (v_3 - v_1)^2] ,$$

$$\frac{1}{2} f_{23}^{(1)} = (v_1 + v_2)(v_1 + v_3) [\ni' + 2\ni'' (v_1 - v_2)(v_1 - v_3)]$$

etc. It is necessary that the three inequalities

$$\ni' + 2\ni'' (v_1 - v_2)^2 \geq 0 , \qquad \ni' + 2\ni'' (v_2 - v_3)^2 \geq 0 , \qquad \ni' + 2\ni'' (v_3 - v_1)^2 \geq 0 \tag{25}$$

should hold and also, for $f_{23}^{(1)} < 0$, say, the inequality

$$f_{22}^{(1)} f_{33}^{(1)} - (f_{23}^{1})^2 > 0 , \tag{26}$$

should be valid. Inequality (26) thus reduces to

$$\ni' \ni'' (v_2 - v_3)^2 > 0 , \qquad \ni'' > 0 .$$

We see that for

$$\ni'' > 0 \tag{27}$$

the three Inequalities (26) hold and so, of course, do the three Inequalities (25).

Assuming that $\ni'' < 0$, and letting $v_1 \geq v_2 \geq v_3$, the Inequalities (25) are satisfied if

$$\ni' \geq 2|\ni''|(v_1 - v_3)^2$$

and this condition is sufficient for the three inequalities

$$f_{23}^{(1)} \geq 0 : \ni' \geq 2|\ni''|(v_1 - v_2)(v_1 - v_3) ,$$

$$f_{31}^{(2)} \geq 0 : \ni' \geq 2|\ni''|(v_2 - v_3)(v_1 - v_2) ,$$

$$f_{12}^{(3)} \geq 0 : \ni' \geq 2|\ni''|(v_1 - v_3)(v_2 - v_3) ,$$

to be fulfilled.

We note that for the material considered Inequality (28) is confirmed by that of (26.30).

CHAPTER 9

THERMODYNAMIC RELATIONS. THE EQUATIONS OF THERMOELASTICITY

1. THE ENERGY BALANCE EQUATION. THE FIRST LAW OF THERMODYNAMICS

The total energy contained in the volume V of a continuous medium is the sum of its internal energy E and its kinetic energy K

$$E + K . \tag{1}$$

The law of conservation of energy is expressed by the energy balance equation

$$\dot{E} + \dot{K} = \frac{\delta' A(e)}{dt} + Q . \tag{2}$$

The left-hand side of this equation represents the change of total energy per unit time, i.e. its material derivative. The first term in the right-hand side has been defined by Equation (2.7.13) as the work done by the external forces (body and contact) per unit time

$$\frac{\delta' A(e)}{dt} = \dot{K} + \iiint_V \underline{T} \cdot\cdot \underline{D}\, dV . \tag{3}$$

The second term Q is the heat supplied to the volume per unit time.

We consider here only mechanical and thermal processes. The energy associated with other processes, e.g. with electromagnetic phenomena and chemical conversions, is not included in our analysis. Q enters equation (2) on an equal footing with the first term in accordance with the Joule-Mayer principle of the equivalence of work and heat and is obviously measured in watts.

By (2) and (3) the energy balance equation is reduced to the form

$$\dot{E} = \iiint_V \underline{T} \cdot\cdot \underline{D}\, dV + Q . \tag{4}$$

In the framework of continuum mechanics, the internal energy is introduced into the energy balance equation, (2), as a term in the left-hand side. The equation $m\underline{b} = \underline{f}$ of mechanics is not an identity ("the definition of force") but rather expresses the equality of two quantities, "information" about the force must be furnished independently. Similarly, Equation (4) asserts the relationship between three quantities but one should not, however, regard this equation as defining $\dot{E}$ by prescription of the right-hand side. Like force in the equation of motion the internal energy should be expressed independently through those quantities which determine the state of the medium.

Just as the magnitude of the stresses (the first term in the right-hand side of (4)) is determined by the external body and contact forces, the heating term Q is composed of the heat supplied to the volume V through its surface (the heat transferred by conduction) and by that distributed throughout the volume (e.g. transferred by radiation from external sources)

$$Q = \iint_O q\, dO + \iiint_V \rho r\, dV \ . \tag{5}$$

The heat flow vector $\underline{h}$ is expressed in terms of q. This is accomplished by the same (but simpler) argument as is used when introducing the stress tensor. At each point of the volume V the oriented area $\underline{N}\, dO$ is associated with $q\, dO$ - the heat passing through the area per unit time. The vector $\underline{N}\, dO$ is thus set into correspondence with the scalar $q\, dO$ and the fact that this relationship is linear can be seen from the theory of heat flow through the faces of an elementary tetrahedron. We have then

$$q = -\underline{N} \cdot \underline{h} \ . \tag{6}$$

The heat leaving the volume is $-q$ and $\underline{N}$ is, as usual, the external normal. In the same way as in Cauchy's fundamental relation (2.2.3), Equation (6) is related to any oriented area with the volume V. Here the field of the heat flow vector $\underline{h}$ is defined and the scalar q which depends linearly upon $\underline{N}$ is represented by the scalar product of $\underline{N}$ and some other vector - see (II.1.2). In exactly the same way, the field of the stress tensor $\underline{T}$ was introduced by a force $\underline{t}_N$ which was a linear function of $\underline{N}$. In the case in which relationship (6) refers to the surface O bounding the volume V it should be interpreted as the boundary condition for $\underline{h}$. Similarly, Equation (2.3.1) represents the boundary condition (the equilibrium condition on the surface) for $\underline{T}$.

The Gauss-Ostrogradskii transformation now enables us to represent Equations (5) and (4) in the following form

$$Q = \iiint (\rho r - \underline{\nabla} \cdot \underline{h})\, dV \ , \tag{7}$$

$$\dot{E} = \iiint_V (\underline{T} \cdot\cdot \underline{D} - \underline{\nabla} \cdot \underline{h} + \rho r)\, dV \ . \tag{8}$$

It is assumed here that the internal energy is an additive function of mass, in other words

$$E = \iiint_V \rho e \, dV = \iiint_v \rho_0 e \, dv \,, \qquad \dot{E} = \iiint_v \rho_0 \dot{e} \, dv = \iiint_V \rho \dot{e} \, dV \tag{9}$$

where e is the internal energy per unit mass, i.e. the specific internal energy. Equation (8) is now reduced to the form

$$\iiint_V (\rho \dot{e} - \underline{T} \cdot\cdot \underline{D} + \underline{\nabla} \cdot \underline{h} - \rho r) dV = 0 \,. \tag{10}$$

This is the first law of thermodynamics in its integral form. Its differential version expresses the vanishing of the integrand since the volume V may be prescribed arbitrarily:

$$\rho \dot{e} - \underline{T} \cdot\cdot \underline{D} + \underline{\nabla} \cdot \underline{h} - \rho r = 0 \,. \tag{11}$$

Representing the heat entering the volume per unit time across boundary surface [see (1.8.5)]

$$\iint_O q \, dO = - \iint_O \underline{N} \cdot \underline{h} \, dO = - \iint_o \sqrt{\frac{G}{g}} \, \underline{n} \cdot \underline{\underline{\nabla r}}^T \cdot \underline{h} \, do =$$

$$= - \iint_o \underline{n} \cdot \overset{0}{\underline{h}} \, do = - \iiint_v \overset{0}{\underline{\nabla}} \cdot \overset{0}{\underline{h}} \, dv \tag{12}$$

we introduce the vector $\overset{0}{\underline{h}}$

$$\overset{0}{\underline{h}} = \sqrt{\frac{G}{g}} \, \underline{\underline{\nabla r}}^T \cdot \underline{h} \,, \tag{13}$$

which is expressed in terms of $\underline{h}$ in exactly the same way as the Piola tensor is expressed in terms of the Cauchy stress tensor [see (2.6.2)]. The vector $\overset{0}{\underline{h}}$ could now be termed the Piola heat flow vector. Turning now to the expressions for the first law of thermodynamics in its integral and differential forms and referring to the representation (2.7.14) of power in terms of the Piola tensor, we arrive at the following forms of this law in the reference configuration

$$\iiint_v (\rho_0 \dot{e} - \underline{P} \cdot\cdot \overset{0}{\underline{\underline{\nabla v}}}^T + \overset{0}{\underline{\nabla}} \cdot \overset{0}{\underline{h}} - \rho_0 r) dv = 0 \,, \tag{14}$$

$$\rho_0 \dot{e} - \underline{P} \cdot\cdot \overset{0}{\underline{\underline{\nabla v}}}^T - \overset{0}{\underline{\nabla}} \cdot \overset{0}{\underline{h}} - \rho_0 r = 0 \,. \tag{15}$$

2. THE SECOND LAW OF THERMODYNAMICS

For continuum mechanics, temperature is a fundamental concept. Its definition in words as an estimate of "how hot a body is", this definition being augmented by an indication of how this quantity may be measured, may be regarded as satisfactory. It must be noted that the lowest temperature defined on the Kelvin Scale is $\theta = 0$, so that $\theta > 0$ always.

The second law of thermodynamics is associated with the concept of enthropy. The material time derivative of the enthropy supplied to the volume V by the external sources of heat is defined (similarly to (1.5)) by the expression

$$\iiint_V \frac{1}{\theta}\rho r\, dV + \iint_O \frac{q}{\theta}\, dO = \iiint_V \frac{1}{\theta}\rho r\, dV - \iint_O \frac{1}{\theta}\underline{N}\cdot\underline{h}\, dO \ . \tag{1}$$

The internal energy of the volume V, denoted by H, an additive function of mass and is expressed in terms of the specific enthropy η (per unit mass) by the formula

$$H = \iiint_V \rho\eta\, dV = \iiint_v \rho_0\eta\, dv \ . \tag{2}$$

From a great number of formulations (not equivalent) of the second law of thermodynamics, we will choose, together with Truesdell, the version based on the Clausius-Duhem inequality

$$\dot{H} \geq \iiint_V \frac{1}{\theta}\rho r\, dV - \iint_O \frac{1}{\theta}\underline{N}\cdot\underline{h}\, dO \ , \tag{3}$$

which expresses the fact that the internal enthropy of the volume does not decrease. The process characterized by the equality sign is termed reversible, and that associated with the inequality irreversible.

Transforming the integral evaluated over the surface into the volume integral

$$\iint_O \frac{1}{\theta}\underline{N}\cdot\underline{h}\, dO = \iiint_V \underline{\nabla}\cdot\frac{h}{\theta}\, dV = \iiint_V \frac{1}{\theta}(\underline{\nabla}\cdot\underline{h} - \underline{h}\cdot\underline{\nabla}\ln\theta)\, dV$$

we obtain an alternative presentation of the Clausius-Duhem inequality

$$\iiint_V \frac{1}{\theta}(\rho\theta\dot{\eta} - \rho r + \underline{\nabla}\cdot\underline{h} - \underline{h}\cdot\underline{\nabla}\ln\theta)\, dV \geq 0 \ , \tag{4}$$

and in the differential form we have

$$\rho\theta\dot{\eta} - (\rho r - \underline{\nabla}\cdot\underline{h}) - \underline{h}\cdot\underline{\nabla}\ln\theta \geq 0 \ . \tag{5}$$

3. FREE ENERGY. THE DISSIPATIVE INEQUALITY

Making use of the differential form of the first law of thermodynamics, we eliminate the term

$$\rho r - \underline{\nabla} \cdot \underline{h} \ ,$$

from the Inequality (2.5) which represents the differential form of the second law. This term represents the external heat supplied to the volume V. We thus arrive at the inequality

$$\rho\theta\dot{\eta} - \rho\dot{e} + \underline{T} \cdot\cdot \underline{D} - \underline{h} \cdot \underline{\nabla} \ln\theta \geq 0 \ , \tag{1}$$

which is transformed on substituting

$$\rho\theta\dot{\eta} = \rho(\theta\eta)^{\cdot} - \rho\dot{\theta}\eta$$

into the following form

$$\underline{T} \cdot\cdot \underline{D} - \rho(e - \theta\eta)^{\cdot} - \rho\dot{\theta}\eta - \underline{h} \cdot \underline{\nabla} \ln\theta \geq 0 \ . \tag{2}$$

We have thus introduced the quantity

$$f = e - \theta\eta \ , \tag{3}$$

called the specific free energy or the Helmholtz thermodynamic potential. The second law of thermodynamics is now expressed in the form of an inequality

$$\underline{T} \cdot\cdot \underline{D} - \rho(\dot{f} + \dot{\theta}\eta) - \underline{h} \cdot \underline{\nabla} \ln\theta \geq 0 \ . \tag{4}$$

The terms

$$\varphi = \underline{T} \cdot\cdot \underline{D} - \rho(\dot{f} + \dot{\theta}\eta) \tag{5}$$

in this relationship define the "specific energy dissipation" or the "specific dissipative function". Materials for which $\varphi = 0$ are called non-dissipative (as will be seen elastic materials belong to this class). For such materials we have by (4)

$$\underline{h} \cdot \underline{\nabla} \ln\theta = \frac{1}{\theta} \underline{h} \cdot \underline{\nabla}\theta \leq 0 \ . \tag{6}$$

The meaning of this relationship will become clear if we recall the inequality $\theta > 0$ and the sign convention of (1.6). This means that the vector $\underline{\nabla}\theta$

directed along the normal to the isothermal surface in the direction of increasing values of θ, forms an angle which is not less than 90° with the heat flow vector $\underline{h}$. In other words, the heat propagates in the direction of decreasing temperature. This conclusion may not hold for positive dissipation ($\varphi > 0$) when the possibility that the inequality in (6) may be reversed cannot be excluded, i.e. the possibility that, under certain conditions, water may flow up the side of a mountain, cannot be excluded.

By virtue of (1.11), (3) and (5) the expression for the first law of thermodynamics in terms of the free energy is reduced to the form

$$\rho\theta\dot{\eta} = \underline{T} \cdot\cdot \underline{D} - \rho(\dot{f} + \dot{\theta}\eta) + \rho r - \underline{\nabla} \cdot \underline{h} = \varphi + \rho r - \underline{\nabla} \cdot \underline{h} \,. \tag{7}$$

For non-dissipative media ($\varphi = 0$), this relationship represents the equation of propagation of heat (heat conduction)

$$\rho\theta\dot{\eta} = \rho r - \underline{\nabla} \cdot \underline{h} \,. \tag{8}$$

In the reference configuration, these formulas take the forms given below. The first and the second laws of thermodynamics are written

$$\rho_0\theta\dot{\eta} = \underline{P} \cdot\cdot \underline{\underline{\overset{0}{\nabla} v}}^{T} - \rho_0(\dot{f} + \dot{\theta}\eta) + \rho_0 r - \underline{\overset{0}{\nabla}} \cdot \underline{\overset{0}{h}} \,, \tag{9}$$

$$\underline{P} \cdot\cdot \underline{\underline{\overset{0}{\nabla} v}}^{T} - \rho_0(\dot{f} + \eta\dot{\theta}) - \underline{\overset{0}{h}} \cdot \underline{\overset{0}{\nabla}} \ln\theta \geq 0 \,, \tag{10}$$

respectively and the dissipative function and the equation of heat propagation in non-dissipative media become

$$\varphi = \underline{P} \cdot\cdot \underline{\underline{\overset{0}{\nabla} v}}^{T} - \rho_0(\dot{f} + \dot{\theta}\eta) \,, \tag{11}$$

$$\rho_0\theta\dot{\eta} = \rho_0 r - \underline{\overset{0}{\nabla}} \cdot \underline{\overset{0}{h}} \,. \tag{12}$$

4. THE THERMODYNAMIC POTENTIALS. THE DEFINING VARIABLES

The application of the laws of thermodynamics to continuum mechanics requires the prescription of one of the thermodynamic potentials, e.g. the free or internal energy, as well as of the heat flow vector as functions of the quantities which define the motion of the medium and the thermal processes taking place in it. There is no hope for any further progress without such a prescription in the same way as the Newton's law $m\underline{b} = \underline{f}$ is useless unless we can specify the dependence of the force $\underline{f}$ on a point in the field of force, the velocity of the point and the time.

Confining our analysis to the case of a simple elastic material and assuming this to be uniform we postulate that its state can be specified by the "defining parameters": the deformation gradient $\overset{0}{\nabla}\underline{\underline{R}}$, the temperature θ and its gradient $\overset{0}{\nabla}\underline{\theta}$. The free energy f and the heat flow vector $\overset{0}{\underline{h}}$ are prescribed as frame-indifferent functions of these parameters

$$f = f(\overset{0}{\nabla}\underline{\underline{R}}\,;\theta,\overset{0}{\nabla}\underline{\theta}) \;, \qquad \overset{0}{\underline{h}} = \overset{0}{\underline{h}}(\overset{0}{\nabla}\underline{\underline{R}}\,;\theta,\overset{0}{\nabla}\underline{\theta}) \;, \tag{1}$$

in other words, these functions are transformed by the law

$$\begin{aligned} f(\overset{0}{\nabla}\underline{\underline{R}}\cdot\underline{O}\,;\theta,\overset{0}{\nabla}\underline{\theta}) &= f(\overset{0}{\nabla}\underline{\underline{R}}\,;\theta,\overset{0}{\nabla}\underline{\theta}) \;; \\ \overset{0}{\underline{h}}(\overset{0}{\nabla}\underline{\underline{R}}\cdot\underline{O}\,;\theta,\overset{0}{\nabla}\underline{\theta}) &= \overset{0}{\underline{h}}(\overset{0}{\nabla}\underline{\underline{R}}\,;\theta,\overset{0}{\nabla}\underline{\theta}) \;. \end{aligned} \tag{2}$$

when we pass from the "non-primed" actual configuration to the "primed" configuration.

In order to simplify subsequent work, these prescriptions are represented here in the reference configuration. The defining parameters are themselves functions of the coordinates and the time though many of the results given below retain their value when f and $\overset{0}{\underline{h}}$ are thought of as functionals over the history of a medium.

The problem amounts to the construction of the state equations using the prescriptions (1), i.e. to the determination of the functional dependences of the Piola tensor $\underline{P}$ and the enthropy η on $\overset{0}{\nabla}\underline{\underline{R}}$, θ, $\overset{0}{\nabla}\underline{\theta}$

$$\underline{P} = \underline{P}(\overset{0}{\nabla}\underline{\underline{R}}\,;\theta,\overset{0}{\nabla}\underline{\theta}) \;, \qquad \eta = \eta(\overset{0}{\nabla}\underline{\underline{R}}\,;\theta,\overset{0}{\nabla}\underline{\theta}) \;, \tag{3}$$

these dependences satisfying the principle of material frame-indifference

$$\begin{aligned} \underline{P}(\overset{0}{\nabla}\underline{\underline{R}}\cdot\underline{O}\,;\theta,\overset{0}{\nabla}\underline{\theta}) &= \underline{P}(\overset{0}{\nabla}\underline{\underline{R}}\,;\theta,\overset{0}{\nabla}\underline{\theta}) \;, \\ \eta(\overset{0}{\nabla}\underline{\underline{R}}\cdot\underline{O}\,;\theta,\overset{0}{\nabla}\underline{\theta}) &= (\overset{0}{\nabla}\underline{\underline{R}}\,;\theta,\overset{0}{\nabla}\underline{\theta}) \;. \end{aligned} \tag{4}$$

This problem can be solved with the aid of the second law of thermodynamics in the form of the dissipative inequality (3.9). The time derivative of the free energy which enters this inequality can, by (1.10.17), be represented as

$$\dot{f} = f_{\overset{0}{\nabla}\underline{\underline{R}}} \cdot\cdot \overset{0}{\nabla}\underline{\underline{v}}^{T} + \frac{\partial f}{\partial\theta}\,\dot{\theta} + \frac{\partial f}{\partial\overset{0}{\nabla}\underline{\theta}} \cdot (\overset{0}{\nabla}\underline{\theta})^{\cdot} \;. \tag{5}$$

The elimination of $\dot{f}$ from (3.10) now leads to the inequality

$$\left(\underline{P} - \rho_0 f_{\overset{0}{\nabla}\underline{\underline{R}}}\right) \cdot\cdot \overset{0}{\nabla}\underline{\underline{v}}^{T} - \rho_0\left(\frac{\partial f}{\partial \theta} + \eta\right)\dot{\theta} + \frac{\partial f}{\partial \overset{0}{\nabla}\underline{\theta}} \cdot (\overset{0}{\nabla}\underline{\theta})^{\cdot} - \overset{0}{\underline{h}} \cdot \overset{0}{\underline{\nabla}} \ln \theta \geq 0 \ . \tag{6}$$

But the inequality

$$Ax + By + Cz + D \geq 0$$

which is linear in x, y, z (the coefficients A, B, C, D are these variables) can only be satisfied provided that

$$A = 0 \ , \qquad B = 0 \ , \qquad C = 0 \ , \qquad D \geq 0$$

conversely, having prescribed suitable values for x, y and z the left-hand side of the inequality may take any sign.

In Inequality (4.6) the variables $\nabla\underline{\underline{v}}^{T}$, $\dot{\theta}$, $(\overset{0}{\nabla}\underline{\theta})^{\cdot}$ play the same role as do x, y, z in the above inequality and the coefficients are independent of these quantities. We thus have

$$\underline{P} = \rho_0 f_{\overset{0}{\nabla}\underline{\underline{R}}} \ , \qquad \eta = -\frac{\partial f}{\partial \theta} \ , \qquad \frac{\partial f}{\partial \overset{0}{\nabla}\underline{\theta}} = 0 \ , \qquad -\overset{0}{\underline{h}} \cdot \overset{0}{\underline{\nabla}} \ln \theta \geq 0 \ . \tag{7}$$

These fundamental results cannot be inferred from the first law (3.9) since it contains the quantity $\rho_0 r - \overset{0}{\underline{\nabla}} \cdot \overset{0}{\underline{h}}$ which is given by the heat supplied to the volume V from outside and this may be of any sign.

From (7) it follows that a simple elastic material is non-dissipative. By (3.11) and (5) we have in fact that

$$\varphi = \rho_0\left(f_{\overset{0}{\nabla}\underline{\underline{R}}} \cdot\cdot \overset{0}{\nabla}\underline{\underline{v}}^{T} + \frac{\partial f}{\partial \theta}\dot{\theta} - \dot{f}\right) \equiv 0 \ . \tag{8}$$

We thus confirm Inequality (3.6) and the form (3.12) of the heat equation. The free energy has been shown to be independent of the temperature gradient

$$\frac{\partial f}{\partial \overset{0}{\nabla}\underline{\theta}} = 0 \ , \qquad f = f(\overset{0}{\nabla}\underline{\underline{R}} \ ; \theta) \ , \tag{9}$$

so that the same holds for $\underline{P}$ and η

$$\underline{P} = \rho_0 f(\overset{0}{\nabla}\underline{\underline{R}} \ ; \theta)_{\overset{0}{\nabla}\underline{\underline{R}}} \ , \tag{10}$$

$$\eta = -\frac{\partial}{\partial \theta} f(\overset{0}{\nabla}\underline{\underline{R}} \ ; \theta) \ . \tag{11}$$

Putting $\overset{0}{\underline{\underline{\nabla R}}}$ into polar coordinates and bearing (2) in mind we have

$$f(\overset{0}{\underline{\underline{\nabla R}}}\,;\theta) = f(\overset{0}{\underline{\underline{\nabla R}}}\cdot\underline{O}\,;\theta) = f(\underline{U}\cdot\underline{O}^{\times}\cdot\underline{O}\,;\theta) = f(\underline{U}\,;\theta) = f(\underline{G}\,;\theta)\ . \tag{12}$$

Here we have assumed that $\underline{O}^{\times}\cdot\underline{O} = \underline{E}$ and used transformation (3.2.9). Retaining the notation f for the free energy, we have now, from (10) and (II.3.5),

$$\underline{P} = \rho_0 f(\underline{G}\,;\theta)_{\overset{0}{\underline{\underline{\nabla R}}}} = 2\rho_0 f(\underline{G}\,;\theta)_{\underline{G}}\cdot\overset{0}{\underline{\underline{\nabla R}}}$$

and by (2.7.3) and (2.7.11) we arrive at the following representations of the Cauchy stress tensor and the tensor derived from the energy equation

$$\underline{T} = 2\rho\overset{0}{\underline{\underline{\nabla R}}}^{T}\cdot f_{\underline{G}}\cdot\overset{0}{\underline{\underline{\nabla R}}}\ , \qquad \overset{\vee}{\underline{T}} = 2\rho f_{\underline{G}}\ , \tag{13}$$

because $\rho_0\sqrt{\frac{g}{G}} = \rho$. With this notation, we have instead of (11)

$$\eta = -\frac{\partial}{\partial\theta}f(\underline{G},\theta)\ . \tag{14}$$

The heat equation (3.12), from (1.13) and (1.7.3), reduces to the form

$$\rho\theta\dot{\eta} = \rho r - \sqrt{\frac{g}{G}}\,\overset{0}{\underline{\nabla}}\cdot\sqrt{\frac{G}{g}}\,\underline{\underline{\nabla r}}^{T}\cdot\underline{h} = \rho r - \underline{r}^{s}\cdot\underline{\underline{\nabla r}}^{T}\frac{\partial\underline{h}}{\partial q^{s}} = \rho r - \underline{R}^{s}\frac{\partial\underline{h}}{\partial q^{s}}$$

and as may be expected

$$\rho\theta\dot{\eta} = \rho r - \underline{\nabla}\cdot\underline{h}\ . \tag{15}$$

Here $\underline{h}$ denotes the heat flow vector. But the Piola heat flow vector, $\overset{0}{\underline{h}}$, satisfies the requirement of frame-indifference, hence, using transformation (12), we have by (1.13)

$$\begin{aligned}\overset{0}{\underline{h}} &= \overset{0}{\underline{h}}(\underline{U}\,;\theta,\overset{0}{\underline{\nabla}}\theta) = \overset{0}{\underline{h}}(\underline{G}^{1/2}\,;\theta,\overset{0}{\underline{\nabla}}\theta)\ ,\\ \underline{h} &= \sqrt{\frac{g}{G}}\,\overset{0}{\underline{\underline{\nabla R}}}^{T}\cdot\overset{0}{\underline{h}} = \underline{h}(\underline{G}\,;\theta,\overset{0}{\underline{\nabla}}\theta)\ .\end{aligned} \tag{16}$$

5. REPRESENTATIONS IN TERMS OF THE SPECIFIC INTERNAL ENERGY

Provided that

$$\frac{\partial^2 f}{\partial\theta^2} \neq 0$$

the Equation (4.11) is uniquely solvable with respect to θ

$$\theta = \theta(\overset{0}{\nabla}\underline{\underline{R}}\ ;\ \eta)\ . \tag{2}$$

At the same time Equation (4.11) allows us to designate f a generating function of the Legendre transformation $\theta \to \eta$. From the rules governing the latter the generating function of the inverse transformation $\eta \to \theta$ is given by the expression

$$\theta(\overset{0}{\nabla}\underline{\underline{R}}\ ;\ \eta) + f(\overset{0}{\nabla}\underline{\underline{R}}\ ;\ \theta(\overset{0}{\nabla}\underline{\underline{R}}\ ;\ \eta)) = e(\overset{0}{\nabla}\underline{\underline{R}}\ ;\ \eta)\ . \tag{3}$$

This expression, from (3.3), determines the internal energy expressed in terms of the deformation gradient and the enthropy. Differentiating (3) with respect to $\overset{0}{\nabla}\underline{\underline{R}}$ and η we obtain

$$e_{\overset{0}{\nabla}\underline{\underline{R}}} = \eta\theta_{\overset{0}{\nabla}\underline{\underline{R}}} + f_{\overset{0}{\nabla}\underline{\underline{R}}} + \frac{\partial f}{\partial \theta}\theta_{\overset{0}{\nabla}\underline{\underline{R}}} = f_{\overset{0}{\nabla}\underline{\underline{R}}} + \eta\left(\theta_{\overset{0}{\nabla}\underline{\underline{R}}} - \theta_{\overset{0}{\nabla}\underline{\underline{R}}}\right) = f_{\overset{0}{\nabla}\underline{\underline{R}}}\ ,$$

$$\frac{\partial e}{\partial \eta} = \theta(\overset{0}{\nabla}\underline{\underline{R}}\ ;\ \eta) + \eta\frac{\partial \theta}{\partial \eta} + \frac{\partial f}{\partial \theta}\frac{\partial \theta}{\partial \eta} = \theta(\overset{0}{\nabla}\underline{\underline{R}}\ ;\ \eta) + \frac{\partial \theta}{\partial \eta}(\eta - \eta) = \theta(\overset{0}{\nabla}\underline{\underline{R}}\ ;\ \eta)$$

here we have made use of Equation (3.11).

We have thus arrived at the relationships

$$\underline{P} = \rho e(\overset{0}{\nabla}\underline{\underline{R}}\ ;\ \eta)_{\overset{0}{\nabla}\underline{\underline{R}}}\ , \qquad \theta = \frac{\partial}{\partial \eta}e(\overset{0}{\nabla}\underline{\underline{R}}\ ;\ \eta)\ , \tag{4}$$

which, of course, confirm the properties of the Legendre transformation.

The prescription of the internal energy should be consistent with the principle of material frame-indifference, i.e. it should satisfy the functional relationship

$$e(\overset{0}{\nabla}\underline{\underline{R}}\ ;\ \eta) = e(\overset{0}{\nabla}\underline{\underline{R}}\cdot\underline{O}\ ;\ \eta)\ . \tag{5}$$

Hence, from the foregoing considerations, we may represent the Cauchy stress tensor, the tensor derived from the energy equation and the temperature by the expressions

$$\underline{T} = 2\rho\overset{0}{\nabla}\underline{\underline{R}}^{T}\cdot e(\underline{G}\ ;\ \eta)_{\underline{G}}\cdot\overset{0}{\nabla}\underline{\underline{R}}\ , \qquad \underline{T}^{\vee} = 2\rho e(\underline{G}\ ;\ \eta)_{\underline{G}}\ ,$$

$$\theta = \frac{\partial}{\partial \eta}e(\underline{G}\ ;\ \eta)\ . \tag{6}$$

In an isotropic medium, when the undistorted state of a material is chosen as its reference configuration, the quantities f and e can be expressed in terms of the Finger strain measure $\underline{F}$

$$f(\underline{G}\,;\,\theta) = f(\underline{O}^{\times}\cdot\underline{F}\cdot\underline{O}^{\times T}\,;\,\theta) = f(\underline{F}\,;\,\theta)\,, \qquad e(\underline{G}\,;\,\eta) = e(\underline{F}\,;\,\eta)\,, \tag{7}$$

this is because f and e then represent isotropic scalars [see (II.5.30)].

However, by (II.3.6) we have

$$f_{\overset{0}{\underline{\nabla R}}} = 2\overset{0}{\underline{\nabla R}}\cdot f_{\underline{F}}\,,$$

so that in view of (4.10), (4) and (1.5.3) we obtain

$$\underline{T} = 2\rho\underline{F}\cdot f_{\underline{F}}\,, \qquad \underline{T} = 2\rho\underline{F}\cdot e_{\underline{F}}\,. \tag{8}$$

Here, f and e denote functions of the invariants $I_k(\underline{F}) = I_k(\underline{G})$.

We have assumed in the expressions for the free and internal energy that the material is uniform and that its properties which are uniform throughout the volume, do not depend upon the material coordinates. We can take the non--homogeneity of the material into account if we introduce into Equations (4.1) and (5) the position vector $\underline{r}(q^1,q^2,q^3)$ in the reference configuration. This would not influence the form of the resultant relationships but would complicate their interpretation.

6. THE HEAT CONDUCTION EQUATION

It is assumed that vector $\overset{0}{\underline{h}}$ may be expressed as the difference of two terms

$$\overset{0}{\underline{h}}(\overset{0}{\underline{\nabla R}}\,;\,\theta,\overset{0}{\underline{\nabla}}\theta) = \underline{h}_0(\overset{0}{\underline{\nabla R}}\,;\,\theta,0) - \overset{0}{\underline{K}}(\overset{0}{\underline{\nabla R}},\theta\,;\,\overset{0}{\underline{\nabla}}\theta)\cdot\overset{0}{\underline{\nabla}}\theta\,, \tag{1}$$

where the first term is independent of the temperature gradient $\overset{0}{\underline{\nabla}}\theta$ and the second is a vector expressed as a product of the second rank tensor $\overset{0}{\underline{K}}$ and the temperature gradient $\overset{0}{\underline{\nabla}}\theta$, this product going to zero with $\overset{0}{\underline{\nabla}}\theta$.

Inequality (4.7) now yields

$$\overset{0}{\underline{h}}\cdot\overset{0}{\underline{\nabla}}\theta = \underline{h}_0\cdot\overset{0}{\underline{\nabla}}\theta - \overset{0}{\underline{\nabla}}\theta\cdot\underline{K}(\overset{0}{\underline{\nabla R}}\,;\,\theta,\overset{0}{\underline{\nabla}}\theta)\cdot\overset{0}{\underline{\nabla}}\theta \le 0 \tag{2}$$

and as has been explained above, the appearance in its left-hand side of a tensor linear in $\overset{0}{\underline{\nabla}}\theta$ enables us to assign an arbitrary sign to the whole expression. We have therefore

$$\overset{0}{\underline{h}}_0 = \underline{0}\ , \qquad \overset{0}{\underline{h}} = -\overset{0}{\underline{K}}(\overset{0}{\underline{\underline{\nabla R}}},\theta,\overset{0}{\underline{\nabla}}\theta)\cdot\overset{0}{\underline{\nabla}}\theta\ , \qquad \overset{0}{\underline{\nabla}}\theta\cdot\overset{0}{\underline{K}}\cdot\overset{0}{\underline{\nabla}}\theta \geq 0\ . \tag{3}$$

The quadratic form associated with the tensor $\overset{0}{\underline{K}}$ is nonnegative, in other words, the tensor $\overset{0}{\underline{K}}$ representing the tensor of thermal conductiviy of a medium, is nonnegative.

For a material which is isotropic with respect to the thermal processes, the tensor $\overset{0}{\underline{K}}$ is spherical when the reference configuration is undistorted

$$\overset{0}{\underline{K}} = \underline{E}k\ , \qquad k \geq 0\ , \qquad \overset{0}{\underline{h}} = -k\overset{0}{\underline{\nabla}}\theta\ , \tag{4}$$

here, k denotes the coefficient of thermal conductivity which is by (3) nonnegative. Below we will confine our analysis to this representation of $\overset{0}{\underline{K}}$.

The heat equation (3.12) now takes the form

$$\rho_0\theta\dot{\eta} = \rho_0 r + \overset{0}{\underline{\nabla}}\cdot k\overset{0}{\underline{\nabla}}\theta\ . \tag{5}$$

The specific heat κ which is defined as the heat required to raise the temperature of unit mass by one degree is usually written

$$\kappa = \left(\frac{\delta q}{d\theta}\right)_\Gamma . \tag{6}$$

The index Γ indicates that the value is related to a certain "thermodynamic path" of a material point. This definition is necessary because the quantity is not a total differential. The concepts of the specific heats of a gas at constant volume and constant pressure are generalized as the specific heats of a solid body at constant strain and constant stress.

Turning to (1.7) and (4.15) we may write down the relationships

$$\delta q = \delta\left(r - \frac{\underline{\nabla}\cdot\underline{h}}{\rho}\right)_\Gamma\ , \qquad \left(\frac{\delta q}{d\theta}\right)_\Gamma = \left(\frac{d\eta}{d\theta}\right)_\Gamma\ . \tag{7}$$

By (4.11) we have

$$\eta = \eta(\overset{0}{\underline{\underline{\nabla R}}}\ ;\ \theta)\ , \qquad \left(\frac{d\eta}{d\theta}\right)_\Gamma = \left(\eta_{0\atop \underline{\underline{\nabla R}}}\cdot\cdot\frac{\partial\overset{0}{\underline{\underline{\nabla R}}}^T}{\partial\theta}\right)_\Gamma + \frac{\partial\eta}{\partial\theta}\ . \tag{8}$$

The first term in the right-hand side of this relationship vanishes when the deformation is constant ($\overset{0}{\underline{\underline{\nabla R}}}$ is constant). By (6) and (4.11) we obtain

$$\kappa = \theta\frac{\partial\eta}{\partial\theta} = -\theta\frac{\partial^2 f}{\partial\theta^2}\ , \tag{9}$$

i.e. the expression for the specific heat at constant deformation. For the left-hand side (5) we have now

$$\dot{\eta} = \frac{\partial\eta}{\partial\theta}\dot{\theta} + \eta_{\overset{0}{\underline{\underline{\nabla R}}}} \cdot\cdot \overset{0}{\underline{\underline{\nabla v}}}^{T} = \frac{1}{\theta}\kappa\dot{\theta} + \frac{1}{\rho_0}\frac{\partial \underline{P}}{\partial\theta} \cdot\cdot \overset{0}{\underline{\underline{\nabla v}}}^{T}$$

and the heat equation takes the form

$$\rho_0\kappa\dot{\theta} = \rho_0 r + \theta\frac{\partial \underline{P}}{\partial\theta} \cdot\cdot \overset{0}{\underline{\underline{\nabla v}}}^{T} + \overset{0}{\underline{\nabla}} \cdot k\overset{0}{\underline{\nabla}}\theta \ . \tag{10}$$

For the steady state ($\overset{0}{\underline{\underline{\nabla v}}}^{T} = \underline{0}$), this becomes the well-known Fourier equation

$$\rho_0\kappa\dot{\theta} = \rho_0 r + \overset{0}{\underline{\nabla}} \cdot k\overset{0}{\underline{\nabla}}\theta \ . \tag{11}$$

Assuming that the Equation (4.10) is solved with respect to $\overset{0}{\underline{\underline{\nabla R}}}$, we have

$$\overset{0}{\underline{\underline{\nabla R}}} = \overset{0}{\underline{\underline{\nabla R}}}(\underline{P}\ ;\theta)\ ,$$

$$\left(\frac{d\eta}{d\theta}\right)_\Gamma = \frac{\partial}{\partial\overset{0}{\underline{\underline{\nabla R}}}} \cdot\cdot \left[\frac{\partial\overset{0}{\underline{\underline{\nabla R}}}^{T}}{\partial\theta} + \overset{0}{\underline{\underline{\nabla R}}}^{T}_{\underline{P}} \cdot\cdot \left(\frac{\partial \underline{P}}{\partial\theta}\right)_\Gamma\right] + \frac{\partial\eta}{\partial\theta} = \eta_{\overset{0}{\underline{\underline{\nabla R}}}} \cdot\cdot \frac{\partial\overset{0}{\underline{\underline{\nabla R}}}^{T}}{\partial\theta} + \frac{\partial\eta}{\partial\theta}\ , \tag{12}$$

this relationship being valid when the process is carried out at constant stress (a constant value of $\underline{P}$). At the same time, $\overset{0}{\underline{\underline{\nabla R}}}$ and θ are the independent variables in the representation of the free energy, and by (4.10) and (4.11), we have

$$\eta_{\overset{0}{\underline{\underline{\nabla R}}}} = -\left(\frac{\partial f}{\partial\theta}\right)_{\overset{0}{\underline{\underline{\nabla R}}}} = -\frac{\partial}{\partial\theta} f_{0\,\overset{0}{\underline{\underline{\nabla R}}}} = -\frac{1}{\rho_0}\frac{\partial \underline{P}}{\partial\theta}\ .$$

From (12) and (9) we arrive at the expression

$$\kappa' = \kappa - \frac{1}{\rho_0}\theta\frac{\partial \underline{P}}{\partial\theta} \cdot\cdot \frac{\partial\overset{0}{\underline{\underline{\nabla R}}}^{T}}{\partial\theta} \tag{13}$$

for the specific heat κ' at constant stress.

7. ISOTHERMAL AND ADIABATIC PROCESSES

A process is termed isothermal if at any instant the temperature is constant throughout the volume

$$\theta = \theta_\times = \text{const}\ . \tag{1}$$

Such an idealized process is realized approximately within a medium of very high thermal conductivity when it is reasonable to assume that any non-homogeneity at any point of the temperature field is eliminated immediately due to the resultant flow of heat.

For an isothermal process, the heat flow equation vanishes and the temperature enters the expressions for the free energy and the Piola tensor (or the Cauchy stress tensor) as a constant parameter θ_x. By (4.10) and (4.11) we have

$$f = f(\overset{0}{\underline{\underline{\nabla R}}} ; \theta_x) , \qquad \underline{P} = \rho_0 f_{0\,\overset{0}{\underline{\underline{\nabla R}}}} (\overset{0}{\underline{\underline{\nabla R}}} ; \theta_x) , \qquad \eta = - \frac{\partial}{\partial \theta_x} (\overset{0}{\underline{\underline{\nabla R}}} ; \theta_x) . \tag{2}$$

The temperature will also enter as a parameter into the equations of motion which may now be solved without any reference to thermal processes.

Another limiting case is that of a thermally isolated volume. From (1.5) and (1.6) we have then

$$r = 0 , \qquad q = 0 , \qquad \underline{h} = \underline{0} . \tag{3}$$

These formulas characterize an adiabatic process. For non-dissipative media the first law of thermodynamics leads, by (4.15), to the equation

$$\dot{\eta} = \frac{\partial \eta}{\partial t} + \underline{v} \cdot \underline{\underline{\nabla}} \eta = 0 , \tag{4}$$

which expresses the constancy of the enthropy along the trajecotry of a material particle. By (2.2) from here it follows that $\dot{H} = 0$, i.e. the internal energy of the volume remains constant, and the Clausius-Duhem inequality becomes an equality. An adiabatic process carried out within a non-dissipative medium particularly within an elastic material, is reversible. The Piola tensor and the temperature are then determined by the value of the internal energy $e(\overset{0}{\underline{\underline{\nabla R}}} ; \eta)$ with the aid of Equations (5.4). The problem is reduced to the analysis of the equations of motion (2.3.5) where the stress tensor $\underline{T}(\overset{0}{\underline{\underline{\nabla R}}} ; \eta)$ is given by Equation (5.6) together with (4). The time dependence of the temperature distribution within the volume V is determined after the problem has been solved. The steady state problem does not introduce any complexities. The mechanical problem is thus separated from the thermal one and the enthropy enters as a constant parameter into the equilibrium equations.

Equation (3) for $r = 0$ and Equation (4) are satisfied by virtue of (6.5) for media which are characterized by vanishingly small values of thermal conductivity $k \to 0$. The idealized processes - isothermal and adiabatic - presuppose completely opposite material properties, i.e. infinitely large infinitely small values of thermal conductivity. This makes acceptable the view that by analysing the strain in these idealized processes one may set limits on the deformation of real materials with finite thermal conductivities.

8. THE EQUATIONS OF THERMOELASTICITY

The properties of an elastic material are prescribed either by the functional dependence of its free energy on the deformation gradient and temperature or by the dependence of its internal energy on the deformation gradient and enthropy. Furthermore, the coefficient (tensor) of thermal conductivity as a function of the deformation gradient, the temperature and the temperature gradient must be specified. The body forces and the heat transferred by radiation from external sources must also be prescribed. The boundary conditions assigned on the surface of a body may be quite dissimilar. It would be pointless to list all their possible combinations. For dynamic problems, the initial conditions must also be formulated.

We thus have in all four independent variables - three material coordinates q^1, q^2, q^3 and the time t. It is required to determine the radius vector $\underline{R}$ and the temperature θ (or the enthropy η) as functions of these variables. If this is the case, the state of stress and the enthropy (or temperature) will also be known.

The above indicates the following sequence of operations:

a) Prescription of the free energy enables us to construct, using Equations (4.10) and (4.11), expressions for the Piola tensor and the enthropy or, from Equation (4.13), the Cauchy stress tensor.

b) Substitution of these expressions into the equations of motion (2.3.5) leads to a system of three differential equations for the components of the displacement vector $\underline{u} = \underline{R} - \underline{r}$. These equations are an analog of the "equations in terms of displacements" of the linear theory. The second derivatives with respect to the coordinates and time are the highest derivatives entering these equations.

c) This system of equations of motion is augmented by the heat flow equation(6.10). Of course, this sequence of operations is preserved if we start from a prescription of the internal energy e when we may make use either of Equations (5.4) or Equations (5.6).

Except for some trivial cases, the dynamic problem thus formulated is intractable and not only from the purely mathematical aspect. The difficulty lies in the lack of reliable experimental data which would enable us either to confirm or reject the accepted concept of thermodynamical potential as a function of the strain measures or the temperature (enthropy), and this is even more true for the coefficient (generally, a tensor) of thermal conductivity as a function of its arguments. These difficulties also persist for static problems of thermoelasticity though the mathematical problem is for that case simplified. The mechanical and thermal problems, however, remain interconnected. The dynamic problem for an isothermal material (an isothermal process) is simplified since the heat equation vanishes and the temperature enters the expression for the free energy and the equations of motion as a constant. For an adiabatic process, this simplification does not arise, and the equations of motion must be considered along with Equation (7.4) which expresses the constancy of the enthropy along the particle trajectory.

For an isothermal process, referring to (4.10) and (2.7.5), we have

$$\underline{P} \cdot\cdot \delta \underline{\underline{\overset{0}{\nabla}R}}^T = \rho_0 f(\underline{\underline{\overset{0}{\nabla}R}} ; \theta_\times)_{\underline{\underline{\overset{0}{\nabla}R}}} \cdot\cdot \delta \underline{\underline{\overset{0}{\nabla}R}}^T = \rho_0 \delta f(\underline{\underline{\overset{0}{\nabla}R}} ; \theta_\times) = \delta' a_{(e)} , \tag{1}$$

i.e. the specific incremental work is equal to the variation in the free energy. The notation

$$\ni(\underline{\underline{\overset{0}{\nabla}R}} ; \theta_\times) = \rho_0 [f(\underline{\underline{\overset{0}{\nabla}R}} ; \theta_\times) - f(\underline{E} ; \theta_\times)] \tag{2}$$

defines the quantity termed as the specific stored free energy of a process which is zero in the reference configuration (this, however, is not essential). Using the above formulas we obtain

$$\begin{aligned} \underline{P} &= \ni(\underline{\underline{\overset{0}{\nabla}R}} ; \theta_\times)_{\underline{\underline{\overset{0}{\nabla}R}}} = 2\ni(\underline{G} ; \theta_\times)_{\underline{G}} \cdot \underline{\underline{\overset{0}{\nabla}R}} , \\ \underline{T} &= 2\sqrt{\frac{g}{G}}\, \underline{\underline{\overset{0}{\nabla}R}}^T \cdot \ni(\underline{G} ; \theta_\times)_{\underline{G}} \cdot \underline{\underline{\overset{0}{\nabla}R}} \end{aligned} \tag{3}$$

or for an isotropic elastic body

$$\underline{T} = 2\sqrt{\frac{g}{G}}\, \underline{F} \cdot \ni(I_1(\underline{F}), I_2(\underline{F}), I_3(\underline{F}) ; \theta_\times)_{\underline{F}} . \tag{4}$$

All the above also holds for an adiabatic process if we introduce, instead of (2), the specific stored internal energy

$$\ni(\underline{\underline{\overset{0}{\nabla}R}} ; \eta_\times) = \rho_0 [e(\underline{\underline{\overset{0}{\nabla}R}} ; \eta_\times) - e(\underline{E} ; \eta_\times)] . \tag{5}$$

Of course, in Equations (3) and (4) we should now substitute $\ni(\underline{\underline{\overset{0}{\nabla}R}} ; \theta_\times)$ by $\ni(\underline{\underline{\overset{0}{\nabla}R}} ; \eta_\times)$. These formulas may also be written

$$\underline{P} = \ni_{\underline{\underline{\overset{0}{\nabla}R}}} , \qquad \underline{T} = 2\sqrt{\frac{g}{G}}\, \underline{\underline{\overset{0}{\nabla}R}}^T \cdot \ni_{\underline{G}} \cdot \underline{\underline{\overset{0}{\nabla}R}} = 2\sqrt{\frac{g}{G}}\, \underline{F} \cdot \ni_{\underline{F}} \tag{6}$$

we should not forget, however, that the same notation $\ni$ is used here to designate of two different quantities. For example in the linear theory of elasticity the adiabatic elastic moduli (obtained with the aid of the representation (5) for $\ni$ expressed in terms of the internal energy) and the isothermal elastic moduli (determined by the free energy) are distinguished. A similar (but more complex) analysis may also be carried out in nonlinear theory.

It is obvious that the states of stress (except for those corresponding to dynamic problems for adiabatic processes) which are determined by the prescription of either the stored free or internal energy, cannot differ since these states are determined by the same equilibrium equations and boundary conditions and both are formulated without any reference to thermal quantities (θ_x or η_x).

An application of the laws of thermodynamics has thus enabled us to prove the existence of "stress potentials" (the "stored energies") for isothermal and adiabatic processes.

The theory of elasticity may, however, be constructed without the introduction of such quantities, i.e. as a purely mechanical discipline which does not consider thermal quantities at all. As it was stated at the beginning of Chapter 4 (Par. 1), all the above has this standpoint as its basis.

APPENDICES

TENSOR ALGEBRA AND TENSOR ANALYSIS

The contents of these Appendices is confined to a presentation of the rules and analytical procedures of tensor calculus in three-dimensional Euclidean space (in E_3) necessary for a study of continuum mechanics. The notation, although somewhat original, is consistent with that of the main text. "Direct", rather than component notation of tensor quantities is mainly used. This results in a brevity and emphasis in the formulas and statements of mechanics which are absent from component representations. Transformation to the latter requires but some skill in elementary algebraic manipulation. Experience in teaching the subject shows that this approach does not give rise to any difficulties.

It is assumed that the reader is familiar with the primary operations of linear algebra and calculus, as well as with operations with matrices. It should also be stressed that the "Appendices" by no means claim to be substitutes for the standard textbooks on tensor calculus, rather they are intended to free the reader from the necessity of continually referring to them.

The Appendices are numbered in Roman numerals and the chapters of the main text in Arabic. They are divided into paragraphs (par.) and the latter into sections (1, 2, etc.). Within each paragraph, the formulas are numbered commencing each time at 1. When a formula is referred to within the same paragraph only its number is given, if the formula referred to belongs to another paragraph of the same Chapter the number of the paragraph is also cited. When the formula referred to belongs to another chapter, the chapter, paragraph and formula numbers are given.

Appendix I. TENSOR ALGEBRA

1. VECTOR BASES

A vector basis is formed by three non-coplanar vectors $\underline{r}_1$, $\underline{r}_2$, $\underline{r}_3$. An orthonormal basis, when it arises in the course of discussion, will be formed by the vectors $\underline{i}_1$, $\underline{i}_2$, $\underline{i}_3$ ($\underline{i}_s \cdot \underline{i}_k = 0$ for $s \neq k$, $\underline{i}_s \cdot \underline{i}_s = 1$).

The volume of a parallelepiped constructed on the basic vectors is represented by the expression*)

*) The scalar and vector products of vectors $\underline{a}$, $\underline{b}$ are designated as $\underline{a} \cdot \underline{b}$, $\underline{a} \times \underline{b}$ respectively.

$$v = \mathbf{r}_1 \cdot (\mathbf{r}_2 \times \mathbf{r}_3) = \mathbf{r}_2 \cdot (\mathbf{r}_3 \times \mathbf{r}_1) = \mathbf{r}_3 \cdot (\mathbf{r}_1 \times \mathbf{r}_2) \; . \tag{1}$$

A dual vector is defined by a triplet of vectors

$$\mathbf{r}^1 = \frac{1}{v} \mathbf{r}_2 \times \mathbf{r}_3 \, , \qquad \mathbf{r}^2 = \frac{1}{v} \mathbf{r}_3 \times \mathbf{r}_1 \, , \qquad \mathbf{r}_3 = \frac{1}{v} \mathbf{r}_1 \times \mathbf{r}_2 \, , \tag{2}$$

which are perpendicular to the planes containing the vectors forming the initial basis. It follows from this definition that

$$\mathbf{r}^s \cdot \mathbf{r}_k = \delta_k^s = \begin{cases} 0 & \text{for } s \neq k \, , \\ 1 & \text{for } s = k \end{cases} \tag{3}$$

(δ_k^s denoting the Kronecker symbol). The following notation for scalar products of vectors from the initial and dual bases is adopted

$$\mathbf{r}_s \cdot \mathbf{r}_k = g_{sk} = g_{ks} \, , \qquad \mathbf{r}^s \cdot \mathbf{r}^k = g^{sk} = g^{ks} \, , \qquad \mathbf{r}^s \cdot \mathbf{r}_k = \delta_k^s = g_k^s \; . \tag{4}$$

It is easily verified that the vectors of the initial basis may be written in terms of those of the dual basis with the aid of the formulas

$$\mathbf{r}_1 = v\mathbf{r}^2 \times \mathbf{r}^3 \, , \qquad \mathbf{r}_2 = v\mathbf{r}^3 \times \mathbf{r}^1 \, , \qquad \mathbf{r}_3 = v\mathbf{r}^1 \times \mathbf{r}^2 \; . \tag{5}$$

Indeed,

$$v(\mathbf{r}^2 \times \mathbf{r}^3) = \frac{1}{v} (\mathbf{r}_3 \times \mathbf{r}_1) \times (\mathbf{r}_1 \times \mathbf{r}_2) = \frac{1}{v} [\mathbf{r}_1 \mathbf{r}_2 \cdot (\mathbf{r}_3 \times \mathbf{r}_1) - \mathbf{r}_2 \mathbf{r}_1 \cdot (\mathbf{r}_3 \times \mathbf{r}_1)] = \mathbf{r}_1 \; .$$

Alternative representations of $\mathbf{r}^s$ through $\mathbf{r}_k$ and vice versa are given by the formulas

$$\mathbf{r}^s = g^{sk} \mathbf{r}_k \, , \qquad \mathbf{r}_s = g_{sk} \mathbf{r}^k \; . \tag{6}$$

In these and below the summation sign over the repeated upper and lower indices ("dummy" indices) is omitted. Equations (6) which may be checked immediately by (3) and (4)

$$\mathbf{r}^s \cdot \mathbf{r}^t = g^{sk} \mathbf{r}^t \cdot \mathbf{r}_k = g^{sk} \delta_k^t = g^{st} \, , \qquad \mathbf{r}_s \cdot \mathbf{r}_t = g_{sk} \mathbf{r}^k \cdot \mathbf{r}_t = g_{st}$$

represent the simplest example of the operation of raising and lowering indices as illustrated by the quantities g_{sk}, g^{sk}.

Naturally, for an orthonormal basis $\mathbf{i}_s$, the difference between the initial and dual bases vanishes. However, in order to retain the summation rule in its generally accepted form, we will sometimes use the notation $\underline{i}^s$ for the dual basis. A vector $\underline{a}$ in an orthonormal basis is sometimes represented in the forms

$$\underline{a} = a^s \underline{i}_s = a_s \underline{i}^s .$$

Here, of course, $a_s = a^s$. Applied to basic vectors, this representation takes the form

$$\underline{r}_s = \rho_s^k \underline{i}_k , \qquad \underline{r}_t = \rho_{tm} \underline{i}^m , \tag{7}$$

so that

$$g_{st} = \underline{r}_s \cdot \underline{r}_t = \rho_s^k \rho_{tm} \underline{i}_k \cdot \underline{i}^m = \rho_s^k \rho_{tk} . \tag{8}$$

Representing now v^2 in the form of a product of determinants

$$\begin{vmatrix} \rho_1^1 & \rho_1^2 & \rho_1^3 \\ \rho_2^1 & \rho_2^2 & \rho_2^3 \\ \rho_3^1 & \rho_3^2 & \rho_3^3 \end{vmatrix} \cdot \begin{vmatrix} \rho_{11} & \rho_{12} & \rho_{13} \\ \rho_{21} & \rho_{22} & \rho_{23} \\ \rho_{31} & \rho_{32} & \rho_{33} \end{vmatrix} = \begin{vmatrix} \rho_1^k \rho_{1k} & \rho_1^k \rho_{2k} & \rho_1^k \rho_{3k} \\ \rho_2^k \rho_{1k} & \rho_2^k \rho_{2k} & \rho_2^k \rho_{3k} \\ \rho_3^k \rho_{1k} & \rho_3^k \rho_{2k} & \rho_3^k \rho_{3k} \end{vmatrix}$$

we obtain, by (8),

$$v^2 = \begin{vmatrix} g_{11} & g_{12} & g_{13} \\ g_{21} & g_{22} & g_{23} \\ g_{31} & g_{32} & g_{33} \end{vmatrix} = |g_{st}| , \qquad v = \underline{r}_1 \cdot (\underline{r}_2 \times \underline{r}_3) = \sqrt{g} . \tag{9}$$

By (6) and (7) we have

$$\underline{r}^s = g^{sk} \underline{r}_k = g^{sk} g_{kt} \underline{r}^t , \qquad \underline{r}^s \cdot \underline{r}_m = \delta_m^s = g^{sk} g_{kt} \delta_m^t = g^{sk} g_{km}$$

the matrices $\| g^{sk} \|$ and $\| g_{km} \|$ are each the inverse of the other, their product is the unit diagonal matrix $\| \delta_m^s \|$

$$\| g^{sk} \| \cdot \| g_{km} \| = \| \delta_m^s \| .$$

In other words, g^{sk} is the adjugate of the element g_{sk} of the determinant $|g_{sk}|$ divided by that determinant. Moreover,

$$|g^{sk}| = g^{-1} \; . \tag{10}$$

2. THE LEVI-CIVITA SYMBOL

Equations (1.1) and their analogues obtained by (1.5) may be represented in the form

$$\sqrt{g}\, e_{skt} = \underline{r}_s \cdot (\underline{r}_k \times \underline{r}_t) \; , \qquad \frac{1}{\sqrt{g}}\, e^{skt} = \underline{r}^s \cdot (\underline{r}^k \times \underline{r}^t) \; . \tag{1}$$

The quantities e_{skt}, e^{skt} are zero if there are repeated indices within the triplet skt and are equal to unity for the sequence 123 and for the sequences 231, 312 obtained from the former by cyclic permutation. These quantities are equal to -1 if the cyclic order of the indices is violated (e.g. for the sequences 213, 132, 321). Following the notation

$$\epsilon_{skt} = \sqrt{g}\, e_{skt} \; , \qquad \epsilon^{skt} = \frac{1}{\sqrt{g}}\, e^{skt} \tag{2}$$

we arrive at the representations

$$\epsilon_{skt} = \underline{r}_s \cdot (\underline{r}_k \times \underline{r}_t) \; , \qquad \epsilon^{skt} = \underline{r}^s \cdot (\underline{r}^k \times \underline{r}^t) \; , \tag{3}$$

$$\underline{r}^k \times \underline{r}^t = \epsilon^{kts}\underline{r}_s \; , \qquad \underline{r}_k \times \underline{r}_t = \epsilon_{kts}\underline{r}^s \; . \tag{4}$$

Turning to (1.7) we have the relationships

$$\underline{r}_s = \rho^k_s \underline{i}_k \; , \qquad \underline{r}_s \cdot \underline{r}^m = \delta^m_s = \rho^k_s \underline{i}_k \cdot \rho^m_t \underline{i}^t = \rho^k_s \rho^m_k \; ,$$

which enable us to represent $\epsilon_{skt}\epsilon^{mnp}$ in the form

$$\epsilon_{skt}\epsilon^{mnp} = \underline{r}_s \cdot (\underline{r}_k \times \underline{r}_t)\, \underline{r}^m \cdot (\underline{r}^n \times \underline{r}^p) =$$

$$= \begin{vmatrix} \rho^1_s & \rho^2_s & \rho^3_s \\ \rho^1_k & \rho^2_k & \rho^3_k \\ \rho^1_t & \rho^2_t & \rho^3_t \end{vmatrix} \cdot \begin{vmatrix} \rho^m_1 & \rho^m_2 & \rho^m_3 \\ \rho^n_1 & \rho^n_2 & \rho^n_3 \\ \rho^p_1 & \rho^p_2 & \rho^p_3 \end{vmatrix} = \begin{vmatrix} \rho^q_s\rho^m_q & \rho^q_s\rho^n_q & \rho^q_s\rho^p_q \\ \rho^q_k\rho^m_q & \rho^q_k\rho^n_q & \rho^q_k\rho^p_q \\ \rho^q_t\rho^m_q & \rho^q_t\rho^n_q & \rho^q_t\rho^p_q \end{vmatrix} ,$$

so that

$$\epsilon_{skt}\epsilon^{mnp} = \begin{vmatrix} \delta_s^m & \delta_s^n & \delta_s^p \\ \delta_k^m & \delta_k^n & \delta_k^p \\ \delta_t^m & \delta_t^n & \delta_t^p \end{vmatrix} . \tag{5}$$

From here we arrive at the "rules of contraction" - i.e. the formulas for sums

$$\epsilon_{skt}\epsilon^{snp} = 3(\delta_k^n\delta_t^p - \delta_k^p\delta_t^n) - \delta_k^s(\delta_s^n\delta_t^p - \delta_s^p\delta_t^n) + \delta_t^s(\delta_s^n\delta_k^p - \delta_s^p\delta_k^n) = \delta_k^n\delta_t^p - \delta_t^n\delta_k^p \tag{6}$$

and furthermore,

$$\epsilon_{skt}\epsilon^{skp} = 2\delta_t^p , \qquad \epsilon_{skt}\epsilon^{skt} = 6 . \tag{7}$$

We now obtain, by (4),

$$\underline{r}^k \times \underline{r}^t\epsilon_{ktm} = \epsilon^{kts}\epsilon_{ktm}\underline{r}_s = 2\delta_m^s\underline{r}_s = 2\underline{r}_m ,$$

which imply the inverse formulas to (4)

$$\underline{r}_m = \frac{1}{2}\epsilon_{ktm}\underline{r}^k \times \underline{r}^t , \qquad \underline{r}^m = \frac{1}{2}\epsilon^{ktm}\underline{r}_k \times \underline{r}_t . \tag{8}$$

3. THE REPRESENTATION OF A VECTOR IN THE ORIGINAL AND DUAL BASES

The term "vector" is nowadays often associated with a set of n numbers called its components - "the vector of generalized coordinates $(q_1,q_2,\ldots,q_n)$". In the context of this book we attach to this term its traditional meaning; i.e. that of a physical quantity defined in a three-dimensional Euclidean space and specified by its numerical value (the vector modulus) and by its direction, for example force, velocity etc. The choice of a basis determines the components of a vector. The numerical values of those components, but certainly not the vector itself, are different in different bases.

The representations of a vector in the original and dual bases have the form

$$\underline{a} = a^s\underline{r}_s = a_s\underline{r}^s . \tag{1}$$

The quantities a_s are called covariant, a^s are the contravariant components of the vector $\underline{a}$. By (1.3)

$$a_s = \underline{a}\cdot\underline{r}_s , \qquad a^s = \underline{a}\cdot\underline{r}^s , \tag{2}$$

so that a_s is the product of $|\underline{r}_s| = \sqrt{g_{ss}}$ and the projection of $\underline{a}$ in the direction $\underline{r}_s$ and similarly a^s is the product of $|\underline{r}^s| = \sqrt{g^{ss}}$ and the component of $\underline{a}$ in the direction $\underline{r}^s$. Alternatively, $\underline{a}$ is the diagonal of a parallelepiped with edges $a_s\sqrt{g^{ss}}$ in the direction $\underline{r}^s$ or the diagonal of a parallelepiped with edges $a^s\sqrt{g_{ss}}$ in the direction $\underline{r}_s$.

The formulas connecting co- and contravariant components have the form

$$a_s = g_{sk}a^k \,, \qquad a^s = g^{sk}a_k \tag{3}$$

here again we see the operations of raising and lowering of indices. These formulas follow from (1) and (2)

$$\underline{a}\cdot\underline{r}^k = a^k = a_s\underline{r}^s\cdot\underline{r}^k = g^{sk}a_s \,, \qquad \underline{a}\cdot\underline{r}_k = a_k = a^s\underline{r}_s\cdot\underline{r}_k = g_{sk}a^s \,.$$

The scalar product of two vectors is given by

$$\underline{a}\cdot\underline{b} = a_s\underline{r}^s\cdot b_k\underline{r}^k = g^{sk}a_sb_k = a^s\underline{r}_s\cdot b^k\underline{r}_k = a^sb^kg_{sk} = a^s\underline{r}_s\cdot b_k\underline{r}^k = a^sb_k\delta^k_s = a^sb_s \,. \tag{4}$$

In particular, the square of the vector modulus is

$$a^2 = \underline{a}\ \underline{a} = a^sa_s = g^{sk}a_sa_k = g_{sk}a^sa^k \,. \tag{5}$$

The vector product is, by (2.4), represented by the expressions

$$\underline{c} = \underline{a}\times\underline{b} = a^sb^k\epsilon_{skt}\underline{r}^t = a_sb_k\epsilon^{skt}\underline{r}_t \,, \qquad c_t = a^sb^k\epsilon_{skt} \,, \qquad c^t = a_sb_k\epsilon^{skt} \,.$$

As an example, consider the transformation

$$\underline{a}\times(\underline{b}\times\underline{c}) = a_tb^sc^k\underline{r}^t\times\epsilon_{skq}\underline{r}^q = a_tb^sc^k\epsilon^{tqm}\epsilon_{skq}\underline{r}_m =$$

$$= a_tb^sc^k\underline{r}_m(\delta^m_s\delta^t_k - \delta^m_k\delta^t_s) = b^s\underline{r}_sa_tc^t - c^k\underline{r}_ka_tb^t = \underline{\underline{ba}}\cdot\underline{c} - \underline{\underline{ca}}\cdot\underline{b} \,,$$

which is the desired result.

We may also check the relationship

$$\underline{r}_s\times\underline{r}^s = \underline{0} \,.$$

Indeed, by (2.8)

$$\underline{r}_s\times\underline{r}^s = \frac{1}{2}\underline{r}_s\times(\underline{r}_k\times\underline{r}_t)\epsilon^{kts} = \frac{1}{2}\epsilon^{kts}(\underline{r}_kg_{ts} - \underline{r}_tg_{sk}) =$$

$$= \frac{1}{2}\,(\epsilon^{tks} - \epsilon^{kts})\,\underline{r}_t g_{sk} = \epsilon^{tks}\underline{r}_t g_{sk} = 0$$

because $g_{sk} = g_{ks}$ and $\epsilon^{tks} = -\epsilon^{kts}$.

4. THE TENSOR OF THE SECOND RANK

We consider the transformation of triplet a^k (a column-matrix) to triplet b_s (a line-matrix) produced by the multiplication of a^k from the left by the square matrix $\|q_{sk}\|$

$$b_s = q_{sk}a^k \qquad (s = 1,2,3)\ . \tag{1}$$

Assuming that a^k are contravariant and b_s the covariant components of the vectors $\underline{a}$ and $\underline{b}$, we may write Equation (1) in the form

$$\underline{b} = q_{sk}a^k\underline{r}^s = q_{sk}\underline{r}^s\underline{r}^k \cdot \underline{a}\ . \tag{2}$$

Alternative expressions maintaining the correspondence between the vectors $\underline{a}$ and $\underline{b}$ may be obtained from other decompositions of their components

$$b^s = q^{sk}a_k\ , \qquad \underline{b} = q^{sk}\underline{r}_s\underline{r}_k \cdot \underline{a}\ ;$$

$$\underline{b} = q^{s}_{\cdot k}\underline{r}_s\underline{r}^k \cdot \underline{a}\ , \qquad \underline{b} = q_s^{\cdot k}\underline{r}^s\underline{r}_k \cdot \underline{a}\ . \tag{3}$$

The quantity

$$\underline{\underline{Q}} = q_{sk}\underline{r}^s\underline{r}^k = q^{sk}\underline{r}_s\underline{r}_k = q^{s}_{\cdot k}\underline{r}_s\underline{r}^k = q_s^{\cdot k}\underline{r}^s\underline{r}_k\ , \tag{4}$$

termed a tensor of the second rank defines the transformation of a vector into vector as a right product of a tensor and a vector

$$\underline{b} = \underline{\underline{Q}} \cdot \underline{a}\ . \tag{5}$$

In this representation, one physical quantity (the vector $\underline{a}$) is transformed into another (the vector $\underline{b}$) and hence $\underline{\underline{Q}}$ must be regarded as having physical meaning. To specify $\underline{\underline{Q}}$ in the initial and dual bases its covariant (q_{sk}), contravariant (q^{sk}) or mixed contraco- ($q^{s}_{\cdot k}$) and cocontravariant ($q_s^{\cdot k}$) components are used. But the tensor $\underline{\underline{Q}}$ is not a matrix of its components, whose values depend on the specified bases, it is rather an invariant physical quantity. The meaning of (5) is invariant, it is totally independent of the choice of basis.

The definitions of a vector and a tensor of the second rank introduced here and in Par. 3 must be augmented by a consideration of the conditions which must be fulfilled in order that the three numbers a_s or a^s or the nine numbers q_{sk}, q^{sk} or $q^s_{.k}$, $q_s^{.k}$ may define a quantity which is independent of the basis specified (a vector or a tensor). In other words it is required to establish the rules governing the transformation which these quantities should undergo when they are transferred from the basis where they are assumed to be known to a new basis (see Par. 7).

Post multiplication of $\underline{\underline{Q}}$ by $\underline{a}$ is defined by an equation similar to (5). This defines the vector

$$\underline{c} = \underline{a} \quad \underline{\underline{Q}} = \underline{a} \cdot q_{st}\underline{r}^s\underline{r}^t = q_{st}\underline{r}^t a^s = q_{st}\underline{r}^t\underline{r}^s \cdot \underline{a} \tag{6}$$

which is different from $\underline{b}$. Interchanging the base vectors in (4) we arrive at the transposed tensor denoted by $\underline{\underline{Q}}^T$

$$\underline{\underline{Q}}^T = q_{st}\underline{r}^t\underline{r}^s = q^{st}\underline{r}_t\underline{r}_s = q^s_{.k}\underline{r}^k\underline{r}_s = q_s^{.k}\underline{r}_k\underline{r}^s \ . \tag{7}$$

Its ks-component equals the sk-component of $\underline{\underline{Q}}$. By (5)

$$\underline{b} = \underline{\underline{Q}} \cdot \underline{a} = \underline{a} \cdot \underline{\underline{Q}}^T \tag{8}$$

this rule furnishes a definition of $\underline{\underline{Q}}^T$ which is independent of the choice of basis.

The formulas which establish the relationships between tensors of different denominations follow from (4) and (1.6). E.g.,

$$\underline{\underline{Q}} = q_{sk}\underline{r}^s\underline{r}^k = q_{sk}g^{sm}\underline{r}_m g^{kn}\underline{r}_n = q_{sk}g^{sm}g^{kn}\underline{r}_m\underline{r}_n = q^{mn}\underline{r}_m\underline{r}_n \ .$$

We thus obtain the following relationships

$$q^{mn} = q_{sk}g^{sm}g^{kn} = q_s^{.n}g^{sm} \ , \qquad q_{mn} = q^{sk}g_{sm}g_{kn} = q^s_{.n}g_{sm} \tag{9}$$

etc., where again we have the familiar "rules of play" for indices.

The components of a tensor can also be represented by the formulas

$$\underline{r}_s \cdot \underline{\underline{Q}} \cdot \underline{r}_t = q_{st} \ , \qquad \underline{r}^s \cdot \underline{\underline{Q}} \cdot \underline{r}^t = q^{st} \ ,$$

$$\underline{r}^s \cdot \underline{\underline{Q}} \cdot \underline{r}_k = q^s_{.k} \ , \qquad \underline{r}_s \cdot \underline{\underline{Q}} \cdot \underline{r}^k = q_s^{.k} \ . \tag{10}$$

A vector dyad. This is the simplest example of a tensor of the second rank formed by two vectors. Their dyad is designated as $\underline{ab}$ (often as $\underline{a} \times \underline{b}$). Post--multiplication by a vector $\underline{c}$ yields, by definition, the vector $\underline{ab} \cdot \underline{c}$ and pre--multiplication the vector $\underline{c} \cdot \underline{ab}$. The representation of a dyad in terms of the base dyads using the transposition rule is given by the formulas

$$\underline{ab} = a_s b_k \underline{r}^s \underline{r}^k = a^s b^k \underline{r}_s \underline{r}_k = a_s b^k \underline{r}^s \underline{r}_k = a^s b_k \underline{r}_s \underline{r}^k , \qquad (\underline{ab})^{\mathrm{T}} = \underline{ba} . \tag{11}$$

Equations (4) represent a tensor as the sum of nine basic dyads multiplied by appropriate components.

The unit tensor $\underline{E}$. This is a tensor whose product (either right or left) by a vector $\underline{a}$ is the vector $\underline{a}$ itself

$$\underline{E} \cdot \underline{a} = \underline{a} \cdot \underline{E} = \underline{a} . \tag{12}$$

This definition together with Equations (8) and (10) yields

$$\underline{E} = \underline{E}^{\mathrm{T}} , \quad \underline{r}^s \cdot \underline{E} \cdot \underline{r}^t = g^{sk} , \quad \underline{r}_s \cdot \underline{E} \cdot \underline{r}_k = g_{sk} , \quad \underline{r}^s \cdot \underline{E} \cdot \underline{r}_k = g^s_k = \delta^s_k . \tag{13}$$

By (4), the dyadic expressions of a unit tensor are written as

$$\underline{E} = g^{sk} \underline{r}_s \underline{r}_k = g_{sk} \underline{r}^s \underline{r}^k = g^s_{.k} \underline{r}_s \underline{r}^k = \delta^s_k \underline{r}_s \underline{r}^k = \underline{r}_s \underline{r}^s = \underline{r}^s \underline{r}_s , \tag{14}$$

so that g^{sk}, g_{sk}, δ^s_k are components of a unit tensor. Formula (3.5) for the square of the length of a line segment allows us to term this tensor a metric tensor.

A tensor which is equal to its transpose is called symmetric

$$\underline{\underline{Q}} = \underline{\underline{Q}}^{\mathrm{T}} , \quad q_{sk} = q_{ks} , \quad q^{sk} = q^{ks} , \quad q^s_{.k} = q_k^{.s} = q^s_k \tag{15}$$

(there is no need to indicate the place from which an index has been raised). For a symmetric tensor

$$\underline{\underline{Q}} = \underline{\underline{Q}}^{\mathrm{T}} : \underline{\underline{Q}} \cdot \underline{a} = \underline{a} \cdot \underline{\underline{Q}} . \tag{16}$$

The simplest examples are the dyad $\underline{aa}$ and the unit tensor $\underline{E}$.

A skew tensor is defined by the condition

$$\underline{\underline{Q}}^{\mathrm{T}} = -\underline{\underline{Q}} , \quad q_{ks} = -q_{sk} , \quad q^{ks} = -q^{sk} , \quad q^s_{.k} = -q_k^{.s} , \tag{17}$$

where the diagonal components are zero. The matrix of the components of a symmetric tensor is specified by six components and that of a skew tensor by three.

A tensor $\underset{\approx}{Q}$ may be represented by its decomposition into symmetric and skew parts

$$\underset{\approx}{Q} = \frac{1}{2}(\underset{\approx}{Q} + \underset{\approx}{Q}^T) + \frac{1}{2}(\underset{\approx}{Q} - \underset{\approx}{Q}^T) = \underline{S} + \underline{\Omega}, \qquad \underline{S} = \underline{S}^T = \frac{1}{2}(\underset{\approx}{Q} + \underset{\approx}{Q}^T),$$

$$\underline{\Omega} = \frac{1}{2}(\underset{\approx}{Q} - \underset{\approx}{Q}^T) = -\underline{\Omega}^T. \qquad (18)$$

In order to save space we have not mentioned here the obvious statements that a product of a tensor by a number (scalar) equals a tensor whose components are the products of the components of that tensor by that number, and that a tensor sum is a tensor whose components are the sums of the components of the individual tensors.

Returning to Equations (4) and (5), we may write the expression for a scalar product as

$$\underline{c} \cdot \underline{b} = \underline{b} \cdot \underline{c} = \underline{c} \cdot \underset{\approx}{Q} \cdot \underline{a} = \underline{a} \cdot \underset{\approx}{Q}^T \cdot \underline{c} = q^{sk} c_s a_k \qquad (19)$$

the bilinear form of the components of two vectors $\underline{a}$ and $\underline{c}$, which is formed by the matrix of the components of $\underset{\approx}{Q}$, is invariant and its numerical value is independent of the choice of basis. The quadratic form of the components of vector $\underline{a}$ is also invariant

$$q^{sk} a_s a_k = \underline{a} \cdot \underset{\approx}{Q} \cdot \underline{a} = \underline{a} \cdot \underset{\approx}{Q}^T \cdot \underline{a}. \qquad (20)$$

This equality contains an alternative definition of a symmetric tensor of the second rank as a physical quantity which relates the vector $\underline{a}$ to an invariant quadratic form of its components. A quadratic form specifies only the symmetric part $\underline{S}$ of a tensor $\underset{\approx}{Q}$ because, by (18)

$$\underline{a} \cdot \underline{\Omega} \cdot \underline{a} = \underline{a} \cdot \underline{\Omega}^T \cdot \underline{a} = -\underline{a} \cdot \underline{\Omega} \cdot \underline{a} = 0.$$

5. THE DETERMINANT OF A TENSOR

From Equations (4.9) we have

$$\|q_{st}\| = \|q_s^{\cdot m} g_{mt}\| = \|q^m_{\cdot t} g_{sm}\| ; \qquad \|q^{st}\| = \|q^s_{\cdot m} g^{mt}\| = \|q_m^{\cdot t} g^{ms}\| ;$$

and by (1.9) and (1.11) and in view of the rule for the multiplication of determinants

$$|q_{st}| = g|q^m_{\cdot t}| = g|q_s^{\cdot m}|, \qquad |q^{st}| = \frac{1}{g}|q^s_{\cdot m}| = \frac{1}{g}|q_m^{\cdot t}|. \qquad (1)$$

The determinant, $\det \underset{=}{Q}$, of a tensor $\underset{=}{Q}$ is defined as the determinant of the matrix of its mixed components

$$\det \underset{=}{Q} = |q^{s}_{\cdot t}| = |q_{s}^{\cdot t}| = \frac{1}{g}\,|q_{st}| = g|q^{st}| = \det \underset{=}{Q}^{\mathrm{T}} . \tag{2}$$

In Equation (2.5) we take one of the six permuations of the indices mnp, e.g., 123 and we arrive at the equality

$$\epsilon_{skt} q_1^{\cdot s} q_2^{\cdot k} q_3^{\cdot t} = \begin{vmatrix} \delta_s^1 & \delta_s^2 & \delta_s^3 \\ \delta_k^1 & \delta_k^2 & \delta_k^3 \\ \delta_t^1 & \delta_t^2 & \delta_t^3 \end{vmatrix} q_1^{\cdot s} q_2^{\cdot k} q_3^{\cdot t} = \begin{vmatrix} q_1^{\cdot 1} & q_1^{\cdot 2} & q_1^{\cdot 3} \\ q_2^{\cdot 1} & q_2^{\cdot 2} & q_2^{\cdot 3} \\ q_3^{\cdot 1} & q_3^{\cdot 2} & q_3^{\cdot 3} \end{vmatrix} = \det \underset{=}{Q} .$$

The same result will follow for any other permutation of the indices mnp. Therefore

$$\det \underset{=}{Q} = \frac{1}{6}\,\epsilon_{skt}\epsilon^{mnp} q_m^{\cdot s} q_n^{\cdot k} q_p^{\cdot t} . \tag{3}$$

A tensor $\underset{=}{Q}$ is called non-singular if $\det \underset{=}{Q} \neq 0$. A skew tensor is singular, the dyad $\underline{\underline{ab}}$ of vectors is also a singular tensor.

The coefficient of $q_r^{\cdot l}$ in (3) represents an adjugate of this element written $A_l^{\cdot r}$. But the quantity $q_r^{\cdot l}$ appears three times in (3), i.e. $(l = s\ , r = m)$, $(l = k\ , r = n)$, $(l = t\ , r = p)$. Therefore

$$A_s^{\cdot m} = \frac{1}{2}\,\epsilon_{skt}\epsilon^{mnp} q_n^{\cdot k} q_p^{\cdot t} . \tag{4}$$

The product of the matrices $\|q_m^{\cdot s}\|$ and $\|A_l^{\cdot q}\|$ is the matrix

$$\|q_m^{\cdot s}\|\|A_l^{\cdot q}\| = \|q_m^{\cdot s} A_s^{\cdot q}\| = \frac{1}{2}\,\|\epsilon_{skt}\epsilon^{qnp} q_m^{\cdot s} q_n^{\cdot k} q_p^{\cdot t}\| .$$

But for $q \neq m$, repeated indices within the triples qnp inevitably appear because there are, by definition, no such repeated indices within the triplet mnp. If, however, we substitute q for m, then the desired result would be tripled because the summation rule would come into play. Returning to (3) we thus have

$$\|q_m^{\cdot s} A_s^{\cdot q}\| = \begin{cases} 0 & \text{for } q \neq m \\ \det \underset{=}{Q} & \text{for } q = m \end{cases} = \delta_m^q \det \underset{=}{Q} . \tag{5}$$

The tensor of the agebraic cofactors of the matrix of a tensor is defined as the tensor (see (4))

$$\underline{\underline{Q}} = A_s^{\cdot m}\underline{r}^s\underline{r}_m = \frac{1}{2}\epsilon_{skt}\underline{r}^s\epsilon^{mnp}\underline{r}_m q_n^{\cdot k} q_p^{\cdot t} =$$

$$= \frac{1}{2}\underline{r}_k \times \underline{r}_t q_n^{\cdot k}\underline{r}^n \times \underline{r}^p q_p^{\cdot t} = \frac{1}{2}(\underline{r}_k \times \underline{r}_t)(\underline{\underline{Q}} \cdot \underline{r}^k) \times (\underline{\underline{Q}} \cdot \underline{r}^t) \,, \qquad (6)$$

this is given in terms of the dyads $(\underline{r}_k \times \underline{r}_t)$, $(\underline{\underline{Q}} \cdot \underline{r}^k) \times (\underline{\underline{Q}} \cdot \underline{r}^t)$.

If a tensor $\underline{\underline{Q}}$ is non-singular then, by (4), the matrix

$$\| p_s^{\cdot q} \| = \frac{\| A_s^{\cdot q} \|}{\det \underline{\underline{Q}}} = \frac{1}{2 \det \underline{\underline{Q}}} \| \epsilon_{skt}\epsilon^{mnp} q_n^{\cdot k} q_p^{\cdot t} \| \,, \qquad (7)$$

may be related to the matrix of the components of $\underline{\underline{Q}}$ and, by (5), it can be seen that (7) is the inverse with respect to the matrix of the components of $\underline{\underline{Q}}$

$$\| q_m^{\cdot s} p_s^{\cdot q} \| = \| \delta_m^q \| \,. \qquad (8)$$

6. THE TENSOR PRODUCT. THE INVERSE TENSOR

The vector $\underline{c} = \underline{\underline{p}} \cdot \underline{b} = \underline{\underline{P}} \cdot (\underline{\underline{Q}} \cdot \underline{a})$ may be formed from two tensors $\underline{\underline{P}}$ and $\underline{\underline{Q}}$ of the second rank and a vector $\underline{a}$. Representing this operation in the form $\underline{c} = (\underline{\underline{P}} \cdot \underline{\underline{Q}}) \cdot \underline{a}$ we introduce the quantity $\underline{\underline{R}} = \underline{\underline{P}} \cdot \underline{\underline{Q}}$ which is a tensor of the second rank because the right product of $\underline{\underline{R}}$ and $\underline{a}$ determines a vector $\underline{c} = \underline{\underline{R}} \cdot \underline{a}$. The components of this tensor are expressed in terms of those of the multipliers by one of the expressions

$$\underline{\underline{P}} \cdot \underline{\underline{Q}} = p_{st}\underline{r}^s\underline{r}^t \cdot q^{mn}\underline{r}_m\underline{r}_n = p_{st}q^{tn}\underline{r}^s\underline{r}_n = p_s^{\cdot t} q_{tp}\underline{r}^s\underline{r}^q \qquad (1)$$

etc. The products of three, four etc. tensors are defined in the same way and, in particular, so too are the positive integral powers of a tensor

$$\underline{\underline{Q}}^2 = \underline{\underline{Q}} \cdot \underline{\underline{Q}} = q^{st}q_{tn}\underline{r}_s\underline{r}^n \,, \qquad \underline{\underline{Q}}^3 = q^{st}q_{tn}q^{nm}\underline{r}_s\underline{r}_m \,. \qquad (2)$$

The transposition rule

$$(\underline{\underline{P}} \cdot \underline{\underline{Q}})^T = \underline{\underline{Q}}^T \cdot \underline{\underline{P}}^T \,, \qquad (\underline{\underline{P}} \cdot \underline{\underline{Q}} \cdot \underline{\underline{R}})^T = \underline{\underline{R}}^T \cdot \underline{\underline{Q}}^T \cdot \underline{\underline{P}}^T \,. \qquad (3)$$

is verified directly. In particular, $(\underline{\underline{Q}} \cdot \underline{\underline{Q}}^T)^T = \underline{\underline{Q}} \cdot \underline{\underline{Q}}^T$ since $\underline{\underline{Q}}^{TT} = \underline{\underline{Q}}$ and tensor $\underline{\underline{Q}} \cdot \underline{\underline{Q}}^T$ is symmetric. The right or left product of a tensor by a unit tensor $\underline{\underline{E}}$ results in the same tensor, any power of a unit tensor is a unit tensor and a determinant of a tensor product is equal to the product of their determinants

$$\underline{\underline{Q}}\cdot\underline{\underline{E}} = \underline{\underline{E}}\cdot\underline{\underline{Q}} = \underline{\underline{Q}}\ , \quad \underline{\underline{E}}\cdot\underline{\underline{E}} = \underline{\underline{E}}^2 = \underline{\underline{E}}\ , \quad \underline{\underline{E}}^n = \underline{\underline{E}}\ , \quad \det\underline{\underline{P}}\cdot\underline{\underline{Q}} = \det\underline{\underline{P}}\cdot\det\underline{\underline{Q}}\ . \tag{4}$$

The inverse tensor. To a non-singular tensor $\underline{\underline{Q}}$ there is related an inverse tensor $\underline{\underline{Q}}^{-1}$ using the relationship

$$\underline{\underline{Q}}\cdot\underline{\underline{Q}}^{-1} = \underline{\underline{E}}\ . \tag{5}$$

The dyad decomposition of $\underline{\underline{Q}}^{-1}$ is by (5.8) and (5.6) given by

$$\underline{\underline{Q}}^{-1} = p_s^{\cdot q}\underline{r}^s\underline{r}_q = \frac{1}{2\det\underline{\underline{Q}}}\,\epsilon_{skt}\epsilon^{qnp}q_n^{\cdot k}q_p^{\cdot t}\underline{r}^s\underline{r}_q = \frac{1}{2\det\underline{\underline{Q}}}\,\underline{r}_k\times\underline{r}_t(\underline{\underline{Q}}\cdot\underline{r}^k)\times(\underline{\underline{Q}}\cdot\underline{r}^t)\ . \tag{6}$$

The solution to system (4.1) may, using the tensor $\underline{\underline{Q}}^{-1}$, be written as

$$\underline{a} = \underline{\underline{Q}}^{-1}\cdot\underline{b}\ , \tag{7}$$

as follows immediately from (4.5) and (5). The equalities

$$\det\underline{\underline{Q}}^{-1} = (\det\underline{\underline{Q}})^{-1}\ , \quad (\underline{\underline{P}}\cdot\underline{\underline{Q}})^{-1} = \underline{\underline{Q}}^{-1}\cdot\underline{\underline{P}}^{-1}\ , \quad (\underline{\underline{Q}}^{\mathrm{T}})^{-1} = (\underline{\underline{Q}}^{-1})^{\mathrm{T}}\ ,$$

$$(\underline{\underline{Q}}^{-1})^{-1} = \underline{\underline{Q}} \tag{8}$$

may be easily verified.

The operation of the vector multiplication of a tensor by a vector $\underline{a}$ from the right and from the left defines the tensors of the second rank

$$\underline{\underline{Q}}\times\underline{a} = q^{st}a^q\epsilon_{tqm}\underline{r}_s\underline{r}^m\ , \quad \underline{a}\times\underline{\underline{Q}} = a^q q^{st}\epsilon_{qsm}\underline{r}^m\underline{r}_t = -(\underline{\underline{Q}}^{\mathrm{T}}\times\underline{a})^{\mathrm{T}}\ . \tag{9}$$

7. TRANSFORMATION OF TENSOR COMPONENTS. THE INVARIANTS

We now consider the new bases $\overset{\vee}{\underline{r}}_s$, $\underline{r}^{\vee s}$ together with the bases $\underline{r}_s$, $\underline{r}^s$. The formulas which relate old and new bases may be written as follows

$$\underline{r}_s = \underline{\underline{E}}\cdot\underline{r}_s = \overset{\vee}{\underline{r}}_k\underline{r}^{\vee k}\cdot\underline{r}_s\ , \quad \overset{\vee}{\underline{r}}_s = \underline{r}_k\underline{r}^k\cdot\overset{\vee}{\underline{r}}_s\ , \quad \underline{r}^s = \underline{r}^{\vee k}\overset{\vee}{\underline{r}}_k\cdot\underline{r}^s\ ,$$

$$\underline{r}^{\vee s} = \underline{r}^k\underline{r}_k\cdot\underline{r}^{\vee s}\ .$$

Applied to a vector $\underline{a}$, these formulas yield

$$\underline{a} = a_s\underline{r}^s = a_s\underline{r}^{\vee k}\overset{\vee}{\underline{r}}_k\cdot\underline{r}^s = \overset{\vee}{a}_k\underline{r}^{\vee k}\ , \quad \underline{a} = a^s\underline{r}_s = a^s\overset{\vee}{\underline{r}}_k\underline{r}^{\vee k}\cdot\underline{r}_s = a^{\vee k}\overset{\vee}{\underline{r}}_k\ , \tag{1}$$

here

$$\overset{\vee}{a}_k = a_s\overset{\vee}{\underline{r}}_k\cdot\underline{r}^s\ , \quad a^{\vee k} = a^s\underline{r}^{\vee k}\cdot\underline{r}_s \tag{2}$$

the covariant components are transformed by the same rule as vectors of the initial basis ("covariantly" with them) whereas the contravariant components follow the rule for the transformation of vectors of the dual bases.

The components of a tensor of the second rank are transformed by a similar rule

$$\check{q}_{sk} = \check{\mathbf{r}}_s \cdot \mathbf{r}^m \check{\mathbf{r}}_k \cdot \mathbf{r}^n q_{mn} \,, \qquad \check{q}^{sk} = \check{\mathbf{r}}^s \cdot \mathbf{r}_m \check{\mathbf{r}}^k \cdot \mathbf{r}_n q^{mn} \,,$$

$$\check{q}_s^{\cdot k} = \check{\mathbf{r}}_s \cdot \mathbf{r}^m \check{\mathbf{r}}^k \cdot \mathbf{r}_n q_m^{\cdot n} \,. \tag{3}$$

<u>The invariants of a tensor</u>. This term is applied to those functions of its components whose values are independent of the choice of basis. A scalar (a tensor of zero rank) is invariant. The widely quoted expression "the projection of vector is a scalar" is inadmissible.

Any function of the invariants is itself an invariant. The square of the length of a vector (a tensor of the first rank) is its invariant. This is confirmed by the calculation

$$a^2 = \check{a}^s \check{a}_s = a_k \mathbf{r}^k \cdot \check{\mathbf{r}}_s \check{\mathbf{r}}^s \cdot \mathbf{r}_m a^m = a_k a^m \mathbf{r}^k \cdot \mathbf{E} \cdot \mathbf{r}_m = a_k a^k \,.$$

A vector has no invariants other than those given by functions of its length. A symmetric tensor of the second rank possesses three independent invariants. We consider here its algebraic invariants which are expressed in terms of the components of a tensor by summation and multiplication. These expressions are linear, quadratic and cubic functions of the tensor's components, and are designated $I_1(\mathbf{Q})$, $I_2(\mathbf{Q})$ and $I_3(\mathbf{Q})$, respectively.

The linear invariant $I_1(\mathbf{Q})$ is formed by double contraction of $\mathbf{Q}$ with the unit tensor $\mathbf{E}$.

$$I_1(\mathbf{Q}) = \mathbf{E} \cdot\cdot \mathbf{Q} \,. \tag{4}$$

This may be written in expanded form as

$$\mathbf{E} \cdot\cdot \mathbf{Q} = \mathbf{r}^s \mathbf{r}_s \cdot\cdot q_n^{\cdot m} \mathbf{r}^n \mathbf{r}_m = \mathbf{r}_s \cdot \mathbf{r}^n \mathbf{r}^s \cdot \mathbf{r}_m q_n^{\cdot m} = \delta_s^n \delta_m^s q_n^{\cdot m} = \delta_m^n q_n^{\cdot m} = q_n^{\cdot n} \tag{5}$$

and we see that the inner and outer vectors are multiplied scalarly. The same final result is, of course, achieved if we use different components of $\mathbf{Q}$, e.g.

$$\mathbf{E} \cdot\cdot \mathbf{Q} = g^{sk} \mathbf{r}_s \mathbf{r}_k \cdot\cdot q^{mn} \mathbf{r}_m \mathbf{r}_n = g^{sk} g_{km} g_{sn} q^{mn} = \delta_m^s g_{sn} q^{mn} = g_{mn} q^{mn} = q_n^{\cdot n}$$

etc. We may also arrive at the same formulas either by substituting in the dyadic representation of the tensor the dyads of the base vectors by their scalar products or by forming the left and right products of the base vectors and the corresponding vectors of the dual basis

$$I_1(\underline{\underline{Q}}) = q^{st}\underline{r}_s\cdot\underline{r}_t = g_{st}q^{st} = q_t^{\cdot t}\,,\qquad I_1(\underline{\underline{Q}}) = \underline{r}_m\cdot\underline{\underline{Q}}\cdot\underline{r}^m = g_{ms}q^{sm} = \underline{r}^m\cdot\underline{\underline{Q}}\cdot\underline{r}_m\,. \tag{6}$$

All these expressions are given by (4), this formula does not include any vector bases and therefore it defines those quantities which preserve their value in any basis. It is sufficient to assume $\underline{\underline{E}} = \overset{\vee}{\underline{r}}{}^s\overset{\vee}{\underline{r}}_s$ in order to obtain immediately

$$I_1(\underline{\underline{Q}}) = \underline{\underline{E}}\cdot\cdot\underline{\underline{Q}} = q_s^{\cdot s} = \overset{\vee}{q}{}_s^{\cdot s}\,.$$

The linear invariants of tensors $\underline{\underline{Q}}^2$ and $\underline{\underline{Q}}^3$ are, by (4) and (6.2), given by the expressions

$$I_1(\underline{\underline{Q}}^2) = \underline{\underline{E}}\cdot\cdot\underline{\underline{Q}}^2 = q_s^{\cdot n}q_n^{\cdot s}\,,\qquad I_1(\underline{\underline{Q}}^3) = q_s^{\cdot n}q_n^{\cdot t}q_t^{\cdot s}\,. \tag{7}$$

The quadratic invariant $I_2(\underline{\underline{Q}})$ may be defined as a linear invariant $I_1(\underline{\underline{Q}}_\times)$ of a tensor of cofactors (5.6)

$$I_2(\underline{\underline{Q}}) = I_1(\underline{\underline{Q}}_\times) = \frac{1}{2}\underline{\underline{E}}\cdot\cdot\epsilon_{skt}\epsilon^{mnp}q_n^{\cdot k}q_p^{\cdot t}\underline{r}^s\underline{r}_m =$$

$$= \frac{1}{2}\delta_m^s\epsilon_{skt}\epsilon^{mnp}q_n^{\cdot k}q_p^{\cdot t} = \frac{1}{2}(\delta_k^n\delta_t^p - \delta_t^n\delta_k^p)q_n^{\cdot k}q_p^{\cdot t} = \frac{1}{2}(q_n^{\cdot n}q_t^{\cdot t} - q_k^{\cdot p}q_p^{\cdot k}) \tag{8}$$

by (5) and (6) this tensor may be expressed as

$$I_2(\underline{\underline{Q}}) = \frac{1}{2}(I_1^2(\underline{\underline{Q}}) - I_1(\underline{\underline{Q}}^2))\,. \tag{9}$$

The third invariant $I_3(\underline{\underline{Q}}) = \det\underline{\underline{Q}}$. By (5.3) we have

$$I_3(\underline{\underline{Q}}) = \det\underline{\underline{Q}} = \frac{1}{6}\begin{vmatrix}\delta_s^m & \delta_s^n & \delta_s^p\\ \delta_k^m & \delta_k^n & \delta_k^p\\ \delta_t^m & \delta_t^n & \delta_t^p\end{vmatrix} q_m^{\cdot s}q_n^{\cdot k}q_p^{\cdot t} = \frac{1}{6}\begin{vmatrix}q_s^{\cdot s} & q_s^{\cdot k} & q_s^{\cdot t}\\ q_k^{\cdot s} & q_k^{\cdot k} & q_k^{\cdot t}\\ q_t^{\cdot s} & q_t^{\cdot k} & q_t^{\cdot t}\end{vmatrix}\,. \tag{10}$$

Expanding this determinant and making use of Equations (5), (7) and (9) we arrive at the formulas

$$I_3(\underset{=}{Q}) = \frac{1}{6}[I_1^3(\underset{=}{Q}) - 3I_1(\underset{=}{Q})I_1(\underset{=}{Q}^2) + 2I_1(\underset{=}{Q}^3)] = \frac{1}{3}[I_1(\underset{=}{Q}^3) - I_1^3(\underset{=}{Q}) + 3I_1(\underset{=}{Q})I_2(\underset{=}{Q})] \ . \tag{11}$$

If $\underset{=}{Q}$ is a non-singular tensor, then by (6.6) and (9)

$$I_1(\underset{=}{Q}^{-1}) = \frac{I_2(\underset{=}{Q})}{I_3(\underset{=}{Q})} \ , \quad I_2(\underset{=}{Q}^{-1}) = \frac{I_1(\underset{=}{Q})}{I_3(\underset{=}{Q})} \ , \quad I_3(\underset{=}{Q}^{-1}) = \det \underset{=}{Q}^{-1} = \frac{1}{I_3(\underset{=}{Q})} \ , \tag{12}$$

where the second equation follows from the first by substitution of $\underset{=}{Q}^{-1}$ for $\underset{=}{Q}$.

It is easy to see that $I_1(\underset{=}{Q}) = I_1(\underset{=}{Q}^{\mathrm{T}})$ and therefore referring to (8) and (10) we have

$$I_k(\underset{=}{Q}) = I_k(\underset{=}{Q}^{\mathrm{T}}) \qquad (k = 1,2,3) \ . \tag{13}$$

The Cayley-Hamilton theorem

$$\underset{=}{Q}^3 = I_1(\underset{=}{Q})\underset{=}{Q}^2 - I_2(\underset{=}{Q})\underset{=}{Q} + \underset{=}{E}I_3(\underset{=}{Q}) \tag{14}$$

which is proved in Par. 9, enables us to represent successively the tensors $\underset{=}{Q}^n$ $(n \geq 3)$ as linear functions of $\underset{=}{Q}^2$, $\underset{=}{Q}$ and $\underset{=}{Q}^0 = \underset{=}{E}$ with the coefficients expressed as functions of $I_1(\underset{=}{Q})$, $I_2(\underset{=}{Q})$ and $I_3(\underset{=}{Q})$. A tensor of the second rank possesses three independent invariants namely $I_k(\underset{=}{Q})$ or $I_1(\underset{=}{Q}^k)$ $(k = 1,2,3)$. Any function of these invariants is, of course, itself invariant.

Contraction of a tensor product. Following the definition (6.1) and the rule for contraction, (6), we have

$$I_1(\underset{=}{P} \cdot \underset{=}{Q}) = p_{st}q^{tn}\underset{-}{r}^s \cdot \underset{-}{r}_n = p_{st}q^{ts} = I_1(\underset{=}{Q} \cdot \underset{=}{P}) \ .$$

When we constructed the product of $\underset{=}{P} = p_{st}\underset{-}{r}^s\underset{-}{r}^t$ and $\underset{=}{Q} = q^{mn}\underset{-}{r}_m\underset{-}{r}_n$, we calculated the contraction of vectors $\underset{-}{r}^t$ and $\underset{-}{r}_m$. The calculation of the invariant $I_1(\underset{=}{P} \cdot \underset{=}{Q})$ is therefore reduced to the operation of double contraction

$$I_1(\underset{=}{P} \cdot \underset{=}{Q}) = \underset{=}{E} \cdot\cdot \underset{=}{P} \cdot \underset{=}{Q} = \underset{=}{P} \cdot\cdot \underset{=}{Q} = p_{st}\underset{-}{r}^s\underset{-}{r}^t \cdot\cdot q^{mn}\underset{-}{r}_m\underset{-}{r}_n = p_{st}q^{mn}\underset{-}{r}^t \cdot \underset{-}{r}_m\underset{-}{r}^s \cdot \underset{-}{r}_n = p_{st}q^{ts} \ . \tag{15}$$

The rule for the double contraction of three tensors

$$I_1(\underset{=}{P} \cdot \underset{=}{Q} \cdot \underset{=}{S}) = \underset{=}{P} \cdot\cdot \underset{=}{Q} \cdot \underset{=}{S} = \underset{=}{P} \cdot \underset{=}{Q} \cdot\cdot \underset{=}{S} = \underset{=}{S} \cdot \underset{=}{P} \cdot\cdot \underset{=}{Q} \tag{16}$$

and of a larger number of tensors is similar to that obtained above. It may also be recalled that

$$I_1(\underline{P}\cdot\underline{Q}) = I_1((\underline{P}\cdot\underline{Q})^T) = I_1(\underline{Q}^T\cdot\underline{P}^T) \ . \tag{17}$$

We shall make frequent use of these rules.

8. THE ORTHOGONAL TENSOR

A tensor $\underline{O}$ is called orthogonal if it is a solution to the equation

$$\underline{O}\cdot\underline{O}^T = \underline{O}^T\cdot\underline{O} = \underline{E} \Rightarrow \underline{O}^T = \underline{O}^{-1} \ . \tag{1}$$

Let $\underline{a}$ and $\underline{b}$ be the vectors transformed by a tensor $\underline{O}$ into $\underline{a}'$ and $\underline{b}'$ according to rule (4.5)

$$\underline{a}' = \underline{a}\cdot\underline{O} = \underline{O}^T\cdot\underline{a} \ , \qquad \underline{b}' = \underline{b}\cdot\underline{O} = \underline{O}^T\cdot\underline{b} \ . \tag{2}$$

Then

$$\underline{a}'\cdot\underline{b}' = \underline{a}\cdot\underline{O}\cdot\underline{O}^T\cdot\underline{b} = \underline{a}\cdot\underline{E}\cdot\underline{b} = \underline{a}\cdot\underline{b} \tag{3}$$

and in particular

$$\underline{a}'\cdot\underline{a}' = a'^2 = \underline{a}\cdot\underline{a} = a^2 \tag{4}$$

an orthogonal transformation produces no change in the length of a vector and preserves the angle between the two vectors.

Setting

$$\underline{O} = o^{s}_{\cdot t}\underline{r}_s\underline{r}^t = o^{\cdot n}_{m}\underline{r}^m\underline{r}_n \ , \tag{5}$$

we have

$$\underline{O}\cdot\underline{O}^T = o^{s}_{\cdot t}o^{\cdot n}_{m}\underline{r}_s\underline{r}^t\cdot\underline{r}_n\underline{r}^m = o^{s}_{\cdot t}o^{\cdot t}_{m}\underline{r}_s\underline{r}^m = \underline{E} = \delta^s_m\underline{r}_s\underline{r}^m$$

we thus arrive at the six conditions of orthogonality given by the forms

$$o^{s}_{\cdot t}o^{\cdot t}_{m} = \delta^s_m \ , \qquad o^{st}o_{mt} = \delta^s_m \ , \qquad g^{ks}g^{tr}o_{kt}o_{mr} = \delta^s_m \ . \tag{6}$$

The nine components of $\underline{O}$ are subject to six constraints - an orthogonal tensor is thus defined by three independent parameters. Furthermore,

$$\det\underline{Q} = \det\underline{Q}^T \ , \qquad (\det\underline{Q})^2 = 1 \ , \qquad \det\underline{O} = 1 \quad \text{and} \quad \det\underline{O} = -1 \ . \tag{7}$$

If $\det \underline{O} = 1$ then the tensor is called properly orthogonal; it effects the rotation of a system of vectors. Tensor $\underline{O}$ is expressed below in terms of the angle of rotation it produces and the unit vector along the axis of rotation. If $\det \underline{O} = -1$ then rotation is accompanied by reflection of the axis. Below we will deal mainly with properly orthogonal tensors and unless stated to the contrary a properly orthogonal tensor will be called an orthogonal tensor (i.e. a rotation tensor).

The vector bases $\underline{r}_s$, $\underline{r}^s$ are rotated under an orthogonal transformation into the rotated bases

$$\underline{r}'_s = \underline{r}_s \cdot \underline{O} = \underline{O}^{\mathrm{T}} \cdot \underline{r}_s \,, \qquad \underline{r}'^s = \underline{r}^s \cdot \underline{O} = \underline{O}^{\mathrm{T}} \cdot \underline{r}^s \tag{8}$$

and conversely,

$$\underline{r}_s = \underline{r}'_s \cdot \underline{O}^{\mathrm{T}} = \underline{O} \cdot \underline{r}_s \,, \qquad \underline{r}^s = \underline{r}'^s \cdot \underline{O}^{\mathrm{T}} = \underline{O} \cdot \underline{r}'^s \,. \tag{9}$$

The components of $\underline{a}'$ in the rotated bases are obviously the same as those of $\underline{a}$ in the initial bases, i.e. the vector is rotated together with the basis

$$\underline{a}' = \underline{a} \cdot \underline{O} = a_s \underline{r}^s \cdot \underline{O} = a_s \underline{r}'^s \,, \qquad a'_s = a_s \,, \qquad a'^s = a^s \,. \tag{10}$$

This is also true for a tensor of the second rank

$$\underline{\underline{Q}}' = \underline{O}^{\mathrm{T}} \cdot \underline{\underline{Q}} \cdot \underline{O} \,, \qquad \underline{\underline{Q}} = \underline{O} \cdot \underline{\underline{Q}}' \cdot \underline{O}^{\mathrm{T}} \,, \qquad q'_{st} = q_{st} \,. \tag{11}$$

The representation of an orthogonal tensor as a dyad can, by (8), be written as follows

$$\underline{O} = \underline{r}_t \underline{r}'^t = \underline{r}^t \underline{r}'_t \,, \qquad \underline{O}^{\mathrm{T}} = \underline{r}'^t \underline{r}_t = \underline{r}'_t \underline{r}^t \,. \tag{12}$$

For example $\underline{a} \cdot \underline{O} = \underline{a} \cdot \underline{r}_t \underline{r}'^t = a_t \underline{r}'^t$ which is the desired result. The components of $\underline{a}'$ in the rotated bases should, of course, be differentiated from the components $\overset{\vee}{a}_s$ of the vector $\underline{a}$ in these bases

$$\underline{a} = \overset{\vee}{a}_s \underline{r}'^s = a_k \underline{r}^k \,, \qquad \overset{\vee}{a}_s = a_k \underline{r}^k \cdot \underline{r}'_s = a_k \underline{r}^k \cdot \underline{O}^{\mathrm{T}} \cdot \underline{r}_s = o_s^{\cdot k} a_k \tag{13}$$

etc.

Similarly, for a tensor of the second rank we have

$$\overset{\vee}{q}_{st} = o_{sk} o_{tr} q^{kr} \,, \qquad \overset{\vee}{q}{}^{s}_{\cdot t} = q^{k}_{\cdot m} o^{s}_{\cdot k} o^{\cdot m}_{t} \tag{14}$$

here the orthogonal transformation has been carried out by repeated application of (7.3).

The invariants of an orthogonal tensor (a rotation tensor) are defined by (7.6), (7.12), (12), (5) and (7) and these are equal to

$$I_1(\underline{O}) = \underline{r}_t \, \underline{r}'^t = o^s_{.s} \,, \quad I_2(\underline{O}) = I_3(\underline{O}) I_1(\underline{O}^{\mathrm{T}}) = I_1(\underline{O}) \,, \quad I_3(\underline{O}) = 1 \,. \tag{15}$$

If a Cartesian system is rotated through angle ω about the axis $\underline{i}_3$, we have

$$\underline{i}_1' = \underline{i}_1 \cos\omega + \underline{i}_2 \sin\omega \,, \qquad \underline{i}_2' = -\underline{i}_1 \sin\omega + \underline{i}_2 \cos\omega \,, \qquad \underline{i}_3' = \underline{i}_3$$

and by (12)

$$\underline{O} = \underline{i}_1(\underline{i}_1 \cos\omega + \underline{i}_2 \sin\omega) + \underline{i}_2(-\underline{i}_1 \sin\omega + \underline{i}_2 \cos\omega) + \underline{i}_3\underline{i}_3 =$$

$$= (\underline{i}_1\underline{i}_1 + \underline{i}_2\underline{i}_2) \cos\omega + (\underline{i}_1\underline{i}_2 - \underline{i}_2\underline{i}_1) \sin\omega + \underline{i}_3\underline{i}_3 =$$

$$= \underline{E} \cos\omega + \underline{i}_3\underline{i}_3(1 - \cos\omega) - \underline{i}_3 \times \underline{E} \sin\omega \,,$$

so that irrespective of the vector basis, we arrive at the invariant definition of a rotation tensor

$$\underline{O} = \underline{E} \cos\omega + \underline{ee}(1 - \cos\omega) - \underline{e} \times \underline{E} \sin\omega \,, \tag{16}$$

where $\underline{e}$ denotes the unit vector along the axis of rotation. Equation (16) contains only specification of that axis and the angle ω. From this definition we infer that

$$I_1(\underline{O}) = I_2(\underline{O}) = 1 + 2\cos\omega \,, \qquad -1 \le I_1(\underline{O}) \le 3 \,. \tag{17}$$

From (14) and (6) we have

$$q'^{s}_{.s} = q^{k}_{.r} o^{s}_{.k} o^{.r}_{s} = q^{k}_{.r} \delta^{r}_{k} = q^{k}_{.k} \,, \qquad I_1(\underline{\underline{Q}}') = I_1(\underline{\underline{Q}}) \,.$$

But the invariants $I_2(\underline{\underline{Q}}')$ and $I_3(\underline{\underline{Q}}')$ may be given in terms of $I_1(\underline{\underline{Q}}'^k)$ so that

$$I_k(\underline{\underline{Q}}') = I_k(\underline{\underline{Q}}) \qquad (k = 1,2,3) \,. \tag{18}$$

The inverse is also true and provided that the three invariants of the two symmetric tensors are equal to each other ($I_k(\underline{P}) = I_k(\underline{Q})$, $k = 1,2,3$), there exists an orthogonal transformation connecting these tensors ($\underline{P} = \underline{Q}'$) (see Par. 10).

9. THE PRINCIPAL AXES AND PRINCIPAL VALUES OF A TENSOR OF THE SECOND RANK

The vectors $\underline{e}$ and $\overset{\times}{\underline{e}}$ whose directions are invariant under both pre- and post-multiplication by $\underline{Q}$ are defined by $\underline{Q}$ where $\underline{Q}$ is a tensor of the second rank

$$\underline{Q}\cdot\underline{e} = \lambda\underline{e}\;, \qquad \overset{\times}{\underline{e}}\cdot\underline{Q} = \overset{\times}{\lambda}\overset{\times}{\underline{e}}\;, \tag{1}$$

here λ and $\overset{\times}{\lambda}$ are scalars to be determined. Alternatively, we may write

$$(\underline{Q} - \lambda\underline{E})\cdot\underline{e} = 0\;, \qquad \overset{\times}{\underline{e}}\cdot(\underline{Q} - \overset{\times}{\lambda}\underline{E}) = 0\;. \tag{2}$$

Substituting $\underline{Q}$ by its dyadic representation containing its mixed components we arrive at the linear systems

$$(q^{s}_{.t} - \delta^{s}_{t}\lambda)e^{t} = 0 \qquad (q^{s}_{.t} - \delta^{s}_{t}\overset{\times}{\lambda})\overset{\times}{e} = 0 \tag{3}$$

for the unknown values of e^{t} and $\overset{\times}{e}_{s}$. Non-trivial solutions to these systems exist whenever the determinants

$$|q^{s}_{.t} - \delta^{s}_{t}\lambda| = 0\;, \qquad |q^{s}_{.t} - \delta^{s}_{t}\overset{\times}{\lambda}| = 0\;, \tag{4}$$

are different from zero from which it follows that $\lambda = \overset{\times}{\lambda}$. The polynomial of the third degree in λ

$$P(\lambda) = \det(\underline{Q} - \lambda\underline{E}) = |q^{s}_{.t} - \delta^{s}_{t}\lambda| \tag{5}$$

is called the defining or characteristic polynomial of a tensor and the equation $P(\lambda) = 0$ is called its defining or characteristic equation. The roots $\lambda_1, \lambda_2, \lambda_3$ are called principal (or eigen-) values of the tensor $\underline{Q}$. An expanded decomposition of the defining polynomial has the form

$$P(\lambda) = \det(\underline{Q} - \lambda\underline{E}) = \begin{vmatrix} q^{1}_{.1} - \lambda & q^{1}_{.2} & q^{1}_{.3} \\ q^{2}_{.1} & q^{2}_{.2} - \lambda & q^{2}_{.3} \\ q^{3}_{.1} & q^{3}_{.2} & q^{3}_{.3} - \lambda \end{vmatrix} = -\lambda^3 + \lambda^2(q^{1}_{.1} + q^{2}_{.2} + q^{3}_{.3}) -$$

$$- \lambda [(q^1_{.1}q^2_{.2} - q^1_{.2}q^2_{.1}) + (q^2_{.2}q^3_{.3} - q^2_{.3}q^3_{.2}) + (q^3_{.3}q^1_{.1} - q^3_{.1}q^1_{.3})] + \det \underline{\underline{Q}} =$$

$$= -\lambda^3 + I_1(\underline{\underline{Q}})\lambda^2 - I_2(\underline{\underline{Q}}) + I_3(\underline{\underline{Q}}) \ . \tag{6}$$

Here we have used (7.6), (7.8) and (7.10) for the representation of the invariants. This demonstrates the invariance of the principal values λ_1, λ_2, λ_3. Just as for the invariants $I_k(\underline{\underline{Q}})$, their values are independent of the choice of vector base. A polynomial may, of course, be written in terms of its roots

$$P(\lambda) = (\lambda_1 - \lambda)(\lambda_2 - \lambda)(\lambda_3 - \lambda) = -\lambda^3 + \lambda^2(\lambda_1 + \lambda_2 + \lambda_3) -$$

$$- \lambda(\lambda_1\lambda_2 + \lambda_2\lambda_3 + \lambda_3\lambda_1) + \lambda_1\lambda_2\lambda_3 \ . \tag{7}$$

Comparison with (6) allows us to express the invariants of a tensor in terms of its eigenvalues

$$I_1(\underline{\underline{Q}}) = \lambda_1 + \lambda_2 + \lambda_3 \ , \quad I_2(\underline{\underline{Q}}) = \lambda_1\lambda_2 + \lambda_2\lambda_3 + \lambda_3\lambda_1 \ , \quad I_3(\underline{\underline{Q}}) = \lambda_1\lambda_2\lambda_3 \ . \tag{8}$$

It is obvious that the λ_s are at the same time the eigenvalues of $\underline{\underline{Q}}^T$. The eigenvectors of the tensor $\underline{\underline{Q}}$ corresponding to the first and second equations of (1) and which are associated with the eigenvalue λ_k and designated $\underline{e}_k$, $\underline{e}^k$ are determined from (1) with the aid of the equations

$$\underline{\underline{Q}} \cdot \underline{e}_k = \lambda_k \underline{e}_k \ , \qquad \underline{e}^k \cdot \underline{\underline{Q}} = \lambda_k \underline{e}^k \ . \tag{9}$$

Here no summation over k is assumed.*) The components of these vectors represent non-trivial solutions of Equations (2) in which λ (or $\overset{\times}{\lambda}$) is substituted for λ_k. The common factor in these expressions may be defined by the condition

$$\underline{e}_s \cdot \underline{e}^s = 1 \qquad (s = 1,2,3) \ . \tag{10}$$

Moreover,

$$\underline{e}^s \cdot \underline{\underline{Q}} \cdot \underline{e}_k = \lambda_k \underline{e}^s \cdot \underline{e}_k \ , \qquad \underline{e}^s \cdot \underline{\underline{Q}} \cdot \underline{e}_k = \lambda_s \underline{e}^s \cdot \underline{e}_k \ , \qquad (\lambda_k - \lambda_s)\underline{e}^s \cdot \underline{e}_k = 0$$

and for $\lambda_s \neq \lambda_k$ (it is assumed that, unless stated to the contrary, the roots of a defining polynomial are simple) we obtain

*) An index which occasions in the left and right-hand sides of (9) is of course not mute.

$$\underline{e}^s \cdot \underline{e}_k = \delta^s_k \tag{11}$$

the vector bases $\underline{e}_s$, $\underline{e}^s$ are dual bases and therefore

$$v\underline{e}^1 = \underline{e}_2 \times \underline{e}_3 , \quad v\underline{e}^2 = \underline{e}_3 \times \underline{e}_1 , \quad v\underline{e}^3 = \underline{e}_1 \times \underline{e}_2 , \quad v = \underline{e}_1 \cdot (\underline{e}_2 \times \underline{e}_3) . \tag{12}$$

The above has only formal meaning because a polynomial $P(\lambda)$ may have a pair of complex conjugate roots λ_1, $\lambda_2 = \bar{\lambda}_1$ and these correspond to pairs of complex conjugate vectors ($\underline{e}_2 = \bar{\underline{e}}_1$, $\underline{e}^2 = \bar{\underline{e}}^1$). This follows from (3) since the components of $\underline{\underline{Q}}$ are real.

In the bases $\underline{e}_s$, $\underline{e}^s$ the tensor $\underline{E}$ is given, in view of (4.14), by the expression

$$\underline{E} = \underline{e}_s\underline{e}^s = \underline{e}^s\underline{e}_s = \underline{e}_1\underline{e}^1 + \underline{e}_2\underline{e}^2 + \underline{e}_3\underline{e}^3 = \underline{e}^1\underline{e}_1 + \underline{e}^2\underline{e}_2 + \underline{e}^3\underline{e}_3 \tag{13}$$

and by (1) we have

$$\begin{aligned} \underline{\underline{Q}} &= \underline{E} \cdot \underline{\underline{Q}} = \underline{\underline{Q}} \cdot \underline{E} = \sum_{s=1}^{3} \lambda_s \underline{e}_s \underline{e}^s = \lambda_1 \underline{e}_1 \underline{e}^1 + \lambda_2 \underline{e}_2 \underline{e}^2 + \lambda_3 \underline{e}_3 \underline{e}^3 , \\ \underline{\underline{Q}}^T &= \sum_{s=1}^{3} \lambda_s \underline{e}^s \underline{e}_s = \lambda_1 \underline{e}^1 \underline{e}_1 + \lambda_2 \underline{e}^2 \underline{e}_2 + \lambda_3 \underline{e}^3 \underline{e}_3 . \end{aligned} \tag{14}$$

It would be inappropriate to use the form $\lambda_s \underline{e}_s \underline{e}^s$ since the rule presupposes summation over two indices.

If det $\underline{\underline{Q}} \neq 0$, the polynomial $P(\lambda)$ has no zero root. This enables us to construct the expression for the inverse tensor $\underline{\underline{Q}}^{-1}$. We have

$$\underline{\underline{Q}}^{-1} \cdot \underline{\underline{Q}} = \underline{E} , \qquad \underline{\underline{Q}}^{-1} \cdot \underline{\underline{Q}} \cdot \underline{e}_s = \underline{E} \cdot \underline{e}_s = \underline{e}_s ,$$

$$\underline{\underline{Q}}^{-1} \cdot \sum_{k=1}^{3} \lambda_k \underline{e}_k \underline{e}^k \cdot \underline{e}_s = \underline{\underline{Q}}^{-1} \cdot \lambda_s \underline{e}_s = \underline{e}_s , \qquad \underline{\underline{Q}}^{-1} \cdot \underline{e}_s = \lambda_s^{-1} \underline{e}_s$$

the eigenvalues of the inverse tensor are the inverses of those of the initial tensor, and the eigenvectors of $\underline{\underline{Q}}$ and $\underline{\underline{Q}}^{-1}$ are identical

$$\underline{\underline{Q}}^{-1} = \frac{1}{\lambda_1} \underline{e}_1 \underline{e}^1 + \frac{1}{\lambda_2} \underline{e}_2 \underline{e}^2 + \frac{1}{\lambda_3} \underline{e}_3 \underline{e}^3 = \sum_{s=1}^{3} \lambda_s^{-1} \underline{e}_s \underline{e}^s . \tag{15}$$

We also have immediately by (14) that

$$\underline{\underline{Q}}^2 = \underline{\underline{Q}} \cdot \underline{\underline{Q}} = \sum_{s=1}^{3} \sum_{k=1}^{3} \lambda_s \underline{e}_s \underline{e}^s \cdot \lambda_k \underline{e}_k \underline{e}^k = \sum_{s=1}^{3} \lambda_s^2 \underline{e}_s \underline{e}^s$$

and in general we have that for positive integral values of n, and also for a non-singular tensor, for negative integral values of n,

$$\underline{\underline{Q}}^n = \sum_{s=1}^{3} \lambda_s^n \underline{e}_s \underline{e}^s = \lambda_1^n \underline{e}_1 \underline{e}^1 + \lambda_2^n \underline{e}_2 \underline{e}^2 + \lambda_3^n \underline{e}_3 \underline{e}^3 \ . \tag{16}$$

The dyads $\underline{e}_1\underline{e}^1$, $\underline{e}_2\underline{e}^2$, $\underline{e}_3\underline{e}^3$ are determined as solutions of the linear system

$$\underline{e}_1\underline{e}^1 + \underline{e}_2\underline{e}^2 + \underline{e}_3\underline{e}^3 = \underline{E} \ ,$$

$$\lambda_1\underline{e}_1\underline{e}^1 + \lambda_2\underline{e}_2\underline{e}^2 + \lambda_3\underline{e}_3\underline{e}^3 = \underline{\underline{Q}} \ ,$$

$$\lambda_1^2\underline{e}_1\underline{e}^1 + \lambda_1^2\underline{e}_2\underline{e}^2 + \lambda_3^2\underline{e}_3\underline{e}^3 = \underline{\underline{Q}}^2$$

with a Vandermond determinant

$$\Delta = (\lambda_1 - \lambda_2)(\lambda_2 - \lambda_3)(\lambda_3 - \lambda_1) \ ,$$

which is different from zero when the roots of the defining equation are simple. We obtain

$$\underline{e}_1\underline{e}^1 = -\frac{1}{P'(\lambda_1)}\left[\underline{\underline{Q}}^2 - (\lambda_2 + \lambda_3)\underline{\underline{Q}} + \underline{E}\lambda_2\lambda_3\right] = -\frac{1}{P'(\lambda_1)}(\underline{\underline{Q}} - \lambda_2\underline{E}) \cdot (\underline{\underline{Q}} - \lambda_3\underline{E})$$

etc. where

$$P'(\lambda_1) = -(\lambda_1 - \lambda_2)(\lambda_1 - \lambda_3) \quad \text{etc.}$$

In an alternative formulation, these expressions may be put into the form

$$\underline{e}_s\underline{e}^s = -\frac{1}{P'(\lambda_s)}\left[\underline{\underline{Q}}^2 - (I_1(\underline{\underline{Q}}) - \lambda_s)\underline{\underline{Q}} + \underline{E}\lambda_s^{-1} I_s(\underline{\underline{Q}})\right] \ . \tag{17}$$

The following refers to the case when $\underline{\underline{Q}}$ is a symmetric tensor. In this case, when there is a multiple root ($\lambda_1 = \lambda_2 \neq \lambda_3$), we have by (14)

$$\underline{\underline{Q}} = (\underline{e}_1\underline{e}^1 + \underline{e}_2\underline{e}^2)\lambda_1 + \lambda_3\underline{e}_3\underline{e}^3 = \underline{E}\lambda_1 + (\lambda_3 - \lambda_1)\underline{e}_3\underline{e}^3 \tag{18}$$

the dyads $\underline{e}_1\underline{e}^1$, $\underline{e}_2\underline{e}^2$ do not enter the definition of $\underline{\underline{Q}}$ and these dyads need not be known. For a triple root ($\lambda_1 = \lambda_2 = \lambda_3$), the tensor $\underline{\underline{Q}}$ is "spherical", it is the product of a scalar by a unit tensor

$$\underset{\approx}{Q} = \lambda \underset{\approx}{E} \,. \tag{19}$$

If the tensor $\underset{\approx}{Q}$ is non-symmetric and polynomial (5) has multiple roots then the decomposition of the tensor into its eigenvectors generally becomes more complicated. This is determined by the structure of the corresponding elementary divisors. This type of decomposition of a non-symmetric tensor is not dealt with in the present context.

The tensor $\underset{\approx}{Q}^{1/2}$. A tensor is called positive if the quadratic form $\underset{-}{a} \cdot \underset{\approx}{Q} \cdot \underset{-}{a}$ associated with it is positive for any non-zero real vector $\underset{-}{a}$. The unit tensor $\underset{\approx}{E}$ is positive since $\underset{-}{a} \cdot \underset{\approx}{E} \cdot \underset{-}{a} = \underset{-}{a} \cdot \underset{-}{a} = a^2$. Another such example is the product $\underset{\approx}{Q} \cdot \underset{\approx}{Q}^T$ where $\underset{\approx}{Q}$ is a non-singular tensor. In fact, we have

$$\underset{-}{a} \cdot \underset{\approx}{Q} \cdot \underset{\approx}{Q}^T \cdot \underset{-}{a} = \underset{-}{b} \cdot \underset{-}{b} = b^2 > 0 \qquad (\underset{-}{b} = \underset{-}{a} \cdot \underset{\approx}{Q} = \underset{\approx}{Q}^T \cdot \underset{-}{a}) \,,$$

because $\underset{-}{b} \neq \underset{-}{0}$ since $\underset{-}{a} = \underset{-}{b} \cdot \underset{\approx}{Q}^{-1} \neq \underset{-}{0}$. A tensor is positive provided that its eigenvalues are positive, because $\underset{-}{e}^s \cdot \underset{\approx}{Q} \cdot \underset{-}{e}^s = \lambda_s |\underset{-}{e}_s|^2$ for $s = 1,2,3$. The matrix of the components of a positive tensor is a Sylvesterian matrix, its determinant and the main diagonal subdeterminants are positive.

By a positive tensor there may be defined a tensor whose eigenvalues are equal to the square roots of the eigenvalues of the initial tensor

$$\underset{\approx}{Q}^{1/2} = \varepsilon_1 \sqrt{\lambda_1}\, \underset{-}{e}_1 \underset{-}{e}^1 + \varepsilon_2 \sqrt{\lambda_2}\, \underset{-}{e}_2 \underset{-}{e}^2 + \varepsilon_3 \sqrt{\lambda_3}\, \underset{-}{e}_3 \underset{-}{e}^3 \,, \qquad \varepsilon_s = \pm 1 \,.$$

Different combinations of signs may be specified but we shall assume, unless stated to the contrary, that the roots $\sqrt{\lambda_s}$ $(s = 1,2,3)$ are assumed positive, in other words, the tensor $\underset{\approx}{Q}^{1/2}$ will be regarded as positive

$$\underset{\approx}{Q}^{1/2} = \sqrt{\lambda_1}\, \underset{-}{e}_1 \underset{-}{e}^1 + \sqrt{\lambda_2}\, \underset{-}{e}_2 \underset{-}{e}^2 + \sqrt{\lambda_3}\, \underset{-}{e}_3 \underset{-}{e}^3 \,. \tag{20}$$

A tensor $\underset{\approx}{Q}^{1/2}$ may be constructed from the components of $\underset{\approx}{Q}$, the procedure being reduced to the solution of the system

$$m^s_{\cdot r} m^r_{\cdot t} = q^s_{\cdot t} \qquad (m^s_{\cdot r} = \underset{-}{r}^s \cdot \underset{\approx}{Q}^{1/2} \cdot \underset{-}{r}_r) \,.$$

This problem is essentially of the same difficulty as that of the solution of the cubic equation which yields the eigenvalues of $\underset{\approx}{Q}$.

The Cayley-Hamilton theorem. As has already been mentioned, this theorem (7.14) enables us to express $\underset{\approx}{Q}^3$ and consequently $\underset{\approx}{Q}^n$ in terms of $\underset{\approx}{Q}^2$, $\underset{\approx}{Q}$, $\underset{\approx}{E} = \underset{\approx}{Q}^0$ for $n > 3$. We start from the equality

$$\underline{\underline{Q}}^3 = \sum_{s=1}^{3} \lambda_s^3 \underline{e}_s \underline{e}^s ,$$

where λ_s^3 may be substituted by its value obtained from the characteristic equation

$$\lambda_s^3 = I_1(\underline{\underline{Q}})\lambda_s^2 - I_2(\underline{\underline{Q}})\lambda_s + I_3(\underline{\underline{Q}}) .$$

We get therefore

$$\underline{\underline{Q}}^3 = I_1(\underline{\underline{Q}}) \sum_{s=1}^{3} \lambda_s^2 \underline{e}_s \underline{e}^s - I_2(\underline{\underline{Q}}) \sum_{s=1}^{3} \lambda_s \underline{e}_s \underline{e}^s + I_3(\underline{\underline{Q}}) \sum_{s=1}^{3} \underline{e}_s \underline{e}^s ,$$

from which relationship we infer by (16) that

$$\underline{\underline{Q}}^3 - I_1(\underline{\underline{Q}})\underline{\underline{Q}}^2 + I_2(\underline{\underline{Q}})\underline{\underline{Q}} - I_3(\underline{\underline{Q}})\underline{\underline{E}} = \underline{\underline{0}} , \tag{21}$$

i.e. that a "tensor satisfies its characteristic equation". For a non-singular tensor this equality implies that

$$\underline{\underline{Q}}^{-1} = \frac{1}{I_3(\underline{\underline{Q}})} [\underline{\underline{Q}}^2 - I_1(\underline{\underline{Q}})\underline{\underline{Q}} + I_2(\underline{\underline{Q}})\underline{\underline{E}}] . \tag{22}$$

Inequalities for invariants of a positive tensor

$$I_1(\underline{\underline{Q}}) \geq 3I_3^{1/3}(\underline{\underline{Q}}) , \quad I_2(\underline{\underline{Q}}) \geq 3I_3^{2/3}(\underline{\underline{Q}}) , \quad I_1^2(\underline{\underline{Q}}) \geq 3I_2(\underline{\underline{Q}}) . \tag{23}$$

The first two follow from the inequalities for the geometric and arithmetic means

$$(\lambda_1\lambda_2\lambda_3)^{1/3} < \frac{1}{3}(\lambda_1 + \lambda_2 + \lambda_3) , \quad \left(\frac{1}{\lambda_1\lambda_2\lambda_3}\right)^{1/3} < \frac{1}{3}\left(\frac{1}{\lambda_1} + \frac{1}{\lambda_2} + \frac{1}{\lambda_3}\right) .$$

The third is a consequence of the formula

$$I_1^2(\underline{\underline{Q}}) - 3I_2(\underline{\underline{Q}}) = \frac{1}{2}[(\lambda_1 - \lambda_2)^2 + (\lambda_2 - \lambda_3)^2 + (\lambda_3 - \lambda_1)^2] .$$

which is easily checked by (8). The Inequalities (23) become equalities only for a spherical tensor. The third Inequality (23) does not require a tensor $\underline{\underline{Q}}$ to be positive.

10. THE SYMMETRIC TENSOR

The essential simplification, introduced by the assumption of symmetry for a tensor, amounts to the fact that the difference between the results of pre- and post-multiplication of a tensor and a vector vanishes. The bases $\underline{e}_k$ and $\underline{e}^k$ are identical. This is a unique orthonormal base and by (9.10) and (9.11) we have

$$\underline{e}_s \cdot \underline{e}_k = 0 \,, \qquad s \neq k \,; \qquad \underline{e}_s \cdot \underline{e}_s = 1 \qquad (s = 1,2,3) \,. \tag{1}$$

The roots of the defining equation of a symmetric tensor are real. In fact, the eigenvectors of the complex conjugate roots λ_1 and $\lambda_2 = \bar{\lambda}_1$ should also be complex conjugate

$$\underline{e}_1 = \frac{\underline{a} + i\underline{b}}{\sqrt{a^2 + b^2}} \,, \qquad \underline{e}_2 = \bar{\underline{e}}_1 = \frac{\underline{a} - i\underline{b}}{\sqrt{a^2 + b^2}} \,.$$

But in that case $\underline{e}_1 \cdot \underline{e}_2 = 1$ which contradicts the condition $\underline{e}_1 \cdot \underline{e}_2 = 0$.

By (9.17) we have

$$\underline{r}_m \cdot \underline{e}_s\underline{e}_s \cdot \underline{r}_m = g_{mm} \cos^2 (\underline{e}_s, \underline{r}_m) =$$

$$= -\frac{1}{P'(\lambda_s)} \left[q_m^{\cdot t} q_{tm} - (I_1(\underline{Q}) - \lambda_s) q_{mm} + I_3(\underline{Q}) \frac{g_{mm}}{\lambda_s} \right] \tag{2}$$

(no summation over m, s!). These formulas determine directions of $\underline{e}_s$, apart from sign, in the vector basis $\underline{r}_m$.

Two orthonormal bases $\underline{e}_s$, $\underline{e}'_s$ can be carried one into another with the aid of an orthogonal transformation. Two symmetric tensors with equal invariants may therefore be connected by the relationship (see the end of Par. 8)

$$\underline{Q} = \underline{O}^T \cdot \underline{P} \cdot \underline{O} \,, \qquad \underline{O} = \sum_{s=1}^{3} \underline{e}_s \underline{e}'_s$$

11. THE SKEW TENSOR AND THE ORTHOGONAL TENSOR

Substituting the dyads $\underline{r}_s\underline{r}_t$ in the expression for a skew tensor by the vector products $\underline{r}_t \times \underline{r}_s$ we obtain

$$\underline{\omega} = \frac{1}{2} \omega^{st} \underline{r}_t \times \underline{r}_s = \frac{1}{2} \omega^{st} \epsilon_{tsq} \underline{r}^q \qquad (\underline{\Omega} = \omega^{st} \underline{r}_s \underline{r}_t) \,, \tag{1}$$

which is called the vector associated with the tensor $\underline{\Omega}$. By (2.1) its covariant components are equal to

$$\omega_1 = \frac{1}{2}\sqrt{g}(\omega^{32} - \omega^{23}) = \sqrt{g}\omega^{32}, \qquad \omega_2 = \sqrt{g}\omega^{13}, \qquad \omega_3 = \sqrt{g}\omega^{21}. \tag{2}$$

Definition (1) implies that

$$\underline{\Omega} \cdot \underline{a} = \underline{\omega} \times \underline{a}, \qquad \underline{a} \cdot \underline{\Omega} = \underline{a} \times \underline{\omega}. \tag{3}$$

Indeed,

$$\underline{\omega} \times \underline{a} = \frac{1}{2} \epsilon_{tsq} \omega^{st} \underline{r}^q \times \underline{a} = \frac{1}{2} \epsilon_{tsq} \epsilon^{qmn} \omega^{st} a_m \underline{r}_n =$$

$$= \frac{1}{2}(\delta^m_t \delta^n_s - \delta^m_s \delta^n_t)\omega^{st} a_m \underline{r}_n = \omega^{nm} a_m \underline{r}_n = \underline{\Omega} \cdot \underline{a},$$

as required. Considering now the quadratic form associated with the symmetric tensor $\underline{\Omega}^2$

$$\underline{a} \cdot \underline{\Omega} \cdot \underline{\Omega} \cdot \underline{a} = (\underline{a} \times \underline{\omega}) \cdot (\underline{\omega} \times \underline{a}) = \underline{a} \cdot [\underline{\omega} \times (\underline{\omega} \times \underline{a})] = \underline{a} \cdot \underline{\omega\omega} \cdot \underline{a} - {}^2\underline{\omega} \cdot \underline{\omega} =$$

$$= \underline{a} \cdot (\underline{\omega\omega} - \underline{E\omega} \cdot \underline{\omega}) \cdot \underline{a},$$

we have

$$\underline{\Omega}^2 = \underline{\omega\omega} - \underline{E\omega} \cdot \underline{\omega}. \tag{4}$$

The invariants of a skew tensor are defined by the formulas

$$I_1(\underline{\Omega}) = 0, \qquad I_2(\underline{\Omega}) = -\frac{1}{2} I_1(\underline{\Omega}^2) = \omega^2, \qquad I_3(\underline{\Omega}) = 0. \tag{5}$$

The defining polynomial (9.6) and its roots turn out to be

$$P(\lambda) = -\lambda(\lambda^2 + \omega^2), \qquad \lambda_1 = -i\omega, \qquad \lambda_2 = i\omega, \qquad \lambda_3 = 0. \tag{6}$$

The right and left vectors $\underline{e}_3$, $\underline{e}^3$ are by (9.1) specified by the relationships

$$\underline{\Omega} \cdot \underline{e}_3 = \omega \times \underline{e}_3 = \lambda_3 \underline{e}_3 = 0, \qquad \underline{e}_3 \cdot \underline{e}^3 = 1, \qquad \underline{e}_3 = \underline{e}^3 = \frac{\underline{\omega}}{\omega}. \tag{7}$$

Therefore

$$\underline{\Omega} \cdot \underline{e}_1 = \underline{\omega} \times \underline{e}_1 = \omega \underline{e}_3 \times \underline{e}_1 = -i\omega \underline{e}_1, \qquad \underline{e}_1 = i\underline{e}_3 \times \underline{e}_1, \qquad \underline{e}_2 = -i\underline{e}_3 \times \underline{e}_2;$$

$$\underline{e}^1 \cdot \underline{\Omega} = \underline{e}^1 \times \underline{\omega} = \omega \underline{e}^1 \times \underline{e}^3 = -i\omega \underline{e}^1, \qquad \underline{e}^1 = -i\underline{e}^3 \times \underline{e}_1, \qquad \underline{e}^2 = i\underline{e}^3 \times \underline{e}_2$$

and comparison of these relationships makes it possible to assume that

$$\underline{e}_1 = \underline{e}^2 , \qquad \underline{e}_2 = \underline{e}^1 , \qquad \underline{E} = \underline{e}_1\underline{e}_2 + \underline{e}_2\underline{e}_1 + \underline{e}_3\underline{e}_3 .$$

The complex conjugate pairs $(\underline{e}_1, \underline{e}_2)$, $(\underline{e}^1, \underline{e}^2)$ of these vectors can be represented as

$$\underline{e}_1 = \underline{e}^2 = \frac{1}{\sqrt{2}} (\underline{c}_1 - i\underline{c}_2) , \qquad \underline{e}_2 = \underline{e}^1 = \frac{1}{\sqrt{2}} (\underline{c}_1 + i\underline{c}_2)$$

and the conditions $\underline{e}_1 \cdot \underline{e}^1 = 1$, $\underline{e}_2 \cdot \underline{e}^2 = 1$ will be satisfied if we assume that $|\underline{c}_1| = |\underline{c}_2| = 1$. We thus obtain

$$\underline{c}_1 - i\underline{c}_2 = i\underline{e}_3 \times (\underline{c}_1 - i\underline{c}_2) , \qquad \underline{c}_1 = \underline{e}_3 \times \underline{c}_2 , \qquad \underline{c}_2 = \underline{c}_1 \times \underline{e}_3$$

the real vectors basis $\underline{c}_1$, $\underline{c}_2$, $\underline{e}_3$ is orthonormal, the vectors $\underline{c}_1$ and $\underline{c}_2$ lie in a plane perpendicular to $\underline{e}_3$ and are specified up to a rotation around $\underline{e}_3$. By (9.14), the tensor $\underline{\Omega}$ is represented by the expression

$$\underline{\Omega} = -i\omega(\underline{e}_1\underline{e}^1 - \underline{e}_2\underline{e}^2) = -i\omega(\underline{e}_1\underline{e}_2 - \underline{e}_2\underline{e}_1) = \omega(\underline{c}_1\underline{c}_2 - \underline{c}_2\underline{c}_1) =$$

$$= \omega\underline{e}_3 \times \underline{E} = \underline{\omega} \times \underline{E} = \underline{E} \times \underline{\omega} . \tag{8}$$

<u>The orthogonal tensor.</u> Its invariants are defined by Equations (8.17), the characteristic equation reduces to the form

$$P(\lambda) = -\lambda^3 + (1 + 2\cos\omega)(\lambda^2 - \lambda) + 1 = (1 - \lambda)(\lambda^2 - 2\lambda\cos\omega + 1) = 0$$

and the eigenvalues turn out to be

$$\lambda_1 = e^{-i\omega} , \qquad \lambda_2 = e^{i\omega} , \qquad \lambda_3 = 1 . \tag{9}$$

For the root λ_3

$$\underline{e}^3 \cdot \underline{O} = \underline{e}^3 , \qquad \underline{O} \cdot \underline{e}_3 = \underline{e}_3$$

and this determines direction $\underline{e}_3 = \underline{e}^3$ which remains constant under an orthogonal transformation - this is the axis of the rotation. By (9.14)

$$\underline{O} = \underline{e}_1\underline{e}^1 e^{-i\omega} + \underline{e}_2\underline{e}^2 e^{i\omega} + \underline{e}_3\underline{e}^3 = (\underline{e}_1\underline{e}^1 + \underline{e}_2\underline{e}^2)\cos\omega - (\underline{e}_1\underline{e}^1 - \underline{e}_2\underline{e}^2) i \sin\omega + \underline{e}_3\underline{e}^3$$

and this expression can be reduced to the real form (8.16) provided that we assume

$$\underline{e}_1 = \underline{e}^2 = \frac{1}{\sqrt{2}} (\underline{c}_1 - i\underline{c}_2) \ , \qquad \underline{e}_2 = \underline{e}^1 = \frac{1}{\sqrt{2}} (\underline{c}_1 + i\underline{c}_2) \ ,$$

and all conditions (9.11) are fulfilled for the orthonormal basis $\underline{c}_1$, $\underline{c}_2$, $\underline{e}^3$.

We obtain

$$\underset{=}{O} = (\underline{c}_1\underline{c}_1 + \underline{c}_2\underline{c}_2) \cos\omega + (\underline{c}_1\underline{c}_2 - \underline{c}_2\underline{c}_1) \sin\omega + \underline{e}\underline{e} =$$

$$= \underset{=}{E} \cos\omega + (1 - \cos\omega)\underline{e}\underline{e} - \underline{e} \times \underset{=}{E} \sin\omega \qquad (10)$$

as required.

12. THE POLAR REPRESENTATION OF A TENSOR

A non-singular tensor $\underset{=}{Q}$ can be represented as a right or left product of a symmetric positive tensor $\underset{=}{U}$ (or $\underset{=}{V}$) by an orthogonal tensor, i.e.

$$\underset{=}{Q} = \underset{=}{U} \cdot \underset{=}{O} \ , \qquad \underset{=}{Q} = \underset{=}{O} \cdot \underset{=}{V} \ . \qquad (1)$$

This follows from the definition of an orthogonal tensor

$$\underset{=}{Q} \cdot \underset{=}{Q}^T = \underset{=}{U} \cdot \underset{=}{O} \cdot \underset{=}{O}^T \cdot \underset{=}{U} = \underset{=}{U}^2 \ , \qquad \underset{=}{Q}^T \cdot \underset{=}{Q} = \underset{=}{V} \cdot \underset{=}{O}^T \cdot \underset{=}{O} \cdot \underset{=}{V} = \underset{=}{V}^2 \ ;$$

$$\underset{=}{U} = (\underset{=}{Q} \cdot \underset{=}{Q}^T)^{1/2} \ , \qquad \underset{=}{V} = (\underset{=}{Q}^T \cdot \underset{=}{Q})^{1/2} \ . \qquad (2)$$

As has been stated in Par. 9 in connection with the definition of the square root of the positive tensor $\underset{=}{Q} \cdot \underset{=}{Q}^T$, the symmetric tensors $\underset{=}{U}$ and $\underset{=}{V}$ are positive.

With this definition of $\underset{=}{U}$ and $\underset{=}{V}$, the orthogonality of $\underset{=}{O}$ follows from the relationships

$$\underset{=}{O} = \underset{=}{U}^{-1} \cdot \underset{=}{Q} \ , \qquad \underset{=}{O}^T = \underset{=}{Q}^T \cdot \underset{=}{U}^{-1} \ , \qquad \underset{=}{O} \cdot \underset{=}{O}^T = (\underset{=}{Q} \cdot \underset{=}{Q}^T)^{-1/2} \cdot \underset{=}{Q} \cdot \underset{=}{Q}^T \cdot (\underset{=}{Q} \cdot \underset{=}{Q}^T)^{-1/2} = \underset{=}{E} \ ,$$

$$\underset{=}{O} = \underset{=}{Q} \cdot \underset{=}{V}^{-1} \ , \qquad \underset{=}{O}^T = \underset{=}{V}^{-1} \cdot \underset{=}{Q}^T \ , \qquad \underset{=}{O}^T \cdot \underset{=}{O} = (\underset{=}{Q}^T \cdot \underset{=}{Q})^{-1/2} \cdot \underset{=}{Q}^T \cdot \underset{=}{Q} \cdot (\underset{=}{Q}^T \cdot \underset{=}{Q})^{-1/2} = \underset{=}{E} \ .$$

The definitions of tensors $\underset{=}{U}$ and $\underset{=}{V}$ are consistent with the relationships

$$\underset{=}{U} \cdot \underset{=}{O} = \underset{=}{O} \cdot \underset{=}{V} \ , \qquad \underset{=}{U} = \underset{=}{O} \cdot \underset{=}{V} \cdot \underset{=}{O}^T \ , \qquad \underset{=}{V} = \underset{=}{O}^T \cdot \underset{=}{U} \cdot \underset{=}{O} \ , \qquad (3)$$

provided that the tensor $\underset{=}{O}$ in Equations (1) is the same. In fact

$$\underline{V}^2 = \underline{O}^T \cdot \underline{U} \cdot \underline{O} \cdot \underline{O}^T \cdot \underline{U} \cdot \underline{O} = \underline{O}^T \cdot \underline{U}^2 \cdot \underline{O} = \underline{Q}^T \cdot \underline{Q}$$

as required.

13. THE REPRESENTATION OF A TENSOR BY A SUM OF A SYMMETRIC TENSOR AND A DEVIATOR

By a tensor $\underline{Q}$ there is given by its spherical part $1/3\, I_1(\underline{Q})\underline{E}$, and the deviation of the tensor from this simplest structure is given by its deviator $\operatorname{dev} \underline{Q}$

$$\underline{Q} = \frac{1}{3} \underline{E} I_1(\underline{Q}) + \operatorname{dev} \underline{Q} \; . \tag{1}$$

It follows from the definition that $I_1(\operatorname{dev} \underline{Q}) = 0$. The first invariant of a skew tensor $\underline{\Omega}$ equals zero, $\operatorname{dev} \underline{\Omega} = \underline{\Omega}$, and consideration of its deviator becomes pointless. Further we assume that $\underline{Q}$ is a symmetric tensor

$$\underline{Q} = \underline{Q}^T \, , \qquad \underline{Q} \cdot \underline{e} = \lambda \underline{e} \, , \qquad \operatorname{dev} \underline{Q} \cdot \underline{e} = \left[\lambda - \frac{1}{3} I_1(\underline{Q}) \right] \underline{e} = -\kappa \underline{e} \; .$$

The principal directions of the tensors $\underline{Q}$ and $\operatorname{dev} \underline{Q}$ are the same, and the principal values κ_s of a deviator are determined by the formulas

$$\kappa_s = \lambda_s - \frac{1}{3} I_1(\underline{Q}) \qquad (s = 1,2,3) \; . \tag{2}$$

By (9.8)

$$\begin{aligned} I_1(\operatorname{dev} \underline{Q}) &= \kappa_1 + \kappa_2 + \kappa_3 = 0 \, , \\ I_2(\operatorname{dev} \underline{Q}) &= \kappa_1\kappa_2 + \kappa_2\kappa_3 + \kappa_3\kappa_1 = I_2(\underline{Q}) - \frac{1}{3} I_1^2(\underline{Q}) \, , \\ I_3(\operatorname{dev} \underline{Q}) &= \kappa_1\kappa_2\kappa_3 = I_3(\underline{Q}) - \frac{1}{3} I_1(\underline{Q}) I_2(\underline{Q}) + \frac{2}{27} I_1^3(\underline{Q}) \; . \end{aligned} \tag{3}$$

Another representation of the second invariant of a deviator follows from (7.8)

$$I_2(\operatorname{dev} \underline{Q}) = -\frac{1}{2} (\kappa_1^2 + \kappa_2^2 + \kappa_3^2) = -\frac{1}{6} [(\lambda_1 - \lambda_2)^2 + (\lambda_2 - \lambda_3)^2 + (\lambda_3 - \lambda_1)^2] \; . \tag{4}$$

Using the notation

$$-I_2(\operatorname{dev} \underline{Q}) = \frac{1}{4} g_2 \, , \qquad I_3(\operatorname{dev} \underline{Q}) = \frac{1}{4} g_3 \tag{5}$$

we reduce the characteristic equation of a deviator to the form

$$4\kappa^3 - g_2\kappa - g_3 = 0 \,. \tag{6}$$

Transformation from a cubic equation to this form reproduces the well-known procedure for the reduction of a complete cubic equation

$$y^3 + ay^2 + by + c = 0$$

to the standard form

$$x^3 + px + q = 0 \,.$$

The discriminant of Equation (6)

$$\Delta = g_2^3 - 27g_3^2 = 64\left[\,|I_2(\operatorname{dev}\underline{\underline{Q}})|^3 - \frac{27}{4}\,I_3^2(\operatorname{dev}\underline{\underline{Q}})\right] > 0 \,, \tag{7}$$

since the eigenvalues κ_s are real. This is an "irreducible case" for which the irrational expressions for the real roots inevitably contain complex quantities. Numerical values of the roots are sought for in the trigonometric form

$$\kappa = \sqrt{\frac{g_2}{3}}\,\sin\psi \,, \tag{8}$$

where ψ is defined by substitution into (6)

$$g_3 = \frac{g_2^{3/2}}{3\sqrt{3}}\,(4\sin^3\psi - 3\sin\psi) = -\left(\frac{g_2}{3}\right)^{3/2}\sin 3\psi \,, \tag{9}$$

so that $\Delta = g_2^3\cos^2 3\psi > 0$ as required. From (9) three values of κ are found and from these the eigenvalues of the deviator

$$\kappa_1 = \sqrt{\frac{g_2}{3}}\,\sin\psi \,, \qquad \kappa_2 = \sqrt{\frac{g_2}{3}}\,\sin\left(\psi + \frac{2\pi}{3}\right) \,, \qquad \kappa_3 = \sqrt{\frac{g_2}{3}}\,\sin\left(\psi + \frac{4\pi}{3}\right) \,,$$
$$|\psi| < \frac{\pi}{6} \,. \tag{10}$$

Applied to a deviator, the Hamilton-Cayley theorem reduces to

$$(\operatorname{dev}\underline{\underline{Q}})^3 + I_2(\operatorname{dev}\underline{\underline{Q}})\operatorname{dev}\underline{\underline{Q}} - \underline{\underline{E}}I_3(\operatorname{dev}\underline{\underline{Q}}) = \underline{\underline{0}}$$

which implies the formulas

$$I_2(\operatorname{dev}\underline{\underline{Q}}) = -\frac{1}{2}\,I_1((\operatorname{dev}\underline{\underline{Q}})^2) \,, \qquad I_3(\operatorname{dev}\underline{\underline{Q}}) = \frac{1}{3}\,I_1((\operatorname{dev}\underline{\underline{Q}})^3) \,,$$

$$I_2^2(\operatorname{dev}\underline{\underline{Q}}) = \frac{1}{2} I_1((\operatorname{dev}\underline{\underline{Q}})^4) \tag{11}$$

etc. Comparing (6) with the differential equation

$$P'^2(z) = 4P^3(z) - g_2 P(z) - g_3$$

for the Weierstrass function $P(z \mid g_2, g_3)$, we infer that the quantities κ_1, κ_2, κ_3 represent the roots e_1, e_2, e_3 of $P'(z \mid g_2, g_3)$.

The formulas (11) are from the theory of elliptic functions

$$\kappa_1^2 + \kappa_2^2 + \kappa_3^2 = \frac{1}{2} g_2 , \qquad \kappa_1^3 + \kappa_2^3 + \kappa_3^3 = \frac{3}{4} g_3 , \qquad \kappa_1^4 + \kappa_2^4 + \kappa_3^4 = \frac{1}{8} g_2^2 . \tag{12}$$

14. ON TENSORS OF HIGHER RANK

The quantity

$$^n\underline{\underline{Q}} = q_{s_{k+1}\ldots s_n}^{s_1 s_2 \ldots s_k} \underline{r}_{s_1} \underline{r}_{s_2} \ldots \underline{r}_{s_k} \underline{r}^{s_{k+1}} \ldots \underline{r}^{s_n} \tag{1}$$

defines a tensor of the n-th rank. The rank decreases by two as a result of contraction over two indices μ and ν. The so obtained tensor may be denoted as

$$^{(n-2)}\underline{\underline{Q}} = \overset{(\mu\cdot\nu)}{^n\underline{\underline{Q}}} = \overset{(\nu\cdot\mu)}{^n\underline{\underline{Q}}} . \tag{2}$$

E.g., the contraction of a tensor $^3\underline{C} = \underline{\underline{Q}}\underline{a}$ of the third rank (a dyadic of a tensor of the second rank and a vector) gives rise to the vectors

$$\overset{(1\cdot 2)}{^3\underline{C}} = I_1(\underline{\underline{Q}})\underline{a} , \qquad \overset{(2\cdot 3)}{^3\underline{C}} = \underline{\underline{Q}} \cdot \underline{a} , \qquad \overset{(3\cdot 1)}{^3\underline{C}} = \underline{\underline{Q}}^T \cdot \underline{a} .$$

An important example of a tensor of the third rank is given by the tensor of Levi-Civita

$$\underline{\epsilon} = \epsilon_{stq} \underline{r}^s \underline{r}^t \underline{r}^q = \epsilon^{stq} \underline{r}_s \underline{r}_t \underline{r}_q . \tag{3}$$

Its components are specified by (2.3); all its contractions lead to zero vectors.

The contraction of tensors of the fourth rank generates tensors of the second rank. We have, e.g.,

$$^4\underline{C} = \underline{\underline{P}}\underline{\underline{Q}} , \qquad \overset{(1\cdot 2)}{^4\underline{C}} = I_1(\underline{\underline{P}})\underline{\underline{Q}} , \qquad \overset{(1\cdot 3)}{^4\underline{C}} = \underline{\underline{P}}^T \cdot \underline{\underline{Q}} \qquad \text{etc} .$$

In particular, all contractions of the tensor $\underline{\underline{E}}\,\underline{\underline{E}}$ reduce to the tensors $3\underline{\underline{E}}$, $\underline{\underline{E}}$. Among the results of contraction of the tensor $\underline{\underline{Q}}\,\underline{\underline{Q}}\,\underline{\underline{Q}}$ of the sixth rank there are tensors $\underline{\underline{Q}}^2\underline{\underline{Q}}$, $\underline{\underline{Q}}\,\underline{\underline{Q}}^2$ of the fourth rank; a double contraction leads to the tensor $\underline{\underline{Q}}^3$ of the second rank.

The operation of relating a tensor of the $(n-1)$-th rank to a tensor of the n-th rank is accomplished by vector multiplication by two base vectors whose positions correspond to indices μ and ν ($\mu > \nu$). The vector product is ascribed either to the index ν (on the left) or to the index μ (on the right). The following notation is possible

$$^{(n-1)\leftarrow}\underline{\underline{Q}} = {}^{n}\overset{\overleftarrow{[\mu\times\nu]}}{\underline{\underline{Q}}} \,, \qquad ^{(n-1)\rightarrow}\underline{\underline{Q}} = {}^{n}\overset{\overrightarrow{[\mu\times\nu]}}{\underline{\underline{Q}}} \,. \tag{4}$$

The vector associated with a symmetric tensor of the second rank is a zero vector

$$q^{st}\underline{r}_t \times \underline{r}_s = \epsilon_{tsq}q^{st}\underline{r}^q = \epsilon_{stq}q^{st}\underline{r}^q = q^{st}\underline{r}_s \times \underline{r}_t = \underline{0}$$

and the vector $2\underline{\omega}$ associated with $\underline{\underline{Q}}$ is defined only by the skew part $\underline{\Omega}$ of that tensor with the aid of Equations (11.1), (11.2).

The tensors of the second rank which are related to the tensors $\underline{\underline{Q}}\,\underline{a}$ and $\underline{a}\,\underline{\underline{Q}}$, include the tensors determined by Equations (6.9) as

$$\underline{\underline{Q}} \times \underline{a} = q^{st}a^m\epsilon_{tmq}\underline{r}_s\underline{r}^q = -(\underline{a}\times\underline{\underline{Q}}^{\mathrm{T}}) \,, \qquad \underline{a}\times\underline{\underline{Q}} = -(\underline{\underline{Q}}^{\mathrm{T}}\times\underline{a})^{\mathrm{T}} \,. \tag{5}$$

In particular, for a symmetric tensor

$$\underline{\underline{Q}} \times \underline{a} = -(\underline{a}\times\underline{\underline{Q}})^{\mathrm{T}} \,, \qquad \underline{a}\times\underline{\underline{Q}} = -(\underline{\underline{Q}}\times\underline{a})^{\mathrm{T}} \tag{6}$$

and it is easily verified that the first invariant of these tensors equals zero

$$\begin{aligned} I_1(\underline{\underline{Q}}\times\underline{a}) &= q^{st}a^m\epsilon_{tmq}\underline{r}_s\cdot\underline{r}^q = q^{st}a^m\epsilon_{tms} = q^{st}a^m\epsilon_{smt} = 0 \,, \\ I_1(\underline{a}\times\underline{\underline{Q}}) &= 0 \,. \end{aligned} \tag{7}$$

One of the tensors of the third rank related to the tensor $\underline{\underline{E}}\,\underline{\underline{E}}$ is the Levi-Civita tensor

$$-\underline{\underline{E}}\times\underline{\underline{E}} = -\underline{r}_s\underline{r}^s \times \underline{r}^k\underline{r}_k = -\underline{r}_s\underline{r}_t\underline{r}_k\epsilon_{skt} = \underline{r}_s\underline{r}_t\underline{r}_k\epsilon_{stk} \,, \qquad \underline{\underline{\epsilon}} = -\underline{\underline{E}}\times\underline{\underline{E}} \,. \tag{8}$$

The representation (11.8) of a tensor $\underline{\Omega}$ in terms of the associated vector

$\underline{\omega}$ is now transformed to

$$\underline{\Omega} = \underline{\omega} \times \underline{E} = \underline{\omega} \cdot \underline{E} \times \underline{E} = -\underline{\omega} \cdot \underline{\epsilon} = -\underline{\epsilon} \cdot \underline{\omega} \ . \tag{9}$$

The relationship (8) also holds for the transformation of vector products. We thus arrive at the familiar relationships (11.3)

$$\underline{a} \times \underline{\omega} = \underline{a} \cdot \underline{E} \times \underline{E} \cdot \underline{\omega} = -\underline{a} \cdot \underline{\epsilon} \cdot \underline{\omega} = \underline{a} \cdot \underline{\Omega} \ , \qquad \underline{\omega} \times \underline{a} = \underline{\Omega} \cdot \underline{a} \ , \tag{10}$$

which are applicable to tensors of any rank

$$\underline{\underline{Q}} \times \underline{\omega} = \underline{\underline{Q}} \cdot \underline{E} \times \underline{E} \cdot \underline{\omega} = -\underline{\underline{Q}} \cdot \underline{\epsilon} \cdot \underline{\omega} = \underline{\underline{Q}} \cdot \underline{\Omega} \ , \qquad \underline{\omega} \times \underline{\underline{Q}} = \underline{\Omega} \cdot \underline{\underline{Q}} \ . \tag{11}$$

Applying this to tensors $\underline{\underline{Q}} \cdot \underline{\underline{Q}}^T$, $\underline{\underline{Q}}^T \cdot \underline{\underline{Q}}$ we have by (4.18) and (1.4)

$$\underline{\underline{Q}} \cdot \underline{\underline{Q}}^T = (\underline{S} + \underline{\Omega}) \cdot (\underline{S} - \underline{\Omega}) = \underline{S}^2 + \underline{\Omega} \cdot \underline{S} - \underline{S} \cdot \underline{\Omega} - \underline{\Omega}^2 = \underline{S}^2 + \underline{\omega} \times \underline{S} + (\underline{\omega} \times \underline{S})^T + $$
$$+ \underline{\underline{E\omega}} \cdot \underline{\omega} - \underline{\underline{\omega\omega}} \ , \tag{12}$$

$$\underline{\underline{Q}}^T \cdot \underline{\underline{Q}} = \underline{S}^2 - \underline{\omega} \times \underline{S} - (\underline{\omega} \times \underline{S})^T + \underline{\underline{E\omega}} \cdot \underline{\omega} - \underline{\underline{\omega\omega}} \ . \tag{13}$$

Contraction of the tensor $\underline{\underline{\epsilon\epsilon}}$ of the sixth rank leads to the fourth rank tensor

$$\underline{\epsilon} \cdot \underline{\epsilon} = (\underline{E} \times \underline{E}) \cdot (\underline{E} \times \underline{E}) = \underline{E} \times \underline{E} \times \underline{E} = \underline{r}_s \underline{r}^s \times \underline{r}^t \underline{r}_t \times \underline{r}_k \underline{r}^k =$$
$$= \epsilon^{stq} \epsilon_{tkm} \underline{r}_s \underline{r}_q \underline{r}^m \underline{r}^k = (\delta^q_k \delta^s_m - \delta^q_m \delta^s_k) \underline{r}_s \underline{r}_q \underline{r}^m \underline{r}^k = \underline{r}_s \underline{r}_k \underline{r}^s \underline{r}^k - \underline{r}_s \underline{r}_m \underline{r}^m \underline{r}^s =$$
$$= \underline{C}_{II} - \underline{C}_{III} \ .$$

We introduced here the notation for isotropic tensors of the fourth rank

$$\underline{C}_{II} = \underline{r}_s \underline{r}_k \underline{r}^s \underline{r}^k \ , \qquad \underline{C}_{III} = \underline{r}_s \underline{r}_k \underline{r}^k \underline{r}^s = \underline{r}_s \underline{\underline{E}} \underline{r}^s \ . \tag{14}$$

We arrive at the relationship

$$\underline{\epsilon} \cdot \underline{\epsilon} = \underline{E} \times \underline{E} \times \underline{E} = \underline{C}_{II} - \underline{C}_{III} \ . \tag{15}$$

By a double contraction

$$\underline{\epsilon} \cdot\cdot \underline{\epsilon} = \underline{r}_s \underline{r}_k \cdot \underline{r}^s \underline{r}^k - \underline{r}_s \underline{r}_m \cdot \underline{r}^m \underline{r}^s = \underline{E} - 3\underline{E} = -2\underline{E} \ . \tag{16}$$

This relationship implies the following inverses of (8) and (11.8), i.e.

$$\underline{\epsilon} \cdot\cdot \underline{\Omega} = -\underline{\epsilon} \cdot\cdot \underline{\epsilon} \cdot \underline{\omega} = 2\underline{E} \cdot \underline{\omega} = 2\underline{\omega} , \qquad \underline{\omega} = \frac{1}{2} \underline{\epsilon} \cdot\cdot \underline{\Omega} . \tag{17}$$

In particular, the vector associated with the dyadic $\underline{ab}$ is by (11.1) equal to

$$\underline{\omega} = \frac{1}{2} \underline{b} \times \underline{a} \tag{18}$$

and by (17) the vector product $\underline{a} \times \underline{b}$ can be represented as

$$\underline{a} \times \underline{b} = \underline{\epsilon} \cdot\cdot (\underline{ba} - \underline{ab}) , \qquad \underline{ba} - \underline{ab} = \underline{E} \times (\underline{a} \times \underline{b}) . \tag{19}$$

Note. The notation of a vector product in the form of a "vector invariant"

$$\frac{1}{2} (\underline{ab} - \underline{ba})_{\times} = \frac{1}{2} (\underline{a} \times \underline{b} - \underline{b} \times \underline{a}) = \underline{a} \times \underline{b}$$

is equally applicable.

15. ISOTROPIC TENSORS

The components of an isotropic tensor remain the same under any orthogonal transformation of the vector basis. The product of an isotropic tensor by a scalar is, of course, also an isotropic tensor.

The unit tensor is the only isotropic tensor of the second rank, e.g. its covariant components in the initial and rotated bases are

$$g_{sk} = \underline{r}_s \cdot \underline{r}_k = \underline{r}'_s \cdot \underline{O}^T \cdot \underline{O} \cdot \underline{r}'_k = \underline{r}'_s \cdot \underline{r}'_k = g'_{sk} .$$

There are no isotropic tensors of the second rank different from $\lambda\underline{E}$.

Those tensors of a higher rank which can be written in terms of $\underline{E}$ are all isotropic. It can be seen from (14.8) that the Levi-Civita tensor of the third rank, $\underline{\epsilon}$, is such a tensor, $\lambda\underline{\epsilon}$ being the only isotropic tensor of the third rank. *)

The tensors of the fourth rank formed by $\underline{E}$ are $\underline{EE}$ and $\underline{C}_{II} - \underline{C}_{III}$ (see (14.15)) and both of the tensors $\underline{C}_{II}$ and $\underline{C}_{III}$ are also isotropic. In fact, by (8.5) and (8.8)

$$\underline{r}'^s = \underline{r}^s \cdot o^k_{.m} \underline{r}_k \underline{r}^m = o^s_{.m} \underline{r}^m , \qquad \underline{r}'_s = \underline{r}_s \cdot o^{.n}_m \underline{r}^m \underline{r}_n = o^{.n}_s \underline{r}_n ,$$

so that

---

*) The components of $\underline{\epsilon}$ change their sign under an improperly orthogonal transformation. Such tensors (of odd rank) are called demitropic.

$$\mathbf{r}'^{s}\mathbf{E}\mathbf{r}'_{s} = o^{s}_{.m}\,o^{.n}_{s}\,\mathbf{r}^{m}\mathbf{E}\mathbf{r}_{n} = \delta^{n}_{m}\mathbf{r}^{m}\mathbf{E}\mathbf{r}_{n} = \mathbf{r}^{m}\mathbf{E}\mathbf{r}_{m} = \mathbf{C}_{III} \,.$$

For the three isotropic tensors of the fourth rank

$$\mathbf{C}_{I} = \mathbf{EE} = \mathbf{r}_{s}\mathbf{r}^{s}\mathbf{r}_{k}\mathbf{r}^{k} \,, \qquad \mathbf{C}_{II} = \mathbf{r}_{s}\mathbf{r}_{k}\mathbf{r}^{s}\mathbf{r}^{k} \,, \qquad \mathbf{C}_{III} = \mathbf{r}_{s}\mathbf{E}\mathbf{r}^{s} = \mathbf{r}_{s}\mathbf{r}_{k}\mathbf{r}^{k}\mathbf{r}^{s} \tag{1}$$

several equivalent forms are possible. In each of the dyads $\mathbf{r}_{s}\mathbf{r}^{s}$, $\mathbf{r}_{k}\mathbf{r}^{k}$ of the tensor $\mathbf{C}_{I}$ the upper and lower indices are interchangeable, the tensor remaining equal to $\mathbf{EE}$. In $\mathbf{C}_{II}$, the dyads $\mathbf{r}_{s}\mathbf{r}_{k}$, $\mathbf{r}^{s}\mathbf{r}^{k}$ may be interchanged. They may also be written as $\mathbf{r}_{s}\mathbf{r}^{k}$, $\mathbf{r}^{s}\mathbf{r}_{k}$. In $\mathbf{C}_{III}$ the first and fourth vectors are interchangeable, as well as the second and third. For example

$$\mathbf{C}_{II} = \mathbf{r}_{s}\mathbf{r}_{k}\mathbf{r}^{s}\mathbf{r}^{k} = g_{sq}\mathbf{r}^{q}\mathbf{r}_{k}g^{sm}\mathbf{r}_{m}\mathbf{r}^{k} = \delta^{m}_{q}\mathbf{r}^{q}\mathbf{r}_{k}\mathbf{r}_{m}\mathbf{r}^{k} = \mathbf{r}^{q}\mathbf{r}_{k}\mathbf{r}_{q}\mathbf{r}^{k}$$

etc. An isomer is defined as a tensor arising from a given tensor under permutation of the indices of the basis vectors. For example, $\mathbf{Q}^{T}$ is an isomer of $\mathbf{Q}$ and tensors $c_{abc}\mathbf{r}^{c}\mathbf{r}^{a}\mathbf{r}^{b}$ are isomers of the tensor $c_{abc}\mathbf{r}^{a}\mathbf{r}^{b}\mathbf{r}^{c}$. There are $n!$ isomers of a tensor of the n-th rank, this number decreasing with symmetry of structure of a tensor. E.g., of the total number of 24 isomers of an isotropic tensor of the fourth rank only three tensors (1) cannot be reduced to each other.

The general expression for an isotropic tensor of the fourth rank is therefore given by

$$^{4}\mathbf{C} = \lambda\mathbf{EE} + \mu(\mathbf{C}_{II} - \mathbf{C}_{III}) + \nu(\mathbf{C}_{III} + \mathbf{C}_{II}) \,. \tag{2}$$

The double contraction $\mathbf{C}_{K}\cdot\cdot\,\mathbf{C}_{L}$ $(K,L = \mathrm{I,II,III})$ leads of course to one of these isotropic tensors as is illustrated below

	$\mathbf{C}_{I}$	$\mathbf{C}_{II}$	$\mathbf{C}_{III}$
$\mathbf{C}_{I}$	$3\mathbf{C}_{I}$	$\mathbf{C}_{I}$	$\mathbf{C}_{I}$
$\mathbf{C}_{II}$	$\mathbf{C}_{I}$	$\mathbf{C}_{III}$	$\mathbf{C}_{II}$
$\mathbf{C}_{III}$	$\mathbf{C}_{I}$	$\mathbf{C}_{II}$	$\mathbf{C}_{II}$

(3)

Double contractions with a tensor of the second rank result in the tensors

$$\mathbf{C}_{I}\cdot\cdot\,\mathbf{Q} = \mathbf{E}I_{1}(\mathbf{Q}) = \mathbf{Q}\cdot\cdot\,\mathbf{C}_{I} \,, \quad \mathbf{C}_{II}\cdot\cdot\,\mathbf{Q} = \mathbf{Q}\cdot\cdot\,\mathbf{C}_{II} = \mathbf{Q}^{T} \,, \quad \mathbf{C}_{III}\cdot\cdot\,\mathbf{Q} = \mathbf{Q}\cdot\cdot\,\mathbf{C}_{III} = \mathbf{Q} \,. \tag{4}$$

The double contraction with $1/2(\mathbf{C}_{II} + \mathbf{C}_{III})$ therefore separates out the symmetric part of $\mathbf{Q}$, and the same operation with $1/2(\mathbf{C}_{III} - \mathbf{C}_{II})$ produces the skew part. We thus arrive at the relationship

$$\left[\lambda \underline{C}_I + \frac{1}{2}\mu(\underline{C}_{II} + \underline{C}_{III}) + \frac{1}{2}\nu(\underline{C}_{III} - \underline{C}_{II})\right] \cdot\cdot\, \underline{Q} = \lambda \underline{E} I_1(\underline{Q}) + \mu \underline{S} + \nu \underline{\Omega} \; . \tag{5}$$

The rules (7.16) imply the relationships

$$(\underline{C}_I \cdot\cdot\, \underline{A}) \cdot \underline{B} = I_1(\underline{A})\underline{B} \, , \qquad \underline{C}_I \cdot\cdot\, \underline{A} \cdot \underline{B} = \underline{E} I_1(\underline{A} \cdot \underline{B}) \, ,$$

$$(\underline{C}_{II} \cdot\cdot\, \underline{A}) \cdot \underline{B} = \underline{A}^T \cdot \underline{B} \, , \qquad \underline{C}_{II} \cdot\cdot\, \underline{A} \cdot \underline{B} = \underline{B}^T \cdot \underline{A}^T \, ,$$

$$(\underline{C}_{III} \cdot\cdot\, \underline{A}) \cdot \underline{B} = \underline{A} \cdot \underline{B} \, , \qquad \underline{C}_{III} \cdot\cdot\, \underline{A} \cdot \underline{B} = \underline{A} \cdot \underline{B} \, ,$$

where $\underline{A}$, $\underline{B}$ denote tensors of the second rank. The formulas

$$\underline{C}_I \cdot \underline{A} = \underline{E}\underline{A} \, , \qquad \underline{C}_{III} \cdot \underline{A} = \underline{A} \cdot \underline{C}_{III} \tag{9}$$

will also prove useful.

The convolutions of tensors (9 in number)

$$\underline{C}_K \cdot \underline{C}_L \, , \qquad K, L = I, II, III \tag{10}$$

are isotropic tensors of the sixth rank. These special tensors do not exhaust, however, the totality of all tensors of that type: to these should be added all those terms contained in the representation of the tensor $\underline{\underline{\epsilon\epsilon}}$ in terms of the basis vectors. We have by (2.5)

$$\underline{\underline{\epsilon\epsilon}} = \underline{r}^s \underline{r}^t \underline{r}^k \underline{r}_m \underline{r}_n \underline{r}_p \begin{vmatrix} \delta_s^m & \delta_s^n & \delta_s^p \\ \delta_t^m & \delta_t^n & \delta_t^p \\ \delta_k^m & \delta_k^n & \delta_k^p \end{vmatrix} =$$

$$= \underline{r}^s \underline{r}^t \underline{r}^k [\underline{r}_s(\underline{r}_t \underline{r}_k - \underline{r}_k \underline{r}_t) + \underline{r}_t(\underline{r}_k \underline{r}_s - \underline{r}_s \underline{r}_k) + \underline{r}_k(\underline{r}_s \underline{r}_t - \underline{r}_t \underline{r}_s)] \; . \tag{11}$$

In Equations (10) and (11) 15 isotropic tensors out of a possible $6! = 720$ such tensors are specified. The formulas for sixfold contraction of these tensors with the sixth rank tensor $\underline{\underline{AAA}}$, where $\underline{A} = \underline{A}^T$, will be found useful. We have for example

$$\underline{\underline{EEE}} \cdots\cdots \underline{\underline{AAA}} = I_1^3(\underline{A}) \, , \qquad \underline{\underline{EC}}_I \cdots\cdots \underline{\underline{AAA}} = I_1(\underline{A}) I_1(\underline{A}^2) \, ,$$

$$\underset{-}{C}_{II} \cdot \underset{-}{C}_{III} \cdots\cdots \underline{AAA} = I_1(\underset{-}{A}^3) \; . \tag{12}$$

There are, however, no invariant scalar structures of the third degree with respect to the components of $\underset{-}{A}$ except for those enumerated in (12). The same invariants would therefore result from the contraction of $\underline{AAA}$ with a general isotropic tensor $^6\underset{-}{C}$

$$^6\underset{-}{C} = \sum_K \sum_L \alpha_{KL} \underset{-}{C}_K \cdot \underset{-}{C}_L + \underset{-}{r}^s \underset{-}{r}^t \underset{-}{r}^k \, [\underset{-}{r}_s(\alpha_{10} \underset{-}{r}_t \underset{-}{r}_k + \alpha_{11} \underset{-}{r}_k \underset{-}{r}_t) +$$

$$+ \underset{-}{r}_t(\alpha_{12} \underset{-}{r}_k \underset{-}{r}_s + \alpha_{13} \underset{-}{r}_s \underset{-}{r}_k) + \underset{-}{r}_k(\alpha_{14} \underset{-}{r}_s \underset{-}{r}_t + \alpha_{15} \underset{-}{r}_t \underset{-}{r}_s)] \; . \tag{13}$$

We obtain

$$^6\underset{-}{C} \cdots\cdots \underline{AAA} = \frac{1}{6} \, [\nu_1 I_1^3(\underset{-}{A}) + 6\nu_2 I_1(\underset{-}{A}) I_1(\underset{-}{A}^2) + 8\nu_3 I_1(\underset{-}{A}^3)] \; , \tag{14}$$

where ν_1, ν_2, ν_3 depend linearly upon α_k. Applied to the specific strain energy of a non-linearly isotropic medium, Bernstein and Toupin [8.13] have termed ν_1, ν_2, ν_3 Lamé moduli of the third order. The right-hand side of (14) may also be rewritten as

$$\frac{1}{3} \, (1 + 2m) I_1^3(\underset{-}{A}) - 2m I_1(\underset{-}{A}) I_2(\underset{-}{A}) + n I_3(\underset{-}{A}) \; ;$$

$$l = \frac{1}{2} \nu_1 + \nu_2 \, , \qquad m = \nu_2 + 2\nu_3 \, , \qquad n = 4\nu_3 \; . \tag{15}$$

We have denoted the Murnaghan constants (F.D. Murnaghan, 1951) by l, m, n.

The fourfold contraction of tensor (2) with that of $\underline{AA}$ (for $\underset{-}{A} = \underset{-}{A}^T$) is represented in a similar, but simpler, form by

$$^4\underset{-}{C} \cdots\cdot \underline{AA} = \frac{1}{2} \, [\lambda I_1^2(\underset{-}{A}) + 2\mu I_1(\underset{-}{A}^2)] \; , \tag{16}$$

here, λ and μ denote the Lamé moduli of the second order.

Contractions of the fourth rank tensors

$$\underline{AB} \, , \; \underline{AB}^2 \, , \; \underset{-}{A}^2\underset{-}{B} \, , \; \underset{-}{A}^2\underset{-}{B}^2 \; ; \qquad \underline{BA} \, , \; \underset{-}{B}^2\underset{-}{A} \, , \; \underline{BA}^2 \, , \; \underset{-}{B}^2\underset{-}{A}^2 \; ,$$

determined by two symmetric tensors of the second rank, result in four invariant quantities

$$I_1(\underset{-}{A} \cdot \underset{-}{B}) \, , \quad I_1(\underset{-}{A} \cdot \underset{-}{B}^2) \, , \quad I_1(\underset{-}{A}^2 \cdot \underset{-}{B}) \, , \quad I_1(\underset{-}{A}^2 \cdot \underset{-}{B}^2) \; .$$

Adding the six invariants $I_1(\underset{-}{A}^k)$, $I_1(\underset{-}{B}^k)$ $(k = 1,2,3)$ to this list, we obtain ten invariants from the two symmetric tensors. The invariants I_2, I_3 can be expressed in terms of those given with the aid of Equations (7.9), (7.10) and the Cayley-Hamilton theorem.*)

*) Spencer A.J.M. Theory of Invariants. In: Continuum Physics, Vol. I, part III. - New York, London, 1971.

Appendix II. TENSOR FUNCTIONS

1. THE LINEAR FUNCTION OF A TENSOR ARGUMENT

We now examine the linear homogeneous functions of the components of a tensor of the first rank (vector $\underline{a}$) and of the second rank (tensor $\underline{Q}$). These functions may be scalars φ, vectors $\underline{c}$ or tensors $\underline{P}$ of the second rank. They are specified by a single relationship in the first case and by three and nine relationships respectively in the second and third (three components c_s and nine components p_{st} in the chosen basis $\underline{r}_s$), i.e.

$$\varphi(\underline{a}) = m_k a^k , \qquad \varphi(\underline{Q}) = m_{st} q^{ts} , \tag{1a}$$

$$c_s(\underline{a}) = m_{st} a^t , \qquad c_s(\underline{Q}) = m_{str} q^{rt} , \tag{1b}$$

$$p_{st}(\underline{a}) = m_{str} a^r , \qquad p_{st}(\underline{Q}) = m_{stmn} q^{nm} . \tag{1c}$$

By definition, on transfer to a new basis $\check{\underline{r}}_s$ a scalar φ remains the same and the quantities c_s, p_{st} should be transformed, respectively, as the components of a vector and of a tensor of the second rank. The quantities m_k, m_{st}, m_{str}, m_{stmn} must be the components in the basis $\underline{r}_s$ of a vector $\underline{m}$, a tensor $\underline{M}$ of the second rank and of tensors ${}^3\underline{M}$, ${}^4\underline{M}$ of the third and fourth ranks, i.e. we have

$$\varphi = \underline{m} \cdot \underline{a} , \qquad \varphi = \underline{M} \cdot\cdot \underline{Q} , \tag{2a}$$

$$\underline{c} = \underline{M} \cdot \underline{a} , \qquad \underline{c} = {}^3\underline{M} \cdot\cdot \underline{Q} , \tag{2b}$$

$$\underline{P} = {}^3\underline{M} \cdot \underline{a} , \qquad \underline{P} = {}^4\underline{M} \cdot\cdot \underline{Q} . \tag{2c}$$

The proof rests on rules (1.7.2) and (1.7.3). For example, for the second relationship

$$q^{ts} = \check{q}^{nm} \underline{r}^t \cdot \check{\underline{r}}_n \underline{r}^s \cdot \check{\underline{r}}_m , \qquad \varphi = m_{st} q^{ts} = \check{q}^{nm} \underline{r}^t \cdot \check{\underline{r}}_n \underline{r}^s \cdot \check{\underline{r}}_m m_{st} .$$

If we define the quantities $\check{m}_{mn}$ by the rules for the transformation of tensor components

$$\check{m}_{mn} = m_{st}\underline{r}_m \cdot \underline{r}^s \check{\underline{r}}_n \cdot \underline{r}^t ,$$

we obtain

$$\varphi = \check{q}^{nm}\check{m}_{mn} = \underline{M} \cdot\cdot \underline{Q}$$

as required. The other relationships are checked in the same way.

The following functions may be regarded as examples of linear tensor functions

$$\varphi(\underline{Q}) = \underline{E} \cdot\cdot \underline{Q} = I_1(\underline{Q}) , \qquad \underline{P}(\underline{Q}) = \underline{C}_I \cdot\cdot \underline{Q} = \underline{E}I_1(\underline{Q}) ,$$

$$\underline{P}(\underline{Q}) = \underline{C}_{II} \cdot\cdot \underline{Q} = \underline{Q}^T , \qquad \underline{P}(\underline{Q}) = \underline{C}_{III} \cdot\cdot \underline{Q} = \underline{Q}$$

the tensor ${}^4\underline{M}$ takes the values of the isotropic tensors $\underline{C}_k$.

A contraction of an isotropic tensor $-\underline{\epsilon}$ with a vector $\underline{\omega}$ defines, from (I.14.9), a skew tensor, and a double contraction with the second rank tensor results, by (I.14.17), in a vector $2\underline{\omega}$.

The quadratic form of the components of $\underline{Q}$ is a scalar defined by the formula (2a) where a linear function of $\underline{Q}$, i.e. $\underline{P}$ (see (2c)) stands for a tensor $\underline{M}$. We have

$$\varphi(\underline{P}(\underline{Q})) = \varphi({}^4\underline{M} \cdot\cdot \underline{Q}) = ({}^4\underline{M} \cdot\cdot \underline{Q}) \cdot\cdot \underline{Q} = m^{tknm} q_{mn} q_{kt} . \tag{4}$$

A specification of this form determines the tensor ${}^4\underline{M}$ provided that the pairs (tk), (nm) are interchangeable

$$m^{tknm} = m^{nmtk} \tag{5}$$

just as the specification of the quadratic form $q^{st}a_s a_t$ prescribes only the symmetric part of a tensor $\underline{Q}$.

For $\underline{Q} = \underline{Q}^T$

$$p^{st} = m^{stmn} q_{nm} , \qquad m^{stmn} = m^{stnm} ,$$

and for $\underline{P} = \underline{P}^T$

$$p^{st} = p^{ts} = m^{stmn} q_{nm} , \qquad m^{stmn} = m^{tsmn} .$$

If both tensors

$$\underline{Q} = \underline{Q}^T , \qquad \underline{P}(\underline{Q}) = [\underline{P}(\underline{Q})]^T$$

are symmetric, then the tensor $\overset{4}{\underline{M}}$ is symmetric with respect to the indices within each pair

$$m^{stmn} = m^{tsmn} = m^{stnm} \tag{6}$$

and the number of its components decreases from 81 to 36 and under transformation (5) to 21.

In the linear theory of elasticity, the role of tensors $\underline{P}$ and $\underline{Q}$ is taken over by a stress tensor $\underline{T}$ and a linear strain tensor $\underline{\varepsilon}$, the tensor $\overset{4}{\underline{M}}$, the tensor of the elastic moduli, being designated $\overset{4}{\underline{C}}$

$$\underline{T} = \overset{4}{\underline{C}} \cdot\cdot\, \underline{\varepsilon} \tag{7}$$

because $\underline{T}$ and $\underline{\varepsilon}$ are symmetric tensors there are 36 elastic moduli ("Cauchy elasticity"). The quadratic form (4) defines twice the potential energy per unit volume

$$2W = \underline{T} \cdot\cdot\, \underline{\varepsilon} = (\overset{4}{\underline{C}} \cdot\cdot\, \underline{\varepsilon}) \cdot\cdot\, \underline{\varepsilon} \ . \tag{8}$$

Its existence, which depends upon whether or not Equations (5) hold

$$\frac{\partial^2 W}{\partial \varepsilon_{st} \partial \varepsilon_{mn}} = c_{stmn} = \frac{\partial^2 W}{\partial \varepsilon_{mn} \partial \varepsilon_{st}} = c_{mnst} \ , \tag{9}$$

reduces the number of elastic moduli to 21 ("Green elasticity").

2. THE SCALAR FUNCTION OF A TENSOR ARGUMENT. THE DERIVATIVE OF A SCALAR WITH RESPECT TO A TENSOR

The function of the components q^{st} of a tensor $\underline{Q}$ defined in the $\underline{r}_s$-basis

$$\varphi = \varphi(q^{11}, q^{22}, \ldots, q^{31}) \ , \tag{1}$$

is a scalar ("scalar valued") provided that it has the same value for any coordinate system

$$\varphi(q^{11}, q^{22}, \ldots, q^{31}) = \bar{\varphi}(\check{q}^{11}, \check{q}^{22}, \ldots, \check{q}^{31}) \ . \tag{2}$$

Such, for example, is the function

$$\varphi = \underline{a} \cdot \underline{Q} \cdot \underline{b} \ .$$

It may be immediately verified by returning to the initial basis that $\bar{\varphi} = \varphi$

$$\varphi = \underline{a} \cdot \underline{\underline{Q}} \cdot \underline{b} = \underline{a} \cdot \check{\underline{r}}_s \check{\underline{r}}^s \cdot \underline{\underline{Q}} \cdot \check{\underline{r}}^t \check{\underline{r}}_t \cdot \underline{b} = \check{a}_s \check{b}_t \check{q}^{st} = a_m b_n q^{mn} \ , \tag{3}$$

as required.

The definition of a derivative

$$f'(x) = \lim_{h \to} \frac{1}{h} [f(x+h) - f(x)] \tag{4}$$

of ordinary calculus, cannot be generalized to functions of a tensor argument. One should proceed from the definition of a derivative as the multiplier of the linear increment δx of the function $f(x)$

$$\delta f = f(x + \delta x) - f(x) = f'(x)\, \delta x \ . \tag{5}$$

The derivative of a scalar (1)

$$\delta\varphi(q^{11}, q^{22}, \ldots, q^{31}) = \frac{\partial \varphi}{\partial q^{st}} \delta q^{st} \tag{6}$$

can be represented as a double contraction of the tensors

$$\varphi_{\underline{\underline{Q}}} = \frac{\partial \varphi}{\partial q^{st}} \underline{r}^s \underline{r}^t \ , \qquad \delta \underline{\underline{Q}}^{\mathrm{T}} = \underline{r}_m \underline{r}_n \delta q^{nm} \ , \qquad \frac{\partial \varphi}{\partial q^{st}} \delta q^{st} = \varphi_{\underline{\underline{Q}}} \cdot\cdot\, \delta \underline{\underline{Q}}^{\mathrm{T}} = \delta \varphi$$

and it is natural, referring to (5), to term $\varphi_{\underline{\underline{Q}}}$ the derivative of φ with respect to the tensor $\underline{\underline{Q}}$. By (1.1a) and (1.2a) the quantity $\varphi_{\underline{\underline{Q}}}$ is a tensor of the second rank

$$\varphi_{\underline{\underline{Q}}} = \frac{\partial \varphi}{\partial \underline{\underline{Q}}} = \frac{\partial \varphi}{\partial q^{st}} \underline{r}^s \underline{r}^t \ , \qquad \delta \varphi = \varphi_{\underline{\underline{Q}}} \cdot\cdot\, \delta \underline{\underline{Q}}^{\mathrm{T}} \ . \tag{7}$$

The differentiation in (6) is carried out with respect to all (nine) components of the tensor $\underline{\underline{Q}}$. In order to avoid mistakes, we must, for a symmetric tensor, make the substitution

$$\varphi(q^{mn}) = \varphi\left(\frac{1}{2}(q^{mn} + q^{nm})\right)$$

and set $q^{mn} = q^{nm}$ after differentiation with respect to any of the nine components. We thus obtain

$$\underline{\underline{Q}} = \underline{\underline{Q}}^{\mathrm{T}} : \varphi_{\underline{\underline{Q}}} = \frac{1}{2} \frac{\partial \varphi}{\partial q^{st}} (\underline{r}^s \underline{r}^t + \underline{r}^t \underline{r}^s) \ , \tag{8}$$

which implies that the derivative of a scalar with respect to a symmetric tensor is itself a symmetric tensor

$$\underset{=}{Q} = \underset{=}{Q}^T : (\varphi_{\underset{=}{Q}})^T = \varphi_{\underset{=}{Q}} . \tag{9}$$

The formal rules for the differentiation of a sum and product are immediately applicable to the differentiation of a scalar by a tensor. We have, for example

$$\delta\varphi\psi = \varphi\delta\psi + \psi\delta\varphi = (\varphi\psi_{\underset{=}{Q}} + \psi\varphi_{\underset{=}{Q}}) \cdot\cdot\, \delta\underset{=}{Q}^T , \qquad (\varphi\psi)_{\underset{=}{Q}} = \varphi\psi_{\underset{=}{Q}} + \psi\varphi_{\underset{=}{Q}} . \tag{10}$$

3. THE FORMULAS FOR THE DIFFERENTIATION OF A SCALAR

The following formulas find numerous applications throughout the main text. Their derivation is based upon the invariant representation (7) of the variation $\delta\varphi$.

1. The derivatives of invariants of a tensor. The first invariant $I_1(\underset{=}{Q})$ is a linear function of $\underset{=}{Q}$. By definition (I.7.4) we have

$$\delta I_1(\underset{=}{Q}) = I_1(\underset{=}{Q} + \delta\underset{=}{Q}) - I_1(\underset{=}{Q}) = \underset{=}{E} \cdot\cdot\, (\underset{=}{Q} + \delta\underset{=}{Q}) - \underset{=}{E} \cdot\cdot\, \underset{=}{Q} = \underset{=}{E} \cdot\cdot\, \delta\underset{=}{Q} = \underset{=}{E} \cdot\cdot\, \delta\underset{=}{Q}^T$$

so that

$$I_1(\underset{=}{Q})_{\underset{=}{Q}} = \underset{=}{E} .$$

Further, recalling rules of contraction, (I.7.16) and (I.7.17), we obtain

$$\delta I_1(\underset{=}{Q}^2) = I_1((\underset{=}{Q} + \delta\underset{=}{Q}) \cdot (\underset{=}{Q} + \delta\underset{=}{Q})) - I_1(\underset{=}{Q}^2) = I_1(\underset{=}{Q} \cdot \delta\underset{=}{Q} + \delta\underset{=}{Q} \cdot \underset{=}{Q}) =$$

$$= \underset{=}{E} \cdot\cdot\, \underset{=}{Q} \cdot \delta\underset{=}{Q} + \underset{=}{E} \cdot\cdot\, \delta\underset{=}{Q} \cdot \underset{=}{Q} = \underset{=}{Q} \cdot\cdot\, \delta\underset{=}{Q} + \underset{=}{Q} \cdot\cdot\, \delta\underset{=}{Q} = 2\underset{=}{Q}^T \cdot\cdot\, \delta\underset{=}{Q}^T ,$$

$$I_1(\underset{=}{Q}^2)_{\underset{=}{Q}} = 2\underset{=}{Q}^T$$

and similarly

$$\delta I_1(\underset{=}{Q}^3) = I_1(\underset{=}{Q}^2 \cdot \delta\underset{=}{Q} + \underset{=}{Q} \cdot \delta\underset{=}{Q} \cdot \underset{=}{Q} + \delta\underset{=}{Q} \cdot \underset{=}{Q}^2) = 3\underset{=}{Q}^2 \cdot\cdot\, \delta\underset{=}{Q} = 3\underset{=}{Q}^{T^2} \cdot\cdot\, \delta\underset{=}{Q}^T ,$$

$$I_1(\underset{=}{Q}^3)_{\underset{=}{Q}} = 3\underset{=}{Q}^{T^2} .$$

We have thus arrived at the expressions for the derivatives of the first invariant of powers of a tensor

$$I_1(\underset{\approx}{Q})_{\underset{\approx}{Q}} = \underset{\approx}{E}\,, \qquad I_1(\underset{\approx}{Q}^2)_{\underset{\approx}{Q}} = 2\underset{\approx}{Q}^T\,, \qquad I_1(\underset{\approx}{Q}^3)_{\underset{\approx}{Q}} = 3\underset{\approx}{Q}^{T^2}\,. \tag{1}$$

From these formulas by virtue of (I.7.9) and (I.7.11) and using the Cayley--Hamilton theorem, we infer that

$$I_2(\underset{\approx}{Q})_{\underset{\approx}{Q}} = \underset{\approx}{E} I_1(\underset{\approx}{Q}) - \underset{\approx}{Q}^T\,, \quad I_3(\underset{\approx}{Q})_{\underset{\approx}{Q}} = \underset{\approx}{Q}^{T^2} - I_1(\underset{\approx}{Q})\underset{\approx}{Q}^T + \underset{\approx}{E} I_2(\underset{\approx}{Q}) = I_3(\underset{\approx}{Q})(\underset{\approx}{Q}^T)^{-1}\,. \tag{2}$$

These fundamental formulas enable us to construct an expression for the derivative of a scalar function of the invariants

$$\varphi(I_1(\underset{\approx}{Q}), I_2(\underset{\approx}{Q}), I_3(\underset{\approx}{Q}))_{\underset{\approx}{Q}} = \left[\frac{\partial\varphi}{\partial I_1} + I_1(\underset{\approx}{Q})\frac{\partial\varphi}{\partial I_2}\right]\underset{\approx}{E} - \frac{\partial\varphi}{\partial I_2}\underset{\approx}{Q}^T + \frac{\partial\varphi}{\partial I_3} I_3(\underset{\approx}{Q})(\underset{\approx}{Q}^T)^{-1} =$$

$$= \left(\frac{\partial\varphi}{\partial I_1} + I_1\frac{\partial\varphi}{\partial I_2} + I_2\frac{\partial\varphi}{\partial I_3}\right)\underset{\approx}{E} - \left(\frac{\partial\varphi}{\partial I_2} + I_1\frac{\partial\varphi}{\partial I_3}\right)\underset{\approx}{Q}^T + \frac{\partial\varphi}{\partial I_3}\underset{\approx}{Q}^{T^2}\,. \tag{3}$$

2. <u>The derivative of $\varphi(\underset{\approx}{Q}\cdot\underset{\approx}{Q}^T)$</u>. If, in the basic definition (2.7), we substitute $\underset{\approx}{Q}$ for $\underset{\approx}{Q}^T$ we have, referring to (I.7.13),

$$\delta\varphi = \varphi_{\underset{\approx}{Q}}\cdot\cdot\,\delta\underset{\approx}{Q}^T = \varphi_{\underset{\approx}{Q}^T}\cdot\cdot\,\delta\underset{\approx}{Q} = \varphi^T_{\underset{\approx}{Q}^T}\cdot\cdot\,\delta\underset{\approx}{Q}^T\,, \qquad \varphi^T_{\underset{\approx}{Q}^T} = \varphi_{\underset{\approx}{Q}}\,, \tag{4}$$

which also implies (2.9). Regarding now φ as a function of the symmetric tensor $\underset{\approx}{S} = \underset{\approx}{Q}\cdot\underset{\approx}{Q}^T$ we have

$$\delta\varphi = \varphi_{\underset{\approx}{Q}}\cdot\cdot\,\delta\underset{\approx}{Q}^T = \varphi_{\underset{\approx}{S}}\cdot\cdot\,\delta\underset{\approx}{S}^T = \varphi_{\underset{\approx}{Q}\cdot\underset{\approx}{Q}^T}\cdot\cdot\,(\underset{\approx}{Q}\cdot\delta\underset{\approx}{Q}^T + \delta\underset{\approx}{Q}\cdot\underset{\approx}{Q}^T) =$$

$$= \varphi_{\underset{\approx}{Q}\cdot\underset{\approx}{Q}^T}\cdot\underset{\approx}{Q}\cdot\cdot\,\delta\underset{\approx}{Q}^T + \underset{\approx}{Q}^T\cdot\varphi_{\underset{\approx}{Q}\cdot\underset{\approx}{Q}^T}\cdot\cdot\,\delta\underset{\approx}{Q} = [\varphi_{\underset{\approx}{Q}\cdot\underset{\approx}{Q}^T}\cdot\underset{\approx}{Q} + (\varphi_{\underset{\approx}{Q}\cdot\underset{\approx}{Q}^T})^T\cdot\underset{\approx}{Q}]\cdot\cdot\,\delta\underset{\approx}{Q}^T$$

and by (2.9)

$$\varphi_{\underset{\approx}{Q}} = 2\varphi_{\underset{\approx}{Q}\cdot\underset{\approx}{Q}^T}\cdot\underset{\approx}{Q}\,, \qquad \varphi_{\underset{\approx}{Q}\cdot\underset{\approx}{Q}^T} = \frac{1}{2}\varphi_{\underset{\approx}{Q}}\cdot\underset{\approx}{Q}^{-1}\,, \tag{5}$$

where the latter equality holds provided that the tensor $\underset{\approx}{Q}$ is non-singular.

After substituting $\underset{\approx}{Q}$ for $\underset{\approx}{Q}^T$ we have in view of (4)

$$\varphi_{\underset{\approx}{Q}^T} = 2\varphi_{\underset{\approx}{Q}^T\cdot\underset{\approx}{Q}}\cdot\underset{\approx}{Q}^T = \varphi^T_{\underset{\approx}{Q}}\,, \qquad \varphi_{\underset{\approx}{Q}} = 2\underset{\approx}{Q}\cdot(\varphi_{\underset{\approx}{Q}^T\cdot\underset{\approx}{Q}})^T\,,$$

so that

$$\varphi_{\underline{\underline{Q}}} = 2\underline{\underline{Q}} \cdot \varphi_{\underline{\underline{Q}}^T \cdot \underline{\underline{Q}}} , \qquad \varphi_{\underline{\underline{Q}}^T \cdot \underline{\underline{Q}}} = \frac{1}{2} \underline{\underline{Q}}^{-1} \cdot \varphi_{\underline{\underline{Q}}} . \tag{6}$$

3. <u>The formulas connecting</u> $\varphi_{\underline{\underline{Q}}}$ <u>and</u> $\varphi_{\underline{\underline{Q}}^{-1}}$. We have

$$\underline{\underline{Q}} \cdot \underline{\underline{Q}}^{-1} = \underline{\underline{E}} , \quad \delta\underline{\underline{Q}} \cdot \underline{\underline{Q}}^{-1} + \underline{\underline{Q}} \cdot \delta\underline{\underline{Q}}^{-1} = 0 , \quad \delta\underline{\underline{Q}}^{-1} = -\underline{\underline{Q}}^{-1} \cdot \delta\underline{\underline{Q}} \cdot \underline{\underline{Q}}^{-1}$$

and therefore

$$\delta\varphi = \varphi_{\underline{\underline{Q}}} \cdot\cdot\, \delta\underline{\underline{Q}}^T = \varphi_{\underline{\underline{Q}}^{-1}} \cdot\cdot\, \delta\underline{\underline{Q}}^{-1^T} = -\varphi_{\underline{\underline{Q}}^{-1}} \cdot\cdot\, \underline{\underline{Q}}^{-1^T} \cdot \delta\underline{\underline{Q}}^T \cdot \underline{\underline{Q}}^{-1^T} =$$

$$= -\underline{\underline{Q}}^{-1^T} \cdot \varphi_{\underline{\underline{Q}}^{-1}} \cdot \underline{\underline{Q}}^{-1^T} \cdot\cdot\, \delta\underline{\underline{Q}}^T ,$$

so that

$$\underline{\underline{Q}}^T \cdot \varphi_{\underline{\underline{Q}}} = -\varphi_{\underline{\underline{Q}}^{-1}} \cdot \underline{\underline{Q}}^{-1^T} . \tag{7}$$

4. <u>The derivatives of the bilinear forms</u> $\underline{a} \cdot \underline{\underline{Q}} \cdot \underline{b}$, $\underline{a} \cdot \underline{\underline{Q}}^2 \cdot \underline{b}$. We have

$$\delta(\underline{a} \cdot \underline{\underline{Q}} \cdot \underline{b}) = \underline{a} \cdot \delta\underline{\underline{Q}} \cdot \underline{b} = \delta\underline{\underline{Q}} \cdot\cdot\, \underline{\underline{ba}} = \underline{\underline{ab}} \cdot\cdot\, \delta\underline{\underline{Q}}^T , \qquad (\underline{a} \cdot \underline{\underline{Q}} \cdot \underline{b})_{\underline{\underline{Q}}} = \underline{\underline{ab}} , \tag{8}$$

$$\delta(\underline{a} \cdot \underline{\underline{Q}}^2 \cdot \underline{b}) = \underline{a} \cdot (\underline{\underline{Q}} \cdot \delta\underline{\underline{Q}} + \delta\underline{\underline{Q}} \cdot \underline{\underline{Q}}) \cdot \underline{b} = (\underline{\underline{Q}} \cdot \delta\underline{\underline{Q}} + \delta\underline{\underline{Q}} \cdot \underline{\underline{Q}}) \cdot\cdot\, \underline{\underline{ba}} =$$

$$= \delta\underline{\underline{Q}} \cdot\cdot\, (\underline{\underline{ba}} \cdot \underline{\underline{Q}} + \underline{\underline{Q}} \cdot \underline{\underline{ba}}) = (\underline{\underline{Q}}^T \cdot \underline{\underline{ab}} + \underline{\underline{ab}} \cdot \underline{\underline{Q}}^T) \cdot\cdot\, \delta\underline{\underline{Q}}^T ,$$

so that

$$(\underline{a} \cdot \underline{\underline{Q}}^2 \cdot \underline{b})_{\underline{\underline{Q}}} = \underline{a} \cdot \underline{\underline{Qb}} + \underline{\underline{aQ}} \cdot \underline{b} , \qquad (\underline{a} \cdot \underline{\underline{Q}}^2 \cdot \underline{b})_{\underline{\underline{Q}}^T} = \underline{\underline{ba}} \cdot \underline{\underline{Q}} + \underline{\underline{Q}} \cdot \underline{\underline{ba}} . \tag{9}$$

5. <u>The derivatives of</u> $I_k(\underline{\underline{Q}})$ <u>with respect to</u> $\underline{\underline{Q}}^2$. We have

$$\delta I_1(\underline{\underline{Q}}) = \underline{\underline{E}} \cdot\cdot\, \delta\underline{\underline{Q}}^T = I_1(\underline{\underline{Q}})_{\underline{\underline{Q}}^2} \cdot\cdot\, (\underline{\underline{Q}}^T \cdot \delta\underline{\underline{Q}}^T + \delta\underline{\underline{Q}}^T \cdot \underline{\underline{Q}}^T) =$$

$$= [I_1(\underline{\underline{Q}})_{\underline{\underline{Q}}^2} \cdot \underline{\underline{Q}}^T + \underline{\underline{Q}}^T \cdot I_1(\underline{\underline{Q}})_{\underline{\underline{Q}}^2}] \cdot\cdot\, \delta\underline{\underline{Q}}^T .$$

The tensor (matrix) equation

$$I_1(\underline{\underline{Q}})_{\underline{\underline{Q}}^2} \cdot \underline{\underline{Q}}^T + \underline{\underline{Q}}^T \cdot I_1(\underline{\underline{Q}})_{\underline{\underline{Q}}^2} = \underline{\underline{E}}$$

is satisfied by the solution

$$I_1(\underline{\underline{Q}})_{\underline{\underline{Q}}^2} = \frac{1}{2}\underline{\underline{Q}}^{\mathrm{T}^{-1}} . \tag{10}$$

In a similar way, we arrive at the tensor equation

$$I_1(\underline{\underline{Q}}^3)_{\underline{\underline{Q}}^2} \cdot \underline{\underline{Q}}^{\mathrm{T}} + \underline{\underline{Q}}^{\mathrm{T}} \cdot I_1(\underline{\underline{Q}}^3)_{\underline{\underline{Q}}^2} = 3\underline{\underline{Q}}^{\mathrm{T}^2} ,$$

which has solution

$$I_1(\underline{\underline{Q}}^3)_{\underline{\underline{Q}}^2} = \frac{3}{2}\underline{\underline{Q}}^{\mathrm{T}} \tag{11}$$

and in which, of course, $I_1(\underline{\underline{Q}}^2)_{\underline{\underline{Q}}^2} = \underline{\underline{E}}$. Turning now to (I.7.9) and (I.7.11) we arrive at the formulas

$$\begin{aligned} I_2(\underline{\underline{Q}})_{\underline{\underline{Q}}^2} &= \frac{1}{2}[I_1(\underline{\underline{Q}})\underline{\underline{Q}}^{\mathrm{T}^{-1}} - \underline{\underline{E}}] , \\ I_3(\underline{\underline{Q}})_{\underline{\underline{Q}}^2} &= \frac{1}{2}[\underline{\underline{Q}}^{\mathrm{T}} - I_1(\underline{\underline{Q}})\underline{\underline{E}} + I_2(\underline{\underline{Q}})\underline{\underline{Q}}^{\mathrm{T}^{-1}}] = \frac{1}{2} I_3(\underline{\underline{Q}})\underline{\underline{Q}}^{\mathrm{T}^{-2}} . \end{aligned} \tag{12}$$

In (10) and (12) it is assumed that tensor $\underline{\underline{Q}}$ is non-singular.

4. THE DERIVATIVE OF A TENSOR WITH RESPECT TO A TENSOR ARGUMENT

In a vector basis $\underline{r}_s$ nine functions p_{mn} of the components of the tensor $\underline{\underline{Q}}$ are specified

$$p_{mn} = p_{mn}(q^{11},\dots,q^{31}) . \tag{1}$$

In a new basis $\check{\underline{r}}_s$ they are given the values

$$\check{p}_{st} = \check{\underline{r}}_s \cdot \underline{r}^{m}\check{\underline{r}}_t \cdot \underline{r}^{n} p_{mn}(\check{q}^{kq}\underline{r}^{1} \cdot \check{\underline{r}}_k\check{\underline{r}}_q \cdot \underline{r}^{1},\dots,\check{q}^{kq}\underline{r}^{3} \cdot \check{\underline{r}}_k\check{\underline{r}}_q \cdot \underline{r}^{1}) \tag{2}$$

and in this way $\underline{\underline{P}}(\underline{\underline{Q}})$, a tensor of the second rank, is defined

$$\underline{\underline{P}}(\underline{\underline{Q}}) = \check{p}_{st}\check{\underline{r}}^{s}\check{\underline{r}}^{t} = \check{\underline{r}}^{s}\check{\underline{r}}_s \cdot \underline{r}^{m}\check{\underline{r}}^{t}\check{\underline{r}}_t \cdot \underline{r}^{n} p_{mn} = \underline{r}^{m}\underline{r}^{n} p_{mn} . \tag{3}$$

By (2.7), the derivative $\underline{P}_{\underline{\underline{Q}}}$ is given by the expression

$$\underline{P}_{\underline{\underline{Q}}} = \underline{r}^{m}\underline{r}^{n}(p_{mn})_{\underline{\underline{Q}}} = \underline{r}^{m}\underline{r}^{n}\underline{r}^{s}\underline{r}^{t}\frac{\partial p_{mn}}{\partial q^{st}} . \tag{4}$$

By (3) and (2.7)

$$\delta \underline{P} = \underline{r}^m \underline{r}^n \delta p_{mn} = \underline{r}^m \underline{r}^n (p_{mn})_{\underline{Q}} \cdot\cdot\, \delta \underline{Q}^T = \underline{r}^m \underline{r}^n \underline{r}^s \underline{r}^t \frac{\partial p_{mn}}{\partial q^{st}} \cdot\cdot\, \delta \underline{Q}^T$$

and in view of (4)

$$\delta \underline{P} = \underline{P}_{\underline{Q}} \cdot\cdot\, \delta \underline{Q}^T . \tag{6}$$

Here, the tensor $\delta \underline{P}$ of the second rank depends linearly on $\delta \underline{Q}^T$, this dependence being expressed by a relationship of the type (1.2c). The quantity (4), called the derivative of $\underline{P}$ with respect to the tensor $\underline{Q}$, is a tensor of the fourth rank. Equation (6) amounts to an invariant definition of the "derivative of a tensor by a tensor", and formula (4) gives the rule for its calculation. The latter can also be put into the forms

$$\underline{P}_{\underline{Q}} = \underline{r}^m \underline{r}^n \underline{r}_s \underline{r}_t \frac{\partial p_{mn}}{\partial q_{st}} = \underline{r}_m \underline{r}_n \underline{r}^s \underline{r}^t \frac{\partial p^{mn}}{\partial q^{st}} = \underline{r}^m \underline{r}_n \underline{r}_s \underline{r}^t \frac{\partial p_m^{\cdot n}}{\partial q_s^{\cdot t}} \tag{7}$$

etc. For a symmetric tensor $\underline{Q} = \underline{Q}^T$ we have by (2.8)

$$\underline{P}_{\underline{Q}} = \frac{1}{2} \underline{r}^m \underline{r}^n (\underline{r}^s \underline{r}^t + \underline{r}^t \underline{r}^s) \frac{\partial p_{mn}}{\partial q^{st}} . \tag{8}$$

1. <u>The derivative of the product of a scalar and a tensor</u> $\varphi(\underline{Q})\underline{P}(\underline{Q})$. We have

$$\delta(\varphi \underline{P}) = \underline{P}\delta\varphi + \varphi\delta\underline{P} = (\underline{P}\varphi_{\underline{Q}} + \varphi \underline{P}_{\underline{Q}}) \cdot\cdot\, \delta\underline{Q}^T , \qquad (\varphi\underline{P})_{\underline{Q}} = \underline{P}\varphi_{\underline{Q}} + \varphi\underline{P}_{\underline{Q}} . \tag{9}$$

Rearranging the terms, we would obtain

$$\delta\varphi\underline{P} = (\delta\varphi)\underline{P} + \varphi\delta\underline{P} = (\varphi_{\underline{Q}} \cdot\cdot\, \delta\underline{Q}^T)\underline{P} + \varphi\underline{P}_{\underline{Q}} \cdot\cdot\, \delta\underline{Q}^T = \underline{P}\varphi_{\underline{Q}} \cdot\cdot\, \delta\underline{Q}^T + \varphi\underline{P}_{\underline{Q}} \cdot\cdot\, \delta\underline{Q}^T ,$$

but since the scalar $\varphi_{\underline{Q}} \cdot\cdot\, \delta\underline{Q}^T$ may also be placed on the right, we would arrive at the same expression (9). The forms $\underline{P}\varphi_{\underline{Q}}$ and $\varphi_{\underline{Q}}\underline{P}$ obviously represent different tensors but in the correct expression, (9), the tensor $\underline{P}$ in the first term should be placed on the left.

2. <u>The derivative of a tensor product</u>

$$\delta\underline{P} \cdot \underline{S} = (\delta\underline{P}) \cdot \underline{S} + \underline{P} \cdot \delta\underline{S} = (\underline{P}_{\underline{Q}} \cdot\cdot\, \delta\underline{Q}^T) \cdot \underline{S} + \underline{P} \cdot \underline{S}_{\underline{Q}} \cdot\cdot\, \delta\underline{Q}^T .$$

By rule (6), the factor $\delta\underline{Q}^T$ in the first term should be removed to the right.

Making use of (I.15.4) we have

$$\delta \underline{\underline{Q}}^{T} = \underline{C}_{III} \cdot\cdot\, \delta \underline{\underline{Q}}^{T} = \underline{r}_s \underline{r}_t (\underline{r}^t \underline{r}^s \cdot\cdot\, \delta \underline{\underline{Q}}^{T}) ,$$

$$(\underline{P}_{\underline{Q}} \cdot\cdot\, \delta \underline{\underline{Q}}^{T}) \cdot \underline{S} = (\underline{P}_{\underline{Q}} \cdot\cdot\, \underline{r}_s \underline{r}_t) \cdot \underline{S}\underline{r}^t \underline{r}^s \cdot\cdot\, \delta \underline{\underline{Q}}^{T} ,$$

because the scalar in parentheses can be removed to the right. We get*)

$$(\underline{P} \cdot \underline{S})_{\underline{Q}} = \underline{P} \cdot \underline{S}_{\underline{Q}} + (\underline{P}_{\underline{Q}} \cdot\cdot\, \underline{r}_s \underline{r}_t) \cdot \underline{S}\underline{r}^t \underline{r}^s = [\underline{P} \cdot (\underline{S}_{\underline{Q}} \cdot\cdot\, \underline{r}_s \underline{r}_t) + (\underline{P}_{\underline{Q}} \cdot\cdot\, \underline{r}_s \underline{r}_t) \cdot \underline{S}] \underline{r}^t \underline{r}^s , \tag{10}$$

since $\underline{S}_{\underline{Q}} \cdot \underline{r}_s \underline{r}_t \cdot \underline{r}^t \underline{r}^s = \underline{S}_{\underline{Q}} \cdot\cdot\, \underline{C}_{III} = \underline{S}_{\underline{Q}}$.

3. Change of independent variable. The tensor $\underline{P}$ is assumed to depend on $\underline{\underline{Q}}$ through $\underline{S}(\underline{\underline{Q}})$. We have for this case

$$\delta \underline{P} = \underline{P}_{\underline{Q}} \cdot\cdot\, \delta \underline{\underline{Q}}^{T} = \underline{P}_{\underline{S}} \cdot\cdot\, \delta \underline{S}^{T} , \quad \delta \underline{S} = \underline{S}_{\underline{Q}} \cdot\cdot\, \delta \underline{\underline{Q}}^{T} , \quad \delta \underline{S}^{T} = \underline{C}_{II} \cdot\cdot\, \delta \underline{S} = \underline{C}_{II} \cdot\cdot\, \underline{S}_{\underline{Q}} \cdot\cdot\, \delta \underline{\underline{Q}}^{T} ,$$

---

*) The formula for a derivative of a tensor product can be reduced to the invariant form

$$(\underline{P} \cdot \underline{S})_{\underline{Q}} = \underline{C}_{II} \cdot\cdot\, (\underline{S}^{T} \cdot (\underline{C}_{II} \cdot\cdot\, \underline{P}_{\underline{Q}})) + \underline{P} \cdot \underline{S}_{\underline{Q}} \tag{*}$$

We start from the relationship

$$\delta(\underline{P} \cdot \underline{S}) = (\delta \underline{P}) \cdot \underline{S} + \underline{P} \cdot \delta \underline{S} = (\underline{P}_{\underline{Q}} \cdot\cdot\, \delta \underline{\underline{Q}}^{T}) \cdot \underline{S} + \underline{P} \cdot \underline{S}_{\underline{Q}} \cdot\cdot\, \delta \underline{\underline{Q}}^{T}$$

The first term in the right-hand side can be transformed with the use of (I.6.3), (I.15.4)

$$(\underline{P}_{\underline{Q}} \cdot\cdot\, \delta \underline{\underline{Q}}^{T}) \cdot \underline{S} = (\underline{S}^{T} \cdot (\underline{P}_{\underline{Q}} \cdot\cdot\, \delta \underline{\underline{Q}}^{T})^{T})^{T} = \underline{C}_{II} \cdot\cdot\, (\underline{S}^{T} \cdot (\underline{P}_{\underline{Q}} \cdot\cdot\, \delta \underline{\underline{Q}}^{T})^{T}) ,$$

$$(\underline{P}_{\underline{Q}} \cdot\cdot\, \delta \underline{\underline{Q}}^{T})^{T} = \underline{C}_{II} \cdot\cdot\, (\underline{P}_{\underline{Q}} \cdot\cdot\, \delta \underline{\underline{Q}}^{T})$$

and

$$(\underline{P}_{\underline{Q}} \cdot\cdot\, \delta \underline{\underline{Q}}^{T}) \cdot \underline{S} = \underline{C}_{II} \cdot\cdot\, (\underline{S}^{T} \cdot (\underline{C}_{II} \cdot\cdot\, (\underline{P}_{\underline{Q}} \cdot\cdot\, \delta \underline{\underline{Q}}^{T}))) =$$

$$= (\underline{C}_{II} \cdot\cdot\, (\underline{S}^{T} \cdot (\underline{C}_{II} \cdot\cdot\, \underline{P}_{\underline{Q}}))) \cdot\cdot\, \delta \underline{\underline{Q}}^{T}$$

the last step can be verified immediately.

With the aid of (*) the relationships (14)-(17) (see below) may also be represented in invariant form. (Suggested by Yu.M. Malinin. - The translator.)

$$\underline{P}_{\underline{\underline{Q}}} = \underline{P}_{\underline{S}} \cdot\cdot \underline{C}_{II} \cdot\cdot \underline{S}_{\underline{\underline{Q}}} = \underline{P}_{\underline{S}} \cdot\cdot \underline{r}_s \underline{r}_t (\underline{r}^s \underline{r}^t \cdot\cdot \underline{S}_{\underline{\underline{Q}}}) \ . \tag{11}$$

Similarly, for a scalar we have

$$\varphi_{\underline{\underline{Q}}} = \varphi_{\underline{S}} \cdot\cdot \underline{C}_{II} \cdot\cdot \underline{S}_{\underline{\underline{Q}}} = \varphi_{\underline{S}}^{T} \cdot\cdot \underline{S}_{\underline{\underline{Q}}} \ , \tag{12}$$

because $\varphi_{\underline{S}} \cdot\cdot \underline{C}_{II} = \varphi_{\underline{S}}^{T}$ by (I.15.4).

4. The derivatives of $\underline{Q}$, $\underline{Q}^2$, $\underline{Q}\cdot\underline{Q}^T$, $\underline{Q}^T\cdot\underline{Q}$ and $\underline{Q}^{-1}$. We have

$$\delta\underline{Q} = \underline{C}_{II} \cdot\cdot \delta\underline{Q}^{T} \ , \qquad \delta\underline{Q}^{T} = \underline{C}_{III} \cdot\cdot \delta\underline{Q}^{T} \ ;$$

$$\underline{Q}_{\underline{\underline{Q}}} = \underline{C}_{II} \ , \quad \underline{Q}^{T}_{\underline{\underline{Q}}} = \underline{C}_{III} \ , \quad \underline{Q}_{\underline{\underline{Q}}^{T}} = \underline{C}_{III} \ . \tag{13}$$

Further, by (10)

$$(\underline{Q}^2)_{\underline{\underline{Q}}} = \underline{Q} \cdot \underline{C}_{II} + \underline{r}_t \underline{r}_s \cdot \underline{\underline{Q}}\underline{r}^t \underline{r}^s = (\underline{Q} \cdot \underline{r}_t \underline{r}_s + \underline{r}_t \underline{r}_s \cdot \underline{Q})\underline{r}^t \underline{r}^s \ , \tag{14}$$

$$(\underline{Q} \cdot \underline{Q}^{T})_{\underline{\underline{Q}}} = (\underline{Q} \cdot \underline{r}_s \underline{r}_t + \underline{r}_t \underline{r}_s \cdot \underline{Q}^{T})\underline{r}^t \underline{r}^s \ , \tag{15}$$

$$(\underline{Q}^{T} \cdot \underline{Q})_{\underline{\underline{Q}}} = (\underline{Q}^{T} \cdot \underline{r}_t \underline{r}_s + \underline{r}_s \underline{r}_t \cdot \underline{Q})\underline{r}^t \underline{r}^s \ . \tag{16}$$

We have also

$$\underline{E}_{\underline{\underline{Q}}} = 0 = (\underline{Q} \cdot \underline{Q}^{-1})_{\underline{\underline{Q}}} = (\underline{C}_{II} \cdot\cdot \underline{r}_s \underline{r}_t) \cdot \underline{Q}^{-1} \underline{r}^t \underline{r}^s + \underline{Q} \cdot (\underline{Q}^{-1})_{\underline{\underline{Q}}} =$$

$$= \underline{r}_t \underline{r}_s \cdot \underline{Q}^{-1} \underline{r}^t \underline{r}^s + \underline{Q} \cdot (\underline{Q}^{-1})_{\underline{\underline{Q}}} \ ,$$

so that

$$(\underline{Q}^{-1})_{\underline{\underline{Q}}} = -\underline{Q}^{-1} \cdot r_t r_s \cdot \underline{Q}^{-1} \underline{r}^t \underline{r}^s \ . \tag{17}$$

For a symmetric $\underline{Q}$, Equations (13), (14) and (17) are reduced to

$$\underline{Q}_{\underline{\underline{Q}}} = \frac{1}{2} (\underline{C}_{II} + \underline{C}_{III}) \ , \tag{18}$$

$$(\underline{Q}^2)_{\underline{\underline{Q}}} = \frac{1}{2} [\underline{Q} \cdot (\underline{r}_t \underline{r}_s + \underline{r}_s \underline{r}_t) + (\underline{r}_t \underline{r}_s + \underline{r}_s \underline{r}_t) \cdot \underline{Q}^{T}] \, \underline{r}^t \underline{r}^s \ , \tag{19}$$

$$(\underline{Q}^{-1})_{\underline{\underline{Q}}} = -\frac{1}{2} \underline{Q}^{-1} \cdot \underline{r}_t \underline{r}_s \cdot \underline{Q}^{-1} (\underline{r}^t \underline{r}^s + \underline{r}^s \underline{r}^t) \ . \tag{20}$$

5. <u>The derivative of the first invariant of a tensor product</u>. We have

$$\delta I_1(\underline{P}\cdot\underline{S}) = \delta(\underline{P}\cdot\cdot\underline{S}) = \underline{P}\cdot\cdot\delta\underline{S} + \underline{S}\cdot\cdot\delta\underline{P} = (\underline{P}\cdot\cdot\underline{S}_{\underline{Q}} + \underline{S}\cdot\cdot\underline{P}_{\underline{Q}})\cdot\cdot\delta\underline{Q}^{T},$$

$$I_1(\underline{P}\cdot\underline{S})_{\underline{Q}} = \underline{P}\cdot\cdot\underline{S}_{\underline{Q}} + \underline{S}\cdot\cdot\underline{P}_{\underline{Q}}\,. \tag{21}$$

6. <u>The second variation of a tensor scalar function.</u> Retaining the quadratic terms in δq^{st} in the Taylor series for the scalar function (2.1), we have

$$\Delta_2\varphi = \varphi(q^{11}+\delta q^{11},\dots,q^{31}+\delta q^{31}) - \varphi(q^{11},\dots,q^{31}) =$$

$$= \frac{\partial\varphi}{\partial q^{st}}\delta q^{st} + \frac{1}{2}\frac{\partial^2\varphi}{\partial q^{mn}\partial q^{st}}\delta q^{mn}\delta q^{st}\,.$$

By (2.7) and definition (4) for the derivative of a tensor with respect to a tensor argument

$$\varphi_{\underline{Q}} = \underline{r}^s\underline{r}^t\frac{\partial\varphi}{\partial q^{st}}, \qquad \varphi_{\underline{QQ}} = \underline{r}^s\underline{r}^t\underline{r}^m\underline{r}^n\frac{\partial^2\varphi}{\partial q^{mn}\partial q^{st}}\,. \tag{22}$$

It may be immediately verified that

$$\frac{\partial^2\varphi}{\partial q^{mn}\partial q^{st}}\delta q^{mn}\delta q^{st} = \delta\underline{Q}^{T}\cdot\cdot\varphi_{\underline{QQ}}\cdot\cdot\delta\underline{Q}^{T}$$

We thus arrive at the expressions

$$\Delta_2\varphi = \delta\varphi + \delta^2\varphi\,, \qquad \delta\varphi = \varphi_{\underline{Q}}\cdot\cdot\delta\underline{Q}^{T}\,, \qquad \delta^2\varphi = \frac{1}{2}\delta\underline{Q}^{T}\cdot\cdot\varphi_{\underline{QQ}}\cdot\cdot\delta\underline{Q}^{T}\,. \tag{23}$$

5. THE ISOTROPIC SCALAR FUNCTION OF A TENSOR

In I, Par. 8, we denoted by q^{mn}, q'^{mn} the components of a tensor $\underline{Q}$ subjected to an orthogonal transformation

$$\underline{Q}' = \underline{O}^{T}\cdot\underline{Q}\cdot\underline{O} \tag{1}$$

respectively, in the "old" and "new" bases $\underline{r}_s$ and $\underline{r}'_s$. Of course, $q^{mn} = q'^{mn}$, and therefore a scalar function of the components of $\underline{Q}$ preserves its form under this change of variables

$$\varphi(q^{11},q^{22},\dots,q^{31}) = \varphi(q'^{11},q'^{22},\dots,q'^{31})\,.$$

If we, however, pass from q^{mn} to any other system $\check{q}^{mn}$ of independent variables (for arbitrary, not necessarily orthogonal transformation, see (2.1)), then

$$\varphi(q^{11},q^{22},\ldots,q^{31}) = \tilde{\varphi}(\check{q}^{11},\check{q}^{22},\ldots,\check{q}^{31})$$

and a scalar (scalar-valued) function retains its numerical value, but not the form of the functional dependence of φ on q^{mn}. This dependence may, however, remain the same if, for example, the function φ depends only upon the invariants.

Confining ourselves to orthogonal transformations, a function of tensor components will be termed isotropic, provided that the form of its dependence on these components is invariant under any orthogonal transformation*)

$$\forall \underline{O} \subset o : \varphi(q^{11},\ldots,q^{31}) = \varphi(\check{q}^{11},\ldots,\check{q}^{31}) . \tag{2}$$

This property of φ may not hold for the totality of orthogonal transformations but for a certain subgroup $\overset{\times}{o} \subset o$. In this subgroup a scalar is termed isotropic, this property, as will be seen, is associated with its dependence on certain combinations of functional arguments. If φ represents the specific potential energy, then its applicability to any particular subgroup is determined by the symmetry properties of the medium.

This definition does not exclude improperly orthogonal transformations because a rotated tensor (1) is invariant under the substitution of $\underline{O}$ for $-\underline{O}$, and $\det(-\underline{O}) = -\det(\underline{O}) = 1$.

1. A triclinic subgroup of orthogonal transformations contains only the unit tensor $\underline{O} = \underline{E}$. All scalars are isotropic in this subgroup.

2. A monoclinic subgroup contains transformations of rotation through about the fixed direction $\underline{c}_1$. By (I.8.16)

$$\underline{O}^{\pi}_{\underline{c}_1} = 2\underline{c}_1\underline{c}_1 - \underline{E} = (\underline{O}^{\pi}_{\underline{c}_1})^{T} . \tag{3}$$

In an orthonormal basis $\underline{c}_1$, $\underline{c}_2$, $\underline{c}_3$, a tensor $\underline{\underline{Q}}$ is represented by the expression

$$\underline{\underline{Q}} = q^{11}\underline{c}_1\underline{c}_1 + q^{1\alpha}(\underline{c}_1\underline{c}_\alpha + \underline{c}_\alpha\underline{c}_1) + q^{\alpha\beta}\underline{c}_\alpha\underline{c}_\beta \qquad (\alpha,\beta = 2,3) . \tag{4}$$

The rotated tensor is

$$\underline{\underline{Q}}' = (\underline{O}^{\pi}_{\underline{c}_1})^{T}\cdot\underline{\underline{Q}}\cdot\underline{O}^{\pi}_{\underline{c}_1} = (2\underline{c}_1\underline{c}_1 - \underline{E})\cdot[q^{11}\underline{c}_1\underline{c}_1 + q^{1\alpha}(\underline{c}_1\underline{c}_\alpha + \underline{c}_\alpha\underline{c}_1) + q^{\alpha\beta}\underline{c}_\alpha\underline{c}_\beta]\cdot$$

*) The symbol $\forall\underline{O} \subset o$ means "for all elements $\underline{O}$ of the group o of orthogonal transformations.

$$\cdot\,(2\underline{c}_1\underline{c}_1 - \underline{E}) = q^{11}\underline{c}_1\underline{c}_1 + q^{\alpha\beta}\underline{c}_\alpha\underline{c}_\beta - q^{1\alpha}(\underline{c}_1\underline{c}_\alpha + \underline{c}_\alpha\underline{c}_1)$$

and a tensor which is isotropic in a monoclinic subgroup retains, as might be expected, the form of its dependence on components of $\underline{Q}$ in an orthonormal system $\underline{c}_1$, $\underline{c}_2$, $\underline{c}_3$ under the following transformation of variables

$$(q^{11},q^{22},q^{33},q^{12},q^{23},q^{31}) \rightarrow (q^{11},q^{22},q^{33},-q^{12},q^{23},-q^{31}) . \tag{5}$$

We will confine ourselves to the consideration of polynomial representations of φ in terms of the components of $\underline{Q}$. There are only three combinations x^2, y^2, xy for which the polynomial $F(x,y)$ is invariant when the signs of the independent variables $x = q^{12}$, $y = q^{31}$ are changed. The representation of an isotropic scalar in this subgroup is thus given by

$$\varphi = \varphi(q^{11},q^{22},q^{33},q^{12^2},q^{23},q^{31^2},q^{12}q^{31}) . \tag{6}$$

3. <u>The orthotropic group</u>. A scalar which is isotropic in this subgroup is invariant under transformation (5), and also under the transformation

$$(q^{11},q^{22},q^{33},q^{12},q^{23},q^{31}) \rightarrow (q^{11},q^{22},q^{33},-q^{12},-q^{23},q^{31}) . \tag{7}$$

This means that the rotation

$$\underline{O}^{\pi}_{\underline{c}_2} = 2\underline{c}_2\underline{c}_2 - \underline{E}$$

is also included in this subgroup. For this reason, an invariant scalar turns out to be a function of the arguments

$$q^{11},q^{22},q^{33},q^{12^2},q^{23^2},q^{31^2},q^{12}q^{23}q^{31} . \tag{8}$$

The expression for $I_3(\underline{Q})$ contains the product $q^{12}q^{23}q^{31}$

$$I_3(\underline{Q}) = q^{11}q^{22}q^{33} - q^{11}q^{23^2} - q^{22}q^{31^2} - q^{33}q^{12^2} + 2q^{12}q^{23}q^{31} , \tag{9}$$

and the other terms can be expressed in terms of the arguments (8). An isotropic scalar can be represented in this subgroup by the expression

$$\varphi = \varphi(q^{11},q^{22},q^{33},q^{12^2},q^{23^2},q^{31^2},I_3(\underline{Q})) . \tag{10}$$

This representation shows that a scalar also is invariant under the rotation

$$\underline{O}^{\pi}_{\underline{c}_3} = \underline{O}^{\pi}_{\underline{c}_1} \cdot \underline{O}^{\pi}_{\underline{c}_2} = -2(\underline{c}_1\underline{c}_1 + \underline{c}_2\underline{c}_2) + \underline{E} = 2\underline{c}_3\underline{c}_3 - \underline{E} \ . \qquad (11)$$

4. The transformation $\underline{O}^{\pi/2}_{\underline{c}_1}$. A tensor $\underline{Q}$ specified by the expression (4) can be reduced to the form

$$\underline{Q}' = (\underline{O}^{\pi/2}_{\underline{c}_1})^T \cdot \underline{Q} \cdot \underline{O}^{\pi/2}_{\underline{c}_1} = (\underline{c}_1\underline{c}_1 + \underline{c}_3\underline{c}_2 - \underline{c}_2\underline{c}_3) \cdot \underline{Q} \cdot (\underline{c}_1\underline{c}_1 + \underline{c}_2\underline{c}_3 - \underline{c}_3\underline{c}_2) =$$

$$= q^{11}\underline{c}_1\underline{c}_1 + q^{33}\underline{c}_2\underline{c}_2 + q^{22}\underline{c}_3\underline{c}_3 + q^{12}(\underline{c}_3\underline{c}_1 + \underline{c}_1\underline{c}_3) -$$

$$- q^{23}(\underline{c}_2\underline{c}_3 + \underline{c}_3\underline{c}_2) - q^{31}(\underline{c}_1\underline{c}_2 + \underline{c}_2\underline{c}_1) \qquad (12)$$

whence the transformation

$$(q^{11},q^{22},q^{33},q^{12},q^{23},q^{31}) \rightarrow (q^{11},q^{33},q^{22},-q^{31},-q^{23},q^{12})$$

may be carried out. The diagonal and non-diagonal elements are transformed according to the rules

$$(q^{11},q^{22},q^{33}) \rightarrow (q^{11},q^{33},q^{22}) \ , \qquad (q^{12},q^{23},q^{31}) \rightarrow (-q^{31},-q^{23},q^{12}) \ . \qquad (14)$$

These requirements will be fulfilled if φ is a function of the diagonal components in terms of the combinations $(q^{11},q^{22}+q^{33},q^{22}q^{33})$ and on non-diagonal elements through the combinations $({q^{12}}^2+{q^{31}}^2,{q^{23}}^2,q^{12}q^{23}q^{31})$. Each of the congruences (13) will be fulfilled if we assume

$$\varphi = \varphi(q^{11},q^{22}+q^{33},q^{22}q^{33},{q^{23}}^2,{q^{12}}^2+{q^{23}}^2+{q^{31}}^2,q^{12}q^{23}q^{31}) \qquad (15)$$

or, alternatively,

$$\varphi = \varphi(I_1(\underline{Q}),I_2(\underline{Q}),q^{11},{q^{23}}^2,{q^{12}}^2+{q^{23}}^2+{q^{31}}^2,q^{12}q^{23}q^{31}) \ , \qquad (16)$$

because the arguments $q^{22}+q^{33}$, $q^{22}q^{33}$ in (15) are reconstructed by the relationships

$$q^{22}+q^{33} = I_1(\underline{Q}) - q^{11} \ , \qquad q^{22}q^{33} = I_2(\underline{Q}) - q^{11}I_1(\underline{Q}) + {q^{12}}^2 + {q^{23}}^2 + {q^{31}}^2 \ . \qquad (17)$$

Equation (16) guarantees the invariance of φ with respect to any transformations given by any group of variables (14). This does not exhaust all possible arguments which might ensure the invariance of φ under transformation (13). To these we must also add $I_3(\underline{Q})$ which would enable us to substitute $q^{12}q^{23}q^{31}$ for the sum $q^{11}q^{23^2}+q^{22}q^{31^2}+q^{33}q^{12^2}$. We thus arrive at the following representation of φ as a function of seven arguments

$$\varphi = \varphi(I_1(\underline{Q}),I_2(\underline{Q}),I_3(\underline{Q}),q^{11},q^{23^2},q^{12^2}+q^{23^2}+q^{31^2},q^{11}q^{23^2}+q^{22}q^{31^2}+q^{33}q^{12^2}) \ . \tag{18}$$

This exhausts all possible representations of φ by a polynomial of degree no higher than three. The list of arguments which leave a polynomial of any degree invariant, should also include $q^{31^2}q^{12^2}$.

5. The cubic symmetry group. To representation (18) should be added two others which are obtained by cyclic permutation of the indices and which are invariant under transformations

$$\underset{-\underline{c}_2}{O}^{\pi/2} = \underline{c}_2\underline{c}_2 + \underline{c}_3\underline{c}_1 - \underline{c}_1\underline{c}_3 \ , \qquad \underset{-\underline{c}_3}{O}^{\pi/2} = \underline{c}_3\underline{c}_3 + \underline{c}_1\underline{c}_2 - \underline{c}_2\underline{c}_1 \ . \tag{19}$$

A cubic polynomial which is invariant with respect to the cubic symmetry group should therefore preserve its form under the substitutions

$$(q^{11},q^{23}) \to (q^{22},q^{31}) \to (q^{33},q^{12})$$

and this requires that list (18) be augmented by three arguments

$$q^{11}q^{22}q^{33},q^{11}q^{23^2} + q^{22}q^{31^2} + q^{33}q^{12^2},q^{23^2} + q^{31^2} + q^{12^2} \ ,$$

because degrees higher than the third are not included. But two of these three arguments are already contained in (18). The invariant polynomial, of the third degree at most, in the symmetry group considered, turns out to be a function of six arguments

$$\varphi = \varphi(I_1(\underline{Q}),I_2(\underline{Q}),I_3(\underline{Q}),q^{11}q^{22}q^{33},q^{12^2}+q^{23^2}+q^{31^2},q^{11}q^{23^2}+q^{22}q^{31^2}+q^{33}q^{12^2}) \ . \tag{20}$$

Note. The representations of a polynomial which is isotropic in three classes of cubic symmetry are given by Green and Adkins [1]. These contain six arguments and are reducible to (20) (provided we restrict our analysis to polynomials of the third degree).

In an isotropic group with cubic symmetry a polynomial of the fourth degree also contains the argument

$$q^{31^2}q^{12^2} + q^{12^2}q^{23^2} + q^{23^2}q^{31^2} \ . \tag{21}$$

6. A transverse isotropy. An isotropic scalar in this group is invariant under

the transformation of rotation through any angle ω about the fixed direction $\underline{c}_3$. A tensor $\underline{\underline{Q}}$ is specified in the orthonormal basis $\underline{c}_1$, $\underline{c}_2$, $\underline{c}_3$

$$\underline{\underline{Q}} = \underline{c}_3\underline{c}_3 q^{33} + q^{\alpha 3}(\underline{c}_\alpha\underline{c}_3 + \underline{c}_3\underline{c}_\alpha) + q^{\alpha\beta}\underline{c}_\alpha\underline{c}_\beta \qquad (\alpha,\beta = 1,2) \ . \tag{22}$$

A rotated tensor is represented in the following way

$$\underline{\underline{Q}}' = [\underline{E}\cos\omega + (1-\cos\omega)\underline{c}_3\underline{c}_3 + \underline{c}_3 \times \underline{E}\sin\omega]\cdot\underline{\underline{Q}}\cdot[\underline{E}\cos\omega + (1-\cos\omega)\underline{c}_3\underline{c}_3 - \underline{c}_3 \times$$

$$\times\ \underline{E}\sin\omega] = \underline{c}_3\underline{c}_3\check{q}^{33} + (\underline{c}_\alpha\underline{c}_3 + \underline{c}_3\underline{c}_\alpha)\check{q}^{\alpha 3} + \check{q}^{\alpha\beta}\underline{c}_\alpha\underline{c}_\beta \ , \tag{23}$$

where $\check{q}^{st}$ denote the components of $\underline{\underline{Q}}'$ in the same basis. These components turn out to be

$$\check{q}^{11} = q^{11}\cos^2\omega + q^{22}\sin^2\omega - 2q^{12}\cos\omega\sin\omega \ ,$$

$$\check{q}^{22} = q^{11}\sin^2\omega + q^{22}\cos^2\omega + 2q^{12}\cos\omega\sin\omega \ , \tag{24}$$

$$\check{q}^{12} = (q^{22} - q^{11})\cos\omega\sin\omega + q^{12}\cos^2\omega \ ,$$

$$\check{q}^{31} = q^{31}\cos\omega + q^{23}\sin\omega \ , \qquad \check{q}^{23} = -q^{31}\sin\omega + q^{23}\cos\omega \ , \qquad \check{q}^{33} = q^{33} \ . \tag{25}$$

Equations (24) defined the transformation of the components $q^{\alpha\beta}$ of a two--dimensional tensor $q^{\alpha\beta}\underline{c}_\alpha\underline{c}_\beta$ which results from the rotation of the coordinate axes $\underline{c}_1$, $\underline{c}_2$ through the angle ω. The invariants of this tensor are

$$q^{11} + q^{22} = \check{q}^{11} + \check{q}^{22} \ , \qquad q^{11}q^{22} - q^{12^2} = \check{q}^{11}\check{q}^{22} - \check{q}^{12^2} \tag{26}$$

this is the result of the elimination of ω from the three Equations (24). The elimination of ω from the two Equations (25) determines the invariant

$$q^{31^2} + q^{23^2} = \check{q}^{31^2} + \check{q}^{23^2} \ .$$

The arguments also include $I_3(\underline{\underline{Q}})$ which is the result of the elimination of ω from the five Equations (24) and (25). The scalar invariant in a group of transverse isotropy turns out to be a function (not necessarily a polynomial) of five arguments

$$\varphi = \varphi(q^{11} + q^{22}, q^{11}q^{22} - q^{12^2}, q^{33}, q^{31^2} + q^{23^2}, I_3(\underline{\underline{Q}})) \ . \tag{27}$$

This can also be represented as a function of the arguments

$$\varphi = \varphi(I_1(\underline{\underline{Q}}), I_2(\underline{\underline{Q}}), I_3(\underline{\underline{Q}}), q^{33}, q^{31^2} + q^{32^2} + q^{33^2}) =$$

$$= \varphi(I_1(\underline{\underline{Q}}), I_2(\underline{\underline{Q}}), I_3(\underline{\underline{Q}}), \underline{c}_3 \cdot \underline{\underline{Q}} \cdot \underline{c}_3, \underline{c}_3 \cdot \underline{\underline{Q}}^2 \cdot \underline{c}_3) \ . \tag{28}$$

7. An isotropic scalar preserves the form of its functional dependence on the components of a tensor $\underline{\underline{Q}}$ under all transformations $\underline{\underline{O}}_{\underline{c}}^{\omega}$ (rotation through an arbitrary angle about an arbitrarily directed axis). The arguments in (28) which depend upon $\underline{c}_3$, vanish

$$\varphi = \varphi(I_1(\underline{\underline{Q}}), I_2(\underline{\underline{Q}}), I_3(\underline{\underline{Q}})) \ . \tag{29}$$

Of course, this representation does not include the basis in which the tensor $\underline{\underline{Q}}$ was defined. The definition of an isotropic scalar is contained in the representation

$$\forall \underline{\underline{O}}_{\underline{c}}^{\omega} : \varphi(\underline{\underline{Q}}) = \varphi(\underline{\underline{O}}^T \cdot \underline{\underline{Q}} \cdot \underline{\underline{O}}) \ . \tag{30}$$

The inverse may also be proved: a scalar function of symmetric tensor represents a scalar invariant if and only if it can be expressed as a function of invariants of this tensor. To this the following statement may be added. An invariant scalar which is a polynomial of the components of a symmetric tensor can be represented as a polynomial of its invariants $I_k(\underline{\underline{Q}})$.

8. The number of terms of polynomial invariants of the second (φ_2) and third (φ_3) degree.

a. *A triclinic group*. The form φ_2 contains 21 terms and the form φ_3 56. For other groups the number of terms is lower.

b. *A monoclinic group*. A quadratic form of n variables contains $(1/2)n(n+1)$ terms. In φ_2 the variables q^{11}, q^{22}, q^{33}, q^{23} would produce 10 terms in all. The variable q^{11} enters φ_3 as q^{11^3}, also participates three times in the formation of the products of q^{11^2} and q^{22}, q^{33}, q^{23}, six times in the products of q^{11} and the quadratic form of these variabels, and three times more in products of q^{11} and q^{12^2}, q^{31^2}, $q^{12}q^{31}$ in all as many as 13 terms are obtained. The variable q^{11} does not come into the subsequent calculation. This then provides 6 terms in q^{22} and three terms containing q^{33}. In total, the form φ_2 contains 13 terms and φ_3 32.

c. *The orthotropic group*. The forms φ_2 and φ_3 contain 9 and 20 terms, respectively.

d. *The cubic symmetry group*. The forms φ_2 and φ_3 contain 6 and 9 terms, respectively.

e. *The transverse isotropy group*. The form φ_2 contains 5 terms and the form φ_3 8 terms.

f. *The isotropic group*. The form φ_2 will contain the terms $I_1^2(\underline{Q})$, $I_2(\underline{Q})$, and φ_3 will contain $I_1^3(\underline{Q})$, $I_1(\underline{Q})I_2(\underline{Q})$, $I_3(\underline{Q})$ - see also (I.15.16) and (I.15.14).

6. THE SCALAR FUNCTION OF VECTOR ARGUMENTS

The scalars defined by two vectors $\underline{a}$ and $\underline{b}$ are either their scalar products $\underline{a}\cdot\underline{a}$, $\underline{a}\cdot\underline{b}$, $\underline{b}\cdot\underline{b}$ or quantities expressible in terms of these products, e.g., $\underline{a}\times\underline{b}^2 = \underline{a}\cdot\underline{a}\underline{b}\cdot\underline{b} - (\underline{a}\cdot\underline{b})^2$. One of the quantities defined by three vectors is the scalar $\underline{a}\cdot(\underline{b}\times\underline{c})$ which can be represented in terms of their scalar products taken in pairs

$$\underline{a}\cdot(\underline{b}\times\underline{c}) = \begin{vmatrix} a\cdot a & a\cdot b & a\cdot c \\ b\cdot a & b\cdot b & b\cdot c \\ c\cdot a & c\cdot b & c\cdot c \end{vmatrix} .$$

Under an orthogonal transformation, $\underline{a}\cdot\underline{b} = \underline{a}'\cdot\underline{b}'$. Therefore, in bases which are connected by an orthogonal transformation, the scalar

$$\varphi(\underline{a}\cdot\underline{b},\underline{b}\cdot\underline{c},\underline{c}\cdot\underline{a},\underline{a}\cdot\underline{a},\underline{b}\cdot\underline{b},\underline{c}\cdot\underline{c}) = \varphi(\underline{a}'\cdot\underline{b}',\underline{b}'\cdot\underline{c}',\underline{c}'\cdot\underline{a}',\underline{a}'\cdot\underline{a}',\underline{b}'\cdot\underline{b}',\underline{c}'\cdot\underline{c}')$$

is the same function of the vector components in those bases. This is an isotropic scalar. The inverse is also true: a scalar function of vectors is isotropic if and only if its expression in terms of the components of these vectors has the form

$$\varphi(a_s b^s, b_s c^s, c_s a^s, a_s a^s, b_s b^s, c_s c^s) =$$

$$= \varphi(g^{sk} a_s b_k, g^{sk} b_s c_k, g^{sk} c_s a_k, g^{sk} a_s a_k, g^{sk} b_s b_k, g^{sk} c_s c_k)$$

etc. The base vectors may enter the representation of an isotropic scalar only in terms of the components of the unit tensor. This, of course, is true for any number of vectors.

7. TENSOR FUNCTIONS OF TENSOR ARGUMENT

The concept of an isotropic function may be generalized to the tensor functions of tensor argument. A function $\underline{F}(\underline{Q},\underline{S},\ldots)$ is termed isotropic if for all tensors $\underline{O}$ of the orthogonal group the relationship

$$\underline{F}' = \underline{O}^T \cdot \underline{F}(\underline{Q},\underline{S},\ldots) \cdot \underline{O} = \underline{F}(\underline{O}^T \cdot \underline{Q} \cdot \underline{O}, \underline{O}^T \cdot \underline{S} \cdot \underline{O},\ldots) = \underline{F}(\underline{Q}',\underline{S}',\ldots) \tag{1}$$

holds.

We now consider an isotropic function of a single tensor such as

$$\underline{F}(\underline{Q}) = \underline{Q}^2 , \qquad \underline{O}^T \cdot \underline{Q}^2 \cdot \underline{O} = \underline{O}^T \cdot \underline{Q} \cdot \underline{O} \cdot \underline{O}^T \cdot \underline{Q} \cdot \underline{O} = (\underline{O}^T \cdot \underline{Q} \cdot \underline{O})^2 = \underline{Q}'^2$$

or, in general, any integral power $\underline{Q}$.

Substituting the tensor $\underline{Q}$ for the variable λ in the expression

$$F(\lambda) = c_0 + c_1\lambda + c_2\lambda^2 + c_3\lambda^3 + \ldots \tag{2}$$

we arrive at the representation of the tensor function $\underline{F}(\underline{Q})$ in the form of a power series

$$\underline{F}(\underline{Q}) = c_0\underline{E} + c_1\underline{Q} + c_2\underline{Q}^2 + c_3\underline{Q}^3 + \ldots = (c_0 + c_1\lambda_1 + c_2\lambda_1^2 + \ldots)\underline{e}_1\underline{e}^1 +$$

$$+ (c_0 + c_1\lambda_2 + c_2\lambda_2^2 + \ldots)\underline{e}_2\underline{e}^2 + (c_0 + c_1\lambda_3 + c_2\lambda_3^2 + \ldots)\underline{e}_3\underline{e}^3 . \tag{3}$$

Here λ_s, $\underline{e}_s$, $\underline{e}^s$ denote the principal values and principal directions of $\underline{Q}$. The tensor $\underline{F}$ has the same principal directions as $\underline{Q}$ (a coaxial tensor), its principal values being $F(\lambda_s)$

$$\underline{F}(\underline{Q}) = F(\lambda_1)\underline{e}_1\underline{e}^1 + F(\lambda_2)\underline{e}_2\underline{e}^2 + F(\lambda_3)\underline{e}_3\underline{e}^3 . \tag{4}$$

This tensor is isotropic because $\underline{O}^T \cdot \underline{e}_s\underline{e}^s \cdot \underline{O} = \underline{e}'_s\underline{e}'^s$ and the principal values are invariant under orthogonal transformation. Equation (I.9.14) implies that

$$\underline{F}(\underline{Q}^T) = \underline{F}(\underline{Q})^T . \tag{5}$$

The tensor $\exp K$ is obtained by substituting tensor K for λ throughout the expression for the function $\exp\lambda$. If K is a skew tensor, then by (5)

$$(\exp K)^T = \exp(-K) , \qquad \exp K \cdot (\exp K)^T = \exp(K - K) = \underline{E} , \tag{6}$$

so that $\exp K = \underline{O}$ is an orthogonal tensor. By (4) and (I.11.9)

$$\exp K = e^{-i\omega}\underline{e}_1\underline{e}^1 + e^{i\omega}\underline{e}_2\underline{e}^2 + \underline{e}_3\underline{e}^3 = \cos\omega(\underline{e}_1\underline{e}^1 + \underline{e}_2\underline{e}^2) -$$

$$- i\sin\omega(\underline{e}_1\underline{e}^1 - \underline{e}_2\underline{e}^2) + \underline{e}_3\underline{e}^3 = (\underline{c}_1\underline{c}_1 + \underline{c}_2\underline{c}_2)\cos\omega +$$

$$+ (\underline{c}_1\underline{c}_2 - \underline{c}_2\underline{c}_1) \sin\omega + \underline{e}_3\underline{e}^3 ,$$

which is required by (I.11.10). The skew-symmetric tensor K generating this orthogonal tensor determines, according to (I.11.8), the angle of rotation, ω, and direction $\underline{e}_3 = \underline{c}_1 \times \underline{c}_2$ of its axis.

In the expression for an entire function of $\underline{Q}$, $\underline{F}(\underline{Q})$, powers of $\underline{Q}$ higher than the second may be substituted using the Cayley-Hamilton theorem (I.9.21) by expressions linear in $\underline{Q}^0 = \underline{E}$, $\underline{Q}$, $\underline{Q}^2$ with coefficients $I_k(\underline{Q})$. The tensor $\underline{F}(\underline{Q})$ will then be represented as a quadratic polynomial

$$\underline{F}(\underline{Q}) = \varphi_0\underline{E} + \varphi_1\underline{Q} + \varphi_2\underline{Q}^2 \tag{7}$$

whose coefficients are functions of the invariants $I_k(\underline{Q})$

$$\varphi_\Gamma = \varphi_\Gamma(I_1(\underline{Q}), I_2(\underline{Q}), I_3(\underline{Q})) \qquad (\Gamma = 0,1,2) . \tag{8}$$

The coefficients of the polynomial in (7) are determined as follows. Eliminating in the representation (2) for the entire function powers of λ higher than the second using the characteristic equation

$$P(\lambda) = -\lambda^3 + I_1(\underline{Q})\lambda^2 - I_2(\underline{Q})\lambda + I_3(\underline{Q}) = (\lambda_1 - \lambda)(\lambda_2 - \lambda)(\lambda_3 - \lambda) = 0 \tag{9}$$

we reduce (2) to the form

$$F(\lambda) = \varphi_0 + \varphi_1\lambda + \varphi_2\lambda^2 , \tag{10}$$

and taking λ equal to λ_1, λ_2, λ_3 we arrive at three equations for the unknowns φ_0, φ_1, φ_2

$$F(\lambda_s) = \varphi_0 + \varphi_1\lambda_s + \varphi_2\lambda_s^2 \qquad (s = 1,2,3) . \tag{11}$$

Their solution can be represented with the notation of (I.9.8) in the following form

$$\varphi_0 = -I_3(\underline{Q}) \sum_{s=1}^{3} \frac{F(\lambda_s)}{P'(\lambda_s)} , \qquad \varphi_1 = \sum_{s=1}^{3} \frac{F(\lambda_s)}{P'(\lambda_s)} [I_1(\underline{Q}) - \lambda_s] , \tag{12}$$

$$\varphi_2 = - \sum_{s=1}^{3} \frac{F(\lambda_s)}{P'(\lambda_s)} .$$

But the roots λ_s are functions of the coefficients of Equation (9) so that φ_Γ

can be rewritten in the form (8). It should be noted in addition that, if $F(\lambda)$ is a polynomial in λ, then the sums

$$\sigma_k = \sum_{s=1}^{3} \frac{\lambda_s^k}{P'(\lambda_s)} \qquad (k = 0,1,2,3,\ldots)$$

may be represented as polynomials*) in terms of the coefficients of Equation (9)

$$\sigma_0 = 0\ ,\ \sigma_1 = 0\ ,\ \sigma_2 = -1\ ,\ \sigma_3 = -I_1(\underline{Q})\ ,\ \sigma_4 = -I_1^2(\underline{Q}) + I_2(\underline{Q}),\ldots\ .$$

We may thus assert that if $\underline{F}(\underline{Q})$ is a polynomial in $\underline{Q}$, in other words, if the components of $\underline{F}$ are isotropic polynomials of the components of $\underline{Q}$, then φ_Γ are polynomials in the invariants $I_k(\underline{Q})$.

For Equation (7) to be valid it is essential only that there should exist a complete generating function (2). The functions (3) which are generated in this way certainly do not exhaust the totality of tensor functions. For example, a linear tensor function $\underline{F}(\underline{Q}) = {}^4\underline{M}\cdot\cdot\underline{Q}$, where ${}^4\underline{M}$ denotes an isotropic tensor of the 4th rank, cannot be represented in the form (2) if tensor $\underline{Q}$ is asymmetric. If, however, we limit the class of tensors to isotropic tensor functions of symmetric tensors then representation (7) will remain valid. The following theorem is true: any isotropic function $\underline{D} = \underline{F}(\underline{Q})$ of a symmetric tensor ($\underline{Q} = \underline{Q}^T$) can be represented in the form (7).

Proof**). It is obvious that any invariant of $\underline{D}$ is at the same time an invariant of $\underline{Q}$. Indeed, a scalar function of a tensor is called its invariant provided that

$$J(\underline{D}) = J(\underline{O}^T\cdot\underline{D}\cdot\underline{O})\ , \qquad \forall\underline{O} : \underline{O}^T\cdot\underline{O} = \underline{E}\ . \tag{13}$$

Denoting by $J(\underline{F}(\underline{Q}))$ an invariant of $\underline{F}(\underline{Q})$, we may write

$$J[\underline{F}(\underline{Q})] = J_1(\underline{Q})\ .$$

---

*) These formulas may be obtained from the decomposition of $[P(\lambda)]^{-1}$ into the sum of elementary fractions

$$[P(\lambda)]^{-1} = \sum_{s=1}^{3} \frac{1}{P'(\lambda_s)(\lambda - \lambda_s)}$$

by equating like powers of λ^{-1} in the expansions for the left- and right-hand sides of this equality.

**) This proof is due to P.A. Zhilin.

If J is an invariant of $\underline{F}$, then J_1 is an invariant of $\underline{Q}$. In fact, by virtue of (13) and (1) we have

$$J[\underline{F}(\underline{Q})] \equiv J_1(\underline{Q}) = J[\underline{O}^T \cdot \underline{F}(\underline{Q}) \cdot \underline{O}] = J[\underline{F}(\underline{O}^T \cdot \underline{Q} \cdot \underline{O})] = J_1(\underline{O}^T \cdot \underline{Q} \cdot \underline{O}) \ ,$$

which means that J_1 is an invariant of $\underline{Q}$.

We will now verify that the set of values admitted by isotropic functions of symmetric tensors is itself a set of symmetric tensors. To this end we introduce a symmetry group $\underline{O}_{\underline{Q}}$ of a tensor $\underline{Q}$. This group is defined as a set of orthogonal solutions to the equation

$$\underline{O}^T \cdot \underline{Q} \cdot \underline{O} = \underline{Q} \ , \qquad \underline{Q} = \underline{Q}^T \ . \tag{14}$$

For symmetric tensors, the group $\underline{O}_{\underline{Q}}$ always contains a subgroup $\underline{O}^{\varepsilon}_{\underline{Q}}$

$$\underline{O}^{\varepsilon}_{\underline{Q}} = \sum_{k=1}^{3} \varepsilon_k \underline{e}_k \underline{e}_k \ , \qquad \varepsilon_k^2 = 1 \ , \tag{15}$$

where $\underline{e}_k$ denote the eigenvectors of $\underline{Q}$, and $\underline{e}_k \cdot \underline{e}_s = \delta_{ks}$. If the eigenvalues of $\underline{Q}$ are all distinct, then $\underline{O}_{\underline{Q}}$ coincides with $\underline{O}^{\varepsilon}_{\underline{Q}}$. We now denote the symmetry group of $\underline{D} = \underline{F}(\underline{Q})$ by $\underline{O}_{\underline{D}}$ and prove that $\underline{O}_{\underline{D}}$ contains $\underline{O}_{\underline{Q}}$. Let us assume that $\underline{O} \in \underline{O}_{\underline{Q}}$. Then we have by (1) and (14)

$$\underline{O}^T \cdot \underline{D} \cdot \underline{O} = F(O^T \cdot Q \cdot O) = F(Q) = D \ , \qquad O \in \underline{O}_{\underline{Q}} \ .$$

From this it follows that the element $\underline{O} \in \underline{O}_{\underline{Q}}$ also belongs to $\underline{O}_{\underline{D}}$, and consequently, $\underline{O}_{\underline{Q}}$ is a subgroup of $\underline{O}_{\underline{D}}$.

Now we represent $\underline{D}$ in the basis of the principal axes of $\underline{Q}$

$$\underline{D} = \sum_{k,s=1}^{3} D^{ks} \underline{e}_k \underline{e}_s$$

and require that the tensors $\underline{O}^{\varepsilon}_{\underline{Q}}$ belong to the symmetry group of $\underline{D}$. Then we get

$$\sum_{k,s=1}^{3} D^{ks} \varepsilon_k \varepsilon_s \underline{e}_k \underline{e}_s = \sum_{s,k=1}^{3} D^{ks} \underline{e}_k \underline{e}_s \ .$$

This yields the equalities

$$D^{ks} \varepsilon_k \varepsilon_s = D^{ks} \qquad \text{(no summation over } k \text{ and } s) \ .$$

Choosing the numbers ε_k so that $\varepsilon_k\varepsilon_s = -1$ $(k \neq s)$, we obtain $D^{ks} = 0$ $(k \neq s)$. The numbers ε_k can always be specified in this way: indeed, selecting from the set of tensors (15) the element for which $\varepsilon_m = 1$ (where m is a number chosen from the set 1,2,3) and other elements $\varepsilon_k = -1$, we obtain $D^{ms} = 0$ $(s \neq m)$. Varying m, i.e. proceeding to other elements of the set (15), we obtain the required property. We have thus arrived at the following representation for $\underline{D}$

$$\underline{D} = \sum_{k=1}^{3} d_k \underline{e}_k \underline{e}_k \,, \qquad d_k = D^{kk} \,, \qquad \underline{D} = \underline{D}^{T} \,. \tag{16}$$

This equality shows that the principal axes of $\underline{\underline{Q}}$ are also the principal axes of $\underline{D}$ and that the coefficients d_k are eigenvalues of $\underline{D}$ and consequently, its invariants. From the above they are also invariants of the tensor $\underline{\underline{Q}}$. The following important property of eigenvalues d_k should also be noted. If any two eigenvalues of $\underline{\underline{Q}}$ coincide, then so do the corresponding eigenvalues of $\underline{D}$. Indeed, if some eigenvalues of $\underline{\underline{Q}}$, say q_1 and q_2, are the same, then the symmetry group of this tensor contains, for example, the orthogonal tensor

$$\underline{O} = \underline{e}_1\underline{e}_2 + \underline{e}_2\underline{e}_1 + \underline{e}_3\underline{e}_3 \,, \qquad \underline{O} \in \underline{O}_{\underline{\underline{Q}}} \,.$$

Consequently, this tensor also belongs to the symmetry group of $\underline{D}$ which can only be possible if $d_1 = d_2$. Extending this argument slightly, it is plain that the number of distinct eigenvalues of $\underline{D}$ does not exceed that of $\underline{\underline{Q}}$. Now the theorem becomes obvious. To prove it, it is sufficient to express the dyads $\underline{e}_k\underline{e}_k$ in terms of the tensor $\underline{\underline{Q}}$ itself. To this end, we write the system of equations

$$\underline{E} = \sum_{k=1}^{3} \underline{e}_k\underline{e}_k \,, \qquad \underline{\underline{Q}} = \sum_{k=1}^{3} q_k \underline{e}_k\underline{e}_k \,, \qquad \underline{\underline{Q}}^2 = \sum_{k=1}^{3} q_k^2 \underline{e}_k\underline{e}_k \,. \tag{17}$$

The solution to this system with respect to dyads $\underline{e}_k\underline{e}_k$ is unique if the eigenvalues q_k are all distinct. Each of the dyads will have the form

$$\underline{e}_k\underline{e}_k = \varphi_0^{(k)} \underline{E} + \varphi_1^{(k)} \underline{\underline{Q}} + \varphi_2^{(k)} \underline{\underline{Q}}^2 \,, \tag{18}$$

where the scalar functions $\varphi_i^{(k)}$ are invariants of $\underline{\underline{Q}}$. Substituting (18) into (16) and bearing in mind that d_k are also invariants of $\underline{\underline{Q}}$, we obtain the proof of the theorem for the case of distinct eigenvalues of the tensor $\underline{\underline{Q}}$. If some of the eigenvalues of $\underline{\underline{Q}}$ are equal, e.g. $q_1 = q_2$, then one should take into account that the system (17) contains only two unknown tensors, namely those of $\underline{e}_1\underline{e}_1 + \underline{e}_2\underline{e}_2$ and the dyad $\underline{e}_3\underline{e}_3$. Neglecting any but the first of Equations (17) we obtain two equations for the two unknown tensors. By the property of the term $\underline{D}$ $(d_1 = d_2)$

proved above, Equation (16) contains just those unknown tensors. Determining these, we again arrive at representation (7). This, of course, is now not unique.

An isotropic tensor (7) with coefficients given by (3.3)

$$\varphi_0 = \frac{\partial\varphi}{\partial I_1} + I_1\frac{\partial\varphi}{\partial I_2} + I_2\frac{\partial\varphi}{\partial I_3}\,, \qquad \varphi_1 = -\left(\frac{\partial\varphi}{\partial I_2} + I_1\frac{\partial\varphi}{\partial I_3}\right), \qquad \varphi_2 = \frac{\partial\varphi}{\partial I_3}\,, \tag{19}$$

where $\varphi(I_1,I_2,I_3)$ denotes an isotropic scalar, represents the derivative of φ with respect to $\underline{\underline{Q}}$

$$\varphi_{\underline{\underline{Q}}} = \varphi_0\underline{\underline{E}} + \varphi_1\underline{\underline{Q}} + \varphi_2\underline{\underline{Q}}^2\,, \tag{20}$$

i.e. φ represents the potential of a tensor $\underline{\underline{F}}(\underline{\underline{Q}})$. The existence of a potential requires the fulfillment of the conditions of integrability obtained by the elimination of φ from (19)

$$\left(\frac{\partial}{\partial I_1} + I_1\frac{\partial}{\partial I_2}\right)(\varphi_1 + I_1\varphi_2) + \frac{\partial}{\partial I_2}(\varphi_0 - I_2\varphi_2) = 0\,,$$

$$\frac{\partial\varphi_2}{\partial I_2} + \frac{\partial}{\partial I_3}(\varphi_1 + \varphi_2 I_1) = 0\,, \qquad \frac{\partial\varphi}{\partial I_1} - I_1\frac{\partial}{\partial I_3}(\varphi_1 + \varphi_2 I_1) - \frac{\partial}{\partial I_3}(\varphi_0 - I_2\varphi_2) = 0\,. \tag{21}$$

It is often preferable to represent $\underline{\underline{F}}(\underline{\underline{Q}})$ in the form

$$\underline{\underline{F}}(\underline{\underline{Q}}) = \psi_0\underline{\underline{E}} + \psi_1\underline{\underline{Q}} + \psi_{-1}\underline{\underline{Q}}^{-1}\,, \qquad \psi_\Gamma = \psi_\Gamma(I_1,I_2,I_3)\,, \qquad \Gamma = -1,0,1\,.$$

Here

$$\psi_0 = \varphi_0 - I_2\varphi_2 = \frac{\partial\varphi}{\partial I_1} + I_1\frac{\partial\varphi}{\partial I_2}\,, \qquad \psi_1 = \varphi_1 + I_1\varphi_2 = -\frac{\partial\varphi}{\partial I_2}\,,$$

$$\psi_{-1} = I_3\varphi_2 = I_3\frac{\partial\varphi}{\partial I_3}\,.$$

8. INVERSION OF A TENSOR RELATION

In the non-linear theory of elasticity tensors $\underline{\underline{F}}$ and $\underline{\underline{Q}}$ are suitably defined tensors of stress and strain. The inverse problem also arises (for an isotropic medium). Starting from the relationship

$$\underline{\underline{F}} = \varphi_0\underline{\underline{E}} + \varphi_1\underline{\underline{Q}} + \varphi_2\underline{\underline{Q}}^2\,, \tag{1}$$

where φ_0, φ_1, φ_2 denote functions of the invariants $I_k(\underline{\underline{Q}})$, $\underline{\underline{Q}}$ may be expressed in terms of $\underline{\underline{F}}$, thus

$$\underset{=}{Q} = \chi_0 \underline{E} + \chi_1 \underline{F} + \chi_2 \underline{F}^2 . \tag{2}$$

The following procedure is proposed. The values of $\underline{F}$ and $\underline{F}^2$ are substituted into (2) and the tensors $\underset{=}{Q}^3$, $\underset{=}{Q}^4$ in the expression for $\underline{F}^2$ are expressed in terms of $\underline{E}$, $\underset{=}{Q}$ and $\underset{=}{Q}^2$ with the aid of the Cayley-Hamilton theorem. The coefficients of $\underline{E}$, $\underset{=}{Q}$ and $\underset{=}{Q}^2$ in the right-hand side of the thus transformed Equation (2) are then set equal to 0,1,0. This leads to a linear system for coefficients χ_0, χ_1, χ_2 which are thus expressed in terms of φ_0, φ_1, φ_2 and the invariants $I_k(\underset{=}{Q})$. The invariants $I_k(\underline{F})$ are expressed in terms of $I_k(\underset{=}{Q})$ by Equation (1) and by the representations for $\underline{F}^2$, $\underline{F}^3$ which follow from that equation. The invariants $I_k(\underline{F})$, so obtained, will turn out to be polynomials in $I_s(\underset{=}{Q})$ if we assume that φ_k as function of $I_s(\underset{=}{Q})$ are also polynomials. The inverse representation of the invariants $I_s(\underset{=}{Q})$ in terms of $I_k(\underline{F})$ is therefore not possible in general algebraic form.

This procedure can be simplified somewhat if we substitute the tensors $\underset{=}{Q}$ and $\underline{F}$ by their decompositions into the apherical and deviatoric parts

$$\underset{=}{Q} = \frac{1}{3} I_1(\underset{=}{Q})\underline{E} + \operatorname{dev}\underset{=}{Q} , \qquad \underline{F} = \frac{1}{3} I_1'(\underline{F})\underline{E} + \operatorname{dev}\underline{F} . \tag{3}$$

Here by (1)

$$I_1(\underline{F}) = 3\varphi_0 + \varphi_1 I_1(\underset{=}{Q}) + \varphi_2 I_1(\underset{=}{Q}^2) , \tag{4}$$

so that

$$\operatorname{dev}\underline{F} = \varphi_1\underset{=}{Q} + \varphi_2\underset{=}{Q}^2 - \frac{1}{3}\underline{E}[\varphi_1 I_1(\underset{=}{Q}) + \varphi_2 I_1(\underset{=}{Q}^2)] = \varphi_1 \operatorname{dev}\underset{=}{Q} + \varphi_2\left[\underset{=}{Q}^2 - \frac{1}{3}\underline{E}I_1(\underset{=}{Q}^2)\right] .$$

Here we should also express $\underset{=}{Q}^2$ through $\operatorname{dev}\underset{=}{Q}$

$$\underset{=}{Q}^2 = \frac{1}{9} I_1^2(\underset{=}{Q})\underline{E} + \frac{2}{3} I_1(\underset{=}{Q}) \operatorname{dev}\underset{=}{Q} + (\operatorname{dev}\underset{=}{Q})^2 .$$

Referring now to (I.13.3) we obtain

$$\operatorname{dev}\underline{F} = \alpha \operatorname{dev}\underset{=}{Q} + \beta\left[(\operatorname{dev}\underset{=}{Q})^2 + \frac{2}{3} I_2(\operatorname{dev}\underset{=}{Q})\underline{E}\right] , \tag{5}$$

$$\alpha = \varphi_1 + \frac{2}{3} I_1(\underset{=}{Q})\varphi_2 , \qquad \beta = \varphi_2 . \tag{6}$$

The expression for $\operatorname{dev}\underset{=}{Q}$ in terms of $\operatorname{dev}\underline{F}$ can be put into a similar form

$$\operatorname{dev}\underset{=}{Q} = \alpha_1 \operatorname{dev}\underline{F} + \beta_1\left[(\operatorname{dev}\underline{F})^2 + \frac{2}{3} I_2(\operatorname{dev}\underline{F})\underline{E}\right] . \tag{7}$$

Following the procedure proposed we now have

$$\operatorname{dev}\underline{Q} = \alpha_1\left\{\alpha \operatorname{dev}\underline{Q} + \beta\left[(\operatorname{dev}\underline{Q})^2 + \frac{2}{3} I_2(\operatorname{dev}\underline{Q})\underline{E}\right]\right\} +$$

$$+ \beta_1\left\{\alpha^2(\operatorname{dev}\underline{Q})^2 + 2\alpha\beta\left[(\operatorname{dev}\underline{Q})^3 + \frac{2}{3} I_2(\operatorname{dev}\underline{Q}) \operatorname{dev}\underline{Q}\right] +\right.$$

$$\left. + \beta^2\left[(\operatorname{dev}\underline{Q})^4 + \frac{4}{3} I_2(\operatorname{dev}\underline{Q})(\operatorname{dev}\underline{Q})^2 + \frac{4}{9}\underline{E}I_2^2(\operatorname{dev}\underline{Q}) + \frac{2}{3}\underline{E}I_2(\operatorname{dev}\underline{F})\right]\right\}, \quad (8)$$

where $I_2(\operatorname{dev}\underline{F})$ is determined from the condition that the first invariant of the expression in the right-hand side is zero

$$I_2(\operatorname{dev}\underline{F}) = \alpha^2 I_2(\operatorname{dev}\underline{Q}) - 3\alpha\beta I_3(\operatorname{dev}\underline{Q}) - \frac{1}{3}\beta^2 I_2^2(\operatorname{dev}\underline{Q}) . \quad (9)$$

Also substituting $(\operatorname{dev}\underline{Q})^3$ and $(\operatorname{dev}\underline{Q})^4$ in (8) using the Cayley-Hamilton theorem we arrive at the two equations

$$\alpha_1\alpha + \beta_1\beta\left[-\frac{2}{3}\alpha I_2(\operatorname{dev}\underline{Q}) + \beta I_3(\operatorname{dev}\underline{Q})\right] = 1 ,$$

$$\alpha_1\beta + \beta_1\left[\alpha^2 + \frac{1}{3}\beta^2 I_2(\operatorname{dev}\underline{Q})\right] = 0 .$$

From these we determine α_1 and β_1 and on substituting into (7) we obtain

$$\operatorname{dev}\underline{Q} = \frac{1}{H}\left\{\left[\alpha^2 + \frac{1}{3}\beta^2 I_2(\operatorname{dev}\underline{Q})\right]\operatorname{dev}\underline{F} - \beta\left[(\operatorname{dev}\underline{F})^2 + \frac{2}{3}\underline{E}I_2(\operatorname{dev}\underline{F})\right]\right\}, \quad (10)$$

$$H = \alpha^3 + I_2(\operatorname{dev}\underline{Q})\alpha\beta^2 - \beta^3 I_3(\operatorname{dev}\underline{Q}) . \quad (11)$$

The problem of inversion is not yet solved because Equation (10) contains $I_k(\underline{Q})$. The two equations (4) and (9) which connect the invariants should be augmented by a third for the determination of $I_3(\operatorname{dev}\underline{F})$. To this end we should first express $(\operatorname{dev}\underline{F})^3$ as in (5). To find $I_3(\operatorname{dev}\underline{F})$, the relationships

$$I_1((\operatorname{dev}\underline{Q})^5) = -5I_2(\operatorname{dev}\underline{Q}) , \qquad I_1((\operatorname{dev}\underline{Q})^6) = -2I_2^3(\operatorname{dev}\underline{Q}) + 3I_3^2(\operatorname{dev}\underline{Q})$$

must be added to (I.13.11).

We arrive at the relationship

$$I_3(\operatorname{dev}\underline{F}) = \alpha^3 I_3(\operatorname{dev}\underline{Q}) + \frac{2}{3}\alpha^2\beta I_2^3(\operatorname{dev}\underline{Q}) -$$

$$- \alpha\beta^2 I_2(\operatorname{dev}\underline{Q}) I_3(\operatorname{dev}\underline{Q}) + \beta^3\left[I_3^2(\operatorname{dev}\underline{Q}) + \frac{2}{27} I_2^3(\operatorname{dev}\underline{Q})\right], \quad (12)$$

here and in (4) and (8), we recall that α and β are functions of $I_k(\underline{\underline{Q}})$. It is impossible, of course, to obtain analytic expressions for $I_k(\underline{\underline{Q}})$ as functions of $I_1(\underline{F})$ from these equations.

If a scalar potential (7.20) exists

$$\underline{F} = \varphi_{\underline{\underline{Q}}} \tag{13}$$

then the formal representation of $\underline{\underline{Q}}$ in terms of $\underline{F}$ can be expressed in terms of a generating function of the inverse transformation - the Legendre transformation

$$\psi(\underline{\underline{Q}}(\underline{F})) = \underline{F} \cdot\cdot \underline{\underline{Q}}^T(\underline{F}) - \varphi(\underline{\underline{Q}}(\underline{F})) . \tag{14}$$

Then

$$\underline{\underline{Q}} = \psi_{\underline{F}} . \tag{15}$$

In fact we have

$$\delta\psi = \psi_{\underline{F}} \cdot\cdot \delta\underline{F}^T = \delta\underline{F} \cdot\cdot \underline{\underline{Q}}^T(\underline{F}) + \underline{F} \cdot\cdot \delta\underline{\underline{Q}}^T - \varphi_{\underline{\underline{Q}}} \cdot\cdot \delta\underline{\underline{Q}}^T = \underline{\underline{Q}} \cdot\cdot \delta\underline{F}^T ,$$

which yields (15).

9. THE TRIGONOMETRIC TRANSFORMATION OF V.V. NOVOZHILOV

In Equations (8.9) and (8.12) we return to the notation of (I.13.5) for the invariants of $\operatorname{dev}\underline{\underline{Q}}$ and introduce a similar notation for the invariants of $\operatorname{dev}\underline{F}$

$$G_2 = -4I_2(\operatorname{dev}\underline{F}) , \qquad G_3 = 4I_3(\operatorname{dev}\underline{F}) . \tag{1}$$

Making use of (I.13.9) we may now rewrite Equation (8.9) in the following form

$$G_2 = \alpha^2 g_2 - 3\alpha\beta\left(\frac{g_2}{3}\right)^{3/2}\sin 3\psi + \frac{1}{12}\beta^2 g_2 =$$

$$= g_2\left[\left(\alpha - \frac{1}{2}\beta\sqrt{\frac{g_2}{3}}\sin 3\psi\right)^2 + \left(\frac{1}{2}\beta\sqrt{\frac{g_2}{3}}\cos 3\psi\right)^2\right] .$$

This equality will be satisfied if we assume

$$\alpha - \frac{1}{2}\beta\sqrt{\frac{g_2}{3}}\sin 3\psi = \sqrt{\frac{G_2}{g_2}}\cos\omega ,$$

$$-\frac{1}{2}\beta\sqrt{\frac{g_2}{3}}\cos 3\psi = \sqrt{\frac{G_2}{g_2}}\sin\omega . \tag{2}$$

These formulas show that

$$\alpha = \sqrt{\frac{G_2}{g_2}}\frac{\cos(\omega+3\psi)}{\cos 3\psi} , \qquad \beta = -2\sqrt{\frac{G_2}{g_2}}\sqrt{\frac{3}{g_2}}\frac{\sin\omega}{\cos 3\psi} \tag{3}$$

and this enables us to write representation (8.5) of tensor dev $\underline{F}$ in the following form

$$\operatorname{dev}\underline{F} = \sqrt{\frac{G_2}{g_2}}(\cos 3\psi)^{-1}\left\{\cos(\omega+3\psi)\operatorname{dev}\underline{Q} - 2\sqrt{\frac{3}{g_2}}\left[(\operatorname{dev}\underline{Q})^2 - \frac{1}{6}g_2\underline{E}\right]\sin\omega\right\} . \tag{4}$$

The expression (8.12) for the third invariant of $\underline{F}$ is now transformed to

$$-3\sqrt{3}\frac{G_3}{G_2^{3/2}} = \frac{1}{\cos^3 3\psi}[\cos^3(\omega+3\psi)\sin 3\psi + 3\cos^2(\omega+3\psi)\sin\omega +$$

$$+ 3\cos(\omega+3\psi)\sin^2\omega + 2\sin^2 3\psi\sin^3\omega - \sin^3\omega] .$$

A cumbersome but quite elementary transformation results in the following form for the right-hand side

$$(3\cos^2\omega - \sin^2\omega)\sin\omega\cos 3\psi + (\cos^2\omega - 3\sin^2\omega)\cos\omega\sin 3\psi = \sin 3(\omega+\psi) ,$$

so that

$$G_3 = -\left(\frac{G_2}{3}\right)^{3/2}\sin 3(\omega+\psi) = -\left(\frac{G_2}{3}\right)^{3/2}\sin 3\chi . \tag{5}$$

The expression (I.13.9) enables us to ascribe to the quantity

$$\chi = \omega + \psi \tag{6}$$

the same role in assigning dev $\underline{F}$ as was played by ψ with respect to dev $\underline{Q}$. The principal values of dev $\underline{F}$ are now determined in terms of χ by (I.13.10)

$$\nu_1 = \sqrt{\frac{G_2}{3}}\sin\chi , \quad \nu_2 = \sqrt{\frac{G_2}{3}}\sin\left(\chi+\frac{2\pi}{3}\right) , \quad \nu_3 = \sqrt{\frac{G_2}{3}}\sin\left(\chi+\frac{4\pi}{3}\right) ,$$

$$|\chi| < \frac{\pi}{6} . \tag{7}$$

Therefore

$$\frac{\nu_s}{\kappa_s} = \sqrt{\frac{G_2}{g_2}}\,\frac{\sin\chi_s}{\sin\psi_s} = \sqrt{\frac{G_2}{g_2}}\,\frac{\sin(\omega+\psi_s)}{\sin\psi_s} \qquad (s = 1,2,3)\ . \tag{8}$$

For $\omega = 0$

$$\frac{\nu_s}{\kappa_s} = \sqrt{\frac{G_2}{g_2}}\,, \qquad \operatorname{dev}\underline{F} = \sqrt{\frac{G_2}{g_2}}\operatorname{dev}\underline{\underline{Q}}\,, \tag{9}$$

which follows from (4). It is thus reasonable to term ω the degree of similarity of deviators - this quantity characterizes the non-similarity of these tensors as well as the non-linearity of the mutual dependence between $\underline{F}$ and $\underline{\underline{Q}}$ because, in the original expression (8.1), the quantity $\beta = \varphi_2$ goes to zero with ω and term $\underline{\underline{Q}}^2$ vanishes.

The quantities ψ, ω and G_2 in (4) are determined by the invariants of tensor $\underline{\underline{Q}}$ itself and by the coefficients φ_1, φ_2 of the expression for $\underline{F}$ in terms of $\underline{\underline{Q}}$. By (I.13.9) and (I.13.7) we have in fact

$$\sin 3\psi = -\frac{3\sqrt{3}}{g_2\sqrt{g_2}}\,g_3\,, \qquad \cos 3\psi = \frac{\sqrt{\Delta}}{g_2\sqrt{g_2}} \qquad \left(|\psi| < \frac{\pi}{6}\right) \tag{10}$$

and furthermore, by (3) and (8.9)

$$\sqrt{\Delta}\,\operatorname{ctg}\omega = -\sqrt{3}\left\{\left[\frac{2\varphi_1}{\varphi_2} + \frac{4}{3}I_1(\underline{\underline{Q}})\right]g_2 + 3g_3\right\}\,, \tag{11}$$

$$\sqrt{\frac{G_2}{g_2}} = \frac{1}{g_2}\left\{\left[\varphi_1 g_2 + \varphi_2\left(\frac{2}{3}\,g_2 I_1(\underline{\underline{Q}}) + \frac{3}{2}\,g_3\right)\right]^2 + \frac{1}{12}\,\varphi_2^2\Delta\right\}^{1/2}\ . \tag{12}$$

The relationship connecting the first invariants of $\underline{F}$ and $\underline{\underline{Q}}$ also includes φ_0.

Relationship (4) allows immediate inversion: it is sufficient to substitute ψ for χ and ω for $-\omega$

$$\operatorname{dev}\underline{\underline{Q}} = \sqrt{\frac{g_2}{G_2}}\,(\cos 3\chi)^{-1}\left\{\cos(3\chi-\omega)\operatorname{dev}\underline{F} + 2\sqrt{\frac{3}{G_2}}\left[(\operatorname{dev}\underline{F})^2 - \frac{1}{6}\,G_2\underline{E}\right]\sin\omega\right\}\ . \tag{13}$$

Appendix III. NOTES ON THE TENSOR CALCULUS

1. THE RADIUS VECTOR. THE UNIT (METRIC) TENSOR

We now introduce a system of curvilinear coordinates q^1, q^2, q^3. The position of the point M is fixed by the radius vector $\underline{r}$, the origin O of which is fixed ($\underline{r} = \overrightarrow{OM}$)

$$\underline{r} = \underline{r}(q^1, q^2, q^3) \ . \tag{1}$$

The Cartesian coordinates of the point M, i.e. projections of its radius-vector on the three axes of the system $OXYZ$ specified by the orthonormal system of vectors $\underline{i}_1$, $\underline{i}_2$, $\underline{i}_3$ are designated*) a^1, a^2, a^3

$$a^s = a^s(q^1, q^2, q^3) \ , \qquad \underline{r} = \underline{i}_s a^s \ . \tag{2}$$

It is assumed that within the range of their definition, these functions are continuously differentiable and the Jacobian

$$\left| \frac{\partial a^s}{\partial q^k} \right| \neq 0 \tag{3}$$

is different from zero. Under this condition, Equations (2) are uniquely solvable with respect to q^k.

In tensor algebra a vector basis is defined by three arbitrarily specified non-coplanar vectors $\underline{r}_1$, $\underline{r}_2$, $\underline{r}_3$. Here these vectors are assumed to be equal to the three partial derivatives of the position vector

$$\underline{r}_s = \frac{\partial \underline{r}}{\partial q^s} \qquad (s = 1,2,3) \tag{4}$$

and in the notation of (I.1.9) we have

$$\sqrt{g} = \underline{r}_1 \cdot (\underline{r}_2 \times \underline{r}_3) = \left| \frac{\partial a^s}{\partial q^k} \right| \ . \tag{5}$$

---

*) In order to preserve the summation rule over the upper and lower dummy indices, we substitute, wherever required, $\underline{i}_s$ for $\underline{i}^s$ and a_s for a^s.

The Jacobian may be regarded as positive (by a suitable numeration of the coordinates).

In algebra, all operations are related to a fixed point, the introduction of curvilinear coordinates and a vector basis (4) corresponds to moving to the examination of a field, i.e. to the comparison of quantities (scalar, vectors, tensors) which are defined at different points in three-dimensional Euclidean space E_3. In this space it is possible to specify the position of any point by a unique Cartesian system $OXYZ$ as has been done above.

A vector $d\underline{r} = \vec{MM'}$ where a point M' lies in an infinitely close neighbourhood of M is defined by the formula

$$d\underline{r} = \frac{\partial \underline{r}}{\partial q^s} dq^s = \underline{r}_s dq^s \ , \tag{6}$$

so that the square of a linear element $(\vec{MM'})^2$ is represented by a positive definite form

$$ds^2 = d\underline{r} \cdot d\underline{r} = \underline{r}_s dq^s \cdot \underline{r}_k dq^k = g_{sk} dq^s dq^k \ . \tag{7}$$

Here and below we retain the notation of Appendix I and $\| g_{sk} \|$ is a matrix of the covariant components of a unit (metric) tensor $\underline{E}$. Its representation in terms of contravariant g^{sk} and mixed $g_k^s = \delta_k^s$ components are given by the formulas (I.4.14). In Cartesian axes,

$$\underline{E} = \underline{i}_s \underline{i}^s \tag{8}$$

and this corresponds to the representation of the square of a linear element in the "Pythagorean" form

$$ds^2 = (da^1)^2 + (da^2)^2 + (da^3)^2 \ . \tag{9}$$

On introducing the dual basis $\underline{r}^s$ we have by (6)

$$dq^s = \underline{r}^s \cdot d\underline{r} = d\underline{r} \cdot \underline{r}^s \ . \tag{10}$$

This simple relationship is made repeated use of below. It obviates the necessity of obtaining expressions which do not contain the base vectors, i.e. which are not associated with any particular choice of curvilinear coordinates. These should be specified at the stage of transfer from general relationships to specific calculations.

2. THE HAMILTON NABLA-OPERATOR

We consider a scalar $\varphi(q^1, q^2, q^3)$ and its differential $d\varphi$ which is an increment defined by a displacement from a point M in the field to a point M' in its immediate neighbourhood. By (1.10)

$$d\varphi = \frac{\partial\varphi}{\partial q^s}\, dq^s = \underline{r}^s \frac{\partial\varphi}{\partial q^s} \cdot d\underline{r} \tag{1}$$

the right-hand side contains a scalar product of the vector $d\underline{r}$ and a quantity which is also a vector since the product is the scalar $d\varphi$ [see (II.1.1a)]. This vector is called the gradient of the scalar and written

$$\underline{\nabla}\varphi = \underline{r}^s \frac{\partial}{\partial q^s}\varphi , \qquad \underline{\nabla} = \underline{r}^s \frac{\partial}{\partial q^s} . \tag{2}$$

This representation may be regarded as the product of a symbolic vector $\underline{\nabla}$, the Hamilton nabla-operator and a scalar. We might equally place the nabla-operator in (2) on the right of the quantity upon which it acts ($\varphi\underline{\nabla}$). It is quite permissible to write this product in this form, but we will not do so here.

The above can also be applied to a vector

$$d\underline{a} = \frac{\partial\underline{a}}{\partial q^s}\, dq^s = d\underline{r} \cdot \underline{r}^s \frac{\partial\underline{a}}{\partial q^s} = d\underline{r} \cdot \underline{\underline{\nabla a}} , \qquad d\underline{a} = \frac{\partial\underline{a}}{\partial q^s}\, \underline{r}^s \cdot d\underline{r} . \tag{3}$$

In the first form the vector $d\underline{a}$ is expressed as the result of the multiplication of $d\underline{r}$ from the left by a quantity which may be visualized, from (II.1.1b) as a tensor of the second rank and designated $\underline{\underline{\nabla a}}$

$$\underline{\underline{\nabla a}} = \underline{r}^s \frac{\partial}{\partial q^s}\underline{a} . \tag{4}$$

This tensor, formally represented by a dyad of vectors $\underline{\nabla}$ and $\underline{a}$, is called the gradient of the vector $\underline{a}$. Alternatively, (3) may be written in terms of $\underline{\underline{a\nabla}}$, but in accordance with the above, we will denote this combination by $\underline{\underline{\nabla a}}^{\mathrm{T}}$, the transposed gradient. Thus*) we have

---

*) Here we make use of the notation of N.E. Kotchine in which $\underline{\underline{\nabla a}}$ is the gradient of the vector $\underline{a}$ and the tensor $\underline{\underline{\nabla a}}^{\mathrm{T}}$ is called by N.E. Kotchine the derivative of vector $\underline{a}$ in the direction of $\underline{r}$

$$\underline{\underline{\nabla a}}^{\mathrm{T}} = \frac{d\underline{a}}{d\underline{r}} , \qquad \underline{\underline{\nabla a}}^{\mathrm{T}} \cdot d\underline{r} = d\underline{a} .$$

In foreign literature, the term gradient is applied to $\underline{\underline{\nabla a}}^{\mathrm{T}}$.

$$d\underline{a} = \underline{\underline{\nabla a}}^{T} \cdot d\underline{r} \ . \tag{5}$$

The extension of these operations to a tensor of the second rank leads to a tensor of the third rank

$$d\underline{\underline{Q}} = \frac{\partial \underline{\underline{Q}}}{\partial q^s} dq^s = d\underline{r} \cdot \underline{r}^s \frac{\partial \underline{\underline{Q}}}{\partial q^s} = d\underline{r} \cdot \underline{\underline{\nabla Q}} \ , \quad \underline{\underline{\nabla Q}} = \underline{r}^s \frac{\partial \underline{\underline{Q}}}{\partial q^s} \ , \quad \frac{\partial \underline{\underline{Q}}}{\partial q^s} = \underline{r}_s \cdot \underline{\underline{\nabla Q}} \ . \tag{6}$$

Just as a scalar $\underline{a} \cdot \underline{b}$ and vector $\underline{a} \times \underline{b}$ correspond to the dyad $\underline{\underline{ab}}$, a tensor $\underline{\underline{\nabla a}}$ determines a scalar

$$\underline{\nabla} \cdot \underline{a} = \underline{r}^s \cdot \frac{\partial \underline{a}}{\partial q^s} = \operatorname{div} \underline{a} \ , \tag{7}$$

called the divergence of $\underline{a}$, and a vector, the curl $\underline{a}$

$$\underline{\nabla} \times \underline{a} = \underline{r}^s \times \frac{\partial \underline{a}}{\partial q^s} = \operatorname{rot} \underline{a} = 2\underline{\omega} \ . \tag{8}$$

The vector $\underline{\omega}$ is called the vorticity of $\underline{a}$, this vector is associated with the skew tensor $\underline{\Omega}$. By (I.14.19)

$$\underline{\Omega} = \underline{E} \times \underline{\omega} = \frac{1}{2} \underline{r}_k \underline{r}^k \times \left(\underline{r}^s \times \frac{\partial \underline{a}}{\partial q^s} \right) = \frac{1}{2} \underline{r}_k \left(\underline{r}^k \cdot \frac{\partial \underline{a}}{\partial q^s} \underline{r}^s - \underline{r}^k \cdot \underline{r}^s \frac{\partial \underline{a}}{\partial q^s} \right) =$$

$$= \frac{1}{2} \left(\underline{\underline{\nabla a}}^T - \underline{\underline{\nabla a}} \right) \ . \tag{9}$$

Conversely, by (I.14.17)

$$\underline{\omega} = \underline{\in} \cdot\cdot \underline{\Omega} \ . \tag{10}$$

The skew tensor $\underline{\Omega}$ is called the spin tensor and the symmetric tensor determined by $\underline{\underline{\nabla a}}$ is called the linear strain tensor

$$\underline{\varepsilon} = \frac{1}{2} \left(\underline{\underline{\nabla a}} + \underline{\underline{\nabla a}}^T \right) \ . \tag{11}$$

With this notation

$$\underline{\underline{\nabla a}} = \underline{\varepsilon} - \underline{\Omega} \ , \qquad \underline{\underline{\nabla a}}^T = \underline{\varepsilon} + \underline{\Omega} \ . \tag{12}$$

Referring to (3), we arrive at the Helmholtz formula

$$d\underline{a} = \underline{\underline{\nabla a}}^T \cdot d\underline{r} = \underline{\varepsilon} \cdot d\underline{r} + \underline{\Omega} \cdot d\underline{r} = \underline{\varepsilon} \cdot d\underline{r} + \underline{\omega} \times d\underline{r} \tag{13}$$

this determines the field of the vector $\underline{a}$ in the neighbourhood of M in terms of strain and spin tensors at the point M.

The contraction $\underline{\nabla} \cdot \underline{Q}$ determines a vector called the divergence of a tensor

$$\underline{\nabla} \cdot \underline{Q} = \operatorname{div} \underline{Q} = \underline{r}^s \cdot \frac{\partial \underline{Q}}{\partial q^s} . \tag{14}$$

The curl of a tensor is a tensor of the second rank defined by the relationship

$$\underline{\nabla} \times \underline{Q} = \operatorname{rot} \underline{Q} = \underline{r}^s \times \frac{\partial \underline{Q}}{\partial q^s} . \tag{15}$$

The effect of the nabla-operator applied to tensors of arbitrary rank is defined in a similar way.

Applied to a radius vector we have

$$\underline{\underline{\nabla r}} = \underline{r}^s \underline{r}_s = \underline{E} , \qquad \underline{\nabla} \cdot \underline{r} = 3 , \qquad \underline{\nabla} \times \underline{r} = \underline{r}^s \times \underline{r}_s = \underline{0} . \tag{16}$$

It is easy to verify the relationships

$$\underline{\underline{\nabla E}} = \underline{0} , \qquad \underline{\underline{\nabla \epsilon}} = -\underline{\underline{\nabla E}} \times \underline{E} = \underline{0} , \tag{17}$$

since in a Cartesian system $\underline{E} = \underline{i}_s \underline{i}^s$ and the tensor $\underline{\underline{\nabla E}}$ represents an invariant which is independent of the basis.

As applied to a skew-symmetric tensor $\underline{\Omega}$, the formula (15) is transformed to

$$\underline{\nabla} \times \underline{\Omega} = \underline{\nabla} \times (\underline{\omega} \times \underline{E}) = \underline{r}^s \times \left(\frac{\partial \underline{\omega}}{\partial q^s} \times \underline{E} \right) = \underline{r}^s \times \left(\frac{\partial \underline{\omega}}{\partial q^s} \times \underline{r}_k \right) \underline{r}^k =$$

$$= \frac{\partial \underline{\omega}}{\partial q^s} \delta^s_k \underline{r}^k - \underline{r}_k \underline{r}^k \underline{r}_s \cdot \frac{\partial \underline{\omega}}{\partial q^s}$$

or

$$\underline{\nabla} \times \underline{\Omega} = \underline{\underline{\nabla \omega}}^T - \underline{\underline{E \nabla}} \cdot \underline{\omega} . \tag{18}$$

From this it follows that

$$I_1 (\underline{\nabla} \times \underline{\Omega}) = -2 \underline{\nabla} \cdot \underline{\omega} . \tag{19}$$

3. EXAMPLES OF OPERATIONS WITH THE NABLA-OPERATOR

We give here the formulas which define the differential operations over certain configurations of the tensor quantities. Our purpose is to express the results of such operations in invariant form in terms of the nabla-operator and the tensor quantities themselves and not the base vectors and components of the tensors. These quantities are sometimes required in analysis.

1. The gradient of the product of two scalars, of a scalar and a vector, of a scalar and a tensor

$$\underline{\nabla}\varphi\psi = \psi\underline{\nabla}\varphi + \varphi\underline{\nabla}\psi , \qquad \underline{\nabla}\varphi\underline{a} = (\underline{\nabla}\varphi)\underline{a} + \varphi\underline{\underline{\nabla a}} , \qquad \underline{\nabla}\varphi\underline{\underline{Q}} = (\underline{\nabla}\varphi)\underline{\underline{Q}} + \varphi\underline{\underline{\nabla Q}} . \tag{1}$$

2. The gradient of scalar product, vector product, the curl of a vector product, the divergence of a dyad and a vector product

$$\underline{\underline{\nabla a}} \cdot \underline{b} = (\underline{\underline{\nabla a}}) \cdot \underline{b} + \underline{a} \cdot \underline{\underline{\nabla b}}^T = (\underline{\underline{\nabla a}}) \cdot \underline{b} + (\underline{\underline{\nabla b}}) \cdot \underline{a} , \tag{2}$$

$$\underline{\underline{\nabla a}} \times \underline{b} = (\underline{\underline{\nabla a}}) \times \underline{b} - (\underline{\underline{\nabla b}}) \times \underline{a} , \tag{3}$$

$$\underline{\nabla} \times (\underline{a} \times \underline{b}) = \underline{b} \cdot \underline{\underline{\nabla a}} - \underline{\underline{b\nabla}} \cdot \underline{a} - \underline{a} \cdot \underline{\underline{\nabla b}} + \underline{\underline{a\nabla}} \cdot \underline{b} , \tag{4}$$

$$\underline{\nabla} \cdot \underline{\underline{ab}} = (\underline{\nabla} \cdot \underline{a})\underline{b} + \underline{a} \cdot \underline{\underline{\nabla b}} , \tag{5}$$

$$\underline{\nabla} \cdot (\underline{a} \times \underline{b}) = \underline{b} \cdot (\underline{\nabla} \times \underline{a}) - \underline{a} \cdot (\underline{\nabla} \times \underline{b}) . \tag{6}$$

These may be written in a number of different ways. For example, from (2.12) and (2.8) we have

$$\underline{\underline{\nabla a}} = \underline{\underline{\nabla a}}^T - 2\underline{\Omega} ,$$

$$(\underline{\underline{\nabla a}}) \cdot \underline{b} = \underline{b} \cdot \underline{\underline{\nabla a}} - 2\underline{\Omega} \cdot \underline{b} = \underline{b} \cdot \underline{\underline{\nabla a}} - 2\underline{\omega} \times \underline{b} = \underline{b} \cdot \underline{\underline{\nabla a}} + \underline{b} \times (\underline{\nabla} \times \underline{a})$$

and Equation (2) may be rewritten as follows

$$\underline{\underline{\nabla a}} \cdot \underline{b} = \underline{b} \cdot \underline{\underline{\nabla a}} + \underline{a} \cdot \underline{\underline{\nabla b}} + \underline{b} \times (\underline{\nabla} \times \underline{a}) + \underline{a} \times (\underline{\nabla} \times \underline{b}) . \tag{2'}$$

3. The divergence of $\underline{\underline{Q}} \times \underline{r}$

$$\underline{\nabla} \cdot (\underline{\underline{Q}} \times \underline{r}) = (\underline{\nabla} \cdot \underline{\underline{Q}}) \times \underline{r} + \underline{r}^s \cdot \underline{\underline{Q}} \times \underline{r}_s .$$

Resolving $\underline{\underline{Q}}$ into its symmetric and skew-symmetric parts, we have

$$(\underline{r}^s \cdot \underline{\underline{S}}) \times \underline{r}_s = S^{st}\underline{r}_t \times \underline{r}_s = S^{ts}\underline{r}_s \times \underline{r}_t = S^{st}\underline{r}_s \times \underline{r}_t = \underline{0} ,$$

$$(\underline{r}^s \cdot \underline{\underline{\Omega}}) \times \underline{r}_s = (\underline{r}^s \times \underline{\omega}) \times \underline{r}_s = \underline{\underline{\omega r}}^s \cdot \underline{r}_s - \underline{r}^s\underline{\omega} \cdot \underline{r}_s = 2\underline{\omega} .$$

We have thus arrived at the equation

$$\underline{\nabla} \cdot (\underline{Q} \times \underline{r}) = (\underline{\nabla} \cdot \underline{Q}) \times \underline{r} + 2\underline{\omega} , \tag{7}$$

where $\underline{\omega}$ is the vector associated with the skew-symmetric part of $\underline{Q}$. If $\underline{Q} = \underline{Q}^{T}$, then

$$\underline{\nabla} \cdot (\underline{Q} \times \underline{r}) = (\underline{\nabla} \cdot \underline{Q}) \times \underline{r} . \tag{8}$$

4. The divergence of $\underline{r} \times \underline{Q}$

$$\underline{\nabla} \cdot (\underline{r} \times \underline{Q}) = \underline{\nabla} \cdot (\underline{r} \times \underline{r}_m \underline{r}_n q^{mn}) = \underline{r}^s \cdot (\underline{r}_s \times \underline{r}_m) \underline{r}_n q^{mn} + \underline{r}^s \cdot (\underline{r} \times \underline{r}_m) \underline{r}_n \nabla_s q^{mn} =$$

$$= (\underline{r}^s \times \underline{r}_s) \cdot \underline{Q} - \underline{r} \cdot (\underline{r}^s \times \underline{r}_m \underline{r}_n) \nabla_s q^{mn} = (\underline{r}^s \times \underline{r}_s) \cdot \underline{Q} - \underline{r} \cdot (\underline{\nabla} \times \underline{Q}) .$$

The first term vanishes by (2.16). We thus obtain

$$\underline{\nabla} \cdot (\underline{r} \times \underline{Q}) = -\underline{r} \cdot (\underline{\nabla} \times \underline{Q}) \tag{9}$$

in the course of this calculation we have employed rule (2.3) for the differentiation of a tensor.

5. The divergence of product of a tensor and vector is given by

$$\underline{\nabla} \cdot (\underline{Q} \cdot \underline{a}) = (\underline{\nabla} \cdot \underline{Q}) \cdot \underline{a} + \underline{r}^s \cdot \underline{Q} \cdot \frac{\partial \underline{a}}{\partial q^s} = (\underline{\nabla} \cdot \underline{Q}) \cdot \underline{a} + \underline{r}^s \cdot (\underline{Q} \cdot \underline{\underline{\nabla a}}^{T}) \cdot \underline{r}_s =$$

$$= (\underline{\nabla} \cdot \underline{Q}) \cdot \underline{a} + I_1 (\underline{Q} \cdot \underline{\underline{\nabla a}}^{T})$$

and we have arrived at the often used transformation

$$\underline{\nabla} \cdot (\underline{Q} \cdot \underline{a}) = (\underline{\nabla} \cdot \underline{Q}) \cdot \underline{a} + \underline{Q} \cdot\cdot \underline{\underline{\nabla a}}^{T} . \tag{10}$$

6. The divergence of the product of two tensors. In this calculation we make use of the differentiation formula for a tensor of the second rank (this formula is given below in Par. 5)

$$\frac{\partial \underline{P}}{\partial q^s} = \underline{r}^m \underline{r}^n \nabla_s p_{mn} , \qquad \underline{\underline{\nabla P}} = \underline{r}^s \underline{r}^m \underline{r}^n \nabla_s p_{mn}$$

where $\underline{\underline{\nabla P}}$ is a tensor of the third rank. We thus obtain

$$\underline{\nabla} \cdot (\underline{Q} \cdot \underline{P}) = \left(\underline{r}^s \cdot \frac{\partial \underline{Q}}{\partial q^s}\right) \cdot \underline{P} + (\underline{r}^s \cdot \underline{Q}) \cdot \frac{\partial \underline{P}}{\partial q^s} =$$

$$= (\underline{\nabla} \cdot \underline{\underline{Q}}) \cdot \underline{P} + (\underline{\underline{Q}}^T \cdot \underline{r}^s) \cdot \underline{r}^m \underline{r}^n \nabla_s p_{mn} = \underline{P}^T \cdot \underline{\nabla} \cdot \underline{\underline{Q}} + \underline{\underline{Q}}^T \cdot\cdot \underline{r}^s \underline{r}^m \underline{r}^n \nabla_s p_{mn} ,$$

so that

$$\underline{\nabla} \cdot (\underline{\underline{Q}} \cdot \underline{P}) = \underline{P}^T \cdot (\underline{\nabla} \cdot \underline{\underline{Q}}) + \underline{\underline{Q}}^T \cdot\cdot \underline{\underline{\nabla P}} . \tag{11}$$

7. The curl of the vector product $\underline{\underline{Q}} \times \underline{r}$

$$\underline{\nabla} \times (\underline{\underline{Q}} \times \underline{r}) = (\underline{\nabla} \times \underline{\underline{Q}}) \times \underline{r} + \underline{r}^s \times (\underline{\underline{Q}} \times \underline{r}_s) ,$$

$$\underline{r}^s \times (\underline{\underline{Q}} \times \underline{r}_s) = \underline{r}^s \times \underline{r}^m \underline{r}_n \times \underline{r}_s q_m^{\cdot n} = \epsilon^{smq} \epsilon_{nst} \underline{r}_q \underline{r}^t q_m^{\cdot n} = (\delta_t^m \delta_n^q - \delta_n^m \delta_t^q) \underline{r}_q \underline{r}^t q_m^{\cdot n} =$$

$$= q_t^{\cdot q} \underline{r}_q \underline{r}^t - \underline{r}_t \underline{r}^t q_n^{\cdot n} = \underline{\underline{Q}}^T - \underline{\underline{E}} I_1(\underline{\underline{Q}}) ,$$

so that

$$\underline{\nabla} \times (\underline{\underline{Q}} \times \underline{r}) = (\underline{\nabla} \times \underline{\underline{Q}}) \times \underline{r} + \underline{\underline{Q}}^T - \underline{\underline{E}} I_1(\underline{\underline{Q}}) . \tag{12}$$

4. THE DERIVATIVES OF THE BASE VECTORS. CHRISROFFEL SYMBOLS

1. We assume the usual notation for the derivatives of the base vectors

$$\frac{\partial \underline{r}_s}{\partial q^t} = \underline{r}_{st} = \underline{r}_{ts} = \frac{\partial \underline{r}_t}{\partial q^s} = \frac{\partial^2 \underline{r}}{\partial q^t \partial q^s} . \tag{1}$$

Their representation in the form of decompositions over the base vectors is given by the expressions

$$\underline{r}_{st} = \left\{ {q \atop st} \right\} \underline{r}_q . \tag{2}$$

The coefficients of this decomposition are called Christoffel symbols of the second kind (and often denoted as Γ_{st}^q). Their definition implies symmetry with respect to the lower indices

$$\left\{ {q \atop st} \right\} = \left\{ {q \atop ts} \right\} = \Gamma_{st}^q = \Gamma_{ts}^q . \tag{3}$$

Equation (2) implies the relationships

$$\left\{ {q \atop st} \right\} g_{qk} = \underline{r}_{st} \cdot \underline{r}_k , \tag{4}$$

their right-hand sides can be expressed through the derivatives of the covariant components of a metric tensor. In fact

$$\frac{\partial}{\partial q^t}\,\underline{r}_s \cdot \underline{r}_k = \frac{\partial}{\partial q^t}\,g_{sk} = \underline{r}_{st} \cdot \underline{r}_k + \underline{r}_s \cdot \underline{r}_k \;,$$

$$\frac{\partial}{\partial q^s}\,g_{kt} = \underline{r}_{ks} \cdot \underline{r}_t + \underline{r}_k \cdot \underline{r}_{ts} \;,$$

$$\frac{\partial}{\partial q^k}\,g_{ts} = \underline{r}_{tk} \cdot \underline{r}_s + \underline{r}_t \cdot \underline{r}_{sk}$$

and the expression at the right-hand side can be obtained by subtraction of the third equality from the sum of the two others

$$\underline{r}_{st} \cdot \underline{r}_k = \frac{1}{2}\left(\frac{\partial g_{sk}}{\partial q^t} + \frac{\partial g_{kt}}{\partial q^s} - \frac{\partial g_{st}}{\partial q^k}\right) = [st,k] \;.$$

The quantities in the right-hand side are called the Christoffel symbols of the first kind

$$[st,k] = [ts,k] = \Gamma_{stk} = \frac{1}{2}\left(\frac{\partial g_{sk}}{\partial q^t} + \frac{\partial g_{tk}}{\partial q^s} - \frac{\partial g_{st}}{\partial q^k}\right) . \tag{5}$$

Equations (4) and their inverse may be put into the form

$$\left\{ {q \atop st} \right\} g_{qk} = \left\{ {q \atop st} \right\} \underline{r}_q \cdot \underline{r}_k = [st,k] \;, \qquad \left\{ {q \atop st} \right\} = g^{qk}\,[st,k] \;. \tag{6}$$

These formulas fully determine the derivatives (2) of the vectors of the initial basis. We may also use them to obtain the derivatives of the vectors of the dual basis:

$$\delta^s_t = \underline{r}^s \cdot \underline{r}_t = 0 \;, \qquad \frac{\partial \underline{r}^s}{\partial q^k} \cdot \underline{r}_t + \underline{r}^s \cdot \underline{r}_{tk} = \frac{\partial \underline{r}^s}{\partial q^k} \cdot \underline{r}_t + \underline{r}^s \cdot \left\{ {q \atop tk} \right\} \underline{r}_q = 0 \;.$$

We thus obtain the formulas

$$\frac{\partial \underline{r}^s}{\partial q^k} \cdot \underline{r}_t + \left\{ {s \atop kt} \right\} = 0 \;, \qquad \frac{\partial \underline{r}^s}{\partial q^k} = -\left\{ {s \atop kt} \right\} \underline{r}^t \;. \tag{7}$$

2. <u>The derivative of $\sqrt{g}$</u>. We have

$$\frac{\partial \sqrt{g}}{\partial q^k} = \frac{\partial}{\partial q^k}\,\underline{r}_1 \cdot (\underline{r}_2 \times \underline{r}_3) = \left\{ {m \atop 1k} \right\} \underline{r}_m \cdot (\underline{r}_2 \times \underline{r}_3) + \left\{ {m \atop 2k} \right\} \underline{r}_m \cdot (\underline{r}_3 \times \underline{r}_1) + $$

$$+ \left\{ {m \atop 3k} \right\} \underline{r}_m \cdot (\underline{r}_1 \times \underline{r}_2) = \sqrt{g}\left(\left\{ {1 \atop 1k} \right\} + \left\{ {2 \atop 2k} \right\} + \left\{ {3 \atop 3k} \right\}\right) ,$$

since, e.g., $\underline{r}_m \cdot (\underline{r}_2 \times \underline{r}_3) \neq 0$ only if $m = 1$. We arrive at the often used relationships

$$\frac{\partial\sqrt{g}}{\partial q^k} = \sqrt{g}\left\{ {s \atop sk} \right\}, \qquad \frac{\partial \ln\sqrt{g}}{\partial q^k} = \left\{ {s \atop sk} \right\}. \tag{8}$$

3. Transformation of Christoffel symbols. In a new coordinate system

$$\check{q}^r = \check{q}^r(q^1, q^2, q^3)$$

the base vectors are tranformed by the law

$$\check{\underline{r}}_s = \frac{\partial \underline{r}}{\partial \check{q}^s} = \frac{\partial \underline{r}}{\partial q^r}\frac{\partial q^r}{\partial \check{q}^s} = \underline{r}_r \frac{\partial q^r}{\partial \check{q}^s} = \underline{r}_r \underline{r}^r \cdot \check{\underline{r}}_s = c^r_s \underline{r}_r ,$$

$$c^r_s = \underline{r}^r \cdot \check{\underline{r}}_s = \frac{\partial q^r}{\partial \check{q}^s}.$$

Therefore

$$\check{\underline{r}}_{st} = \frac{\partial \check{\underline{r}}_s}{\partial \check{q}^t} = \left\{ {\check{q} \atop st} \right\} \check{\underline{r}}_q = c^m_t \frac{\partial c^r_s}{\partial q^m} \underline{r}_r + c^m_t c^r_s \left\{ {k \atop rm} \right\} \underline{r}_k ,$$

so that

$$\left\{ {q \atop st} \right\} = \underline{c^m_t \frac{\partial c^r_s}{\partial q^m} \check{c}^q_r} + c^m_t c^r_s \check{c}^q_k \left\{ {k \atop rm} \right\}, \qquad \check{c}^q_k = \check{\underline{r}}^q \cdot \underline{r}_k = \frac{\partial \check{q}^q}{\partial q^k}. \tag{9}$$

The transformation formuls for the components of a tensor of the third rank (I, Par. 7) have the form

$$\check{P}^q_{st.} = P^k_{rm.} c^r_s c^m_t \check{c}^q_k , \tag{10}$$

whereas Equation (9) contains the extra (underlined) term. For this reason, the Christoffel symbols do not represent components of a tensor.

5. COVARIANT DIFFERENTIATION

An operation with tensors requires introduction of a coordinate basis and a consideration of components possessing a certain structure in this basis. A change in an invariant (scalar, vector or tensor) is governed by its properties. It is determined specified using the nabla-operator i.e., symbolically, $d(\ldots) = d\underline{r} \cdot \underline{\nabla}(\ldots)$. A change in components is calculated differently. This depends upon the base considered. For example, if $\underline{a}$ is a constant vector, $d\underline{a} = \underline{0}$, but a_k or a^k are by no means constant because of the change of basis.

On the other hand, if the components are constant, vector $\underline{a}$ does not remain constant in either magnitude or direction. It is therefore necessary to introduce those features of the variation of a tensor which can account for both the change due to the variability of its components and the contribution due to the non-constancy of the basis. This is achieved by covariant differentiation.

Having specified a vector in terms of its contravariant components, we have, referring to the differentiation formula (4.2) for the base vectors

$$\frac{\partial \underline{a}}{\partial q^s} = \frac{\partial}{\partial q^s} a^k \underline{r}_k = \frac{\partial a^k}{\partial q^s} \underline{r}_k + a^k \left\{ {q \atop sk} \right\} \underline{r}_q = \underline{r}_k \left(\frac{\partial a^k}{\partial q^s} + \left\{ {k \atop sq} \right\} a^q \right) .$$

If a vector is specified by its covariant components then we have by (4.7)

$$\frac{\partial \underline{a}}{\partial q^s} = \frac{\partial}{\partial q^s} a_k \underline{r}^k = \frac{\partial a_k}{\partial q^s} \underline{r}^k - a_k \left\{ {k \atop sq} \right\} \underline{r}^q = \underline{r}^k \left(\frac{\partial a_k}{\partial q^s} - \left\{ {q \atop sk} \right\} a_q \right) .$$

The quantities

$$\nabla_s a^k = \frac{\partial a^k}{\partial q^s} + \left\{ {k \atop sq} \right\} a^q , \qquad \nabla_s a_k = \frac{\partial a_k}{\partial q^s} - \left\{ {q \atop sk} \right\} a_q \tag{1}$$

are called the covariant derivatives of the components (contra- and covariant) of a vector. Introducing these, the derivative of a vector is given by the expressions

$$\frac{\partial \underline{a}}{\partial q^s} = \underline{r}_k \nabla_s a^k , \qquad \frac{\partial \underline{a}}{\partial q^s} = \underline{r}^k \nabla_s a_k . \tag{2}$$

A similar calculation for a tensor of the second rank yields

$$\frac{\partial \underline{Q}}{\partial q^s} = \frac{\partial}{\partial q^s} q^{mn} \underline{r}_m \underline{r}_n = \frac{\partial q_{mn}}{\partial q^s} \underline{r}_m \underline{r}_n + q^{mn} \left(\left\{ {q \atop sm} \right\} \underline{r}_q \underline{r}_n + \left\{ {q \atop sn} \right\} \underline{r}_m \underline{r}_q \right)$$

or after renaming the dummy indices

$$\frac{\partial \underline{Q}}{\partial q^s} = \underline{r}_m \underline{r}_n \nabla_s q^{mn} , \qquad \nabla_s q^{mn} = \frac{\partial q^{mn}}{\partial q^s} + \left\{ {m \atop sq} \right\} q^{qn} + \left\{ {n \atop sq} \right\} q^{mq} . \tag{3}$$

For covariant and mixed components we have in the same way

$$\frac{\partial \underline{Q}}{\partial q^s} = \underline{r}^m \underline{r}^n \nabla_s q_{mn} , \qquad \nabla_s q_{mn} = \frac{\partial q_{mn}}{\partial q^s} - \left\{ {q \atop sm} \right\} q_{qn} - \left\{ {q \atop sn} \right\} q_{mq} , \tag{4}$$

$$\frac{\partial \underline{Q}}{\partial q^s} = \underline{r}_m \underline{r}^n \nabla_s q^m_{\cdot n} , \qquad \nabla_s q^m_{\cdot n} = \frac{\partial q^m_{\cdot n}}{\partial q^s} + \left\{ {m \atop sq} \right\} q^q_{\cdot n} - \left\{ {q \atop sn} \right\} q^m_{\cdot q} , \tag{5}$$

$$\frac{\partial \underline{Q}}{\partial q^s} = \underline{r}^m \underline{r}_n \nabla_s q^{\cdot n}_m , \qquad \nabla_s q^{\cdot n}_m = \frac{\partial q^{\cdot n}_m}{\partial q^s} - \left\{ {q \atop sm} \right\} q^{\cdot n}_q + \left\{ {n \atop sq} \right\} q^{\cdot q}_m . \tag{6}$$

Examining the structure of these formulas it is easy to understand precisely how they are formulated. The rules for the arrangement of the upper and lower indices and the summation over dummy indices are so apt and explicit as to eliminate the possibility of error and the formulas "write themselves down". The expanded expressions, in which the values of the Christoffel symbols are given explicitly, are, naturally, very cumbersome, if written in this form the formulas would lose their simplicity and elegance.

In covariant differentiation, the formal rules for the differentiation of sum and product are retained. E.g.,

$$\nabla_r a^s b_t = (\nabla_r a^s) b_t + a^s \nabla_r b_t . \tag{7}$$

By definition (2.2) of the nabla-operator

$$\underline{\underline{\nabla a}} = \underline{r}^s \frac{\partial \underline{a}}{\partial q^s} = \underline{r}^s \underline{r}^k \nabla_s a_k = \underline{r}^s \underline{r}_k \nabla_s a^k , \tag{8}$$

so that $\nabla_s a_k$, $\nabla_s a^k$ are the covariant and cocontravariant components of $\underline{\underline{\nabla a}}$. Of course,

$$\underline{\underline{\nabla a}}^T = \underline{r}^k \underline{r}^s \nabla_s a_k = \underline{r}_k \underline{r}^s \nabla_s a^k . \tag{9}$$

Similarly

$$\underline{\underline{\nabla Q}} = \underline{r}^s \underline{r}^m \underline{r}^n \nabla_s q_{mn} = \underline{r}^s \underline{r}_m \underline{r}_n \nabla_s q^{mn} \tag{10}$$

etc. Here $\nabla_s q_{mn}$, $\nabla_s q^{mn}$ are various components of a tensor $\underline{\underline{\nabla Q}}$ of the third rank.

Equation (2.17) implies the vanishing of the covariant derivatives of the components of a unit tensor (Ricci's theorem)

$$\nabla_s g_{mn} = 0 , \qquad \nabla_s g^{mn} = 0 , \qquad \nabla_s g^m_n = 0 . \tag{11}$$

This is easily verified by direct calculation. For example, by (4) we have

$$\nabla_s g_{mn} = \frac{\partial g_{mn}}{\partial q^s} - \left\{ {q \atop sm} \right\} g_{qn} - \left\{ {q \atop sn} \right\} g_{mq} = \frac{\partial g_{mn}}{\partial q^s} - [sm,n] - [sn,m] = 0 \ , \qquad (12)$$

which follows from definition (4.5) of the Christoffel symbols of the first kind.

The components of the metric tensor behave like constants under covariant differentiation, they may be placed either before or after the operator, e.g.,

$$\nabla_s a^k = \nabla_s g^{kt} a_t = g^{kt} \nabla_s a_t \ , \qquad \nabla_s a_k = g_{kt} \nabla_s a^t \ .$$

This is in agreement with the definition of a metric tensor

$$\underline{\nabla a} = \underline{\nabla E} \cdot \underline{a} = \underline{E} \cdot \underline{\nabla a} \ .$$

Ricci's theorem can be extended to isotropic tensors of arbitrary rank (I, Par. 15), to the tensor $\underline{\epsilon} = -\underline{E} \times \underline{E}$ of Levi-Civita and to the tensors (I.15.1) of the fourth rank, etc.

The purpose of this Appendix is to explain the rules of operation with tensors in E_3. We have therefore given no space here to a geometric interpretation of covariant differentiation, i.e. to the concepts of parallel translation, geodesics etc.

6. DIFFERENTIAL OPERATIONS WITH TENSORS

The invariant representations of differential operations with tensors (gradient, divergence, curl) were defined in Par. 2. Here we consider the rules for obtaining them.

1. From (5.8) for the gradient of a vector its curl is defined as

$$\underline{\nabla} \times \underline{a} = \underline{r}^s \times \underline{r}^k \nabla_s a_k = \epsilon^{skt} \underline{r}_t \nabla_s a_k = \epsilon^{skt} \underline{r}_t \left(\frac{\partial a_k}{\partial q^s} - \left\{ {q \atop sk} \right\} a_q \right)$$

and its divergence as

$$\underline{\nabla} \cdot \underline{a} = \underline{r}^s \cdot \underline{r}^k \nabla_s a_k = g^{sk} \nabla_s a_k = \nabla_s a^s = \frac{\partial a^s}{\partial q^s} + \left\{ {s \atop sm} \right\} a^m \ .$$

By (4.3) and (4.8)

$$\epsilon^{skt} \left\{ {q \atop sk} \right\} = \epsilon^{kst} \left\{ {q \atop sk} \right\} = -\epsilon^{skt} \left\{ {q \atop sk} \right\} = 0 \ , \qquad \left\{ {s \atop sm} \right\} a^m = \frac{\partial \ln \sqrt{g}}{\partial q^m} a^m \ ,$$

so that

$$\underline{\nabla} \times \underline{a} = \underline{r}_t \epsilon^{skt} \frac{\partial a_k}{\partial q^s}\,, \quad 2\omega' = \frac{\partial a^3}{\partial q^2} - \frac{\partial a^2}{\partial q^3}\,, \quad 2\omega^2 = \frac{\partial a^1}{\partial q^3} - \frac{\partial a^3}{\partial q^1}\,, \quad 2\omega^3 = \frac{\partial a^2}{\partial q^1} - \frac{\partial a^1}{\partial q^2}\,, \tag{1}$$

$$\underline{\nabla} \cdot \underline{a} = \frac{1}{\sqrt{g}} \frac{\partial}{\partial q^s} \sqrt{g}\, a^s\,. \tag{2}$$

2. <u>The curl and divergence of tensor of the second rank</u> are given by

$$\underline{\nabla} \times \underline{\underline{Q}} = \underline{r}^s \times \underline{r}^m \underline{r}^n \nabla_s q_{mn} = \epsilon^{smt} \underline{r}_t \underline{r}^n \left(\frac{\partial q_{mn}}{\partial q^s} - \left\{ {q \atop sm} \right\} q_{qn} - \left\{ {q \atop sn} \right\} q_{mq} \right),$$

$$\underline{\nabla} \cdot \underline{\underline{Q}} = \underline{r}^s \cdot \underline{r}_m \underline{r}_n \nabla_s q^{mn} = \underline{r}_n \nabla_s q^{sn} = \underline{r}_n \left(\frac{\partial q^{sn}}{\partial q^s} + \left\{ {s \atop sm} \right\} q^{mn} + \left\{ {n \atop sm} \right\} q^{sm} \right).$$

The relationships

$$\epsilon^{smt} \left\{ {q \atop sm} \right\} = 0\,, \qquad -\underline{r}^n \left\{ {q \atop sn} \right\} = -\frac{\partial \underline{r}^q}{\partial q^s}\,, \qquad \underline{r}_n \left\{ {n \atop sm} \right\} = \frac{\partial \underline{r}_m}{\partial q^s}$$

enable us to transform these expressions to

$$\underline{\nabla} \times \underline{\underline{Q}} = \epsilon^{smt} \underline{r}_t \frac{\partial}{\partial q^s} \underline{r}^q q_{mq} = \underline{r}^s \times \underline{r}^m \underline{r}^q \nabla_s q_{mq}\,, \tag{3}$$

$$\underline{\nabla} \cdot \underline{\underline{Q}} = \frac{1}{\sqrt{g}} \frac{\partial}{\partial q^s} \sqrt{g}\, q^{st} \underline{r}_t\,. \tag{4}$$

The first invariant of the curl of a symmetric tensor turns out to be zero

$$\underline{\underline{Q}} = \underline{\underline{Q}}^{\mathrm{T}} : I_1(\underline{\nabla} \times \underline{\underline{Q}}) = (\underline{r}^s \times \underline{r}^m) \cdot \underline{r}^q \nabla_s q_{mq} = \epsilon^{smq} \nabla_s q_{mq} = 0\,. \tag{5}$$

The first invariant of the curl of a skew-symmetric tensor is defined by (2.19).

3. <u>Double differentiation</u>. The nabla-operator applied to the gradient defines a symmetric tensor of the second rank

$$\underline{\nabla\nabla}\varphi = \underline{r}^s \frac{\partial}{\partial q^s} \underline{r}^k \frac{\partial}{\partial q^k} = \underline{r}^s \underline{r}^k \left(\frac{\partial^2 \varphi}{\partial q^s \partial q^k} - \left\{ {m \atop sk} \right\} \frac{\partial \varphi}{\partial q^m} \right) = (\underline{\nabla\nabla}\varphi)^{\mathrm{T}}\,. \tag{6}$$

The scalar defined by this tensor is a Laplacian in

$$\underline{\nabla} \cdot \underline{\nabla}\varphi = \nabla^2 \varphi = g^{sk} \left(\frac{\partial^2 \varphi}{\partial q^s \partial q^k} - \left\{ {m \atop sk} \right\} \frac{\partial \varphi}{\partial q^m} \right)\,, \tag{7}$$

and the vector (the curl of the gradient) is zero

$$\underline{\nabla} \times \underline{\nabla}\varphi = \epsilon^{skt}\underline{r}_t\left(\frac{\partial^2\varphi}{\partial q^s \partial q^k} - \left\{ {m \atop sk} \right\} \frac{\partial \varphi}{\partial q^m}\right) = 0 \ . \tag{8}$$

In the alternative representation of Equation (6) we take into account that the operations of differentiation and covariant differentiation are indistinguishable when applied to a scalar

$$\underline{\underline{\nabla\nabla}}\varphi = (\underline{\underline{\nabla\nabla}}\varphi)^{\mathrm{T}} = \underline{r}^k \frac{\partial}{\partial q^k} \underline{r}^s \nabla_s \varphi = \underline{r}^k \underline{r}^s \nabla_k \nabla_s \varphi = \underline{r}^s \underline{r}^k \nabla_s \nabla_k \varphi = \underline{r}^k \underline{r}^s \nabla_s \nabla_k \varphi \ . \tag{9}$$

The relationship $\nabla_k \nabla_s \varphi = \nabla_s \nabla_k \varphi$ is obvious for a scalar. However, it remains true (as shown in Par. 10) for components of a tensor of arbitrary rank

$$\nabla_k \nabla_s (\ldots) = \nabla_s \nabla_k (\ldots) \ . \tag{10}$$

The expression for a Laplacian now takes the form

$$\underline{\nabla}^2 \varphi = \underline{r}^k \cdot \underline{r}^s \nabla_k \nabla_s \varphi = g^{ks} \nabla_k \nabla_s \varphi = \nabla^s \nabla_s \varphi \ . \tag{11}$$

Here we have introduced the notation for "contravariant" differentiation

$$\nabla^s = g^{sk} \nabla_k \ . \tag{12}$$

4. The expanded expressions for the second order operations applied to a vector are cumbersome. The tensor $\underline{\underline{\nabla a}}$ of the third rank allows the following contractions decreasing its rank by two, namely the formation of the Laplacian and of the gradient of the divergence

$$\nabla^2 \underline{a} = \underline{\nabla} \cdot \underline{\underline{\nabla a}} = \underline{r}^s \cdot \frac{\partial}{\partial q^s} \underline{r}^t \underline{r}_k \nabla_t a^k = g^{st} \underline{r}_k \nabla_s \nabla_t a^k = \underline{r}_k \nabla^t \nabla_t a^k \ , \tag{13}$$

$$\underline{\underline{\nabla\nabla}} \cdot \underline{a} = \underline{r}^s \frac{\partial}{\partial q^s} \frac{1}{\sqrt{g}} \frac{\partial}{\partial q^k} \sqrt{g}\, a^k = \underline{r}^s \nabla_s \nabla_k a^k \ . \tag{14}$$

These two are related by the vector

$$\underline{\nabla} \times (\underline{\nabla} \times \underline{a}) = \underline{\underline{\nabla\nabla}} \cdot \underline{a} - \nabla^2 \underline{a} \ . \tag{15}$$

The divergence of a tensor $\underline{\underline{\nabla a}}^{\mathrm{T}}$ is given by

$$\underline{\nabla} \cdot \underline{\underline{\nabla a}}^{\mathrm{T}} = \underline{r}^s \cdot \frac{\partial}{\partial q^s} \underline{r}_k \underline{r}^t \nabla_t a^k = \delta^s_k \underline{r}^t \nabla_k \nabla_t a^k = \underline{r}^t \nabla_t \nabla_k a^k$$

and in view of (14)

$$\underline{\nabla} \cdot \underline{\nabla a}^T = \underline{\nabla\nabla} \cdot \underline{a} \; . \tag{16}$$

Reducing the rank of a tensor $\underline{\nabla\nabla a}$ by one we obtain tensors of the second rank, namely the curl of the gradient and the gradient of the curl

$$\underline{\nabla} \times \underline{\nabla a} = \underline{0} \; , \qquad \underline{\nabla\nabla} \times \underline{a} = \underline{r}^q \underline{r}^s \times \underline{r}^t \nabla_q \nabla_s a_t = \epsilon^{stk} \underline{r}^q \underline{r}_k \nabla_q \nabla_s a_t \; . \tag{17}$$

5. The gradient of the divergence, the divergence of the gradient (the Laplacian) and the curl of the curl, which are tensors of the second rank, are formed from the fourth rank tensor $\underline{\nabla\nabla Q}$

$$\underline{\nabla\nabla} \cdot \underline{Q} = \underline{r}^s \underline{r}_n \nabla_s \nabla_k q^{kn} \; , \qquad \underline{\nabla} \cdot \underline{\nabla Q} = \nabla^2 \underline{Q} = \underline{r}_k \underline{r}_t \nabla^s \nabla_s q^{kt} \; , \tag{18}$$

$$\underline{\nabla} \times (\underline{\nabla} \times \underline{Q}) = \underline{\nabla\nabla} \cdot \underline{Q} - \nabla^2 \underline{Q} \tag{19}$$

as well as the tensor formed from (9)

$$\underline{\nabla\nabla} I_1(\underline{Q}) = \underline{r}^k \underline{r}^s \nabla_k \nabla_s I_1(\underline{Q}) \; . \tag{20}$$

By $\underline{\nabla\nabla Q}$ we can also form the vectors

$$\underline{\nabla} \times \underline{\nabla} I_1(\underline{Q}) = \underline{0} \; , \qquad \underline{\nabla} \cdot (\underline{\nabla} \times \underline{Q}) = \underline{0} \; , \qquad \underline{\nabla} \times (\underline{\nabla} \cdot \underline{Q}) = \epsilon^{snm} \underline{r}_m \nabla_s \nabla_t q^t_{\cdot n} \; , \tag{21}$$

and the scalar

$$\underline{\nabla} \cdot \underline{\nabla} \cdot \underline{Q} = \nabla_k \nabla_s q^{ks} \; . \tag{22}$$

6. <u>The tensor</u> Ink $\underline{Q}$. This tensor, called "incompatibility tensor of $\underline{Q}$" (Inkompatibilität), has substantial applications in continuum mechanics. It represents the curl of the transposed curl of $\underline{Q}$

$$\mathrm{Ink}\, \underline{Q} = \underline{\nabla} \times (\underline{\nabla} \times \underline{Q})^T \; . \tag{23}$$

We assume below that tensor $\underline{Q}$ is symmetric. By (19), the expression (23) can be put into the form

$$\mathrm{Ink}\, \underline{Q} = -\nabla^2 \underline{Q} + \nabla\nabla \cdot \underline{Q} - \underline{\nabla} \times [(\underline{\nabla} \times \underline{Q}) - (\underline{\nabla} \times \underline{Q})^T] \tag{24}$$

in which the expressions in brackets are to be transformed. Referring to (I.14.9) we have

$$\underline{\nabla} \times \underline{\underline{Q}} - (\underline{\nabla} \times \underline{\underline{Q}})^{\mathrm{T}} = (\underline{r}^s \times \underline{r}_m \underline{r}_n - \underline{r}_n \underline{r}^s \times \underline{r}_m) \nabla_s q^{mn} = \underline{\underline{E}} \times [\underline{r}_n \times (\underline{r}^s \times \underline{r}_m)] \nabla_s q^{mn} =$$

$$= \underline{\underline{E}} \times (\underline{r}^s g_{mn} - \underline{r}_m \delta_n^s) \nabla_s q^{mn} = \underline{\underline{E}} \times (\underline{r}^s \nabla_s g_{mn} q^{mn} - \underline{r}_m \nabla_n q^{mn}) \ ,$$

so that

$$\underline{\nabla} \times \underline{\underline{Q}} - (\underline{\nabla} \times \underline{\underline{Q}})^{\mathrm{T}} = \underline{\underline{E}} \times [\underline{\nabla} I_1(\underline{\underline{Q}}) - \underline{\nabla} \cdot \underline{\underline{Q}}] = \underline{\underline{E}} \times \underline{\omega} \ . \tag{25}$$

Here $\underline{\omega}$ denotes by (I.11.8) the vector associated with the skew-symmetric tensor $\underline{\Omega}$, and from (2.18) we have

$$\underline{\nabla} \times \underline{\Omega} = \underline{\underline{\nabla\omega}}^{\mathrm{T}} - \underline{\underline{E\nabla}} \cdot \underline{\omega} = \underline{\underline{\nabla\nabla}} I_1(\underline{\underline{Q}}) - (\underline{\underline{\nabla\nabla}} \cdot \underline{\underline{Q}})^{\mathrm{T}} - \underline{\underline{E}} [\nabla^2 I_1(\underline{\underline{Q}}) - \underline{\nabla} \cdot \underline{\nabla} \cdot \underline{\underline{Q}}] \ .$$

Substitution to (24) leads to the required representation

$$\mathrm{Ink}\,\underline{\underline{Q}} = -\underline{\nabla}^2 \underline{\underline{Q}} + \underline{\underline{\nabla\nabla}} \cdot \underline{\underline{Q}} + (\underline{\underline{\nabla\nabla}} \cdot \underline{\underline{Q}})^{\mathrm{T}} + (\underline{\underline{E\nabla}}^2 - \underline{\underline{\nabla\nabla}}) I_1(\underline{\underline{Q}}) - \underline{\underline{E\nabla}} \cdot \underline{\nabla} \cdot \underline{\underline{Q}} \ . \tag{26}$$

This immediately implies that $\mathrm{Ink}\,\underline{\underline{Q}}$ is symmetric if $\underline{\underline{Q}} = \underline{\underline{Q}}^{\mathrm{T}}$

$$\underline{\underline{Q}} = \underline{\underline{Q}}^{\mathrm{T}} : \mathrm{Ink}\,\underline{\underline{Q}} = (\mathrm{Ink}\,\underline{\underline{Q}})^{\mathrm{T}} \ . \tag{27}$$

Another representation of $\mathrm{Ink}\,\underline{\underline{Q}}$ follows directly from the definition (23)

$$\mathrm{Ink}\,\underline{\underline{Q}} = \epsilon^{tnq} \epsilon^{smr} \underline{r}_q \underline{r}_r \nabla_t \nabla_s q_{nm} = \underline{r}^t \times \underline{r}^n \underline{r}^s \times \underline{r}^m \nabla_t \nabla_s q_{nm} \ , \tag{28}$$

and taking into account the interchangeability of the operators of covariant differentiation, we arrive at the following expressions for its contravariant components

$$\begin{aligned} (\mathrm{Ink}\,\underline{\underline{Q}})^{11} &= \frac{1}{g} (\nabla_2^2 q_{33} + \nabla_3^2 q_{22} - 2\nabla_2 \nabla_3 q_{23}) \ , \\ (\mathrm{Ink}\,\underline{\underline{Q}})^{12} &= \frac{1}{g} [-\nabla_1 \nabla_2 q_{33} + \nabla_3 (\nabla_1 q_{23} + \nabla_2 q_{31} - \nabla_3 q_{12})] \end{aligned} \tag{29}$$

(the remaining four components are obtained by a cyclic permutation of the indices).

Let $\underline{\underline{Q}}$ be a linear strain tensor (2.11). In Cartesian coordinates the expressions (29) take the form

$$(\mathrm{Ink}\,\underline{\underline{\varepsilon}})^{11} = \frac{\partial^2 \varepsilon_{33}}{\partial a^{2^2}} + \frac{\partial^2 \varepsilon_{32}}{\partial a^{3^2}} - 2 \frac{\partial^2 \varepsilon_{23}}{\partial a^2 \partial a^3} \ ,$$

$$(\text{Ink}\,\underline{\varepsilon})^{12} = -\frac{\partial^2 \varepsilon_{33}}{\partial a^1 \partial a^2} + \frac{\partial}{\partial a^3}\left(\frac{\partial \varepsilon_{23}}{\partial a^1} + \frac{\partial \varepsilon_{31}}{\partial a^2} - \frac{\partial \varepsilon_{12}}{\partial a^3}\right) \tag{30}$$

and the conditions $\text{Ink}\,\underline{\varepsilon} = \underline{0}$ lead to the six well-known Saint-Venant equations of continuity of the linear theory of elasticity.

7. ORTHOGONAL CURVILINEAR COORDINATES

1. The choice of curvilinear coordinates q^1, q^2, q^3 is subject to the requirement of the orthogonality of the base vectors

$$\underline{r}_s \cdot \underline{r}_k = g_{sk} = 0\,, \quad s \neq k\,; \qquad \underline{r}_s \cdot \underline{r}_s = g_{ss} = H_s^2 \qquad (s = 1,2,3)\,. \tag{1}$$

The quantities $H_s = |\underline{r}_s|$ are called the Lamé coefficients. We introduce here the orthonormal basis

$$\underline{e}_s = \frac{\underline{r}_s}{H_s}\,, \quad s = 1,2,3\,; \qquad |\underline{e}_s| = 1\,; \qquad \underline{e}_s \cdot \underline{e}_k = 0\,, \quad s \neq k\,. \tag{2}$$

The vector $\underline{e}$ is the unit vector of the normal to the surface $q^s = q_0^s = \text{const}$, it is directed along the tangent to a curve $[q^s]$ along which the coordinate q^s varies in the direction of increasing q^s, and the two others are constant.

The matrix of the covariant components of the metric tensor $\|g_{sk}\|$ is diagonal as is the inverse matrix $\|g^{sk}\|$

$$\|g_{sk}\| = \text{diag}(H_1^2, H_2^2, H_3^2)\,, \qquad \|g^{sk}\| = \text{diag}\left(\frac{1}{H_1^2}, \frac{1}{H_2^2}, \frac{1}{H_3^2}\right). \tag{3}$$

Therefore

$$\underline{r}^s = g^{sk}\underline{r}_k = \frac{\underline{r}_s}{H_s^2}\,, \qquad \underline{r}_s = H_s \underline{e}_s\,, \qquad \underline{r}^s = \frac{\underline{e}_s}{H_s}\,,$$

$$\underline{\nabla} = \frac{\underline{e}_s}{H_s}\frac{\partial}{\partial q^s}\,, \qquad \underline{\underline{E}} = \underline{e}_s \underline{e}_s\,. \tag{4}$$

The index appearing in the left- and right-hand sides of this formula is, of course, not a summy index and no summation is implied over it. Summation is still carried out over the dummy indices and in contrast with the rule adopted earlier, it is also carried out over the three repeated indices regardless of their order. Examples of this are provided by the expressions for $\underline{\nabla}$ and $\underline{\underline{E}}$ in (4).

2. Cylindrical and spherical coordinates are well-known examples of orthogonal curvilinear coordinates.

For cylindrical coordinates we have

$$q^1 = r\,,\ q^2 = \varphi\,,\ q^3 = z\,;\qquad 0 < r < \infty\,,\ 0 \le \varphi \le 2\pi\,,\ -\infty < z < \infty \tag{5}$$

these are the radius, the azimuthal angle and the height. The coordinate surfaces are right circular cylinders $r = r_0$ with the axis OZ (defined by a unit vector $\underline{k} = \underline{e}_3$), halfplanes $\varphi = \varphi_0$ passing through the axis and planes $z = z_0$ perpendicular to this axis. The coordinate lines, i.e. the intersections of the corresponding pairs of surfaces, are the straight lines z, parallel to the axis OZ, the radially directed straight lines r, and circles φ. The tangents to these lines are the vectors $\underline{e}_3 = \underline{k}$, $\underline{e}_1 = \underline{e}_r$, $\underline{e}_2 = \underline{e}_\varphi$. The Jacobian $J = \sqrt{g} = r$ vanishes on the axis OZ and so this is not included in the range of definition of coordinates. The radius vector of a point is given in cylindrical coordinates by the expression

$$\underline{r} = r\underline{e}_r + \underline{k}z\ . \tag{6}$$

For spherical coordinates

$$q^1 = R\,,\ q^2 = \vartheta\,,\ q^3 = \lambda\,;\qquad 0 < R < \infty\,,\ 0 < \vartheta < \pi\,,\ 0 \le \lambda \le 2\pi \tag{7}$$

these being the radius of the sphere, the angle measured along the meridian from the north pole and the longitude (from the east to the west). The coordinate surfaces are spheres $R = R_0$ about the origin O, circular cones $\vartheta = \vartheta_0$ with vertex O, and halfplanes $\lambda = \lambda_0$ cutting the axis OZ. The coordinate lines are parallel circles $[\lambda]$ lying in the intersection of the spherical and conical surfaces, the radially divergent straight lines $[R]$ which are the intersections of the cones ϑ_0 and halfplanes λ_0, and meridians $[\vartheta]$ which are intersections of the halfplanes λ_0 and spheres R_0. The tangents to these lines are the vectors $\underline{e}_3 = \underline{e}_\lambda$, $\underline{e}_1 = \underline{e}_R$, $\underline{e}_2 = \underline{e}_\vartheta$. The vector $\underline{k} = \underline{e}_R \cos\vartheta - \underline{e}_\vartheta \sin\vartheta$ is a unit vector of the axis OZ.

The Jacobian $\sqrt{g} = R^2 \sin\vartheta$. The radius vector is defined by the formula

$$\underline{r} = \underline{e}_R R\ . \tag{8}$$

The square of a linear element in the orthogonal coordinates is given by the expression

$$d\underline{r}\cdot d\underline{r} = H_1^2 {dq^1}^2 + H_2^2 {dq^2}^2 + H_3^2 {dq^3}^2 = ds^2\ , \tag{9}$$

and the elements of arc along the coordinate lines $[q^k]$ are

$$d_k s = H_k dq^k = |\underline{r}_k| dq^k \ . \tag{10}$$

For cylindrical and spherical coordinates

$$d_1 s = dr\ ,\ H_1 = H_r = 1\ ;\quad d_2 s = r\, d\varphi\ ,\ H_2 = H_\varphi = r\ ;\quad d_3 s = dz\ ,\ H_3 = H_z = 1\ , \tag{11}$$

$$d_1 s = dR\ ,\ H_1 = H_R = 1\ ;\quad d_2 s = R\, d\vartheta\ ,\ H_2 = H_\vartheta = R\ ;$$

$$d_3 s = R \sin\vartheta\, d\lambda\ ,\ H_3 = H_\lambda = R \sin\vartheta\ . \tag{12}$$

3. The differentiation of the vectors of the orthonormal basis. Although we might use the formulas of Par. 4, it is preferable to avoid the calculation of Christoffel symbols and to resort to a kinematical method of a moving basis (G. Darboux).

Let the vertex M of a moving trehedron $\underline{e}_s$ slide along the coordinate line $[q^m]$ with a unit velocity $v = 1$, so that

$$d_m s = H_m dq^m = dt$$

(where t is the time). In any instanteneous position of a basis, the vectors $\underline{e}_s$ must be directed along the tangents to the coordinate lines at the point M. The motion of this point is therefore accompanied by a rotation of the basis about M, the angular velocity of this rotation being denoted by $\overset{m}{\underline{\omega}}$. By the usual formula of kinematics, the end points of the unit vectors will have the velocities

$$\frac{d\underline{e}_s}{dt} = \frac{\partial \underline{e}_s}{H_m \partial q^m} = \overset{m}{\underline{\omega}} \times \underline{e}_s$$

with respect to the origin. The required formulas take the form

$$\frac{\partial \underline{e}_s}{\partial q^m} = \overset{m}{\underline{o}} \times \underline{e}_s\ ,\qquad \overset{m}{\underline{o}} = H_m \overset{m}{\underline{\omega}}\ . \tag{13}$$

We now introduce a skew-symmetric tensor $\overset{m}{\underline{\Omega}}$ for which $\overset{m}{\underline{o}}$ is the associated vector

$$\overset{m}{\underline{\Omega}} = \overset{m}{\underline{o}} \times \underline{E} = \overset{m}{\underline{o}} \times \underline{e}_s \underline{e}_s = \frac{\partial \underline{e}_q}{\partial q^m}\, \underline{e}_s = \overset{m}{o}_{pq}\, \underline{e}_p \underline{e}_q \tag{14}$$

with components

$$\overset{m}{o}_{pq} = \frac{\partial \underline{e}_q}{\partial q^m} \cdot \underline{e}_p$$

and by (I.11.1) the vector $\overset{m}{\underline{o}}$ is given by the following expression in the orthonormal system $\underline{e}_s$

$$\overset{m}{\underline{o}} = \frac{1}{2}\overset{m}{o}_{pq}\epsilon_{qpt}\underline{e}_t = \frac{1}{2}\,\frac{\partial \underline{e}_q}{\partial q^m}\cdot\underline{e}_p\epsilon_{qpt}\underline{e}_t\ . \tag{15}$$

By (4.3), (1) and (2) we have

$$\underline{r}_{st}\cdot\underline{r}_k = \frac{1}{2}\left(\frac{\partial g_{sk}}{\partial q^t} + \frac{\partial g_{tk}}{\partial q^s} - \frac{\partial g_{st}}{\partial q^k}\right) = H_s\frac{\partial H_k}{\partial q^t}\delta_{sk} + H_t\frac{\partial H_k}{\partial q^s}\delta_{kt} - H_t\frac{\partial H_s}{\partial q^k}\delta_{st} =$$

$$= \frac{\partial}{\partial q^t}(H_s\underline{e}_s)\cdot H_k\underline{e}_k = \frac{\partial H_s}{\partial q^t}H_k\delta_{sk} + H_sH_k\frac{\partial \underline{e}_s}{\partial q^t}\cdot\underline{e}_k\ ,$$

so that

$$\frac{\partial \underline{e}_s}{\partial q^t}\cdot\underline{e}_k = \frac{H_t}{H_sH_k}\left(\frac{\partial H_k}{\partial q^s}\delta_{kt} - \frac{\partial H_s}{\partial q^k}\delta_{st}\right) = \frac{\partial H_k}{H_s\partial q^s}\delta_{kt} - \frac{\partial H_s}{H_k\partial q^k}\delta_{st}\ .$$

Substitution into (15) now leads to the relationship

$$\overset{m}{\underline{o}} = \frac{1}{2}\left(\frac{\partial H_p}{H_q\partial q^q}\delta_{pm} - \frac{\partial H_q}{H_p\partial q^p}\delta_{qm}\right)\epsilon_{qpt}\underline{e}_t =$$

$$= \frac{1}{2}\left(\frac{\partial H_m}{H_q\partial q^q}\epsilon_{qmt} - \frac{\partial H_m}{H_p\partial q^p}\epsilon_{mpt}\right)\underline{e}_t = \frac{\partial H_m}{H_q\partial q^q}\underline{e}_q\times\underline{e}_m = \underline{\nabla}H_m\times\underline{e}_m\ .$$

We have thus

$$\overset{m}{\underline{o}} = \underline{\nabla}H_m\times\underline{e}_m\ ,\qquad \frac{\partial \underline{e}_s}{\partial q^m} = (\underline{\nabla}H_m\times\underline{e}_n)\times\underline{e}_s = \underline{e}_m\frac{\partial H_m}{H_s\partial q^s} - \delta_{ms}\underline{\nabla}H_m\ . \tag{16}$$

These are the required derivation formulas for the base vectors of an orthonormal trihedron. Note that the vectors $\overset{m}{\underline{o}}$ may often be determined without calculation, by appealing to a kinematical interpretation.

For cylindrical coordinates

$$H_r = 1\ ,\ H_\varphi = r\ ,\ H_3 = 1\ ;\qquad \overset{1}{\underline{o}} = \underline{0}\ ,\ \overset{\varphi}{\underline{o}} = \underline{\nabla r}\times\underline{e}_\varphi = \underline{k}\ ,\ \overset{3}{\underline{o}} = \underline{0}\ , \tag{17}$$

as might be expected the motion of a trihedron along the coordinate lines [z] and [r] is progressive and the motion along the circle [φ] is a rotation with angular velocity $\overset{\varphi}{\underline{\omega}} = \underline{k}/\underline{r}$.

The following derivatives are different from zero

$$\frac{\partial \mathbf{e}_r}{\partial \varphi} = \overset{\varphi}{\mathbf{o}} \times \mathbf{e}_r = \mathbf{e}_\varphi \,, \qquad \frac{\partial \mathbf{e}_\varphi}{\partial \varphi} = \overset{\varphi}{\mathbf{o}} \times \mathbf{e}_\varphi = -\mathbf{e}_r \,. \tag{18}$$

In spherical coordinates

$$H_R = 1 \,, \quad \overset{R}{\mathbf{o}} = \mathbf{0} \,; \quad H_\vartheta = R \,, \quad \overset{\vartheta}{\mathbf{o}} = \nabla R \times \mathbf{e}_\vartheta = \mathbf{e}_R \times \mathbf{e}_\vartheta = \mathbf{e}_\lambda \,; \quad H_\lambda = R \sin\vartheta \,,$$

$$\overset{\lambda}{\mathbf{o}} = \nabla H_\lambda \times \mathbf{e}_\lambda = (\mathbf{e}_R \sin\vartheta + \mathbf{e}_\vartheta \cos\vartheta) \times \mathbf{e}_\lambda = -\mathbf{e}_\vartheta \sin\vartheta + \mathbf{e}_R \cos\vartheta = \mathbf{k} \,. \tag{19}$$

The following derivatives are different from zero

$$\frac{\partial \mathbf{e}_R}{\partial \vartheta} = \mathbf{e}_\lambda \times \mathbf{e}_R = \mathbf{e}_\vartheta \,, \qquad \frac{\partial \mathbf{e}_\vartheta}{\partial \vartheta} = \mathbf{e}_\lambda \times \mathbf{e}_\vartheta = -\mathbf{e}_R \,, \qquad \frac{\partial \mathbf{e}_R}{\partial \lambda} = \mathbf{k} \times \mathbf{e}_R = \mathbf{e}_\lambda \sin\vartheta \,,$$

$$\frac{\partial \mathbf{e}_\vartheta}{\partial \lambda} = \mathbf{k} \times \mathbf{e}_\vartheta = \mathbf{e}_\lambda \cos\vartheta \,, \qquad \frac{\partial \mathbf{e}_\lambda}{\partial \lambda} = \mathbf{k} \times \mathbf{e}_\lambda = -(\mathbf{e}_R \sin\vartheta + \mathbf{e}_\vartheta \cos\vartheta) \,.$$

Substitution of the expressions (13) for the derivatives into the identities

$$\frac{\partial}{\partial q^s} \frac{\partial \mathbf{e}_k}{\partial q^t} - \frac{\partial}{\partial q^t} \frac{\partial \mathbf{e}_k}{\partial q^s} = \mathbf{0} \tag{21}$$

leads to the equality

$$\frac{\partial}{\partial q^s} \overset{t}{\mathbf{o}} \times \mathbf{e}_k - \frac{\partial}{\partial q^t} (\overset{s}{\mathbf{o}} \times \mathbf{e}_k) = \left(\frac{\partial \overset{t}{\mathbf{o}}}{\partial q^s} - \frac{\partial \overset{s}{\mathbf{o}}}{\partial q^t} \right) \times \mathbf{e}_k + \overset{t}{\mathbf{o}} \times (\overset{s}{\mathbf{o}} \times \mathbf{e}_k) - \overset{s}{\mathbf{o}} \times (\overset{t}{\mathbf{o}} \times \mathbf{e}_k) =$$

$$= \left[\frac{\partial \overset{t}{\mathbf{o}}}{\partial q^s} - \frac{\partial \overset{s}{\mathbf{o}}}{\partial q^t} + (\overset{t}{\mathbf{o}} \times \overset{s}{\mathbf{o}}) \right] \times \mathbf{e}_k = \mathbf{0} \qquad (k = 1,2,3) \,,$$

which implies that the following relationship exists between the vectors $\overset{s}{\mathbf{o}}$

$$\frac{\partial \overset{t}{\mathbf{o}}}{\partial q^s} - \frac{\partial \overset{s}{\mathbf{o}}}{\partial q^t} + \overset{t}{\mathbf{o}} \times \overset{s}{\mathbf{o}} = \mathbf{0} \,. \tag{22}$$

4. Differential operations in orthogonal coordinates. The following relationships are based on representation (4) of the nabla-operator, the derivation formulas and on the definition (16) of the vectors $\overset{s}{\mathbf{o}}$.

a. *The gradient of a vector*

$$\nabla \mathbf{a} = \frac{\mathbf{e}_s}{H_s} \frac{\partial}{\partial q^s} \mathbf{e}_k a_k = \mathbf{e}_s \mathbf{e}_k \frac{\partial a_k}{H_s \partial q^s} + \mathbf{e}_s \frac{a_m}{H_s} \overset{s}{\mathbf{o}} \times \mathbf{e}_m = \mathbf{e}_s \mathbf{e}_k \frac{\partial a_k}{H_s \partial q^s} + \frac{a_m}{H_s} \mathbf{e}_k \cdot (\overset{s}{\mathbf{o}} \times \mathbf{e}_m)$$

and furthermore

$$\underline{e}_k \cdot (\overset{s}{\underline{o}} \times \underline{e}_m) = \overset{s}{\underline{o}} \cdot (\underline{e}_m \times \underline{e}_k) = \underline{\nabla} H_s \cdot [\underline{e}_s \times (\underline{e}_m \times \underline{e}_k)] = \frac{\partial H_s}{H_m \partial q^m} \delta_{sk} - \frac{\partial H_s}{H_k \partial q^k} \delta_{sm} .$$

We thus obtain

$$\underline{\underline{\nabla a}} = \underline{e}_s \underline{e}_k \left(\frac{\partial a_k}{H_s \partial q^s} - \frac{a_s}{H_s H_k} \frac{\partial H_s}{\partial q^k} + \delta_{sk} \frac{a_m}{H_m} \frac{\partial H_s}{H_s \partial q^m} \right) . \tag{23}$$

b. *The linear strain tensor*

$$\underline{\varepsilon} = \frac{1}{2} \underline{e}_s \underline{e}_k \left(\frac{\partial a_k}{H_s \partial q^s} + \frac{\partial a_s}{H_k \partial q^k} - \frac{a_s}{H_s H_k} \frac{\partial H_s}{\partial q^k} - \frac{a_k}{H_k H_s} \frac{\partial H_k}{\partial q^k} + 2\delta_{sk} \frac{a_m}{H_m H_s} \frac{\partial H_s}{\partial q^m} \right) . \tag{24}$$

c. *The divergence of a vector*

$$\underline{\nabla} \cdot \underline{a} = \frac{\partial}{\partial q^s} \frac{a_s}{H_s} + \frac{a_m}{H_m} \frac{\partial \ln H_s}{\partial q^m} \delta_{ss} ,$$

$$\frac{\partial \ln H_s}{\partial q^m} \delta_{ss} = \frac{\partial}{\partial q^m} (\ln H_1 + \ln H_2 + \ln H_3) = \frac{\partial}{\partial q^m} \ln \sqrt{g} .$$

We get

$$\underline{\nabla} \cdot \underline{a} = \frac{1}{\sqrt{g}} \frac{\partial}{\partial q^s} \sqrt{g} \frac{a_s}{H_s} = \frac{1}{H_1 H_2 H_3} \left(\frac{\partial}{\partial q^1} H_2 H_3 a_1 + \frac{\partial}{\partial q^2} H_3 H_1 a_2 + \frac{\partial}{\partial q^3} H_1 H_2 a_3 \right) . \tag{25}$$

d. *The curl of a vector.* On substituting $\underline{e}_s \times \underline{e}_k = (1/2)(\underline{e}_s \times \underline{e}_k - \underline{e}_k \times \underline{e}_s)$ we obtain

$$\underline{\nabla} \times \underline{a} = \frac{1}{2} \frac{\underline{e}_s \times \underline{e}_k}{H_s H_k} \left(\frac{\partial}{\partial q^s} H_k a_k - \frac{\partial}{\partial q^k} H_s a_s \right) . \tag{26}$$

e. *The Laplacian of a scalar.* Setting $\underline{a} = \underline{\nabla}\psi$ in (25) we have

$$\nabla^2 \psi = \frac{1}{H_1 H_2 H_3} \left(\frac{\partial}{\partial q^1} \frac{H_2 H_3}{H_1} \frac{\partial \psi}{\partial q^1} + \frac{\partial}{\partial q^2} \frac{H_3 H_1}{H_2} \frac{\partial \psi}{\partial q^2} + \frac{\partial}{\partial q^3} \frac{H_1 H_2}{H_3} \frac{\partial \psi}{\partial q^3} \right) . \tag{27}$$

f. *The divergence of a tensor of the second rank.*

$$\underline{\nabla} \cdot \underline{\underline{Q}} = \frac{\underline{e}_r}{H_r} \cdot \frac{\partial}{\partial q^r} \underline{e}_s \underline{e}_t q_{st} =$$

$$= \frac{1}{H_1H_2H_3}\left(\frac{\partial}{\partial q^1} H_2H_3q_{1t}\underline{e}_t + \frac{\partial}{\partial q^2} H_3H_1q_{2t}\underline{e}_t + \frac{\partial}{\partial q^3} H_1H_2q_{3t}\underline{e}_t\right) . \tag{28}$$

Below we will make use of Equations (16).

The formulas given are well suited to calculations in those problems where orthogonal coordinates are commonly used. However, they lack the precision and symmetry inherent in the relationships of Par. 6. The latter are preferable in general when detailed calculations are not of paramount importance.

8. THE GAUSS-OSTROGRADSKII TRANSFORMATION. STOKES' TRANSFORMATION

1. We assume here that the following form of the Gauss-Ostrogradskii theorem is familiar

$$\iiint\limits_{v} \frac{\partial\varphi}{\partial a^s}\, dv = \iint\limits_{o} n_s\varphi\, do \; . \tag{1}$$

Here $dv = da^1da^2da^3$ denotes the element of volume v, do the element of the boundary surface o, and n_s is the projection along the s-axis of a Cartesian coordinate system of the exterior normal $\underline{n}$. The function $\varphi(a^1,a^2,a^3)$ is assumed, together with its first partial derivatives, to be continuous and bounded in the closed domain $v + o$.

The volume v may contain internal cavities, and the direction of $\underline{n}$ may change suddenly at some curves lying on the surface o, e.g. at the edges of a parallelepiped.*)

The simplest generalization of Equation (1) is given by the relationship

$$\iiint\limits_{v} \underline{\nabla}\cdot\underline{c}\, dv = \iiint\limits_{v}\left(\frac{\partial c_1}{\partial a^1} + \frac{\partial c_2}{\partial a^2} + \frac{\partial c_3}{\partial a^3}\right) dv =$$

$$= \iint\limits_{o} (n_1c_1 + n_2c_2 + n_3c_3)\, do = \iint\limits_{o} \underline{n}\cdot\underline{c}\, do \; , \tag{2}$$

connecting the volume integral of the divergence $\underline{\nabla}\cdot\underline{c}$ of a vector with the flux $\underline{n}\cdot\underline{c}$ of the same vector across the boundary surface of the volume.

The rule for the substitution of the nabla-operator in a volume integral by a vector $\underline{n}$ in a surface integral which is embodied in (2), can be applied to other relationships since the matter is reduced, finally, to the initial transformation (1). For example,

---

*) Exact formulations of sufficient conditions of validity of the Gauss-Ostrogradskii transformation are given in textbooks on potential theory.

$$\iiint_v \nabla \times \underline{c}\, dv = \iint_o \underline{n} \times \underline{c}\, do\,, \quad \iiint_v \nabla \underline{c}\, dv = \iint_o \underline{n}\underline{c}\, do\,, \quad \iiint_v \nabla \underline{c}^T do = \iint_o \underline{c}\underline{n}\, do\,. \tag{3}$$

Applying this to a tensor, we have similarly that

$$\iiint_v \nabla \cdot \underline{\underline{Q}}\, dv = \iint_o \underline{n} \cdot \underline{\underline{Q}}\, do\,, \quad \iiint_v \nabla \times \underline{\underline{Q}}\, dv = \iint_o \underline{n} \times \underline{\underline{Q}}\, do\,. \tag{4}$$

The formulas based on Equations (3.7), (3.10)

$$\iint_o (\underline{r} \times \underline{n} \cdot \underline{\underline{Q}}) do = - \iint_o (\underline{n} \cdot \underline{\underline{Q}}) \times \underline{r}\, do = - \iiint_v [(\nabla \cdot \underline{\underline{Q}}) \times \underline{r} + 2\underline{\omega}]\, dv =$$

$$= \iiint_v [(\underline{r} \times \nabla \cdot \underline{\underline{Q}}) - 2\underline{\omega}]\, dv\,, \tag{5}$$

$$\iint_o \underline{n} \cdot \underline{\underline{Q}} \cdot \underline{a}\, do = \iiint_v \nabla \cdot (\underline{\underline{Q}} \cdot \underline{a}) dv = \iiint_v [(\nabla \cdot \underline{\underline{Q}}) \cdot \underline{a} + \underline{\underline{Q}} \cdot\cdot\, \nabla \underline{a}^T]\, dv\,. \tag{6}$$

enjoy wide application. In Equation (5) we have denoted the vector associated with the skew-symmetric part of $\underline{\underline{Q}}$ by $\underline{\omega}$. For a symmetric tensor we have

$$\underline{\underline{Q}} = \underline{\underline{Q}}^T : \iint_o (\underline{r} \times \underline{n} \cdot \underline{\underline{Q}}) do = \iiint_v (\underline{r} \times \nabla \cdot \underline{\underline{Q}}) dv\,. \tag{7}$$

2. Stokes' transformation. In a volume v let there be a closed contour Γ, which is reducible to a point by continuous transformation so long as it does not intersect the bounding surface of the volume. Let us consider for example a region bounded both inside and out by spherical surfaces. On the contour let there be constructed a surface o ("a cap") which is contained entirely within v.

Let $\underline{a}$ be a vector field which, together with its first derivatives, is continuous in v. Consider the circulation (a line integral) evaluated over Γ

$$C = \oint_\Gamma d\underline{r} \cdot \underline{a}$$

($d\underline{r}$ denotes a vector element on Γ and $|d\underline{r}| = ds$). We assume that the motion along the contour takes place in a positive direction around the given direction of the normal to the "cap's" surface.*)

Stokes' transformation establishes the equality

*) In a right-handed system a motion is deemed positive if it is counterclockwise as seen by an observer situated at the end of $\underline{n}$ and negative if it is clockwise.

$$\oint_{\Gamma} d\underline{r} \cdot \underline{a} = \oint_{\Gamma} \underline{a} \cdot d\underline{r} = \iint_{O} \underline{n} \cdot (\underline{\nabla} \times \underline{a}) do = \iint_{O} (\underline{n} \times \underline{\nabla}) \cdot \underline{a}\, do \tag{8}$$

i.e. the circulation of a vector equals the flux of its curl across the surface of the "cap".

When considering the circulation of a tensor we must distinguish between the left and right products with $d\underline{r}$

$$\oint_{\Gamma} d\underline{r} \cdot \underline{Q}\,, \qquad \oint \underline{Q} \cdot d\underline{r} = \oint d\underline{r} \cdot \underline{Q}^{T}$$

the Stokes' transformation now takes the form

$$\oint_{\Gamma} d\underline{r} \cdot \underline{Q} = \iint_{O} (\underline{n} \times \underline{\nabla}) \cdot \underline{Q}\, dv\,, \qquad \oint_{\Gamma} \underline{Q} \cdot d\underline{r} = \iint_{O} (\underline{n} \cdot \underline{\nabla}) \cdot \underline{Q}^{T} do\,. \tag{9}$$

These expressions give the rule for the substitution of the vector $d\underline{r}$ in the line integral at the left-hand side for the vector $\underline{n} \times \underline{\nabla}\, do$ in the surface integral. This rule is a consequence of the equality

$$\oint_{\Gamma} da^{s} \varphi(a^{1},a^{2},a^{3}) = e^{smn} \iint_{O} n_{m} \frac{\partial \varphi}{\partial a^{n}} do\,, \tag{10}$$

which is the basis for the derivation of Stokes' transformation. We have for example

$$\oint_{\Gamma} \underline{i}_{s} da^{s} \varphi = \oint_{\Gamma} d\underline{r}\, \varphi = \underline{i}_{s} e^{smn} \iint_{O} n_{m} \frac{\partial \varphi}{\partial a^{n}} do = \iint_{O} \underline{i}_{m} n_{m} \times \underline{i}_{n} \frac{\partial \varphi}{\partial a^{n}} do = \iint_{O} \underline{n} \times \underline{\nabla} \varphi\, do\,,$$

which is the desired result. Application of this rule enables Equations (8) and (9) to be augmented by the relationships

$$\oint_{\Gamma} d\underline{r}\,\underline{a} = \iint_{O} (\underline{n} \times \underline{\nabla}) \underline{a}\, do\,, \qquad \oint_{\Gamma} d\underline{r} \times \underline{a} = \iint_{O} (\underline{n} \times \underline{\nabla}) \times \underline{a}\, do = \iint_{O} (\underline{n} \cdot \underline{\nabla}\underline{a}^{T} - \underline{n}\underline{\nabla} \cdot \underline{a}) do\,.$$

For instance, as is to be expected

$$\frac{1}{2} \oint_{\Gamma} \underline{r} \times d\underline{r} = -\frac{1}{2} \oint_{\Gamma} d\underline{r} \times \underline{r} = \frac{1}{2} \iint_{O} (\underline{n}\underline{\nabla} \cdot \underline{r} - \underline{n} \cdot \underline{\nabla}\underline{r}^{T}) do =$$

$$= \frac{1}{2} \iint_{O} (3\underline{n} - \underline{n}) do = \iint_{O} \underline{n}\, do\,.$$

The surface of the "cap" is assigned arbitrarily. Therefore, if the circulation equals zero,

$$\oint_{\Gamma} d\underline{r} \cdot \underline{a} = 0 : \underline{\nabla} \times \underline{a} = \underline{0} , \quad \underline{a} = \underline{\nabla}\varphi , \quad \int_{M_0}^{M} d\underline{r} \cdot \underline{a} = \int_{M_0}^{M} d\underline{r} \cdot \underline{\nabla}\varphi = \int_{M_0}^{M} d\varphi = $$

$$= \varphi_M - \varphi_{M_0} \qquad (12)$$

then the line integral along the curve does not depend on the curve itself but is determined only by the coordinates of its initial and end points, i.e. by the potential difference between these points.

The analogous relationships for a tensor of the second rank have the form

$$\oint_{\Gamma} d\underline{r} \cdot \underline{Q} = \underline{0} , \quad \underline{\nabla} \times \underline{Q} = \underline{0} , \quad \underline{Q} = \underline{\nabla a} , \quad \int_{M_0}^{M} d\underline{r} \cdot \underline{Q} = \int_{M_0}^{M} d\underline{r} \cdot \underline{\nabla a} = \int_{M_0}^{M} d\underline{a} = $$

$$= \underline{a}_M - \underline{a}_{M_0} , \qquad (13)$$

$$\oint_{\Gamma} \underline{Q} \cdot \underline{r} = \underline{0} , \quad \underline{\nabla} \times \underline{Q}^T = \underline{0} , \quad \underline{Q} = \underline{\nabla a}^T , \quad \int_{M_0}^{M} \underline{Q} \cdot d\underline{r} = \int_{M_0}^{M} \underline{\nabla a}^T \cdot d\underline{r} = \int_{M_0}^{M} d\underline{a} = $$

$$= \underline{a}_M - \underline{a}_{M_0} . \qquad (14)$$

3. The doubly connected volume. In a doubly connected volume there are closed contours (which we will call K-contours) which can be reduced to each other by continuous transformation but cannot be reduced to a point. Such a contour is the axis of a torus as also is the contour between two coaxial cylinders which encloses the inner cylindrical surface. A doubly connected volume can be transformed into a simply connected volume with the aid of a barrier without loss of connectedness, i.e. the volume does not split into two pieces. For example, for coaxial cylinders the barrier may be a plane passing through their common axis.

Assume that, just as for (12), the condition of integrability $\underline{a} = \underline{\nabla}$ is fulfilled. Stokes' theorem cannot, however, be applied to a K-contour since one cannot construct a "cap" o lying entirely in the volume v and therefore there is no reason to regard the circulation over a K-contour as zero, i.e. the function is not a single valued function of the coordinates

$$\underline{a} = \underline{\nabla}\varphi , \qquad \oint_K d\underline{r} \cdot \underline{a} = \oint_K d\underline{r} \cdot \underline{\nabla}\varphi = \oint_K d\varphi = \chi . \qquad (15)$$

The "cyclic constant" χ is the same for all K-contours. This is confirmed by consideration of an integral evaluated over the contour $N_1M_1M_2N_2M_2M_1N_1$ which is reducible to a point and which consists of two contours $K_1(N_1M_1N_1)$ and

$K_2(M_2N_2M_2)$ and the barrier M_1M_2. Stokes' theorem can be applied to this contour, and the condition that the circulation be zero for this is expressed by the equality

$$\int_{N_1}^{M_1} d\underline{r}\cdot\underline{a} + \int_{M_1}^{M_2} d\underline{r}\cdot\underline{a} + \oint_{K_2} d\underline{r}\cdot\underline{a} + \int_{M_2}^{M_1} d\underline{r}\cdot\underline{a} + \oint_{K_1} d\underline{r}\cdot\underline{a} + \int_{M_1}^{N_1} d\underline{r}\cdot\underline{a} = 0$$

after cancelling similar terms we obtain

$$\oint_{K_2} d\underline{r}\cdot\underline{a} + \oint_{K_1} d\underline{r}\cdot\underline{a} = 0\,, \qquad \oint_{K_1} d\underline{r}\cdot\underline{a} = \oint_{K_2} d\underline{r}\cdot\underline{a}\,,$$

as required. Of course, the same also holds for circulation of a tensor

$$\underline{\underline{Q}} = \underline{\underline{\nabla a}}\,, \qquad \oint_K d\underline{r}\cdot\underline{\underline{Q}} = \oint_K d\underline{r}\cdot\underline{\underline{\nabla a}} = \oint_K d\underline{a} = \underline{b}\,,$$

$$\underline{\underline{Q}} = \underline{\underline{\nabla a}}^T\,, \qquad \int_K \underline{\underline{Q}}\cdot d\underline{r} = \oint_K \underline{\underline{\nabla a}}^T\cdot d\underline{r} = \oint_K d\underline{a} = \underline{c}\,.$$

9. THE EXPRESSION OF A VECTOR IN TERMS OF A LINEAR STRAIN TENSOR

In the Helmholtz equation (2.13)

$$d\underline{a} = (\underline{\varepsilon} + \underline{\Omega})\cdot d\underline{r} = \underline{\varepsilon}\cdot d\underline{r} + \underline{\omega}\times d\underline{r} \tag{1}$$

the first term in the right-hand side is assumed known and the second must be determined from the integrability condition, i.e. from the existence of a vector $\underline{a}$ for which the prescribed tensor $\underline{\varepsilon}$ is a linear strain tensor. The role of $\underline{\underline{Q}}$ in Equation (8.14) is assumed by the tensor $\underline{\varepsilon} + \underline{\Omega}$ and the integrability condition takes the form

$$\underline{\nabla}\times(\underline{\varepsilon} + \underline{\Omega})^T = \underline{\nabla}\times\underline{\varepsilon} - \underline{\nabla}\times\underline{\Omega} = \underline{0}\,, \tag{2}$$

this leading by (2.18) to the relationship

$$\underline{\nabla}\times\underline{\varepsilon} = \underline{\underline{\nabla\omega}}^T - \underline{\underline{E\nabla}}\cdot\underline{\omega}\,. \tag{3}$$

But by (6.5) and (2.19)

$$I_1(\underline{\nabla}\times\underline{\varepsilon}) = -2\underline{\nabla}\cdot\underline{\omega} = 0\,, \qquad \underline{\nabla}\cdot\underline{\omega} = 0\,, \qquad \underline{\nabla}\times\underline{\varepsilon} = \underline{\underline{\nabla\omega}}^T\,, \tag{4}$$

so that

$$(\underline{\nabla}\times\underline{\varepsilon})\cdot d\underline{r} = \underline{\underline{\nabla\omega}}^T\cdot d\underline{r} = d\underline{\omega}\,. \tag{5}$$

Again, by (8.14) and from the definition (6.23) of an incompatibility tensor, the condition of integrability for this expression takes the form

$$\underline{\nabla} \times (\underline{\nabla} \times \underline{\varepsilon})^{\mathrm{T}} = \mathrm{Ink}\,\underline{\varepsilon} = \underline{0} \qquad (6)$$

the specified tensor $\underline{\varepsilon}$ should satisfy this condition, i.e. the six Saint-Venant continuity equations.

Returning to (5) we obtain

$$\underline{\omega} = \underline{\omega}_0 + \int_{M_0}^{M} (\underline{\nabla} \times \underline{\varepsilon}) \cdot d\underline{r} \qquad (7)$$

and the Helmholtz equation (1) takes the form

$$d\underline{a} = \underline{\omega}_0 \times d\underline{r} - d\underline{r} \times \int_{M_0}^{M} (\underline{\nabla} \times \underline{\varepsilon}) \cdot d\underline{r} + \underline{\varepsilon} \cdot d\underline{r} \ . \qquad (8)$$

Subsequent integration yields

$$\underline{a} = \underline{a}_0 + \underline{\omega}_0 \times (\underline{r} \quad \underline{r}_0) - \int_{M_0}^{M(s)} d\underline{r}(\sigma) \times \int_{M_0}^{M(\sigma)} (\underline{\nabla} \times \underline{\varepsilon}(\sigma')) \cdot d\underline{r}(\sigma') + \int_{M_0}^{M(s)} \underline{\varepsilon}(\sigma) \cdot d\underline{r}(\sigma) \ . \qquad (9)$$

Here $\underline{a}_0$, $\underline{\omega}_0$, $\underline{r}_0$ denote the constant values of vectors $\underline{a}$, $\underline{\omega}$, $\underline{r}$ at the initial point $M_0(s_0)$ of the path of integration, $\underline{r} = \underline{r}(s)$ is the position vector of the point $M(s)$ and σ, σ' are the variables of integration along the paths (M_0M) and $(M_0, M(\sigma))$.

Changing the order of integration using

$$\int_0^a dx \int_0^x f(x,y)\,dy = \int_0^a dy \int_y^a f(x,y)\,dx \ ,$$

we get

$$\int_{M_0}^{M} d\underline{r}(\sigma) \times \int_{M_0}^{M(\sigma)} (\underline{\nabla} \times \underline{\varepsilon}(\sigma')) \cdot d\underline{r}(\sigma') = - \int_{M_0}^{M(s)} (\underline{\nabla} \times \underline{\varepsilon}(\sigma')) \cdot d\underline{r}(\sigma') \times$$

$$\times \int_{M(\sigma')}^{M} d\underline{r}(\sigma) = \int_{M_0}^{M} [\underline{r}(s) - \underline{r}(\sigma')] \times (\underline{\nabla} \times \underline{\varepsilon}(\sigma')) \cdot d\underline{r}(\sigma') \ .$$

Expression (9) is now transformed to

$$\underline{a}(s) = \underline{a}_0 + \underline{\omega}_0 \times (\underline{r}(s) - \underline{r}_0) + \int_{M_0}^{M(s)} \{\underline{\varepsilon}(\sigma) + [\underline{r}(\sigma) - \underline{r}(s)] \times (\underline{\nabla} \times \underline{\varepsilon}(\sigma))\} \cdot d\underline{r}(\sigma) \tag{10}$$

this is Cesaro's formula which defines a vector by its linear strain tensor.

It remains only to verify that the integral in (10) is independent of the path of integration. Denoting the tensor under the integral sign by $\underline{\Pi}$, we have, referring to (I.14.5),

$$\underline{\Pi}^{\mathrm{T}} = \{\underline{\varepsilon}(\sigma) + [\underline{r}(\sigma) - \underline{r}(s)] \times (\underline{\nabla} \times \underline{\varepsilon}(\sigma)\}^{\mathrm{T}} = \underline{\varepsilon}(\sigma) - (\underline{\nabla} \times \underline{\varepsilon}(\sigma))^{\mathrm{T}} \times [\underline{r}(\sigma) - \underline{r}(s)]$$

and by (8.14) it may be verified that the curl of this tensor equals zero. Setting $\underline{Q} = (\underline{\nabla} \times \underline{\varepsilon})^{\mathrm{T}}$ in Equation (3.12), we have, referring also to (I.14.7)

$$\underline{\nabla} \times \{[\underline{\nabla} \times \underline{\varepsilon}(\sigma)]^{\mathrm{T}} \times [\underline{r}(\sigma) - \underline{r}(s)]\} = \{\underline{\nabla} \times [\underline{\nabla} \times \underline{\varepsilon}(\sigma)]^{\mathrm{T}}\} \times [\underline{r}(\sigma) - \underline{r}(s)] +$$

$$+ [\nabla \times \varepsilon(\sigma)]^{\mathrm{TT}} - \underline{E} I_1(\underline{\nabla} \times \underline{\varepsilon}(\sigma)) = \mathrm{Ink}\,\underline{\varepsilon}(\sigma) \times [\underline{r}(\sigma) - \underline{r}(s)] + \underline{\nabla} \times \underline{\varepsilon}(\sigma) \;.$$

The condition

$$\underline{\nabla} \times \underline{\Pi}^{\mathrm{T}} = \underline{\nabla} \times \underline{\varepsilon}(\sigma) - \mathrm{Ink}\,\underline{\varepsilon}(\sigma) \times [\underline{r}(\sigma) - \underline{r}(s)] - \underline{\nabla} \times \underline{\varepsilon}(\sigma) = \underline{0}$$

thus amounts to the same requirement of (6) which is now fulfilled because the vector $\underline{r}(\sigma) - \underline{r}(s)$ is arbitrary.

10. THE RIEMANN-CHRISTOFFEL TENSOR AND THE RICCI TENSOR

The square of a linear element in Euclidean space can be represented in the Pythagorean form

$$ds^2 = (da^1)^2 + (da^2)^2 + (da^3)^2 \;. \tag{1}$$

Introducing the curvilinear coordinates

$$a^k = a^k(q^1, q^2, q^3) \qquad (k = 1,2,3) \;, \tag{2}$$

this takes the form of a quadratic

$$ds^2 = g_{sk}\,dq^s\,dq^k \;, \tag{3}$$

its coefficients being the covariant components of a metric (unit) tensor $\underline{E}$ determined by mapping (2) with the aid of the formulas

$$g^{sk} = \sum_{m=1}^{3} \frac{\partial a^m}{\partial q^s} \frac{\partial a^m}{\partial q^k} \qquad (s,k = 1,2,3) \ . \tag{4}$$

We may state the problem differently. We assume that the quadratic form (3) is specified by its coefficients $g_{sk}(q^1,q^2,q^3)$ and that this form is positive definite. With certain reservations, it defines a Riemannian space R_3. The mapping (2) is itself not specified, its determination is reduced to the problem of the integration of six differential equations (4) with three unknowns a^1, a^2, a^3. This mapping exists only provided that the conditions of integrability of this system are fulfilled. If g_{sk} are so assigned that these conditions are satisfied then the Riemannian space R_3 becomes Euclidean and in this space, the position of a point may be determined in a unique Cartesian system of axes and the square of a linear element can be reduced to the Pythagorean form (1).

Using a linear transformation, the positive definite form (3) may be transformed to a sum of squares

$$ds^2 = b_1(z^1,z^2,z^3)dz^{1^2} + b_2(z^1,z^2,z^3)dz^{2^2} + b_3(z^1,z^2,z^3)dz^{3^2} \ , \tag{5}$$

where z^k denote the new variables, and the quantities b_k are positive. The quantity $\sqrt{b_k(z^1,z^2,z^3)}dz^k$ is not a differential of any particular function, and only by specifying particular values of z^1, z^2, z^3 may we write $da^k =$ $= \sqrt{b_k(z^1,z^2,z^3)}dz^k$ and in this way define, in the neighbourhood of a point in R_3, a Cartesian system of axes, i.e. the Pythagorean line element. In other words, it is possible to insert the metric of E_3 into a space R_3 locally. The fulfillment of the integrability conditions causes the degeneration of R_3 to E_3 throughout the entire range of definition of functions $g_{sk}(q^1,q^2,q^3)$, not only in the "space E_3 tangential to R_3".

These conditions of integrability thus ensure the existence of three functions (2) or, equivalently, they amount to the possibility of the specification of position of any point by its radius vector

$$\underline{r} = \underline{r}(q^1,q^2,q^3) \tag{6}$$

in a unique Cartesian system with axes $OXYZ$. This specification enables us to determine the base vectors $\underline{r}_s$ and, by virtue of (4.2), their derivatives

$$\frac{\partial \underline{r}_s}{\partial q^t} = \left\{ {m \atop st} \right\} \underline{r}_m \ . \tag{7}$$

The coefficients of this system of differential equations are the Christoffel symbols evaluated for a specified quadratic form (4) by the rules (4.5), (4.6).

The relationships

$$\frac{\partial \underline{r}_s}{\partial q^t} = \frac{\partial \underline{r}_t}{\partial q^s}, \tag{8}$$

are the consequence of the symmetry of these coefficients with respect to the indices s, t. They guarantee the integrability of the expression

$$d\underline{r} = \underline{r}_s dq^s, \tag{9}$$

in other words, the existence of the radius vector (6). The problem now reduces to that of the existence of vectors $\underline{r}_s$ and it is necessary to determine what conditions should be imposed on the coefficients of (7) (essentially, on the specified values (4) of g_{sk}) in order that this system be integrable. The forms

$$d\underline{r}_s = \left\{ {m \atop st} \right\} \underline{r}_m dq^t \tag{10}$$

must be perfect differentials, i.e. the derivative of the coefficient of dq^t with respect to q^k must be equal to that of dq^k with respect to q^t

$$\frac{\partial}{\partial q^k} \left\{ {m \atop st} \right\} \underline{r}_m = \frac{\partial}{\partial q^t} \left\{ {m \atop sk} \right\} \underline{r}_m ;$$

$$\underline{r}_m \left(\frac{\partial}{\partial q^k} \left\{ {m \atop st} \right\} + \left\{ {q \atop st} \right\} \left\{ {m \atop qk} \right\} \right) = \underline{r}_m \left(\frac{\partial}{\partial q^t} \left\{ {m \atop sk} \right\} + \left\{ {q \atop sk} \right\} \left\{ {m \atop qt} \right\} \right) .$$

The integrability conditions may be written

$$R^m_{kts.} = \frac{\partial}{\partial q^k} \left\{ {m \atop st} \right\} - \frac{\partial}{\partial q^t} \left\{ {m \atop sk} \right\} + \left\{ {q \atop st} \right\} \left\{ {m \atop qk} \right\} - \left\{ {q \atop sk} \right\} \left\{ {m \atop qt} \right\} = 0 . \tag{11}$$

The tensor character of these quantities is found by examining the difference between the components of two tensors of the third rank, i.e. those of $\nabla_k \nabla_t a^q$ and $\nabla_t \nabla_k a^q$ which are tensors representing the second covariant derivatives of the contravariant components of the vector $\underline{a}$. We have

$$\nabla_t a^q = \frac{\partial a^q}{\partial q^t} + \left\{ {q \atop ts} \right\} a^s ,$$

$$\nabla_k \nabla_t a^q = \frac{\partial}{\partial q^k} \nabla_t a^q + \left\{ {q \atop km} \right\} \nabla_t a^m - \left\{ {m \atop kt} \right\} \nabla_m a^q =$$

$$= \underline{\frac{\partial^2 a^q}{\partial q^k \partial q^t}} + \underline{\left\{ {q \atop ts} \right\} \frac{\partial a^s}{\partial q^k}} + a^s \frac{\partial}{\partial q^k} \left\{ {q \atop ts} \right\} + \underline{\left\{ {q \atop ks} \right\} \frac{\partial a^s}{\partial q^t}} + \left\{ {q \atop km} \right\} \left\{ {m \atop st} \right\} a^s - \underline{\left\{ {m \atop kt} \right\} \nabla_m a^q} \ .$$

On forming the differences, the underlined terms cancel out and we have

$$\nabla_k \nabla_t a^q - \nabla_t \nabla_k a^q = a^s \left[\frac{\partial}{\partial q^k} \left\{ {q \atop ts} \right\} - \frac{\partial}{\partial q^t} \left\{ {q \atop ks} \right\} + \left\{ {q \atop km} \right\} \left\{ {m \atop st} \right\} - \left\{ {q \atop tm} \right\} \left\{ {m \atop sk} \right\} \right] =$$

$$= R_{kts.}{}^q a^s \ . \qquad (12)$$

The widely used rule for the interchangeability of the indices of covariant differentiation for operations in E_3, follows from (11) and (12), i.e. we have

$$(\nabla_k \nabla_t - \nabla_t \nabla_k) a^q = 0 \ . \qquad (13)$$

Though proved for vectors, this rule holds equally for tensors of arbitrary rank but the above calculation becomes somewhat more complex.

According to (12), a contraction of the quantities $R_{kts.}{}^q$ with a vector results in a tensor of the third rank. These quantities therefore represent a tensor of the fourth rank - the Riemann-Christoffel tensor, also called the tensor of curvature in R_3

$$\overset{4}{\underline{\mathrm{R}}} = \underline{\mathrm{r}}^k \underline{\mathrm{r}}^t \underline{\mathrm{r}}^s \underline{\mathrm{r}}_m R_{kts.}{}^m = \underline{\mathrm{r}}^k \underline{\mathrm{r}}^t \underline{\mathrm{r}}^s \underline{\mathrm{r}}^r R_{ktsr} \ . \qquad (14)$$

In E_3, this is a zero tensor. The fourth order covariant components of $\overset{4}{\underline{\mathrm{R}}}$ are, from (14) and (11), determined by the expressions

$$R_{ktsm} = g_{mr} R_{kts.}{}^m = g_{mr} \left(\frac{\partial}{\partial q^k} g^{mq} [st,q] - \frac{\partial}{\partial q^t} g^{mq} [sk,q] \right) +$$

$$+ g^{qm} ([kq,r][st,m] - [tq,r][sk,m])$$

in which we have also used Equations (4.6). Using the relationships

$$g_{mr} \frac{\partial}{\partial q^k} g^{mq} = -g^{mq} \frac{\partial}{\partial q^k} g_{mr} = -g^{mq} ([rk,m] + [mk,r]) \ ,$$

we arrive at the expressions

$$R_{ktsr} = \frac{\partial}{\partial q^k} [st,r] - \frac{\partial}{\partial q^t} [sk,r] - g^{mq} \{ ([rk,m] + [mk,r]) [st,q] - ([rt,m] +$$

$$+ [mt,r]) [sk,q] - [km,r][st,q] + [tm,r][sk,q] \} \ .$$

Expanding the Christoffel symbols of the first kind (4.5) we arrive at the following representation of the covariant components of the tensor of curvature

$$R_{ktsr} = \frac{1}{2}\left(\frac{\partial^2 g_{tr}}{\partial q^k \partial q^s} - \frac{\partial^2 g_{kr}}{\partial q^t \partial q^s} + \frac{\partial^2 g_{sk}}{\partial q^t \partial q^r} - \frac{\partial^2 g_{st}}{\partial q^k \partial q^r}\right) + g^{mq}([sk,q][rt,m] - [st,q] \times$$

$$\times [rk,m]) = \frac{1}{2}\left(\frac{\partial^2 g_{tr}}{\partial q^k \partial q^s} - \frac{\partial^2 g_{kr}}{\partial q^t \partial q^s} + \frac{\partial^2 g_{sk}}{\partial q^t \partial q^r} - \frac{\partial^2 g_{st}}{\partial q^k \partial q^r}\right) + \left\{ {q \atop rt} \right\}[sk,q] - \left\{ {q \atop rk} \right\}[st,q] \, . \tag{15}$$

Their structure if symmetric with respect to the pairs of indices (kt) and (sr)

$$R_{ktsr} = R_{srkt} \tag{16}$$

and skew-symmetric with respect to the indices k, t and s, r

$$R_{ktrs} = -R_{tkrs} \, , \qquad R_{ktsr} = -R_{ktrs} \, . \tag{17}$$

For this reason, the pairs of indices kt, sr should be ascribed the values 12, 23, 31 only, and rearranging these pairs, one has to consider only the six components

$$R_{1212} \, , \; R_{1223} \, , \; R_{1231} \, , \; R_{2323} \, , \; R_{2331} \, , \; R_{3131} \tag{18}$$

of the total of 81. The other components are either zero or are expressible in terms of those already given.

The Ricci identities hold here and having fixed the fourth index and carried out a cyclic permutation of the other three, we get

$$R_{ktsr} + R_{tskr} + R_{sktr} = 0 \, . \tag{19}$$

This follows because one of the three indices kts is certain to be equal to r. Indeed, assuming that $k = r$ we have

$$R_{ktsk} + R_{tskk} + R_{sktk} = 0 \, ,$$

because the second term and the sum of the other two are zero.

The quantities $g_{sk} = \underline{r}_s \cdot \underline{r}_k$ are evaluated at those values of the position vector given by (6) and which make the tensor of curvature zero identically. If, on the other hand, we start from the specified positive symmetric matrix $\| g_{sk} \|$ and calculate the inverse matrix $\| g^{sk} \|$ and the quantities (18), which all turn out to be zero, then it would indicate that the quadratic form (3) is reducible to the Pythagorean form (1), and that g_{sk} represent the covariant components of a metric tensor. In the opposite case, ds^2 is the square of a linear element in R_3.

Using tensor $^4\underline{R}$ we may construct a tensor $\underline{R}$ of the second rank by substituting the dyads $\underline{r}^k\underline{r}^t$, $\underline{r}^s\underline{r}^r$ in (14) by the vector products $\underline{r}^k \times \underline{r}^t$, $\underline{r}^s \times \underline{r}^r$

$$\underline{R} = \frac{1}{4}\underline{r}^k \times \underline{r}^t\underline{r}^s \times \underline{r}^r R_{ktsr} = \frac{1}{4}\epsilon^{ktm}\epsilon^{srn}\underline{r}_m\underline{r}_n R_{ktsr} \tag{20}$$

(where the factor 1/4 has been introduced purely for convenience). By (14), this tensor, called the Ricci tensor, is symmetric. Its contravariant components are

$$\begin{aligned} &R^{11} = \frac{1}{g}R_{2323}\,, \quad R^{12} = \frac{1}{g}R_{2331}\,, \quad R^{13} = \frac{1}{g}R_{2312}\,,\\ &R^{22} = \frac{1}{g}R_{3131}\,, \quad R^{23} = \frac{1}{g}R_{3112}\,,\\ &R^{33} = \frac{1}{g}R_{1212}\,. \end{aligned} \tag{21}$$

Of course, the Ricci tensor vanishes together with the tensor of curvature.

In the orthogonal curvilinear coordinates, relationships (10) are substituted for the conditions

$$d\underline{e}_s = \frac{\partial \underline{e}_s}{\partial q^k}dq^k \qquad (s = 1,2,3) \tag{22}$$

and the requirement of integrability, i.e. of the existence of an orthonormal base $\underline{e}_1$, $\underline{e}_2$, $\underline{e}_3$, is reduced to that of (7.22). Eliminating the vectors $\overset{t}{\underline{o}}$ with the aid of (7.16) we arrive at the six Lamé equations

$$\begin{aligned} &\frac{\partial}{\partial q^1}\frac{1}{H_1}\frac{\partial H_2}{\partial q^1} + \frac{\partial}{\partial q^2}\frac{1}{H_2}\frac{\partial H_1}{\partial q^3}\frac{\partial H_2}{\partial q^3} = 0\,,\\ &\frac{\partial^2 H_1}{\partial q^2\partial q^3} - \frac{1}{H_3}\frac{\partial H_3}{\partial q^2}\frac{\partial H_1}{\partial q^3} - \frac{1}{H_2}\frac{\partial H_1}{\partial q^2}\frac{\partial H_2}{\partial q^3} = 0 \end{aligned} \tag{23}$$

(the remaining Equations are obtained by cyclic permutation of the indices). The fulfillment of these relationships guarantees the reducibility of a square of a linear element (7.9) to a Pythagorean form. Equations (23) represent the requirement (expressed in orthogonal coordinates) that the Ricci tensor be zero in E_3.

11. NOTES ON THE THEORY OF SURFACES

The following data are employed in the main text. The contents of Par. 11 cannot, of course, replace any of the many monographs on the theory of surfaces.

1. The first and the second quadratic forms of a surface. A surface is defined by the prescription of its radius vector $\underline{\rho}$ as a function of Gaussian coordinates q^1, q^2

$$\underline{\rho} = \underline{\rho}(q^1, q^2) \ . \tag{1}$$

The square of a linear element $d\sigma^2$ on the surface is defined as a positive definite quadratic form (the first quadratic form)

$$I = d\sigma^2 = d\underline{\rho} \cdot d\underline{\rho} = \underline{\rho}_\alpha dq^\alpha \cdot \underline{\rho}_\alpha dq^\beta = a_{\alpha\beta} dq^\alpha dq^\beta \ , \qquad a_{\alpha\beta} = \underline{\rho}_\alpha \cdot \underline{\rho}_\beta \tag{2}$$

(the Greek indices take the values 1, 2). Here

$$\underline{\rho}_a = \frac{\partial \underline{\rho}}{\partial q^\alpha} \qquad (\alpha = 1,2) \tag{3}$$

are the base vectors of the surface. The matrix $\| a_{\alpha\beta} \|$ has a corresponding inverse $\| a^{\alpha\beta} \|$

$$a^{11} = \frac{a_{22}}{a} \ , \quad a^{22} = \frac{a_{11}}{a} \ , \quad a^{12} = -\frac{a_{12}}{a} \ , \quad a = |a_{\alpha\beta}| = a_{11} a_{22} - a_{12}^2 \ ,$$

$$a^{\alpha\beta} a_{\beta\gamma} = \delta^\alpha_\gamma \ . \tag{4}$$

In this way we determine the metric tensor on the surface

$$\underline{E}_2 = a_{\alpha\beta} \underline{\rho}^\alpha \underline{\rho}^\beta = a^{\alpha\beta} \underline{\rho}_\alpha \underline{\rho}_\beta = \underline{\rho}_\alpha \underline{\rho}^\alpha \ , \qquad \underline{\rho}^\alpha = a^{\alpha\beta} \underline{\rho}_\beta \ , \qquad \underline{\rho}^\alpha \cdot \underline{\rho}_\beta = \delta^\alpha_\beta \ . \tag{5}$$

Here $\underline{\rho}^\alpha$ denote these vectors of the dual basis which also belong to the surface.

The surface represents a Riemannian manifold R_2. Its geometry may be constructed on the basis of the quadratic form (2), i.e. of its coefficients $a_{\alpha\beta}(q^1, q^2)$. These coefficients may be determined by measurements of the lengths of the segments carried out by beings confined to a surface and which are totally unaware of the third dimension. We leave the surface as soon as we introduce the unit vector of its normal

$$\underline{n} = \frac{\underline{\rho}_1 \times \underline{\rho}_2}{|\underline{\rho}_1 \times \underline{\rho}_2|} = \frac{1}{\sqrt{a}} \underline{\rho}_1 \times \underline{\rho}_2 \ . \tag{6}$$

Differentiating the relationships $\underline{\rho}_\alpha \cdot \underline{n} = 0$ we arrive at the formulas

$$\underline{\rho}_{\alpha\beta} \cdot \underline{n} = -\underline{\rho}_\alpha \cdot \underline{n}_\beta = b_{\alpha\beta}\,, \qquad \underline{\rho}_{\alpha\beta} \cdot \underline{n} = -\underline{\rho}_\beta \cdot \underline{n}_\alpha\,, \qquad b_{\alpha\beta} = b_{\beta\alpha}\,, \tag{7}$$

which determine the quantities $b_{\alpha\beta} = b_{\beta\alpha}$. From these we construct the second quadratic form of a surface

$$\mathrm{II} = b_{\alpha\beta} dq^\alpha dq^\beta\,, \tag{8}$$

which takes the form

$$\mathrm{II} = b_{\alpha\beta} \frac{\partial q^\alpha}{\partial \check{q}^\gamma} \frac{\partial q^\beta}{\partial \check{q}^\delta} d\check{q}^\gamma d\check{q}^\delta = \check{b}_{\gamma\delta} d\check{q}^\gamma d\check{q}^\delta\,, \qquad \check{b}_{\gamma\delta} = b_{\alpha\beta} \frac{\partial q^\alpha}{\partial \check{q}^\gamma} \frac{\partial q^\beta}{\partial \check{q}^\delta} \tag{9}$$

on transferring to the new variables $q^\alpha = q^\alpha(\check{q}^1, \check{q}^2)$. This means that $b_{\alpha\beta}$ are components of a symmetric tensor in two dimensions

$$\underline{B} = b_{\alpha\beta} \underline{\rho}^\alpha \underline{\rho}^\beta = b^\beta_\alpha \underline{\rho}^\alpha \underline{\rho}_\beta = b^{\alpha\beta} \underline{\rho}_\alpha \underline{\rho}_\beta\,, \qquad b^\beta_\alpha = a^{\beta\gamma} b_{\alpha\gamma}\,, \tag{10}$$

and the vectors $\underline{n}_\alpha$, $\underline{n}_\beta$ can be represented by the formulas

$$\underline{n}_\alpha = -b^\beta_\alpha \underline{\rho}_\beta\,, \qquad \underline{n}_\beta = -b^\alpha_\beta \underline{\rho}_\alpha\,. \tag{11}$$

The invariants of the tensor $\underline{B}$ define the quantities

$$I_1(\underline{B}) = b^1_1 + b^2_2 = 2H\,, \qquad I_2(\underline{B}) = b^1_1 b^2_2 - b^2_1 b^1_2 = K = \frac{1}{a}(b_{11} b_{22} - b^2_{12}) \tag{12}$$

which are termed the mean (H) and Gaussian (K) curvatures of a surface.

2. The derivation formulas. The second derivatives $\underline{\rho}_{\alpha\beta}$ of a radius vector of a surface are not the surface vectors themselves. It can be seen from (7) that their normal components are equal to $b_{\alpha\beta}$. Following the definition of coefficients $a_{\alpha\beta} = \underline{\rho}_\alpha \cdot \underline{\rho}_\beta$ of the first quadratic form and reproducing in detail the derivation of Equations (4.5) we arrive at the equalities

$$\underline{\rho}_{\alpha\beta} \cdot \underline{\rho}_\gamma = \frac{1}{2}\left(\frac{\partial a_{\alpha\gamma}}{\partial q^\beta} + \frac{\partial a_{\beta\gamma}}{\partial q^\alpha} - \frac{\partial a_{\alpha\beta}}{\partial q^\gamma} \right) = [\alpha\beta, \gamma] = a_{\gamma\delta} \left\{ {\delta \atop \alpha\beta} \right\}\,. \tag{13}$$

Here we introduced the surface Christoffel symbols.

The differentiation formulas for the basic vectors are put into the form

$$\underline{\rho}_{\alpha\beta} = \left\{ {\gamma \atop \alpha\beta} \right\} \underline{\rho}^{\gamma} + b_{\alpha\beta}\underline{n} \tag{14}$$

(these are the derivation formulas of Weingarten). For vectors of the dual basis, these formulas take the form

$$\frac{\partial \underline{\rho}^{\alpha}}{\partial q^{\beta}} = - \left\{ {\alpha \atop \beta\gamma} \right\} \underline{\rho}^{\gamma} + b^{\alpha}_{\beta}\underline{n} \ , \tag{15}$$

which follows from Equations (4.7) and the relationships $\underline{\rho}_{\alpha} \cdot \underline{\rho}^{\beta} = \delta^{\beta}_{\alpha}$, $a^{\alpha\gamma} b_{\gamma\delta} = \delta^{\alpha}_{\delta}$.

3. Geometry in the neighbourhood of a surface. The radius vector of a point in Euclidean space E_3 in the neighbourhood of a surface is represented by the relationship

$$\underline{r} = \underline{\rho} + \underline{n}\zeta \ . \tag{16}$$

Here $q^3 = \zeta$ denotes the third coordinate measured along the normal to the surface. The base vectors, the covariant components of the metric tensor and determinant of these components are defined by the formulas

$$\underline{r}_{\alpha} = \underline{\rho}_{\alpha} + \zeta\underline{n}_{\alpha} = \underline{\rho}_{\alpha} - \zeta b^{\gamma}_{\alpha}\underline{\rho}_{\gamma} \ , \qquad \underline{r}_3 = \underline{n} \ , \tag{17}$$

$$g_{\alpha\beta} = a_{\alpha\beta} - 2\zeta b_{\alpha\beta} + \zeta^2 b_{\alpha\gamma} b^{\gamma}_{\beta} \ , \qquad g_{\alpha 3} = \underline{\rho}_{\alpha} \cdot \underline{n} = 0 \ , \qquad g_{33} = 1 \ , \tag{18}$$

$$g = g_{11}g_{22} - g_{12}^2 \ , \qquad \sqrt{g} = (\underline{r}_1 \times \underline{r}_2) \cdot \underline{n} = \sqrt{a}\,(1 - 2H\zeta + K\zeta^2) \tag{19}$$

here we have made use of Equations (7), (10), (11) and (12).

The contravariant components of the metric tensor are determined from the expressions

$$g^{\alpha\beta} = a^{\alpha\beta} + 2\zeta b^{\alpha\beta} \ , \tag{20}$$

retaining only terms linear with respect to ζ. This is easily verified by immediate calculation

$$g_{\alpha\beta}g^{\beta\gamma} = (a_{\alpha\beta} - 2\zeta b_{\alpha\beta})(a^{\beta\gamma} + 2\zeta b^{\beta\gamma}) = \delta^{\gamma}_{\alpha} + 2\zeta(a_{\alpha\beta}b^{\beta\gamma} - a^{\beta\gamma}b_{\alpha\beta}) = \delta^{\gamma}_{\alpha} \ ,$$

as required. The vectors of the dual basis to this approximation, the nabla-operator and the metric tensor are now determined by the formulas

$$\underline{r}^{\alpha} = g^{\alpha\beta}\underline{r}_{\beta} = \underline{\rho}^{\alpha} + \zeta b^{\alpha}_{\beta}\underline{\rho}^{\beta} \ , \qquad \underline{r}^{3} = \underline{n} \ , \tag{21}$$

$$\underline{\nabla} = \underline{r}^{\alpha} \frac{\partial}{\partial q^{\alpha}} + \underline{n} \frac{\partial}{\partial \zeta} = (\underline{\rho}^{\alpha} + \zeta b^{\alpha}_{\beta} \underline{\rho}^{\beta}) \frac{\partial}{\partial q^{\alpha}} + \underline{n} \frac{\partial}{\partial \zeta} , \tag{22}$$

$$\underline{E} = \underline{r}_{\alpha} \underline{r}^{\alpha} + \underline{nn} = \underline{\rho}_{\alpha} \underline{\rho}^{\alpha} + \zeta (b^{\alpha}_{\beta} \underline{\rho}_{\alpha} \underline{\rho}^{\beta} + b^{\beta}_{\alpha} \underline{\rho}_{\beta} \underline{\rho}^{\alpha}) + \underline{nn} = \underline{E}_2 + 2\zeta \underline{B} + \underline{nn} . \tag{23}$$

These formulas enable us to formulate rules for differential operations up to terms linear in ζ with respect to tensors defined in E_3 in the neighbourhood of the surface considered. In particular, for $\zeta = 0$ these formulas determine the quantities defined on this surface itself. Let us consider here an example of a tensor $(\underline{\nabla\nabla}\varphi)_{\zeta=0}$. We have

$$(\underline{\nabla\nabla}\varphi)_{\zeta=0} = \left[\left(\underline{r}^{\alpha} \frac{\partial}{\partial q^{\alpha}} + \underline{n} \frac{\partial}{\partial \zeta}\right)\left(\underline{r}^{\beta} \frac{\partial \varphi}{\partial q^{\beta}} + \underline{n} \frac{\partial \varphi}{\partial \zeta}\right)\right]_{\zeta=0} = \left[\underline{r}^{\alpha} \underline{r}^{\beta} \frac{\partial^2 \varphi}{\partial q^{\alpha} \partial q^{\beta}} + \right.$$

$$\left. + (\underline{r}^{\alpha} \underline{n} + \underline{n} \underline{r}^{\alpha}) \frac{\partial^2 \varphi}{\partial \zeta \partial q^{\alpha}} + \underline{nn} \frac{\partial^2 \varphi}{\partial \zeta^2} + \underline{r}^{\alpha} \frac{\partial \underline{r}^{\beta}}{\partial q^{\alpha}} \frac{\partial \varphi}{\partial q^{\beta}} + \underline{r}^{\alpha} \underline{n} \frac{\partial \varphi}{\partial \zeta} + \underline{n} \frac{\partial \underline{r}^{\beta}}{\partial \zeta} \frac{\partial \varphi}{\partial \zeta}\right]_{\zeta=0}$$

and by (17), (21)

$$(\underline{\nabla\nabla}\varphi)_{\zeta=0} = \left[\underline{\rho}^{\alpha} \underline{\rho}^{\beta} \left(\frac{\partial^2 \varphi}{\partial q^{\alpha} \partial q^{\beta}} - \left\{ {\gamma \atop \alpha\beta} \right\} \frac{\partial \varphi}{\partial q^{\gamma}} - b_{\alpha\beta} \frac{\partial \varphi}{\partial \zeta}\right) + \right.$$

$$\left. + (\underline{\rho}^{\alpha} \underline{n} + \underline{n} \underline{\rho}^{\alpha}) \left(\frac{\partial^2 \varphi}{\partial q^{\alpha} \partial \zeta} + b^{\gamma}_{\alpha} \frac{\partial \varphi}{\partial q^{\gamma}}\right) + \underline{nn} \frac{\partial^2 \varphi}{\partial \zeta^2}\right]_{\zeta=0} . \tag{24}$$

Assuming that the functions

$$\varphi , \quad \frac{\partial \varphi}{\partial q^{\alpha}} , \quad \frac{\partial \varphi}{\partial \zeta}$$

are continuously differentiable with respect to coordinates q^{α} on this surface*)

$$\left[\frac{\partial^2 \varphi}{\partial q^{\alpha} \partial q^{\gamma}}\right]_{\zeta=0} = 0 , \quad \left[\frac{\partial^2 \varphi}{\partial q^{\alpha} \partial \zeta}\right]_{\zeta=0} = 0 , \tag{25}$$

we obtain

$$\underline{nn} \left[\frac{\partial^2 \varphi}{\partial \zeta^2}\right]_{\zeta=0} = \underline{nn} (\underline{\nabla\nabla}\varphi)_{\zeta=0} , \quad \left[\frac{\partial^2 \varphi}{\partial \zeta^2}\right]_{\zeta=0} = \underline{n} \cdot [\underline{\nabla\nabla}\varphi]_{\zeta=0} \cdot \underline{n} . \tag{26}$$

---

*) We denote here by brackets the difference between the limiting values for $\zeta \to +0$ and $\zeta \to -0$:

$$[\psi(\zeta)]_{\zeta=0} = \lim_{\zeta \to +0} \psi(\zeta) - \lim_{\zeta \to -0} \psi(\zeta) .$$

4. The Gauss and Codazzi formulas. The second derivatives of a radius vector $\underline{r}$ in the neighbourhood of a surface, evaluated using (17) and (14), are defined by the formulas

$$\underline{r}_{\alpha\beta} = \frac{\partial}{\partial q^\beta}(\underline{\rho}_\alpha - \zeta b^\gamma_\alpha \underline{\rho}_\gamma) = \left\{ {\gamma \atop \alpha\beta} \right\} \underline{\rho}_\gamma + b_{\alpha\beta}\underline{n} - \zeta\left(b^\gamma_\alpha \left\{ {\delta \atop \beta\gamma} \right\} \underline{\rho}_\delta + b^\gamma_\alpha b_{\gamma\beta}\underline{n} + \underline{\rho}_\delta \frac{\partial}{\partial q^\beta} b^\delta_\alpha \right) =$$

$$= \left(\left\{ {\delta \atop \alpha\beta} \right\} - \zeta b^\gamma_\alpha \left\{ {\delta \atop \beta\gamma} \right\} - \zeta \frac{\partial}{\partial q^\beta} b^\delta_\alpha \right) \underline{\rho}_\delta + (b_{\alpha\beta} - \zeta b^\gamma_\alpha b_{\gamma\beta})\underline{n} \,, \tag{27}$$

$$\underline{r}_{3\alpha} = \underline{n}_\alpha = -b^\gamma_\alpha \underline{\rho}_\gamma \,, \qquad \underline{r}_{33} = \underline{0} \,. \tag{28}$$

The use of the formulas (10.15) for the representation of the components of the tensor of curvature (equal to zero in E_3) leads to a cumbersome calculation. Direct analysis of the integrability conditions of the system (10.10) which is reduced here to the form

$$d\underline{r}_\alpha = \underline{r}_{\alpha\beta} dq^\beta + \underline{r}_{\alpha 3} d\zeta \,, \qquad d\underline{r}_3 = \underline{r}_{3\alpha} dq^\alpha + \underline{r}_{33} d\zeta = \underline{n}_\alpha dq^\alpha \tag{29}$$

is preferable.

The second group is plainly integrable $\underline{r}_3 = \underline{n}$. The first group is divided into two systems

$$\text{a)}\ \frac{\partial \underline{r}_{\alpha\beta}}{\partial \zeta} = \frac{\partial \underline{r}_{\alpha 3}}{\partial q^\beta} \,;$$

$$\text{b)}\ \frac{\partial \underline{r}_{\alpha\beta}}{\partial q^\gamma} = \frac{\partial \underline{r}_{\alpha\gamma}}{\partial q^\beta} \quad \text{or} \quad \frac{\partial \underline{r}_{\alpha 1}}{\partial q^2} - \frac{\partial \underline{r}_{\alpha 2}}{\partial q^1} = \underline{0} \,. \tag{30}$$

Equations (30a) are satisfied identically, this being easily verified by Equations (27, (28), (14) and (11). The Equations (30b) will be of interest for us only on the basic surface

$$\left(\frac{\partial \underline{r}_{\alpha 2}}{\partial q^1} - \frac{\partial \underline{r}_{\alpha 1}}{\partial q^2} \right)_{\zeta=0} = \frac{\partial}{\partial q^1}\left(\left\{ {\delta \atop \alpha 2} \right\} \underline{\rho}_\delta + b_{\alpha 2}\underline{n} \right) - \frac{\partial}{\partial q^2}\left(\left\{ {\delta \atop \alpha 1} \right\} \underline{\rho}_\delta + b_{\alpha 1}\underline{n} \right) =$$

$$= \left[\frac{\partial}{\partial q^1} \left\{ {\delta \atop \alpha 2} \right\} - \frac{\partial}{\partial q^2} \left\{ {\delta \atop \alpha 1} \right\} + \left\{ {\gamma \atop \alpha 2} \right\} \left\{ {\delta \atop \gamma 1} \right\} - \left\{ {\gamma \atop \alpha 1} \right\} \left\{ {\delta \atop \gamma 2} \right\} \right] \underline{\rho}_\delta +$$

$$+ \left[\left\{ {\delta \atop \alpha 2} \right\} b_{\delta 1} - \left\{ {\delta \atop \alpha 1} \right\} b_{\delta 2} + \frac{\partial b_{\alpha 2}}{\partial q^1} - \frac{\partial b_{\alpha 1}}{\partial q^2} \right] \underline{n} - (b_{\alpha 2} b^\delta_1 - b_{\alpha 1} b^\delta_2)\underline{\rho}_\delta \,.$$

Equating coefficients of $\underline{n}$ and $\underline{\rho}_\delta$ to zero we arrive at the two Codazzi formulas

$$\frac{\partial b_{\alpha 1}}{\partial q^2} - \frac{\partial b_{\alpha 2}}{\partial q^1} = \left\{ {\delta \atop \alpha 2} \right\} b_{\delta 1} - \left\{ {\delta \atop \alpha 1} \right\} b_{\delta 2} \qquad (\alpha = 1,2) \tag{31}$$

and the Gauss formula

$$\frac{\partial}{\partial q^1}\left\{ {\delta \atop \alpha 2} \right\} - \frac{\partial}{\partial q^2}\left\{ {\delta \atop \alpha 1} \right\} + \left\{ {\gamma \atop \alpha 2} \right\}\left\{ {\delta \atop \gamma 1} \right\} - \left\{ {\gamma \atop \alpha 1} \right\}\left\{ {\delta \atop \gamma 2} \right\} = -(b_{\alpha 1} b_2^{\delta} - b_{\alpha 2} b_1^{\delta}) . \tag{32}$$

Its left-hand side parallels the expression (10.11) for $k = 1$, $m = \delta$, $s = \alpha$, $t = 2$ and this is the component $R_{12\alpha .}^{\ \ \ \ \delta}$ of the tensor of curvature in R_2, i.e. on a surface. Proceeding to its quadruply covariant representation we have

$$a_{\delta\lambda} R_{12\alpha .}^{\ \ \ \ \delta} = R_{12\alpha\lambda} = -(b_{\alpha 1} b_{2\lambda} - b_{\alpha 2} b_{1\lambda})$$

and by (10.18) it may be assumed that $\alpha = 1$, $\lambda = 2$. Referring to (10.15) we arrive at the representation of the Gaussian curvature in terms of the derivatives of the coefficients of the first quadratic form

$$R_{1212} = \frac{1}{2}\left(\frac{\partial^2 a_{11}}{\partial q^{2^2}} - \frac{\partial^2 a_{22}}{\partial q^{1^2}} - 2\frac{\partial^2 a_{12}}{\partial q^1 \partial q^2} \right) + a^{\alpha\beta}([11,\alpha][22,\beta] -$$

$$- [12,\alpha][12,\beta]) = -(b_{11} b_{22} - b_{12}^2) = -aK . \tag{33}$$

5. Representations in terms of lines of curvature. In an orthonormal base we have

$$\underline{e}_\alpha = \frac{\underline{\rho}_\alpha}{|\underline{\rho}_\alpha|} = \frac{\underline{\rho}_\alpha}{H_\alpha} , \quad \underline{e}_3 = \underline{n} ; \qquad a_{11} = H_1^2 , \ a_{22} = H_2^2 , \ a_{12} = 0 , \ \sqrt{a} = H_1 H_2 . \tag{34}$$

We now introduce the plane curve Γ_α which is the intersection of a surface with the plane of vectors $\underline{e}_\alpha$, $\underline{n}$. The principal normal to that curve (i.e. the normal directed towards the centre of curvature on its concave side) is denoted by $\underline{m}$ and obviously, $\underline{m} = \varepsilon\underline{n}$, $\varepsilon = \pm 1$. By the formulas describing the curve we have

$$\frac{\partial \underline{m}}{\partial \sigma_\alpha} = -\frac{\underline{e}_\alpha}{p_\alpha} , \qquad \frac{\partial \underline{e}_\alpha}{\partial \sigma_\alpha} = \frac{\underline{m}}{p_\alpha} . \tag{35}$$

Here $d\sigma_\alpha = H_\alpha dq^\alpha$ is the element of arc of Γ_α and p_α its radius of curvature. It would be erroneous to identify the principal normal $\underline{m}'$ at the infinitely close point M' on Γ_α with $\varepsilon\underline{n}'$ where $\underline{n}'$ denotes the normal to the surface at M'. Such a property is inherent not in any orthogonal set on the surface but in the set of its lines of curvature. We assume below that the curves $[q^\alpha]$ are the lines of curvature. By their definition, the three vectors $\underline{e}_1$, $\underline{n}$ and $\underline{n} + \underline{n}_1 dq^1$ lie in the same plane

$$(\underline{e}_1 \times \underline{n}) \cdot (\underline{n} + \underline{n}_1 dq^1) = 0 , \qquad -\underline{e}_2 \cdot \underline{n}_1 = -\frac{1}{H_2} \underline{\rho}_2 \cdot \underline{n}_1 = \frac{1}{H_2} b_1^2 = 0 .$$

Along the lines of curvature we have thus

$$b_1^2 = b_2^1 = 0 , \qquad b_{12} = 0 . \tag{36}$$

By (35), (36) and (11) we now have

$$\underline{n}_1 = -\frac{\varepsilon}{p_1} \underline{\rho}_1 = -b_1^1 \underline{\rho}_1 , \qquad \frac{\varepsilon}{p_1} = b_1^1 , \qquad \frac{\varepsilon}{p_2} = b_2^2 .$$

The quantities $\varepsilon p_\alpha^{-1}$, designated as R_α^{-1}, are termed the principal curvatures of a surface - these are the curvatures of principal normal intersections of Γ_α taken with appropriate signs. For the lines of curvature we have thus

$$b_1^1 = \frac{1}{R_1} , \quad b_2^2 = \frac{1}{R_2} , \quad b_{11} = \frac{H_1^2}{R_1} , \quad b_{22} = \frac{H_2^2}{R_2} , \quad H = \frac{1}{2}\left(\frac{1}{R_1} + \frac{1}{R_2}\right) , \tag{37}$$

$$K = \frac{1}{R_1 R_2} .$$

For coordinate lines forming an orthogonal set on a surface the Christoffel symbols are represented by the expressions

$$[11,1] = H_1 \frac{\partial H_1}{\partial q^1} , \qquad [11,2] = -H_1 \frac{\partial H_1}{\partial q^2} , \qquad [12,1] = H_1 \frac{\partial H_1}{\partial q^2} ,$$
$$[12,2] = H_2 \frac{\partial H_2}{\partial q_1} , \qquad [22,1] = -H_2 \frac{\partial H_2}{\partial q^1} , \qquad [22,2] = H_2 \frac{\partial H_2}{\partial q^2} \tag{38}$$

$$\left\{ {1 \atop 11} \right\} = \frac{\partial \ln H_1}{\partial q^1} , \qquad \left\{ {2 \atop 11} \right\} = -\frac{H_1}{H_2^2} \frac{\partial H_1}{\partial q^2} , \qquad \left\{ {1 \atop 12} \right\} = \frac{\partial \ln H_1}{\partial q^2} ,$$
$$\left\{ {2 \atop 12} \right\} = \frac{\partial \ln H_2}{\partial q^1} , \qquad \left\{ {1 \atop 22} \right\} = -\frac{H_2}{H_1^2} \frac{\partial H_2}{\partial q^1} , \qquad \left\{ {2 \atop 22} \right\} = \frac{\partial \ln H_2}{\partial q^2} . \tag{39}$$

These formulas together with (37) enable us to represent the formulas of Codazzi and Gauss in the forms

$$\frac{\partial}{\partial q^2}\frac{H_1}{R_1} = \frac{\partial H_1}{\partial q^2}\frac{1}{R^2}, \qquad \frac{\partial}{\partial q^1}\frac{H_2}{R_2} = \frac{\partial H_2}{\partial q^1}\frac{1}{R^1}, \tag{40}$$

$$\frac{H_1 H_2}{R_1 R_2} = -\frac{\partial}{\partial q^1}\frac{1}{H_1}\frac{\partial H_2}{\partial q^1} - \frac{\partial}{\partial q^2}\frac{1}{H_2}\frac{\partial H_1}{\partial q^2}. \tag{41}$$

For an observer on a surface, it is impossible, in principle, to determine the curvatures of the lines lying on it. But measurements performed on a surface (i.e. knowledge of its first quadratic form) allow beings on it to discover the presence or absence of Gaussian curvature. The formulas (30) and, particularly, (41) embody the theorem of Gauss - one of his greatest achievements. In Par. 10 we already stated that knowledge of the quadratic form (10.3) of three-dimensional space enables us to assert whether the space is Euclidean E_3 or Riemannian R_3 - for the first case ${}^4\underline{R} = \underline{0}$, for the second ${}^4\underline{R} \neq \underline{0}$. Analogy with the geometry of surfaces shows why this tensor is called the tensor of curvature of space.

The deformation of a surface is called bending if it preserves distances between points, in other words, if it leaves the same the first quadratic form of the surface. The Gaussian curvature is an invariant of a bending deformation.

BIBLIOGRAPHY AND BIBLIOGRAPHICAL NOTES

MONOGRAPHS

1. Green, A.E. and J.E. Adkins, Large Elastic Deformations and Nonlinear Continuum Mechanics, Clarendon Press, Oxford, 1960.
2. Lurie, A.I., Theory of Elasticity (in Russian Nauka, Moscow, 1970.
3. Novozhilov, V.V., Foundations of Nonlinear Theory of Elasticity (in Russian), Gostekhizdat, Moscow, 1948.
4. Sedov, L.V., Introduction to Continuun Mechanics (in Russian), Fizmatgiz, Moscow, 1962.
5. Truesdell, C., A First Course in Rational Continuum Mechanics, The John Hopkins University, Baltimore, Maryland, 1972.
6. Eringen, A.C., Nonlinear Theory of Continuous Media, McGraw-Hill, New York, London, 1962.
7. Green, A.E. and W. Zerna, Theoretical elasticity, Oxford, Clarendon Press, 1954.
8. Grioli, S., Mathematical theory of elastic equilibrium (recent results), Ergebnisse angew. Math., No. 7, Springer-Verlag, 1962.
9. Leigh, D.C., Nonlinear Continuum Mechanics, McGraw-Hill, New York, London, 1968.
10. Truesdell, C., The mechanical foundations of elasticity and fluid dynamics, J. Rational Mech. and Anal., 1952, v. 1, p. 125-300; Corrections and additions, J. Rational Mech. and Anal., 1953, v. 2, p. 593-616.
11. Truesdell, C. and W. Noll, The nonlinear field theories of mechanics, Handbuch der Physik, Encyclopedia of Physics, Vol. 3, Part 3, Springer-Verlag, 1965.

This book contains a detailed bibliography on the nonlinear mechanics of continua published up to 1965. The growth of interest in this area is illustrated by the increasing number of publications: 1938 - 4, 1949 - 28, 1960 - 60, 1962 - 79.

12. Truesdell, C. and R.A. Toupin, The classical field theories, Handbuch der Physik, Encyclopedia of Physics, Vol. 3, Part 1, Springer-Verlag, 1960.
13. Wang, C.C. and C. Truesdell, Introduction to rational elasticity, Wolters-Noordhoff, Groningen, 1972.

TO CHAPTER 1

This chapter covers the geometrical concepts and forms a natural preface to a study of the mechanics of a deformable solid body.

The basic concepts of material and spatial coordinates, reference and actual configurations and their vector bases, tensors, deformation gradients are illustrated in Pars. 1-3.

In Pars. 4-5 we define measures of deformation and the inverse tensors. The names ascribed to them (Cauchy-Green, Almansi, Finger) do not pretend to be historically precise.

Par. 6 introduces the polar decompositions of the place gradients and discusses the stretch tensors (6.4) and orthogonal tensors accompanying deformation (6.5)-(6.7). The transition from strain measures to tensors is dealt with in Par. 7.

The mapping of an oriented area from the reference to the actual configuration given by Equations (8.5)-(8.8) finds wide application in the subsequent text.

The Pars. 9-12 are included in order to prepare operational techniques employed in Chapters 4 and 8.

The fundamental kinematical quantities, the concept of a material derivative and its calculation are given in Pars. 13-14.

The concept of a frame-indifferent quantity is introduced in Par. 15. The theorem of Zorawski (15.10) which asserts the frame-indifference of a rate tensor os strain and the non-frame-indifference of vorticity is used in Par. 16 to formulate the operation of objective differentiation.

In Par. 18 the method of solving the problem of the determination of a position vector by prescribing the strain measure is outlined. The affine strain tensors of the third rank introduced by A.P. Norden have been used in Par. 19. For example, the problem of Par. 18 reduces to the linear system (19.12) of the deformation gradient, the coefficients of this system being components for the affine strain tensor (19.9); the differential operations over functions of the gradient or strain measures are compared with derivatives calculated with respect to those measures; see Equations (19.20), (19.23).

The background material for this chapter may be found in the books [3], [4], [6], [10-12] together with the following papers:

1.1. Norden, A.P., On the geometrical theory of finite deformations (in Russian), Proceedings of the Kazan Branch of the Academy of Sciences of the USSR, Ser. Phys.-Math. and Eng. Sci., 1950, No. 2, pp. 53-61.

1.2. Zubov, L.M., On Jaumann's derivative for a tensor of the second rank (in Russian), Proceedings of the North Caucasian High School Science Centre, Natural sciences, 1976, No. 2, pp. 27-30.

TO CHAPTER 2

In Par. 1 body and surface forces are introduced and the concept of a force tensor is illustrated (1.16). In Pars. 2-3 the stress tensor is introduced and also the equations of motion of continuous medium are put forward. The contents of Pars. 1-3 are explained with greater precision in the book [5].

Equations (3.7) and (3.14) define the mean value and moments of the first order of a stress tensor. This material is given in more detail in the books [2] and [8].

In its most complete form, the concept of a tensor of stress functions is examined in the book

2.1. Krutkov, Yu.A., Tensor of Stress Functions and General Solutions in Statics of the Theory of Elasticity (in Russian), Published by the USSR Academy of Sciences, Moscow, 1949.

We have only touched upon the concept of a polar medium since this is not used in the subsequent text. See [6], [8], [10].

In Pars. 6-7 the Piola tensor (6.2) and an "energetic" stress tensor (6.11) are defined. The use of the Piola tensor simplifies many of the calculations and relationships of nonlinear theory. This may be explained by the simplicity of representation (7.5) of the specific incremental work in terms of this tensor. It is also essential that the equilibrium equations (6.4) and (6.5) expressed in terms of the Piola tensor are associated with the reference configuration. For various definitions of a stress tensor see also [3].

TO CHAPTER 3

In Pars. 1-3 we discuss the definitions of a simple body and a simple elastic body (material). The exposition rests upon the fundamental principles of causality, local action and material frame-indifference formulated by Noll in his earliest publication.

3.1. Noll, W., On the continuity of the solid and fluid states, J. Rational Mech. and Anal., 1955, v. 4, p. 13-81.

Here we introduce the concept of a simple material (2.4) and establish its fundamental property (2.9). We also introduce the definitions (3.9) and (3.10) of an elastic material and give the various forms of its equations of state.

A change in the reference configuration may result in a change in the form of the state equation under the same actual state of stress but one cannot rule out in advance the possibility that this form may remain the same. This is associated with the concept of a material "symmetry group" (Par. 4) and with the correct definition of its property of isotropy (Par. 5). Consequently, Equation (5.7), the stress tensor, turns out to be an isotropic tensor function of the right stretch tensor and may therefore be represented as a quadratic trinomial (5.9).

The definitions and illustrations of the concepts of a solid body (isotropic and anisotropic) and of elastic fluid are given in Pars. 6-8.

The papers [5], [9] and [11] are referred to throughout the chapter.

TO CHAPTER 4

In this chapter and in the subsequent Chapters 5-8, a hyperelastic medium (Green elasticity) is considered. To the definition of an elastic body is added the requirement of the existence of stored strain energy, i.e. the existence of a function of the deformation gradient whose variation equals the incremental work done by the external forces (the "stored" energy). The stress tensors are expressed in terms of derivatives of this function, see (1.4)-(1.9).

In Par. 2 the equations of state for an orthotropic and transversally isotropic elastic body are introduced. Applications of these equations are to be found in the book [1].

The state equations of an isotropic elastic body are given in Par. 3 in the form of representations of the Cauchy stress tensor in terms of the Finger and Almansi measures of deformation. The formulas (3.9) for principal stresses and (3.11) for principal forces are often used. The material properties of a nonlinearly elastic isotropic body are thus described by three functions of the strain invariants, e.g. by those of (3.5), connected by the differential relationships (3.6). Linear theory disposes of only two Lamé constants and so, naturally, many effects cannot be explained within its framework.

In Pars. 5-6 we introduce the tensor $\underline{\theta}$ which plays the role of a stress tensor in the changed state of stress (see (6.2)-(6.4)). The relation of $\underline{\theta}$ to the convected derivatives of the Cauchy and Piola tensors is given by Equations (5.9), (5.10). The latter formula was given in the paper

4.1. Zubov, L.M., Variational principles of the nonlinear theory of elasticity. The case of a small deformation superposed on a finite deformation (in Russian), Prikladnaja Matematika i Mekhanika, 1971, Vol. 35, No. 5.

The tensor $\underline{\theta}$ is calculated by Equation (5.8), this formula is given in [2] but less precisely.

In Pars. 7-9 there is given the definition of a relaxation tensor, i.e. of a tensor of the fourth rank which is the derivative of a stress tensor with respect to the strain measure in terms of which this tensor is expressed. The formula (7.9) provides an invariant representation of the tensor of elasticities and Equation (8.4) gives expressions for its components. Attention should be paid to Equations (7.12)-(7.14) which give Lamé coefficients in terms of the values of the coefficients of the state equation and their derivatives in the natural reference configuration.

In a vector basis consisting of the principal directions of the Finger measure the components of a relaxation tensor are evaluated by Equations (8.6)-(8.8); their content is illustrated by representations (8.9)-(8.10) in terms of the principal stresses.

The equations of equilibrium which are expressed in terms of a position vector in the actual configuration (the "equations in terms of displacements") are given in various forms in Par. 10, the simplest being that of (10.3). This is a system of partial differential equations of complex structure which is linear with respect to the highest (second) derivatives of the position vector.

The notion of ellipticity (Par. 11) is associated with smoothness of solutions. Singularities are to be expected outside the particular set of parameters characterizing the deformation. See the papers

4.2. Knowles, J.K. and E. Sternberg, On the ellipticity of the equations of nonlinear elastostatics for a special material, J. Elasticity, 1975, v. 5, No. 3-4, p. 341-361.

4.3. Knowles, J.K. and E. Sternberg, On the failure of ellipticity of the equations for finite elastostatic plane strain, Arch. Rational Mech. and Anal., 1977, v. 63, No. 4, p. 321-336.

A criterion of strong ellipticity is embodied in Inequalities (11.6) and (11.8). This criterion reduces to the requirement that the acoustical tensor (11.14) should be positive definite. A general representation of this tensor and its specialization for certain assignments of the specific strain energy are given in the paper

4.4. Lurie, A.I., A criterion of ellipticity of the equations of equilibrium in the nonlinear theory of elasticity (in Russian), Mekhanika tverdogo tela, 1979, No. 2, pp. 23-34.

The acoustical tensor has also been expressed in terms of the principal directions of the Finger measure, see Equations (12.9), (12.10) and also [5].

The formulation and methods of solution of static boundary value problems in nonlinear elasticity theory are given in Pars. 13-14 and the concept of "universal" solutions is also illustrated. Ericksen's theorem (which is cited) asserts the non-existence of such solutions for transformations different from affine.

4.5. Ericksen, J.L., Deformations possible in every compressible isotropic, perfectly elastic material, J. Math. Phys., 1955, v. 34, p. 126-128.

The proof of this theorem given in the text is due to Shield:

4.6. Shield, R.T., Deformations possible in every compressible, isotropic, perfectly elastic material, J. Elasticity, 1971, v. 1, No. 1, p. 71-72.

The variational principles are analysed in Pars. 16-19. For a complete account (seven formulations) see paper

4.7. Zubov, L.M., Variational principles of the nonlinear theory of elasticity (in Russian), Prikladnaya Matematika i Mekhanika, 1971, Vol. 35, No. 3.

Of special interest is the principle of stationary complementary energy because it proves possible to express this in terms of a functional not containing quantities determined by the strain measures. See

4.8. Zubov, L.M., The principle of stationary complementary energy in the nonlinear theory of elasticity (in Russian), Prikladnaya Matematika i Mekhanika, 1970, Vol. 34, No. 2.

This paper is discussed in

4.9. Koiter, W.T., On the principle of stationary complementary energy in the nonlinear theory of elasticity, SIAM J. Appl. Math., 1973, v. 25, No. 3.

4.10. Koiter, W.T., On the complementary energy theorem in nonlinear elasticity theory, Technische Hogeschool Delft, 1975, No. 72.

4.11. Christoffersen, J., On Zubov's principle of stationary complementary energy and a related principle, Danish Center for Applied Math. and Mech., 1973, Rep. 44

(the principle applied to an isotropic body).

The non-uniqueness of representation of a deformation gradient in terms of the Piola tensor mentioned in [4.9]-[4.11] is discussed in the paper

4.12. Zubov, L.M., On the representation of a displacement gradient of an isotropic elastic body in terms of the Piola tensor (in Russian), Prikladnaya Matematika i Mekhanika, 1976, Vol. 40, No. 6.

TO CHAPTER 5

The state equation acquires concrete content once a description of the specific strain energy is given in terms of the invariants of strain measures.

The substitution of a linear strain tensor for that of finite deformation in Hooke's law as suggested by Seth, is not successful since the conditions for the existence of specific strain energy are not fulfilled.

5.1. Seth, B.R., Finite strain in elastic problems, Phil. Trans. Roy. Soc., London, 1935, v. A234, p. 231-264.

A general representation of the quadratic law of dependence of a stress tensor on a tensor of deformation is embodied in the state equation (2.9) of Signorini consistent with the existence of a specific potential energy.

5.2. Signorini, A., Transformazioni termoelastiche finite, Mem. 1[a]. Ann. di Mat. (4), 1943, v. 22, pp. 33-143; Mem. 2[a]. Ann. di Mat. Pura Appl. (4), 1949, v. 30, p. 1-72.

See also [8] and the paper

5.3. Zvolinsky, N.V. and P.M. Riz, On certain problems of the nonlinear theory of elasticity (in Russian), Prikladnaya Matematika i Mekhanika, 1939, Vol. 2, No. 4.

Murnaghan has suggested starting with the polynomial representation (3.1) or (3.2) of the specific stored energy of deformation. His papers have been collected in the monograph

5.4. Murnaghan, F.D., Finite deformation of an elastic solid, Second Edition, J. Wiley, New York, 1967.

The constants of the third order in Murnaghan's law are determined by acoustical measurements and extensive information on these constants has now been accumulated. Data relating to 1975 are contained in Tables 1-3.

The formulas connecting the Murnaghan constants with the bulk and shear moduli have been obtained by Wang.

5.5. Wang, C.C., Second order change of volume in isotropic materials free from applied loads, Zeitsch. angew. Math. und Mech., 1966, v. 46, No. 2, p. 141-144.

The formulas (4.5)-(4.8) determining the transformation of the coefficients of the laws of Signorini and Murnaghan in a similarity transformation of the reference configuration, in particular for isothermal states of different temperatures, can be found in the book

5.6. Brillouin, L., Les Tenseurs en Mechanique et en Elasticité, Paris, Masson, 1938.

The specific strain energy of a material which John has called harmonic and which is termed below quasi-linear, can be represented by a quadratic form (5.2) of stretches, this form being reducible to that of (5.4).

5.7. John, F., On the finite deformation of an elastic isotropic material, Inst. Math. Sci. New York Univ. Report IMM-NYU, 1958, No. 250.

The representations (5.9) and (5.11) of the specific potential energy and complementary energy in terms of the Piola tensor and the formulas for the position vector were obtained in [4.8] and [2]. The conditions of strong ellipticity for a quasi-linear material were communicated to the author by E.L. Gurvich; see also [4.4].

The representation (6.1) of the specific strain energy was proposed in the paper

5.8. Blatz, P.J. and W.L. Ko, Application of finite elasticity theory to the deformation of rubbery materials, Tran. Soc. Rheol., 1962, v. 6, p. 223-251.

A simplified version has been used in [4.2]. The acoustical tensor is defined by Equation (6.14) which is reducible to the form (6.15) used in [4.2]. In an alternative version of the expression (6.1) the material turns out to be strongly elliptic for arbitrary deformations.

In Par. 7 the specific strain energy is represented as the sum of the volume strain energy and the shear strain energy. The state equation is reduced to the form (7.11). See

5.9. Pal'mov, V.A., Vibrations of Elastic-Plastic Bodies (in Russian), Nauka, Moscow, 1976.

The trigonometric representation of the state equation suggested by V.V. Novozhilov is expressed by Equation (8.18); the specific strain energy is represented by (8.16) in terms of the first invariant of the logarithmic measure of deformation (the Hencky measure), the second invariant of its deviator and the phase difference of the deviators of the stress tensor and the Hencky tensor.

The problem of the formulation of the general requirements necessary for the prescription of the specific strain energy is formulated in Pars. 9-12. It should not lead to results which are incompatible with intuitive ideas concerning the behaviour of an elastic body. Much space is devoted in [5], [12] and [13] to a discussion of the proposed criteria.

A criterion of monotonicity for the state of stress is put forward in the paper

5.10. Coleman, B.D. and W. Noll, Material symmetry and thermostatic inequalities in finite elastic deformation, Arch. Rational Mech. and Anal., 1964, v. 15, p. 87-111.

Provided that the polar representations of the actual and admissible deformation gradients are expressed in terms of one and the same orthogonal tensor, we arrive at Inequality (9.7) and at its differential form (9.10). For an admissible configuration of a more general type the Inequalities (9.10) are complemented by those of (10.14).

The consequences inferred from the criteria of monotonicity and strong ellipticity are reviewed in Par. 13.

The values of the Murnaghan constants are given in Tables 1-3 which summarize the data of papers [5.11]-[5.19].

5.11. Crecraft, D.I., Ultrasonic wave velocities in stressed nickel steel, Nature, 1962, v. 195, No. 4847, p. 1193-1194.

5.12. Smith, R.T., R. Stern and R.W.B. Stephens, Third-order elastic moduli of polycrystalline metals from ultrasonic velocity measurements, J. Acoust. Soc. Amer., 1966, v. 40, No. 5, p. 1002-1008.

5.13. Savin, G.N., A.A. Lukashov, E.M. Lysko, S.V. Veremeenko and S.M. Vozhevskaya, The propagation of elastic waves in a solid for the case of a nonlinearly elastic model of a continuous medium (in Russian), Prikladnaya Mekhanika, 1970, Vol. 36, No. 2, pp. 38-42.

5.14. Subbotina, E.K. and S.S. Sekoyan, On the determination of the pressure dependence for the velocities of propagation of elastic waves in solids from the results of ultrasonic measurements under uniaxial loading of samples (in Russian), In the book: A Collection of Papers of the First All-Union Meeting on Physics and Technology of High Pressures (in Russian), Donetsk, 1973.

5.15. Seeger, A. and O. Buck, Die experimentalle Ermittelung der elastischen Konstanten höherer Ordnung, Z. Naturf., 1960, 15a, p. 1056-1067.

5.16. Guz, A.N., F.G. Makhort, O.I. Guscha and V.K. Lebedev, On the theory of propagation of waves in an elastic isotropic body with initial deformations (in Russian), Prikladnaya Mekhanika, 1970, Vol. 6, No. 12, p. 42-49.

5.17. Bogardus, E.H., Third-order elastic constants of Ge, MgO and fused SiO_2, J. Appl. Phys., 1965, v. 36, p. 2504.

5.18. Hughes, D.S. and J.L. Keily, Second order elastic deformation of solids, Phys. Rev. 1953, v. 92, No. 5, p. 1145-1149.

Detailed data for ЭП-56 steel are quoted in the paper

5.19. Sekoyan, S.S., E.K. Subbotina and I.I. Averbukh, An influence of thermal work upon the values of elastic constants of the third order for ЭП-56 steel (in Russian), Proceedings of the All-Union Research Institute of Measurements, No. 5 (35), Moscow, 1971.

TO CHAPTER 6

An affine mapping from the reference to the actual configuration does not require the specification of the specific strain energy for the determination of the state of stress.

In Pars. 1-3 similarity transformations and the uniaxial state of stress are considered. For the problem of simple shear (Par. 4) normal stresses which are not predictable from linear theory are revealed. The problem of pure shear (Par. 5) is more complex, here the principal values of the strain measure are sought using the state of shear stress specified by (5.1). The shear stress is given by (5.16) with the aid of function (5.6) of shear measure but the specific law for this dependence can only be formulated in an approximate form. The problem has also been considered in the paper

6.1. Moon, H. and C. Truesdell, Interpretation of adscititious inequalities through the effects pure shear stress produces upon an isotropic elastic solid, Arch. Rational Mech., Math., 1974, v. 55, No. 1, p. 1-17.

In Pars. 6-9 we consider a quasi-linear material. Solutions to the Lamé problem for a hollow cylinder and sphere are given in Par. 6. The problem of the bending of a circular membrane (Par. 7) is more complicated. The principle of stationary complementary energy is used and the equations obtained in [4.10] are given with greater precision.

A statement of the plane problem is considered in Par. 8. The presentation is similar to that proposed in the paper

6.2. Lurie, A.I., The plane problem for a quasi-linear material (in Russian), In the book: Continuum Mechanics and Related Problems of Calculus (in Russian), Nauka, Moscow, 1972, pp. 289-296.

The plane problem for quasi-linear material has also been considered in the paper

6.3. Ovsyannikov, A.V., On the solution of problems of the plane deformation of a quasi-linear elastic medium (in Russian), Prikladnaya Mekhanika, 1972, Vol. 8, No. 10, pp. 47-54.

The plane problem for a Murnaghan material has been examined in many papers by Yu. I. Koifman, some of whose papers are cited in the bibliography of the review.

6.4. Savin, G.N. and Yu.I. Koifman, General nonlinear theory of elasticity (a survey) (in Russian), Prikladnaya Mekhanika, 1970, Vol. 6, No. 12, pp. 3-26.

The general equations of the plane problem (material not specified) are given in the review

6.5. Chernykh, K.F., Generalized plane deformation in the nonlinear theory of elasticity (in Russian), Prikladnaya Mekhanika, 1977, Vol. 13, No. 1, pp. 3-30.

The relationships of Par. 8 are illustrated by the problem of the bending of a plate into a right cylindrical form (Par. 9) and by that of the torsion of a hollow cylinder, both problems yielding productive formulas.

Second order effects are considered in Pars. 10-12. Equations (10.7) and (10.10) provide representations of the Cauchy and Piola stress tensors with all terms quadratic with respect to the components of the gradient of the displacement vector taken into account. In particular expression (10.9) is given for a Murnaghan material.

The formulation of a solution and the problem of its existence are discussed in Par. 11. The problem reduces to the system (11.12) but difficulties arise when its determinant is zero. These are analysed at length in [8], [11], [5.2].

An approach to the analysis of the problem of second order effects which is somewhat different from that just described has been proposed in the paper

6.6. Rivlin, R.S. and C. Topakoglu, A theorem in the theory of finite elastic deformation, J. Rational Mech. and Anal., 1954, v. 3, p. 581-589.

A survey of methods for the calculation of second order effects is given in the paper

6.7. Truesdell, C., Second order effects in the mechanics of materials, Proc. Int. Symp. Second-Order Effects. Haifa, 1962, p. 1-47.

The formula (12.5) determining change of volume of a body subjected to distortion was obtained in the paper

6.8. Zener, C., Theory of lattice expansion introduced by coldwork, Trans. Amer. Soc. Mining Engrs., 1942, v. 147, p. 361-364.

More general results (also for an isotropic body) can be found in

6.9. Toupin, R.A. and R.S. Rivlin, Dimensional changes in crystals caused by dislocations, J. Math. Phys., 1960, v. 1, p. 8-15.

The second order effects in the problem of torsion of a rod (Par. 13) are considered in the paper

6.10. Rivlin, R.S., The solution of problems in second order elasticity theory, J. Rational Mech. and Anal., 1953, v. 2, p. 53-81.

An example of a twisted rod compressed over its lateral surface is given in [8]. The problems of the plane deformation of compressible and incompressible materials are examined in monographs [1] and [7].

The plane deformation of compressible and incompressible materials is also considered in the paper

6.11. Adkins, J.E. and A.E. Green, Plane problems in second order elasticity theory, Proc. Roy. Soc. London, 1958, v. A239, p. 557-576.

The poorly found premises of the "physically nonlinear" theory of elasticity presented in the book

6.12. Kauderer, H., Nichtlineare Mechanik, Springer-Verlag, Berlin, Göttingen, Heidelberg, 1958

have been discussed in the paper

6.13. Bharatha, S. and M. Levinson, On physically nonlinear elasticity, Journ. of Elasticity, 1977, v. 7, No. 3, p. 307-324.

TO CHAPTER 7

The concept of superposition of constraints (Equations (1.1)-(1.4)) is explained in Par. 1 together with the state equations for a material with superposed constraints.

In Par. 2 we consider the case of an incompressible material, formulate the equations of static equilibrium (2.11) and (2.19), and put forward the principle of stationary potential energy and the Hamilton-Ostrogradskii principle.

In Par. 3, equations of static equilibrium are formulated for internal (3.13) and boundary (3.10) points of the volume. These equations, together with the Equation (3.12) of constraint, serve for the determination of second order effects.

Expressions for the specific strain energy are given in Par. 4 in the form of the Mooney potential together with its simplified version - the neo-Hookean potential of Treloar, and with its more complex version - the Rivlin-Saunders potential.

7.1. Mooney, M., A theory of large elastic deformation, J. Appl. Phys., 1940, v. 11, p. 582-592.

7.2. Rivlin, R.S. and D.W. Saunders, Large elastic deformations of isotropic materials. VII. Experiments on the deformation of rubber, Phil. Trans. Roy. Soc. London, 1951, v. A234, p. 251-288.

The experimental data for several rubbers provide the basis for the consideration of the potential (4.8) in the paper

7.3. Chernykh, K.F. and I.M. Shubina, The laws of elasticity for isotropic compressible materials (in Russian), Mekhanika elastomerov, Krasnodar, 1977, Vol. I, pp. 54-65.

A plane problem is considered in Pars. 5-6. See also the paper [6.11]. Much space is devoted to this point in [1], [2], [7].

Adkins' inverse theorem, mentioned in Par. 5, is stated in the paper

7.4. Adkins, J.E., A reciprocal property of the finite plane strain equations, J. Mech. Phys. Solids, 1958, v. 6, p. 267-275.

This investigation continued in

7.5. Hill, J.M., On the duality of stress and deformation fields in finite elasticity, J. Elasticity, 1973, v. 3, No. 1, p. 51-56.

7.6. Hill, J.M. and R.T. Shield, Notes on the duality of stress and deformation fields in plane finite elasticity, J. Elasticity, 1974, v. 4, No. 2, p. 167.

The problems of tension and simple shear for an incompressible material are considered in Par. 7. Expressions for principal forces are given for a plane state of stress. These expressions were used for the interpretation of the results of measurements in the experiments by Rivlin and Saunders, see [7.2] and [11].

Ericksen's theorem for general solutions of the equations of the theory of elasticity for an incompressible body is formulated in Par. 8. Its proof, divided into several stages, is given in Par. 21.

7.7. Ericksen, J.L., Deformations possible in every isotropic incompressible perfectly elastic body, Zeitsch. angew. Math. Phys., 1954, No. 5, p. 466-489.

A list of general mappings is given in Par. 9:

1) the bending of a plate into a right cylindrical form, Equations (9.1)-(9.10);

2) the unbending of a cylindrical panel, Equations (9.11)-(9.13);

3) the transformation of a cylindrical panel, Equations (9.14)-(9.21);

4) the axisymmetric transformation of a hollow sphere, Equations (9.22)-(9.25);

5) transformation of a right cylindrical form, Equations (9.26)-(9.28).

A fifth family of transformations has been proposed in the paper

7.8. Klingbeil, W.W. and R.T. Shield, On a class of solutions in plane finite elasticity, Z. angew. Math. Phys., 1966, v. 17, p. 489-511.

A particular choice of the constants of the general solutions enables solustions to the problems of Pars. 10-14 to be found. These exact solutions to the equations of the nonlinear theory of elasticity were found before Ericksen's work in the fundamental papers

7.9. Rivlin, R.S., Large elastic deformations of isotropic materials. IV. Further developments of the general theory, Phil. Trans. Roy. Soc. London, 1948, v. A241, p. 379-397.

7.10. Rivlin, R.S., Large elastic deformations of isotropic materials. V. The problem of flexure, Proc. Roy. Soc. London, 1949, v. A195, p. 463-473.

7.11. Rivlin, R.S., Large elastic deformations of isotropic materials. VI. Further results in the theory of torsion, shear and flexure, Phil. Trans. Roy. Soc. London, 1949, v. 242, p. 173-195.

The problem of the state of stress within a rotating cylinder has been considered in

7.12. Green, A.E. and R.T. Shield, Finite elastic deformation of incompressible isotropic bodies, Proc. Roy. Soc. London, 1950, v. A202, p. 407-417.

Considerable space is devoted in [11] and [13] to the formulation of universal solutions to the equations of motion (Pars. 15-16). Examples of radial oscillations of a cylindrical tube (Par. 18) and of axisymmetric oscillations of a hollow sphere (Par. 19) are also considered. See also the papers

7.13. Knowles, J.K., Large amplitude oscillations of a tube of incompressible elastic material, Quart. Appl. Math., 1960, v. 18, p. 71-77.

7.14. Guo Zhong Heng and R. Solecki, Free and forced finite amplitude oscillations of an elastic thick-walled hollow sphere of incompressible material, Arch. Mech. Stosow., Warszawa, 1963, v. 15, p. 427-433.

The procedure for solving these problems suggested in Pars. 18-19 is different from those recommended in [11], [13] and in the cited papers.

"Anti-plane" deformation is considered in

7.15. Knowles, J.K., On finite anti-plane shear for incompressible elastic materials, Journ. Aust. Math. Soc., 1976, XIX, ser. B, part 4, p. 400-415.

7.16. Knowles, J.K., A note on anti-plane shear for compressible materials in finite elastostatics, Journ. Aust. Math. Soc., 1977, XX, ser. B, part 1, p. 1-7.

7.17. Adkins, J.E., Some generalizations of the shear problem of isotropic incompressible materials, Proc. Camb. Phil. Soc., 1954, v. 50, p. 334-345.

TO CHAPTER 8

The basic relationships, i.e. the state equations, the equations of neutral equilibrium and the stationary value principles for a small deformation superposed on a finite deformation are discussed in Pars. 1-3. The results of [2] and [10] are used and the duality theorems (1.15) and (1.18) are formulated in [4.1].

Applications of these relationships are given in Pars. 4-6. In Par. 4 is considered the superposition of a small deformation on a hydrostatically prestressed elastic body and its equilibrium equations are shown to be reducible to the form of those of linear theory if the Lamé constants are defined by Equations (4.4), (4.10). The problem of the torsion of an initially compressed rod is analysed in [7] and in the paper

8.1. Green, A.E. and R.T. Shield, Finite extension and torsion of cylinders, Phil. Trans. Roy. Soc. London, 1951, v. A244, p. 47-86.

Wave propagation in an uniformly prestressed elastic medium (Pars. 4-9) has been examined in the papers

8.2. Truesdell, C., General and exact theory of waves in finite elastic strain, Arch. Rational Mech. and Anal., 1961, v. 8, p. 263-296.
8.3. Hayes, M. and R.S. Rivlin, Propagation of a plane wave in an isotropic elastic material subjected to pure homogeneous deformation, Arch. Rational Mech. and Anal., 1961, v. 8, p. 15-22.

Stability of the state of equilibrium is characterized in Par. 10 by the condition that the potential energy of a system be minimal at this state. The problem is simplified when this condition is expressed by Inequality (10.5) but difficulties which arise when the inequality degenerates into a strict equality are left aside.

The equations of neutral equilibrium for a quasi-linear material (11.11)-(11.12) have been obtained in [2] by a different procedure. These equations turn out to parallel those found long before the view-points and methods of the nonlinear theory became prevalent.

8.4. Southwell, R.V., On the general theory of elastic stability, Phil. Trans. Roy. Soc. London, 1913, v. A213, p. 187-244.
8.5. Biezeno, C.B. and H. Hencky, On the general theory of elastic stability, K. Akad. Wet. Amsterdam Proc., 1930, v. 31, p. 569-592; v. 32, p. 444-456.

This situation arises because the tensor of additional stress which causes the particles of a medium to be displaced by an amount $\eta\underline{w}$ from their actual state is connected with the linear strain tensor $\varepsilon(\underline{w})$ by the state equation of the linear theory of elasticity. This turns out to be valid for quasi-linear material. In [8.4] and [8.5], no mention is made of material properties.

The equations of neutral equilibrium are applied in [2] to the problem of a cylinder compressed by a longitudinal force and to that of a sphere compressed by a pressure distributed over its surface; see also the paper

8.6. Sensenig, C.B., Instability of thick elastic solids, Comm. Pure and Appl. Math., 1964, v. 17, p. 451-491.

The problem of the buckling of a plate is examined in the paper

8.7. Zubov, L.M., An approximate theory of the buckling of thin plates of a quasi-linear material under affine initial deformation (in Russian), Prikladnaya Matematika i Mekhanika, 1969, Vol. 33, No. 4, pp. 668-675.

The problem of the stability of a sphere made of a Murnaghan material is considered in the paper

8.8. Volokitin, G.I., The influence of physical and geometrical nonlinearity on the values of upper critical pressure in the buckling of a hollow sphere (in Russian), Prikladnaya Matematika i Mekhanika, 1978, Vol. 42, No. 3, pp. 504-510.

A different procedure for the construction of the equations of neutral equilibrium has been suggested in

8.9. Balabukh, L.I. and M.G. Yakovenko, The equations of the bifurcation of an equilibrated elastic isotropic body in terms of the rates of change of Lagrangian variables (in Russian), Prikladnaya Matematika i Mekhanika, 1974, Vol. 38, No. 4, pp. 694-702.

8.10. Balabukh, L.I. and G.M. Yakovenko, On the calculation of deformational anisotropy in problems of the stability of isotropic elastic bodies (in Russian), In the book: "Mechanics of Deformable Bodies and Constructions", Mashinostroenie, Moscow, 1975, pp. 52-61.

A procedure for the approximate determination of the critical load of a column for the case of quasi-linear material is considered in Par. 12. The results parallel those presented in the paper

8.11. Pearson, C.E., General theory of elastic stability, Quart. Appl. Math., 1956, v. 14, p. 133-144.

Pars. 17-19 are based on the paper

8.12. Holden, J.T., Estimation of critical loads in elastic stability theory, Arch. Rational Mech. and Anal., 1964, v. 17, p. 171-183.

In Par. 14 there is a summary of the required data on Korn's constant.

8.13. Bernstein, B. and R.A. Toupin, Korn inequalities for the sphere and circle, Arch. Rational Mech. and Anal., 1960, v. 6, p. 51-64.

8.14. Payne, L.E. and H.F. Weinberger, On Korn's inequality, Arch. Rational Mech. and Anal., 1961, v. 8, p. 89-98.

Holden's investigation has been continued by Beatty in the paper

8.15. Beatty, M.F. Estimation of ultimate safe loads in elastic stability theory, J. Elasticity, 1971, v. 1, No. 2, p. 95-120, this paper contains references to earlier publications.

The Beatty criterion is applied in Par. 16 to the determination of a longitudinal force compressing a rod. In spite of the roughness of the estimates carried out, this force turns out to be 0.4-0.6 of the Eulerian critical value for a rod clamped at its ends.

In Par. 17 we examine a quasi-linear material. Particular importance is attached to the state equation (17.7) for the case in which the specific strain energy is given as a function of the invariants of the stretch tensor and in particular as a function of (17.9) for a quasi-linear material. The inequality determining safe loading is reduced to the form (17.14).

In Pars. 18-21 the condition of stability of equilibrium is examined together with the equations of neutral equilibrium for an incompressible material (18.12)-(18.14) and an expression is given for the Beatty-criterion as applied to problems of the stability of equilibrium of the undistorted state (20.4) and of a compressed rod (21.4). See also [8.15] and [1].

The concept of the convexity of the specific strain energy with respect to a deformation gradient is defined by Inequality (22.6) and in particular by Hadamard's inequality (22.8) which is equivalent to the property of strong ellipticity. The examples given in Par. 22 are designed to demonstrate the superiority of Hadamard's criterion over other convexity criteria.

The author is indebted for the contents of Pars. 22-25 to his talks with E.L. Gurvich.

In Par. 23 it is shown that stability of equilibrium necessitates Hadamard's condition to hold in its strengthened form, in other words, it brings strong ellipticity. The inverse proposition is stated in Par. 24 for the first boundary value problem. An example is given of the deformation of a disc confined by a rigid band (Par. 25).

Wave propagation in an incompressible elastic medium (Par. 26) has been examined in the papers

8.16. Ericksen, J.L., On the propagation of waves in isotropic incompressible perfectly elastic materials, J. Rational Mech. Anal., 1953, v. 2, p. 329-337.

8.17. Sawyers, K.N. and R.S. Rivlin, Instability of an elastic material, Int. J. of Solids and Structures, 1973, v. 9, p. 607-613.

8.18. Sawyers, K.N. and R.S. Rivlin, On the speed of propagation of waves in a deformed elastic material, Zeitsch. angew. Math. Phys., 1977, v. 28. p. 1045-1057.

8.19. Sawyers, K.N. and R.S. Rivlin, A note on the Hadamard criterion for an incompressible isotropic elastic material, Mech. Res. Comm., 1978, v. 5 (4), p. 211-214.

Problems associated with Hadamard's criterion (Par. 27) have been analysed in the paper

8.20. Gurvich, E.L., Hadamard's condition in the nonlinear theory of elasticity (in Russian), Mekhanika Tverdogo Tela, 1979, No. 1, pp. 45-51.
See also [4.4] and
8.21. Gurvich, E.L. and A.I. Lurie, On the theory of wave propagation in a non-linearly elastic medium (effective verification of Hadamard's condition) (in Russian), Mekhanika Tverdogo Tela, 1980, No. 6, pp. 110-116. (Note by the translator).

TO CHAPTER 9

The books [5], [8], [11], [12] and [5.9] have been referred to as well as

9.1. Truesdell, C., Rational Thermodynamics, McGraw-Hill, 1969.

BIBLIOGRAPHY TO APPENDICES

A.1. Kotchine, N.E., Vector Calculus and Foundations of Tensor Calculus (in Russian), Nauka, Moscow, 1965.
A.2. Sokolnikoff, I.S., Tensor Analysis. Theory and Applications to Geometry and Mechanics of Continua, Wiley, New York, 1964.
A.3. McConnel, A.J., Applications of Tensor Analysis, Dover Publications, Inc. New York, 1957.

TO APPENDIX I

Pars. 1-11 contain illustrations of the fundamental notions of the vector bases, symbols of Levi-Civita, a tensor of the second rank, its invariants, principal axes and directions and polar representation. For additional information see A.3 .

In the well-known textbooks, e.g. [A.1] and [A.2] the principal values and directions of a tensor of the second rank are discussed only as applied to the case of a symmetric tensor. Following [4], we consider the general case in Par. 9.

The contents of Par. 15 find numerous applications in the main text. The concept of isotropic tensors of the fourth rank is illustrated in the book

A.I.1. Jeffreys, H., Cartesian Tensors, Cambridge University Press, 1931.

Isotropic tensors of the sixth rank were used in the paper

A.I.2. Toupin, R.A. and B. Bernstein, Sound waves in deformed perfectly elastic materials; the acoustoelastic effect, J. Acoustic Soc. Amer., 1961, v. 33, p. 216-225.

The reader will facilitate his studies if he acquaints himself fully with the following widely used relationships:

1. The contraction formula for components of a tensor of Levi-Civita (2.5)-(2.7).
2. The rules (7.15)-(7.17) for the double contraction of a tensor product.
3. The representations (7.4), (7.9), (7.11) and (9.8) of the invariants of a second rank tensor and the Cayley-Hamilton theorem (9.21) and (9.22).
4. The rules for operation with the isotropic tensors (15.3)-(15.9).
5. The formulas defining the connections between a skew tensor and its associated vector (14.9), (14.17).
6. The definitions (8.1), (8.12) and (8.16) of an orthogonal tensor and the representations (8.10) of a vector and (8.11) of a tensor of the second rank in rotated bases.

TO APPENDIX II

This deals mainly with the operations of the differentiation of a scalar and tensor by a tensor argument. The formulas (2.7) and (4.6) provide invariant definitions of these operations and Equalities (3.1), (3.2) and (3.3) define the derivatives of tensor invariants and of their scalar functions. Rules for the differentiation of tensor products (4.10) and of change of variabels (4.12) are given. The use of tensors of the fourth rank leads to brevity of presentation of the relationships (4.10)-(4.16) obtained.

The contents of Pars. 2-4 is reproduced in the paper

A.II.1. Lurie, A.I., Differentiation by a tensor argument (in Russian), In the book: "Problems in Mathematical Physics" (in Russian), Nauka, Leningrad, 1976, pp. 48-57.

See also

A.II.2. Pietraszkiewicz, W., Stress in an isotropic elastic solid under superposed deformations, Archiwum Mechaniki Stosowanej, 1974, v. 26, p. 871-884.

A.II.3. Lurie, A.I., The differentiation of tensors with respect to time (in Russian), Trudy LPI (Proceedings of the Leningrad Polytechnical Institute), Mashinostroenie, Leningrad, 1971, No. 318.

An explanation of the concept of a tensor isotropic in a subgroup of orthogonal tensors is given in Par. 5. Examples of monoclinic, cubic and transversally isotropic symmetries are considered. The treatment leads directly to the desired result but is less general than that given in [1].

The theorem of the quadratic dependence of an isotropic tensor function on a symmetric tensor is given in Par. 7. In Pars. 8-9 the problem of inversion of this dependence is considered. Here we have made considerable use of the paper

A.II.4. Novozhilov, V.V., On the dependence between stress and strain for nonlinearly elastic bodies (in Russian), Prikladnaya Matematika i Mekhanika, 1951, Vol. 15, No. 2.

See also

A.II.5. Truesdell, C. and H. Moon, Inequalities sufficient to ensure semi-invertibility of isotropic functions, J. of Elasticity, 1975, v. 5, No. 3-4, p. 183-189.

TO APPENDIX III

The reader familiar with tensor calculus from the books [A.1]-[A.3] will not find anything substantially new in this appendix.

The presentation is designed to train the reader in the use of the nabla-operator and in covariant differentiation. The concepts of parallel translation, geodesics etc. not used in the main text are omitted.

It is recommended that the reader pay careful attention to Equations (3.7), (3.10), (2.19), (6.4) and (6.5). The Gauss-Ostrogradskii transformation (Par. 8) is of universal application.

Par. 11 was included in view of the statement substantially used in Chapter 4, Par. 11 (Equations (4.11.26)). The notions of the mean and Gaussian curvatures of a surface are employed in Chapter 7 for finding the general solutions of Ericksen.

SUBJECT INDEX

Printed in Dunstable, United Kingdom

91338437R00358